# Springer Texts in Statistics

*Series Editors:*
G. Casella
S. Fienberg
I. Olkin

T0180039

For other titles published in this series, go to
http://www.springer.com/series/417

Jiming Jiang

# Large Sample Techniques for Statistics

 Springer

Jiming Jiang
University of California
Department of Statistics
1 Shields Avenue
Davis, California 95616
USA
jiang@wald.ucdavis.edu

ISBN 978-1-4614-2623-3          ISBN 978-1-4419-6827-2(eBook)
DOI 10.1007/978-1-4419-6827-2
Springer New York Dordrecht Heidelberg London

For my parents, Huifen and Haoliang,
and my sisters, Qiuming and Dongming,
with love

# Preface

In a way, the world is made up of approximations, and surely there is no exception in the world of statistics. In fact, approximations, especially large sample approximations, are very important parts of both theoretical and applied statistics. The Gaussian distribution, also known as the normal distribution, is merely one such example, due to the well-known central limit theorem. Large-sample techniques provide solutions to many practical problems; they simplify our solutions to difficult, sometimes intractable problems; they justify our solutions; and they guide us to directions of improvements. On the other hand, just because large-sample approximations are used everywhere, and every day, it does not guarantee that they are used properly, and, when the techniques are misused, there may be serious consequences.

*Example 1* (Asymptotic $\chi^2$ distribution). Likelihood ratio test (LRT) is one of the fundamental techniques in statistics. It is well known that, in the "standard" situation, the asymptotic null distribution of the LRT is $\chi^2$, with the degrees of freedom equal to the difference between the dimensions, defined as the numbers of free parameters, of the two nested models being compared (e.g., Rice 1995, pp. 310). This might lead to a wrong impression that the asymptotic (null) distribution of the LRT is always $\chi^2$. A similar mistake might take place when dealing with Pearson's $\chi^2$-test—the asymptotic distribution of Pearson's $\chi^2$-test is not always $\chi^2$ (e.g., Moore 1978).

*Example 2* (Approximation to a mean). It might be thought that, in a large sample, one could always approximate the mean of a random quantity by the quantity itself. In some cases this technique works. For example, suppose $X_1, \ldots, X_n$ are observations that are independent and identically distributed (i.i.d.) such that $\mu = \mathrm{E}(X_1) \neq 0$. Then one can approximate $\mathrm{E}(\sum_{i=1}^{n} X_i) = n\mu$ by simply removing the expectation sign, that is, by $\sum_{i=1}^{n} X_i$. This is because the difference $\sum_{i=1}^{n} X_i - n\mu = \sum_{i=1}^{n}(X_i - \mu)$ is of the order $O(\sqrt{n})$, which is lower than the order of the mean of $\sum_{i=1}^{n} X_i$. However, this technique completely fails if one considers $(\sum_{i=1}^{n} X_i)^2$ instead. To see this, let us assume for simplicity that $X_i \sim N(0, 1)$. Then $\mathrm{E}(\sum_{i=1}^{n} X_i)^2 = n$. On the other hand, since $\sum_{i=1}^{n} X_i \sim N(0, n)$, $(\sum_{i=1}^{n} X_i)^2 = n\{(1/\sqrt{n}) \sum_{i=1}^{n} X_i\}^2 \sim n\chi_1^2$, where

$\chi_1^2$ is a random variable with a $\chi^2$ distribution with one degree of freedom. Therefore, $(\sum_{i=1}^n X_i)^2 - \mathrm{E}(\sum_{i=1}^n X_i)^2 = n(\chi_1^2 - 1)$, which is of the same order of $E(\sum_{i=1}^n X_i)^2$. Thus, $(\sum_{i=1}^n X_i)^2$ is not a good approximation to its mean.

*Example 3* (Maximum likelihood estimation). Here is another example of the so-called large-sample paradox. Because of the popularity of the maximum likelihood and its well-known large-sample theory in the classical situation, one might expect that the maximum likelihood estimator is always consistent. However, this is not true in some fairly simple, and practical, situations. For example, Neyman and Scott (1948) gave the following example. Suppose that two measurements are taken from each of the $n$ patients. Let $y_{ij}$ denote the $j$th measurement from the $i$th patient, $i = 1, \ldots, n, j = 1, 2$. Suppose that the measurements are independent and $y_{ij}$ is normally distributed with unknown mean $\mu_i$ and variance $\sigma^2$. Then, as $n \to \infty$, the maximum likelihood estimator of $\sigma^2$ is inconsistent.

The above are only a few examples out of many, but the message is just as clear: It is time to unravel such confusions.

This book deals with large-sample techniques in statistics. More importantly, we show how to argue with large-sample techniques and how to use these techniques the right way. It should be pointed out that there is an extensive literature on large-sample theory, including books and published papers, some of which are highly mathematical. Traditionally, there have been several approaches to introducing these materials. The first is the theorem/proof approach, which provides rigorous proofs for all or most of the theoretical results (e.g., Petrov 1975). The second is the method/application approach, which focuses on using the results without paying attention to any of the proofs (e.g., Barndorff-Nielsen and Cox 1989). Our approach is somewhere in between. Instead of giving a formal, technical proof for every result, we focus on the ideas of asymptotic arguments and how to use the methods developed by these arguments in various less-than-textbook situations.

We begin by reviewing some of the very simple and fundamental concepts that most of us have learned, say, from a calculus book. More specifically, Chapters 1–5 are devoted to a comprehensive review of the basic tools for large-sample approximations, such as the $\epsilon$-$\delta$ arguments, Taylor expansion, different types of convergence, and inequalities. Chapters 6–10 discuss limit theorems in specific situations of observational data. These include the classical case of i.i.d. observations, independent but not identically distributed observations such as those encountered in linear regression, empirical processes, martingales, time series, stochastic processes, and random fields. Each of the first 10 chapters contains at least one section of case study as applications of the methods or techniques covered in the chapter. Some more extensive applications of the large-sample techniques are discussed in Chapters 11–15. The areas of applications include nonparametric statistics, linear and generalized linear mixed models, small-area estimation, jackknife and bootstrap, and Markov-chain Monte Carlo methods.

As mentioned, there have been several major texts on similar topics. These include, in the order of year published: [1] Hall & Heyde (1980), *Martingale Limit Theory and Its Application*, Academic Press; [2] Barndorff-Nielsen & Cox (1989), *Asymptotic Techniques for Use in Statistics*, Chapman & Hall; [3] Ferguson (1996), *A Course in Large Sample Theory*, Chapman & Hall; [4] Lehmann (1999), *Elements of Large-Sample Theory*, Springer; and [5] Das-Gupta (2008), *Asymptotic Theory of Statistics and Probability*, Springer. A comparison with these existing texts would help to highlight some of the features of the current book. Text [2] deals with the case of independent observations. In practice, however, the observations are often correlated. A main purpose of the current book is to introduce large-sample theory and methods for correlated observations, such as those in time series, mixed models, and spatial statistics. Furthermore, the approach of [2] is more like "use this formula," rather than "why?" and "what's the trick?." In contrast, the current text focuses more on the way of thinking. For example, the current text covers basic elements in asymptotic theory, such as $\epsilon$-$\delta$, $O_P$, and $o_P$, in addition to the asymptotic results, such as a formula of asymptotic expansion. This reflects the current author's belief that methodology is more important and applicable to a broader range of problems than formulas.

Text [3] provides an account of large-sample theory for independent random variables (mostly in the i.i.d. case) with applications to efficient estimation and testing problems. Several classical cases of dependent random variables are also considered, such as $m$-dependent sequences, rank, and order statistics, but the basic method was to convert these to the case of independent observations plus some extra terms that are asymptotically negligible. The chapters are written in a theorem–proof style which is what the author intended to do.

Like [2] and [3], text [4] deals with independent observations, mostly the i.i.d. case. However, the approach of [4] has motivated the current author. For example, [4] begins with very simple and fundamental concepts and eventually gets to a much advanced level. It might be worth mentioning that the current author assisted Professor E. L. Lehmann in the mid-1990s during his writing of book [4].

Text [5] provides a very comprehensive account of asymptotic theory in statistics and probability. However, similar to books [2]–[4], the focus of [5] is mainly on independent observations. Also, since a large number of topics need to be covered, it is unavoidable that the coverage is a little sketchy.

Unlike books [2]–[5], text [1] deals with one special case of dependent observations—the martingales. Whereas the martingale limit theory applies to a broad ranges of problems, such as linear mixed models and some cases of time series, it does not cover many other cases encountered in practice. Furthermore, the book starts at a relatively high level, assuming that the reader has taken an advanced course in probability theory. As mentioned, the current book begins with very basic concepts in asymptotic arguments, such as $\epsilon$-$\delta$ and Taylor expansion, which requires nothing more than a course in calculus, and eventually covers much more than the martingale limit theory.

We realize that there have been other books covering similar or related topics, for example, Serfling (1980), van der Vaart and Wellner (1996), and van der Vaart (1998), to mention just a few; however, space does not allow us to make comparisons here.

The current book is supplemented by a large number of exercises. The exercises are attached to each chapter and closely related to the materials covered, giving the readers plenty of opportunities to practice the large-sample techniques that they have learned. The book is mostly self-contained with the appendixes providing some backgrounds for matrix algebra and mathematical statistics. A list of notation is also provided in the appendixes for the readers' convenience. The book is intended for a wide audience, ranging from senior undergraduate students to researchers with Ph.D. degrees. More specifically, Chapters 1–5 and parts of Chapters 10–15 are intended for senior undergraduate and M.S. students. For Ph.D. students and researchers, all chapters are suitable. A first course in mathematical statistics and a course in calculus are prerequisites. As it is unlikely that all 15 chapters will be covered in a single-semester or quarter-course, the following combinations of chapters are recommended for a single-semester course, depending on the focus of interest (for a single-quarter course some adjustment is necessary):

For a senior undergraduate or M.S.-level course on large sample techniques, Chapters 1–6.

For those interested in linear models, generalized linear models, mixed effects models, and their applications, Chapters 1–6, 8, and 12.

For those interested in time series, stochastic processes, and their applications, Chapters 1–6 and 8–10.

For those interested in semiparametric, nonparametric statistics, and their applications, Chapters 1–7 and 11.

For those interested in empirical Bayes methods, small-area estimation, and related fields, Chapters 1-6, 12, and 13.

For those interested in resampling methods, Chapters 10–7, 11, and 14.

For those interested in Monte Carlo methods and their applications in Bayesian inference, Chapters 1–6, 10, and 15.

For those interested in spatial statistics, Chapters 1–6, 9, and 10.

Thus, in particular, Chapters 1–6 are vital to any sequence recommended.

The book is motivated by the author's research work, who has used large-sample techniques throughout his career. The author wishes to give his sincere thanks to Professor Peter J. Bickel for guiding the author in his Ph.D. dissertation that led to one of his best theoretical work on REML asymptotics (see Section 12.2) and for the many helpful discussions afterwards including those regarding the bootstrap method that is covered in Chapter 14; to Professor David Aldous for communications regarding an example in Chapter 10; to Professor Samuel Kou for helpful discussion on Markov-chain Monte Carlo methods; to Professor Jun Liu for kindly providing a plot to be included in

Chapter 15 of this book; and to the author's long-time collaborator and friend, Professor Partha Lahiri, for leading the author to some of the important application areas of large-sample techniques, such as small-area estimation and resampling methods. In addition, a number of anonymous reviewers have made valuable comments regarding earlier versions of the book chapters. For example, several reviewers have suggested inclusion of a chapter on nonparametric methods; one reviewer suggested another case study regarding Chapter 8. The author appreciates their valuable suggestions. The author also wishes to express his gratefulness to Dr. Thuan Nguyen for computational and graphic assistance and to Mr. Peter Scully for reading and improving the English presentation of the Preface. Finally, the author has grown up reading Professor Erich Lehmann's classical texts in Statistics, from whom he learned to write his first paper in America (Jiang 1997b) and his first book on mixed models (Jiang 2007). While the author is heartfeltly grateful to Professor Lehmann's lifetime inspiration, he had wished to show his appreciation by sending him the first copy of this book. (Professor Lehmann died on September 12, 2009.)

Jiming Jiang
Davis, California
December, 2009

# Contents

# 1

## The $\epsilon$-$\delta$ Arguments

Let's start at the very beginning
A very good place to start
When you read you begin with A-B-C
When you sing you begin with do-re-mi

**Rodgers and Hammerstein (1959)**
*The Sound of Music*

### 1.1 Introduction

Every subject has its A, B, and C. The A-B-C for large-sample techniques is $\epsilon$, $\delta$, and a line of arguments. For the most part, this line of arguments tells how large the sample size, $n$, has to be in order to achieve an arbitrary accuracy that is characterized by $\epsilon$ and $\delta$. It should be pointed out that, sometimes, the arguments may involve no $\delta$ ($\epsilon$), or more than one $\delta$ ($\epsilon$), but the basic lines of the arguments are all similar. Here is a simple example.

*Example 1.1.* Suppose that one needs to show $\log(n + 1) - \log(n) \to 0$ as $n \to \infty$. The arguments on one's scratch paper (before it is printed nicely in a book) might look something like the following. To show

$$\log(n + 1) - \log(n) \longrightarrow 0$$

means to show

$$\log\left(1 + \frac{1}{n}\right) < \epsilon. \tag{1.1}$$

What is $\epsilon$? $\epsilon$ is a (small) positive number. Okay. Go on. This means $1 + 1/n < e^\epsilon$, or $n > (e^\epsilon - 1)^{-1}$. If we take $N = [(e^\epsilon - 1)^{-1}] + 1$, where $[x]$ represents the integer part of $x$ (i.e., the largest integer less than or equal to $x$), then when $n \geq N$, we have (1.1). Now grab a nice piece of paper, or a computer file, and write the following proof:

J. Jiang, *Large Sample Techniques for Statistics*,
DOI 10.1007/978-1-4419-6827-2_1, © Springer Science+Business Media, LLC 2010

For any $\epsilon > 0$, let $N = [(e^\epsilon - 1)^{-1}] + 1$. Then, for all $n \geq N$, we have (1.1). This proves that $\log(n + 1) - \log(n) \to 0$ as $n \to \infty$.

The above proof looks nice and short, but it is the arguments on the scratch paper (which probably gets thrown out to the trash basket after the proof is written) that is more useful in training the way that one thinks.

## 1.2 Getting used to the $\epsilon$-$\delta$ arguments

In this argument, the order of choosing $\epsilon$ and $N$ is critically important; $\epsilon$ has to be chosen (or given) first before $N$ is chosen. For example, in Example 1.1, if one were allowed to choose $N$ first and then $\epsilon$, the same "argument" can be used to show that $\log(n + 1) - \log(n) \to 1$ (or any other constant) as $n \to \infty$. This is because for any $n \geq N$, one can always find $\epsilon > 0$ such that

$$| \log(n + 1) - \log(n) - 1| = \left| \log\left( 1 + \frac{1}{n} \right) - 1 \right|$$
$$< \epsilon.$$

Here is another example.

*Example 1.2.* Define $f(x) = x^2$ if $x < 1$, and $f(x) = 2$ if $x \geq 1$. A plot of $f(x)$ is shown in Figure 1.1. Show that $f(x)$ is continuous at $x = 0$. Once again, first work on the scratch paper. To show that $f(x)$ is continuous at $x = 0$ is to show that $|f(x) - f(0)| < \epsilon$, if $|x - 0| = |x| < \delta$. What is $\epsilon$? $\epsilon$ is a given (small) positive number. Okay. Go on. What is $\delta$? $\delta$ is another (small) positive number chosen after $\epsilon$ and therefore depending on $\epsilon$. Okay and go on. Since $f(0) = 0$, this means to choose $\delta$ such that $|f(x)| < \epsilon$, if $|x| < \delta$. Because $f(x) = x^2$ when $x$ is close to 0, we need to show that $x^2 < \epsilon$, or $|x| < \sqrt{\epsilon}$. Now, it is clear how $\delta$ should be chosen: $\delta = \sqrt{\epsilon}$.

Note that if the order in which $\epsilon$ and $\delta$ are chosen is reversed, the same "argument" can be used to show that $f(x)$ is continuous at $x = 1$, which is obviously not true from Figure 1.1. To see this, note that for any $0 < \delta < 1$, when $1 - \delta < x < 1$, we have $|f(x) - f(1)| = |x^2 - 2| < 2 - (1 - \delta)^2$. Thus, if one lets $\epsilon = 2 - (1 - \delta)^2$, one has $|f(x) - f(1)| < \epsilon$, but this is wrong! The choice of $\epsilon$ should be arbitrary, and it cannot depend on the value of $\delta$. In fact, it is the other way around; $\delta$ typically depends on the value of $\epsilon$, such as in Example 1.2.

One of the important concepts in large-sample theory is called convergence in probability. This is closely related to another important concept in statistics—namely, the consistency of an estimator. To show that an estimator is consistent is to show that it converges in probability to the quantity (e.g., parameter) that it intends to estimate. Convergence in probability is defined

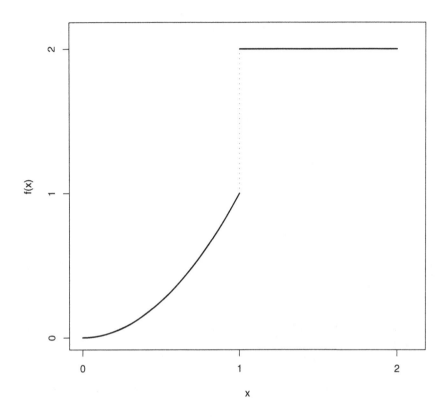

**Fig. 1.1.** *A plot of the function in Example 1.2*

through an $\epsilon$-$\delta$ argument as follows. Let $\xi_n$, $n = 1, 2, \ldots$, be a sequence of random variables. The sequence converges in probability to a random variable $\xi$, denoted by $\xi_n \xrightarrow{P} \xi$, if for any $\epsilon > 0$, the probability $P(|\xi_n - \xi| > \epsilon) \to 0$ as $n \to \infty$. In other words, for any $\epsilon > 0$ and (then) for any $\delta > 0$, there is $N \geq 1$ such that $P(|\xi_n - \xi| > \epsilon) < \delta$ if $n \geq N$.

In particular, the random variable $\xi$ can be a constant, which is often the case in the context of consistent estimators. We consider some examples.

*Example 1.3* (Consistency of the sample mean). Let $X_1, \ldots, X_n$ be observations that are independent and identically distributed (i.i.d.). Then the sample mean

$$\bar{X} = \frac{X_1 + \cdots + X_n}{n} \tag{1.2}$$

is a consistent estimator of the population mean $\mu = \mathrm{E}(X_i)$ (note that the latter does not depend on $i$ due to the i.i.d. assumption), provided that it is finite. This result is also known as the *weak law of large numbers, WLLN* (WLLN) in probability theory. To prove this result, we will make a stronger assumption, for now, that $\mathrm{E}(X_i^2) < \infty$. Then, for any $\epsilon > 0$ and for any $\delta > 0$, by Chebyshev's inequality, we have

$$\mathrm{P}(|\bar{X} - \mu| > \epsilon) \leq \frac{\mathrm{E}(|\bar{X} - \mu|^2)}{\epsilon^2}$$

$$= \frac{1}{\epsilon^2 n^2} \mathrm{E}\left\{\sum_{i=1}^{n}(X_i - \mu)\right\}^2$$

$$= \frac{1}{\epsilon^2 n^2} \sum_{i=1}^{n} \mathrm{E}(X_i - \mu)^2$$

$$\leq \frac{c}{\epsilon^2 n} \tag{1.3}$$

for some constant $c > 0$. Therefore, for any $\delta > 0$, as long as $c/\epsilon^2 n < \delta$ or $n > c/\epsilon^2 \delta$, we have, by (1.3), $\mathrm{P}(|\bar{X} - \mu| > \epsilon) < \delta$. Thus, by letting $N = [c/\epsilon^2\delta] + 1$, we have $\mathrm{P}(|\bar{X} - \mu| > \epsilon) < \delta$ if $n \geq N$.

*Example 1.4* (Consistency of MLE in the Uniform distribution). Let $X_1$, ..., $X_n$ be i.i.d. observations from the Uniform$[0, \theta]$ distribution, where $\theta$ is an unknown (positive) parameter. It can be shown that the maximum likelihood estimator (MLE) of $\theta$ is $\hat{\theta} = X_{(n)} = \max_{1 \leq i \leq n} X_i$. We show that $\hat{\theta}$ is a consistent estimator of $\theta$.

For any $\epsilon > 0$, since, by the definition, $\hat{\theta} \leq \theta$ with probability 1, we have

$$\mathrm{P}(|\hat{\theta} - \theta| > \epsilon) = \mathrm{P}(\hat{\theta} < \theta - \epsilon)$$

$$= \mathrm{P}(X_1 < \theta - \epsilon, \ldots, X_n < \theta - \epsilon)$$

$$= \mathrm{P}(X_1 < \theta - \epsilon) \cdots \mathrm{P}(X_n < \theta - \epsilon)$$

$$= \left(1 - \frac{\epsilon}{\theta}\right)^n. \tag{1.4}$$

Here, we may assume, without loss of generality, that $\epsilon < \theta$. Now, for any $\theta > 0$, if we want the left side of (1.4) to be less than $\delta$, we need $n > \log(\delta)/\log(1 - \epsilon/\theta)$. This gives the choice of $N$ (e.g., $N = [\log(\delta)/\log(1 - \epsilon/\theta)] + 1$), so that $\mathrm{P}(|\hat{\theta} - \theta| > \epsilon) < \delta$, if $n \geq N$.

It should be pointed out that the right end of the interval $[0, \theta]$ of the Uniform distribution is closed, which is critically important here. For example, if the Uniform$[0, \theta]$ distribution is replaced by the Uniform$[0, \theta)$ distribution, then it can be shown that the MLE of $\theta$ does not even exist. Of course, in this case, the MLE is inconsistent (Exercise 1.8).

## 1.3 More examples

Sometimes, the $\epsilon$-$\delta$ arguments may involve several steps that have to be put together at the end. The arguments in the following examples are somehow more complicated than the previous ones, but the way of thought is very similar.

*Example 1.5* (Consistency of the sample median). Let $X$ be a random variable. The median of $X$ is defined as any number $a$ such that $P(X \leq a) \geq 1/2$ and $P(X \geq a) \geq 1/2$. In general, the median may be an interval instead of a single number. Here, we assume, for simplicity, that $X$ is a continuous random variable with a unique median $a$. It follows that $P(X \leq x) < 1/2$, $x < a$, $P(X \leq a) = 1/2$, and $P(X \leq x) > 1/2$, $x > a$.

The sample median is defined in terms of the order statistics, $X_{(1)} < \cdots < X_{(n)}$, which are the observations $X_1, \ldots, X_n$ listed in the increasing order. Here, we assume that $X_1, \ldots, X_n$ are independent with the same distribution as the random variable $X$ above. If $n$ is an odd number, say, $n = 2m + 1$, the sample median is defined as $X_{(m+1)}$; otherwise, if $n$ is an even number, say, $n = 2m$, the sample median is defined as $\{X_{(m)} + X_{(m+1)}\}/2$.

We consider the case $n = 2m+1$ here. The case $n = 2m$ is left to the readers as an exercise. For any $\epsilon > 0$, we need to show that $P\{|X_{(m+1)} - a| > \epsilon\} \to 0$ as $n \to \infty$. Note that

$$P\{|X_{(m+1)} - a| > \epsilon\} = P\{X_{(m+1)} > a + \epsilon\} + P\{X_{(m+1)} < a - \epsilon\}$$
$$\leq P\{X_{(m+1)} > a + \epsilon\} + P\{X_{(m+1)} \leq a - \epsilon\}. \quad (1.5)$$

For any $x$, define the random variable $Y_n$ as the total number of $X_i$'s that are less than or equal to $x$. Then $X_{(m+1)} \leq x$ if and only if $Y_n \geq m+1$. Notice that $Y_n$ has a Binomial$(n, p)$ distribution, where $p = F(x) = P(X \leq x)$. Therefore,

$$P\{X_{(m+1)} \leq x\} = P(Y_n \geq m + 1)$$
$$= P\left\{\frac{Y_n - np}{\sqrt{np(1-p)}} \geq \frac{m + 1 - np}{\sqrt{np(1-p)}}\right\}. \quad (1.6)$$

We now use the (classical) central limit theorem (CLT), which will be further discussed in the sequel. It follows that $(Y_n - np)/\sqrt{np(1-p)}$ converges in distribution to $N(0,1)$. On the other hand, note that $m + 1 = (n + 1)/2$. Thus, we have

$$\frac{m + 1 - np}{\sqrt{np(1-p)}} = \left\{\left(\frac{1}{2} - p\right)n + \frac{1}{2}\right\}/\sqrt{np(1-p)}. \quad (1.7)$$

It follows that $(1.7) \to \infty$ if $p < 1/2$ and $\to -\infty$ if $p > 1/2$. If we combine $(1.5)$–$(1.7)$ with $x = a - \epsilon$ (and hence $p < 1/2$), we come up with the following argument. For any $\delta > 0$, choose $B > 0$ such that $\Phi(B) > 1 - \delta$, where $\Phi(\cdot)$

denotes the cumulative distribution function (cdf) of $N(0,1)$. Then, according to the CLT, there is $N_1 \geq 1$ such that when $n \geq N_1$, we have

$$\left| P\left\{ \frac{Y_n - np}{\sqrt{np(1-p)}} \leq B \right\} - \Phi(B) \right| < \delta.$$

Furthermore, there is $N_2 \geq 1$ such that when $n \geq N_2$, the left side of (1.7) is greater than $B$. Thus, when $n \geq N_1 \vee N_2$, we have [by (1.6)]

$$
\begin{aligned}
P\{X_{(m+1)} \leq a - \epsilon\} &\leq P\left\{ \frac{Y_n - np}{\sqrt{np(1-p)}} > B \right\} \\
&= 1 - P\left\{ \frac{Y_n - np}{\sqrt{np(1-p)}} \leq B \right\} \\
&\leq 1 - \Phi(B) + \left| P\left\{ \frac{Y_n - np}{\sqrt{np(1-p)}} \leq B \right\} - \Phi(B) \right| \\
&< 2\delta. \qquad\qquad\qquad\qquad\qquad\qquad\qquad\qquad (1.8)
\end{aligned}
$$

By a similar argument, it can be shown that there are $N_3$, $N_4 \geq 1$ such that when $n \geq N_3 \vee N_4$, we have $P\{X_{(m+1)} > a + \epsilon\} < 2\delta$. Therefore, by (1.5), when $n \geq N_1 \vee N_2 \vee N_3 \vee N_4$, we have $P\{|X_{(m+1)} - u| > \epsilon\} < 4\delta$.

*Note 1.* The role of $B$ in this argument is called *a bridge*. It helps to connect the $\epsilon$-$\delta$ arguments; once the connection is made, the role of $B$ is finished (and thus resembles the role of a bridge). For example, in (1.8), all one needs are the left and right ends of these inequalities to hold when $n \geq N_1 \vee N_2$, but $B$ helps to make the connections. Such usages of bridges are fairly common in asymptotic arguments (see the Exercises at the end of the chapter).

*Note 2.* Unlike in Examples 1.1–1.4, here several $N$'s were chosen in different pieces of the arguments. Typically, one needs to take the maximum of those $N$'s at the end.

*Note 3.* Also note that, at the end, we showed that the probability is less than $4\delta$, not $\delta$. However, this makes no difference in the asymptotic arguments because $\delta$ is arbitrary. If one wishes, one could replace $\delta$ by $\delta/4$ at the intermediate steps where $N_1, \ldots, N_4$ were chosen and repeat the argument so that, at the end, one has the probability less than $\delta$.

Convergence in distribution is another important concept in large-sample theory. In particular, it is closely related to the CLT that was used in the previous example. A sequence of distributions, represented by their cdf's $F_1, F_2, \ldots$, converges weakly to a distribution with cdf $F$, denoted by $F_n \xrightarrow{w} F$ if $F_n(x) \to F(x)$ as $n \to \infty$ for every $x$ at which $F(x)$ is continuous. Note that as a cdf, $F$ can only have countably many discontinuity points (Exercise 1.12). A sequence of random variables $\xi_1, \xi_2, \ldots$ converges in distribution to a random variable $\xi$, denoted by $\xi_n \xrightarrow{d} \xi$, if $F_n \xrightarrow{w} F$, where $F_n$ is the cdf of $\xi_n$,

$n = 1, 2, \ldots$, and $F$ is the cdf of $\xi$. One of the striking results of convergence in distribution is the following.

*Example 1.6* (Pólya's theorem). Suppose that $F$ is continuous. If $F_n \xrightarrow{w} F$, then the convergence is uniform in that

$$\sup_x |F_n(x) - F(x)| \longrightarrow 0, \tag{1.9}$$

as $n \to \infty$. The result is striking because weak convergence is defined point-wisely, and, in general, pointwise convergence does not necessarily imply uniform convergence. However, a cdf is monotone and has limits at $-\infty$ and $\infty$. Such nice properties make it possible to derive uniform convergence from pointwise convergence. Result (1.9) actually holds for multivariate cdf's as well, but here, for simplicity, we consider the univariate case only.

To show (1.9) we need to show that for any $\epsilon > 0$, there is $N \geq 1$ such that the left side of (1.9) is less than $\epsilon$ if $n \geq N$. First, choose $A < 0$ and $B > 0$ such that $F(A) < \epsilon$ and $F(B) > 1 - \epsilon$. Because $F(x)$ is continuous over $[A, B]$, there are points $A < x_1 < \cdots < x_k < B$ such that $F(x_{j+1}) - F(x_j) < \epsilon$, $0 \leq j \leq k$, where $x_0 = A$ and $x_{k+1} = B$. Now, because $F_n \xrightarrow{w} F$, for each $0 \leq j \leq k+1$, there is $N_j \geq 1$ such that

$$|F_n(x_j) - F(x_j)| < \epsilon \quad \text{if } n \geq N_j.$$

Let $N = N_0 \vee N_1 \vee \cdots \vee N_{k+1}$ and suppose $n \geq N$. If $x \leq A$, we have

$$
\begin{aligned}
F_n(x) - F(x) &\leq F_n(A) \\
&= F(A) + F_n(A) - F(A) \\
&\leq F(A) + |F_n(A) - F(A)| \\
&= F(A) + |F_n(x_0) - F(x_0)| \\
&< 2\epsilon
\end{aligned}
$$

and $F_n(x) - F(x) \geq -F(A) > -\epsilon$, so $|F_n(x) - F(x)| < 2\epsilon$. If $x \geq B$, then $F_n(x) - F(x) \leq 1 - F(x) \leq 1 - F(B) < \epsilon$, and

$$
\begin{aligned}
F_n(x) - F(x) &\geq F_n(B) - 1 \\
&= F(B) - 1 + F_n(B) - F(B) \\
&\geq F(B) - 1 - |F_n(B) - F(B)| \\
&= F(B) - 1 - |F_n(x_{k+1}) - F(x_{k+1})| \\
&> -2\epsilon.
\end{aligned}
$$

Thus, $|F_n(x) - F(x)| < 2\epsilon$. Finally, if $A < x < B$, then there is $0 \leq j \leq k$ such that $x \in [x_j, x_{j+1}]$. It follows that

$$
\begin{aligned}
F_n(x) - F(x) &\leq F_n(x_{j+1}) - F(x_j) \\
&= F_n(x_{j+1}) - F(x_{j+1}) + F(x_{j+1}) - F(x_j) \\
&\leq |F_n(x_{j+1}) - F(x_{j+1})| + F(x_{j+1}) - F(x_j) \\
&< 2\epsilon,
\end{aligned}
$$

and

$$\begin{aligned}
F_n(x) - F(x) &\geq F_n(x_j) - F(x_{j+1}) \\
&= F_n(x_j) - F(x_j) + F(x_j) - F(x_{j+1}) \\
&\geq -|F_n(x_j) - F(x_j)| + F(x_j) - F(x_{j+1}) \\
&> -2\epsilon.
\end{aligned}$$

Thus, once again, we have $|F_n(x) - F(x)| < 2\epsilon$. In conclusion, we have $|F_n(x) - F(x)| < 2\epsilon$ for all $x$, as long as $n \geq N$. Thus, for $n \geq N$, we have the left side of (1.9) $\leq 2\epsilon$. This completes the proof.

This example has the same flavor as the previous one in that (i) $A$ and $B$ were used as the bridge(s); (ii) a number of $N_j$'s were chosen at intermediate steps and the final $N$ was the maximum of those; and (iii) at the end we showed the left side of (1.9) is $\leq 2\epsilon$ rather than $< \epsilon$, but this did not matter since $\epsilon$ was arbitrary.

## 1.4 Case study: Consistency of MLE in the i.i.d. case

One of the fundamental results regarding the MLE is its consistency under regularity conditions. It should be pointed out that there are two types of consistency so long as the MLE is concerned. The first type of consistency is called the Cramér consistency (Cramér 1946), which states that there exists a root to the likelihood equation that is consistent. Thus, the result does not explicitly imply that the (global) maximum of the likelihood function (i.e., the MLE) is consistent. Furthermore, in case there are multiple roots to the likelihood equation, the theorem does not tell which root is consistent. Nevertheless, the Cramér consistency is a fundamental result because typically the MLE is typically a solution to the likelihood equation, and in some cases, the solution is unique. Another type of consistency is called the Wald consistency (Wald 1949). It states that the global maximum of the likelihood function (i.e., the MLE) is consistent. From a theoretical point of view the Wald consistency is a more desirable asymptotic property, although it is usually more difficult to prove than the Cramér consistency.

In the following we present a proof of the Cramér consistency due to Lehmann (1983) and a proof of the Wald consistency given by Wolfowitz (1949) in a note following Wald's paper. Both proofs involve the $\epsilon$-$\delta$ arguments, which is why they are presented here. The difference between Wald (1949) and Wolfowitz (1949) is that Wald proved strong consistency defined in terms of almost sure convergence, whereas Wolfowitz established consistency, which is defined in terms of convergence in probability. See Chapter 2 for a detailed account of different types of convergence. It should be pointed out that both proofs require some regularity conditions, which we will discuss later in Chapter 10, as our main goal here is to demonstrate the use of the $\epsilon$-$\delta$ argument.

We assume that $X_1, \ldots, X_n$ are i.i.d. observations that have the common probability density function (pdf) $f(x|\theta)$, where, for simplicity, we assume that $\theta$ is a real-valued unknown parameter, with the parameter space $\Theta = (-\infty, \infty)$ (see Exercise 1.13 for an extension of the proof to the more general case). Here, the pdf is with respect to a $\sigma$-finite measure $\mu$ (see Appendix A.2).

*Cramér consistency.* Denote the likelihood function by

$$L(\theta) = \prod_{i=1}^{n} f(X_i|\theta).$$

We need to show that there is a sequence of roots to the likelihood equation

$$\frac{\partial}{\partial\theta} \log\{L(\theta)\} = \sum_{i=1}^{n} \frac{1}{f(X_i|\theta)} \frac{\partial}{\partial\theta} f(X_i|\theta)$$

$$= 0, \qquad (1.10)$$

say, $\tilde{\theta}_n$, such that $\tilde{\theta}_n \xrightarrow{P} \theta$, where $\theta$ is the true parameter. For any $\epsilon > 0$, consider the sequence of random variables $Y_i = \log\{f(X_i|\theta)/f(X_i|\theta - \epsilon)\}$, $i = 1, 2, \ldots$. By Jensen's inequality (see Chapter 5), we have

$$\begin{aligned}
E_\theta(Y_i) &= E_\theta[-\log\{f(X_i|\theta - \epsilon)/f(X_i|\theta)\}] \\
&> -\log[E_\theta\{f(X_i|\theta - \epsilon)/f(X_i|\theta)\}] \\
&= -\log\left\{\int \frac{f(x|\theta - \epsilon)}{f(x|\theta)} f(x|\theta)\, d\mu\right\} \\
&= -\log\left\{\int f(x|\theta - \epsilon)\, d\mu\right\} \\
&= 0.
\end{aligned}$$

Hereafter, $E_\theta$ denotes expectation under the pdf $f(x|\theta)$. Similarly, let $P_\theta$ denote probability under the pdf $f(x|\theta)$. Then we have, by the WLLN, $n^{-1}\sum_{i=1}^{n} Y_i \xrightarrow{P} E_\theta(Y_1) > 0$; hence $P_\theta(\sum_{i=1}^{n} Y_i > 0) \to 1$ (Exercise 1.14). Therefore, for any $\delta > 0$, there is $N_1 \geq 1$ such that when $n \geq N_1$, $P_\theta(\sum_{i=1}^{n} Y_i > 0) > 1 - \delta$. Similarly, consider the sequence of random variables $Z_i = \log\{f(X_i|\theta)/f(X_i|\theta + \epsilon)\}$, $i = 1, 2, \ldots$. It can be shown that $P_\theta(\sum_{i=1}^{n} Z_i > 0) \to 1$. Therefore, there is $N_2 \geq 1$ such that when $n \geq N_2$, $P_\theta(\sum_{i=1}^{n} Z_i > 0) > 1 - \delta$.

Define $\tilde{\theta}_n$ as the root to the likelihood equation (1.10) that is closest to $\theta$. Note that $\tilde{\theta}_n$ exist as long as a root to (1.10) exists. In particular, the limit of a sequence of roots is also a root, provided that the left side of (1.10) is continuous. Also note that $\sum_{i=1}^{n} Y_i > 0$ and $\sum_{i=1}^{n} Z_i > 0$ imply that the value of the log-likelihood is higher at $\theta$ than at $\theta - \epsilon$ and $\theta + \epsilon$ and, hence, the existence of a root inside the interval $(\theta - \epsilon, \theta + \epsilon)$. Since the latter event implies that $|\tilde{\theta}_n - \theta| < \epsilon$, we have

$$P_\theta(|\tilde{\theta}_n - \theta| < \epsilon) \geq P_\theta\left(\sum_{i=1}^n Y_i > 0, \sum_{i=1}^n Z_i > 0\right)$$
$$> 1 - 2\delta,$$

or $P_\theta(|\tilde{\theta}_n - \theta| \geq \epsilon) < 2\delta$.

*Wald consistency.* The ingenious proof of Wald (1949) originated from the following simple property of the pdf, which is proved above by Jensen's inequality. Let $\theta$ be the true parameter. Then for any $\theta_1 \neq \theta$, we have

$$E_\theta[\log\{f(X_1|\theta)\}] \geq E_\theta[\log\{f(X_1|\theta_1)\}].$$

In fact, Wald proved the following stronger result. For every $\theta_1 \neq \theta$, there is $\rho = \rho(\theta_1) > 0$ such that

$$E_\theta\left[\log\left\{\sup_{|\theta_2 - \theta_1| \leq \rho} f(X_1|\theta_2)\right\}\right] < E_\theta[\log\{f(X_1|\theta)\}].$$

Furthermore, there is a positive number $a$ such that

$$E_\theta\left[\log\left\{\sup_{|\theta_2| > a} f(X_1|\theta_2)\right\}\right] < E_\theta[\log\{f(X_1|\theta)\}].$$

For any $\epsilon > 0$, the collection of open sets $S(\theta_1, \rho) = \{\theta_2 : |\theta_2 - \theta_1| < \rho\}$, $\theta_1 \in \Theta$, form an open cover of the compact set $(\theta - \epsilon, \theta + \epsilon)^c \cap [-a, a]$. By the Heine–Borel theorem (see the next subsection), there exist a finite subcover, say, $S(\theta_{1,1}, \rho_1), \ldots, S(\theta_{1,K}, \rho_K)$, of $(\theta - \epsilon, \theta + \epsilon)^c \cap [-a, a]$. Define the sequences of i.i.d. random variables as follows:

$$Y_{k,i} = \log\left\{\sup_{|\theta_2 - \theta_{1,k}| \leq \rho_k} f(X_i|\theta_2)\right\} - \log\{f(X_i|\theta)\}, \quad i = 1, 2, \ldots,$$

$1 \leq k \leq K$, and

$$Y_{K+1,i} = \log\left\{\sup_{|\theta_2| > a} f(X_i|\theta_2)\right\} - \log\{f(X_i|\theta)\}, \quad i = 1, 2, \ldots.$$

By the WLLN, we have $n^{-1} \sum_{i=1}^n Y_{k,i} \xrightarrow{P} E_\theta(Y_{k,1}) < 0$; hence,

$$P_\theta\left\{\sum_{i=1}^n Y_{k,i} < \frac{n}{2} E_\theta(Y_{k,1})\right\} \longrightarrow 1,$$

as $n \to \infty$, $1 \leq k \leq K + 1$.

Let $A_k$ be the event that $\sum_{i=1}^n Y_{k,i} < (n/2)\mathrm{E}_\theta(Y_{k,1})$, $1 \le k \le K+1$. Then for any $\delta > 0$, there is $N_k \ge 1$ such that when $n \ge N_k$, we have $\mathrm{P}_\theta(A_k) > 1 - \delta$, $1 \le k \le K+1$. Let $\eta = (1/2)\max_{1 \le k \le K+1} \mathrm{E}_\theta(Y_{k,1})$ and $N = \max_{1 \le k \le K+1} N_k$. Then when when $n \ge N$, we have

$$\mathrm{P}_\theta\left(\sum_{i=1}^n Y_{k,i} < n\eta, \ 1 \le k \le K+1\right) \ge \mathrm{P}_\theta\left(\cap_{k=1}^{K+1} A_k\right)$$
$$= 1 - \mathrm{P}_\theta\left(\cup_{k=1}^{K+1} A_k^c\right)$$
$$> 1 - (K+1)\delta. \tag{1.11}$$

Let $h = e^\eta$. Since $\eta < 0$, we have $0 < h < 1$. Furthermore, it can be shown (Exercise 1.15) that $\sum_{i=1}^n Y_{k,i} \le n\eta$, $1 \le k \le K+1$, imply

$$\sup_{|\theta_2 - \theta| \ge \epsilon} \left\{ \frac{\prod_{i=1}^n f(X_i|\theta_2)}{\prod_{i=1}^n f(X_i|\theta)} \right\} \le h^n. \tag{1.12}$$

Thus, by (1.11), the probability of event (1.12) is greater than $1 - (K+1)\delta$. Note that (1.12), in turn, implies that $|\hat\theta - \theta| < \epsilon$, where $\hat\theta$ is the MLE of $\theta$. In other words, the maximum of the likelihood function must lie within the interval $(\theta - \epsilon, \theta + \epsilon)$ (because the ratio is less than 1 for any $\theta$ outside the interval). It follows that $\mathrm{P}_\theta(|\hat\theta - \theta| < \epsilon) > 1 - (K+1)\delta$ if $n \ge N$. Since $\delta$ is arbitrary, we must have $\mathrm{P}_\theta(|\hat\theta - \theta| \ge \epsilon) \to 0$, as $n \to \infty$; hence, $\hat\theta$ is consistent.

## 1.5 Some useful results

In this section we present a list of useful results in mathematical analysis that involve the $\epsilon$-$\delta$ arguments or are often used in such arguments.

### 1.5.1 Infinite sequence

1. *Limit of a sequence.* A sequence $a_n$, $n = 1, 2, \ldots$, converges to a limit $a$, denoted by $a_n \to a$ or $\lim_{n\to\infty} a_n = a$, if for any $\epsilon > 0$, there is $N \ge 1$ such that $|a_n - a| < \epsilon$ if $n \ge N$. Note that this definition applies to both a real-valued sequence and a vector-valued sequence, where for a vector $v = (v_k)_{1 \le k \le d}$, $|v|$ is defined as its Euclidean norm; that is, $|v| = (\sum_{k=1}^d v_k^2)^{1/2}$.

The above definition of convergence of a sequence involves the limit of the sequence. Sometimes the limit is unknown, and it would be nice if one could judge the convergence by the sequence itself [i.e., without relying on the (unknown) limit]. A well-known criterion for convergence is the following.

2. *Cauchy criterion.* The sequence $a_n$, $n = 1, 2, \ldots$, is a *Cauchy sequence* if for any $\epsilon > 0$, there is $N \ge 1$ such that $|a_{n+k} - a_n| < \epsilon$ for any $n \ge N$ and $k \ge 1$. The sequence $a_n$, $n = 1, 2, \ldots$, converges if and only if it is a Cauchy sequence.

The following results can be established using either the definition of convergence or the Cauchy criterion (Exercises 1.16 and 1.17).

*3. Monotone sequence.* A sequence $a_n$, $n = 1, 2, \ldots$, is increasing if $a_n \leq a_{n+1}$ for any $n$; it is decreasing if $a_n \geq a_{n+1}$ for any $n$. Increasing or decreasing sequences are called monotone sequences. Every monotone sequence is convergent, provided that it is bounded. More specifically, if the sequence $a_n$ is increasing, then $\lim_{n \to \infty} a_n = \sup_{n \geq 1} a_n$, provided that the latter is finite; if $a_n$ is decreasing, then $\lim_{n \to \infty} a_n = \inf_{n \geq 1} a_n$, provided that the latter is finite.

*4. Convergent subsquence.* Every bounded sequence has a convergent subsequence.

*5. Upper and lower limits.* Let $a_n$, $n = 1, 2, \ldots$, be a sequence. The upper limit of the sequence, denoted by $\lim \sup a_n$, is defined as the largest limit point of $a_n$. Note that $\lim \sup a_n$ is always well defined according to the above result on convergent subsequence, provided that $a_n$ is bounded. In such a case, since the supremum of all the limit points of $a_n$ is also a limit point, the largest limit point always exists. Similarly, the lower limit of the sequence, denoted by $\lim \inf a_n$, is defined as the smallest limit point of $a_n$. The following are some properties of the upper and lower limits.

*5.1.* $\lim_{n \to \infty} a_n = a$ if and only if $\lim \sup a_n = \lim \inf a_n = a$.

*5.2.* Let the seqeuence $a_n$ be bounded. Then we have

$$\lim \sup(-a_n) = -\lim \inf a_n,$$
$$\lim \inf(-a_n) = -\lim \sup a_n.$$

*5.3.* Suppose that $a_n$ and $b_n$ are two sequences that are bounded. Then the following inequalities hold:

$$\lim \inf a_n + \lim \inf b_n \leq \lim \inf(a_n + b_n)$$
$$\leq \lim \inf a_n + \lim \sup b_n$$
$$\leq \lim \sup(a_n + b_n)$$
$$\leq \lim \sup a_n + \lim \sup b_n.$$

*6 (The argument of subsequences).* $\lim_{n \to \infty} a_n = a$ if and only if for any subsequence $a_{n_k}$, $k = 1, 2, \ldots$, there is a further subsequence $a_{n_{k_l}}$, $l = 1, 2, \ldots$, such that $\lim_{l \to \infty} a_{n_{k_l}} = a$.

### 1.5.2 Infinite series

*7. Convergence of a series.* The notation $\sum_{i=1}^{\infty} x_i$ represents an infinite series if it converges. The latter is defined as the existence of the limit $\lim_{n \to \infty} \sum_{i=1}^{n} x_i$. In other words, the infinite series $\sum_{i=1}^{\infty} x_i$ converges to $s$, denoted by $\sum_{i=1}^{\infty} x_i = s$, if for any $\epsilon > 0$, there is $N \geq 1$ such that $|\sum_{i=1}^{n} x_i - s| < \epsilon$ if $n \geq N$. Once again, this definition applies to both real-valued and vector-valued infinite series.

8. *Cauchy criterion for convergence of infinite series.* The infinite series $\sum_{i=1}^{\infty} x_i$ converges if and only if for any $\epsilon > 0$, there is $N \geq 1$ such that $\left| \sum_{i=n+1}^{n+k} x_i \right| < \epsilon$ for any $n \geq N$ and $k \geq 1$.

A test for convergence is a (sufficient) condition that ensures convergence of the infinite series. There are various tests for convergence. The following are some of them involving positive series. A series $\sum_{i=1}^{\infty} x_i$ is positive if $x_i > 0$, $i \geq 1$. These tests can be established, for example, using the Cauchy criterion (Exercises 1.18–1.20).

9. Suppose $\sum_{i=1}^{\infty} x_i$ and $\sum_{i=1}^{\infty} y_i$ are positive series such that $x_i \leq y_i$, $i \geq 1$. (i) if $\sum_{i=1}^{\infty} y_i$ is convergent, so is $\sum_{i=1}^{\infty} x_i$; (ii) conversely, if $\sum_{i=1}^{\infty} x_i$ is divergent (i.e., it is not convergent), so is $\sum_{i=1}^{\infty} y_i$.

10. Suppose $\sum_{i=1}^{\infty} x_i$ and $\sum_{i=1}^{\infty} y_i$ are positive series. If $\lim_{n \to \infty} (x_n/y_n) = r$, where $r \in (0, \infty)$, the two series are both convergent or both divergent.

11. Let $\sum_{i=1}^{\infty} x_i$ be a positive series. (i) If $\limsup_{n \to \infty} (x_{n+1}/x_n) < 1$, the series is convergent; (ii) if $\liminf_{n \to \infty} (x_{n+1}/x_n) > 1$, the series is divergent.

Note that no conclusion can be made if $\liminf_{n \to \infty} (x_{n+1}/x_n) \leq 1$ and $\limsup_{n \to \infty} (x_{n+1}/x_n) \geq 1$.

12. Let $\sum_{i=1}^{\infty} x_i$ be a positive series and $\rho = \limsup_{n \to \infty} (x_n^{1/n})$. (i) If $\rho < 1$, the series is convergent; (ii) if $\rho > 1$, the series is divergent.

Note that no conclusion can be made if $\rho = 1$.

For infinite series with positive and negative terms, we have the following result.

13. *Absolute convergence.* The infinite series $\sum_{i=1}^{\infty} x_i$ is absolutely convergent if $\sum_{i=1}^{\infty} |x_i|$ is convergent. Absolute convergence of an infinite series implies convergence of the infinite series.

### 1.5.3 Topology

14. *Neighborhood.* A neighborhood of $x \in R^d$ is defined as a subset of $R^d$ of the form $S(x, \epsilon) = \{y \in R^d : |y - x| < \epsilon\}$ for some $\epsilon > 0$.

15. *Open sets.* A subset $S \subset R^d$ is an open set if for every $x \in S$, there is $\epsilon > 0$ such that $S(x, \epsilon) \subset S$.

16. *Limit point of a set.* A point $x \in R^d$ is a limit point of a set $S \subset R^d$ if $S(x, \epsilon) \cap S \setminus \{x\} \neq \emptyset$ for every $\epsilon > 0$. In other words, every neighborhood of $x$ contains at least one point of $S$ that is different from $x$ (if $x \in S$).

17. *Closed sets.* A subset $S \subset R^d$ is a closed set if it contains every limit point of it.

The following fact can be used as an equivalent definition of a closed set.

18. A set $S$ is closed if and only if its complement, $S^c$, is open.

The following theorems, which are equivalent, are fundamental results in real analysis. An open cover of $S \subset R$ is a collection of open sets $\mathcal{S} = \{S_\alpha, \alpha \in A\}$ such that $S \subset \cup_{\alpha \in A} S_\alpha$. If a subcollection of $\mathcal{S}$, $\mathcal{S}_1$, is also an open cover of $S$, $\mathcal{S}_1$ is called a subcover. In particular, if $\mathcal{S}_1$ is a finite collection, it is called a finite subcover. Finally, a set $S \subset R$ is compact if every open cover of $S$ has a finite subcover.

*19. Heine–Borel theorem.* Every bounded closed subset of $R$ is compact.

*20. Bolzano–Weierstrass theorem.* Every bounded infinite subset of $R$ has a limit point.

For a proof of the equivalence of the Heine–Borel and Bolzano–Weierstrass theorems, see, for example, Khan and Thaheem (2000).

### 1.5.4 Continuity, differentiation, and intergration

For simplicity we consider real-valued functions defined on $R$.

*21.* A function $f(x)$ is continuous at $x = x_0$ if for every $\epsilon > 0$, there is $\delta > 0$ such that $|f(x) - f(x_0)| < \epsilon$ if $|x - x_0| < \delta$. The function $f(x)$ is continuous on $S \subset R$ if it is continuous at every $x \in S$.

Some important properties of continuous functions are the following.

*22.* If $f(x)$ is continuous at $x = x_0$ and $f(x_0) > 0$, there is a neighborhood $S(x_0, \delta)$ for some $\delta > 0$ such that $f(x) > 0$, $x \in S(x_0, \delta)$.

*23. The intermediate-value theorem.* If $f(x)$ is continuous on $[a, b]$, then for any $\lambda \in (A, B)$, where $A = f(a) \wedge f(b)$ and $B = f(a) \vee f(b)$, there is $c \in (a, b)$ such that $f(c) = \lambda$.

*24.* If $f(x)$ is continuous on $S$ and $S$ is closed and bounded (or compact according to the Heine–Borel theorem), then $f(x)$ is bounded on $S$. Furthermore, let $A = \inf_{x \in S} f(x)$ and $B = \sup_{x \in S} f(x)$. There are $x_1, x_2 \in S$, such that $f(x_1) = A$ and $f(x_2) = B$. In other words, $f(x)$ achieves its infimum and supremum on $S$.

*25. Uniform continuity.* The function $f(x)$ is uniformly continuous on $S$ if for any $\epsilon > 0$, there is $\delta > 0$ such that $|f(x_1) - f(x_2)| < \epsilon$ for any $x_1, x_2 \in S$, such that $|x_1 - x_2| < \delta$. If $f(x)$ is continuous on $S$ and $S$ is closed and bounded, then $f(x)$ is uniformly continuous on $S$.

*26. Differentiability of a function.* Let $f(x)$ be defined in a neighborhood of $x_0$, $S(x_0, \delta)$, for some $\delta > 0$. If the limit of

$$\frac{f(x_0 + h) - f(x_0)}{h}$$

exists as $h \to 0$, where $h \neq 0$ and $|h| < \delta$, $f(x)$ is differentiable at $x_0$ and its derivative at $x_0$ is denoted by

$$f'(x_0) = \lim_{h \to 0} \frac{f(x_0 + h) - f(x_0)}{h}.$$

If $f(x)$ is differentiable at every $x \in S$, then $f(x)$ is differentiable on $S$.

Some important properties of differentiable functions are the following.

*27.* If $f(x)$ is differentiable at $x_0$, $f(x)$ is continuous at $x_0$. In other words, differentiability implies continuity.

*28. Rolle's theorem.* Suppose that $f(x)$ is continuous on $[a, b]$ and differentiable on $(a, b)$, and $f(a) = f(b)$; then there is $c \in (a, b)$ such that $f'(c) = 0$.

A consequence of Rolle's theorem is the following theorem.

*29. The mean value theorem.* If $f(x)$ is continuous on $[a, b]$ and differentiable on $(a, b)$, there is $c \in (a, b)$ such that

$$f'(c) = \frac{f(b) - f(a)}{b - a}.$$

*30.* If $f(x)$ is increasing (decreasing) and differentiable on $(a, b)$, then $f'(x) \geq 0$ ($f'(x) \leq 0$), $x \in (a, b)$.

*31. Maxima and minima of a differentiable function.* The function $f(x)$ has a local maximum (minimum) at $x_0 \in S$ if there is $\delta > 0$ such that $f(x) \leq f(x_0)$ ($f(x) \geq f(x_0)$) for every $x \in S(x_0, \delta)$. If $f(x)$ is differentiable on $(a, b)$ and has a local maximum or local minimum at $x_* \in (a, b)$, then $f'(x_*) = 0$. In particular, if $f(x)$ is continuous on $[a, b]$ and differentiable on $(a, b)$ and there is $x_0 \in (a, b)$ such that $f(x_0) > f(a) \vee f(b)$ or $f(x_0) < f(a) \wedge f(b)$, then there is $x_* \in (a, b)$ such that $f'(x_*) = 0$ (Exercise 1.22).

For an extension of the above results to partial and higher order derivatives, see Chapter 4.

*32. Riemann integral.* Let $f(x)$ be a function defined on $[a, b]$. For any sequence $a < x_1 < \cdots < x_{n-1} < b$, let $m_i$ and $M_i$ denote the infimum and maximum of $f(x)$ on $[x_{i-1}, x_i]$, $1 \leq i \leq n$, where $x_0 = a$ and $x_n = b$. Then $f(x)$ is Riemann integrable on $[a, b]$ if for any $\epsilon > 0$, there is $\delta > 0$ such that

$$\sum_{i=1}^{n} (M_i - m_i)(x_i - x_{i-1}) < \epsilon$$

whenever $\max_{1 \leq i \leq n}(x_i - x_{i-1}) < \delta$. In this case, the integral $\int_a^b f(x) \, dx$ is defined as the limit of

$$\sum_{i=1}^{n} f(t_i)(x_i - x_{i-1})$$

as $\max_{1 \leq i \leq n}(x_i - x_{i-1}) \to 0$, where $t_i$ is any point in $[x_{i-1}, x_i]$, $1 \leq i \leq n$.

Some important properties of Riemann integrals are given below.

*33.* Any continuous function $f(x)$ on $[a, b]$ is Riemann integrable on $[a, b]$.

*34. The mean value theorem for integrals.* If $f(x)$ is continuous on $[a, b]$, then there is $c \in [a, b]$ such that

$$\frac{\int_a^b f(x) \, dx}{b - a} = f(c).$$

*35.* Let $f(x)$ be Riemann integrable on $[a, b]$. The following hold for

$$F(x) = \int_a^x f(t) \, dt.$$

(i) $F(x)$ is uniformly continuous on $[a, b]$;

(ii) if $f(x)$ is continuous on $[a, b]$, then $F(x)$ is differentiable on $(a, b)$ and $F'(x) = f(x)$, $x \in (a, b)$.

Result (ii) can actually be extended to $[a, b]$, provided that the derivatives of $F(x)$ at $a$ and $b$ are understood as the right and left derivatives, respectively, defined as follows:

$$F'_+(a) = \lim_{h>0, h\to 0} \frac{F(a+h) - F(a)}{h},$$

$$F'_-(b) = \lim_{h<0, h\to 0} \frac{F(b) - F(b-h)}{h}.$$

Two other types of integrals are also frequently used in mathematical statistics. The first is the Riemann–Stieltjes integral, which may be regarded as an extension of the Riemann integral. The second is the Lebesgue integral, which is defined in terms of measure theory. The latter sets up the foundation of probability and mathematical statistics.

*36. Riemann–Stieltjes integral.* An extension of the Riemann integral involving another function, $g(x)$, is the following. Let $g(x)$ be an increasing function on $[a, b]$. If we replace $x_i - x_{i-1}$ in the Riemann integral by $g(x_i) - g(x_{i-1})$, we have the definition of the Riemann–Stieltjes integral. Suppose that for any $\epsilon > 0$, there is $\delta > 0$ such that

$$\sum_{i=1}^{n} (M_i - m_i)\{g(x_i) - g(x_{i-1})\} < \epsilon$$

whenever $\max_{1\leq i\leq n}(x_i - x_{i-1}) < \delta$. $f(x)$ is said to be Riemann–Stieltjes integrable with respect to $g(x)$ on $[a, b]$. In this case, the Riemann-Stieltjes integral, denoted by $\int_a^b f(x)\, dg(x)$, is defined as the limit of

$$\sum_{i=1}^{n} f(t_i)\{g(x_i) - g(x_{i-1})\}$$

as $\max_{1\leq i\leq n}(x_i - x_{i-1}) \to 0$, where $t_i$ is any point in $[x_{i-1}, x_i]$, $1 \leq i \leq n$.

The definition of the Lebesgue integral through measure theory is deferred to Appendix A.2, so are those of Lebesgue measure and measurable functions used below. We conclude this section by pointing out an important connection between the Riemann integral and the Lebesgue integral.

*37.* A bounded measurable function $f(x)$ on $[a, b]$ is Riemann integrable if and only if the set of points at which $f(x)$ is discontinuous has Lebesgue measure zero, and in that case, the Riemann integral of $f(x)$ on $[a, b]$ is equal in value to its Lebesgue integral on $[a, b]$.

## 1.6 Exercises

*1.1.* Use the $\epsilon$-$\delta$ argument to show that for any $a \in (-\infty, \infty)$,

$$\left(1 + \frac{1}{n}\right)^a \longrightarrow 1$$

as $n \to \infty$.

1.2. Use the $\epsilon$-$\delta$ argument to show that

$$\left(1 + \frac{1}{n}\right)^n \longrightarrow e$$

as $n \to \infty$. (Hint: First prove the inequality $x - x^2/2 \le \log(1+x) \le x$, $x > 0$.)

1.3. Use the $\epsilon$-$\delta$ argument to show the following:

(a) $(1 + 1/n)^{\sqrt{n}} \to 1$ as $n \to \infty$.

(b) $(1 + 1/\sqrt{n})^n \to \infty$ as $n \to \infty$; in other words, $(1 + 1/\sqrt{n})^{-n} \to 0$, as $n \to \infty$.

1.4. The Student's $t$-distribution has extensive statistical applications. It is defined as a continuous distribution with the following pdf:

$$\phi(x|\nu) = \frac{\Gamma\{(\nu+1)/2\}}{\sqrt{\nu\pi}\,\Gamma(\nu/2)}\left(1 + \frac{x^2}{\nu}\right)^{-(\nu+1)/2}, \qquad -\infty < x < \infty,$$

where $\nu$ is the degrees of freedom (d.f.) of the $t$-distribution. Show that the pdf of the t-distribution converges to that of the standard normal distribution as the d.f. goes to infinity; that is,

$$\phi(x|\nu) \longrightarrow \phi(x) = \frac{1}{\sqrt{2\pi}}e^{-x^2/2}, \qquad -\infty < x < \infty$$

as $\nu \to \infty$.

1.5. A sequence $a_n$, $n = 0, 1, \ldots$, is defined as follows. Starting with initial values $a_0$ and $a_1$, let

$$a_{n+1} = \frac{3}{2}a_n - \frac{1}{2}a_{n-1}, \quad n = 1, 2, \ldots.$$

(a) Use Cauchy's criterion to show that the sequence converges.

(b) Find the limit of the sequence. Does the limit depend on the initial values $a_0$ and $a_1$?

1.6. Determine the ranges of $x$ for which each of the following infinite series converges, absolutely converges, and uniformly converges.

(a) $\sum_{n=1}^{\infty}\{(-1)^n/n4^n\}x^n$.

(b) $\sum_{n=0}^{\infty}(\log x)^n/n!$.

(c) $\sum_{n=1}^{\infty}\sin(n\pi x)/n(n+1)$.

1.7. The Riemann's $\zeta$-function is defined as the infinite series

$$\zeta(x) = \sum_{n=1}^{\infty}\frac{1}{n^x}.$$

(a) Show that $\zeta(x)$ is uniformly convergent for $x \in [a, \infty)$, where $a$ is any number greater than 1.

(b) Show that $\zeta(x)$ is continuous on $[a, \infty)$ for the same $a$.

(c) Is $\zeta(x)$ differentiable on $[a, \infty)$? If so, find an expression of $\zeta'(x)$ in terms of an infinite series.

*1.8.* Suppose that $X_1, \ldots, X_n$ are i.i.d. observations from the Uniform$[0, \theta)$ distribution, where $\theta > 0$ is an unknown parameter.

(a) Show that the MLE of $\theta$ does not exist.

(b) Find an estimator of $\theta$ that is consistent.

*1.9.* Suppose that $X$ is a continuous random variable with a unique median $a$ (see Example 1.5). Show that $P(X \leq x) < 1/2$, $x < a$, $P(X \leq a) = 1/2$, and $P(X \leq x) > 1/2$, $x > a$.

*1.10.* Using a similar argument as in Example 1.5 that led to (1.8), show that there are $N_3$ and $N_4$ such that when $n \geq N_3 \vee N_4$, we have $P\{X_{(m+1)} > a + \epsilon\} < 2\delta$.

*1.11.* Complete the second half of Example 1.5; that is, prove the consistency of the sample median for the case $n = 2m$.

*1.12.* Prove the following property of a cdf: A cdf $F$ can only have countably many discontinuity points.

*1.13.* Extend the proof of the Cramér consistency given in Section 1.4 to the case of multivariate observations and parameters; that is, $X_1, \ldots, X_n$ are i.i.d. vector-valued observations that have the common joint pdf $f(x|\theta)$, where $\theta$ is a vector-valued parameter with the parameter space $\Theta \in R^p$ $(p \geq 1)$.

*1.14.* In the proof of the Cramér consistency given in Section 1.4, show that $P_\theta(\sum_{i=1}^n Y_i > 0) \to 1$.

*1.15.* In the proof of the Wald consistency given in Section 1.4, show that $\sum_{i=1}^n Y_{k,i} \leq n\eta$, $1 \leq k \leq K + 1$, imply (1.12).

*1.16.* Use the $\epsilon$-$\delta$ argument to prove the monotone convergence criterion of §1.5.1.3.

*1.17.* Use the $\epsilon$-$\delta$ argument to prove the result on convergent subsequence of §1.5.1.4.

*1.18.* Establish the test for convergence §1.5.2.9.

*1.19.* Establish the test for convergence §1.5.2.10.

*1.20.* Establish the test for convergence §1.5.2.11.

*1.21.* Show that if $f(x)$ is continuous on $[a, b]$ and differentiable on $(a, b)$ and there is $x_0 \in (a, b)$ such that $f(x_0) > f(a) \vee f(b)$, or $f(x_0) < f(a) \wedge f(b)$, then there is $x_* \in (a, b)$ such that $f'(x_*) = 0$.

# 2

# Modes of Convergence

## 2.1 Introduction

In this chapter we discuss different types of convergence in probability and statistics. Types of convergence have already been introduced. They are convergence in probability and convergence in distribution. In addition, we introduce other types of convergence, such as almost sure convergence and $L^p$ convergence. We discuss properties of different types of convergence, the connections between them, and how to establish these properties. The discussion will mainly focus on the case of univariate random variables. However, most of the results presented here can be easily extended to the multivariate situation.

The concept of different types of convergence is critically important in mathematical statistics. In fact, misusage of such concepts often leads to confusions, even errors. The following is a simple example.

*Example 2.1* (Asymptotic variance). The well-known result of CLT states that, under regularity conditions, we have $\sqrt{n}(\bar{X} - \mu) \xrightarrow{\text{d}} N(0, \sigma^2)$, where $\sigma^2$ is called the asymptotic variance. The definition seems to be clear enough: $\sigma^2$ is the variance of the limiting normal distribution. Even so, some confusion still arises, and the following are some of them.

(a) $\sigma^2$ is the asymptotic variance of $\bar{X}$.
(b) $\lim_{n\to\infty} n\text{var}(\bar{X}) \to \sigma^2$ as $n \to \infty$.
(c) $n(\bar{X} - \mu)^2 \to \sigma^2$ as $n \to \infty$.

Statement (a) is clearly incorrect. It would be more appropriate to say that $\sigma^2$ is the asymptotic variance of $\sqrt{n}\bar{X}$; however, this does not mean that $\lim_{n\to\infty} \text{var}(\sqrt{n}\bar{X}) \to \sigma^2$, as $n \to \infty$, or Statement (b). In fact, convergence in distribution and convergence of the variance (which is essentially the convergence of moments) are two different concepts, and they do not imply each other. In some cases, even if the variance does not exist, the CLT still holds (e.g., Ibragimov and Linnik 1971, pp. 85, Theorem 2.6.3). As for Statement (c), it is not clear in what sense the convergence is. Even if the latter is made

J. Jiang, *Large Sample Techniques for Statistics*,
DOI 10.1007/978-1-4419-6827-2_2, © Springer Science+Business Media, LLC 2010

clear, say, in probability, it is still a wrong statement because, according to the CLT, $n(\bar{X} - \mu)^2 = \{\sqrt{n}(\bar{X} - \mu)\}^2$ converges in distribution to $\sigma^2 \chi_1^2$, where $\chi_1^2$ is the $\chi^2$-distribution with one degree of freedom. Since the latter is a random variable, not a constant, Statement (c) is incorrect even in the sense of convergence in probability.

In a way, the problem associated with Statement (c) is very similar to the second example in the Preface regarding approximation to a mean.

## 2.2 Convergence in probability

For the sake of completeness, here is the definition once again. A sequence of random variables $\xi_1, \xi_2, \ldots$ converges in probability to a random variable $\xi$, denoted by $\xi_n \xrightarrow{P} \xi$, if for any $\epsilon > 0$, we have $P(|\xi_n - \xi| > \epsilon) \to 0$ as $n \to \infty$.

It should be pointed out that, more precisely, convergence in probability is a property about the distributions of the random variables $\xi_1, \xi_2, \ldots, \xi$ rather than the random variables themselves. In particular, convergence in probability does not imply that the sequence $\xi_1, \xi_2, \ldots$ converges pointwisely at all. For example, consider the following.

*Example 2.2.* Define the sequence of random variables $\xi_n = \xi_n(x)$, $x \in [0, 1]$, which is the probability space with the probability being the Lebesgue measure (see Appendix A.2), as follows.

$$\xi_1(x) = \begin{cases} 1, & x \in [0, 1/2) \\ 0, & x \in [1/2, 1]; \end{cases}$$

$$\xi_2(x) = \begin{cases} 0, & x \in [0, 1/2) \\ 1, & x \in [1/2, 1]; \end{cases}$$

$$\xi_3(x) = \begin{cases} 1, & x \in [0, 1/4) \\ 0, & x \in [0, 1] \setminus [0, 1/4); \end{cases}$$

$$\xi_4(x) = \begin{cases} 1, & x \in [1/4, 1/2) \\ 0, & x \in [0, 1] \setminus [1/4, 1/2); \end{cases}$$

$$\xi_5(x) = \begin{cases} 1, & x \in [1/2, 3/4) \\ 0, & x \in [0, 1] \setminus [1/2, 3/4); \end{cases}$$

$$\xi_6(x) = \begin{cases} 1, & x \in [3/4, 1] \\ 0, & x \in [0, 1] \setminus [3/4, 1], \end{cases}$$

and so forth (see Figure 2.1). It can be shown that $\xi_n \xrightarrow{P} 0$ as $n \to \infty$; however, $\xi_n(x)$ does not converge pointwisely at any $x \in [0, 1]$ (Exercise 2.1).

So, what does convergence in probability really mean after all? It means that the overall probability that $\xi_n$ is not close to $\xi$ goes to zero as $n$ increases, and nothing more than that. We consider another example.

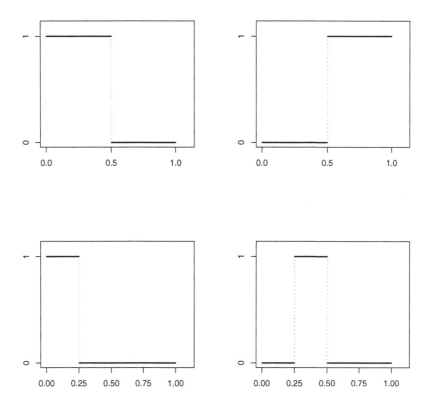

**Fig. 2.1.** *A plot of the random variables in Example 2.2*

*Example 2.3.* Suppose that $\xi_n$ is uniformly distributed over the intervals

$$\left[ i - \frac{1}{2n^2}, i + \frac{1}{2n^2} \right], \quad i = 1, \ldots, n.$$

Then the sequence $\xi_n$, $n \geq 1$, converges in probability to zero. To see this, note that the pdf of $\xi_n$ is given by

$$f_n(x) = \begin{cases} n, x \in [i - 1/2n^2, i + 1/2n^2], \ 1 \leq i \leq n \\ 0, \text{ elsewhere.} \end{cases}$$

It follows that for any $\epsilon > 0$, $\mathrm{P}(|\xi_n| > \epsilon) = 1/n \to 0$, as $n \to \infty$; hence, $\xi_n \xrightarrow{\mathrm{P}} 0$. The striking thing about this example is that, as $n \to \infty$, the height of the density function actually approaches infinity. Meanwhile, the total area in which the density is nonzero approaches zero as $n \to \infty$, which is what counts in the convergence in probability of the sequence (see Figure 2.2).

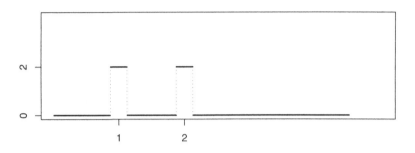

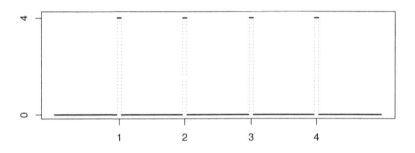

**Fig. 2.2.** *A plot of the pdfs of the random variables in Example 2.3*

The follow theorems provide useful sufficient conditions for convergence in probability.

**Theorem 2.1.** Suppose that $E(|\xi_n - \xi|^p) \to 0$ as $n \to \infty$ for some $p > 0$. Then $\xi_n \xrightarrow{P} \xi$ as $n \to \infty$.

The proof follows from the Chebyshev's inequality (Exercise 2.2).

**Theorem 2.2.** Suppose that $\xi_n = a_n \eta_n + b_n$, where $a_n$ and $b_n$ are non-random sequences such that $a_n \to a$, $b_n \to b$ as $n \to \infty$, and $\eta_n$ is a sequence of random variables such that $\eta_n \xrightarrow{P} \eta$ as $n \to \infty$. Then $\xi_n \xrightarrow{P} \xi = a\eta + b$ as $n \to \infty$.

**Theorem 2.3.** Suppose that $\xi_n \xrightarrow{P} \xi$ and $\eta_n \xrightarrow{P} \eta$ as $n \to \infty$. Then $\xi_n + \eta_n \xrightarrow{P} \xi + \eta$ as $n \to \infty$.

**Theorem 2.4.** Suppose that $\xi_n \xrightarrow{P} \xi$ and $\xi$ is positive with probability 1. Then $\xi_n^{-1} \xrightarrow{P} \xi^{-1}$ as $n \to \infty$.

The proofs of Theorems 2.2–2.4 are left to the readers as exercises (Exercises 2.3 2.5).

An important property of convergence in probability is the following. The sequence $\xi_n$, $n = 1, 2, \ldots$, is called bounded in probability if for any $\epsilon > 0$, there is $M > 0$ such that $P(|\xi_n| \leq M) > 1 - \epsilon$ for any $n \geq 1$.

**Theorem 2.5.** If $\xi_n$, $n = 1, 2, \ldots$, converges in probability, then the sequence is bounded in probability.

*Proof.* Suppose that $\xi_n \xrightarrow{P} \xi$ for some random variable $\xi$. Then for any $\epsilon > 0$, there is $B > 0$ such that $P(|\xi| \leq B) > 1 - \epsilon$ (see Example A.5). On the other hand, by convergence in probability, there is $N \geq 1$ such that when $n \geq N$, we have $P(|\xi_n - \xi| \leq 1) > 1 - \epsilon$. It follows that

$$P(|\xi_n| \leq B + 1) \geq P(|\xi_n - \xi| \leq 1, |\xi| \leq B)$$
$$> 1 - 2\epsilon, \quad n \geq N.$$

Now, let $\eta$ be the random variable $\max_{1 \leq n \leq N-1} |\xi_n|$. According to Example A.5, there is a constant $A > 0$ such that $P(\eta \leq A) > 1 - 2\epsilon$. Let $M = A \vee (B + 1)$. Then we have $P(|\xi_n| \leq M) > 1 - 2\epsilon$, $n \geq 1$. Since $\epsilon$ is arbitrary, this completes the proof. Q.E.D.

With the help of Theorem 2.5 it is easy to establish the following result (Exercise 2.6).

**Theorem 2.6.** Suppose that $\xi_n \xrightarrow{P} \xi$ and $\eta_n \xrightarrow{P} \eta$ as $n \to \infty$. Then $\xi_n \eta_n \xrightarrow{P} \xi\eta$ as $n \to \infty$.

## 2.3 Almost sure convergence

A sequence of random variables $\xi_n$, $n = 1, 2, \ldots$, converges almost surely to a random variable $\xi$, denoted by $\xi_n \xrightarrow{a.s.} \xi$ if $P(\lim_{n \to \infty} \xi_n = \xi) = 1$.

Almost sure convergence is a stronger property than convergence in probability, as the following theorem shows.

**Theorem 2.7.** $\xi_n \xrightarrow{a.s.} \xi$ implies $\xi_n \xrightarrow{P} \xi$.

The proof follows from the following lemma whose proof is a good exercise of the $\epsilon$-$\delta$ argument discussed in Chapter 1 (Exercise 2.11).

**Lemma 2.1.** $\xi_n \xrightarrow{\text{a.s.}} \xi$ if and only if for every $\epsilon > 0$,

$$\lim_{N \to \infty} P\left(\cup_{n=N}^{\infty}\{|\xi_n - \xi| \geq \epsilon\}\right) = 0. \tag{2.1}$$

On the other hand, Example 2.2 shows that there are sequences of random variables that converge in probability but not almost surely. We consider some more examples.

*Example 2.4.* Consider the same probability space $[0, 1]$ as in Example 2.2 but a different sequence of random variables $\xi_n$, $n = 1, 2, \ldots$, defined as follows: $\xi_n(i/n) = i$, $1 \leq i \leq n$, and $\xi_n(x) = 0$, if $x \in [0, 1] \setminus \{i/n, 1 \leq i \leq n\}$. Then $\xi_n \xrightarrow{\text{a.s.}} 0$ as $n \to \infty$. To see this, let $\mathcal{A} = \{i/n, i = 1, \ldots, n, n = 1, 2, \ldots\}$. Then $P(\mathcal{A}) = 0$ (note that P is the Lebesgue measure on $[0, 1]$). Furthermore, for any $x \in [0, 1] \setminus \mathcal{A}$, we have $\xi_n(x) = 0$ for any $n$; hence, $\xi_n(x) \to 0$ as $n \to \infty$. Therefore, $P(\lim_{n \to \infty} \xi_n = 0) \geq P([0, 1] \setminus \mathcal{A}) = 1$.

*Example 2.5.* Suppose that $X_i$ is a random variable with a Binomial$(i, p)$ distribution, $i = 1, 2, \ldots$, where $p \in [0, 1]$. Define

$$\xi_n = \sum_{i=1}^{n} \frac{X_i}{i^{2+\delta}}, \quad n = 1, 2, \ldots,$$

where $\delta > 0$. Then

$$\xi_n \xrightarrow{\text{a.s.}} \xi = \sum_{i=1}^{\infty} \frac{X_i}{i^{2+\delta}} \quad \text{as } n \to \infty. \tag{2.2}$$

To see this, note that $0 \leq X_i/i^{2+\delta} \leq i/i^{2+\delta} = 1/i^{1+\delta}$, and the infinite series $\sum_{i=1}^{\infty} 1/i^{1+\delta}$ converges. Therefore, by the result of §1.5.2.9 (i), the infinite series $\sum_{i=1}^{\infty} X_i/i^{2+\delta}$ always converges, which implies (2.2).

The following result is often useful in proving almost sure convergence.

**Theorem 2.8.** If, for every $\epsilon > 0$, we have $\sum_{n=1}^{\infty} P(|\xi_n - \xi| \geq \epsilon) < \infty$, then $\xi_n \xrightarrow{\text{a.s.}} \xi$ as $n \to \infty$.

*Proof.* By Lemma 2.1 we need to show (2.1). Since

$$P(\cup_{n=N}^{\infty}\{|\xi_n - \xi| \geq \epsilon\}) \leq \sum_{n=N}^{\infty} P(|\xi_n - \xi| \geq \epsilon),$$

and the latter converges to zero as $N \to \infty$, because the sequence $\sum_{n=1}^{\infty} P(|\xi_n - \xi| \geq \epsilon)$ is convergent, the result follows. Q.E.D.

*Example 2.6.* In Example 1.4 we showed consistency of the MLE in the Uniform distribution, that is, $\hat{\theta} \xrightarrow{P} \theta$ as $n \to \infty$, where $\hat{\theta} = X_{(n)}$ and $X_1, \ldots, X_n$ are i.i.d. observations from the Uniform$[0, \theta]$ distribution. We now show that, in fact, $\hat{\theta} \xrightarrow{a.s.} \theta$ as $n \to \infty$. For any $\epsilon > 0$, we have

$$
\begin{aligned}
P\{|X_{(n)} - \theta| \geq \epsilon\} &= P\{X_{(n)} \leq \theta - \epsilon\} \\
&= P(X_1 < \theta - \epsilon, \ldots, X_n \leq \theta - \epsilon) \\
&= [P(X_1 \leq \theta - \epsilon\}^n \\
&= \left(1 - \frac{\epsilon}{\theta}\right)^n.
\end{aligned}
$$

Thus, we have

$$
\begin{aligned}
\sum_{n=1}^{\infty} P\{|X_{(n)} - \theta| \geq \epsilon\} &= \sum_{n=1}^{\infty} \left(1 - \frac{\epsilon}{\theta}\right)^n \\
&= \frac{\theta - \epsilon}{\epsilon} < \infty.
\end{aligned}
$$

Here, we assume, without loss of generality, that $\epsilon < \theta$. It follows by Theorem 2.8 that $X_{(n)} \xrightarrow{a.s.} \theta$ as $n \to \infty$.

The following example is known as the bounded strong law of large numbers, which is a special case of the strong law of large numbers (SLLN; see Chapter 6).

*Example 2.7.* Suppose that $X_1, \ldots, X_n$ are i.i.d. and $|X_i| \leq b$ for some constant $b$. Then

$$
\xi_n = \frac{1}{n} \sum_{i=1}^{n} X_i \xrightarrow{a.s.} E(X_1) \tag{2.3}
$$

as $n \to \infty$. To show (2.3), note that, for any $\epsilon > 0$,

$$
\begin{aligned}
P\{|\xi_n - E(X_1)| \geq \epsilon\} &= P\left\{\frac{1}{n}\sum_{i=1}^{n} X_i - E(X_1) \geq \epsilon\right\} \\
&\quad + P\left\{\frac{1}{n}\sum_{i=1}^{n} X_i - E(X_1) \leq -\epsilon\right\} \\
&= I_1 + I_2. \tag{2.4}
\end{aligned}
$$

Furthermore, we have, by Chebyshev's inequality (see Section 5.2),

$$
\begin{aligned}
I_1 &= P\left[\sum_{i=1}^{n}\left\{\frac{X_i - E(X_1)}{\sqrt{n}}\right\} \geq \epsilon\sqrt{n}\right] \\
&= P\left(\exp\left[\sum_{i=1}^{n}\left\{\frac{X_i - E(X_1)}{\sqrt{n}}\right\}\right] \geq e^{\epsilon\sqrt{n}}\right)
\end{aligned}
$$

$$\leq e^{-\epsilon\sqrt{n}}\mathrm{E}\left(\exp\left[\sum_{i=1}^{n}\left\{\frac{X_i - \mathrm{E}(X_1)}{\sqrt{n}}\right\}\right]\right)$$

$$= e^{-\epsilon\sqrt{n}}\mathrm{E}\left[\prod_{i=1}^{n}\exp\left\{\frac{X_i - \mathrm{E}(X_1)}{\sqrt{n}}\right\}\right]$$

$$= e^{-\epsilon\sqrt{n}}\left(\mathrm{E}\left[\exp\left\{\frac{X_1 - \mathrm{E}(X_1)}{\sqrt{n}}\right\}\right]\right)^{n}. \tag{2.5}$$

By Taylor's expansion (see Section 4.1), we have, for any $x \in R$,

$$e^x = 1 + x + \frac{e^{\lambda x}}{2}x^2$$

for some $0 \leq \lambda \leq 1$. It follows that $e^x \leq 1 + x + (e^c/2)x^2$ if $|x| \leq c$. Since $|\{X_1 - \mathrm{E}(X_1)\}/\sqrt{n}| \leq 2b/\sqrt{n} \leq 2b$, by letting $c = 2b$ we have

$$\exp\left\{\frac{X_1 - \mathrm{E}(X_1)}{\sqrt{n}}\right\} \leq 1 + \frac{X_1 - \mathrm{E}(X_1)}{\sqrt{n}} + \frac{e^{2b}}{2}\left\{\frac{X_1 - \mathrm{E}(X_1)}{\sqrt{n}}\right\}^2$$

$$\leq 1 + \frac{X_1 - \mathrm{E}(X_1)}{\sqrt{n}} + \frac{2b^2 e^{2b}}{n}$$

(because $|X_1 - \mathrm{E}(X_1)| \leq 2b$); hence,

$$\mathrm{E}\left[\exp\left\{\frac{X_1 - \mathrm{E}(X_1)}{\sqrt{n}}\right\}\right] \leq 1 + \frac{2b^2 e^{2b}}{n}$$

$$\leq \exp\left(\frac{2b^2 e^{2b}}{n}\right) \tag{2.6}$$

using the inequality $e^x \geq 1 + x$ for all $x \geq 0$. By (2.5) and (2.6), we have $I_1 \leq ce^{-\epsilon\sqrt{n}}$, where $c = \exp(2b^2 e^{2b})$. By similar arguments, it can be shown that $I_2 \leq ce^{-\epsilon\sqrt{n}}$ (Exercise 2.12). Therefore, by (2.4), we have $\mathrm{P}(|\xi_n - \mathrm{E}(X_1)| \geq \epsilon) \leq 2ce^{-\epsilon\sqrt{n}}$. The almost sure convergence of $\xi_n$ to $\mathrm{E}(X_1)$ then follows from Theorem 1.8, because $\sum_{i=1}^{\infty} e^{-\epsilon\sqrt{n}} < \infty$ (Exercise 2.13).

## 2.4 Convergence in distribution

Convergence in distribution is another concept that was introduced earlier. Again, for the sake of completeness we repeat the definition here. A sequence of random variables $\xi_1, \xi_2, \ldots$ converges in distribution to a random variable $\xi$, denoted by $\xi_n \xrightarrow{d} \xi$, if $F_n \xrightarrow{w} F$, where $F_n$ is the cdf of $\xi_n$ and $F$ is the cdf of $\xi$. The latter means that $F_n(x) \to F(x)$ as $n \to \infty$ for every $x$ at which $F(x)$ is continuous.

Note that convergence in distribution is a property of the distribution of $\xi_n$ rather than $\xi_n$ itself. In particular, convergence in distribution does not imply almost sure convergence or even convergence in probability.

*Example 2.8.* Let $\xi$ be a random variable that has the standard normal distribution, $N(0,1)$. Let $\xi_1 = \xi$, $\xi_2 = -\xi$, $\xi_3 = \xi$, $\xi_4 = -\xi$, and so forth. Then, clearly, $\xi_n \xrightarrow{d} \xi$ (because $\xi$ and $-\xi$ have the same distribution). On the other hand, $\xi_n$ does not converge in probability to $\xi$ or any other random variable $\eta$. To see this, suppose that $\xi_n \xrightarrow{P} \eta$ for some random variable $\eta$. Then we must have $P(|\xi_n - \eta| > 1) \to 0$ as $n \to \infty$. Therefore, we have

$$P(|\xi - \eta| > 1) = P(|\xi_{2k-1} - \eta| > 1) \to 0, \tag{2.7}$$
$$P(|\xi + \eta| > 1) = P(|\xi_{2k} - \eta| > 1) \to 0 \tag{2.8}$$

as $k \to \infty$. Because the left sides of (2.7) and (2.8) do not depend on $k$, we must have $P(|\xi - \eta| > 1) = 0$ and $P(|\xi + \eta| > 1) = 0$. Then because $|2\xi| \le |\xi - \eta| + |\xi + \eta|$, $|\xi| > 1$ implies $|2\xi| > 2$, which, in turn, implies that either $|\xi - \eta| > 1$ or $|\xi + \eta| > 1$. It follows that $P(|\xi| > 1) \le P(|\xi - \eta| > 1) + P(|\xi + \eta| > 1) = 0$, which is, of course, not true.

Since almost sure convergence implies convergence in probability (Theorem 2.7), the sequence $\xi_n$ in Example 2.8 also does not converge almost surely. Nevertheless, the fact that the distribution of $\xi_n$ is the same for any $n$ is enough to imply convergence in distribution.

On the other hand, the following theorem shows that convergence in probability indeed implies convergence in distribution, so the former is a stronger convergent property than the latter.

**Theorem 2.9.** $\xi_n \xrightarrow{P} \xi$ implies $\xi_n \xrightarrow{d} \xi$.

*Proof.* Let $x$ be a continuity point of $F(x)$. We need to show that $P(\xi_n \le x) = F_n(x) \to F(x) = P(\xi \le x)$. For any $\epsilon > 0$, we have

$$\begin{aligned}
F(x - \epsilon) &= P(\xi \le x - \epsilon) \\
&= P(\xi \le x - \epsilon, \xi_n \le x) + P(\xi \le x - \epsilon, \xi_n > x) \\
&\le P(\xi_n \le x) + P(|\xi_n - \xi| > \epsilon) \\
&= F_n(x) + P(|\xi_n - \xi| > \epsilon).
\end{aligned}$$

It follows by the results of §1.5.1.5 that

$$\begin{aligned}
F(x - \epsilon) &\le \liminf F_n(x) + \limsup P(|\xi_n - \xi| > \epsilon) \\
&= \liminf F_n(x).
\end{aligned}$$

By a similar argument, it can be shown that (Exercise 2.18)

$$F(x + \epsilon) \ge \limsup F_n(x).$$

Since $\epsilon$ is arbitrary and $F(x)$ is continuous at $x$, we have

$$\limsup F_n(x) \le F(x) \le \liminf F_n(x).$$

On the other hand, we always have $\liminf F_n(x) \leq \limsup F_n(x)$. Therefore, we have $\liminf F_n(x) = \limsup F_n(x) = F(x)$; hence, $\lim_{n\to\infty} F_n(x) = F(x)$ by the results of §1.5.1.2. This completes the proof. Q.E.D.

Although convergence in distribution can often be verified by the definition, the following theorems sometimes offer more convenient tools for establishing convergence in distribution.

Let $\xi$ be a random variable. The moment generating function (mgf) of $\xi$ is defined as

$$m_\xi(t) = \mathrm{E}(e^{t\xi}), \tag{2.9}$$

provided that the expectation exists; the characteristic function (cf) of $\xi$ is defined as

$$c_\xi(t) = \mathrm{E}(e^{it\xi}), \tag{2.10}$$

where $i = \sqrt{-1}$. Note that the mgf is defined at $t \in R$ for which the expectation (2.9) exists (i.e., finite). It is possible, however, that the expectation does not exist for any $t$ except $t = 0$ (Exercise 2.19). The latter is the one particular value of $t$ at which the mgf is always well defined. On the other hand, the cf is well defined for any $t \in R$. This is because $|e^{it\xi}| \leq 1$ by the properties of complex numbers (Exercise 2.20).

**Theorem 2.10.** Let $m_n(t)$ be the mgf of $\xi_n$, $n = 1, 2, \ldots$. Suppose that there is $\delta > 0$ such that $m_n(t) \to m(t)$ as $n \to \infty$ for all $t$ such that $|t| < \delta$, where $m(t)$ is the mgf of a random variable $\xi$; then $\xi_n \overset{d}{\longrightarrow} \xi$ as $n \to \infty$.

In other words, convergence of the mgf in a neighborhood of zero implies convergence in distribution. The following example shows that the converse of Theorem 2.10 is not true; that is, convergence in distribution does not necessarily imply convergence of the mgf in a neighborhood of zero.

*Example 2.9.* Suppose that $\xi_n$ has a $t$-distribution with $n$ degrees of freedom (i.e., $\xi_n \sim t_n$). Then it can be shown that $\xi_n \overset{d}{\longrightarrow} \xi \sim N(0, 1)$ as $n \to \infty$. However, $m_n(t) = \mathrm{E}(e^{t\xi_n}) = \infty$ for any $t \neq 0$, whereas the mgf of $\xi$ is given by $m(t) = e^{t^2/2}$, $t \in R$ (Exercise 2.21). Therefore, $m_n(t)$ does not converge to $m(t)$ for any $t \neq 0$.

On the other hand, convergence of the cf is indeed equivalent to convergence in distribution, as the following theorem shows.

**Theorem 2.11** (Lévy-Cramér continuity theorem). Let $c_n(t)$ be the cf of $\xi_n$, $n = 1, 2, \ldots$, and $c(t)$ be the cf of $\xi$. Then $\xi_n \overset{d}{\longrightarrow} \xi$ as $n \to \infty$ if and only if $c_n(t) \to c(t)$ as $n \to \infty$ for every $t \in R$.

The proof of Theorem 2.10 is based on the theory of Laplace transformation. Consider, for example, the case that $\xi$ is a continuous random variable that has the pdf $f_\xi(x)$ with respect to the Lebesgue measure (see Appendix A.2). Then

$$m_\xi(t) = \int_{-\infty}^{\infty} e^{tx} f_\xi(x)\, dx, \tag{2.11}$$

which is the Laplace transformation of $f(x)$. A nice property of the Laplace transformation is its uniqueness. This means that if (2.11) holds for all $t$ such that $|t| < \delta$, where $\delta > 0$, then there is one and only one $f_\xi(x)$ that satisfies (2.11). Given this property, it is not surprising that Theorem 2.10 holds, and this actually outlines the main idea of the proof. The idea behind the proof of Theorem 2.11 is similar. We omit the details of both proofs, which are technical in nature (e.g., Feller 1971).

The following properties of the mgf and cf are often useful. The proofs are left as exercises (Exercises 2.22, 2.23).

**Lemma 2.2.** (i) Let $\xi$ be a random variable. Then, for any constants $a$ and $b$, we have

$$m_{a\xi+b}(t) = e^{bt} m_\xi(at),$$

provided that the $m_\xi(at)$ eixsts. (ii) Let $\xi$, $\eta$ be independent random variables. Then we have

$$m_{\xi+\eta}(t) = m_\xi(t) m_\eta(t), \quad |t| \leq \delta,$$

provided that both $m_\xi(t)$ and $m_\eta(t)$ exist.

**Lemma 2.3.** (i) Let $\xi$ be a random variable. Then, for any constants $a$ and $b$, we have

$$c_{a\xi+b}(t) = e^{ibt} c_\xi(at), \quad t \in R.$$

(ii) Let $\xi$ and $\eta$ be independent random variables. Then we have

$$c_{\xi+\eta}(t) = c_\xi(t) c_\eta(t), \quad t \in R.$$

We consider some examples.

*Example 2.10* (Poisson approximation to Binomial). Suppose that $\xi_n$ has a Binomial$(n, p_n)$ distribution such that as $n \to \infty$, $np_n \to \lambda$. It can be shown that the mgf of $\xi_n$ is given by

$$m_n(t) = (p_n e^t + 1 - p_n)^n,$$

which converges to $\exp\{\lambda(e^t - 1)\}$ as $n \to \infty$ for any $t \in R$ (Exercise 2.24). On the other hand, $\exp\{\lambda(e^t - 1)\}$ is the mgf of $\xi \sim \text{Poisson}(\lambda)$. Therefore, by

Theorem 2.10, we have $\xi_n \xrightarrow{d} \xi$ as $n \to \infty$. This justifies an approximation that is often taught in elementary statistics courses; that is, the Binomial$(n, p)$ distribution can be approximated by the Poisson$(\lambda)$ distribution, provided that $n$ is large, $p$ is small, and $np$ is approximately equal to $\lambda$.

*Example 2.11.* The classical CLT may be interpreted as, under regularity conditions, the sample mean of i.i.d. observations, $X_1, \ldots, X_n$, is asymptotically normal. This sometimes leads to the impression that as $n \to \infty$ (and with a suitable normalization), the limiting distribution of

$$\bar{X} = \frac{X_1 + \cdots + X_n}{n}$$

is always normal. However, this is not true. To see a counterexample, suppose that $X_1, \ldots, X_n$ are i.i.d. with the pdf

$$f(x) = \frac{1 - \cos(x)}{\pi x^2}, \quad -\infty < x < \infty.$$

Note that the mgf of $X_i$ does not exist for any $t \neq 0$. However, the cf of $X_i$ is given by $\max(1 - |t|, 0)$, $t \in R$ (Exercise 2.25). Furthermore, by Lemma 2.3 it can be shown that the cf of $\bar{X}$ is given by

$$\left\{ \max\left(1 - \frac{|t|}{n}, 0\right) \right\}^n,$$

which converges to $e^{-|t|}$ as $n \to \infty$ (Exercise 2.25). However, the latter is the cf of the Cauchy$(0, 1)$ distribution. Therefore, in this case, the sample mean is asymptotically Cauchy instead of asymptotically normal. The violation of the CLT is due to the failure of the regularity conditions—namely, that $X_i$ has finite expectation (and variance; see Section 6.4 for details).

In many cases, convergence in distribution of a sequence can be derived from the convergence in distribution of another sequence. We conclude this section with some useful results of this type.

**Theorem 2.12** (Continuous mapping theorem). Suppose that $\xi_n \xrightarrow{d} \xi$ as $n \to \infty$ and that $g$ is a continuous function. Then $g(\xi_n) \xrightarrow{d} g(\xi)$ as $n \to \infty$.

The proof is omitted (e.g., Billingsley 1995, §5). Alternatively, Theorem 2.12 can be derived from Theorem 2.18 given in Section 2.7 (Exercise 2.27).

**Theorem 2.13** (Slutsky's theorem). Suppose that $\xi_n \xrightarrow{d} \xi$ and $\eta_n \xrightarrow{P} c$, as $n \to \infty$, where $c$ is a constant. Then (i) $\xi_n + \eta_n \xrightarrow{d} \xi + c$, and (ii) $\xi_n \eta_n \xrightarrow{d} c\xi$ as $n \to \infty$.

The proof is left as an exercises (Exercise 2.26).

The next result involves an extension of convergence in distribution to the multivariate case. Let $\xi = (\xi_1, \ldots, \xi_k)$ be a random vector. The cdf of $\xi$ is defined as

$$F(x_1, \ldots, x_k) = P(\xi_1 \leq x_1, \ldots, \xi_k \leq x_k), \quad x_1, \ldots, x_k \in R.$$

A sequence of random vectors $\xi_n$, $n = 1, 2, \ldots$, converges in distribution to a random vector $\xi$, denoted by $\xi_n \xrightarrow{d} \xi$, if the cdf of $\xi_n$ converges to the cdf of $\xi$, denoted by $F$, at every continuity point of $F$.

**Theorem 2.14.** Let $\xi_n$, $n = 1, 2, \ldots$, be a sequence of $d$-dimensional random vectors. Then $\xi_n \xrightarrow{d} \xi$ as $n \to \infty$ if and only if $a'\xi_n \xrightarrow{d} a'\xi$ as $n \to \infty$ for every $a \in R^d$.

## 2.5 $L^p$ convergence and related topics

Let $p$ be a positive number. A sequence of random variables $\xi_n$, $n = 1, 2, \ldots$ converges in $L^p$, to a random variable $\xi$, denoted by $\xi_n \xrightarrow{L^p} \xi$, if $E(|\xi_n - \xi|^p) \to 0$ as $n \to \infty$. $L^p$ convergence (for any $p > 0$) implies convergence in probability, as the following theorem states, which can be proved by applying Chebyshev's inequality (Exercise 2.30).

**Theorem 2.15.** $\xi_n \xrightarrow{L^p} \xi$ implies $\xi_n \xrightarrow{P} \xi$.

The converse, however, is not true, as the following example shows.

*Example 2.12.* Let $X$ be a random variable that has the following pdf with respect to the Lebesgue measure

$$f(x) = \frac{\log a}{x(\log x)^2}, \quad x \geq a,$$

where $a$ is a constant such that $a > 1$. Let $\xi_n = X/n$, $n = 1, 2, \ldots$. Then we have $\xi_n \xrightarrow{P} 0$, as $n \to \infty$. In fact, for any $\epsilon > 0$, we have

$$\begin{aligned}
P(|\xi_n| > \epsilon) &= P(X > n\epsilon) \\
&= \int_{n\epsilon}^{\infty} \frac{\log a}{x(\log x)^2} \, dx \\
&= \frac{\log a}{\log(n\epsilon)} \\
&\longrightarrow 0
\end{aligned}$$

as $n \to \infty$. On the other hand, for any $p > 0$, we have

$$E(|\xi|^p) = E\left(\frac{X}{n}\right)^p$$

$$= \frac{\log a}{n^p} \int_a^\infty \frac{dx}{x^{1-p}(\log x)^2}$$

$$= \infty;$$

so it is not true that $\xi_n \xrightarrow{L^p} 0$ as $n \to \infty$.

Note that in the above example the sequence $\xi_n$ converges in probability; yet it does not converge in $L^p$ for any $p > 0$. However, the following theorem states that, under an additional assumption, convergence in probability indeed implies $L^p$ convergence.

**Theorem 2.16** (Dominated convergence theorem). Suppose that $\xi_n \xrightarrow{P} \xi$ as $n \to \infty$, and there is a nonnegative random variable $\eta$ such that $E(\eta^p) < \infty$, and $|\xi_n| \le \eta$ for all $n$. Then $\xi_n \xrightarrow{L^p} \xi$ as $n \to \infty$.

The proof is based on the following lemma whose proof is omitted (e.g., Chow and Teicher 1988, §4.2).

**Lemma 2.4** (Fatou's lemma). Let $\eta_n$, $n = 1, 2, \ldots$, be a sequence of random variables such that $\eta_n \ge 0$, a.s.. Then

$$E(\liminf \eta_n) \le \liminf E(\eta_n).$$

*Proof of Theorem 2.16.* First, we consider a special case so that $\xi_n \xrightarrow{a.s.} \xi$. Then, $|\xi| = \lim_{n\to\infty} |\xi_n| \le \eta$, a.s. Consider $\eta_n = (2\eta)^p - |\xi_n - \xi|^p$. Since $|\xi_n - \xi| \le |\xi_n| + |\xi| \le 2\eta$, a.s., we have $\eta_n \ge 0$, a.s. Thus, by Lemma 2.4 and the results of §1.5.1.5, we have

$$(2\eta)^p = E(\liminf \eta_n)$$
$$\le \liminf E(\eta_n)$$
$$= \liminf\{(2\eta)^p - E(|\xi_n - \xi|^p)\}$$
$$\le (2\eta)^p - \limsup E(|\xi_n - \xi|^p),$$

which implies $\limsup E(|\xi_n - \xi|^p) \le 0$; hence, $E(|\xi_n - \xi|^p) \to 0$ as $n \to \infty$.

Now, we drop the assumption that $\xi_n \xrightarrow{a.s.} \xi$. We use the argument of subsequences (see §1.5.1.6). It suffices to show that for any subsequence $n_k$, $k = 1, 2, \ldots$, there is a further subsequence $n_{k_l}$, $l = 1, 2, \ldots$, such that

$$E\left(\left|\xi_{n_{k_l}} - \xi\right|^p\right) \longrightarrow 0 \tag{2.12}$$

as $l \to \infty$. Since $\xi_n \xrightarrow{P} \xi$, so does the subsequence $\xi_{n_k}$. Then, according to a result given later in Section 2.7 (see §2.7.2), there is a further subsequence

$n_{k_l}$ such that $\xi_{n_{k_l}} \xrightarrow{\text{a.s.}} \xi$ as $l \to \infty$. Result (2.12) then follows from the proof given above assuming a.s. convergence. This completes the proof. Q.E.D.

The dominated convergence theorem is a useful result that is often used to establish $L^p$ convergence given convergence in probability or a.s. convergence. We consider some examples.

*Example 2.13.* Let $X_1, \ldots, X_n$ be i.i.d. Bernoulli($p$) observations. The sample proportion (or binomial proportion)

$$\hat{p} = \frac{X_1 + \cdots + X_n}{n}$$

converges in probability to $p$ (it also converges a.s. according to the bounded SLLN; see Example 2.7). Since $|X_i| \le 1$, by Theorem 2.16, $\hat{p}$ converges to $p$ in $L^p$ for any $p > 0$.

*Example 2.14.* In Example 1.4 we showed that if $X_1, \ldots, X_n$ are i.i.d. observations from the Uniform$[0, \theta]$ distribution, then the MLE of $\theta$, $\hat{\theta} = X_{(n)}$, is consistent; that is $\hat{\theta} \xrightarrow{\text{P}} \theta$ as $n \to \infty$. Because $0 \le \hat{\theta} \le \theta$, Theorem 2.16 implies that $\hat{\theta}$ converges in $L^p$ to $\theta$ for any $p > 0$.

Another concept that is closely related to $L^p$ convergence is called uniform integrability. The sequence $\xi_n$, $n = 1, 2, \ldots$, is uniformly integrable in $L^p$ if

$$\lim_{a \to \infty} \sup_{n \ge 1} E\{|\xi_n|^p 1_{(|\xi_n|>a)}\} = 0. \tag{2.13}$$

**Theorem 2.17.** Suppose that $E(|\xi_n|^p) < \infty$, $n = 1, 2, \ldots$, and $\xi_n \xrightarrow{\text{P}} \xi$ as $n \to \infty$. Then the following are equivalent:
  (i) $\xi_n$, $n = 1, 2, \ldots$, is uniformly integrable in $L^p$;
  (ii) $\xi_n \xrightarrow{L^p} \xi$ as $n \to \infty$ with $E(|\xi|^p) < \infty$;
  (iii) $E(|\xi_n|^p) \to E(|\xi|^p) < \infty$, as $n \to \infty$.

*Proof.* (i) $\Rightarrow$ (ii): First, assume that $\xi_n \xrightarrow{\text{a.s.}} \xi$. Then, for any $a > 0$, the following equality holds almost surely:

$$|\xi|^p 1_{(|\xi|>a)} = \left\{ \lim_{n \to \infty} |\xi_n|^p 1_{(|\xi_n|>a)} \right\} 1_{(|\xi|>a)}.$$

To see this, note that if $|\xi| \le a$, both sides of the equation are zero; and if $|\xi| > a$, then $\xi_n \to \xi$ implies that $|\xi_n| > a$ for large $n$; hence, $|\xi_n|^p 1_{(|\xi_n|>a)} = |\xi_n|^p \to |\xi|^p$, which is the left side. Thus, by Fatou's lemma, we have

$$\begin{aligned}
E\{|\xi|^p 1_{(|\xi|>a)}\} &\le E\left\{ \lim_{n \to \infty} |\xi_n|^p 1_{(|\xi_n|>a)} \right\} \\
&\le \liminf E\{|\xi_n|^p 1_{(|\xi_n|>a)}\} \\
&\le \sup_{n \ge 1}\{|\xi_n|^p 1_{(|\xi_n|>a)}\}.
\end{aligned} \tag{2.14}$$

For any $\epsilon > 0$, choose $a$ such that (2.13) holds; hence, $\mathrm{E}\{|\xi|^p 1_{(|\xi|>a)}\} < \epsilon$ by (2.14). It follows that $\mathrm{E}(|\xi|^p) = \mathrm{E}\{|\xi|^p 1_{(|\xi|\le a)}\} + \mathrm{E}\{|\xi|^p 1_{(|\xi|>a)}\} \le |a|^p + \epsilon < \infty$. Furthermore, we have

$$
\begin{aligned}
|\xi_n - \xi|^p = {} & |\xi_n - \xi|^p 1_{(|\xi_n|\le a)} \\
& + |\xi_n - \xi|^p 1_{(|\xi_n|>a,|\xi|\le a)} \\
& + |\xi_n - \xi|^p 1_{(|\xi_n|>a,|\xi|>a)}.
\end{aligned} \tag{2.15}
$$

If $|\xi| \le a < |\xi_n|$, then $|\xi_n - \xi| \le |\xi_n| + |\xi| < 2|\xi_n|$; hence, the second term on the right side of (2.15) is bounded by $2^p |\xi_n|^p 1_{(|\xi_n|>a)}$. On the other hand, by the inequality

$$|u - v|^p \le 2^p(|u|^p + |v|^p), \quad u, v \in R \tag{2.16}$$

(Exercise 2.32), the third term on the right side of (2.15) is bounded by $2^p\{|\xi_n|^p 1_{(|\xi_n|>a)} + |\xi|^p 1_{(|\xi|>a)}\}$. Therefore, by (2.15), we have

$$
\begin{aligned}
\mathrm{E}(|\xi_n - \xi|^p) \le {} & \mathrm{E}\{|\xi_n - \xi|^p 1_{(|\xi_n|\le a)}\} \\
& + 2^{p+1}\mathrm{E}\{|\xi_n|^p 1_{(|\xi_n|>a)}\} + 2^p \mathrm{E}\{|\xi|^p 1_{(|\xi|>a)}\} \\
\le {} & \mathrm{E}\{|\xi_n - \xi|^p 1_{(|\xi_n|\le a)}\} + 3 \cdot 2^p \epsilon.
\end{aligned} \tag{2.17}
$$

Finally, we $|\xi_n - \xi|^p 1_{(|\xi_n|\le a)} \xrightarrow{\text{a.s.}} 0$ and $|\xi_n - \xi|^p 1_{(|\xi_n|\le u)} \le 2^p(a^p + |\xi|^p)$ by (2.16), and $\mathrm{E}(|\xi|^p) < \infty$ as is proved above. Thus, by the dominated convergence theorem, we have $\mathrm{E}\{|\xi_n - \xi|^p 1_{(|\xi_n|\le a)}\} \to 0$ as $n \to \infty$. It follows, by (2.17) and the results of §1.5.1.5, that

$$\limsup \mathrm{E}(|\xi_n - \xi|^p) \le 3 \cdot 2^p \epsilon.$$

Since $\epsilon$ is arbitrary, we have $\mathrm{E}(|\xi_n - \xi|^p) \to 0$ as $n \to \infty$.

We now drop the assumption that $\xi_n \xrightarrow{\text{a.s.}} \xi$. The result then follows by the argument of subsequences (Exercise 2.33).

(ii) $\Rightarrow$ (iii): For any $a > 0$, we have

$$
\begin{aligned}
|\xi_n|^p - |\xi|^p &= (|\xi_n|^p - |\xi|^p)1_{(|\xi_n|\le a)} + (|\xi_n|^p - |\xi|^p)1_{(|\xi_n|>a)} \\
&= \eta_n + \zeta_n.
\end{aligned} \tag{2.18}
$$

By (2.16), we have

$$
\begin{aligned}
|\zeta_n| &\le |\xi_n|^p 1_{(|\xi_n|>a)} + |\xi|^p 1_{(|\xi_n|>a)} \\
&\le 2^p(|\xi|^p + |\xi_n - \xi|^p)1_{(|\xi_n|>a)} + |\xi|^p 1_{(|\xi_n|>a)} \\
&\le (2^p + 1)|\xi|^p 1_{(|\xi|>a)} + 2^p|\xi_n - \xi|^p.
\end{aligned} \tag{2.19}
$$

Combining (2.18) and (2.19), we have

$$
\begin{aligned}
\mathrm{E}\left(||\xi_n|^p - |\xi|^p|\right) &\le \mathrm{E}(|\eta_n|) + (2^p + 1)\mathrm{E}\{|\xi|^p 1_{(|\xi|>a)}\} + 2^p \mathrm{E}(|\xi_n - \xi|^p) \\
&= I_1 + I_2 + I_3
\end{aligned}
$$

By Theorem 2.15, we have $\eta_n \xrightarrow{\text{P}} 0$; hence, by Theorem 2.16, we have $I_1 \to 0$ as $n \to \infty$. Also (ii) implies $I_3 \to 0$, as $n \to \infty$. Thus, we have (see §1.5.1.5)

$$\limsup \mathrm{E}\left(\left||\xi_n|^p - |\xi|^p\right|\right) \le (2^p + 1)\mathrm{E}\{|\xi|^p 1_{(|\xi|>a)}\}.$$

Note that $a$ is arbitrary and, by Theorem 2.16, it can be shown that

$$\mathrm{E}\{|\xi|^p 1_{(|\xi|>a)}\} \to 0 \quad \text{as } a \to \infty. \tag{2.20}$$

This implies $\mathrm{E}(||\xi_n|^p - |\xi|^p|) \to 0$, which implies $\mathrm{E}(|\xi_n|^p) \to \mathrm{E}(|\xi|^p)$ as $n \to \infty$ (Exercise 2.34).

(iii) $\Rightarrow$ (i): For any $a > 0$, we have

$$\mathrm{E}\{|\xi_n|^p 1_{(|\xi_n|>a)}\} = \mathrm{E}(|\xi_n|^p) - \mathrm{E}\{|\xi_n|^p 1_{(|\xi_n|\le a)}\}$$
$$\le \mathrm{E}(|\xi_n|^p) - \mathrm{E}\{|\xi_n|^p 1_{(|\xi_n|\le a, |\xi|<a)}\} = I_1 - I_2.$$

(iii) implies $I_1 \to \mathrm{E}(|\xi|^p)$. Furthermore, let $\eta_n = |\xi_n|^p 1_{(|\xi_n|\le a, |\xi|<a)}$. It can be shown (Exercise 2.35) that $\eta_n \xrightarrow{\text{P}} \eta = |\xi|^p 1_{(|\xi|<a)}$ as $n \to \infty$. In addition, we have $0 \le \eta_n \le a^p$. Thus, by Theorem 2.16, we have $\eta_n \xrightarrow{L^1} \eta$, which implies $I_2 = \mathrm{E}(\eta_n) \to \mathrm{E}(\eta)$. We now use the arguments of §1.5.1.5 to conclude that

$$\limsup \mathrm{E}\{|\xi_n|^p 1_{(|\xi_n|>a)}\} \le \mathrm{E}(|\xi|^p) - \mathrm{E}\{|\xi|^p 1_{(|\xi|<a)}\}$$
$$= \mathrm{E}\{|\xi|^p 1_{(|\xi|\ge a)}\}. \tag{2.21}$$

For any $\epsilon > 0$, by (2.21) and the definition of $\limsup$, there is $N \ge 1$ such that $\mathrm{E}\{|\xi_n|^p 1_{(|\xi_n|>a)}\} \le \mathrm{E}\{|\xi|^p 1_{(|\xi|\ge a)}\} + \epsilon$ if $n \ge N$. It follows that

$$\sup_{n\ge 1} \mathrm{E}\{|\xi_n|^p 1_{(|\xi_n|>a)}\} \le \left[\max_{1\le n\le N-1} \mathrm{E}\{|\xi_n|^p 1_{(|\xi_n|>a)}\}\right]$$
$$\vee \left[\mathrm{E}\{|\xi|^p 1_{(|\xi|\ge a)}\} + \epsilon\right].$$

Furthermore, by the dominated convergence theorem it can be shown that $\mathrm{E}\{|\xi_n|^p 1_{(|\xi_n|>a)}\} \to 0$, $1 \le n \le N-1$, and $\mathrm{E}\{|\xi|^p 1_{(|\xi|\ge a)}\} \to 0$ as $a \to \infty$ (see Exercise 2.34). Therefore, we have

$$\limsup_{} \sup_{n\ge 1} \mathrm{E}\{|\xi_n|^p 1_{(|\xi_n|>a)}\} \le \epsilon,$$

where the $\limsup$ is with respect to $a$. Since $\epsilon$ is arbitrary, we conclude that $\sup_{n\ge 1} \mathrm{E}\{|\xi_n|^p 1_{(|\xi_n|>a)}\} \to 0$ as $a \to \infty$. this completes the proof. Q.E.D.

*Example 2.15.* Suppose that $\xi_n \xrightarrow{\text{P}} \xi$ as $n \to \infty$, and that $\mathrm{E}(|\xi_n|^q)$, $n \ge 1$, is bounded for some $q > 0$. Then $\xi_n \xrightarrow{L^p} \xi$ as $n \to \infty$ for any $0 < p < q$. To see this, note that for any $a > 0$, $|\xi_n| > a$ implies $|\xi_n|^{p-q} < a^{p-q}$. Thus,

$$\mathrm{E}\{|\xi_n|^p 1_{|\xi_n|>a}\} \le a^{p-q}\mathrm{E}(|\xi_n|^q)$$
$$\le Ba^{p-q},$$

where $B = \sup_{n\geq 1} E(|\xi_n|^q) < \infty$. Because $p - q < 0$, we have

$$\sup_{n\geq 1} E\{|\xi_n|^p 1_{(|\xi_n|>a)}\} \to 0$$

as $a \to \infty$. In other words, $\xi_n$, $n = 1, 2, \ldots$, is uniformly integrable. The result then follows by Theorem 2.17.

*Example 2.16.* Let $X$ be a random variable that has a pdf $f(x)$ with respect to a $\sigma$-finite measure $\mu$ (see Appendix A.2). Suppose that $f_n(x)$, $n = 1, 2, \ldots$, is a sequence of pdf's with respect to $\mu$ such that $f_n(x) \to f(x)$, $x \in R$, as $n \to \infty$. Consider the sequence of random variables

$$\xi_n = \frac{f_n(X)}{f(X)}, \tag{2.22}$$

$n = 1, 2, \ldots$. Then we have $\xi_n \xrightarrow{L^1} 1$ as $n \to \infty$. To see this, note that $f_n(x) \to f(x)$, $x \in R$ implies $\xi_n \xrightarrow{a.s.} 1$. This is because $f_n(x)/f(x) \to 1$ as long as $f(x) > 0$; hence, $P(\xi_n \to 1) \geq P\{f(X) > 0\} = 1 - P\{f(X) = 0\}$ and

$$P\{f(X) = 0\} = \int_{f(x)=0} f(x)\, d\mu$$
$$= 0.$$

It follows by Theorem 2.7 that $\xi_n \xrightarrow{P} 1$. On the other hand, we have

$$E(|\xi_n|) = E\left\{\frac{f_n(X)}{f(X)}\right\}$$
$$= \int \frac{f_n(x)}{f(x)} f(x)\, d\mu$$
$$= \int f_n(x)\, d\mu$$
$$= 1.$$

Thus, by Theorem 2.17, we have $\xi_n \xrightarrow{L^1} 1$ as $n \to \infty$.

When $X$ is a vector of observations, (2.22) corresponds to a likelihood ratio, which may be thought as the probability of observing $X$ under $f_n$ divided by that under $f$. Thus, the above example indicates that if $f_n$ converges to $f$ pointwisely, then the likelihood ratio converges to 1 in $L^1$, provided that $f(x)$ is the true pdf of $X$. To see a specific example, suppose that $X$ has a standard normal distribution; that is, $X \sim f(x)$, where

$$f(x) = \frac{1}{\sqrt{2\pi}} e^{-x^2/2}, \quad -\infty < x < \infty.$$

Let $f_n(x)$ be the pdf of the $t$-distribution with $n$ degrees of freedom; that is,

$$f_n(x) = \frac{\Gamma\{(n+1)/2\}}{\sqrt{n\pi}\,\Gamma(n/2)}\left(1 + \frac{x^2}{n}\right)^{-(n+1)/2}, \quad -\infty < x < \infty.$$

Then, by Exercise 1.4, we have $f_n(x) \to f(x)$, $x \in R$, as $n \to \infty$. It follows that $f_n(X)/f(X) \xrightarrow{L^1} 1$ as $n \to \infty$. It should be pointed out that the $L^1$ convergence may not hold if $f(x)$ is not the true distribution of $X$, even if $f_n(x) \to f(x)$ for every $x$. For example, suppose that in the Example 2.16 involving the $t$-distribution, the distribution of $X$ is $N(0,2)$ instead of $N(0,1)$; then, clearly, we still have $f_n(x) \to f(x)$, $x \in R$ [$f(x)$ has not changed; only that $X \sim f(x)$ no longer holds]. However, it is not true that $f_n(X)/f(X) \xrightarrow{L^1} 1$. This is because, otherwise, by the inequality

$$\frac{f_n(X)}{f(X)} \le 1 + \left|\frac{f_n(X)}{f(X)} - 1\right|,$$

we would have

$$\mathrm{E}\left\{\frac{f_n(X)}{f(X)}\right\} \le 1 + \mathrm{E}\left(\left|\frac{f_n(X)}{f(X)} - 1\right|\right)$$
$$\le 2$$

for large $n$. However,

$$\mathrm{E}\left\{\frac{f_n(X)}{f(X)}\right\} = \int \frac{f_n(x)}{f(x)}\frac{1}{\sqrt{4\pi}}e^{-x^2/4}\,dx$$
$$= \frac{1}{\sqrt{2}}\int \frac{\Gamma\{(n+1)/2\}}{\sqrt{n\pi}\,\Gamma(n/2)}\left(1 + \frac{x^2}{n}\right)^{-(n+1)/2}e^{x^2/4}\,dx$$
$$= \infty.$$

We conclude this section by revisiting the example that began the section.

*Example 2.1 (continued).* It is clear now that CLT means convergence in distribution—that is, $\xi_n = \sqrt{n}(\bar{X} - \mu) \xrightarrow{d} \xi \sim N(0, \sigma^2)$—but this does not necessarily imply $\mathrm{var}(\sqrt{n}\bar{X}) = \mathrm{E}(\xi_n^2) \to \mathrm{E}(\xi^2) = \sigma^2$ (see an extension of parts of Theorem 2.17 in Section 2.7, where the convergence in probability condition is weakened to convergence in distribution). In fact, the CLT even holds in some situations where the variance of the $X_i$'s do not exist (see Chapter 6).

## 2.6 Case study: $\chi^2$-test

One of the celebrated results in classical statistics is Pearson's $\chi^2$ goodness-of-fit test, or simply $\chi^2$-test (Pearson 1900). The test statistic is given by

$$\chi^2 = \sum_{k=1}^{M} \frac{(O_k - E_k)^2}{E_k}, \tag{2.23}$$

where $M$ is the number of cells into which $n$ observations are grouped, $O_k$ and $E_k$ are the observed and expected frequencies of the $k$th cell, $1 \leq k \leq M$, respectively. The expected frequency of the $k$th cell is given by $E_k = np_k$, where $p_k$ is the known cell probability of the $k$th cell evaluated under the assumed model. The asymptotic theory associated with this test is simple: Under the null hypothesis of the assumed model, $\chi^2 \xrightarrow{d} \chi^2_{M-1}$ as $n \to \infty$.

One good feature of Pearson's $\chi^2$-test is that it can be used to test an arbitrary probability distribution, provided that the cell probabilities are completely known. However, the latter actually is a serious constraint, because in practice the cell probabilities often depend on certain unknown parameters of the probability distribution specified by the null hypothesis. For example, under the normal null hypothesis, the cell probabilities depend on the mean and variance of the normal distribution, which may be unknown. In such a case, intuitively one would replace the unknown parameters by their estimators and thus obtain the estimated $E_k$, say $\hat{E}_k$, $1 \leq k \leq M$. The test statistic (2.23) then becomes

$$\hat{\chi}^2 = \sum_{k=1}^{M} \frac{(O_k - \hat{E}_k)^2}{\hat{E}_k}. \tag{2.24}$$

However, this test statistic may no longer have an asymptotic $\chi^2$-distribution.

In a simple problem of assessing the goodness-of-fit to a Poisson or Multinomial distribution, it is known that the asymptotic null-distribution of (2.24) is $\chi^2_{M-p-1}$, where $p$ is the number of parameters estimated by the maximum likelihood method. This is the famous "subtract one degree of freedom for each parameter estimated" rule taught in many elementary statistics books (e.g., Rice 1995, pp. 242). However, the rule may not be generalizable to other probability distributions. For example, this rule does not even apply to testing normality with unknown mean and variance, as mentioned above. Note that here we are talking about MLE based on the original data, not the MLE based on cell frequencies. It is known that the rule applies in general to MLE based on cell frequencies. However, the latter are less efficient than the MLE based on the original data except for special cases where the two are the same, such as the above Poisson and Multinomial cases.

R. A. Fisher was the first to note that the asymptotic null-distribution of (2.24) is not necessarily $\chi^2$ (Fisher 1922a). He showed that if the unknown parameters are estimated by the so-called minimum chi-square method, the asymptotic null-distribution of (2.24) is still $\chi^2_{M-p-1}$, but this conclusion may be false if other methods of estimation (including the ML) are used. Note that there is no contradiction of Fisher's result with the above results related to Poisson and Multinomial distributions, because the minimum chi-square estimators and the MLE are asymptotically equivalent when both are based on cell frequencies. A more thorough result was obtained by Chernoff and Lehmann (1954), who showed that when the MLE based on the original observations are used, the asymptotic null-distribution of (2.24) is not necessarily

$\chi^2$, but instead a "weighted" $\chi^2$, where the weights are eigenvalues of certain nonnegative definite matrix. Note that the problem is closely related to the first example given in the Preface of this book. See Moore (1978) for a nice historical review of the $\chi^2$-test.

There are two components in Pearson's $\chi^2$-test: the (observed) cell frequencies, $O_k$, $1 \le k \le M$, and the cell probabilities, $p_k$, $1 \le k \le M$. Although considerable attention has been given to address the issue associated with the $\chi^2$-test with estimated cell probabilities, there are situations in practice where the cell frequencies also need to be estimated. The following is an example.

*Example 2.17* (Nested-error regression). Consider a situation of clustered observations. Let $Y_{ij}$ denote the $j$th observation in the $i$th cluster. Suppose that $Y_{ij}$ satisfies the following nested-error regression model:

$$Y_{ij} = x'_{ij}\beta + u_i + e_{ij},$$

$i = 1, \ldots, n$, $j = 1, \ldots, b$, where $x_{ij}$ is a known vector of covariates, $\beta$ is an unknown vector of regression coefficients, $u_i$ is a random effect, and $e_{ij}$ is an additional error term. It is assumed that the $u_i$'s are i.i.d. with distribution $F$ that has mean 0, the $e_{ij}$'s are i.i.d. with distribution $G$ that has mean 0, and the $u_i$'s and $e_{ij}$'s are independent. Here, both $F$ and $G$ are unknown. Note that this is a special case of the (non-Gaussian) linear mixed models, which we will further discuss in Chapter 12. The problem of interest here is to test certain distributional assumptions about $F$ and $G$; that is, $H_0$: $F = F_0$ and $G = G_0$, where $F_0$ and $G_0$ are known up to some dispersion parameters. Let $\bar{Y}_{i\cdot} = b^{-1}\sum_{j=1}^{b} Y_{ij}$, $\bar{x}_{i\cdot} = b^{-1}\sum_{j=1}^{b} x_{ij}$, and $\bar{e}_{i\cdot} = b^{-1}\sum_{j=1}^{b} e_{ij}$. Consider $X_i = \bar{Y}_{i\cdot} - \bar{x}'_{i\cdot}\beta = u_i + \bar{e}_{i\cdot}$, $1 \le i \le n$, where $\beta$ is the vector of true regression coefficients. It is easy to show (Exercise 2.36) that $X_1, \ldots, X_n$ are i.i.d. with a distribution whose cf is given by

$$c(t) = c_1(t) \left\{ c_2 \left( \frac{t}{b} \right) \right\}^b, \tag{2.25}$$

where $c_1$ and $c_2$ represent the cf of $F$ and $G$, respectively. If $\beta$ were known, one would consider the $X_i$'s as i.i.d. observations, based on which one could compute the cell frequencies and then apply Pearson's $\chi^2$-test (with estimated cell probabilities). However, because $\beta$ is unknown, the cell frequencies are not observable. In such a case, it is natural to consider $\hat{X}_i = \bar{Y}_{i\cdot} - \bar{x}'_{i\cdot}\hat{\beta}$, where $\hat{\beta}$ is an estimator of $\beta$, and compute the cell frequencies based on the $\hat{X}_i$'s. This leads to a situation where the cell frequencies are estimated.

Jiang, Lahiri, and Wu (2001) extended Pearson's $\chi^2$-test to situations where both the cell frequencies and cell probabilities have to be estimated. In the remaining part of this section we describe their approach without giving all of the details. The details are referred to the reference above. Let $Y$ be a vector of observations whose joint distribution depends on an unknown vector of parameters, $\theta$. Suppose that $X_i(\theta) = X_i(y, \theta)$ satisfy the following conditions: (i) for any fixed $\theta$, $X_1(\theta), \ldots, X_n(\theta)$ are independent; and (ii) if $\theta$ is the true parameter vector, $X_1(\theta), \ldots, X_n(\theta)$ are i.i.d.

*Example 2.17 (continued).* If we let $\theta = \beta$ and $X_i(\theta) = \bar{Y}_i. - \bar{x}'_i.\beta$, $1 \leq i \leq n$, then conditions (i) and (ii) are satisfied (Exercise 2.36).

Let $C_k$, $1 \leq k \leq M$ be disjoint subsets of $R$ such that $\cup_{k=1}^M C_k$ covers the range of $X_i(\theta)$, $1 \leq i \leq n$. Define $p_{i,k}(\theta, \tilde{\theta}) = P_\theta\{X_i(\tilde{\theta}) \in C_k\}$, $1 \leq k \leq M$, and $p_i(\theta, \tilde{\theta}) = [p_{i,k}(\theta, \tilde{\theta})]_{1 \leq k \leq M}$. Here, $P_\theta$ denotes the probability given that $\theta$ is the true parameter vector. Note that under assumption (ii), $p_i(\theta, \theta)$ does not depend on $i$ (why?). Therefore, it will be denoted by $p(\theta) = [p_k(\theta)]_{1 \leq k \leq M}$.

If $\theta$ were known, one would have observed $X_i(\theta)$ and hence compute the $\chi^2$ statistic (2.24); that is,

$$\hat{\chi}_0^2 = \sum_{k=1}^M \frac{\{O_k(\theta) - np_k(\theta)\}^2}{np_k(\theta)}, \tag{2.26}$$

where $O_k(\theta) = \sum_{i=1}^n 1_{\{X_i(\theta) \in C_k\}}$. Here, $p_k(\theta)$ is computed under the null hypothesis. However, $O_k(\theta)$ is not observable, because $\theta$ is unknown. Instead, we compute an estimated cell frequency, $O_k(\hat{\theta}) = \sum_{i=1}^n 1_{\{X_i(\hat{\theta}) \in C_k\}}$, where $\hat{\theta}$ is an estimator of $\theta$. If we replace $O_k(\theta)$ by $O_k(\hat{\theta})$ and $p_k(\theta)$ by $p_k(\hat{\theta})$ in (2.26), we come up with the following $\chi^2$ statistic:

$$\hat{\chi}_e^2 = \sum_{k=1}^M \frac{\{O_k(\hat{\theta}) - np_k(\hat{\theta})\}^2}{np_k(\hat{\theta})}. \tag{2.27}$$

Here, the subscript e represents "estimated" (frequencies).

Our goal is to obtain the asymptotic distribution of $\chi_e^2$. In order to do so, we need some regularity conditions, including assumptions about $\hat{\theta}$. We assume that $p_i(\theta, \tilde{\theta})$ is two times continuously differentiable with respect to $\theta$ and $\tilde{\theta}$. Let $\theta$ denotes the true parameter vector. We assume that $p_k(\theta) > 0$, $1 \leq k \leq M$, and there is $\delta > 0$ such that the following are bounded:

$$\sup_{|\tilde{\theta}-\theta|<\delta} \left\| \frac{\partial}{\partial \tilde{\theta}'} p_i(\theta, \tilde{\theta}) \right\|,$$

$$\sup_{|\tilde{\theta}-\theta|<\delta} \left\| \frac{\partial^2}{\partial\theta\partial\tilde{\theta}'} p_{i,k}(\theta, \tilde{\theta}) \right\|,$$

$$\sup_{|\theta_1-\theta|<\delta, |\theta_2-\theta|<\delta} \left\| \frac{\partial^2}{\partial\theta_1\partial\theta_2'} p_{i,k}(\theta_1, \theta_2) \right\|,$$

$1 \leq k \leq M$, $1 \leq i \leq n$ (see Appendix A.1 for notation of matrix norms and differentiation). Furthermore, we assume that for fixed $\tilde{\theta}$, $X_i(\tilde{\theta})$, $1 \leq i \leq n$, are independent of $\hat{\theta}$, and $\hat{\theta}$ satisfies

$$\sqrt{n}(\hat{\theta} - \theta) \xrightarrow{\text{d}} N\{0, A(\theta)\}, \qquad (2.28)$$

where the covariance matrix $A(\theta)$ may be singular. Then the asymptotic distribution of $\chi_e^2$ is the same as the distribution of

$$\sum_{j=1}^{r}(1 + \lambda_j)Z_j^2 + \sum_{j=r+1}^{M-1} Z_j^2, \qquad (2.29)$$

where $r = \text{rank}\{B(\theta)\}$ with

$$B(\theta) = \text{diag}\{p(\theta)\}^{-1/2}Q(\theta)A(\theta)Q(\theta)' \, \text{diag}\{p(\theta)\}^{-1/2}, \qquad (2.30)$$

$$Q(\theta) = \lim_{n\to\infty} \frac{1}{n} \sum_{i=1}^{n} \frac{\partial}{\partial \theta'} p_i(\theta, \tilde{\theta}) \Big|_{\tilde{\theta}=\theta} \qquad (2.31)$$

(see Appendix A.4 for notation), $\lambda_j$, $1 \leq j \leq r$ are the positive eigenvalues of $B(\theta)$, and $Z_j$, $1 \leq j \leq M - 1$ are independent $N(0,1)$ random variables.

Note that in spite of the fact that $p_i(\theta, \theta) = p(\theta)$, $(\partial/\partial\theta')p_i(\theta, \tilde{\theta})|_{\tilde{\theta}=\theta}$ is not necessarily equal to $(\partial/\partial\theta')p(\theta)$ (Exercise 2.37). Therefore, the right side of (2.31) is not necessarily equal to $(\partial/\partial\theta')p(\theta)$.

Comparing the above result with the well-known results about the $\chi^2$-test (e.g., Chernoff and Lehmann 1954), we observe the following:

(i) If no parameter is estimated, the asymptotic distribution of $\hat{\chi}_0^2$, defined by (2.26), is the same as that of $\sum_{j=1}^{M-1} Z_j^2$.

(ii) If the parameters are estimated by the MLE based on the cell frequencies, the asymptotic distribution of the resulting $\chi^2$ statistic, say, $\hat{\chi}_2^2$, is the same as that of $\sum_{j=1}^{M-s-1} Z_j^2$, where $s$ is the number of (independent) parameters estimated.

(iii) If the parameters are estimated by the MLE based on the original data, the asymptotic distribution of the resulting $\chi^2$ statistic, say, $\hat{\chi}_1^2$, is the same as that of $\sum_{j=1}^{M-s-1} Z_j^2 + \sum_{j=M-s}^{M-1} \mu_j Z_j^2$, where $0 \leq \mu_j \leq 1$, $M - s \leq j \leq M - 1$.

It is interesting to note that, stochastically, we have

$$\sum_{j=1}^{M-s-1} Z_j^2 \leq \sum_{j=1}^{M-s-1} Z_j^2 + \sum_{j=M-s}^{M-1} \mu_j Z_j^2$$

$$\leq \sum_{j=1}^{M-1} Z_j^2 \leq \sum_{j=1}^{r}(1 + \lambda_j)Z_j^2 + \sum_{j=r+1}^{M-1} Z_j^2. \qquad (2.32)$$

The interpretation is the following. In $\hat{\chi}_1^2$ and $\hat{\chi}_2^2$, $\hat{\theta}$ is computed from the same data, whereas in $\chi_e^2$, $\hat{\theta}$ is obtained from an independent source. When using the same data to compute the cell frequencies and estimate $\theta$, the overall variation tends to reduce. To see this, consider a simple example in which $X_1, \ldots, X_n$ are i.i.d. $\sim$ Bernoulli$(p)$, where $p$ is unknown. The observed frequency for $X_i = 1$ is $O_1 = \sum_{i=1}^{n} X_i$; the expected frequency is $E_1 = np$, so $(O_1 - E_1)^2 = (O_1 - np)^2$.

However, if one estimates $p$ by its MLE, $\hat{p} = O_1/n$, one has $\hat{E}_1 = n\hat{p} = O_1$; therefore, $(O_1 - \hat{E}_1)^2 = 0$ (i.e., there is no variation). On the other hand, if $\hat{\theta}$ is obtained from an independent source, it introduces additional variation, which is the implication of (2.32).

The assumption that $\hat{\theta}$ is independent with $X_i(\tilde{\theta})$, $1 \leq i \leq n$, for fixed $\tilde{\theta}$ may seem a bit restrictive. On the other hand, in some cases, information obtained from previous studies can be used to obtain $\hat{\theta}$. In such a case, it may be reasonable to assume that $\hat{\theta}$ is independent with $X_i(\tilde{\theta})$, $1 \leq i \leq n$, if the latter are computed from the current data. Another situation that would satisfy the independence requirement is when $\hat{\theta}$ is obtained by data-splitting. See Jiang, Lahiri, and Wu (2001).

We give an outline of the derivation of the asymptotic distribution of $\hat{\chi}_e^2$. The detail can be found in Jiang, Lahiri, and Wu (2001). First, note that $\hat{\chi}_e^2 = |\xi_n|^2$, where $\xi_n$ is the random vector $\mathrm{diag}\{np(\hat{\theta})\}^{-1/2}\{O(\hat{\theta}) - np(\hat{\theta})\}$ with $O(\theta) = [Q_k(\theta)]_{1 \leq k \leq M}$. The first step is to show

$$\xi_n \xrightarrow{\ \mathrm{d}\ } N\{0, \Sigma(\theta) + B(\theta)\} \tag{2.33}$$

as $n \to \infty$, where

$$\Sigma(\theta) = I_M - \left\{p(\theta)^{1/2}\right\}\left\{p(\theta)^{1/2}\right\}',$$

with $p(\theta)^{1/2} = [p_k(\theta)^{1/2}]_{1 \leq k \leq M}$. By Theorem 2.14, (2.33) is equivalent to

$$\lambda'\xi_n \xrightarrow{\ \mathrm{d}\ } N[0, \lambda'\{\Sigma(\theta) + B(\theta)\}\lambda] \tag{2.34}$$

as $n \to \infty$ for any $\lambda \in R^M$. According to Theorem 2.11, (2.34) is, in turn, equivalent to that the cf of $\lambda'\xi_n$ converges to the cf of the right side of (2.34). However, this is equivalent to

$$E\{\exp(i\lambda'\xi_n)\} \longrightarrow \exp\left[-\frac{1}{2}\lambda'\{\Sigma(\theta) + B(\theta)\}\lambda\right] \tag{2.35}$$

as $n \to \infty$ for any $\lambda \in R^M$ (Exercise 2.38). To show (2.35), we express $\xi_n$ as

$$\xi_n = \eta_n + \zeta_n + \gamma_n, \tag{2.36}$$

where

$$\eta_n = \mathrm{diag}\{np(\hat{\theta})\}^{-1/2}\left\{O(\hat{\theta}) - \sum_{i=1}^{n} p_i(\theta, \hat{\theta})\right\},$$

$$\zeta_n = -\mathrm{diag}\{p(\hat{\theta})\}^{-1/2}\left\{\frac{1}{n}\sum_{i=1}^{n}\frac{\partial}{\partial\theta'}p_i(\theta, \tilde{\theta})\Big|_{\tilde{\theta}=\theta}\right\}n^{1/2}(\hat{\theta} - \theta),$$

and $\gamma_n$ satisfies the following: There is a constant $c$ such that for any $\epsilon > 0$,

$$|\gamma_n| \le c(n|\hat{\theta} - \theta|^2)n^{-1/2} \quad \text{if } |\hat{\theta} - \theta| \le \epsilon.$$

The idea for the proof of (2.35) is therefore to show that, as $n \to \infty$, $\eta_n$ and $\zeta_n$ are the leading terms in (2.36) and $\gamma_n$ is negligible. In fact, the contribution of $\Sigma(\theta)$ in the asymptotic covariance matrix, $\Sigma(\theta) + B(\theta)$ in (2.33), comes from $\eta_n$ and the contribution of $B(\theta)$ from $\zeta_n$ (and $\gamma_n$ has no contribution). In other words, we have

$$\lambda'(\eta_n + \zeta_n)\lambda \xrightarrow{d} N[0, \lambda'\{\Sigma(\theta) + B(\theta)\}\lambda]$$

and $\lambda'\gamma_n \xrightarrow{P} 0$ as $n \to \infty$. Result (2.35) then follows from Slutsky's theorem (Theorem 2.13).

Given that (2.33) holds, we apply Theorem 2.12 (note that a multivariate version of the result also stands—that is, when $\xi_n$, $\xi$ are random vectors) to conclude

$$\hat{\chi}_e^2 = |\xi_n|^2 \xrightarrow{d} |\xi|^2, \tag{2.37}$$

where $\xi \sim N\{0, \Sigma(\theta) + B(\theta)\}$. It remains to determine the distribution of $|\xi^2| = \xi'\xi$. Write $\Sigma = \Sigma(\theta)$ and $B = B(\theta)$ and let $P = \{p(\theta)^{1/2}\}\{p(\theta)^{1/2}\}'$. It can be shown (Exercise 2.39) that $PB = BP = 0$. Thus (see Appendix A.1), there is an orthogonal matrix $T$ such that

$$B = T \operatorname{diag}(\lambda_1, \ldots, \lambda_r, 0, \ldots, 0)T',$$
$$P = T \operatorname{diag}(\rho_1, \ldots, \rho_M)T',$$

where $\lambda_j > 0$, $1 \le j \le r = \operatorname{rank}(B)$, and $\rho_1, \ldots, \rho_M$ are the eigenvalues of $P$. Note that the latter is a projection matrix with rank 1. Therefore, $\rho_1, \ldots, \rho_M$ are zero except for one of them, which is 1. It follows that the distribution of $\xi'\xi$ is the same as that of (2.29) (Exercise 2.39).

## 2.7 Summary and additional results

This section provides a summary of some of the main results in this chapter as well as some additional results. The summary focuses on the connection between different types of convergence.

1. Almost sure (a.s.) convergence implies convergence in probability, which, in turn, implies convergence in distribution.

2. If $\xi_n \xrightarrow{P} \xi$ as $n \to \infty$, then there is a subsequence $n_k$, $k = 1, 2, \ldots$, such that $\xi_{n_k} \xrightarrow{a.s.} \xi$ as $k \to \infty$.

3. $\xi_n \xrightarrow{P} \xi$ as $n \to \infty$ if and only if for every subsequence $n_k$, $k = 1, 2, \ldots$, there is a further subsequence $n_{k_l}$, $l = 1, 2, \ldots$, such that

$$\xi_{n_{k_l}} \xrightarrow{a.s.} \xi \quad \text{as } l \to \infty.$$

4. If for every $\epsilon > 0$ we have $\sum_{n=1}^{\infty} P(|\xi_n - \xi| \geq \epsilon) < \infty$, then $\xi_n \xrightarrow{\text{a.s.}} \xi$ as $n \to \infty$. Intuitively, this result states that convergence in probability at a certain rate implies a.s. convergence.

The proof of the above result follows from the following lemma, which is often useful in establishing a.s. convergence (Exercise 2.40). Let $A_1, A_2, \ldots$ be a sequence of events. Define $\limsup A_n = \cap_{N=1}^{\infty} \cup_{n=N}^{\infty} A_n$.

**Lemma 2.5.** (Borel–Cantelli lemma)
(i) If $\sum_{n=1}^{\infty} P(A_n) < \infty$, then $P(\limsup A_n) = 0$.
(ii) If $A_1, A_2, \ldots$ are pairwise independent and $\sum_{n=1}^{\infty} P(A_n) = \infty$, then $P(\limsup A_n) = 1$.

5. $L^p$ convergence for any $p > 0$ implies convergence in probability.

6. (Dominated convergence theorem) If $\xi_n \xrightarrow{\text{P}} \xi$ as $n \to \infty$ and there is a random variable $\eta$ such that $E(\eta^p) < \infty$ and $|\xi_n| \leq \eta$, $n \geq 1$, then $\xi_n \xrightarrow{L^p} \xi$ as $n \to \infty$ and $E(|\xi|^p) < \infty$.

Let $a_n$, $n = 1, 2, \ldots$, be a sequence of constants. The sequence converges increasingly to $a$, denoted by $a_n \uparrow a$, if $a_n \leq a_{n+1}$, $n \geq 1$ and $\lim_{n \to \infty} a_n = a$. Similarly, let $\xi_n$, $n = 1, 2, \ldots$, be a sequence of random variables. The sequence converges increasingly a.s. to $\xi$, denoted by $\xi_n \uparrow \xi$ a.s., if $\xi_n \leq \xi_{n+1}$ a.s., $n \geq 1$, and $\lim_{n \to \infty} \xi_n = \xi$ a.s.

7. (Monotone convergence theorem) If $\xi_n \uparrow \xi$ a.s. and $\xi_n \geq \eta$ a.s. with $E(|\eta|) < \infty$, then $E(\xi_n) \uparrow E(\xi)$. The result does not imply, however, that $E(\xi)$ is finite. So, if $E(\xi) = \infty$, then $E(\xi_n) \uparrow \infty$. On the other hand, we must have $E(\xi) > -\infty$ (why?).

8. If $\sum_{n=1}^{\infty} E(|\xi_n - \xi|^p) < \infty$ for some $p > 0$, then $\xi_n \xrightarrow{\text{a.s.}} \xi$ as $n \to \infty$. Intuitively, this means that $L^p$ convergence at a certain rate implies a.s. convergence (Exercise 2.40).

The following theorem is useful in establishing the connection between convergence in distribution and other types of convergence.

**Theorem 2.18** (Skorokhod representation theorem). If $\xi_n \xrightarrow{\text{d}} \xi$ as $n \to \infty$, then there are random variables $\eta_n$, $n = 1, 2, \ldots$, and $\eta$ defined on a common probability space such that $\eta_n$ has the same distribution as $\xi_n$, $n = 1, 2, \ldots$, and $\eta$ has the same distribution as $\xi$, and $\eta_n \xrightarrow{\text{a.s.}} \eta$ as $n \to \infty$.

With Skorokhod's theorem we can extend part of Theorem 2.17 as follows.
9. If $\xi_n \xrightarrow{\text{d}} \xi$ as $n \to \infty$, then the following are equivalent:
(i) $\xi_n$, $n = 1, 2, \ldots$, is uniformly integrable in $L^p$.
(ii) $E(|\xi_n|^p) \to E(|\xi|^p) < \infty$ as $n \to \infty$.

10. $\xi_n \xrightarrow{\text{d}} \xi$ as $n \to \infty$ is equivalent to $c_n(t) \to c(t)$ as $n \to \infty$ for every $t \in R$, where $c_n(t)$ is the cf of $\xi_n$, $n = 1, 2, \ldots$, and $c(t)$ the cf of $\xi$.

11. If there is $\delta > 0$ such that the mgf of $\xi_n$, $m_n(t)$, converges to $m(t)$ as $n \to \infty$ for all $t$ such that $|t| < \delta$, where $m(t)$ is the mgf of $\xi$, then $\xi_n \xrightarrow{d} \xi$ as $n \to \infty$.

12. $\xi_n \xrightarrow{d} \xi$ is equivalent to any of the following:
(i) $\lim_{n\to\infty} E\{h(\xi_n)\} = E\{h(\xi)\}$ for every bounded continuous function $h$.
(ii) $\limsup P(\xi_n \in C) \leq P(\xi \in C)$ for any closed set $C$.
(iii) $\liminf P(\xi_n \in O) \geq P(\xi \in O)$ for any open set $O$.

13. Let $f_n(x)$ and $f(x)$ be the pdfs of $\xi_n$ and $\xi$, respectively, with respect to a $\sigma$-finite measure $\mu$ (see Appendix A.2). If $f_n(x) \to f(x)$ a.e. $\mu$ as $n \to \infty$, then $\xi_n \xrightarrow{d} \xi$ as $n \to \infty$.

14. Let $g$ be a continuous function. Then we have the following:
(i) $\xi_n \xrightarrow{a.s.} \xi$ implies $g(\xi_n) \xrightarrow{a.s.} g(\xi)$ as $n \to \infty$;
(ii) $\xi_n \xrightarrow{P} \xi$ implies $g(\xi_n) \xrightarrow{P} g(\xi)$ as $n \to \infty$;
(iii) $\xi_n \xrightarrow{d} \xi$ implies $g(\xi_n) \xrightarrow{d} g(\xi)$ as $n \to \infty$.

15. (Slutsky's theorem) If $\xi_n \xrightarrow{d} \xi$ and $\eta_n \xrightarrow{P} c$, where $c$ is a constant, then the following hold:
(i) $\xi_n + \eta_n \xrightarrow{d} \xi + c$;
(ii) $\eta_n \xi_n \xrightarrow{d} c\xi$;
(iii) $\xi_n / \eta_n \xrightarrow{d} \xi/c$, if $c \neq 0$.

## 2.8 Exercises

2.1. Complete the definition of the sequence of random variables $\xi_n$, $n = 1, 2, \ldots$, in Example 2.1 (i.e., define $\xi_n$ for a general index $n$). Show that $\xi_n \xrightarrow{P} 0$ as $n \to \infty$; however, $\xi_n(x)$ does not converge pointwisely at any $x \in [0, 1]$.

2.2. Use Chebyshev's inequality (see Section 5.2) to prove Theorem 2.1.

2.3. Use the $\epsilon$-$\delta$ argument to prove Theorem 2.2.

2.4. Use the $\epsilon$-$\delta$ argument to prove Theorem 2.3.

2.5. Use the $\epsilon$-$\delta$ argument to prove Theorem 2.4.

2.6. Use Theorem 2.5 and the $\epsilon$-$\delta$ argument to prove Theorem 2.6.

2.7. Let $X_1, \ldots, X_n$ be independent random variables with a common distribution $F$. Define

$$\xi_n = \frac{\max_{1 \leq i \leq n} |X_i|}{a_n}, \quad n \geq 1,$$

where $a_n$, $n = 1, 2, \ldots$, is a sequence of positive constants. Determine $a_n$ for the following cases such that $\xi_n \xrightarrow{P} 0$ as $n \to \infty$:
(i) $F$ is the Uniform$[0, 1]$ distribution.
(ii) $F$ is the Exponential$(1)$ distribution.
(iii) $F$ is the $N(0, 1)$ distribution.

(iv) $F$ is the Cauchy$(0, 1)$ distribution.

*2.8.* Continue with Problem 2.7 with $a_n = n$. Show the following:

(i) If $E(|X_1|) < \infty$, then $\xi_n \xrightarrow{L^1} 0$ as $n \to \infty$.

(ii) If $E(X_1^2) < \infty$, then $\xi_n \xrightarrow{\text{a.s.}} 0$ as $n \to \infty$.

Hint: For (i), first show that for any $a > 0$,

$$\max_{1 \le i \le n} |X_i| \le a + \sum_{i=1}^{n} |X_i| 1_{(|X_i| > a)}.$$

For (ii), use Theorem 2.8 and also note that by exchanging the order of summation and expectation, one can show for any $\epsilon > 0$,

$$\sum_{n=1}^{\infty} n P(|X_1| > \epsilon n) < \infty.$$

*2.9.* Suppose that for each $1 \le j \le k$, $\xi_{n,j}$, $n = 1, 2, \ldots$, is a sequence of random variables such that $\xi_{n,j} \xrightarrow{P} 0$ as $n \to \infty$. Define $\xi_n = \max_{1 \le j \le k} |\xi_{n,j}|$.

(i) Show that if $k$ is fixed, then $\xi_n \xrightarrow{P} 0$ as $n \to \infty$.

(ii) Give an example to show that if $k$ increases with $n$ (i.e., $k = k_n \to \infty$ as $n \to \infty$), the conclusion of (i) may not be true.

*2.10.* Let $\xi_1, \xi_2, \ldots$ be a sequence of random variables. Show that $\xi_n \xrightarrow{P} 0$ as $n \to \infty$ if and only if

$$E\left(\frac{|\xi_n|}{1 + |\xi_n|}\right) \longrightarrow 0 \quad \text{as } n \to \infty.$$

*2.11.* Prove Lemma 2.1 using the $\epsilon$-$\delta$ argument. Then use Lemma 2.1 to establish Theorem 2.7.

*2.12.* Show by similar arguments as in Example 2.7 that $I_2 \le c e^{-\epsilon \sqrt{n}}$, where the notations refer to Example 2.7.

*2.13.* Verify that the infinite series $\sum_{i=1}^{\infty} e^{-\epsilon \sqrt{n}}$ converges. This result was used at the end of Example 2.7.

*2.14.* Suppose that $X_1, \ldots, X_n$ are i.i.d. observations with finite expectation. Show that in the following cases the sample mean $\bar{X} = (X_1 + \cdots + X_n)/n$ is a strongly consistent estimator of the population mean, $\mu = E(X_1)$—that is, $\bar{X} \xrightarrow{\text{a.s.}} \mu$ as $n \to \infty$.

(i) $X_1 \sim \text{Binomial}(m, p)$, where $m$ is fixed and $p$ is an unknown proportion.

(ii) $X_1 \sim \text{Uniform}[a, b]$, where $a$ and $b$ are unknown constants.

(iii) $X_1 \sim N(\mu, \sigma^2)$, where $\mu$ and $\sigma^2$ are unknown parameters.

*2.15.* Suppose that $X_1, X_2, \ldots$ are i.i.d. with a Cauchy$(0, 1)$ distribution; that is, the pdf of $X_i$ is given by

$$f(x) = \frac{1}{\pi(1 + x^2)}, \quad -\infty < x < \infty.$$

Find a positive number $\delta$ such that $n^{-\delta}X_{(n)}$ converges in distribution to a nondegenerate distribution, where $X_{(n)} = \max_{1\leq i\leq n} X_i$. What is the limiting distribution?

2.16. Suppose that $X_1,\ldots,X_n$ are i.i.d. Exponential(1) random variables. Define $X_{(n)}$ as in Exercise 2.15. Show that

$$X_{(n)} - \log(n) \xrightarrow{d} \xi$$

as $n \to \infty$, where the cdf of $\xi$ is given by

$$F(x) = \exp\{-\exp(-x)\}, \quad -\infty < x < \infty.$$

2.17. Let $X_1, X_2,\ldots$ be i.i.d. Uniform$(0,1]$ random variables and $\xi_n = (\prod_{i=1}^n X_i)^{-1/n}$. Show that

$$\sqrt{n}(\xi_n - e) \xrightarrow{d} \xi$$

as $n \to \infty$, where $\xi \sim N(0, e^2)$. (Hint: The result can be established as an application of the CLT; see Chapter 6.)

2.18. Complete the second half of the proof of Theorem 2.9; that is, $\limsup F_n(x) \leq F(x + \epsilon)$ for any $\epsilon > 0$.

2.19. Give examples of a random variable $\xi$ such that the following hold:
(i) The mgf of $\xi$ does not exist for any $t$ except $t = 0$.
(ii) The mgf of $\xi$ exists for $|t| < 1$ but does not exist for $|t| \geq 1$.
(iii) The mgf of $\xi$ exists for any $t \in R$.

2.20. Show that the integrand in (2.10) is bounded in absolute value, and therefore the expectation exists for any $t \in R$.

2.21. Suppose that $\xi_n \sim t_n$, $n = 1,2,\ldots$. Show that the following hold:
(i) $\xi_n \xrightarrow{d} \xi \sim N(0,1)$.
(ii) $m_n(t) = E(e^{t\xi_n}) = \infty$, $\forall t \neq 0$.
(iii) $m(t) = E(e^{t\xi}) = e^{t^2/2}$, $t \in R$.

2.22. Derive the results of Lemma 2.2.

2.23. Derive the results of Lemma 2.3.

2.24. (i) Suppose that $\xi \sim$ Binomial$(n, p)$. Show that $m_\xi(t) = (pe^t+1-p)^n$.
(ii) Show that $(p_n e^t + 1 - p_n)^n \to \exp\{\lambda(e^t - 1)\}$ as $n \to \infty$, $t \in R$, provided that $np_n \to \lambda$ as $n \to \infty$.

2.25. Suppose that $X_1,\ldots,X_n$ are i.i.d. with the pdf

$$f(x) = \frac{1 - \cos(x)}{\pi x^2}, \quad -\infty < x < \infty.$$

(i) Show that the mgf of $X_i$ does not exist.
(ii) Show that the cf of $X_i$ is given by $\max(1 - |t|, 0)$, $t \in R$.
(iii) Show that the cf of $\bar{X} = n^{-1}\sum_{i=1}^n X_i$ is given by

$$\left\{\max\left(1 - \frac{|t|}{n}, 0\right)\right\}^n,$$

which converges to $e^{-|t|}$ as $n \to \infty$.

(iv) Show that the cf of $\xi \sim$ Cauchy$(0, 1)$ is $e^{-|t|}$, $t \in R$. Therefore, $\bar{X} \xrightarrow{d} \xi$ as $n \to \infty$.

*2.26.* Prove Theorem 2.13.

*2.27.* Let $X_1, \ldots, X_n$ be i.i.d Bernoulli$(p)$ observations. Show that

$$\left\{ \frac{n}{p(1-p)} \right\}^{1/2} (\hat{p} - p) \xrightarrow{d} N(0, 1) \quad \text{as } n \to \infty,$$

where $\hat{p}$ is the sample proportion which is equal to $(X_1 + \cdots + X_n)/n$. This result is also known as normal approximation to binomial distribution. (Of course, the result follows from the CLT, but here you are asked to show it directly—without using the CLT.)

*2.28.* Suppose that $\xi_n \xrightarrow{P} \xi$ as $n \to \infty$ and $g$ is a bounded continuous function. Show that $g(\xi_n) \xrightarrow{L^p} g(\xi)$ as $n \to \infty$ for every $p > 0$.

*2.29.* Let $X \sim$ Uniform$(0, 1)$. Define $\xi_n = 2^{n-1} 1_{(0 < X < 1/n)}$, $n = 1, 2, \ldots$.

(i) Show that $\xi_n \xrightarrow{a.s.} 0$ as $n \to \infty$.

(ii) Show that $\xi_n$, $n = 1, 2, \ldots$, does not converge to zero in $L^p$ for any $p > 0$.

*2.30.* Prove Theorem 2.15 using Chebyshev's inequality (see Section 5.2).

*2.31.* Use Skorokhod's theorem (Theorem 2.18) to prove the first half of Theorem 2.11; that is, convergence in distribution implies convergence of the characteristic function.

*2.32.* Prove that the inequality (2.16) holds for any $p > 0$. Note that for $p \geq 1$, this follows from the convex function inequality, but the inequality holds for $0 < p < 1$ as well.

*2.33.* Complete the proof of Theorem 2.17 (i) $\Rightarrow$ (ii) using the argument of subsequences (see §1.5.1.6).

*2.34.* Use the dominated convergence theorem (Theorem 2.16) to show (2.20). Also show that $E(\|\xi_n\|^p - |\xi|^p|) \to 0$ implies $E(|\xi_n|^p) \to E(|\xi|^p)$ as $n \to 0$.

*2.35.* Refer to the (iii) $\Rightarrow$ (i) part of the proof of Theorem 2.17.

(i) Show that $\eta_n \xrightarrow{P} \eta$ as $n \to \infty$.

(ii) Show that it is not necessarily true that $|\xi_n|^p 1_{(|\xi_n| \leq a)} \xrightarrow{P} |\xi|^p 1_{(|\xi| \leq a)}$ as $n \to \infty$.

*2.36.* This exercise refers to Example 2.17.

(i) Show that $X_1, \ldots, X_n$ are i.i.d. with a distribution whose cf is given by (2.25).

(ii) If we define $X_i(\theta) = \bar{Y}_i. - \bar{x}'_i.\beta$ for an arbitrary $\theta = \beta$ (not necessarily the true parameter vector), then conditions (i) and (ii) are satisfied.

*2.37.* Consider the function $f(x, y) = x^2 + y^2$. Show that

$$\frac{\partial}{\partial x} f(x, y) \bigg|_{y=x} \neq \frac{\partial}{\partial x} f(x, x).$$

*2.38.* Show that (2.33) is equivalent to that (2.35) holds for every $\lambda \in R^M$.

*2.39.* Regarding the distribution of $|\xi|^2 = \xi'\xi$ in (2.37), show the following [see the notation below (2.37)]:

(i) $PB = BP = 0$.

(ii) $P$ is a projection matrix with rank 1.

(iii) The distribution of $\xi'\xi$ is the same as (2.29), where $Z_1, \ldots, Z_{M-1}$ are independent standard normal random variables.

*2.40.* Use the Borel–Cantelli lemma (Lemma 2.5) to prove the following:

(i) If for every $\epsilon > 0$ we have $\sum_{n=1}^{\infty} P(|\xi_n - \xi| \geq \epsilon) < \infty$, then $\xi_n \xrightarrow{\text{a.s.}} \xi$ as $n \to \infty$.

(ii) If $\sum_{n=1}^{\infty} E(|\xi_n - \xi|^p) < \infty$ for some $p > 0$, then $\xi_n \xrightarrow{\text{a.s.}} \xi$ as $n \to \infty$.

# 3

# Big $O$, Small $o$, and the Unspecified $c$

## 3.1 Introduction

One of the benefits of using large-sample techniques is that it allows us to separate important factors from those that have minor impact and to replace quantities by equivalents that are of simpler form. For example, recall Example 2 in the Preface. Here, the problem is to estimate the mean of a random variable. In the first case, the mean can be expressed as $\mathrm{E}(\sum_{i=1}^{n} X_i)$, where $X_1, \ldots, X_n$ are i.i.d. observations with $\mathrm{E}(X_i) = \mu \neq 0$. In this case, as mentioned, one could estimate the mean by simply removing the expectation sign, that is, by $\sum_{i=1}^{n} X_i$. The reason can be seen easily because, according to the WLLN, $\bar{X} = n^{-1} \sum_{i=1}^{n} X_i$ is a consistent estimator of $\mu$; therefore, it makes sense to estimate $\mathrm{E}(\sum_{i=1}^{n} X_i) = n\mu$ by $n\bar{X} = \sum_{i=1}^{n} X_i$. Here is another look at this method, which may be easier to generalize to cases where the $X_i$'s are not i.i.d. Note that we can write

$$\mathrm{E}\left(\sum_{i=1}^{n} X_i\right) = \sum_{i=1}^{n} X_i - \left\{\sum_{i=1}^{n} X_i - \mathrm{E}\left(\sum_{i=1}^{n} X_i\right)\right\}$$
$$= I_1 - I_2. \tag{3.1}$$

Now, compare the orders of $I_1$ and $I_2$. Suppose that the $X_i$'s have finite variance, say, $0 < \sigma^2 < \infty$. Then the order of $I_1$ is $O(n)$, and that of $I_2$ is $O(\sqrt{n})$. To see this, note that, by the WLLN, we have $n^{-1} \sum_{i=1}^{n} X_i \xrightarrow{\mathrm{P}} \mu \neq 0$, which explains $I_1 = O(n)$. On the other hand, it is easy to show that $\mathrm{E}(I_2^2) = n\sigma^2$, or $\mathrm{E}(I_2/\sqrt{n})^2 = \sigma^2$. This implies that $I_2/\sqrt{n}$ is bounded in $L^2$; hence, $I_2/\sqrt{n} = O(1)$, or $I_2 = O(\sqrt{n})$. Here we are using the notation big $O$ and small $o$ for random variables (note that both $I_1$ and $I_2$ are random variables), which will be carefully defined in the sequel. Given the orders of $I_1$ and $I_2$, it is easy to see why it is reasonable to approximate the left side of (3.1) by $I_1$—because it captures the main part of it.

Now, consider the second case of Example 2 in the Preface, where the interest is to estimate $\mathrm{E}(\sum_{i=1}^{n} X_i)^2$, assuming $\mu = 0$. Does the previous technique

J. Jiang, *Large Sample Techniques for Statistics*,
DOI 10.1007/978-1-4419-6827-2_3, © Springer Science+Business Media, LLC 2010

still work? Well, formally one can still write

$$\mathrm{E}\left(\sum_{i=1}^{n} X_i\right)^2 = \left(\sum_{i=1}^{n} X_i\right)^2 - \left\{\left(\sum_{i=1}^{n} X_i\right)^2 - \mathrm{E}\left(\sum_{i=1}^{n} X_i\right)^2\right\}$$

$$= I_1 - I_2. \qquad (3.2)$$

The question is whether it works the same way. To answer this question, we, once again, compare the orders of $I_1$ and $I_2$. According to the CLT, we have $(\sqrt{n}\sigma)^{-1}\sum_{i=1}^{n} X_i \xrightarrow{d} N(0,1)$; hence, by Theorem 2.12,

$$\frac{\left(\sum_{i=1}^{n} X_i\right)^2}{n\sigma^2} \xrightarrow{d} \chi_1^2, \qquad (3.3)$$

$$\frac{\left(\sum_{i=1}^{n} X_i\right)^2}{n\sigma^2} - 1 \xrightarrow{d} \chi_1^2 - 1 \qquad (3.4)$$

as $n \to \infty$. Result (3.3) implies that $I_1 = O(n)$. Furthermore, since

$$\frac{\left(\sum_{i=1}^{n} X_i\right)^2 - \mathrm{E}\left(\sum_{i=1}^{n} X_i\right)^2}{n\sigma^2} = \frac{\left(\sum_{i=1}^{n} X_i\right)^2}{n\sigma^2} - 1,$$

(3.4) implies that $I_2 = O(n)$. Thus, the two terms on the right side of (3.2) are of the same order; hence, it is not a good idea to simply approximate the left side by the first term (because then one ignores a major part of it). In conclusion, the previous technique no longer works.

As mentioned, the techniques used here involve the notation big $O$ and small $o$, but please keep in mind that they are more than just notation. The operation of big $O$ and small $o$, and later an unspecified constant $c$, is an art in large-sample techniques.

## 3.2 Big $O$ and small $o$ for sequences and functions

We begin with constant infinite sequences. An infinite sequence $a_n$, $n = 1, 2, \ldots$, is $O(1)$ if it is bounded; that is, there is a constant $c$ such that $|a_n| \leq c$, $n \geq 1$. The concept can be generalized. Let $b_n$, $n = 1, 2, \ldots$, be a positive infinite sequence. We say $a_n = O(b_n)$ if the sequence $a_n/b_n$, $n = 1, 2, \ldots$, is bounded. The simple lemma below gives an alternative expression.

**Lemma 3.1.** $a_n = O(b_n)$ if and only if $a_n = b_n O(1)$.

The proof is straightforward from the definition.

Now, the definition of $o$. A sequence $a_n$, $n = 1, 2, \ldots$ is, $o(1)$ if $a_n \to 0$ as $n \to \infty$. More generally, let $b_n$, $n = 1, 2, \ldots$, be a positive infinite sequence. We say $a_n = o(b_n)$ if $a_n/b_n \to 0$ as $n \to \infty$. Similar to Lemma 3.1, we have the following.

**Lemma 3.2.** $a_n = o(b_n)$ if and only if $a_n = b_n o(1)$.

Below are some simple facts and rules of operation for $O$ and $o$ (Exercises 3.1 and 3.2).

**Lemma 3.3.** If $a_n = o(b_n)$, then $a_n = O(b_n)$.

**Lemma 3.4** (Properties of $O$ and $o$).
(i) If $a_n = O(b_n)$ and $b_n = O(c_n)$, then $a_n = O(c_n)$.
(ii) If $a_n = O(b_n)$ and $b_n = o(c_n)$, then $a_n = o(c_n)$.
(iii) If $a_n = o(b_n)$ and $b_n = O(c_n)$, then $a_n = o(c_n)$.
(iv) If $a_n = o(b_n)$ and $b_n = o(c_n)$, then $a_n = o(c_n)$.

**Lemma 3.5** (Properties of $O$ and $o$).
(i) If $a_n = O(b_n)$, then for any $p > 0$, $|a_n|^p = O(b_n^p)$.
(ii) If $a_n = o(b_n)$, then for any $p > 0$, $|a_n|^p = o(b_n^p)$.
In particular, if $a_n = O(b_n)$ $[o(b_n)]$, then $|a_n| = O(b_n)$ $[o(b_n)]$.

However, the properties of Lemma 3.5 cannot be generalized without caution. This means that $a_n = O(b_n)$ does not imply $g(a_n) = O\{g(b_n)\}$ for any (increasing) function $g$; likewise, $a_n = o(b_n)$ does not imply $g(a_n) = o\{g(b_n)\}$ for any (increasing) function $g$.

*Example 3.1.* Consider $a_n = n$ and $b_n = 2n$. Then, clearly, we have $a_n = O(b_n)$. However, $e^{a_n}/e^{b_n} = e^n/e^{2n} = e^{-n} \to 0$ as $n \to \infty$. Therefore, $e^{a_n} = o(e^{b_n})$ instead of $O(e^{b_n})$.

*Example 3.2.* This time consider $a_n = n$ and $b_n = n^2$, then $a_n = o(b_n)$. However, $\log(a_n) = \log(n)$ and $\log(b_n) = 2\log(n)$, so $\log(a_n) = O\{\log(b_n)\}$ instead of $o\{\log(b_n)\}$.

Among the infinite sequences that are commonly in use, we have the following results, where $0 < p < 1 < q < \infty$. For each of the sequences below we have $a_n = o(b_n)$ if $b_n$ is a sequence to the right of $a_n$ [e.g., $n^p = o(n^q)$].

$$\ldots, \log\log(n), \ldots,$$
$$\ldots, \{\log(n)\}^p, \ldots, \log(n), \ldots, \{\log(n)\}^q, \ldots,$$
$$\ldots, n^p, \ldots, n, \ldots, n^q, \ldots$$
$$\ldots, e^{n^p}, \ldots, e^n, \ldots, e^{n^q}, \ldots,$$
$$\ldots, n!, \ldots, n^n, \ldots. \tag{3.5}$$

By the lemma below, if we take the reciprocals of the sequences in (3.5), we get the small $o$ relationships in the reversed order. For example, $n^{-q} = o(n^{-p})$.

**Lemma 3.6.** If $a_n$ and $b_n$ are nonzero and $a_n = o(b_n)$, then $b_n^{-1} = o(a_n^{-1})$.

The concepts of $O$ and $o$ can be extended to functions of a real variable, $x$. Let $f(x)$ be a function of $x$. First, consider the case $x \to 0$. We say $f(x) = O(x)$ if $f(x)/x$ is bounded as $x \to 0$ (but $x \neq 0$) and $f(x) = o(x)$ if $f(x)/x \to 0$ as $x \to 0$ (but $x \neq 0$). Similarly, for the case $x \to \infty$, we say $f(x) = O(x)$ if $f(x)/x$ is bounded and $f(x) = o(x)$ if $f(x)/x \to 0$. More generally, for any $x_0$ and $p \geq 0$, we say, as $x \to x_0$ (but $x \neq x_0$), $f(x) = O(|x - x_0|^p)$ if $f(x)/|x - x_0|^p$ is bounded and $f(x) = o(|x - x_0|^p)$ if $f(x)/|x - x_0|^p \to 0$; also, as $|x| \to \infty$, $f(x) = O(|x|^p)$ if $f(x)/|x|^p$ is bounded and $f(x) = o(|x|^p)$ if $f(x)/|x|^p \to 0$.

*Example 3.3.* Let $p$ and $q$ be positive integers. Consider

$$f(x) = \frac{a_p x^p + a_{p-1} x^{p-1} + \cdots + a_1 x + a_0}{b_q x^q + b_{q-1} x^{q-1} + \cdots + b_1 x + b_0}.$$

First, assume that $a_p$ and $b_q$ are nonzero. Then, as $|x| \to \infty$, $f(x) = o(1)$ if $p < q$; $f(x) = O(1)$ if $p = q$ and $1/f(x) = o(1)$ if $p > q$. Now, assume $b_0 \neq 0$. Then as $x \to 0$, $f(x) = O(1)$ regardless of $p$ and $q$ (Exercise 3.3).

In the above example, if $a_0 = 0$ and $b_0 \neq 0$, then $f(x) = o(1)$ as $x \to 0$. Nevertheless, there is no contradition with the last conclusion of Example 3.3, because $f(x) = o(1)$ implies that $f(x) = O(1)$ (see Lemma 3.3). On the other hand, in order to characterize the orders of sequences or functions more precisely, we need the following definitions. Two sequences, $a_n$ and $b_n$, are of the same order, denoted by $a_n \propto b_n$, if both $a_n/b_n$ and $b_n/a_n$ are bounded; the two sequences are asymptotically equivalent, denoted by $a_n \sim b_n$, if $a_n/b_n \to 1$ as $n \to \infty$. It is clear from the definition that $a_n \propto b_n$ if and only if $b_n \propto a_n$. The definitions can be easily extended to functions. For example, $f(x) \sim x^p$ as $x \to \infty$ means that $f(x)/x^p \to 1$ as $x \to \infty$.

*Example 3.4* (Stirling's formula). Stirling's approximation, also known as Stirling's formula, states that

$$\frac{n!}{\sqrt{2\pi n}(n/e)^n} \longrightarrow 1 \qquad (3.6)$$

as $n \to \infty$, or, using the notation just introduced,

$$n! \sim \sqrt{2\pi n}\left(\frac{n}{e}\right)^n.$$

Table 3.1 shows astonishing accuracy of this approximation even for small $n$, where the ratio is the left side of (3.6). It is not that straightforward to prove Stirling's formula, especially the exact limit in (3.6) (see Exercise 3.5). However, it is fairly easy to show that the limit of the left side of (3.6) exists. In fact, Table 3.1 suggests that the left side of (3.6) is decreasing in $n$, which is indeed true (Exercise 3.4). Since the sequence is bounded from below (by 0), by the result of §1.5.1.3 it must have a limit.

## Table 3.1. Stirling's approximation

| $n$ | Exact | Approximation | Ratio |
|---|---|---|---|
| 1 | 1 | 0.922 | 1.084 |
| 2 | 2 | 1.919 | 1.042 |
| 3 | 6 | 5.836 | 1.028 |
| 4 | 24 | 23.506 | 1.021 |
| 5 | 120 | 118.019 | 1.017 |
| 6 | 720 | 710.078 | 1.014 |
| 7 | 5040 | 4980.396 | 1.012 |
| 8 | 40320 | 39902.40 | 1.010 |
| 9 | 362880 | 359536.9 | 1.009 |
| 10 | 3628800 | 3598696 | 1.008 |

## 3.3 Big $O$ and small $o$ for vectors and matrices

To extend the concepts of big $O$ and small $o$ to sequences of vectors and matrices we first introduce some notation. Let $v$ denote a $k$-dimensional vector and $A$ a $k \times l$ matrix. The Euclidean norm (or 2-norm) of $v$ is defined as $|v| = \sqrt{\sum_{j=1}^{k} v_j^2}$, where $v_j$, $1 \le j \le k$, are the components of $v$. The spectral norm of $A$ is defined as $\|A\| = \{\lambda_{\max}(A'A)\}^{1/2}$. The 2-norm of $A$ is defined as $\|A\|_2 = \{\text{tr}(A'A)\}^{1/2}$. It is easy to establish the following relationships between the two norms (Exercise 3.6).

**Lemma 3.7.** $\|A\| \le \|A\|_2 \le \sqrt{k \wedge l}\|A\|$, where $k \wedge l = \min(k, l)$.

Due to this result, working on any of the matrix norms would be equivalent as far as the order is concerned, provided that the dimension of the matrix does not increase with $n$. For example, consider the following.

*Example 3.5.* Let $A_n = n^{-1/2}I_l$, where $I_l$ denotes the $l$-dimensional identity matrix. Then we have $\|A_n\| = n^{-1/2}$ and $\|A_n\|_2 = (l/n)^{1/2}$. If $l$ is fixed, then both $\|A_n\|$ and $\|A_n\|_2 \to 0$ as $n \to \infty$. However, if $l = n$, we have $\|A_n\| \to 0$ and $\|A_n\|_2 = 1$ as $n \to \infty$.

Throughout this book we mainly use $\|\cdot\|$ as the norm for matrices, but keep in mind that most of the results can be extended to $\|\cdot\|_2$ if the dimension of the matrix is fixed or bounded. Let $a_n$, $n = 1, 2, \ldots$ be a sequence of positive numbers. Let $v_n$, $n = 1, 2, \ldots$, be a sequence of vectors. We say $v_n = O(a_n)$ if $|v_n|/a_n$ is bounded and $v_n = o(a_n)$ if $|v_n|/a_n \to 0$ as $n \to \infty$. Similarly, let $A_n$, $n = 1, 2, \ldots$, be a sequence of matrices. We say $A_n = O(a_n)$ if $\|A_n\|/a_n$ is bounded and $A_n = o(a_n)$ if $\|A_n\|/a_n \to 0$ as $n \to \infty$. Clearly, $v_n = O(a_n)$ $[o(a_n)]$ if and only if $|v_n| = O(a_n)$ $[o(a_n)]$ and $A_n = O(a_n)$ $[o(a_n)]$ if and only if $\|A_n\| = O(a_n)$ $[o(a_n)]$.

To establish further properties we need to introduce a partial order among matrices so that different matrices may be compared. Let $A$ and $B$ be $k \times k$ matrices. The notation $A \geq B$ means that $A - B$ is nonnegative definite. Similarly, the notation $A > B$ means that $A - B$ is positive definite; $A \geq 0$ means that $A$ is nonnegative definite and $A > 0$ means that $A$ is positive definite. Likewise, the notation $A \leq B$ means that $B \geq A$, and so forth. If $A \geq 0$, the square root of $A$, $A^{1/2}$, is defined as follows. Let $T$ be the orthogonal matrix such that $A = T \operatorname{diag}(\lambda_1, \ldots, \lambda_k)T'$, where $\lambda_j$, $1 \leq j \leq k$, are the eigenvalues of $A$ which are nonnegative. Then $A^{1/2}$ is defined as $T \operatorname{diag}(\sqrt{\lambda_1}, \ldots, \sqrt{\lambda_k})T'$ .

The above definition of "$\geq$" introduces a partial order among matrices. This means that some, but not all, pairs of matrices are comparable. Nevertheless, in many ways such a partial order resembles the (complete) order of real numbers. For example, the following results hold.

**Lemma 3.8.** Suppose that $A \geq B \geq 0$. Then we have the following:
(i) $A^{1/2} \geq B^{1/2}$;
(ii) $A^{-1} \leq B^{-1}$, if $B$ is nonsingular.

See, for example, Chan and Kwong (1985) for the proofs of the above results. However, an easy mistake can be made if one tries too aggresively to extend the properties of real numbers to matrices. For example, it is not even true that $A \geq B$ implies $A^2 \geq B^2$.

*Example 3.6.* Consider $A = \begin{pmatrix} 2 & 1 \\ 1 & 1 \end{pmatrix}$ and $B = \begin{pmatrix} 1 & 0 \\ 0 & 0 \end{pmatrix}$. Then we have $A \geq B$. However, it is not true that $A^2 \geq B^2$. To see this, note that

$$A^2 - B^2 = \begin{pmatrix} 5 & 3 \\ 3 & 2 \end{pmatrix} - \begin{pmatrix} 1 & 0 \\ 0 & 0 \end{pmatrix} = \begin{pmatrix} 4 & 3 \\ 3 & 2 \end{pmatrix},$$

which has determinant $-1$; hence, the difference is not nonnegative definite.

Therefore, it is important to know what are the correct results regarding matrix comparisons and not to assume that every result in real numbers has its matrix analogue. The following are some useful results in this regard.

**Lemma 3.9.** Let $A$ and $B$ be $k \times k$ matrices. The following statements are equivalent:
(i) $A \geq B$;
(ii) $v'Av \geq v'Bv$ for any $k \times 1$ vector $v$;
(iii) $C'AC \geq C'BC$ for any $k \times l$ matrix $C$.

**Lemma 3.10.** Let $A$ be a $k \times k$ symmetric matrix. Then, for any $k \times l$ matrix $C$, we have $C'AC \leq \lambda_{\max}(A)C'C$.

**Lemma 3.11.** $A \geq B$ implies the following:

(i) $\lambda_{\max}(A) \geq \lambda_{\max}(B)$;
(ii) $\lambda_{\min}(A) \geq \lambda_{\min}(B)$;
(iii) $\operatorname{tr}(A) \geq \operatorname{tr}(B)$;
(iv) $\operatorname{tr}(A^2) \geq \operatorname{tr}(B^2)$.

Result (iv) deserves some attention, especially in view of Example 3.6. A proof can be given as follows:

$$
\begin{aligned}
\operatorname{tr}(A^2) &= \operatorname{tr}(A^{1/2} A A^{1/2}) \\
&\geq \operatorname{tr}(A^{1/2} B A^{1/2}) \text{ [Lemma 3.8 and Lemma 3.10(iii)]} \\
&= \operatorname{tr}(B^{1/2} A B^{1/2}) \text{ (property of trace; see Appendix A.1)} \\
&\geq \operatorname{tr}(B^{1/2} B B^{1/2}) \text{ [Lemma 3.8 and Lemma 3.10(iii)]} \\
&= \operatorname{tr}(B^2).
\end{aligned}
$$

The proofs of (i), (ii) and (iii) are left as exercises (Exercise 3.7).

**Corollary 3.1.** For any $j \times k$ matrix $A$, $k \times l$ matrix $B$, and $k \times 1$ vector $v$, we have the following:
(i) $|Av| \leq \|A\| \cdot |v|$.
(ii) $\|AB\| \leq \|A\| \cdot \|B\|$.
(iii) $\|A + B\| \leq \|A\| + \|B\|$.

The proof is left as an exercise (Exercise 3.8). Using the above results, it is easy to establish the following properties of $O$ and $o$, where $a_n$, $b_n$, and $c_n$ denote sequences of positive constants, $A_n$ and $B_n$ are sequences of matrices, and $v_n$ is a sequence of vectors.

**Lemma 3.12.** If $A_n = O(a_n)$, $B_n = O(b_n)$, and $v_n = O(c_n)$, then we have the following:
(i) $A_n v_n = O(a_n c_n)$;
(ii) $A_n B_n = O(a_n b_n)$;
(iii) $A_n + B_n = O(a_n \vee b_n)$.

*Proof.* (i), (ii) and (iii) follow from (i), (ii) and (iii) of Corollary 3.11, respectively. Note that $a_n \vee b_n \leq a_n + b_n \leq 2(a_n \vee b_n)$. Q.E.D.

**Lemma 3.13.** If $A_n = O(a_n)$, $B_n = o(b_n)$ and $v_n = o(c_n)$, then the following hold:
(i) $A_n v_n = o(a_n c_n)$;
(ii) $A_n B_n = o(a_n b_n)$.

The proof is as straightforward as the previous one. The following definitions are often used in operation of sequences of matrices and vectors. A sequence of square matrices $A_n$ is bounded from above if $A_n = O(1)$; it is bounded from below if $A_n^{-1} = O(1)$. Here, by square matrix it means that $A_n$

is $k \times k$ for some $k$, which may depend on $n$. From the definition it is clear that $A_n$ is bounded from above if and only if $\lambda_{\max}(A'_n A_n) = O(1)$. As for boundedness from below, we have the following.

**Lemma 3.14.** $A_n$ is bounded from below if and only if there is a constant $c > 0$ such that $\lambda_{\min}(A'_n A_n) \geq c$, $n \geq 1$.

*Proof.* ($\Rightarrow$) The definition implies that $A_n$ is nonsingular and so are $A'_n A_n$ and $A_n A'_n$. It follows that

$$
\begin{aligned}
\|A_n^{-1}\| &= \lambda_{\max}\{(A_n^{-1})' A_n^{-1}\} \\
&= \lambda_{\max}\{(A'_n)^{-1} A_n^{-1}\} \\
&= \lambda_{\max}\{(A_n A'_n)^{-1}\} \\
&= \frac{1}{\lambda_{\min}(A_n A'_n)} \\
&= \frac{1}{\lambda_{\min}(A'_n A_n)},
\end{aligned}
\tag{3.7}
$$

and the result immediately follows. ($\Leftarrow$) Note that (3.7) holds as long as $A'_n A_n$ is nonsingular (note that $A_n$ is required to be a square matrix). The result thus follows. Q.E.D.

Finally, we have the following results that associate the orders of the vectors and matrices to those of their components and elements.

**Lemma 3.15.** Let $A_n$ be a $k_n \times l_n$ matrix and $v_n$ be a $k_n \times 1$ vector, $n = 1, 2, \ldots$. Furtheremore, let $a_n$ and $b_n$ be sequences of positive numbers. Suppose that both $k_n$ and $l_n$ are bounded. Then we have the following:
(i) $A_n = O(a_n)$ $[o(a_n)]$ if and only if $a_{n,ij} = O(a_n)$ $[o(a_n)]$ for any $1 \leq i \leq k_n$ and $1 \leq j \leq l_n$, where $a_{n,ij}$ is the $(i, j)$ element of $A_n$;
(ii) $v_n = O(b_n)$ $[o(b_n)]$ if and only if $v_{n,i} = O(b_n)$ $[o(b_n)]$ for any $1 \leq i \leq k_n$, where $v_{n,i}$ is the $i$th component of $v_n$.

The proofs are left as an exercise (Exercise 3.9).

## 3.4 Big $O$ and small $o$ for random quantities

A sequence of random variables, $\xi_n$, $n = 1, 2, \ldots$, is bounded in probability, denoted by $\xi_n = O_P(1)$, if for any $\epsilon > 0$, there is $M > 0$ and $N \geq 1$ such that

$$
P(|\xi_n| \leq M) > 1 - \epsilon, \quad n \geq N.
\tag{3.8}
$$

**Lemma 3.16.** $\xi_n = O_P(1)$ if and only if for any $\epsilon > 0$, there is $M > 0$ such that

$$
P(|\xi_n| \leq M) > 1 - \epsilon, \quad n \geq 1.
\tag{3.9}
$$

*Proof.* ($\Rightarrow$) For any $\epsilon > 0$, by the definition there is $M > 0$ and $N \geq 1$ such that (3.8) holds. On the other hand, for each $1 \leq n \leq N - 1$, since $\xi_n$ is a random variable, there is an $M_n > 0$ such that $\mathrm{P}(|\xi_n| \leq M_n) > 1 - \epsilon$ (see Example A.5). Let $M' = M \vee M_1 \vee \cdots \vee M_{N-1}$. Then we have

$$\mathrm{P}(|\xi_n| \leq M') \geq \mathrm{P}(|\xi_n| \leq M_n) > 1 - \epsilon$$

if $1 \leq n \leq N - 1$ and

$$\mathrm{P}(|\xi_n| \leq M') \geq \mathrm{P}(|\xi_n| \leq M) > 1 - \epsilon$$

if $n \geq N$; hence (3.9), holds with $M$ replaced by $M'$. The proof of ($\Leftarrow$) is trivial. Q.E.D.

More generally, let $a_n$ be a sequence of positive numbers. We say $\xi_n = O_{\mathrm{P}}(a_n)$ if $\xi_n / a_n = O_{\mathrm{P}}(1)$ or, equivalently, $\xi_n = a_n O_{\mathrm{P}}(1)$. We say $\xi_n = o_{\mathrm{P}}(a_n)$ if $\xi_n / a_n \xrightarrow{\mathrm{P}} 0$ as $n \to \infty$. Similarly, let $\xi_n$ be a sequence of random vectors (random matrices). Then $\xi_n = O_{\mathrm{P}}(a_n)$ if $|\xi_n| = O_{\mathrm{P}}(a_n)$ $[\|\xi_n\| = O_{\mathrm{P}}(a_n)]$ and $\xi_n = o_{\mathrm{P}}(a_n)$ if $|\xi_n| = o_{\mathrm{P}}(a_n)$ $[\|\xi_n\| = o_{\mathrm{P}}(a_n)]$.

In the following we mainly consider sequences of random variables, but keep in mind that, by the definition, all of the results can be easily extended to sequences of random vectors or random matrices. The properties of $o_{\mathrm{P}}$ are essentially those about convergence in probability (to zero), which we discussed in Chapter 2. Thus, we mainly focus $O_{\mathrm{P}}$ and, because of the definition, it suffices to consider $O_{\mathrm{P}}(1)$ (i.e., $a_n = 1$).

**Theorem 3.1.** $\xi_n = O_{\mathrm{P}}(1)$ if one of the following holds:
(i) There is $p > 0$ such that $\mathrm{E}(|\xi_n|^p)$, $n \geq 1$ is bounded.
(ii) $\xi_n \xrightarrow{\mathrm{P}} \xi$ as $n \to \infty$ for some random variable $\xi$.
(iii) $\xi_n \xrightarrow{\mathrm{d}} \xi$ as $n \to \infty$ for some random variable $\xi$.

*Proof.* In view of Theorem 2.9, it suffices to show that either (i) or (iii) implies $\xi_n = O_{\mathrm{P}}(1)$. Suppose that (i) holds. For any $\epsilon > 0$, we have, by Chebyshev's inequality,

$$\mathrm{P}(|\xi_n| > M) = \mathrm{P}(|\xi_n|^p > M^p)$$
$$\leq \frac{\mathrm{E}(|\xi_n|^p)}{M^p} \leq \frac{c}{M^p},$$

where $c = \sup_{n \geq 1} \mathrm{E}(|\xi_n|^p) < \infty$. Thus, if we choose $M$ such that $M > (c/\epsilon)^{1/p}$, we have $\mathrm{P}(|\xi_n| > M) < \epsilon$; hence, $\mathrm{P}(|\xi_n| \leq M) > 1 - \epsilon$ for any $n \geq 1$. It follows by Lemma 3.16 that $\xi_n = O_{\mathrm{P}}(1)$.

Now suppose that (iii) holds. For any $\epsilon > 0$, there is $M > 0$ such that $\mathrm{P}(|\xi| < M) > 1 - \epsilon/2$. Note that $O = (-M, M)$ is an open set. Thus, by (iii) of §2.7.12, we have

$$\liminf \mathrm{P}(|\xi_n| \le M) \ge \liminf \mathrm{P}(|\xi_n| < M)$$
$$\ge \mathrm{P}(|\xi| < M)$$
$$> 1 - \frac{\epsilon}{2}.$$

Therefore, there is $N \ge 1$ such that $\mathrm{P}(|\xi_n| \le M) > 1 - \epsilon$, $n \ge N$; that is, $\xi_n = O_\mathrm{P}(1)$. Q.E.D.

We consider some examples.

*Example 3.7* (Sample mean). Suppose that $X_1, \ldots, X_n$ are i.i.d. observations from a distribution that has a finite expectation; that is, $\mathrm{E}(|X_1|) < \infty$. Then the sample mean $\bar{X} = n^{-1} \sum_{i=1}^n X_i = O_\mathrm{P}(1)$. This is because

$$\mathrm{E}\left(|\bar{X}|\right) \le \frac{1}{n} \sum_{i=1}^n \mathrm{E}(|X_i|)$$
$$= \mathrm{E}(|X_1|) < \infty.$$

Therefore, by (i) of Theorem 3.1 (with $p = 1$), we have $\bar{X} = O_\mathrm{P}(1)$. It should be pointed out that the condition that $\mathrm{E}(|X_1|) < \infty$ is sufficient for $\bar{X} = O_\mathrm{P}(1)$, but not necessary. For example, suppose that $X_1, \ldots, X_n$ are i.i.d. with the distribution defined in Example 2.11. Then we have $\mathrm{E}(|X_1|) = \infty$. However, according to Exercise 2.25, we have $\bar{X} \xrightarrow{d}$ Cauchy $(0, 1)$; therefore, by (iii) of Theorem 3.1, $\bar{X} = O_\mathrm{P}(1)$.

It should be pointed out that although either (ii) or (iii) of Theorem 3.1 implies $\xi_n = O_\mathrm{P}(1)$, it is often easier to show the latter directly than establishing (ii) or (iii).

*Example 3.8.* Suppose that $X_1, \ldots, X_n$ are i.i.d. observations from the Exponential($\lambda$) distribution with pdf

$$f(x|\lambda) = \frac{1}{\lambda} e^{-x/\lambda}, \quad x \ge 0$$

where $\lambda > 0$ is an unknown parameter. It can be shown that

$$\frac{X_{(n)}}{\log(n)} \xrightarrow{\mathrm{P}} \lambda, \tag{3.10}$$

as $n \to \infty$, where $X_{(n)} = \max_{1 \le i \le n} X_i$. In other words, $X_{(n)}/\log(n)$ is a consistent estimator of $\lambda$ (Exercise 3.11). However, it is easier to show directly that $X_{(n)} = O_\mathrm{P}\{\log(n)\}$. To see this, note that

$$\mathrm{P}\left\{\frac{X_{(n)}}{\log(n)} \le 2\lambda\right\} = \mathrm{P}\{X_{(n)} \le 2\lambda \log(n)\}$$

$$
\begin{aligned}
&= \mathrm{P}\{X_1 \le 2\lambda \log(n), \dots, X_n \le 2\lambda \log(n)\} \\
&= [\mathrm{P}\{X_1 \le 2\lambda \log(n)\}]^n \\
&= (1 - n^{-2})^n \\
&\longrightarrow 1
\end{aligned}
$$

as $n \to \infty$. For any $\epsilon > 0$, there is $N \ge 1$ such that $(1 - n^{-2})^n > 1 - \epsilon, n \ge N$. It follows that (3.8) holds with $\xi_n = X_{(n)}/\log(n)$ and $M = 2\lambda$. Note that this $M$ does not depend on $\epsilon$.

A concept that is often used in large-sample statistics is called $\sqrt{n}$-consistent. Let $\theta$ be a population parameter. A sequence of estimators $\hat{\theta}$ (here we suppress the subscript $n$ in $\hat{\theta}$, as is often done in applied statistics) is called $\sqrt{n}$-consistent if $\sqrt{n}(\hat{\theta} - \theta) = O_\mathrm{P}(1)$.

*Example 3.7 (continued).* Now, suppose the variance of the $X_i$'s exists or, equivalently, $\mathrm{E}(X_1^2) < \infty$. Then the sample mean $\bar{X}$ is a $\sqrt{n}$-consistent estimator of the population mean, $\mu = \mathrm{E}(X_1)$. To see this, note that

$$
\begin{aligned}
\mathrm{E}\left\{\sqrt{n}(\bar{X} - \mu)\right\}^2 &= n\mathrm{E}\left\{\frac{1}{n}\sum_{i=1}^n (X_i - \mu)\right\}^2 \\
&= \frac{1}{n}\sum_{i=1}^n \mathrm{var}(X_i) \\
&= \mathrm{var}(X_1) < \infty.
\end{aligned}
$$

Therefore, by (i) of Theorem 3.1 (with $p = 2$), $\sqrt{n}(\bar{X} - \mu) = O_\mathrm{P}(1)$.

The following are some useful results involving $O_\mathrm{P}(1)$ and $o_\mathrm{P}(1)$.

**Theorem 3.2.** The following hold:
(i) If $\xi_n = O_\mathrm{P}(1)$ and $\eta_n = O_\mathrm{P}(1)$ $[o_\mathrm{P}(1)]$, then $\xi_n \eta_n = O_\mathrm{P}(1)$ $[o_\mathrm{P}(1)]$.
(ii) If $\xi_n$, $n = 1, 2, \dots$, is a sequence of $k \times l$ random matrices, where $k$ and $l$ are fixed, then $\xi_n = O_\mathrm{P}(1)$ $[o_\mathrm{P}(1)]$ if and only if $\xi_{n,ij} = O_\mathrm{P}(1)$ $[o_\mathrm{P}(1)]$, $1 \le i \le k, 1 \le j \le l$.
(iii) If $\xi_n$, $n = 1, 2, \dots$, is a sequence of $k \times k$ random matrices such that $\xi_n \overset{\mathrm{P}}{\longrightarrow} \xi$, where $\xi$ is nonsingular with probability 1. Then $\xi_n^{-1} = O_\mathrm{P}(1)$ and $\xi_n^{-1} - \xi^{-1} = o_\mathrm{P}(1)$.

*Proof.* (i) The proof for the part that $\xi_n = O_\mathrm{P}(1)$ and $\eta_n = O_\mathrm{P}(1)$ imply $\xi_n \eta_n = O_\mathrm{P}(1)$ is left to the reader (Exercise 3.12).
    For any $\epsilon > 0$ and for any $\delta > 0$, since $\xi_n = O_\mathrm{P}(1)$, there is $M > 0$ and $N_1 \ge 1$ such that $\mathrm{P}(|\xi_n \le M) > 1 - \delta, n \ge N_1$. On the other hand, since $\eta_n = o_\mathrm{P}(1)$, there is $N_2 \ge 1$ such that $\mathrm{P}(|\eta_n| > \epsilon/M) < \delta$ if $n \ge N_2$. It follows that when $N \ge N_1 \vee N_2$, we have

$$P(|\xi_n \eta_n| > \epsilon) = P(|\xi_n \eta_n| > \epsilon, |\xi_n| \leq M) + P(|\xi_n \eta_n| > \epsilon, |\xi_n| > M)$$
$$\leq P(|\eta_n| > \epsilon/M) + P(|\xi_n| > M)$$
$$< 2\delta;$$

hence, $P(|\xi_n \eta_n| > \epsilon) \to 0$ as $n \to \infty$, and therefore $\xi_n \eta_n = o_P(1)$.

The proof of (ii) is left to the reader (Exercise 3.12).

(iii) First consider the special case of $k = 1$. According to the results in Appendix A.2, we have $1 = P(|\xi| > 0) = \lim_{k \to \infty} P(|\xi| > 1/k)$. Therefore, for any $\epsilon > 0$, there is $k$ such that $P(|\xi| > 1/k) > 1 - \epsilon$. On the other hand, there is $N \geq 1$ such that $P(|\xi_n - \xi| > 1/2k) < \epsilon$, $n \geq N$. Since $|\xi| \leq |\xi_n| + |\xi_n - \xi|$, $|\xi| > 1/k$ implies that either $|\xi_n| > 1/2k$ or $|\xi_n - \xi| > 1/2k$. Thus, we have

$$1 - \epsilon < P(|\xi| > 1/k)$$
$$\leq P(|\xi_n| > 1/2k) + P(|\xi_n - \xi| > 1/2k)$$
$$< P(|\xi_n| > 1/2k) + \epsilon$$
$$\leq P(|\xi_n^{-1}| \leq 2k) + \epsilon$$

or $P(|\xi_n^{-1}| \leq 2k) > 1 - 2\epsilon$ if $n \geq N$. Therefore, $\xi_n^{-1} = O_P(1)$.

Now, consider the general case $k \geq 1$. We have $\xi_n^{-1} = |\xi_n|^{-1} \xi_n^*$, where $|\xi_n|$ is the determinant of $\xi_n$ and $\xi_n^*$, the adjoint matrix of $\xi_n$. Theorem 2.6 implies that $|\xi_n| \xrightarrow{P} |\xi|$, which is nonzero with probability 1. It follows by part (ii) of Theorem 3.1 and the above result for the $k = 1$ case that each element of $\xi_n^{-1}$ is $O_P(1)$. Thus, once again by part (ii), we have $\xi_n^{-1} = O_P(1)$.

Finally, by the identity $\xi_n^{-1} - \xi^{-1} = \xi^{-1}(\xi - \xi_n)\xi_n^{-1}$ and Corollary 3.1, we have $\|\xi_n^{-1} - \xi^{-1}\| \leq \|\xi^{-1}\| \cdot \|\xi_n^{-1}\| \cdot \|\xi_n - \xi\| = O(1)O_P(1)o_P(1) = o_P(1)$, using, once again, part (ii). Q.E.D.

## 3.5 The unspecified $c$ and other similar methods

Near the end of the proof of Theorem 3.2, we simplified the arguments by writing $O(1)O_P(1)o_P(1) = o_P(1)$. This is actually a useful technique in that although the big $O$'s and small $o$'s are different in their values, there is no need to distinguish them and hence use different notation every time they appear, as far as asymptotics are concerned. A similar technique will be explored in this section.

In many cases, the asymptotic arguments involve a series of inequalities and bounds, but the actual values of the constants involved are not important. For example, if the goal is to derive $a_n = O(b_n)$, then it does not matter whether $a_n \leq b_n$ or $a_n \leq 2b_n$. In other words, as long as one shows $a_n \leq cb_n$ for some constant $c$, it does not make a difference whether $c = 1$ or $c = 2$ as far as the order is concerned. Therefore, in those arguments, we let $c$ represent a positive generic constant whose value may be different at different places (e.g., Shao and Wu 1987, pp. 1566). We illustrate the use of such an unspecified $c$ by some examples.

*Example 3.9.* Suppose that $X_1, X_2, \ldots$ is a sequence of martingale differences with respect to the $\sigma$-fields $\mathcal{F}_i = \sigma(X_1, \ldots, X_i)$, $i \geq 1$. This means that $E(X_1) = 0$ and $E(X_i | X_1, \ldots, X_{i-1}) = 0$ a.s. for any $i \geq 2$. See Chapter 8 for more detail. Furthermore, each $X_i$ has a Uniform$[-1/2, 1/2]$ distribution. For example, if $Y_1, Y_2, \ldots$ are independent and distributed as Uniform$[0,1]$, then $X_i = Y_i - 1/2$, $i \geq 1$, satisfy the above condition of martingale differences as well as the distributional assumption.

Now suppose that one wishes is to obtain the order of $E(\bar{X}^4)$, where $\bar{X} = n^{-1} \sum_{i=1}^{n} X_i$. A formal derivation with specific values of all the constants involved may be given as follows. First, by Burkholder's inequality (see Section 5.4), we have

$$E(\bar{X}^4) = \frac{1}{n^4} E \left( \sum_{i=1}^{n} X_i \right)^4$$

$$\leq \frac{\left( 18 \times 4 \times \sqrt{4/3} \right)^4}{n^4} E \left( \sum_{i=1}^{n} X_i^2 \right)^2$$

$$= \frac{47775744}{n^4} E \left( \sum_{i=1}^{n} X_i^2 \right)^2.$$

Next, by the convex function inequality (see Section 5.1), we have

$$\left( \frac{1}{n} \sum_{i=1}^{n} X_i^2 \right)^2 \leq \frac{1}{n} \sum_{i=1}^{n} X_i^4,$$

which implies

$$\left( \sum_{i=1}^{n} X_i^2 \right)^2 \leq n \sum_{i=1}^{n} X_i^4.$$

It follows that

$$E \left( \sum_{i=1}^{n} X_i^2 \right)^2 \leq n \sum_{i=1}^{n} E(X_i^4)$$

$$= n^2 E(X_1^4).$$

Finally, a simple calculation gives

$$E(X_1^4) = \int_{-1/2}^{1/2} x^4 \, dx = \frac{1}{80}.$$

Therefore, by combining the pieces we get

$$E(\bar{X}^4) \leq \frac{47775744}{n^4} \times \frac{n^2}{80}$$
$$= \frac{597196.8}{n^2}.$$

However, if we use the unspecified $c$, the derivation can be simplified as follows:

$$E(\bar{X}^4) = \frac{1}{n^4}E\left(\sum_{i=1}^{n} X_i\right)^4$$
$$\leq \frac{c}{n^4}E\left(\sum_{i=1}^{n} X_i^2\right)^2$$
$$\leq \frac{c}{n^3}\sum_{i=1}^{n}E(X_i^4)$$
$$\leq \frac{c}{n^2}.$$

In the series of inequalities above, $c$ represents possibly a different constant at each step. So, mathematically speaking, some of these inequalities might not hold if $c$ were to represent the same constant. However, there is no need to work out the specific value of $c$ at each step or to use different notation such as $c_1, c_2, \ldots$ at different steps. In other words, $c$ is a notation just like the big $O$ and small $o$. The end result is all that matters; for example, in Example 3.9, $E(\bar{X}^4) \leq cn^{-2}$ for some constant $c$.

Here is another reason why the specific value of $c$ may not be important. Consider Example 3.9. At the end, we obtained the value of the constant as 597196.8, but do you believe that the constant really has to be this large? In fact, the constant in Burkholder's inequality is for the general situations of martingale differences. In any specific case (such as the i.i.d. Uniform case mentioned in Exampe 3.9), the constant may be improved (i.e., reduced). This is why the actual value of $c$ is not so important (because it may not be so accurate). Here is another example.

*Example 3.10* (Finite sample correction). It is not unusual that a well-known statistic is slightly modified for improved finite-sample performance. For example, the sample proportion, defined as

$$\hat{p} = \frac{Y}{n},$$

is a well-known estimator of the population proportion $p$. Here, $Y = Y_1 + \cdots + Y_n$ and $Y_1, \ldots, Y_n$ are i.i.d. Bernoulli($p$) observations. In some cases, the following alternative estimator of $p$ is considered:

$$\tilde{p} = \frac{Y + a}{n + b},$$

where $a$ and $b$ are some constants. Among different choices of $a$ and $b$ are $a = 2$ and $b = 4$ for constructing a 95% confidence interval for $p$, or, more precisely and generally, $a = 0.5Z^2_{\alpha/2}$ and $b = Z^2_{\alpha/2}$ for constructing a $100(1 - \alpha)\%$ confidence interval for $p$ (e.g., Samuels and Witmer 2003, pp. 209–210), where $Z_{\alpha/2}$ is the $\alpha/2$ critical value of the standard normal distributions [i.e., $P(Z > Z_{\alpha/2}) = \alpha/2$, where $Z \sim N(0,1)$]. Such a modification is often called a finite-sample correction, with the implication that it would maintain the same large-sample behavior of $\hat{p}$ (and, meanwhile, improve the finite-sample performance in some sence). But does it?

To verify this, we consider the difference $d_n = \sqrt{n}(\tilde{p} - p) - \sqrt{n}(\hat{p} - p)$. The motivation is that, according to the CLT, $\sqrt{n}(\hat{p} - p) \xrightarrow{d} N\{0, p(1 - p)\}$ as $n \to \infty$. So, if one can show $d_n \xrightarrow{P} 0$ as $n \to \infty$, then, by Theorem 2.13, we have $\sqrt{n}(\tilde{p} - p) \xrightarrow{d} N\{0, p(1 - p)\}$ as $n \to \infty$. In other words, $\tilde{p}$ has the same large-sample property in terms of asymptotic distribution as $\hat{p}$. By using an unspecified $c$, a simple argument can be given as follows. By Theorem 2.15, it suffices to show that $E(d_n^2) \to 0$ as $n \to \infty$. We have $E(d_n^2) = nE(\tilde{p} - \hat{p})^2$. On the other hand, we have

$$\tilde{p} - \hat{p} = \frac{an - by}{n(n + b)}$$

$$= \frac{a}{n + b} - \left(\frac{b}{n + b}\right)\hat{p}.$$

It follows that

$$E(\tilde{p} - \hat{p})^2 = \left(\frac{a}{n + b}\right)^2 - 2\left(\frac{a}{n + b}\right)\left(\frac{b}{n + b}\right)E(\hat{p}) + \left(\frac{b}{n + b}\right)^2 E(\hat{p}^2)$$

$$\leq \frac{c}{n^2} + \frac{c}{n^2} \times p + \frac{c}{n^2} \times 1$$

$$\leq \frac{c}{n^2}.$$

Thus, we have $E(d_n^2) \leq cn^{-1}$ and, hence, $\to 0$ as $n \to \infty$. Note that not only have we shown $E(d_n^2) \to 0$, we also obtained its convergence rate as $O(n^{-1})$.

As mentioned, notationwise $c$ is very similar to big $O$ and small $o$. In fact, the latter can be operated in very much the same way. For example, we have $O(1)O(1) = O(1)$, $O(1)o(1) = o(1)$, $O(1) + o(1) = O(1)$, and so forth, even though the actual values of $O(1)$s and $o(1)$s may be different at different places. We demonstrate this with a simple example.

*Example 3.11* (Finite population proportion). In Example 3.10 we assumed that $Y_1, \ldots, Y_n$ are i.i.d. Bernoulli observations. Such an assumption holds only if the population from which the $Y_i$'s are sampled is infinite. In real life, however, the population is usually finite, no matter how large. What would happen if one samples from a finite population?

Consider a finite population with $N$ items, of which $D$ are defective and $N - D$ are not. Suppose that a sample of $n$ items are drawn at random so that all $\binom{N}{n}$ possible samples of size $n$ are equally likely. Let $Y_i = 1$ if the $i$th item drawn is defective and $Y_i = 0$ otherwise. Then we have

$$P(Y_i = 1) = \frac{D}{N}, \quad 1 \leq i \leq N. \tag{3.11}$$

Thus, the $Y_i$'s are identically distributed, even though they are not independent (Exercise 3.13). It follows that

$$E(Y_i) = \frac{D}{N}, \tag{3.12}$$

$$\mathrm{var}(Y_i) = \frac{D}{N}\left(1 - \frac{D}{N}\right), \tag{3.13}$$

and it can be shown that (Exercise 3.13)

$$\mathrm{cov}(Y_i, Y_j) = -\frac{D(N - D)}{N^2(N - 1)}, \quad i \neq j. \tag{3.14}$$

Now, consider $Y = \sum_{i=1}^{n} Y_i$, the total number of defective items in the sample. It is known that $Y$ has a hypergeometric distribution (e.g., Casella and Berger 2002, p. 622). The sample proportion of defective items is therefore $\hat{p} = Y/n$. By (3.12)–(3.14), it can be shown that

$$E(\hat{p}) = \frac{D}{N}, \tag{3.15}$$

$$\mathrm{var}(\hat{p}) = \frac{1}{n}\frac{N - n}{N - 1}\frac{D}{N}\left(1 - \frac{D}{N}\right) \tag{3.16}$$

(Exercise 3.13).

Although in real life the population is usually finite, the population size can be huge, so the infinite population model of Example 3.10 may be used as an approximation. More precisely, consider the following asymptotic framework in which the population size, $N$, is increasing such that

$$\frac{D}{N} \longrightarrow p,$$

where $p \in (0, 1)$. Furthermore, we assume that $n = o(N)$; that is, the sample size is negligible compared to the population size. It follows by (3.15) that

$$E(\hat{p}) \sim p. \tag{3.17}$$

As for the variance, we can write, by (3.16),

$$\text{var}(\hat{p}) = \frac{1}{n} \frac{N - o(N)}{N - 1} \{p + o(1)\}\{1 - p - o(1)\}$$

$$= \frac{1}{n} \frac{1 - o(1)}{1 - o(1)} \{p + o(1)\}\{1 - p + o(1)\}$$

$$= \frac{p(1 - p)}{n} \{1 + o(1)\}\{1 + o(1)\}\{1 + o(1)\}$$

$$= \frac{p(1 - p)}{n} \{1 + o(1)\}$$

$$\sim \frac{p(1 - p)}{n}. \tag{3.18}$$

Note that the right sides of (3.17) and (3.18) are exactly the mean and variance, respectively, of $\hat{p}$ under the infinite population sampling (Example 3.10).

## 3.6 Case study: The baseball problem

Efron and Morris (1973) considered the problem of predicting batting averages of 18 major league baseball players during the 1970 season. The authors used this problem as an example to demonstrate the performance of their empirical Bayes method. The dataset has since been analyzed by several authors, including Morris (1983), Gelman et al. (1995), Datta and Lahiri (2000), and Jiang and Lahiri (2006). Efron and Morris first obtained the batting average of Roberto Clemente, an extremely good hitter, from the *New York Times* dated April 26, 1970 when he had already batted 45 times. The batting average of a player is the proportion of hits among the number at-bats. They then selected 17 other major league baseball players who had also batted 45 times from the April 26 and May 2, 1970 issues of the *New York Times*. They considered the problem of predicting the batting averages of all the 18 players for the remainder of the 1970 season based on their batting averages for first 45 at-bats. The authors used the following simple model for the prediction problem:

$$Y_i = \mu + v_i + e_i, \quad i = 1, \ldots, n,$$

where $\mu$ is an unknown mean, $v_i$ is a player-specific random effect, and $e_i$ is the sampling error. It is assumed that the $v_i$'s are independent and distributed as $N(0, A)$, where $A$ is an unknown variance; the $e_i$'s are independent standard normal random variables; and the $v_i$'s and $e_i$'s are independent. The true batting average of a particular player $i$ is $\theta_i = \mu + v_i$, whose prediction is of main interest. Without loss of generality, let $i = 1$.

For the sake of simplicity we assume for the rest of this section that $\mu = 0$. In this case, the best predictor (BP) of $\theta_1 = v_1$ is

$$\tilde{\theta}_1 = \frac{A}{A + 1} Y_1. \tag{3.19}$$

See Chapter 13 for more details about the prediction problem. Because $A$ is unknown, the BP is not computable. In such a case, it is customary to replace $A$ in (3.19) by an estimator, say the MLE, which is given by $\hat{A} = n^{-1} \sum_{i=1}^{n} Y_i^2 - 1$. This leads to the so-called empirical best predictor (EBP),

$$\hat{\theta}_1 = \frac{\hat{A}}{\hat{A}+1} Y_1.$$

The question is how large the difference is between the EBP and BP in terms of the prediction performance.

To answer this question, we first introduce a lemma, which was used in the proofs of Jiang, Lahiri, and Wan (2002b) to establish the asymptotic un-biasedness of their jackknife estimator of the mean squared error (MSE) of an empirical predictor, such as the EBP. The jackknife method will be discussed in detail in Chapter 14. We use this lemma (and its proof) to demonstrate the use of the unspecified $c$ discussed in the previous section. The $c$ in the following lemma and its proof therefore represents the unspecified constant.

**Lemma 3.17.** Let $\xi_n$, $\eta_n$, and $\zeta_n$ be sequences of random variables and let $\mathcal{A}_n$ be a sequence of events. Suppose that $\xi_n = \eta_n + \zeta_n$ on $\mathcal{A}_n$ and the following hold: $\mathrm{E}(\xi_n^2 1_{\mathcal{A}_n^c}) \leq cn^{-a_1}$, $\mathrm{E}(\eta_n^2 1_{\mathcal{A}_n^c}) \leq cn^{-a_2}$, $\mathrm{E}(\eta_n^2) \leq c$, and $|\zeta_n| \leq n^{-a_3} \nu_n$ with $\mathrm{E}(\nu_n^2) \leq c$, where the $a$'s are positive constants. Then, for any $0 < \epsilon \leq a_1 \wedge a_2 \wedge a_3$, we have

$$\left| \mathrm{E}(\xi_n^2) - \mathrm{E}(\eta_n^2) \right| \leq cn^{-\epsilon},$$

where $c$ depends only on the $a$'s and the (unspecified) $c$'s.

*Proof.* We have

$$\mathrm{E}(\xi_n^2) - \mathrm{E}(\eta_n^2) = \mathrm{E}(\xi_n^2 - \eta_n^2) 1_{\mathcal{A}_n} + \mathrm{E}(\xi_n^2 1_{\mathcal{A}_n^c}) - \mathrm{E}(\eta_n^2 1_{\mathcal{A}_n^c})$$
$$= \mathrm{E}(2\eta_n \zeta_n + \zeta_n^2) 1_{\mathcal{A}_n} + \mathrm{E}(\xi_n^2 1_{\mathcal{A}_n^c}) - \mathrm{E}(\eta_n^2 1_{\mathcal{A}_n^c}).$$

Thus, we have

$$\left| \mathrm{E}(\xi_n^2) - \mathrm{E}(\eta_n^2) \right| \leq cn^{-a_3} \mathrm{E}(|\eta_n| \nu_n) + cn^{-2a_3} \mathrm{E}(\nu_n^2) + cn^{-a_1} + cn^{-a_2}$$
$$\leq cn^{-a_1} + cn^{-a_2} + cn^{-a_3}$$
$$\leq cn^{-\epsilon}.$$

Note that, by the Cauchy–Schwarz inequality (see Chapter 5), we have $\mathrm{E}(|\eta_n| \nu_n) \leq (\mathrm{E}\eta_n^2)^{1/2} (\mathrm{E}\nu_n^2)^{1/2} \leq c$. Q.E.D.

Now, return to the baseball prediction problem. Let $\xi_n = \hat{\theta}_1 - \theta_1$, $\eta_n = \tilde{\theta}_1 - \theta_1$, and $\hat{B} = (\hat{A}+1) \vee 0.5$. Then it is easy to show that $\xi_n = \eta_n + \zeta_n$ on $\mathcal{A}_n = \{\hat{A} \geq -0.5\}$, where

$$\zeta_n = \frac{\hat{A} - A}{(A+1)\hat{B}} Y_1.$$

Furthermore, we have, by the Cauchy–Schwarz inequality,

$$E\left(\xi_n^2 1_{\mathcal{A}_n^c}\right) \le \{E(\xi_n^4)\}^{1/2}\{P(\mathcal{A}_n^c)\}^{1/2}$$
$$\le c\{P(\mathcal{A}_n^c)\}^{1/2}.$$

Note that $|\xi_n| \le |Y_1| + |\theta_1| \le 2|v_1| + |e_1|$, whose $k$th moment is finite for any $k > 0$. On the other hand, let $X_i = Y_i^2 - A - 1$. By Chebyshev, Burkholder, and the convex function inequalities (see Chapter 5), we have, for any $k > 2$,

$$
\begin{aligned}
P\left(\mathcal{A}_n^c\right) &= P\left(\frac{1}{n}\sum_{i=1}^n X_i < -A - 0.5\right)\\
&\le P\left(\left|\frac{1}{n}\sum_{i=1}^n X_i\right| > A + 0.5\right)\\
&\le \frac{c}{n^k}E\left(\left|\sum_{i=1}^n X_i\right|^k\right)\\
&\le \frac{c}{n^k}E\left\{\left(\sum_{i=1}^n X_i^2\right)^{k/2}\right\}\\
&= \frac{c}{n^{k/2}}E\left\{\left(\frac{1}{n}\sum_{i=1}^n X_i^2\right)^{k/2}\right\}\\
&\le \frac{c}{n^{k/2}}E\left(\frac{1}{n}\sum_{i=1}^n |X_i|^k\right)\\
&\le \frac{c}{n^{k/2}}.
\end{aligned}
$$

Thus, we have $E(\xi_n^2 1_{\mathcal{A}_n^c}) \le cn^{-k/4}$. By the same argument, it can be shown that $E(\eta_n^2 1_{\mathcal{A}_n^c}) \le cn^{-k/4}$. Furthermore, it is easy to show that $E(\eta_n^2) \le c$. Finally, we have $|\zeta_n| \le cn^{-1/2}|\sqrt{n}(\hat{A} - A)| \cdot |Y_1| = n^{-1/2}\nu_n$ with

$$
\begin{aligned}
E(\nu_n^2) &\le c \cdot nE\{(\hat{A} - A)^2 Y_1^2\}\\
&\le c \cdot n\{E(\hat{A} - A)^4\}^{1/2}\{E(Y_1^4)\}^{1/2}\\
&\le c \cdot n\{E(\hat{A} - A)^4\}^{1/2}\\
&\le c \cdot n \cdot n^{-1}\\
&\le c,
\end{aligned}
$$

using the same inequalities as above. Now, apply Lemma 3.17 with $a_1 = a_2 = k/4$ and $a_3 = 1/2$ to obtain

$$
\begin{aligned}
\left|\text{MSE}(\hat{\theta}_1) - \text{MSE}(\tilde{\theta}_1)\right| &= \left|E(\hat{\theta}_1 - \theta_1)^2 - E(\tilde{\theta}_1 - \theta_1)^2\right|\\
&\le cn^{-1/2};
\end{aligned}
$$

that is, the difference between the MSE of the EBP and that of the BP is $O(n^{-1/2})$. It will be shown later in Chapter 13 that the difference is, in fact, $O(n^{-1})$. By the way, it is easy to show that $\mathrm{MSE}(\tilde{\theta}_1) = A/(A+1) = O(1)$.

## 3.7 Case study: Likelihood ratio for a clustering problem

In community ecology, the term "clustering" is synonymous with what is commonly known as "classification." However, in statistics, there is a major difference between the two. The difference lies in the existence of a training dataset for classification, whereas no such data are available for clustering. For example, suppose that a group of individuals are labeled as men and women, and information about their heights and weights is available. This information provides the training data. Now, suppose that a new individual comes in with an unknown label (i.e., the gender of the individual is unknown) but known height and weight, and we wish to classify this new individual into one of the two classes, men or women, based on his/her height and weight. This is a classification problem. If, instead, the group of individuals are unlabeled (i.e., their genders are unknown) and we wish to classify them into an unknown number of classes based on their heights and weights, we have a clustering problem. Due to such a difference, classification is often associated with the so-called supervised learning (via the training data), whereas clustering is associated with the unsupervised one.

An important problem in cluster analysis is to test the existence of clusters. Consider perhaps the simplest case in which a standard normal distribution is tested against a mixture of the standard normal with another normal distribution with the same variance but a different mean. Let $X_1, \ldots, X_n$ be i.i.d. observations from a normal mixture distribution

$$(1 - p)N(0, 1) + pN(\theta, 1), \tag{3.20}$$

where $\theta$ is an unknown parameter and $p$ is an unknown proportion. We are interested in testing $H_0: \theta = 0$ against $H_a: \theta \neq 0$. Note that the null hypothesis indicates that there is only one cluster in the population distribution (or there is no clustering), whereas the alternative implies that there may be two clusters (or there may be a clustering). The reason that the alternative does not imply for sure that there is clustering is because, when $p = 0$, the distribution of (3.20) becomes $N(0, 1)$ regardless of $\theta$. Therefore, the test result is more decisive when the null hypothesis is accepted than it is rejected. This seemingly unpleasant phenomenon is due to the fact that the distribution of $X_i$ is unidentifiable when $p = 0$.

An alternative testing problem is also often considered; that is, $H_0: p = 0$ against $H_a: p \neq 0$. Note that this is equivalent to the above testing problem in the null hypothesis (i.e., $N(0, 1)$), which implies no clustering. However, there is no escape from the identifiability problem—when $\theta = 0$, the distribution of $X_i$ is $N(0, 1)$ regardless of $p$.

Now suppose that one wishes to test the null hypothesis (in either formation) using likelihood ratio test (LRT). To be more specific, let us focus on the first formation of the testing problem. Standard asymptotic theory (see Chapter 6) asserts that, under regularity conditions, the asymptotic null distribution of the LRT statistic, which is

$$\mathcal{L}^* = 2 \left[ \log \left\{ \sup_{\theta, p} L(\theta, p | X) \right\} - \log \left\{ \sup_{p} L(0, p | X) \right\} \right]$$
$$= 2 \sup_{\theta, p} l^*(\theta, p | X), \tag{3.21}$$

where $L(\theta, p | X)$ is the likelihood function and

$$l^*(\theta, p | X) = \sum_{i=1}^{n} \log \left\{ 1 - p + p \, \exp \left( X_i \theta - \frac{1}{2} \theta^2 \right) \right\} \tag{3.22}$$

(Exercise 3.14), is $\chi^2$ with two degrees of freedom. Here, by asymptotic null distribution we mean the asymptotic distribution under the null hypothesis $\theta = 0$, and the two degrees of freedom corresponds to the number of unknown parameters ($\theta$ and $p$) that have to be estimated. However, one of the regularity conditions requires that the distribution of $X_i$ be identifiable. As mentioned, this condition is not satisfied in this case. The question then is: Does the LRT still have the asymptotic $\chi^2$-distribution under the null hypothesis? The answer is no. In fact, Hartigan (1985) showed that, under the null hypothesis, the LRT statistic (3.21) $\to \infty$ in probability as $n \to \infty$. Hereafter, a sequence of random variable $\xi_n \to \infty$ in probability if for any $M > 0$ the probability $P(\xi_n > M) \to 1$ as $n \to \infty$. Note that this problem is closely related to Example 1 in the Preface. In this case, the asymptotic null distribution of the LRT does not even exist.

Hartigan's proof showed that the divergence of the LRT statistic was an example of $O_P$ and $o_P$ in action. The arguments given below are similar in spirit. For any fixed $\theta \neq 0$, write $l^* = l^*(\theta, p | X)$ for notation simplicity. Also, write $Y_i = \exp(X_i \theta - 0.5\theta^2)$ and $Z_i = Y_i - 1$. Then $Z_1, \ldots Z_n$ are i.i.d. with $E(Z_i) = 0$ and $\text{var}(Z_i) = e^{\theta^2} - 1$ (Exercise 3.14). Furthermore, we have $l^* = \sum_{i=1}^{n} \log(1 + p Z_i)$, and

$$\frac{\partial^2 l^*}{\partial p^2} = - \sum_{i=1}^{n} \frac{Z_i^2}{(1 + p Z_i)^2} < 0$$

with probability 1. It follows that, with probability 1, $l^*$ is strictly concave in $p$, and therefore there is a unique maximum for $p \in [0, 1]$. Denote this maximum by $\hat{p}$.

Next, we show that $\hat{p} = o_P(1)$. For any $p > 0$, we have

$$\frac{\partial l^*}{\partial p} = \sum_{i=1}^{n} \frac{Z_i}{1 + p Z_i}$$

$$= \sum_{i=1}^{n} \frac{1+pZ_i - 1}{1+pZ_i} p^{-1}$$

$$= p^{-1} \sum_{i=1}^{n} \left(1 - \frac{1}{1+pZ_i}\right)$$

$$= p^{-1} \sum_{i=1}^{n} \psi(Z_i),$$

where $\psi(x) = 1-(1+px)^{-1}$. Since $\psi''(x) = -2p^2(1+px)^{-3} < 0$, $\psi(x)$ is strictly concave. It follows by Jensen's inequality (see Chapter 5) that $E\{\psi(Z_i)\} < \psi\{E(Z_i)\} = 1 - \{1 + pE(Z_i)\}^{-1} = 0$. Thus, by the WLLN, we have

$$\frac{p}{n} \cdot \frac{\partial l^*}{\partial p} = \frac{1}{n} \sum_{i=1}^{n} \psi(Z_i)$$

$$= E\{\psi(Z_1)\} + \frac{1}{n} \sum_{i=1}^{n} [\psi(Z_i) - E\{\psi(Z_1)\}]$$

$$= E\{\psi(Z_1)\} + o_P(1).$$

By the properties of a concave function, we have with probability 1 that $\partial l^*/\partial p < 0$ implies $\hat{p} < p$ (why?). It follows that

$$P(\hat{p} \geq p) \leq P\left(\frac{\partial l^*}{\partial p} \geq 0\right)$$

$$= P\left(\frac{p}{n}\frac{\partial l^*}{\partial p} \geq 0\right)$$

$$= P[E\{\psi(Z_1)\} + o_P(1) \geq 0]$$

$$= P[o_P(1) \geq -E\{\psi(Z_1)\}]$$

$$\to 0$$

as $n \to \infty$. Since $p$ is arbitrary and $\hat{p} \geq 0$, we have $\hat{p} = o_P(1)$ by the definition.

We now go one step further to obtain an asymptotic expansion for $\hat{p}$. Let $A_n$ and $B_n$ be two sequences of events. We say $B_n$ holds with probability tending to 1 on $A_n$ or $B_n$ w.p. $\to 1$ on $A_n$ if $P(A_n \setminus B_n) = P(A_n \cap B_n^c) \to 0$ as $n \to \infty$. In the special case of $A_n = \Omega$, the entire sample space, this is the same as that $B_n$ holds with probability tending to 1 or $B_n$ w.p. $\to 1$. First, note that if $\sum_{i=1}^{n} Z_i \leq 0$, then, by the properties of a concave function, we have $\partial l^*/\partial p \leq \partial l^*/\partial p|_{p=0} = \sum_{i=1}^{n} Z_i \leq 0$; hence, $l^* \leq l^*(0,0|X) = 0$ (i.e., $\hat{p} = 0$). Now, suppose that $\sum_{i=1}^{n} Z_i > 0$. Then $\partial l^*/\partial p|_{p=0} > 0$; hence, $\hat{p} > 0$. On the other hand, the argument above shows that $\hat{p} < 1$ w.p. $\to 1$ (Exercise 3.14). It follows that $\hat{p} \in (0,1)$ w.p. $\to 1$ on $\sum_{i=1}^{n} Z_i > 0$; hence, $\partial l^*/\partial p|_{\hat{p}} = 0$ w. p. $\to 1$ on $\sum_{i=1}^{n} Z_i > 0$. Write $g_i(p) = \log(1+pZ_i)$. Then we have $g_i(0) = 0$, $g_i'(0) = Z_i$, $g_i''(0) = -Z_i^2$, and $g_i'''(p) = 2Z_i^3(1+pZ_i)^{-3}$. By the Taylor expansion (see the next chapter), we have, w.p. $\to 1$ on $\sum_{i=1}^{n} Z_i > 0$,

$$0 = \left.\frac{\partial l^*}{\partial p}\right|_{\hat{p}}$$

$$= \sum_{i=1}^{n} g_i'(\hat{p})$$

$$= \sum_{i=1}^{n} \left\{ g_i'(0) + g_i''(0)\hat{p} + \frac{1}{2}g_i'''(p_i)\hat{p}^2 \right\}$$

$$= \sum_{i=1}^{n} Z_i - \hat{p}\sum_{i=1}^{n} Z_i^2 + \frac{\hat{p}^2}{2}\sum_{i=1}^{n} g_i'''(p_i), \tag{3.23}$$

where $0 \le p_i \le \hat{p}$. Since $Z_i = Y_i - 1 \ge -1$, we have $1 + p_i Z_i \ge 1 - p_i \ge 1 - \hat{p} = 1 + o_P(1)$; hence, $|g_i'''(p_i)| \le 2\{1 + o_P(1)\}^{-3}|Z_i|^3$, where $o_P(1)$ does not depend on $i$. It follows that

$$\frac{1}{2}\left|\sum_{i=1}^{n} g_i'''(p_i)\right| \le \{1 + o_P(1)\}^{-3}\sum_{i=1}^{n} |Z_i|^3$$

$$= \{1 + o_P(1)\}^{-3}O_P(n)$$

$$= O_P(n),$$

using the WLLN. Therefore, we have, by (3.23),

$$\sum_{i=1}^{n} Z_i = \hat{p}\left\{ \sum_{i=1}^{n} Z_i^2 - \frac{\hat{p}}{2}\sum_{i=1}^{n} g_i'''(p_i) \right\}$$

$$= \hat{p}\left\{ \sum_{i=1}^{n} Z_i^2 - \hat{p}O_P(n) \right\}$$

$$= n\hat{p}\left\{ \frac{1}{n}\sum_{i=1}^{n} Z_i^2 - \hat{p}O_P(1) \right\}$$

$$= n\hat{p}\{E(Z_1^2) + o_P(1)\},$$

again using the WLLN. Thus, we obtain the following asymptotic expansion:

$$\hat{p} = \frac{1}{E(Z_1^2) + o_P(1)}\frac{1}{n}\sum_{i=1}^{n} Z_i$$

$$= \left\{ \frac{1}{E(Z_1^2)} + o_P(1) \right\}\frac{1}{n}\sum_{i=1}^{n} Z_i$$

$$= \frac{\sum_{i=1}^{n} Z_i}{nE(Z_1^2)} + o_P(n^{-1/2}), \tag{3.24}$$

using the results of Theorem 3.2(iii) and Example 3.7. In conclusion, we have $\hat{p} = 0$ if $\sum_{i=1}^{n} Z_i \le 0$ and (3.24) w.p. $\to 1$ on $\sum_{i=1}^{n} Z_i > 0$.

We now use (3.24) to obtain an asymptotic expansion of $\hat{l}^* = l^*(\theta, \hat{p}|X)$. If $\sum_{i=1}^{n} Z_i \leq 0$, we have $\hat{l}^* = l^*(\theta, 0|X) = 0$. On the other hand, we have, again by the Taylor expansion,

$$
\begin{aligned}
\hat{l}^* &= \sum_{i=1}^{n} g_i(\hat{p}) \\
&= \sum_{i=1}^{n} \left\{ g_i(0) + g_i'(0)\hat{p} + \frac{1}{2}g_i''(0)\hat{p}^2 + \frac{1}{6}g_i'''(\tilde{p}_i)\hat{p}^3 \right\} \\
&= \hat{p}\sum_{i=1}^{n} Z_i - \frac{\hat{p}^2}{2}\sum_{i=1}^{n} Z_i^2 + \frac{\hat{p}^3}{6}\sum_{i=1}^{n} g_i'''(\tilde{p}_i),
\end{aligned}
\tag{3.25}
$$

where $0 \leq \tilde{p}_i \leq \hat{p}$. Now, suppose that (3.24) holds. It follows that $\hat{p} = O_P(n^{-1/2})$. Thus, by an argument similar to the above, it can be shown that the last term on the right side of (3.25) is $\hat{p}^3 O_P(n)$, which is $o_P(1)$. Now, combine (3.24) and (3.25) to get

$$
\begin{aligned}
\hat{l}^* &= \left\{ \frac{\sum_{i=1}^{n} Z_i}{nE(Z_1^2)} + o_P(n^{-1/2}) \right\} \sum_{i=1}^{n} Z_i \\
&\quad - \frac{1}{2}\left\{ \frac{\sum_{i=1}^{n} Z_i}{nE(Z_1^2)} + o_P(n^{-1/2}) \right\}^2 n\{E(Z_1^2) + o_P(1)\} + o_P(1) \\
&= \frac{(\sum_{i=1}^{n} Z_i)^2}{nE(Z_1^2)} + o_P(1) \\
&\quad - \frac{1}{2}\left[ \frac{(\sum_{i=1}^{n} Z_i)^2}{n^2\{E(Z_1^2)\}^2} + o_P(n^{-1}) \right] n\{E(Z_1^2) + o_P(1)\} + o_P(1) \\
&= \frac{(\sum_{i=1}^{n} Z_i)^2}{nE(Z_1^2)} - \frac{1}{2}\left\{ \frac{(\sum_{i=1}^{n} Z_i)^2}{nE(Z_1^2)} + o_P(1) \right\} + o_P(1) \\
&= \frac{(\sum_{i=1}^{n} Z_i)^2}{2nE(Z_1^2)} + o_P(1),
\end{aligned}
\tag{3.26}
$$

using the facts that $\sum_{i=1}^{n} Z_i = O_P(n^{1/2})$ (Example 3.7) and $\sum_{i=1}^{n} Z_i^2 = nn^{-1}\sum_{i=1}^{n} Z_i^2 = n\{E(Z_1^2) + o_P(1)\}$ by the WLLN. Thus, in conclusion, we have $\hat{l}^* = 0$ if $\sum_{i=1}^{n} Z_i \leq 0$ and (3.26) w.p. $\to 1$ on $\sum_{i=1}^{n} Z_i > 0$. It follows that the following holds w.p. $\to 1$:

$$
\begin{aligned}
\hat{l}^* &= \left\{ \frac{(\sum_{i=1}^{n} Z_i)^2}{2nE(Z_1^2)} + o_P(1) \right\} 1_{(\sum_{i=1}^{n} Z_i > 0)} \\
&= \frac{(\sum_{i=1}^{n} Z_i)^2}{2nE(Z_1^2)} 1_{(\sum_{i=1}^{n} Z_i > 0)} + o_P(1).
\end{aligned}
\tag{3.27}
$$

The rest of the arguments is the same as those in Hartigan (1985). Write

$$
U_n(\theta) = \frac{\sum_{i=1}^{n} Z_i}{\sqrt{nE(Z_1^2)}}.
$$

Note that the quantity depends on $\theta$ because the $Z_i$'s do. By the CLT, we have $U_n \xrightarrow{d} N(0,1)$ as $n \to \infty$ for each fixed $\theta$. Furthermore, for any collection of (positive) $\theta$'s, the corresponding $U_n(\theta)$'s are asymptotically jointly normal with mean 0, variance 1, and correlation between $U_n(\theta_j)$ and $U_n(\theta_k)$ given by

$$\frac{e^{\theta_j \theta_k} - 1}{\left\{\left(e^{\theta_j^2} - 1\right)\left(e^{\theta_k^2} - 1\right)\right\}^{1/2}}. \tag{3.28}$$

For any $M > 0$ and for any $\epsilon > 0$, choose an integer $m \geq 1$ such that $\Phi(M)^m < \epsilon/2$, where $\Phi(x)$ is the cdf of $N(0,1)$. Now, choose $\theta_1, \ldots, \theta_m > 0$ such that all pairwise correlations (3.28) are sufficiently small so that

$$\lim_{n \to \infty} \mathrm{P}\left\{\max_{1 \leq j \leq m} U_n(\theta_j) \leq M\right\} < \epsilon \tag{3.29}$$

(Exercise 3.14). Note that when the correlations between random variables $U_1, \ldots, U_m$, which are jointly normal and each distributed as $N(0,1)$, are very small, the $U_j$'s are nearly independent; hence, $\mathrm{P}(\max_{1 \leq j \leq m} U_j \leq M) \approx \Phi(M)^m < \epsilon/2$. Thus, (3.29) is possible.

Note that the $\hat{l}^*$ in (3.27) depends on $\theta$ [i.e., $\hat{l}^* = \hat{l}^*(\theta)$]. For the $\theta_1, \ldots, \theta_m$ chosen above, we have, by (3.27), that w.p. $\to 1$,

$$2\mathcal{L}^* \geq \max_{1 \leq j \leq m} 2\hat{l}^*(\theta_j)$$
$$\geq \max_{1 \leq j \leq m} \left\{U_n^2(\theta_j)1_{(U_n(\theta_j)>0)}\right\} + o_\mathrm{P}(1). \tag{3.30}$$

Thus, w.p. $\to 1$, $2\mathcal{L}^* \leq M^2 - 1$ implies $\max_{1 \leq j \leq m} U_n(\theta_j) \leq M$ (Exercise 3.14). It follows by (3.29) that

$$\limsup \mathrm{P}\left(\mathcal{L}^* \leq \frac{M^2 - 1}{2}\right) \leq \epsilon.$$

Because $\epsilon$ is arbitrary, this proves that $\mathcal{L}^* \to \infty$ in probability.

To add a little bit of drama (even further) to the story, Hartigan's theoretical result was not supported by the results of a series of empirical studies. For example, Wolfe (1971) suggested that the asymptotic distribution of $2\{(n-3)/n\}\mathcal{L}^*$ was $\chi_2^2$, although Wolfe's study was based only on 100 replications of sample size $n = 100$. A much more extensive simulation study was carried out later by Atwood et al. (1996). The authors generated 90,000 replications of each sample size from 50 to 500 in increments of 25 in order to find an empirical distribution of $2\mathcal{L}^*$. Furthermore, to explore the asymptotic distribution of $2\mathcal{L}^*$ the authors generated 10,000 replications for each of the sample size 1000, 2000, 4000, 8000, 16,000, 32,000 and 64,000. In addition, 3211 replications were generated for the sample size 256,000. Yet, the authors found no trace of $2\mathcal{L}^*$ going to infinity. For example, the simulated mean and variance of $2\mathcal{L}^*$ were found to be approximately 2.11 and 4.27, respectively, for the sample size $n = 500$ and 2.02 and 4.21, respectively, for the sample size $n = 64,000$. These values are very close to the mean and variance of a $\chi^2$-distribution with two degrees of freedom, which are 2 and 4, respectively. Based on their simulation results, the authors concluded that the asymptotic distribution of $2\mathcal{L}^*$ could well be $\chi_2^2$.

There is at least one explanation for the seeming contradiction between theoretical and empirical results, which happens, but not surprisingly, to have something to do with the order. At the end of Hartigan's paper, the author has a remark on how fast $2\mathcal{L}^*$ goes to infinity. He estimated the rate of divergence as $\log\log(n)$. To see what this means, suppose that $n$ is one million (1,000,000), which is much larger than any of the sample sizes considered above. Then $\log\log(n)$ is approximately 2.6, which is well within the range of $\chi_2^2$!

## 3.8 Exercises

*3.1.* Verify the properties of Lemma 3.4.

*3.2.* Verify the properties of Lemma 3.5.

*3.3.* Consider the function $f(x)$ in Example 3.3.

(i) Suppose that $a_p, b_q$ are nonzero. Show that as $|x| \to \infty$, $f(x) = o(1)$ if $p < q$, $f(x) = O(1)$ if $p = q$, and $1/f(x) = o(1)$ if $p > q$.

(ii) Suppose that $b_0 \neq 0$. Show that as $x \to 0$, $f(x) = O(1)$ regardless of $p$ and $q$.

*3.4.* Recall Stirling's formula (Example 3.4). Define

$$d_n = \log\left\{ \frac{n!}{\sqrt{n}(n/e)^n} \right\}$$

$$= \log(n!) - \left(n + \frac{1}{2}\right)\log(n) + n.$$

Show that the sequence $d_n$ is decreasing.

*3.5.* Complete the proof of Stirling's formula. Note that there are various proofs of this famous approximation. For example, a standard proof involves the use of the Wallis formula:

$$\prod_{k=1}^{\infty} \frac{(2k)^2}{(2k-1)(2k+1)} = \frac{\pi}{2};$$

an alternative proof can be given via the CLT (e.g., Casella and Berger 2002, pp. 261). You are asked to find at least one complete proof of Stirling's formula.

*3.6.* Prove Lemma 3.7.

*3.7.* Prove parts (i), (ii), and (iii) of Lemma 3.11. (Hint: Use Lemma 3.9.)

*3.8.* Prove Corollary 3.1. [Hint: Use Lemma 3.10 for parts (i) and (ii) and Lemma 3.9 for part (iii).]

*3.9.* Let $A$ be a $k \times l$ matrix whose $(i, j)$ element is $a_{ij}$, $1 \leq i \leq k, 1 \leq j \leq l$. Show that

$$\max_{i,j} |a_{ij}| \leq \|A\| \leq \sqrt{kl} \max_{i,j} |a_{ij}|.$$

Use this result to prove (i) and (ii) of Lemma 3.15.

*3.10.* Show that if $\xi_n \overset{d}{\longrightarrow} \xi$, where $\xi$ is a degenerate random variable [i.e., there is a constant $c$ such that $P(\xi = c) = 1$], then $\xi_n \overset{P}{\longrightarrow} \xi$.

*3.11.* Consider the observations $X_1, \ldots, X_n$ in Example 3.8.
(i) Show that

$$P\left\{ \frac{X_{(n)}}{\log(n)} \leq x \right\} \longrightarrow \begin{cases} 0, & x < \lambda \\ e^{-1}, & x = \lambda \\ 1, & x > \lambda. \end{cases}$$

(ii) Use (i) and the result of the previous exercise to show (3.10).

*3.12.* (i) Complete the proof of the first part of part (i) of Theorem 3.2; that is, $\xi_n = O_P(1)$ and $\eta_n = O_P(1)$ imply $\xi_n \eta_n = O_P(1)$.
(ii) Prove part (ii) of Theorem 3.2.

*3.13.* Consider the $Y_i$'s defined in Example 3.11.
(i) Show that the $Y_i$'s are identically distributed [i.e., (3.11)].
(ii) Show that the $Y_i$'s are not independent.
(iii) Verify (3.14).
(iv) Verify (3.15) and (3.16).

*3.14.* This problem is associated with Section 3.7.
(i) Verify that the LRT statistic is given by (3.21) and (3.22).
(ii) For fixed $\theta$, consider the random variable $Y_i$ defined therein. Show that for any real number $k$,

$$E(Y_i^k) = \exp\left\{ \frac{k(k-1)}{2} \theta^2 \right\}.$$

It follows that $E(Z_i) = 0$ and $\text{var}(Z_i) = e^{\theta^2} - 1$, where $Z_i = Y_i - 1$.
(iii) Show that $\hat{p} < 1$ with probability tending to 1.
(iv) Show by the inequality (3.30) that w.p. $\to 1$, and $2\mathcal{L}^* \leq M^2 - 1$ implies that $\max_{1 \leq j \leq m} U_n(\theta_j) \leq M$.
(v) Show that for any $\delta > 0$ and any $l \geq 1$, one can choose $\theta_1, \ldots, \theta_l > 0$ such that all pairwise correlations (3.28) are less than $\delta$.
(vi) Furthermore, let $U_1, \ldots, U_l$ be jointly normal, each have $N(0, 1)$ distribution, and the correlations between $U_j$ and $U_k$ be given by (3.28). Show that as $\delta \to 0$, $P(\max_{1 \leq j \leq l} U_j \leq x) \to \Phi(x)^l$ for every $x$, where $\delta$ is the maximum absolute value of the correlations between the $U_j$'s.

*3.15.* Determine the order relation of the following sequences $a_n$ and $b_n$:
(i) $a_n = c_0 + c_1 n + \cdots + c_k n^k$, $b_n = a^n$ for any positive integer $k$ and $a > 1$, where $c_1, \ldots, c_k$ are constants.

(ii) $a_n = \{\log(n)\}^{1-\epsilon}$, $b_n = \log(n^\delta)$ for any $0 < \epsilon < 1$ and $\delta > 0$.

(iii) $a_n = \exp[\{\log(n)\}^\epsilon]$, $b_n = n^\delta$ for any $0 < \epsilon < 1$ and $\delta > 0$.

(iv) $a_n = (n/a)^n$, $b_n = n!$, where $a > 0$. (Note: Depending on the value of $a$, the conclusion may be different.)

3.16. Determine the order relation of the following sequences $a_n$ and $b_n$.

(i) $a_n = (n+c)^n$, $b_n = n^n$, where $c$ is any constant.

(ii) $a_n = \sum_{i=1}^{n} i^{-1}$, $b_n = \log(n)$.

(iii) $a_n = c_0 + c_1 n + \cdots + c_k n^k$, $b_n = d_1 + d_2 n + \cdots + d_l n^l$, where $c_1, \ldots, c_k$ and $d_1, \ldots, d_l$ are constants such that $c_k d_l \neq 0$. (Note: The answer depends on the values of $k$ and $l$.)

(iv) $a_n = c_0 + c_1 n^{-1} + \cdots + c_k n^{-k}$, $b_n = d_1 + d_2 n^{-1} + \cdots + d_l n^{-1}$, where $c_1, \ldots, c_k$ and $d_1, \ldots, d_l$ are constants such that $c_0 d_0 \neq 0$. Does the answer depend on the values of $k$ and $l$?

3.17. What sequence is this: $1, 1, 2, 3, 5, 8, 13, 21, 34, 55, \ldots$? If you examine the numbers carefully, you will realize that these are the famous Fibonacci numbers, or Fibonacci sequence, defined by $F_1 = 1$, $F_2 = 1$, and $F_n = F_{n-1} + F_{n-2}$, $n = 3, 4, \ldots$. Fibonacci (Leonardo Pisano) posed the following problem in his treatise *Liber Abaci* published in 1202:

> How many pairs of rabbits will be produced in a year, beginning with a single pair, if in every month each pair bears a new pair which becomes productive from the second month in?

Not surprisingly, the answer is the Fibonacci sequence. Note that the number grows quickly (so did the population of rabbits—it once happened in Australia!).

(i) Show that $\log(F_n) = O(n)$.

(ii) Find the limit $\lim_{n \to} F_n/n$.

(iii) Show that $F_n \log(F_n) \sim nF_{n+1}$ [the definition of $a_n \sim b_n$ is in Section 3.2 above (3.6)].

3.18. The distribution of a continuous random variable $X$ is symmetric if the pdf of $X$, $f(x)$, satisfies $f(-x) = f(x)$ for all $x$; that is, $f(x)$ is symmetric about zero. Suppose that $X_1, \ldots, X_n$ are i.i.d. random variables. In each of the following cases show that the distribution of $X_1$ is symmetric and also obtain the order, in terms of $O_P$, of $(\sum_{i=1}^{n} X_i)^4$:

(i) $X_1 \sim N(0,1)$.

(ii) $X_1 \sim t_5$, the $t$-distribution with five degrees of freedom.

(iii) $X_1$ has the pdf $f(x)$ given in Example 2.11. What do you think is the reason for the order in this case to be different from the previous two cases?

3.19. Let $X_1, \ldots, X_n$ be i.i.d. random variables such that $E(X_i) = 0$ and $\mathrm{var}(X_i) = \sigma^2$, where $\sigma^2 \in (0, \infty)$.

(i) Show that $X. = \sum_{i=1}^{n} X_i = O_P(\sqrt{n})$.

(ii) If you think (i) is straightforward, show that $e^{X.}$ is not $O_P\left(e^{a\sqrt{n}}\right)$ for any constant $a > 0$ (no matter how how large). In other words, for any $a > 0$, $e^{X.}/e^{a\sqrt{n}}$ is not bounded in probability.

(iii) Show that $e^{X} = o\left(e^{bn^{0.5+\delta}}\right)$ for any constants $\delta, b > 0$ (no matter how small).

*3.20.* Let $X_1, \ldots, X_n$ be independent standard normal random variables. Determine the orders of the following sequences of random variables $\xi_n$:

$$\text{(i)} \qquad \xi_n = \left(\sum_{i-1}^{n} X_i\right)\left(\sum_{i=1}^{n} X_i^2\right)\left(\sum_{i=1}^{n} X_i^3\right).$$

$$\text{(ii)} \qquad \xi_n = \frac{\left(\sum_{i=1}^{n} X_i\right)\left(\sum_{i=1}^{n} X_i^3\right)}{\left(1 + \sum_{i=1}^{n} X_i^2\right)\left(1 + \sum_{i=1}^{n} X_i^4\right)}.$$

$$\text{(iii)} \qquad \xi_n = \frac{\left(\sum_{i=1}^{n} X_i^3\right)^2}{1 + \sum_{i=1}^{n} X_i^6}.$$

$$\text{(iv)} \qquad \xi_n = \frac{\left(\sum_{i=1}^{n} X_i^2\right)^4}{1 + \sum_{i=1}^{n} X_i^8}.$$

*3.21.* Let $X_1, \ldots, X_n$ be independent Exponential(1) random variables.
(i) Prove the identity

$$\frac{1}{\sum_{i=1}^{n} X_i} = \frac{1}{n} - \frac{\sum_{i=1}^{n} X_i - n}{n^2} + \frac{\left(\sum_{i=1}^{n} X_i - n\right)^2}{n^3} - \frac{\left(\sum_{i=1}^{n} X_i - n\right)^3}{n^3 \sum_{i=1}^{n} X_i}.$$

(ii) Show that

$$\frac{\sum_{i=1}^{n} X_i^2}{\sum_{i=1}^{n} X_i} = \frac{1}{n}\sum_{i=1}^{n} X_i^2 - \frac{1}{n^2}\left(\sum_{i=1}^{n} X_i^2\right)\left(\sum_{i=1}^{n} X_i - n\right)$$

$$+ \frac{1}{n^3}\left(\sum_{i=1}^{n} X_i^2\right)\left(\sum_{i=1}^{n} X_i - n\right)^2 + O_P(n^{-3/2}).$$

(iii) What are the orders of the first three terms on the right side of the equation in (ii)?

*3.22.* Suppose that $X_1, \ldots, X_n$ are i.i.d. observations whose mgf [defined by (2.9)] exists for some $t > 0$. Show that $X_{(n)} = O_P\{\log(n)\}$, where $X_{(n)}$ is the largest order-statistic (defined below Example 1.5).

# 4

# Asymptotic Expansions

## 4.1 Introduction

One of the techniques used in the latest case study (Section 3.7) is an asymptotic expansion of an estimator as well as that for a log-likelihood function. The most well-known asymptotic expansion is the Taylor expansion, which is a mathematical tool, rather than a statistical method. However, the method is used so extensively in both theoretical and applied statistics that its role in statistics can hardly be overstated. Several other expansions, including the Edgeworth expansion and Laplace approximation, can be derived from the Taylor expansion. It should be pointed out that some elementary expansions can also be very useful (see Section 4.5).

Asymptotic expansions are extremely helpful in cases where the quantities of interest do not have closed-form expressions. Sometimes, even when the quantity does have a closed form, an asymptotic expansion may still be useful in simplifying the expression and revealing the dominant factor(s). For example, consider the following.

*Example 4.1* (Variance estimation in linear regression). A multiple linear regression model may be expressed as

$$Y_i = \beta_0 + \beta_1 x_{i1} + \cdots + \beta_p x_{ip} + \epsilon_i, \tag{4.1}$$

$i = 1, \ldots, n$, where $x_{i1}, \ldots, x_{ip}$ are known covariate, $\beta_0, \ldots, \beta_p$ are unknown regression coefficients, and $\epsilon_i$ is a random error. Here, we assume that $\epsilon_1, \ldots, \epsilon_n$ are independent and distributed as $N(0, \sigma^2)$, where $\sigma^2$ is an unknown variance.

The error variance is typically estimated by the unbiased estimator

$$\hat{\sigma}^2 = \frac{\text{RSS}}{n - p - 1},$$

where RSS represents the residual sum of squares, $\sum_{i=1}^{n} \hat{\epsilon}_i^2$, where $\hat{\epsilon}_i = Y_i - \hat{Y}_i$, $\hat{Y}_i$ is the fitted value given by $\hat{Y}_i = \hat{\beta}_0 + \hat{\beta}_1 x_{i1} + \cdots + \hat{\beta}_p x_{ip}$, and $\hat{\beta} = (\hat{\beta}_0, \ldots, \hat{\beta}_p)'$

J. Jiang, *Large Sample Techniques for Statistics*,
DOI 10.1007/978-1-4419-6827-2_4, © Springer Science+Business Media, LLC 2010

is the least squares estimator of $\beta = (\beta_0, \ldots, \beta_p)'$ given by

$$\hat{\beta} = (X'X)^{-1}X'Y. \tag{4.2}$$

In (4.2), $Y$ is the vector of observations, $Y = (Y_i)_{1 \le i \le n}$, and $X$ the matrix of covariates, $X = (x_{ij})_{1 \le i \le n, 0 \le j \le p}$ with $x_{i0} = 1$, $1 \le i \le n$, corresponding to the intercept. Using this notation, (4.1) can be expressed as:

$$Y = X\beta + \epsilon, \tag{4.3}$$

where $\epsilon = (\epsilon_i)_{1 \le i \le n}$. Furthermore, the residuals and RSS can be expressed in terms of a project matrix $P_{X^\perp} = I_n - P_X$, where $P_X = X(X'X)^{-1}X'$; that is, $\hat{e} = (\hat{e}_i)_{1 \le i \le n} = P_{X^\perp}y$ and RSS $= |\hat{e}|^2 = Y'P_{X^\perp}Y$. Here, we assume, for simplicity, that $X$ is of full (column) rank, but similar expressions can be obtained even without this restriction. See Appendix A.1 for the definition and properties of projection matrices. It follows that

$$\hat{\sigma}^2 = \frac{Y'P_{X^\perp}Y}{n - p - 1}.$$

Alternatively, since normality is assumed, $\sigma^2$ may be estimated by the MLE, which can be expressed as

$$\tilde{\sigma}^2 = \frac{Y'P_{X^\perp}Y}{n}.$$

Both estimators have closed-form expressions, although the expression for $\tilde{\sigma}^2$ is even simpler. On the other hand, an asymptotic expansion shows the close relation between the two estimators in terms of decreasing orders. Note that

$$\frac{1}{1 - x} = 1 + x + x^2 + \cdots \tag{4.4}$$

for any $0 \le x < 1$. Then, for $n > p + 1$, we have

$$\left(1 - \frac{p+1}{n}\right)^{-1} = 1 + \frac{p+1}{n} + \left(\frac{p+1}{n}\right)^2 + \cdots$$

$$= 1 + \frac{p+1}{n} + O\left(\frac{1}{n^2}\right), \tag{4.5}$$

provided that $n \to \infty$ and $p$ is bounded.

Expansion (4.5) implies the following connection between $\hat{\sigma}^2$ and $\tilde{\sigma}^2$:

$$\hat{\sigma}^2 = \left(1 - \frac{p+1}{n}\right)^{-1}\tilde{\sigma}^2$$

$$= \tilde{\sigma}^2 + \left(\frac{p+1}{n}\right)\tilde{\sigma}^2 + O_P\left(\frac{1}{n^2}\right). \tag{4.6}$$

The reason that the remaining terms is $O_P(n^{-2})$ is because $E(|\hat{\epsilon}|^2) = \sigma^2(n - p - 1)$. This follows from the unbiasedness of $\hat{\sigma}^2$; alternatively, it can also be derived using the following simple arguments:

$$
\begin{aligned}
E(|\hat{\epsilon}|^2) &= E(Y'P_{X^\perp}Y) \\
&= E\{(Y - X\beta)'P_{X^\perp}(Y - X\beta)\} \\
&= E[\text{tr}\{(Y - X\beta)'P_{X^\perp}(Y - X\beta)\}] \\
&= E[\text{tr}\{P_{X^\perp}(Y - X\beta)(Y - X\beta)'\}] \\
&= \text{tr}[E\{P_{X^\perp}(Y - X\beta)(Y - X\beta)'\}] \\
&= \text{tr}[P_{X^\perp}E\{(Y - X\beta)(Y - X\beta)'\}] \\
&= \sigma^2 \text{tr}(P_{X^\perp}) \\
&= \sigma^2(n - p - 1).
\end{aligned}
$$

Here, we used the facts that one can exchange the order of trace and expectation and that $E\{(Y - X\beta)(Y - X\beta)'\} = \text{Var}(Y) = \sigma^2 I_n$. It follows from Theorem 3.1 that $|\hat{\epsilon}|^2 = O_P(n)$; hence, $\tilde{\sigma}^2 = O_P(1)$. Equation (4.6) shows that $\tilde{\sigma}^2$ is the leading $[O_P(1)]$ term in an expansion of $\hat{\sigma}^2$. It also shows that the next term in the expansion is $\{(p+1)/n\}\tilde{\sigma}^2$, which is $O_P(n^{-1})$, and the next term is $O_P(n^{-2})$, and so forth. Even though (4.6) is derived as a large-sample approximation, assuming that $n \to \infty$ while $p$ remains fixed or bounded, it can also be useful under a finite-sample consideration. For example, it shows that if the number of covariates, $p$, is comparable to the sample size [i.e., if the ratio $(p+1)/n$ is not very small], the difference between the two methods of variance estimation can be nontrivial. In fact, the latter is the reason for the failure of consistency of the MLE in Example 3 of the Preface. See Chapter 12 for a further discussion.

## 4.2 Taylor expansion

It is the author's view that the Taylor expansion is the single most useful mathematical tool for a statistician. We begin by revisiting (4.4). There is more than one way to derive this identity, one of which is to use the Taylor expansion. First, compute the derivatives of $f(x) = (1-x)^{-1}$. We have $f'(x) = (1-x)^{-2}$, $f''(x) = 2(1-x)^{-3}$, $f'''(x) = 6(1-x)^{-4}$, and so on. Thus, we obtain (4.4) as the Taylor series at $a = 0$

$$
\begin{aligned}
f(x) &= f(a) + f'(a)(x - a) + \frac{f''(a)}{2}(x - a)^2 + \frac{f'''(a)}{6}(x - a)^3 + \cdots \\
&= \sum_{k=0}^{\infty} \frac{f^{(k)}(a)}{k!}(x - a)^k,
\end{aligned}
\tag{4.7}
$$

where $f^{(k)}(x)$ represents the $k$th derivative with $f^{(0)}(x) = f(x)$.

The formal statement of the Taylor expansion is the following.

**Theorem 4.1** (Taylor's theorem). Suppose that the $l$th derivative of $f(x)$ is continuous on $[a, b]$ and the $(l+1)$st derivative of $f(x)$ exists on $(a, b)$. Then for any $x \in [a, b]$, we have

$$f(x) = f(a) + f'(a)(x - a) + \cdots + \frac{f^{(l)}(a)}{l!}(x - a)^l + \frac{f^{(l+1)}(c)}{(l + 1)!}(x - a)^{l+1}$$

$$= \sum_{k=0}^{l} \frac{f^{(k)}(a)}{k!}(x - a)^k + \frac{f^{(l+1)}(z)}{(l + 1)!}(x - a)^{l+1}, \tag{4.8}$$

where $z$ lies between $a$ and $x$; that is, $z = (1 - t)a + tx$ for some $t \in [0, 1]$.

If, in particular, $f^{(k)}(x)$ exists for all $k$ and the last term on the right side of (4.8) $\to 0$ as $n \to \infty$, then (4.7) holds, which is called the Taylor series. In the special case of $a = 0$, the Taylor series is also known as the Maclaurin's series. For example, in addition to (4.4), we have

$$\frac{1}{1 + x} = 1 - x + x^2 - x^3 + x^4 - \cdots,$$

$$e^x = 1 + x + \frac{x^2}{2!} + \frac{x^3}{3!} + \frac{x^4}{4!} + \cdots,$$

$$\log(1 + x) = x - \frac{x^2}{2} + \frac{x^3}{3} - \frac{x^4}{4} + \cdots,$$

$$\sin(x) = x - \frac{x^3}{3!} + \frac{x^5}{5!} - \frac{x^7}{7!} + \cdots,$$

$$\cos(x) = 1 - \frac{x^2}{2!} + \frac{x^4}{4!} - \frac{x^6}{6!} + \cdots.$$

English mathematician Brook Taylor (1685–1731) published a general method for constructing the Taylor series (which are now named after him) in 1715, although various forms of special cases were known much earlier. Colin Maclaurin, a Scottish mathematician who was once a professor in Edinburgh, published the special case of the Taylor series in the 18th century.

It should be pointed out that the Taylor expansion is a local property of a function. This means that the closer $x$ is to $a$ the more accurate is the approximation. We illustrate this with an example.

*Example 4.2.* Consider the accuracy of the Maclaurin expansion for the function $f(x) = e^x$. The Taylor (Maclaurin) series for $e^x$ is given above. Table 4.1 shows the approximations using the first $n$ terms in the series, where the relative error is computed as the absolute value of the approximation error divided by the true value. It is clear that the approximation is much more accurate for $x = 1$ than for $x = 5$. This is because the expansion is at $x = 0$,

**Table 4.1. Maclaurin expansion for $f(x) = e^x$**

| $n$ | $e \approx 2.718$ | Relative Error | $e^5 \approx 148.4$ | Relative Error |
|---|---|---|---|---|
| 1 | 1.000 | 0.632 | 1.0 | 0.993 |
| 2 | 2.000 | 0.264 | 6.0 | 0.960 |
| 3 | 2.500 | 0.080 | 18.5 | 0.875 |
| 4 | 2.667 | 0.019 | 39.3 | 0.735 |
| 5 | 2.708 | 0.004 | 65.4 | 0.560 |
| 6 | 2.717 | 0.001 | 91.4 | 0.384 |

and $x = 1$ is much closer to zero than $x = 5$. However, as long as $n$ is large enough, the same accuracy will be achieved for any $x$ (Exercise 4.1).

A multivariate extension of the Taylor expansion is perhaps even more useful in practice. To illustrate the results, we define a linear differential operator as follows. Let $x = (x_1, \ldots, x_s) \in R^s$, and $\nabla$ denote the gradient operator, or vector differential operator, defined by $\nabla = (\partial/\partial x_1, \ldots, \partial/\partial x_s)'$. Consider the linear differential operator

$$x'\nabla = \sum_{i=1}^{s} x_i \frac{\partial}{\partial x_i}.$$

Note that $(x'\nabla)^k$ can be operated in a similar way as the $k$th power of a sum. For example, with $s = 2$, we have

$$(x'\nabla)^2 = \left( x_1 \frac{\partial}{\partial x_1} + x_2 \frac{\partial}{\partial x_2} \right)^2$$
$$= x_1^2 \frac{\partial^2}{\partial x_1^2} + 2x_1 x_2 \frac{\partial^2}{\partial x_1 \partial x_2} + x_2^2 \frac{\partial^2}{\partial x_2^2};$$
$$(x'\nabla)^3 = \left( x_1 \frac{\partial}{\partial x_1} + x_2 \frac{\partial}{\partial x_2} \right)^3$$
$$= x_1^3 \frac{\partial^3}{\partial x_1^3} + 3x_1^2 x_2 \frac{\partial^3}{\partial x_1^2 x_2} + 3x_1 x_2^2 \frac{\partial^3}{\partial x_1 x_2^2} + x_2^3 \frac{\partial^3}{\partial x_2^3};$$

and so on. The multivariate Taylor expansion, or the Taylor expansion in several variables, can be stated as follows.

**Theorem 4.2** (Multivariate Taylor expansion). Let $f : D \to R$, where $R \subset R^s$. Suppose that there is a neighborhood of $a$, $S_\delta(a) \subset D$ such that $f$ and its up to $(l+1)$st partial derivatives are continuous in $S_\delta(a)$. Then, for any $x \in S_\delta(a)$, we have

$$f(x) = f(a) + \sum_{k=1}^{l} \frac{1}{k!} \{(x-a)'\nabla\}^k f(a)$$

$$+\frac{1}{(l+1)!}\{(x-a)'\nabla\}^{l+1}f(z), \tag{4.9}$$

where $z = (1-t)a + tx$ for some $t \in [0,1]$.

In the author's experience, the (multivariate) Taylor expansion of second or third orders are most useful in practice. For such expansions, there is an alternative expression that may be more interpretable and easier to use. Let

$$\frac{\partial f(x)}{\partial x'} = \left[\frac{\partial f(x)}{\partial x_1}, \dots, \frac{\partial f(x)}{\partial x_s}\right], \tag{4.10}$$

$$\frac{\partial^2 f(x)}{\partial x \partial x'} = \left[\frac{\partial^2 f(x)}{\partial x_i \partial x_j}\right]_{1\le i,j\le s}. \tag{4.11}$$

Note that (4.10) is the transpose of the gradient vector, or $\{\nabla f(x)\}'$, and (4.11) is the matrix of second derivatives, or Hessian matrix. Then the second-order Taylor expansion can be expressed as

$$f(x) = f(a) + \frac{\partial f(a)}{\partial x'}(x-a) + \frac{1}{2}(x-a)'\frac{\partial^2 f(z)}{\partial x \partial x'}(x-a), \tag{4.12}$$

where $z = (1-t)a + tx$ for some $t \in [0,1]$. Equation (4.12) shows that, locally (i.e., in a neighborhood of $a$), $f(x)$ can be approximated by a quadratic function. For example, suppose that $a$ is a point such that $\partial f(a)/\partial x = \{\partial f(a)/\partial x'\}' = 0$. Furthermore, suppose that, locally, the Hessian matrix of $f(x)$ is positive definite. It follows from (4.12) that $f(x) > f(a)$ near $a$ and, hence, has a unique local minimum at $x = a$. Similarly, the third-order Taylor expansion can be expressed as

$$f(x) = f(a) + \frac{\partial f(a)}{\partial x'}(x-a) + \frac{1}{2}(x-a)'\frac{\partial^2 f(a)}{\partial x \partial x'}(x-a)$$
$$+\frac{1}{6}\left[(x-a)'\frac{\partial^3 f(z)}{\partial x_i \partial x \partial x'}(x-a)\right]'_{1\le i\le s}(x-a), \tag{4.13}$$

where $z = (1-t)a + tx$ for some $t \in [0,1]$. Note that in a small neighborhood of $a$, the third-order term (i.e., the last term) in (4.13) is dominated by the leading quadratic function.

The last term in the Taylor expansion (i.e., the term that involves $z$), is called the remaining term. This term is sometimes expressed in terms of a small $o$ or big $O$. For example, suppose that all of the $(l+1)$st partial derivatives of $f(x)$ are bounded in the neighborhood $S_\delta(a)$. Then (4.9) can be written as

$$f(x) = f(a) + \sum_{k=1}^{l}\frac{1}{k!}\{(x-a)'\nabla\}^k f(a) + o(|x-a|^l), \tag{4.14}$$

where $|x-a| = \{\sum_{i=1}^{s}(x_i - a_i)^2\}^{1/2}$, or

$$f(x) = f(a) + \sum_{k=1}^{l} \frac{1}{k!} \{(x-a)'\nabla\}^k f(a) + O(|x-a|^{l+1}). \qquad (4.15)$$

In particular, (4.13) can be expressed as

$$f(x) = f(a) + \frac{\partial f(a)}{\partial x'}(x-a) + \frac{1}{2}(x-a)'\frac{\partial^2 f(a)}{\partial x \partial x'}(x-a)$$
$$+ o(|x-a|^2), \qquad (4.16)$$

and the $o(|x-a|^2)$ can be replaced by $O(|x-a|^3)$. However, caution should be paid when using such an expression for a large-sample approximation, in which the function $f(x)$ may depend on the sample size $n$.

*Example 4.3.* Suppose that $X_1, \ldots, X_n$ are i.i.d. observations from the Logistic($\theta$) distribution whose pdf is given by

$$f(x|\theta) = \frac{e^{\theta-x}}{(1+e^{\theta-x})^2}, \quad -\infty < x < \infty.$$

Consider the second-order Taylor expansion of the log-likelihood function,

$$l(\theta) = \sum_{i=1}^{n} \log\{f(X_i|\theta)\}$$

at the true $\theta$, which we assume, for simplicity, to be zero. Then the Taylor expansion can be expressed as

$$l(\theta) = l(0) + l'(0)\theta + \frac{1}{2}l''(\tilde{\theta})\theta^2, \qquad (4.17)$$

where $\tilde{\theta}$ lies between zero and $\theta$. It can be shown (Exercise 4.2) that

$$l(0) = -\sum_{i=1}^{n}\{X_i + 2\log(1 + e^{-X_i})\},$$

$$l'(0) = \sum_{i=1}^{n} \frac{1 - e^{-X_i}}{1 + e^{-X_i}},$$

$$l''(\theta) = -2\sum_{i=1}^{n} \frac{e^{\theta-X_i}}{(1+e^{\theta-X_i})^2}.$$

Furthermore, it can be shown that $l(0) = O_P(n)$, $l'(0) = O_P(\sqrt{n})$, and $\sup_\theta |l''(\theta)| = O_P(n)$. Now, suppose that one wishes to study the behavior of the log-likelihood near the true value $\theta = 0$ by considering a sequence $\theta_n = t/\sqrt{n}$, where $t$ is a constant known as the local deviation (e.g., Bickel et al. 1993, p. 17). If one blindly uses (4.15) (with $s = 1$ and $l = 1$), one would have $l(\theta) = l(0) + l'(0)\theta + O_P(\theta^2)$ [here we use $O_P(\theta^2)$ instead of $O(\theta^2)$ because $l(\theta)$ is random]; hence,

$$l(\theta_n) = l(0) + l'(0)\theta_n + O_P(\theta_n^2). \tag{4.18}$$

The first term on the right side of (4.18) is $O_P(n)$, the second term is $O_P(n^{1/2})(t/\sqrt{n}) = O_P(1)$, and the third terms appears to be $O_P(n^{-1})$. This suggests that the third term is negligible because it is of lower order than the second term. However, this is not true because the more accurate expression (4.17) (with $\theta = \theta_n$) shows that the third term is $O_P(n)(t^2/n) = O_P(1)$, which is of the same order as the second term.

We conclude this section with an example of a well-known application of the Taylor expansion. More examples will be discussed in the sequel.

*Example 4.4* (The delta-method). Let $\xi_n$, $n = 1, 2, \ldots$ be a sequence of $s$-dimensional random vectors such that

$$a_n(\xi_n - c) \xrightarrow{d} \eta \tag{4.19}$$

as $n \to \infty$, where $c$ is a constant vector, $a_n$ is a sequence of positive constants such that $a_n \to \infty$ as $n \to \infty$, and $\eta$ is an $s$-dimensional random vector. Then for any continuously differentiable function $g(x)\colon R^s \to R$, we have

$$a_n\{g(\xi_n) - g(c)\} \xrightarrow{d} \frac{\partial g(c)}{\partial x'}\eta \tag{4.20}$$

as $n \to \infty$. To establish (4.20), use the first-order Taylor expansion to get

$$g(\xi_n) = g(c) + \frac{\partial g(\zeta_n)}{\partial x'}(\xi_n - c),$$

where $\zeta_n$ lies between $c$ and $\xi_n$. It follows that $|\zeta_n - c| \leq |\xi_n - c|$. (4.19) and the fact that $a_n \to \infty$ implies that $\xi_n - c = o_P(1)$; hence, $\zeta_n - c = o_P(1)$. It follows that (Exercise 4.3)

$$\frac{\partial g(\zeta_n)}{\partial x'} - \frac{\partial g(c)}{\partial x'} = o_P(1).$$

Therefore, we have, by Theorem 2.13,

$$
\begin{aligned}
a_n\{g(\xi_n) - g(c)\} &= \frac{\partial g(c)}{\partial x'}a_n(\xi_n - c) + \left\{\frac{\partial g(\zeta_n)}{\partial x'} - \frac{\partial g(c)}{\partial x'}\right\}a_n(\xi_n - c) \\
&= \frac{\partial g(c)}{\partial x'}a_n(\xi_n - c) + o_P(1)O_P(1) \\
&= \frac{\partial g(c)}{\partial x'}a_n(\xi_n - c) + o_P(1) \\
&\xrightarrow{d} \frac{\partial g(c)}{\partial x'}\eta.
\end{aligned}
$$

In particular, let $\theta$ be a $s$-dimensional parameter vector and $\hat{\theta}$ be an estimator of $\theta$ based on i.i.d. observations $X_1, \ldots, X_n$. We say the estimator $\hat{\theta}$ is asymptotically normal if

$$\sqrt{n}(\hat{\theta} - \theta) \xrightarrow{d} N(0, \Sigma), \tag{4.21}$$

where $\Sigma$ is called the asymptotic covariance matrix. It follows that for any differentiable function $g(x)$: $R^s \rightarrow R$, we have

$$\sqrt{n}\{g(\hat{\theta}) - g(\theta)\} \xrightarrow{d} N(0, \sigma^2), \tag{4.22}$$

where

$$\sigma^2 = \frac{\partial g(\theta)}{\partial x'} \Sigma \frac{\partial g(\theta)}{\partial x},$$

where $\partial g(\theta)/\partial x = (\partial g(\theta)/\partial x')'$. For example, suppose that $X_1, \ldots, X_n$ are i.i.d. with $E(X_i) = \mu$ and $\text{var}(X_i) = \sigma^2 \in (0, \infty)$. Then, according to the CLT, we have $\sqrt{n}(\bar{X} - \mu) \xrightarrow{d} N(0, \sigma^2)$, where $\bar{X}$ is the sample mean. It follows that the following hold (Exercise 4.4):

$$\sqrt{n}(e^{\bar{X}} - e^{\mu}) \xrightarrow{d} N(0, e^{2\mu}\sigma^2),$$

$$\sqrt{n}\{\log(1 + \bar{X}^2) - \log(1 + \mu^2)\} \xrightarrow{d} N\left\{0, \frac{4\mu^2\sigma^2}{(1 + \mu^2)^2}\right\},$$

$$\sqrt{n}\left(\frac{\bar{X}}{1 + \bar{X}^2} - \frac{\mu}{1 + \mu^2}\right) \xrightarrow{d} N\left\{0, \frac{(1 - \mu^2)^2\sigma^2}{(1 + \mu^2)^4}\right\}.$$

Obviously, many more such results can be derived.

## 4.3 Edgeworth expansion; method of formal derivation

The central limit theorem (CLT) states that, subject to a mild condition (that the second moment is finite), the sample mean $\bar{X}$ of i.i.d. observations $X_1, \ldots, X_n$ is asymptotically normal in the sense that

$$\frac{\sqrt{n}}{\sigma}(\bar{X} - \mu) \xrightarrow{d} N(0, 1) \tag{4.23}$$

as $n \rightarrow \infty$, where $\mu = E(X_1)$ and $\sigma^2 = \text{var}(X_1)$. Over the years, this astonishing result has amazed, surprised, or even confused its users. For example, it says that no matter what the population distribution (of $X_i$) is, the limiting distribution on the right side of (4.23) is alway the same: the standard normal distribution. Imagine how different the population distribution can be in terms of its shape: symmetric, skewed, bimodal, continuous, discrete, and so forth. Yet, they do not make a difference as long as the CLT is concerned.

Nevertheless, the CLT is correct from a theoretical point of view—and this has been confirmed by countless empirical studies. Here, from a theoretical point of view it means that $n \to \infty$ or at least is very large. However, in a finite-sample situation, it can well be a different story. For example, suppose that $n = 30$. It can be shown that in this case the shape of the population distribution makes a difference (Exercise 4.5). This raises an issue about the convergence rate of the CLT. In particular, two characteristic measures of the shape of the population distribution are the skewness and kurtosis, defined as

$$\kappa_3 = \frac{E(X_1 - \mu)^3}{\sigma^3}, \tag{4.24}$$

$$\kappa_4 = \frac{E(X_1 - \mu)^4}{\sigma^4} - 3, \tag{4.25}$$

respectively. One would expect these characteristics to have some impact on the convergence rate of the CLT. For example, the celebrated Berry–Esseen theorem, discoverd by Berry (1941) and Esseen (1942), states that if the third moment of $X_1$ is bounded, then

$$\sup_x |F_n(x) - \Phi(x)| \le \frac{cE(|X_1|^3)}{\sqrt{n}}, \tag{4.26}$$

where $F_n(x)$ is the cdf of $\xi_n = (\sqrt{n}/\sigma)(\bar{X} - \mu)$, $\Phi(x)$ is the cdf of $N(0,1)$, and $c$ is an absolute constant (i.e., a constant that does not depend on the distribution of $X_1$). The Edgeworth expansion, named in honor of the Irish mathematician Francis Ysidro Edgeworth (1845–1926), carries the approximation in (4.26) to higher orders.

Like the Taylor expansion, the Edgeworth expansion can be expressed up to $k + 1$ terms plus a remaining term. The difference is that, in the Taylor expansion the terms are in decreasing orders of $|x - a|$ [see (4.9)]; and in the Edgeworth expansion, the terms are in decreasing orders of $n^{-1/2}$. For the sake of simplicity, we mainly focus on the case $k = 2$, which can be expressed as

$$F_n(x) = \Phi(x) + \frac{\kappa_3 p_1(x)}{6\sqrt{n}}\phi(x) + \frac{\kappa_4 p_2(x) - \kappa_3^2 p_3(x)}{24n}\phi(x)$$
$$+ O(n^{-3/2}), \tag{4.27}$$

where $\phi(x)$ is the pdf of $N(0,1)$; that is, $\phi(x) = (1/\sqrt{2\pi})e^{-x^2/2}$,

$$p_1(x) = 1 - x^2,$$
$$p_2(x) = x(3 - x^2),$$
$$p_3(x) = \frac{x}{3}(15 - 10x^2 + x^4).$$

Expansion (4.27) is known as the two-term Edgeworth expansion (rather than three-term Edgeworth expansion). Note that $\Phi(x)$ does not count as a "term" (or it may be counted as the zeroth term), so the first term of the expansion is $O(n^{-1/2})$, the second term is $O(n^{-1})$, and so on. We see that the first and second terms of the Edgeworth expansion involve the skewness, $\kappa_3$, and kurtosis, $\kappa_4$, confirming our earlier speculation that these quantities may influence the convergence rate of the CLT.

To derive the Edgeworth expansion we introduce a method called formal derivation, which will be used repeatedly in this book. Note that the validity of the Taylor expansion is not with conditions. For example, for (4.14) to hold, it is necessary that the remaining term is really $o(|x - a|^l)$, which requires certain conditions. Furthermore, according to the results of Chapter 2, convergence in probability does not necessarily imply convergence in expectation. So, for example, it is not necessarily true that $E\{o_P(1)\} = o(1)$. However, such arguments as the above will be used in the derivation of the Edgeworth expansion as well as many other asymptotic results in the sequel. So what should we do? Should we verify the necessary conditions for every single step of the derivation or should we go ahead with the derivation without having to worry about the conditions? The answer depends on at what stage you are in during the development of a method. Science will not advance if we have to watch our steps for every tiny little move. In the development of most statistical methods there is an important first step—that is, to propose the method. After the method is proposed, the next step is to study the performance of the method, which includes theoretical justification, empirical studies, and applications. At the first stage of the development (i.e., propose the method), one may not need to worry about the conditions. In other words, the first step does not have to wait for the second step to follow immediately. This is what we called formal derivation.

More specifically, in the first step, one derives the formula (or procedure), assuming that all of the necessary conditions are satisfied or that the formula or procedure will hold under certain conditions. Quite often, the first step is done by some researcher(s) and later justified by others. For example, Efron (1979) proposed the bootstrap method without establishing its theoretical properties. General accounts of theory for the bootstrap were latter given by Bickel and Freedman (1981) and Beran (1984), among others. In conclusion, the conditions are important, but they should not tie our hands. This is a moral we learned, among other things, from the development of *bootstrap* (see Chapter 14) and many other statistical methods.

Going back to the Edgeworth expansion, we use the method of formal derivation. Recall $\xi_n = (\sqrt{n}/\sigma)(\bar{X}-\mu) = \sum_{j=1}^{n} Z_j$, where $Z_j = (X_j-\mu)/\sigma\sqrt{n}$. Then the cf of $\xi_n$ can be expressed as

$$c_n(t) = E\{\exp(it\xi_n)\}$$
$$= \prod_{j=1}^{n} E\{\exp(itZ_j)\}$$
$$= [E\{\exp(itZ_1\}]^n. \tag{4.28}$$

Furthermore, by the Taylor expansion, we have

$$\exp(itZ_1) = \sum_{k=0}^{4} \frac{(itZ_1)^k}{k!} + O_P(n^{-5/2}).$$

Note that the Taylor expansion also holds for functions of complex variables ($i = \sqrt{-1}$ is a complex number). Also note that $Z_1 = O_P(n^{-1/2})$. Thus,

$$E\{\exp(itZ_1)\} = \sum_{k=0}^{4} \frac{(it)^k}{k!} E(Z_1^k) + O(n^{-5/2})$$

$$= 1 - \frac{t^2}{2n} + \sum_{k=3}^{4} \frac{(it)^k}{k!} E(Z_1^k) + O(n^{-5/2})$$

because $E(Z_1) = 0$ and $E(Z_1^2) = 1/n$. Another Taylor expansion gives

$$\log[E\{\exp(itZ_1)\}] = -\frac{t^2}{2n} + \sum_{k=3}^{4} \frac{(it)^k}{k!} E(Z_1^k) + O(n^{-5/2})$$

$$-\frac{1}{2}\left\{-\frac{t^2}{2n} + \cdots\right\}^2 + O(n^{-3})$$

$$= -\frac{t^2}{2n} + \frac{(it)^3}{3!}\frac{\kappa_3}{n^{3/2}} + \frac{(it)^4}{4!}\frac{\kappa_4}{n^2} + O(n^{-5/2})$$

because $E(Z_1^3) = \kappa_3/n^{3/2}$ and $E(Z_1^4) = (\kappa_4 + 3)/n^2$. Therefore, we have

$$n \log E\{\exp(itZ_1)\} = -\frac{t^2}{2} + \frac{(it)^3}{3!}\frac{\kappa_3}{n^{1/2}} + \frac{(it)^4}{4!}\frac{\kappa_4}{n} + O(n^{-3/2});$$

hence, by (4.28) and the Taylor expansion of $f(x) = e^x$ at $x = -t^2/2$,

$$c_n(t) = \exp\left\{-\frac{t^2}{2} + \frac{(it)^3}{3!}\frac{\kappa_3}{n^{1/2}} + \frac{(it)^4}{4!}\frac{\kappa_4}{n} + O(n^{-3/2})\right\}$$

$$= e^{-t^2/2} + e^{-t^2/2}\left\{\frac{(it)^3}{3!}\frac{\kappa_3}{n^{1/2}} + \frac{(it)^4}{4!}\frac{\kappa_4}{n} + O(n^{-3/2})\right\}$$

$$+ \frac{e^{-t^2/2}}{2!}\left\{\frac{(it)^3}{3!}\frac{\kappa_3}{n^{1/2}} + \cdots\right\}^2 + O(n^{-3/2})$$

$$= e^{-t^2/2}\left[1 + \frac{(it)^3\kappa_3}{6}n^{-1/2} + \left\{\frac{(it)^4\kappa_4}{24} + \frac{(it)^6\kappa_3^2}{72}\right\}n^{-1} + O(n^{-3/2})\right]$$

$$= e^{-t^2/2} + n^{-1/2}r_1(it)e^{-t^2/2} + n^{-1}r_2(it)e^{-t^2/2} + O(n^{-3/2}), \qquad (4.29)$$

where $r_1(z) = (\kappa_3/6)z^3$ and $r_2(z) = (\kappa_4/24)z^4 + (\kappa_3^2/72)z^6$. Note that

$$c_n(t) = \int e^{itx} d\, F_n(x)$$

is the Fourier–Stieltjes transform of $F_n(x) = \mathrm{P}(\xi_n \leq x)$, which has the asymptotic expansion (4.29). An inversion of the transform transform then gives

$$F_n(x) = \Phi(x) + n^{-1/2}R_1(x) + n^{-1}R_2(x) + O(n^{-3/2}), \qquad (4.30)$$

where $R_j(x)$ is a function that satisfies

$$\int e^{itx}\, dR_j(x) = r_j(it)e^{-t^2/2}, \quad j = 1, 2, \ldots$$

Note that $\int e^{itx}\, d\Phi(x) = e^{-t^2/2}$. It can be shown that (e.g., Hall 1992, p. 44)

$$R_1(x) = -\frac{\kappa_3}{6}(x^2 - 1)\phi(x),$$

$$R_2(x) = -\left\{\frac{\kappa_4}{24}(x^2 - 3) + \frac{\kappa_3^2}{72}(x^4 - 10x^2 + 15)\right\}x\phi(x).$$

Thus, by (4.30) we obtain the expansion (4.27). We consider some examples.

*Example 4.5.* Let $X_1, \ldots, X_n$ be independent with the Beta$(\alpha, \beta)$ distribution. We consider two special cases: (i) $\alpha = \beta = 2$ and (ii) $\alpha = 2$, $\beta = 6$. The skewness and kurtosis of the Beta$(\alpha, \beta)$ distribution are given by

$$\kappa_3 = \frac{2(\beta - \alpha)\sqrt{\alpha + \beta + 1}}{(\alpha + \beta + 2)\sqrt{\alpha\beta}},$$

$$\kappa_4 = 6\frac{\alpha^3 - \alpha^2(2\beta - 1) + \beta^2(\beta + 1) - 2\alpha\beta(\beta + 2)}{\alpha\beta(\alpha + \beta + 2)(\alpha + \beta + 3)},$$

respectively. Therefore, in case (i), we have $\kappa_3 = 0$ and $\kappa_4 = -6/7$. Thus, the Edgeworth expansion (4.27) becomes

$$F_n(x) = \Phi(x) - \frac{p_2(x)\phi(x)}{28n} + O(n^{-3/2}). \qquad (4.31)$$

Note that in case (i), the distribution of $X_i$ is symmetric. As a result, the Edgeworth expansion has a simpler form. On the other hand, in case (ii), we have $\kappa_3 = 2\sqrt{3}/5$ and $\kappa_4 = 6/55$. Thus, (4.27) becomes

$$F_n(x) = \Phi(x) + \frac{\sqrt{3}p_1(x)\phi(x)}{15\sqrt{n}} + \frac{\{5p_2(x) - 22p_3(x)\}\phi(x)}{11000n}$$

$$+O(n^{-3/2}). \qquad (4.32)$$

Comparing (4.31) and (4.32), one would expect the convergence in CLT to be faster for case (i) than for case (ii) (Exercise 4.6).

*Example 4.6.* Suppose that $X_1, \ldots, X_n$ are independent and distributed as Exponential(1). Then it is easy to verify that $\kappa_3 = 2$ and $\kappa_4 = 6$. Thus, the Edgeworth expansion (4.27) becomes

$$F_n(x) = \Phi(x) + \frac{p_1(x)\phi(x)}{3\sqrt{n}} + \frac{3p_2(x) - 2p_3(x)}{12n}\phi(x)$$
$$+ O(n^{-3/2}).$$

Now comes the second step, the justification part. A rigorous treatment of the Edgeworth expansion, including sufficient conditions, can be found in Hall (1992, Section 2.4). One of the key conditions is known as Cramér's condition, which states the following:

$$\limsup_{t \to \infty} |E(e^{itX_1})| < 1. \tag{4.33}$$

In other words, the cf of $X_1$ is bounded strictly by 1. It holds, in particular, if $X_1$ has a pdf with respect to the Lebesgue measure (see Appendix A.2).

In fact, the Edgeworth expansion is not limited to the sample mean $\bar{X}$, as has been discussed so far. Let $\hat{\theta}$ be an estimator of $\theta$ such that $\sqrt{n}(\hat{\theta} - \theta)$ is asymptotically normal with mean 0 and variance $\sigma^2 > 0$. Then the cdf of $\xi_n = (\sqrt{n}/\sigma)(\hat{\theta} - \theta)$, $F_n(x)$, may be expanded as (4.30), where $R_j(x) = p_j(x)\phi(x)$ and $p_j(x)$ is a polynomial of degree no more than $3j - 1$. We consider an example below and refer more details to Hall (1992, Section 2.3).

*Example 4.7.* The $t$-test and confidence interval are associated with the random variable $\xi_n = \sqrt{n}(\bar{X} - \mu)/\hat{\sigma}$, where $\hat{\sigma}^2 = n^{-1}\sum_{i=1}^n (X_i - \bar{X})^2$. Here, $X_1, \dots, X_n$ are assumed to be i.i.d. with a finite fourth moment. The Edgeworth expansion for $F_n(x) = P(\xi_n \le x)$ is given by

$$F_n(x) = \Phi(x) + \frac{P_1(x)\phi(x)}{\sqrt{n}} + \frac{P_2(x)\phi(x)}{n} + o(n^{-1}),$$

where $P_1(x) = (\kappa_3/6)(2x^2 + 1)$ and

$$P_2(x) = x \left\{ \frac{\kappa_4}{12}(x^2 - 3) - \frac{\kappa_3^2}{18}(x^4 + 2x^2 - 3) - \frac{x^2 + 3}{4} \right\}.$$

## 4.4 Other related expansions

### 4.4.1 Fourier series expansion

The Fourier–Stieltjes transform in the previous section is an extension of the Fourier transform, which has had a profound impact in the mathematical world. The Fourier series may be regarded as a discrete version of the inversion of Fourier transform, which is defined as

$$\hat{f}(k) = \frac{1}{2\pi} \int_{-\pi}^{\pi} f(t)e^{-ikt} \, dt, \tag{4.34}$$

for $k = 0, \pm 1, \pm 2, \ldots$, where $i = \sqrt{-1}$. Here, $f$ denotes an integrable function. Given the Fourier transform, one may recover $f$ via the Fourier series

$$\sum_{k=-\infty}^{\infty} \hat{f}(k)e^{ikt}. \tag{4.35}$$

Here, the series is understood as convergent in some sense (see below). Note that one may express (4.35) as

$$\int_Z \hat{f}(x)e^{itx}\mu(dx), \tag{4.36}$$

where $Z = \{0, \pm 1, \pm 2, \ldots\}$ and $\mu$ represents the counting measure on $Z$ [i.e., $\mu(\{k\}) = 1$ for any $k \in Z$]. Comparing (4.36) with (4.35), we see that the Fourier series (4.35) actually corresponds to an inversion formula of the Fourier transform. Note that we have not written (4.35) as

$$f(t) = \sum_{k=-\infty}^{\infty} \hat{f}(k)e^{ikt}. \tag{4.37}$$

The question is whether (4.37) actually holds, or holds in some sense. Before we answer this question, let us point out the following facts.

First, if (4.37) does hold, say, at a certain point $t$, then by truncating the series after a given number of terms, one obtains the Fourier expansion

$$f(t) = \sum_{k=-N}^{N} \hat{f}(k)e^{ikt} + o(1), \tag{4.38}$$

where $N$ is a positive integer and the remaining term is $o(1)$ as $N \to \infty$. Expansion (4.38) is more useful from a practical point of view because, realistically, one can only evaluate a finite number of terms. Unlike the Taylor expansion, there is no general result on the order of the remaining term, so $o(1)$ is all one can say at this point. This is because the Fourier series applies to a much broader class of functions than the Taylor expansion. For a function to have a Taylor series, it must be infinitely differentiable, or at least have some order(s) of derivatives in an interval, if one uses Taylor expansion that involves a finite number of terms. On the other hand, the Fourier series may be used to approximate not only nondifferentiable functions; "they even do a good job in the wilderness of the wildly discontinuous" (Bachman et al. 2000, p. 139). Therefore, it is difficult to evaluate the order of the remaining term because it depends on, among other things, the degree of "smoothness" of the function. For example, for a noncontinuous function, the convergence of the Fourier series may not be in the sense of (4.37) (see below).

Second, the Fourier series (4.35) is expressed in the form of an *exponential series* or, more generally,

$$\sum_{k=-\infty}^{\infty} c_k e^{ikt}. \tag{4.39}$$

Alternatively, it may be expressed in the form of a *trigonometric series*,

$$\frac{a_0}{2} + \sum_{k=1}^{\infty} a_k \cos(kt) + \sum_{k=1}^{\infty} b_k \sin(kt), \tag{4.40}$$

through the simple transformation $a_k = c_k + c_{-k}$ and $b_k = i(c_k - c_{-k})$.

The following theorem, known as Dirichlet's pointwise convergence theorem, states sufficient conditions for the convergence of the Fourier series as well as to what value the series converges. For any function $f$ defined on $[-\pi, \pi]$, its $2\pi$-periodic extension is defined by $f(t+2k\pi) = f(t)$, $k \in Z, t \in R$. Furthermore, the left (right) limit of a function $g$ at a point $t$ is defined as $g(t^-) = \lim_{s \to t-} g(s)$ [$g(t^+) = \lim_{u \to t+} g(u)$]. Here, $s \to t-$ ($u \to t+$) means that $s$ ($u$) approaches $t$ from the left (i.e., $s < t$) [right (i.e., $u > t$)].

**Theorem 4.3.** Let $f$ be the $2\pi$-periodic extension of an integrable function on $[-\pi, \pi]$. If $f'(t^-)$ and $f'(t^+)$ exist for all $t$, then the Fourier series (4.39) or (4.40), where $c_k = \hat{f}(k)$, $k \in Z$, converges to

$$\frac{f(t^-) + f(t^+)}{2}$$

at every $t$. In particular, at any continuity point $t$, the series converges to $f(t)$.

An alternative to pointwise convergence is $L^2$-convergence. A function $f \in L^2[-\pi, \pi]$ if $\int_{-\pi}^{\pi} |f(t)|^2 \, dt < \infty$. For any $f \in L^2[-\pi, \pi]$, its $n$th-order Fourier approximation is defined as

$$S_n f(t) = \sum_{k=-n}^{n} \hat{f}(k) e^{ikt}.$$

Here, we use the term "Fourier approximation" instead of "Fourier expansion," the difference between the two being that the latter is the former plus a remaining term, which may be expressed in terms of big $O$ or small $o$. In fact, $L^2[-\pi, \pi]$ constitutes a Hilbert space if we define the inner product of any $f, g \in L^2[-\pi, \pi]$ by $< f, g > = (2\pi)^{-1} \int_{-\pi}^{\pi} f(t)\overline{g(t)} \, dt$, where the bar denotes complex conjugation. It follows that the $n$th-order Fourier approximation is simply the projection of $f$ onto the subspace of $L^2[-\pi, \pi]$ spanned by $\{e_k, |k| \le n\}$, where $e_k(t) = e^{ikt}$ and the coefficients in the approximation are the inner products, $\hat{f}(k) = < f, e_k >$, $|k| \le n$. A sequence $f_n$ in $L^2[-\pi, \pi]$ converges in $L^2$ to a limit $f \in L^2[-\pi, \pi]$ if

$$\int_{-\pi}^{\pi} |f_n(t) - f(t)|^2 \, dt \longrightarrow 0$$

as $n \to \infty$. We have the following result.

**Theorem 4.4.** For any $f \in L^2[-\pi, \pi]$, its Fourier approximation $S_n f$ converges in $L^2$ to $f$ as $n \to \infty$.

Modern harmonic analysis treats Fourier series as a special case of orthonormal series for the representation or approximation of functions or signals. Let $S$ be a subset of $R$, the real line. An orthonormal system on $S$ is defined as a sequence of functions $\phi_k$, $k \in I$, on $S$ such that

$$\int_S \phi_k(t)\phi_l(t)\, dt = 0, \quad k \neq l, \tag{4.41}$$

$$\int_S |\phi_k(t)|^2\, dt = 1. \tag{4.42}$$

Here, $I$ represents an index set. Give an orthonormal system $\phi_k$, $k \in I$, one may consider the following series expansion of a function $f$:

$$f(t) = \sum_{k \in I} c_k \phi_k(t), \tag{4.43}$$

where $c_k = \int_a^b f(t)\overline{\phi_k(t)}\, dt$. Again, the series expansion may be interpreted in terms of projection in a Hilbert space, with the coefficients being the inner products. Below are some examples.

*Example 4.8.* If we let $\phi_k(t) = e^{ikt}/\sqrt{2\pi}$, then it is easy to verify that $\phi_k$, $k \in Z$, is an orthonormal system (Exercise 4.8). This orthonormal system on $[-\pi, \pi]$ corresponds to the Fourier series (4.39) with $c_k = \hat{f}(k)$, $k \in Z$.

*Example 4.9.* Similarly, the sequence

$$\frac{1}{\sqrt{2\pi}}, \quad \frac{\sin(kt)}{\sqrt{2\pi}}, \quad k = 1, 2, \ldots, \quad \frac{\cos(kt)}{\sqrt{2\pi}}, \quad k = 1, 2, \ldots,$$

defines an orthonormal system (Exercise 4.9). This orthonormal system on $[-\pi, \pi]$ corresponds to the Fourier series (4.40) with $c_k = \hat{f}(k)$, $k \in Z$.

*Example 4.10* (Orthonormal polynomials). Consider polynomial approximation to a function $f$ on $[0, 1]$. A polynomial is a linear combination of the powers $1, x, x^2, \ldots$. Unfortunately, the power functions themselves are not orthonormal. A general procedure for constructing an orthonormal system is called the Gram–Schmidt orthonormalization. The procedure is described as follows. Starting with $\phi_0(x) = 1$, let

$$\phi_1(x) = \frac{x - \int_0^1 u\, du}{\{\int_0^1 (v - \int_0^1 u\, du)^2\, dv\}^{1/2}}$$
$$= \sqrt{3}(2x - 1).$$

In general, the sequence $\phi_k(x)$ is defined recursively by

$$\phi_k(x) = \frac{x^k - \sum_{j=0}^{k-1}\{\int_0^1 u^k \phi_j(u)\,du\}\phi_j(x)}{(\int_0^1 [v^k - \sum_{j=0}^{k-1}\{\int_0^1 u^k \phi_j(u)\,du\}\phi_j(v)]^2\,dv)^{1/2}},$$

$k = 1, 2, \ldots$. This defines an orthonormal system on $[0, 1]$. In particular, it is fairly straightforward to compute the first few orthonormal polynomials and verify that they are orthonormal (Exercise 4.10).

*Example 4.11* (Haar functions). This system is a special case of wavelets. It is a sequence of discontinuous functions defined through transformations of the indicator function of $[0, 1)$: $I_{[0,1)}(t) = 1$ if $0 \le t < 1$ and $0$ otherwise. Let $\phi_0(t) = I_{[0,1)}(2t) - I_{[0,1)}(2t - 1)$. $\phi_0$ is called the Haar mother wavelet and it can be expressed more explicitly as

$$\phi_0(t) = \begin{cases} 1, & 0 \le t < 1/2 \\ -1, & 1/2 \le t < 1 \\ 0, & \text{otherwise} \end{cases}$$

(see Figure 4.1). The subsequent Haar functions are defined as

$$\phi_{j,k}(t) = 2^{j/2}\phi_0(2^j t - k), \quad j = 0, 1, 2, \ldots, \quad k - 0, 1, \ldots, 2^j - 1, \quad (4.44)$$

where $\phi_{0,0} = \phi_0$, the mother wavelet. It can be shown that the Haar functions defined by (4.44) together with $I_{[0,1)}$ constitute an orthonormal system on $(-\infty, \infty)$ (Exercise 4.11).

### 4.4.2 Cornish–Fisher expansion

The Edgeworth expansion discussed in Section 4.3 can be inverted, leading to a useful expansion for the quantiles of $\xi_n$. This is known as the Cornish–Fisher expansion. For any $\alpha \in (0, 1)$, define $q_n(\alpha) = \inf\{x : F_n(x) \ge \alpha\}$, which is called the upper $\alpha$th quantile of $F_n$. Here, as in Section 4.3, $F_n$ denotes the cdf of $\xi_n = (\sqrt{n}/\sigma)(\bar{X} - \mu)$ and $\bar{X}$ is the sample mean of i.i.d. observations $X_1, \ldots, X_n$. Let $z_\alpha$ denote the upper $\alpha$th quantile of $N(0, 1)$ [i.e., $\Phi(z_\alpha) = \alpha$]. Then the two-term Cornish–Fisher expansion may be expressed as

$$q_n(\alpha) = z_\alpha + \frac{(z_\alpha^2 - 1)\kappa_3}{6\sqrt{n}} + \frac{1}{12n}\left\{\frac{(z_\alpha^3 - 3z_\alpha)\kappa_4}{2} - \frac{(2z_\alpha^3 - 5z_\alpha)\kappa_3^2}{3}\right\}$$
$$+ O(n^{-3/2}), \qquad (4.45)$$

where $\kappa_3$ and $\kappa_4$ are defined by (4.24) and (4.25), respectively.

The Cornish–Fisher expansion is useful in obtaining more accurate confidence intervals and critical values for tests. Note that the CLT approximation to $q_n(\alpha)$ would be $z_\alpha$, which is the leading term on the right side of (4.45)

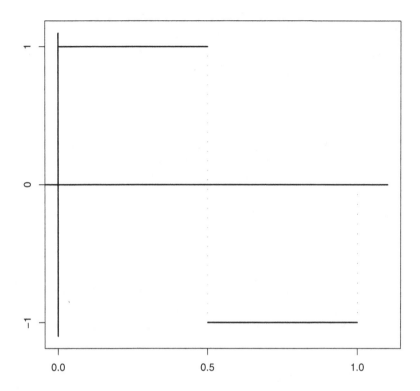

**Fig. 4.1.** *The Haar mother wavelet*

(i.e., $z_\alpha$). The following example shows how much more accuracy the expansion (4.45) may bring compared to the CLT approximation.

*Example 4.12.* Barndorff-Nielsen and Cox (1989, p. 119) reported the results of the Cornish–Fisher approximation in the situation where the $X_i$'s are distributed as $\chi_1^2$. In this case, we have $\kappa_3 = 2\sqrt{2}$ and $\kappa_4 = 12$, so the two-term expansion (4.45) becomes

$$q_n(\alpha) = z_\alpha + \frac{\sqrt{2}(z_\alpha^2 - 1)}{3\sqrt{n}} + \frac{z_\alpha^3 - 7z_\alpha}{18n} + O(n^{-3/2}).$$

Note that the mean and vaiance of $\chi_n^2 = \sum_{i=1}^n X_i$ are $n$ and $2n$, respectively. Thus, the $\alpha$th quantile of $\chi_n^2$ is

$$n + \sqrt{2n}q_n(\alpha) = n + z_\alpha\sqrt{2n} + \frac{2(z_\alpha^2 - 1)}{3} + \frac{z_\alpha^3 - 7z_\alpha}{9\sqrt{2n}} + O(n^{-1}).$$

Of course, the quantiles of $\chi_n^2$ can be calculated exactly. Table 4.2, extracted from Table 4.5 of Barndorff-Nielsen and Cox (1989), compares the approximations by the two-term Cornish–Fisher expansion as well as by the CLT with the exact quantiles for $\alpha = 0.1$, where C-F refers to the two-term Cornish–Fisher expansion. The results showed astonishing accuracy of the C-F approximation even with very small sample size $(n = 5)$.

### Table 4.2. Approximation of quantiles

| n | Exact | CLT | C-F |
|---|---|---|---|
| 5 | 9.24 | 9.65 | 9.24 |
| 10 | 15.99 | 15.73 | 15.99 |
| 50 | 63.17 | 62.82 | 63.16 |
| 100 | 118.50 | 118.12 | 118.50 |

It should be pointed out that, like the Edgeworth expansion, the Cornish–Fisher expansion requires certain regularity conditions in order to hold, and one of the key conditions is (4.33). If the condition fails, the Cornish-Fisher expansion may not improve over the CLT. The following is an example.

*Example 4.13.* Suppose that $X_1, \ldots, X_n$ are i.i.d. from the Bernoulli($p$) distribution with $p = 0.5$. It is easy to show that the distribution does not satisfy (4.33) (Exercise 4.12). If one blindly applies the Cornish–Fisher expansion (4.45), then since in this case $\kappa_3 = 0$ and $\kappa_4 = -2$, one would get

$$q_n(\alpha) = z_\alpha + \frac{3z_\alpha - z_\alpha^3}{12n} + O(n^{-3/2}). \tag{4.46}$$

Despite the simple form, (4.46) may not give a better approximation than the CLT. To see this, note that $X. \sim \text{Binomial}(n, p)$, so the exact $\alpha$th quantile of $X.$ can be calculated. On the other hand, the $\alpha$th quantile of $X.$ is given by

$$n\left\{ p + q_n(\alpha)\sqrt{\frac{p(1-p)}{n}} \right\}. \tag{4.47}$$

Table 4.3 compares the approximations to the $\alpha$th quantiles, where $\alpha = P(X. \leq k)$ for $n = 15$ and $k = 3, 6, 9, 12$, by C-F [i.e., (4.47) with $q_n(\alpha)$ given by (4.46)] as well as by the CLT [i.e., (4.47) with $q_n(\alpha) = z_\alpha$], with the exact quantiles. It is seen that inappropriate use of C-F sometimes makes things worse.

**Table 4.3. Approximation of quantiles**

| $k$ | 3 | 6 | 9 | 12 |
|-----|---|---|---|-----|
| $\alpha$ | 0.0176 | 0.3036 | 0.8491 | 0.9963 |
| CLT | 3.4207 | 6.5046 | 9.4998 | 12.6878 |
| C-F | 3.4533 | 6.4895 | 9.5212 | 12.5674 |

### 4.4.3 Two time series expansions

A time series is a set of observations, each recorded at a specified time $t$. In this subsection we consider a stationary (complex-valued) time series, denoted by $\{X_t, t = 0, \pm 1, \pm 2, \ldots\}$, or simply $X_t$. This means that $E(|X_t|^2) < \infty$ and $E(X_t)$ and $E(X_{t+k}\overline{X_t})$ do not depend on $t$. We can then define the autocovariance function of $X_t$ as

$$\gamma(k) = \text{cov}(X_{t+k}, X_t)$$
$$= E\{(X_{t+k} - \mu)\overline{(X_t - \mu)}\}, \tag{4.48}$$

$k = 0, \pm 1, \pm 2, \ldots$, where $\mu = E(X_t)$. One special stationary time series is called a white noise, for which $\mu = 0$ and $\gamma(k) = \sigma^2 1_{(k=0)}$. In other words, $W_t$ is a white noise if $E(W_t) = 0$, $E(W_t^2) = \sigma^2$, and $E(W_s W_t) = 0$, $s \neq t$. The following conditions (i) and (ii) are both necessary and sufficient for $\gamma$ to be the autocovariance function of a stationary time series $X_t$.

(i) $\gamma(k) = \int_{-\pi}^{\pi} e^{ik\lambda} \, dF(\lambda)$, where $F$ is a right-continuous, nondecreasing and bounded function on $[-\pi, \pi]$ with $F(-\pi) = 0$.

(ii) $\sum_{i,j=1}^{n} \gamma(i-j)a_i\overline{a_j} \geq 0$ for any positive integer $n$ and $a_1, \ldots, a_n \in C$.

Here, $C$ denotes the set of complex numbers and $F$ is right-continuous at $\lambda$ if $F(\nu) \to F(\lambda)$ as $\nu$ approaches $\lambda$ from the right (i.e., $\nu > \lambda$). The functiom $F$ is called the spectral distribution function of $\gamma$ or $X_t$. In particular, if $F$ is absolutely continuous such that $F(\lambda) = \int_{-\pi}^{\lambda} f(\nu) \, d\nu$, $-\pi \leq \lambda \leq \pi$, $f$ is called the spetral density of $\gamma$ or $X_t$. Note that the properties of $F$ imply that $f(\lambda) \geq 0$, $\lambda \in [-\pi, \pi]$. If $\gamma$ is the autocovariance function of a stationary time series $X_t$ that is absolutely summable, that is,

$$\sum_{k=-\infty}^{\infty} |\gamma(k)| < \infty. \tag{4.49}$$

then there exists a function $f$ such that $f(\lambda) \geq 0$, $\lambda \in [-\pi, \pi]$ and

$$\gamma(k) = \int_{-\pi}^{\pi} e^{ik\lambda} f(\lambda) \, d\lambda, \quad k = 0, \pm 1, \pm 2, \ldots. \tag{4.50}$$

In other words, $f$ is the spectral density of $X_t$. Furthermore, we have

$$f(\lambda) = \frac{1}{2\pi} \sum_{k=-\infty}^{\infty} \gamma(k)e^{-ik\lambda}, \tag{4.51}$$

$\lambda \in [-\pi, \pi]$. In other words, we have the asymptotic expansion

$$f(\lambda) = \frac{1}{2\pi} \sum_{k=-n}^{n} \gamma(k)e^{-ik\lambda} + o(1), \qquad (4.52)$$

$\lambda \in [-\pi, \pi]$, where $o(1) \to 0$ as $n \to \infty$. Equation (4.51) or (4.52) can be established using the results of Fourier expansion (see Section 4.4.1) or verified directly using Fubini's theorem (see Exercise 4.12).

Another well-known expansion in time series is called the Wold decomposition. For simplicity, assume that $X_t$ is real-valued. Consider the space $\mathcal{H}$ of all random variables $X$ satisfying $E(X^2) < \infty$. Then $\mathcal{H}$ is a Hilbert space with the inner product $< X, Y >= E(XY)$. Let $\mathcal{H}_t$ denote the subspace of $\mathcal{H}$ spanned by $\{X_s, s \leq t\}$. Let $P_{\mathcal{H}_{t-1}} X_t$ denote the projection of $X_t$ onto $\mathcal{H}_{t-1}$ (called the one-step predictor). See Chapter 9 for more details. Also, define $\mathcal{H}_{-\infty} = \cap_{t=-\infty}^{\infty} \mathcal{H}_t$. A time series $X_t$ is said to be deterministic of $X_t \in \mathcal{H}_{t-1}$ for all $t$. The Wold decomposition states that if $\sigma^2 = E(X_t - P_{\mathcal{H}_{t-1}} X_t)^2 > 0$, then $X_t$ can be expressed as

$$X_t = \sum_{k=0}^{\infty} \psi_k Z_{t-k} + V_t, \qquad (4.53)$$

where $\psi_0 = 1$ and $\sum_{k=0}^{\infty} \psi_k^2 < \infty$; $Z_t$ is a white noise with variance $\sigma^2$ and $Z_t \in \mathcal{H}_t$; $V_t$ and $Z_t$ are uncorrelated [i.e., $E(Z_t V_u) = 0$, $\forall t, u$] and $V_t \in \mathcal{H}_{-\infty}$ and is deterministic. In fact, (4.53) and the above properties uniquely determine $\psi_k$, $Z_t$, and $V_t$. We consider an example.

*Example 4.14.* Consider the real-valued function

$$\gamma(k) = \begin{cases} 1, & k = 0 \\ \rho, & k = \pm 1 \\ 0, & \text{otherwise.} \end{cases}$$

It is easy to show that $\gamma$ is an autocovariance function if $|\rho| \leq 1/2$ (Exercise 4.13). Since (4.49) is obviously satisfied, it follows by the spectral representation (4.51) that $f(\lambda) = (2\pi)^{-1} \sum_{k=-\infty}^{\infty} \gamma(k)e^{-ik\lambda} = (2\pi)^{-1}\{1 + 2\rho\cos(\lambda)\}$. Clearly, we have $f(\lambda) \geq 0$, $\lambda \in [-\pi, \pi]$ provided that $|\rho| \leq 1/2$. In fact, this is the spectral density of an MA(1) process defined by $X_t = Z_t + \theta Z_{t-1}$, where $Z_t$ is a white noice with variance $\sigma^2 > 0$ and $\theta = \rho/\sigma^2$. See Chapter 9 for more details. Clearly, the Wold decomposition holds for this $X_t$ with $\psi_0 = 1$, $\phi_1 = \theta$, $\psi_k = 0$, $k > 1$, and $V_t = 0$.

## 4.5 Some elementary expansions

The asymptotic expansions encountered so far are well known in the mathematical or statistical literature, and their derivations involve (much) more than just a few lines of algebra. However, these are not the only ways to come up with an asymptotic expansion. In this section, we show that one can derive some useful asymptotic expansions oneself using some elementary approaches that involve nothing more than a few lines of simple algebra.

Let us begin with a simple problem. Suppose that one wishes to expand the function $f(x) = x^{-1}$ at $x = a$. Most people would immediately think that, well, let's try the Taylor expansion. Surely one can do so without a problem. However, here is an alternative approach. Write

$$\frac{1}{x} = \frac{1}{a} + \frac{1}{x} - \frac{1}{a}$$
$$= \frac{1}{a} - \frac{x-a}{a}\frac{1}{x}. \tag{4.54}$$

Equation (4.54) suggests an iterative procedure so that we have

$$\frac{1}{x} = \frac{1}{a} - \frac{x-a}{a}\left(\frac{1}{a} - \frac{x-a}{a}\frac{1}{x}\right)$$
$$= \frac{1}{a} - \frac{x-a}{a^2} + \frac{(x-a)^2}{a^2}\frac{1}{x}$$
$$= \frac{1}{a} - \frac{x-a}{a^2} + \frac{(x-a)^2}{a^2}\left(\frac{1}{a} - \frac{x-a}{a}\frac{1}{x}\right)$$
$$= \cdots.$$

In general, one has the asymptotic expansion

$$\frac{1}{x} = \sum_{k=1}^{l}(-1)^{k-1}\frac{(x-a)^{k-1}}{a^k} + (-1)^l\frac{(x-a)^l}{a^l x} \tag{4.55}$$

for $l = 1, 2, \ldots$. If one instead uses the Taylor expansion, then since $f^{(k)}(x) = (-1)^k k! x^{-(k+1)}$, we have, by (4.8),

$$\frac{1}{x} = \sum_{k=1}^{l}(-1)^{k-1}\frac{(x-a)^{k-1}}{a^k} + (-1)^l\frac{(x-a)^l}{\xi^{l+1}}, \tag{4.56}$$

where $\xi$ lies between $a$ and $x$. Comparing the Taylor expansion with (4.55), which we call elementary expansion, the only difference is that (4.55) is more precise in terms of the remaining term than (4.56). In other words, in the Taylor expansion, we only know that $\xi$ is somewhere between $a$ and $x$, whereas in the elementary expansion there is no such uncertainty. Here is another look at the difference. If we drop the remaining term in the Taylor expansion (4.8) with $l$ replaced by $l+1$, we can write

$$\frac{1}{x} \approx \sum_{k=1}^{l}(-1)^{k-1}\frac{(x-a)^{k-1}}{a^k} + (-1)^l\frac{(x-a)^l}{a^{l+1}}. \tag{4.57}$$

Comparing (4.55) with (4.57), the difference is that the elementary expansion is *exact* (characterized by =), whereas the Taylor expansion is *approximate* (characterized by ≈).

In fact, it is not just that the results are (slightly) different. The elementary expansion is derived using very simple algebras—no results of calculus such as derivatives are involved. This is important because such an elementary expansion is easier to extend to situations beyond real numbers, such as matrices. For example, suppose that one wishes to approximate the inverse of matrix $B$ by that of matrix $A$. Then, by a similar derivation, we have

$$
\begin{aligned}
B^{-1} &= A^{-1} + B^{-1} - A^{-1} \\
&= A^{-1} + A^{-1}(A - B)B^{-1} \\
&= A^{-1} + A^{-1}(A - B)\{A^{-1} + A^{-1}(A - B)B^{-1}\} \\
&= A^{-1} + A^{-1}(A - B)A^{-1} + \{A^{-1}(A - B)\}^2 B^{-1} \\
&= A^{-1} + A^{-1}(A - B)A^{-1} \\
&\quad + \{A^{-1}(A - B)\}^2\{A^{-1} + A^{-1}(A - B)B^{-1}\} \\
&= A^{-1} + A^{-1}(A - B)A^{-1} + \{A^{-1}(A - B)\}^2 A^{-1} \\
&\quad + \{A^{-1}(A - B)\}^3 B^{-1} \\
&= \cdots.
\end{aligned}
$$

In general, we have the matrix asymptotic expansion

$$
B^{-1} = \left[\sum_{k=0}^{l}\{A^{-1}(A - B)\}^k\right]A^{-1} + \{A^{-1}(A - B)\}^{l+1}B^{-1} \quad (4.58)
$$

for $l = 0, 1, 2, \ldots$ (e.g., Das et al. 2004, Lemma 5.4).

For example, expansions such as (4.55) and (4.58) are useful in situations where $x$ $(B)$ is a random variable (matrix) and $a$ $(A)$ is its expectation. We consider an example.

*Example 4.15.* Suppose that $X_1, \ldots, X_n$ are i.i.d. $p$-dimensional standard normal random vectors; that is, the $X_i$'s are independent $\sim N(0, I_p)$, where $I_p$ is the $p$-dimensional identity matrix. Let $B = I_P + \bar{X}\bar{X}'$, where $\bar{X} = n^{-1}\sum_{i=1}^{n} X_i$, and suppose that one wishes to evaluate $E(B^{-1})$. Note that $\bar{X} \sim N(0, n^{-1}I_p)$; hence, $n^{-1/2}\bar{X} \sim N(0, I_p)$. It follows that $\bar{X} = O_P(n^{-1/2})$. Let $A = E(B) = \{(n+1)/n\}I_p$. Then we have $E\{(A - B)^2\} = E(\bar{X}\bar{X}'\bar{X}\bar{X}') - n^{-2}I_p$. Write $\xi = n^{1/2}\bar{X} = (\xi_1, \ldots, \xi_p)' \sim N(0, I_p)$. Then the $(i, j)$ element of $\xi\xi'\xi\xi'$ is $\eta_{ij} = \xi_i\xi_j \sum_{k=1}^{p} \xi_k^2$. It is easy to show (Exercise 4.15) that $E(\eta_{ij}) = (p + 2)1_{(i=j)}$. It follows that $E(\xi\xi'\xi\xi') = (p + 2)I_p$; hence, $E\{(A - B)^2\} = n^{-2}E(\xi\xi'\xi\xi') - n^{-2}I_p = n^{-2}(p + 1)I_p$. Now, by (4.58) with $l = 2$, we have

$$
B^{-1} = [I_p + A^{-1}(A - B) + \{A^{-1}(A - B)\}^2]A^{-1} + O_P(n^{-3}).
$$

Here, we used Theorem 3.2 to argue that $B^{-1} = O_P(1)$, and note that $A^{-1} = \{n/(n+1)\}I_p = O(1)$, and $A - B = n^{-1}I_p - \bar{X}\bar{X}' = O(n^{-1}) + O_P(n^{-1}) =$

$O_P(n^{-1})$. Thus, by the method of formal derivation (see Section 4.3), we have

$$E(B^{-1}) = A^{-1} + A^{-1}E\{(A - B)A^{-1}(A - B)\}A^{-1} + O(n^{-3})$$

$$= \frac{n}{n+1}I_p + \left(\frac{n}{n+1}\right)^3 E\{(A - B)^2\} + O(n^{-3})$$

$$= \frac{n}{n+1}\left\{1 + \frac{p+1}{(n+1)^2}\right\}I_p + O(n^{-3})$$

$$= \left\{1 - \frac{1}{n} + \frac{p+2}{n^2}\right\}I_p + O(n^{-3}). \tag{4.59}$$

The last equality in (4.59) is because, by (4.55) with $l = 3$, we have $(n+1)^{-1} = n^{-1} - n^{-2} + n^{-3} + O(n^{-4})$; hence, $n/(n+1) = 1 - n^{-1} + n^{-2} + O(n^{-3})$ and $(n+1)^{-2} = n^{-2} + O(n^{-3})$. Therefore, $\{n/(n+1)\}\{1 + (p+1)/(n+1)^2\} = \{1 - n^{-1} + n^{-2} + O(n^{-3})\}\{1 + (p+1)n^{-2} + O(n^{-3})\} = 1 - n^{-1} + (p+2)n^{-2} + O(n^{-3})$.

We now derive (4.59) using a different method—this time by the Taylor expansion. To do so, we first make use of the following matrix identity (e.g., Sen and Srivastava 1990, p. 275): For any $p \times p$ matrix $P$, $p \times q$ matrix $U$ and $q \times p$ matrix $V$, we have

$$(P + UV)^{-1} = P^{-1} - P^{-1}U(I_q + VP^{-1}U)^{-1}VP^{-1}, \tag{4.60}$$

provided that the inverses involved exist. By letting $P = I_p$, $U = \bar{X}$, and $V = \bar{X}'$ in (4.60), we have

$$B^{-1} = (I_p + \bar{X}\bar{X}')^{-1}$$

$$= I_p - \frac{\bar{X}\bar{X}'}{1 + \bar{X}'\bar{X}}$$

$$= I_p - \frac{\xi\xi'}{n + \xi'\xi},$$

where $\xi$ is defined as above. Note that the $(i, j)$ element of $\zeta = (n + \xi'\xi)^{-1}\xi\xi'$ is $(n + \xi'\xi)\xi_i\xi_j$. If $i \neq j$, then $E(\zeta_{ij}) = 0$; if $i = j$, then $E(\zeta_{ii}) = E\{(n + \sum_{k=1}^p \xi_k^2)^{-1}\xi_i^2\}$ does not depend on $i$ (Exercise 4.16). Thus,

$$E(\zeta_{ii}) = \frac{1}{p}\sum_{i=1}^p E\left(\frac{\xi_i^2}{n + \sum_{k=1}^p \xi_k^2}\right)$$

$$= \frac{1}{p}E\left(\frac{\sum_{k=1}^p \xi_k^2}{n + \sum_{k=1}^p \xi_k^2}\right)$$

$$= \frac{1}{p}E\left(\frac{\chi_p^2}{n + \chi_p^2}\right), \tag{4.61}$$

where $\chi_p^2$ represents a random variable with a $\chi_p^2$-distribution. By the Taylor expansion, we have

$$\frac{\chi_p^2}{n + \chi_p^2} = \frac{\chi_p^2}{n} \frac{1}{1 + n^{-1}\chi_p^2}$$

$$= \frac{\chi_p^2}{n} \left\{ 1 - \frac{\chi_p^2}{n} + O_P(n^{-2}) \right\}$$

$$= \frac{\chi_p^2}{n} - \frac{\chi_p^4}{n^2} + O_P(n^{-3}).$$

Now, again, use the method of formal derivation (Section 4.3), the facts that $E(\chi_p^2) = p$ and $E(\chi_p^4) = p(p+2)$, and (4.61) to get

$$E(\xi_{ii}) = \frac{1}{p} \left\{ \frac{p}{n} - \frac{p(p+2)}{n^2} + O(n^{-3}) \right\}$$

$$= \frac{1}{n} - \frac{p+2}{n^2} + O(n^{-3}).$$

Therefore, in conclusion, we have

$$E(B^{-1}) = I_p - E(\zeta)$$

$$= I_p - \left\{ \frac{1}{n} - \frac{p+2}{n^2} + O(n^{-3}) \right\} I_p$$

$$= \left\{ 1 - \frac{1}{n} + \frac{p+2}{n^2} \right\} I_p + O(n^{-3}),$$

which is the same as (4.59).

Note that in the latest derivation using the Taylor expansion we actually benefited from the identity (4.60) of matrix inversion and results on moments of the $\chi^2$-distribution (otherwise, the derivation could be even more tedious).

## 4.6 Laplace approximation

Suppose that one wishes to approximate an integral of the form

$$\int e^{-q(x)} dx, \tag{4.62}$$

where $q(\cdot)$ is a "well-behaved" function in the sense that it achieves its minimum value at $x = \tilde{x}$ with $q'(\tilde{x}) = 0$ and $q''(\tilde{x}) > 0$. Then we have, by the Taylor expansion,

$$q(x) = q(\tilde{x}) + \frac{1}{2}q''(\tilde{x})(x - \tilde{x})^2 + \cdots,$$

which yields the following approximation (Exercise 4.18):

$$\int e^{-q(x)} dx \approx \sqrt{\frac{2\pi}{q''(\tilde{x})}} e^{-q(\tilde{x})}. \tag{4.63}$$

Approximations such as (4.63) are known as the Laplace approximation, named after the French mathematician and astronomer Pierre-Simon Laplace. There is a multivariate extension of (4.63), which is often more useful in practice. Let $q(x)$ be a well-behaved function that attains its minimum value at $x = \tilde{x}$ with $q'(\tilde{x}) = 0$ and $q''(\tilde{x}) > 0$ (positive definite), where $q'$ and $q''$ denote the gradient vector and Hessian matrix, respectively. Then we have

$$\int e^{-q(x)} dx \approx c|q''(\tilde{x})|^{-1/2} e^{-q(\tilde{x})}, \tag{4.64}$$

where $c$ is a constant depending only on the dimension of the integral (Exercise 4.19) and $|A|$ denotes the determinant of matrix $A$.

Approximations (4.63) or (4.64) are derived using the second-order Taylor expansion. This is called the first-order Laplace approximation. If one uses the higher order Taylor expansions, the results are the higher order Laplace approximations, which are more complicated in their forms (e.g., Barndorff-Nielsen and Cox 1989, Section 3.3; Lin and Breslow 1996). For a fixed-order (e.g., first order) Laplace approximation, its accuracy depends on the behavior of the function $q$. Roughly speaking, the more "concentrate" the function is near $\tilde{x}$ the more accurate; and the more normal-look-like the function is the more accurate. For example, consider the following.

*Example 4.16 (t-distribution).* Consider the function

$$q(x) = \frac{\nu + 1}{2} \log\left(1 + \frac{x^2}{\nu}\right), \quad -\infty < x < \infty,$$

where $\nu$ is a positive integer. Note that, subject to a normalizing constant, $e^{-q(x)}$ corresponds to the pdf of the $t$-distribution with $\nu$ degrees of freedom. It is easy to verify (Exercise 4.20) that, in this case, the exact value of (4.62) is given by $\sqrt{\nu\pi}\,\Gamma(\nu/2)/\Gamma\{(\nu + 1)/2\}$, where $\Gamma$ is the gamma function; the Laplace approximation (4.63) is $\sqrt{2\nu\pi/(\nu + 1)}$. Table 4.4 shows the numerical values (up to the fourth decimal) for a number of different $\nu$'s, where Relative Error is defined as Exact minus Approximate divided by Exact. It is seen that the accuracy improves as $\nu$ increases. This is because as $\nu$ increases, the $t$-distribution becomes more and more concentrate at $x = 0$. In the extreme case where $\nu \to \infty$, the $t$-distribution becomes the standard normal distribution, for which the Laplace approximation is exact (Exercise 4.20).

So, if $q$ is a fixed function, there is a limit for how accurate one can approximate (4.62) with a fixed-order Laplace approximation. Note that, in practice, the first-order Laplace approximation is by far the most frequently used and the higher than the second order Laplace approximation is rarely even considered. This is because as the order increases, the formula for the approximation quickly becomes complicated, especially in the multivariate case. Therefore, practically, increasing the order of Laplace approximation may not be an op-

**Table 4.4. Accuracy of Laplace approximation**

| $\nu$ | Exact | Approximate | Relative Error |
|---|---|---|---|
| 1 | 3.1416 | 1.7725 | 0.4358 |
| 2 | 2.8284 | 2.0467 | 0.2764 |
| 3 | 2.7207 | 2.1708 | 0.2021 |
| 4 | 2.6667 | 2.2420 | 0.1593 |
| 5 | 2.6343 | 2.2882 | 0.1314 |
| 10 | 2.5700 | 2.3900 | 0.0700 |
| 50 | 2.5192 | 2.4819 | 0.0148 |
| 100 | 2.5129 | 2.4942 | 0.0074 |
| 250 | 2.5091 | 2.5016 | 0.0030 |

tion on the table to improve the accuracy of approximation. What else (option) does one have on the table?

In many applications, the function $q$ in (4.62) is not a fixed function but rather depends on $n$, the sample size. In other words, the sample size $n$ may play a role in the accuracy of Laplace approximation, which so far has not been taken into account. To see why the sample size may help, let us consider a simple example. Suppose that the function $q$ in (4.62), or, more precisely, $e^{-q(x)}$, corresponds to the pdf of a sample mean $\bar{X}$ of i.i.d. random variables $X_1, \ldots, X_n$. According to the law of large numbers (LLN), as $n$ increases, $\bar{X}$ becomes more and more concentrated near the population mean $\tilde{x} = \mathrm{E}(X_1)$. Therefore, the Laplace approximation is expected to become more accurate as $n$ increases. To show this more precisely, let us first consider a simple case.

Suppose that in (4.62), $q(x) = nx$, and another function $p(x)$ is added in front of $dx$. More specifically, we consider

$$I_n = \int_0^\infty e^{-nx} p(x) \, dx. \tag{4.65}$$

Suppose that $p^{(k)}(x)e^{-nx} \to 0$ as $x \to \infty$ for $k = 0, 1, 2, \ldots$. Then, by integration by parts, we have

$$
\begin{aligned}
I_n &= \frac{p(0)}{n} + \frac{1}{n}\int_0^\infty e^{-nx} p'(x) \, dx \\
&= \frac{p(0)}{n} + \frac{p'(0)}{n^2} + \frac{1}{n^2}\int_0^\infty e^{-nx} p''(x) \, dx \\
&= \frac{p(0)}{n} + \frac{p'(0)}{n^2} + \frac{p''(0)}{n^3} + \cdots.
\end{aligned}
$$

In other words, we have an asymptotic expansion in terms of increasing powers of $n^{-1}$. Now, let us consider a more general case by replacing the function $x$ in (4.65) by $g(x)$ and assuming that $g'(0) \neq 0$. It is also assumed that as $x \to \infty$, $e^{-ng(x)}p(x)/g'(x) \to 0$, $e^{-ng(x)}\{p'(x)g'(x) - p(x)g''(x)\}/\{g'(x)\}^3 \to 0$, and so forth, so that we get, by integration by parts,

$$I_n = \int_0^\infty e^{-ng(x)} p(x)\, dx$$

$$= e^{-ng(0)} \frac{p(0)}{g'(0)} n^{-1} + \frac{1}{n} \int_0^\infty e^{-ng(x)} \left\{ \frac{p(x)}{g'(x)} \right\}' dx$$

$$= e^{-ng(0)} \frac{p(0)}{g'(0)} n^{-1} + e^{-ng(0)} \frac{p'(0)g'(0) - p(0)g''(0)}{\{g'(0)\}^3} n^{-2} + \cdots. \quad (4.66)$$

So, again, the expansion is in increasing powers of $n^{-1}$.

The assumption that $g'(0) \neq 0$ makes a big difference in the approximation. To see this, consider the integral

$$I_n = \int_{-\infty}^\infty e^{-ng(x)} p(x)\, dx. \quad (4.67)$$

Suppose that $g(x)$ attains its minimum at $\tilde{x}$ such that $g'(\tilde{x}) = 0$, $g''(\tilde{x}) > 0$ and $p(\tilde{x}) \neq 0$. Then, under regularity conditions, we have, by the Taylor expansion,

$$g(x) = g(\tilde{x}) + g'(\tilde{x})(x - \tilde{x}) + \frac{1}{2} g''(\tilde{x})(x - \tilde{x})^2 + \cdots$$

$$= g(\tilde{x}) + \frac{1}{2} g''(\tilde{x})(x - \tilde{x})^2 + \cdots.$$

So, if we make a change of variable $y = \sqrt{ng''(\tilde{x})}(x - \tilde{x})$, we have

$$ng(x) = ng(\tilde{x}) + \frac{n}{2} g''(\tilde{x})(x - \tilde{x})^2 + \cdots$$

$$= ng(\tilde{x}) + \frac{y^2}{2} + \cdots.$$

On the other hand, again by the Taylor expansion, we have

$$p(x) = p\left\{ \tilde{x} + \frac{y}{\sqrt{ng''(\tilde{x})}} \right\}$$

$$= p(\tilde{x}) + p'(\tilde{x}) \frac{y}{\sqrt{ng''(\tilde{x})}} + \frac{1}{2} p''(\tilde{x}) \frac{y^2}{ng''(\tilde{x})} + \cdots.$$

If we ignore the $\cdots$ in both expansions, we obtain the following Laplace approximation of $I_n$ in (4.67):

$$I_n = \int_{-\infty}^\infty \exp\left\{ -ng(\tilde{x}) - \frac{y^2}{2} - \cdots \right\}$$

$$\times \left\{ p(\tilde{x}) + \frac{p'(\tilde{x})}{\sqrt{ng''(\tilde{x})}} y + \frac{p''(\tilde{x})}{2ng''(\tilde{x})} y^2 + \cdots \right\} \frac{dy}{\sqrt{ng''(\tilde{x})}}$$

$$\approx \frac{e^{-ng(\tilde{x})}}{\sqrt{ng''(\tilde{x})}} \int_{-\infty}^\infty e^{-y^2/2} \left\{ p(\tilde{x}) + \frac{p'(\tilde{x})}{\sqrt{ng''(\tilde{x})}} y + \frac{p''(\tilde{x})}{2ng''(\tilde{x})} y^2 \right\} dy$$

$$= \sqrt{\frac{2\pi}{ng''(\tilde{x})}} e^{-ng(\tilde{x})} \left\{ p(\tilde{x}) + \frac{p''(\tilde{x})}{2ng''(\tilde{x})} \right\}$$

$$= \sqrt{\frac{2\pi}{ng''(\tilde{x})}} e^{-ng(\tilde{x})} p(\tilde{x}) \{ 1 + O(n^{-1}) \}. \tag{4.68}$$

Note that, unlike (4.66), expansion (4.68) is in increasing powers of $n^{-1/2}$. We may compare (4.68) with $p(x) = 1$ with (4.63), where $q(x) = ng(x)$. According to (4.63), we have $I_n \approx \sqrt{2\pi/ng''(\tilde{x})} e^{-ng''(\tilde{x})}$, whereas, according to (4.68), we have $I_n \approx \sqrt{2\pi/ng''(\tilde{x})} e^{-ng''(\tilde{x})} \{ 1 + O(n^{-1}) \}$. The leading terms of the two approximations are the same, but (4.68) also ascertains that the next term in the approximation is $O(n^{-3/2})$.

The seemingly nice results might lead an unwary mind to wrong conclusions that expansions such as (4.66) and (4.68) always hold. This is because, in many cases, the function $g$ also depends on $n$, so that as $n$ increases, the remaining term in the Laplace expansion may not have the same order as what we have seen so far. To add a further complication, in some cases the dimension of the integral also depends on $n$. Below is an example.

*Example 4.17.* Suppose that, given the random variables $u_1, \ldots, u_{m_1}$ and $v_1, \ldots, v_{m_2}$, binary responses $Y_{ij}$, $i = 1, \ldots, m_1$, $j = 1, \ldots, m_2$, are conditionally independent such that $p_{ij} = P(Y_{ij} = 1 | u, v)$ and

$$\text{logit}(p_{ij}) = \mu + u_i + v_j,$$

where $\mu$ is an unknown parameter, $u = (u_i)_{1 \le i \le m_1}$, and $v = (v_j)_{1 \le j \le m_2}$. Furthermore, assume the $u_i$'s and $v_j$'s are independent such that $u_i \sim N(0, \sigma_1^2)$ and $v_j \sim N(0, \sigma_2^2)$, where the variances $\sigma_1^2$ and $\sigma_2^2$ are unknown. Here, the $u_i$'s and $v_j$'s are called random effects and the above model is a special case of the generalized linear mixed model (GLMM). See Chapter 12 for more details. Suppose that one wishes to estimate the unknown parameters $\mu$, $\sigma_1^2$, and $\sigma_2^2$ by the maximum likelihood method. It can be shown that the likelihood function can be expressed as

$$c - \frac{m_1}{2} \log(\sigma_1^2) - \frac{m_2}{2} \log(\sigma_2^2) + \mu Y_{..}$$

$$+ \log \int \cdots \int \left[ \prod_{i=1}^{m_1} \prod_{j=1}^{m_2} \{ 1 + \exp(\mu + u_i + v_j) \}^{-1} \right]$$

$$\times \exp \left( \sum_{i=1}^{m_1} u_i Y_{i.} + \sum_{j=1}^{m_2} v_j Y_{.j} - \frac{1}{2\sigma_1^2} \sum_{i=1}^{m_1} u_i^2 - \frac{1}{2\sigma_2^2} \sum_{j=1}^{m_2} v_j^2 \right)$$

$$du_1 \cdots du_{m_1} dv_1 \cdots dv_{m_2}, \tag{4.69}$$

where $c$ is a constant, $Y_{..} = \sum_{i=1}^{m_1} \sum_{j=1}^{m_2} Y_{ij}$, $Y_{i.} = \sum_{j=1}^{m_2} Y_{ij}$, and $Y_{.j} = \sum_{i=1}^{m_1} Y_{ij}$ (Exercise 4.21). The multidimensional integral involved in (4.69) has no closed-form expression, and it cannot be further simplified. Furthermore, the dimension of the integral is $m_1 + m_2$, which increases with the total sample size $n = m_1 m_2$ (in fact, unlike the classical i.i.d. case, here the total sample size $n$ is no longer meaningful if the interest is to estimate the variances $\sigma_1^2$ and $\sigma_2^2$).

Such a high-dimensional integral is difficult to evaluate even numerically. In particular, a fixed-order Laplace approximation no longer provides a good approximation unless $\sigma_1^2$ and $\sigma_2^2$ are very small. In fact, Jiang (1998a) showed that if one approximates a likelihood function such as (4.69) using the Laplace approximation and then estimates the parameters by maximizing the approximate likelihood function, the resulting estimators are inconsistent.

As a final remark, so far the derivations of the Laplace approximation may be viewed as a method of formal derivation (see Section 4.3). As it turns out, this is one of the cases that the second step in the development of a method (see the fourth paragraph in Section 4.3) may reject the first step. Our general recommendation is that the Laplace approximation is useful in many cases, but it should be used with caution.

## 4.7 Case study: Asymptotic distribution of the MLE

A classical application of Taylor series expansion is the derivation of the asymptotic distribution of the MLE. Let us begin with the i.i.d. case with the same set up as in Section 1.4; that is, $X_1, \ldots, X_n$ are i.i.d. observations with pdf $f(x|\theta)$, where $\theta$ is a real-valued unknown parameter with the parameter space $\Theta = (-\infty, \infty)$. Let $\hat{\theta}$ denote the MLE of $\theta$. We assume that $\hat{\theta}$ is consistent (see Section 1.4). Let $l(\theta|X)$ denote the log-likelihood function; that is, $l(\theta|X) = \sum_{i=1}^{n} \log\{f(X_i|\theta)\}$. Here, $X = (X_1, \ldots, X_n)'$ represents the vector of observations. Then under regularity conditions we have, by the Taylor expansion,

$$
\begin{aligned}
0 &= \frac{\partial}{\partial \theta} l(\hat{\theta}|X) \\
&= \frac{\partial}{\partial \theta} l(\theta|X) + \left\{ \frac{\partial^2}{\partial \theta^2} l(\theta|X) \right\} (\hat{\theta} - \theta) \\
&\quad + \frac{1}{2} \left\{ \frac{\partial^3}{\partial \theta^3} l(\tilde{\theta}|X) \right\} (\hat{\theta} - \theta)^2,
\end{aligned}
\tag{4.70}
$$

where $\tilde{\theta}$ lies between $\theta$ and $\hat{\theta}$. Before we continue, let us note the following facts:

(i) We have

$$
\frac{\partial^2}{\partial \theta^2} l(\theta|X) = E\left\{ \frac{\partial^2}{\partial \theta^2} l(\theta|X) \right\} + \frac{\partial^2}{\partial \theta^2} l(\theta|X) - E\left\{ \frac{\partial^2}{\partial \theta^2} l(\theta|X) \right\}
$$

$$= n\mathrm{E}(Y_1) + \sum_{i=1}^{n} \{Y_i - \mathrm{E}(Y_i)\},$$

where $Y_i = (\partial^2/\partial\theta^2)\log\{f(X_i|\theta)\}$. It follows by the WLLN that under a suitable moment condition (see Chapter 6),

$$\frac{\partial^2}{\partial\theta^2} l(\theta|X) = n\mathrm{E}(Y_1) + o_\mathrm{P}(n). \tag{4.71}$$

(ii) Under some regularity conditions, we have

$$\frac{\partial^3}{\partial\theta^3} l(\tilde{\theta}|X) = \sum_{i=1}^{n} \frac{\partial^3}{\partial\theta^3} f(X_i|\tilde{\theta}) = O_\mathrm{P}(n). \tag{4.72}$$

(iii) Write $Z_i = (\partial/\partial\theta)\log\{f(X_i|\theta)\}$. Then under some regularity conditions (see below), we have, by the CLT,

$$\frac{1}{\sqrt{n}} \frac{\partial}{\partial\theta} l(\theta|X) = \frac{1}{\sqrt{n}} \sum_{i=1}^{n} Z_i \xrightarrow{\mathrm{d}} N\{0, \mathrm{E}(Z_1^2)\}. \tag{4.73}$$

One of the regularity conditions makes sure that is legal to interchange the order of differentiation and integration in the following calculation:

$$0 = \frac{\partial}{\partial\theta} \int f(x|\theta)\, dx$$

$$= \int \frac{\partial}{\partial\theta} f(x|\theta)\, dx$$

$$= \int \frac{\partial}{\partial\theta} \log\{f(x|\theta)\} f(x|\theta)\, dx$$

$$= \mathrm{E}(Z_1);$$

$$0 = \frac{\partial^2}{\partial\theta^2} \int f(x|\theta)\, dx$$

$$= \frac{\partial}{\partial\theta} \int \frac{\partial}{\partial\theta} \log\{f(x|\theta)\} f(x|\theta)\, dx$$

$$= \int \left[ \frac{\partial^2}{\partial\theta^2} \log\{f(x|\theta)\} f(x|\theta) + \frac{\partial}{\partial\theta} \log\{f(x|\theta)\} \frac{\partial}{\partial\theta} f(x|\theta) \right] dx$$

$$= \int \frac{\partial^2}{\partial\theta^2} \log\{f(x|\theta)\} f(x|\theta)\, dx + \int \left[ \frac{\partial}{\partial\theta} \log\{f(x|\theta)\} \right]^2 f(x|\theta)\, dx$$

$$= \mathrm{E}(Y_1) + \mathrm{E}(Z_1^2).$$

Thus, in particular, $\mathrm{E}(Y_1) = -\mathrm{E}(Z_1)$. Combining (4.70)–(4.74), we have

$$0 = \frac{\partial}{\partial\theta} l(\theta|X) + \{n\mathrm{E}(Z_1^2) + o_\mathrm{P}(n) + O_\mathrm{P}(n)(\hat{\theta} - \theta)\}(\hat{\theta} - \theta)$$

$$= \frac{\partial}{\partial\theta} l(\theta|X) + n\{\mathrm{E}(Z_1^2) + o_\mathrm{P}(1)\}(\hat{\theta} - \theta)$$

using the consistency of $\hat{\theta}$. Thus, we have

$$\sqrt{n}(\hat{\theta} - \theta) = -\frac{1}{\mathrm{E}(Z_1^2) + o_\mathrm{P}(1)} \frac{1}{\sqrt{n}} \frac{\partial}{\partial \theta} l(\theta|X)$$

$$\xrightarrow{\mathrm{d}} N\left\{0, \frac{1}{\mathrm{E}(Z_1^2)}\right\} \tag{4.74}$$

using Slutsky's theorem (Theorem 2.13).

The quantity $\mathrm{E}(Z_1^2)$ is known as the Fisher information, denoted by

$$I(\theta) = \mathrm{E}\left[\frac{\partial}{\partial \theta} \log\{f(X_1|\theta)\}\right]^2. \tag{4.75}$$

In a suitable sense, $I(\theta)$ represents the amount of information about $\theta$ contained in $X_1$. The concept can be extended to multiple observations; that is, the amount of information contained in $X_1, \ldots, X_n$ is

$$\mathcal{I}(\theta) = \mathrm{E}\left[\frac{\partial}{\partial \theta} \log\{f(X_1, \ldots, X_n|\theta)\}\right]^2. \tag{4.76}$$

Here, with a little abuse of the notation, $f(x_1, \ldots, x_n|\theta)$ represents the joint pdf of $X_1, \ldots, X_n$. Since, in the i.i.d. case, we have $f(x_1, \ldots, x_n|\theta) = \prod_{i=1}^{n} f(x_i|\theta)$, it follows that, under regularity conditions, $\mathcal{I}(\theta) = nI(\theta)$ (Exercise 4.22); that is, the amount of information contained in $X_1, \ldots, X_n$ is $n$ times that contained in $X_1$.

The result (4.74) on asymptotic distribution of the MLE may be generalized in many ways. First, the parameter $\theta$ does not have to be univariate. Second, the observations do not have to be i.i.d. Let $\theta$ be a multi-dimensional vector of parameters; that is, $\theta \in \Theta \subset R^p$ ($p \geq 1$). Let $X_1, \ldots, X_n$ be observations whose joint pdf with respect to a measure $\mu$ depends on $\theta$, denoted by $f(x|\theta)$, where $x = (x_1, \ldots, x_n)'$. Then under some regularity conditions, the MLE of $\theta$, $\hat{\theta}$, satisfies the likelihood equation $\partial l/\partial \theta = 0$, where $l = l(\theta|X) = \log\{f(X|\theta)\}$ is the log-likelihood function with $X = (X_1, \ldots, X_n)'$. By the multivariate Taylor expansion (4.12), we have

$$0 = \frac{\partial l(\hat{\theta}|X)}{\partial \theta_i}$$

$$= \frac{\partial l(\theta|X)}{\partial \theta_i} + \left\{\frac{\partial^2 l(\theta|X)}{\partial \theta' \partial \theta_i}\right\}(\hat{\theta} - \theta)$$

$$+ \frac{1}{2}(\hat{\theta} - \theta)'\left\{\frac{\partial^3 l(\tilde{\theta}^{(i)}|X)}{\partial \theta \partial \theta' \partial \theta_i}\right\}(\hat{\theta} - \theta),$$

$1 \leq i \leq p$, where $\theta_i$ is the $i$th component of $\theta$ and $\tilde{\theta}^{(i)}$ lies between $\theta$ and $\hat{\theta}$. Note that $\tilde{\theta}^{(i)}$ depends on $i$ (which is something that one might overlook).

Here, of course, we assume the existence of all partial derivatives involved. It follows that, as a vector, we can write

$$
\begin{aligned}
0 &= \frac{\partial l(\hat{\theta}|X)}{\partial \theta} \\
&= \frac{\partial l(\theta|X)}{\partial \theta} + \left\{ \frac{\partial^2 l(\theta|X)}{\partial \theta \partial \theta'} \right\} (\hat{\theta} - \theta) \\
&\quad + \frac{1}{2} \left[ (\hat{\theta} - \theta)' \frac{\partial^3 l(\tilde{\theta}^{(i)}|X)}{\partial \theta \partial \theta' \partial \theta_i} \right]_{1 \le i \le p} (\hat{\theta} - \theta) \\
&= \frac{\partial l(\theta|X)}{\partial \theta} + \left\{ \frac{\partial^2 l(\theta|X)}{\partial \theta \partial \theta'} \right. \\
&\quad \left. + \frac{1}{2} \left[ (\hat{\theta} - \theta)' \frac{\partial^3 l(\tilde{\theta}^{(i)}|X)}{\partial \theta \partial \theta' \partial \theta_i} \right]_{1 \le i \le p} \right\} (\hat{\theta} - \theta).
\end{aligned}
\tag{4.77}
$$

Note that $\partial^2 l / \partial \theta' \partial \theta_i = \partial^2 l / \partial \theta_i \partial \theta'$. In order to derive the asymptotic distribution of $\hat{\theta}$, we need to make more assumptions. Basically, these assumptions replace the i.i.d assumption by some weaker conditions so that some kind of WLLN and CLT hold (note that for these results to hold, some distributional assumptions on $X_1, \ldots, X_n$ are necessary). First define

$$
\mathcal{I}_n(\theta) = -\mathrm{E} \left\{ \frac{\partial^2 l(\theta|X)}{\partial \theta \partial \theta'} \right\},
\tag{4.78}
$$

which is called the Fisher information matrix. This may be regarded as an extension of $n \mathcal{I}(\theta)$ in the i.i.d. case. We assume that $\mathcal{I}_n(\theta)$ is positive definite. Furthermore, we assume that

$$
\mathcal{I}_n^{-1/2}(\theta) \left\{ \frac{\partial^2 l(\theta|X)}{\partial \theta \partial \theta'} - \mathcal{I}_n(\theta) \right\} \mathcal{I}_n^{-1/2}(\theta) = o_P(1),
\tag{4.79}
$$

$$
\mathcal{I}_n^{-1/2}(\theta) \left[ (\hat{\theta} - \theta)' \frac{\partial^3 l(\tilde{\theta}^{(i)}|X)}{\partial \theta \partial \theta' \partial \theta_i} \right]_{1 \le i \le p} \mathcal{I}_n^{-1/2}(\theta) = o_P(1)
\tag{4.80}
$$

(see Appendix A.1 for the definition of $A^{-1/2}$, where $A > 0$). Note that since all we know (by Taylor expansion) is that $\tilde{\theta}^{(i)} = (1-t)\theta + t\hat{\theta}$ for some $t \in [0,1]$, for (4.79) to hold we need some kind of uniform convergence in probability for $t \in [0,1]$. Finally, we assume that

$$
\mathcal{I}_n^{-1/2}(\theta) \frac{\partial l(\theta|X)}{\partial \theta} \xrightarrow{\mathrm{d}} N(0, \Sigma)
\tag{4.81}
$$

as $n \to \infty$. Under assumptions (4.78)–(4.81) we have, by (4.77),

$$
-\mathcal{I}_n^{-1/2}(\theta) \frac{\partial l(\theta|X)}{\partial \theta} = \{ I_p + o_P(1) \} \mathcal{I}_n^{1/2}(\theta) (\hat{\theta} - \theta).
$$

Therefore, as $n \to \infty$, we have

$$\mathcal{I}_n^{1/2}(\theta)(\hat{\theta} - \theta) = -\{I_p + o_P(1)\}^{-1} \mathcal{I}_n^{-1/2}(\theta) \frac{\partial l(\theta|X)}{\partial \theta}$$

$$\xrightarrow{\mathrm{d}} N(0, \Sigma). \tag{4.82}$$

Note that in most cases where (4.81) holds, we actually have $\Sigma = I_p$. For example, under some regularity conditions, we have, similar to the previous i.i.d. and univariate $\theta$ case,

$$\mathrm{E}\left\{\frac{\partial l(\theta|X)}{\partial \theta}\right\} = 0, \tag{4.83}$$

$$\mathrm{Var}\left\{\frac{\partial l(\theta|X)}{\partial \theta}\right\} = \mathcal{I}_n(\theta). \tag{4.84}$$

Then, the left side of (4.81) is simply the standardization of $(\partial/\partial\theta)l(\theta|X)$, which in many cases converges in distribution to the standard $p$-variate normal distribution. Therefore, the result of (4.82) may be interpreted as that as $n \to \infty$, $\hat{\theta}$ is asymptotically ($p$-variate) normal with mean vector $\theta$ and covariance matrix equal to $\mathcal{I}_n^{-1}(\theta)$, the inverse of the Fisher information matrix.

## 4.8 Case study: The Prasad–Rao method

Surveys are usually designed to produce reliable estimates of various characteristics of interest for large geographic areas. However, for effective planning of health, social, and other services and for apportioning government funds, there is a growing demand to produce similar estimates for small geographic areas and subpopulations. The usual design-based estimator, which uses only the sample survey data for the particular small area of interest, is unreliable due to relatively small samples available from the area. In the absence of a reliable small-area design-based estimator, one may alternatively use a synthetic estimator (Rao 2003, Section 4.2), which utilizes data from censuses or administrative records to obtain estimates for small geographical areas or subpopulations. Although the synthetic estimators are known to have smaller variances compared to the direct survey estimators, they tend to be biased as they do not make use of the information on the characteristic of interest directly obtainable from sample surveys.

A compromise between the direct survey and the synthetic estimations is the method of composite estimation which uses sample survey data in conjunction with different census and administrative data. Implicit or explicit models, which "borrow strength" from related sources, have been used in this latter approach. Research in this and related areas are usually called small area estimation. See Rao (2003) for a detailed account of different composite estimation and other techniques in small area estimation.

An explicit (linear) model for composite small area estimation may be expressed as follows:

$$Y_i = X_i\beta + Z_i v_i + e_i, \quad i = 1,\ldots,m, \qquad (4.85)$$

where $m$ is the number of small areas; $Y_i$ represents the vector of observations from the $i$th small area; $X_i$ is a matrix of known covariates for the $i$th small area, and $\beta$ is a vector of unknown regression coefficients (the fixed effects); $Z_i$ is a known matrix, and $v_i$ is a vector of small-area specific random effects; and $e_i$ represents a vector of sampling errors. It is assumed that $Y_i$ is $n_i \times 1$, $X_i$ is $n_i \times p$, $\beta$ is $p \times 1$, $Z_i$ is $n_i \times b_i$, $v_i$ is $b_i \times 1$ and $e_i$ is $n_i \times 1$. Also assumed is that the $v_i$'s and $e_i$'s are independent such that $\mathrm{E}(v_i) = 0$, $\mathrm{Var}(v_i) = G_i$; $\mathrm{E}(e_i) = 0$ and $\mathrm{Var}(e_i) = R_i$. Here the matrices $G_i$ and $R_i$ usually depend on a vector $\psi$ of unknown parameters known as variance components. Two special cases of the above small area model are the following.

*Example 4.18* (The Fay–Herriot model). Fay and Herriot (1979) proposed the following model for the estimation of per-capita income of small places with population sizes less than 1000:

$$Y_i = x_i'\beta + v_i + e_i, \qquad (4.86)$$

$i = 1,\ldots,m$, where $x_i$ is a vector of known covariates, $\beta$ is a vector of unknown regression coefficients, $v_i$'s are area-specific random effects, and $e_i$'s represent sampling errors. It is assumed that the $v_i$'s and $e_i$'s are independent with $v_i \sim N(0, A)$ and $e_i \sim N(0, D_i)$. The variance $A$ is unknown, but the sampling variances $D_i$'s are assumed known. It is easy to show that the Fay–Herriot model is a special case of the general small-area model (4.85) (Exercise 4.25).

*Example 4.19* (The nested-error regression model). Battese, Harter, and Fuller (1988) presented data from 12 Iowa counties obtained from the 1978 June Enumerative Survey of the U.S. Department of Agriculture as well as data obtained from land observatory satellites on crop areas involving corn and soybeans. The objective was to predict the mean hectares of corn and soybeans per segment for the 12 counties using the satellite information. The authors introduced the following model, known as the nested-error regression model, for the prediction problem:

$$Y_{ij} = x_{ij}'\beta + v_i + e_{ij}, \qquad (4.87)$$

$i = 1,\ldots,m$, $j = 1,\ldots,n_i$, where $x_{ij}$ is a known vector of covariates, $\beta$ is an unknown vector of regression coefficients, $v_i$ is a random effect associated with the $i$th small area, and $e_{ij}$ is the sampling error. It is assumed that the random effects are independent and distributed as $N(0, \sigma_v^2)$, the sampling errors are independent and distributed as $N(0, \sigma_e^2)$, and the random effects and sampling errors are uncorrelated. It can be shown that this is, again, a special case of the general small-area model (4.85) (Exercise 4.26).

The problem of main interest in the small-area estimation is usually the estimation, or prediction, of small-area means. A small-area mean may be expressed, at least approximately, as a mixed effect, $\eta = b'\beta + a'v$, where $a$ and $b$ are known vectors and $\beta$ and $v = (v_i)_{1 \leq i \leq m}$ are the vectors of fixed and random effectsi, respectively, in (4.85) (it is called a mixed effect because it is a combination of fixed and random effects). If $\beta$ and $\psi$ are both known, the best predictor (BP) for $\eta$, is the conditional expectation $E(\eta|Y)$. Furthermore, if the random effects $v_i$ and errors $e_i$ are normally distributed, this conditional expectation is given by

$$\eta^* = b'\beta + a'E(\alpha|Y)$$
$$= b'\beta + a'GZ'V^{-1}(Y - X\beta),$$

where $X = (X_i)_{1 \leq i \leq m}$, $Y = (Y_i)_{1 \leq i \leq m}$, $G = \text{diag}(G_1, \ldots, G_m)$, $Z_i = \text{diag}(Z_1, \ldots, Z_m)$, and $V = \text{Var}(Y) = \text{diag}(V_1, \ldots, V_m)$ with $V_i = Z_i G_i Z_i' + R_i$. In the absence of the normality assumption, $\eta^*$ is the best linear predictor (BLP) of $\eta$ in the sense that it minimizes the mean squared prediction error (MSPE) of a predictor that is linear in $Y$ (e.g., Jiang 2007, Section 2.3). Of course, $\beta$ is unknown in practice. It is then customary to replace $\beta$ by

$$\tilde{\beta} = (X'V^{-1}X)^{-1}X'V^{-1}Y, \tag{4.88}$$

which is the MLE of $\beta$ under the normality assumption, provided that $\psi$ is known. The result is called the best linear unbiased predictor, or BLUP, denoted by $\tilde{\eta}$. In other words, $\tilde{\eta}$ is given by $\eta^*$ with $\beta$ replaced by $\tilde{\beta}$.

The expression of BLUP involves $\psi$, the vector of variance components, which is typically unknown in practice. It is then customary to replace $\psi$ by a consistent estimator, $\hat{\psi}$. The resulting predictor is often called the empirical BLUP, or EBLUP, denoted by $\hat{\eta}$. To illustrate the EBLUP procedure, we consider a previous example.

*Example 4.18 (continued).* Consider the Fay–Herriot model. Let $\eta$ denote the small-area mean for the $i$th area; that is, $\eta = x_i'\beta + v_i$. Then the BP for $\eta$ is given by (Exercise 4.25)

$$\eta^* = (1 - B_i)Y_i + B_i x_i'\beta,$$

where $B_i = D_i/(A + D_i)$. The BLUP is given by $\tilde{\eta} = \eta^*$ with $\beta$ replaced by

$$\tilde{\beta} = \left(\sum_{i=1}^m \frac{x_i x_i'}{A + D_i}\right)^{-1} \left(\sum_{i=1}^m \frac{x_i y_i}{A + D_i}\right).$$

Finally, the EBLUP is given by

$$\hat{\eta} = (1 - \hat{B}_i)Y_i + \hat{B}_i x_i'\hat{\beta},$$

where $\hat{B}_i$ and $\hat{\beta}$ are $B_i$ and $\tilde{\beta}$, respectively, with $A$ replaced by $\hat{A}$, a consistent estimator of $A$. One example of a consistent estimator of $A$ is the method of moments (MoM) estimator proposed by Prasad and Rao (1990), given by

$$\hat{A} = \frac{Y'P_{X\perp}Y - \operatorname{tr}(P_{X\perp}D)}{m - p},$$

where $P_{X\perp} = I - P_X$, with $P_X = X(X'X)^{-1}X'$, and $D = \operatorname{diag}(D_1, \ldots, D_m)$.

Although the EBLUP is fairly easy to obtain, assessing its uncertainty is quite a challenging problem. As mentioned, a measure of the uncertainty that is commonly used is the MSPE. However, unlike the BLUP, the MSPE of the EBLUP does not, in general, have a closed-form expression. This is because once the variance components $\psi$ are replaced by their (consistent) estimators, the predictor is no longer linear in $Y$. A naive approach to estimation of the MSPE of EBLUP would be to first obtain the MSPE of BLUP, which can be expressed in closed-form as a function of $\psi$ (see below), and then replace $\psi$ by $\hat{\psi}$ in the expression of the MSPE of BLUP, where $\hat{\psi}$ is the consistent estimator of $\psi$. However, as will be seen, this approach underestimates the MSPE of EBLUP, as it does not take into account the additional variation associated with the estimation of $\psi$.

Prasad and Rao (1990) proposed a method based on the Taylor series expansion to produce the second-order unbiased MSPE estimator for EBLUP. Here, the term "second-order unbiased" is with respect to the above naive MSPE estimator, which is first-order unbiased. The latter property is because, roughly speaking, the difference between the BLUP and EBLUP is of the order $O(m^{-1/2})$. To see this, note that the BLUP can be expressed as

$$\begin{aligned} \tilde{\eta} &= \tilde{\eta}(\psi) \\ &= b'\tilde{\beta} + a'GZ'V^{-1}(Y - X\tilde{\beta}), \end{aligned} \tag{4.89}$$

where $\tilde{\beta}$ is given by (4.88). It follows that the EBLUP is simply $\hat{\eta} = \tilde{\eta}(\hat{\psi})$. By the Taylor expansion, we have $\tilde{\eta}(\hat{\psi}) - \tilde{\eta}(\psi) \approx (\partial\tilde{\eta}/\partial\psi')(\hat{\psi} - \psi) = O_P(m^{-1/2})$ under some regularity conditions. Therefore, $\operatorname{E}(\hat{\eta} - \tilde{\eta})^2$ is typically of the order $O(m^{-1})$. On the other hand, Kackar and Harville (1984) showed that under the normality assumption,

$$\begin{aligned} \operatorname{MSPE}(\hat{\eta}) &= \operatorname{E}(\hat{\eta} - \eta)^2 \\ &= \operatorname{E}(\tilde{\eta} - \eta)^2 + \operatorname{E}(\hat{\eta} - \tilde{\eta})^2 \\ &= \operatorname{MSPE}(\tilde{\eta}) + \operatorname{E}(\hat{\eta} - \tilde{\eta})^2. \end{aligned} \tag{4.90}$$

Equation (4.90) clearly suggests that the naive MSPE estimator underestimates the true MSPE, because it only takes into account the first term on the right side. Furthermore, if one replaces $\psi$ by $\hat{\psi}$ in the expression of $\operatorname{MSPE}(\tilde{\eta})$, it introduces a bias of the order $O(m^{-1})$ [not $O(m^{-1/2})$]. Thus, the bias of the naive MSPE estimator is $O(m^{-1})$. By second-order unbiasedness of the Prasad–Rao method, it means that

$$\mathrm{E}\left(\widehat{\mathrm{MSPE}} - \mathrm{MSPE}\right) = o(m^{-1}), \tag{4.91}$$

where $\mathrm{MSPE} = \mathrm{MSPE}(\hat{\eta})$ and $\widehat{\mathrm{MSPE}}$ represents the Prasad–Rao estimator of MSPE. Furthermore, the following closed-form expression can be obtained (Exercise 4.27):

$$\mathrm{MSPE}(\tilde{\eta}) = a'(G - GZ'V^{-1}ZG)a + d'(X'V^{-1}X)^{-1}d, \tag{4.92}$$

where $d = b - X'V^{-1}ZGa$. Here, we assume that $X$ is of full rank $p$. Note that, typically, the first term on the right side of (4.92) is $O(1)$ and the second-term is $O(m^{-1})$. An implication is that $\mathrm{MSPE}(\tilde{\eta}) = O(1)$. In view of (4.90) and (4.92), a main part of the Prasad–Rao method is therefore to derive an approximation to $\mathrm{E}(\hat{\eta} - \tilde{\eta})^2$. Assume that suitable regularity conditions are satisfied. Then we have, by the Taylor expansion and (4.89),

$$\begin{aligned}\hat{\eta} - \tilde{\eta} &= \tilde{\eta}(\hat{\psi}) - \tilde{\eta}(\psi) \\ &= \frac{\partial \tilde{\eta}}{\partial \psi'}(\hat{\psi} - \psi) + \frac{1}{2}(\hat{\psi} - \psi)'\frac{\partial^2 \tilde{\eta}(\check{\psi})}{\partial \psi \partial \psi'}(\hat{\psi} - \psi),\end{aligned}$$

where $\check{\psi}$ lies between $\psi$ and $\hat{\psi}$. Suppose that $\hat{\psi}$ is a $\sqrt{m}$-consistent estimator in the sense that $\sqrt{m}(\hat{\psi} - \psi) = O_\mathrm{P}(1)$ (see Section 3.4, above Example 3.7), and the following hold:

$$\left|\frac{\partial \tilde{\eta}}{\partial \psi}\right| = O_\mathrm{P}(1), \qquad \sup_{|\check{\psi}-\psi|\le|\hat{\psi}-\psi|}\left\|\frac{\partial^2 \tilde{\eta}(\check{\psi})}{\partial \psi \partial \psi'}\right\| = O_\mathrm{P}(1).$$

Then by the method of formal derivation (see Section 4.3), we have

$$\mathrm{E}(\hat{\eta} - \tilde{\eta})^2 = \mathrm{E}\left\{\frac{\partial \tilde{\eta}}{\partial \psi'}(\hat{\psi} - \psi)\right\}^2 + o(m^{-1}). \tag{4.93}$$

Now, suppose the first term on the right side of (4.93) can be expressed as

$$\mathrm{E}\left\{\frac{\partial \tilde{\eta}}{\partial \psi'}(\hat{\psi} - \psi)\right\}^2 = \frac{a(\psi)}{m} + o(m^{-1}), \tag{4.94}$$

where $a(\cdot)$ is a known differentiable function. Also, let $b(\psi)$ denote the right side of (4.92). By (4.93) and (4.94), to obtain a second-order unbiased estima-tor of $\mathrm{E}(\hat{\eta} - \tilde{\eta})^2$, all one needs to do is to replace $\psi$ in $a(\psi)$ by $\hat{\psi}$ because the resulting bias is $o(m^{-1})$ (why?). However, one cannot use the same strategy to estimate $\mathrm{MSPE}(\tilde{\eta}) = b(\psi)$, because the resulting bias is $O(m^{-1})$ rather than $o(m^{-1})$. In order to reduce the latter bias to $o(m^{-1})$, we use the following bias correction procedure. Note that, by the Taylor expansion, we have

$$b(\hat{\psi}) = b(\psi) + \frac{\partial b}{\partial \psi'}(\hat{\psi} - \psi) + \frac{1}{2}(\hat{\psi} - \psi)'\frac{\partial^2 b}{\partial \psi \partial \psi'}(\hat{\psi} - \psi) + o(m^{-1});$$

hence, by the method of formal derivation (Section 4.3),

$$E\{b(\hat{\psi})\} = b(\psi) + \frac{\partial b}{\partial \psi'}E(\hat{\psi} - \psi) + \frac{1}{2}E\left\{(\hat{\psi} - \psi)'\frac{\partial^2 b}{\partial \psi \partial \psi'}(\hat{\psi} - \psi)\right\}$$

$$+o(m^{-1})$$

$$= b(\psi) + \frac{c(\psi)}{m} + o(m^{-1}).$$

Here, we make the assumption that $E(\hat{\psi} - \psi) = O(m^{-1})$, which holds under regularity conditions. Now, we can apply the same plug-in technique used above for estimating $a(\psi)$ to the estimation of $c(\psi)$. In other words, we estimate $b(\psi)$ by $b(\hat{\psi}) - c(\hat{\psi})/m$ because the bias of this estimator is

$$E\left\{b(\hat{\psi}) - \frac{c(\hat{\psi})}{m}\right\} - b(\psi) = b(\psi) + \frac{c(\psi)}{m} + o(m^{-1}) - \frac{E\{c(\hat{\psi})\}}{m} - c(\psi)$$

$$= \frac{E\{c(\psi) - c(\hat{\psi})\}}{m} + o(m^{-1})$$

$$= o(m^{-1}),$$

provided that $c(\cdot)$ is a smooth (e.g., differentiable) function.

In conclusion, if we define the Prasad–Rao estimator as

$$\widehat{\text{MSPE}} = b(\hat{\psi}) + \frac{a(\hat{\psi}) - c(\hat{\psi})}{m}, \tag{4.95}$$

then we have

$$E(\widehat{\text{MSPE}}) = E\{b(\hat{\psi})\} + \frac{E\{a(\hat{\psi})\}}{m} - \frac{E\{c(\hat{\psi})\}}{m}$$

$$= b(\psi) + \frac{c(\psi)}{m} + o(m^{-1})$$

$$+\frac{a(\psi)}{m} + \frac{E\{a(\hat{\psi}) - a(\psi)\}}{m}$$

$$-\frac{c(\psi)}{m} - \frac{E\{c(\hat{\psi}) - c(\psi)\}}{m}$$

$$= b(\psi) + \frac{a(\psi)}{m} + o(m^{-1})$$

$$= \text{MSPE}(\tilde{\eta}) + E(\hat{\eta} - \tilde{\eta})^2 + o(m^{-1})$$

$$= \text{MSPE} + o(m^{-1}),$$

using (4.90), (4.92), and (4.93) near the end. Therefore, (4.91) holds.

Prasad and Rao (1990) obtained a detailed expression of (4.94) and, hence, (4.95) for the two special cases discussed earlier—that is, the Fay–Herriot model (Example 4.18) and the nested-error regression model (Example 4.19), assuming normality and using the MoM estimators of $\psi$. Extensions of the Prasad–Rao method will be discussed in Chapters 12 and 13.

## 4.9 Exercises

*4.1.* Regarding Table 4.1, how large should $n$ be in order to achieve the same accuracy (in terms of the relative error) for $x = 5$?

*4.2.* This is regarding Example 4.3.

(a) Show that

$$l(0) = -\sum_{i=1}^{n}\{X_i + 2\log(1 + e^{-X_i})\},$$

$$l'(0) = \sum_{i=1}^{n}\frac{1 - e^{-X_i}}{1 + e^{-X_i}},$$

$$l''(\theta) = -2\sum_{i=1}^{n}\frac{e^{\theta-X_i}}{(1 + e^{\theta-X_i})^2}.$$

(b) Show that $n^{-1}l(0) \xrightarrow{P} a$ as $n \to \infty$, where $a$ is a positive constant.

(c) Show that $n^{-1/2}l'(0) \xrightarrow{d} N(0, \sigma^2)$ as $n \to \infty$, and determine $\sigma^2$.

(d) Show that there is a sequence of positive random variables $\xi_n$ and a constant $c > 0$ such that $\xi_n \xrightarrow{P} b$, where $b$ is a positive constant, and

$$\xi_n n \leq \sup_{\theta}|l''(\theta)| \leq cn.$$

*4.3.* In Example 4.4, show that

$$\frac{\partial g(\zeta_n)}{\partial x} - \frac{\partial g(c)}{\partial x} = o_P(1),$$

where $\partial g/\partial x = (\partial g/\partial x')'$.

*4.4.* Let $X_1, \ldots, X_n$ be i.i.d. observations such that $E(X_i) = \mu$ and $\mathrm{var}(X_i) = \sigma^2$, where $0 < \sigma^2 < \infty$. Derive the (three) results at the end of Section 4.2.

*4.5.* Let $X_1, \ldots, X_n$ be i.i.d. observations generated from the following distributions, where $n = 30$. Construct the histograms of the empirical distribution of $\bar{X}$ based on 10,000 simulated values. Does the population distribution of the $X_i$'s make a difference?

(i) $N(0, 1)$;

(ii) Uniform$[0, 1]$;

(iii) Exponential(1);

(iv) Bernoulli($p$), where $p = 0.1$.

*4.6.* In this exercise you are asked to study empirically the convergence of CLT in regard to Example 4.5.

(i) Generate two sequences of random variables. The first sequence is generated independently from the Beta($\alpha, \beta$) distribution with $\alpha = \beta = 2$ [case (i)]; the second sequence is generated independently from the Beta($\alpha, \beta$) distribution with $\alpha = 2$ and $\beta = 6$ [case (ii)]. Based on each sequence, compute $\xi_n = (\sqrt{n}/\sigma)(\bar{X} - \mu)$, where $n$ is the sample size (i.e., the number of random variables in the sequence, which is the same for both sequences),

$$\mu = \frac{\alpha}{\alpha + \beta},$$

$$\sigma^2 = \frac{\alpha\beta}{(\alpha + \beta)^2(\alpha + \beta + 1)}.$$

(ii) For each of sample sizes $n = 15, 30, 60, 150$, and $400$, repeat (i) 1000 times. Make a histogram for case (i) and case (ii).

(iii) In addition to the histograms, obtain the 5th and 95th percentiles based on the 1000 values of $\xi_n$ for each case and sample size and compare the percentiles with the corresponding standard normal percentiles.

(iv) Make a nice plot that compares the histograms and a nice table that compares the percentiles for the increasing sample size. What do you conclude?

*4.7.* Obtain the two-term Edgeworth expansion [i.e., (4.27)] for the following distributions of $X_i$:

(i) $X_i \sim$ the double exponential distribution DE$(0, 1)$, where the pdf of DE$(\mu, \sigma)$ is given by

$$f(x|\mu, \sigma) = \frac{1}{2\sigma} \exp\left(-\frac{|x - \mu|}{\sigma}\right), \quad -\infty < x < \infty.$$

(ii) $X_i \sim \chi_4^2$.

*4.8.* Show that the sequence of functions $\phi_k$, $k \in Z$, of Example 4.8 is an orthonormal system.

*4.9.* Show that the sequence of functions defined in Example 4.9 is an orthonormal system.

*4.10.* Compute $\phi_k(x)$ for $k = 2, 3, 4$ in Example 4.10. Also verify that $\phi_k$, $k = 0, 1, 2, 3, 4$, are orthonormal; that is, they satisfy (4.41) and (4.42) with $S = [0, 1]$.

*4.11.* Show that the Haar functions defined in Example 4.11 [i.e., (4.44) plus $I_{[0,1)}$] constitute an orthonormal system on $(-\infty, \infty)$.

*4.12.* Use Fubini's theorem (see Appendix A.2) to establish (4.51) given the condition (4.49).

*4.13.* Show that the function $\gamma$ defined in Example 4.14 is an autocovariance function.

*4.14.* Prove the (identity) expansions (4.55) and (4.58) by mathematical induction.

*4.15.* Show that in Example 4.15 we have $E(\eta_{ij}) = (p+2)1_{(i=j)}$, $1 \leq i, j \leq p$.

*4.16.* Show that in Example 4.15 we have $E(\zeta_{ij}) = 0$ if $i \neq j$ and $E(\zeta_{ii})$ does not depend on $i$.

*4.17.* Suppose that $X$ has a $\chi_\nu^2$-distribution, where $\nu > 2$. Use the elementary expansion (4.55) with $l = 4$ and without the remaining term to approximate $E(X^{-1})$. Note that closed-form expressions of moments of $X$, including $E(X^{-1})$, can be obtained, so that one can directly compare the accuracy of the approximation. Does the approximation improve as $\nu \to \infty$? (Hint: Consider the relative error of the approximation defined as $|\text{approximate} - \text{exact}|/\text{exact}$.)

*4.18.* Derive the approximation (4.63) using the Taylor expansion

$$q(x) = q(\tilde{x}) + \frac{1}{2}q''(\tilde{x})(x - \tilde{x})^2 + \cdots.$$

*4.19.* Derive the Laplace approximation (4.64). What is the constant $c$?

*4.20.* This exercise is related to Example 4.16.

(i) Show that in this case the exact value of (4.62) is given by

$$\frac{\sqrt{\nu\pi}\,\Gamma(\frac{\nu}{2})}{\Gamma(\frac{\nu+1}{2})},$$

and the Laplace approximation (4.63) is

$$\sqrt{\frac{2\nu\pi}{\nu + 1}}.$$

(ii) Show that if $q(x) = (x - \mu)^2/2\sigma^2$ for some $\mu \in R$ and $\sigma^2 > 0$, the Laplace approximation (4.63) is exact.

*4.21.* Show that the likelihood function in Example 4.17 can be expressed as (4.69).

*4.22.* Show that in the i.i.d. case, the amount of information contained in $X_1, \ldots, X_n$ is $n$ times that contained in $X_1$; that is, $\mathcal{I}(\theta) = nI(\theta)$ [see (4.75) and (4.76)]. The result requires some regularity conditions to hold. What regularity conditons?

*4.23.* Let $X_1, \ldots, X_n$ be i.i.d. observations with the pdf or pmf $f(x|\theta)$, where $\theta$ is a univariate parameter. Here, the pdf is with respect to the Lebesgue measure, whereas the pmf may be regarded as a pdf with respect to the counting measure [see below (4.36)]. Obtain the Fisher information (4.75) for the following cases:

(i) $X_1 \sim$ Bernoulli($\theta$), so that

$$f(x|\theta) = \theta^x(1 - \theta)^{1-x}, \quad x = 0, 1,$$

where $\theta \in (0, 1)$.

(ii) $X_1 \sim$ Poisson($\theta$), so that

$$f(x|\theta) = e^{-\theta} \frac{\theta^x}{x!}, \quad x = 0, 1, \ldots,$$

where $\theta > 0$.

(iii) $X_1 \sim$ Exponential($\theta$), so that

$$f(x|\theta) = \frac{1}{\theta} e^{-x/\theta}, \quad x \geq 0,$$

where $\theta > 0$.

(iv) $X_1 \sim N(\theta, \theta^2)$, so that

$$f(x|\theta) = \frac{1}{\sqrt{2\pi\theta^2}} \exp\left\{ -\frac{(x-\theta)^2}{2\theta^2} \right\}, \quad -\infty < x < \infty,$$

where $\theta \in (-\infty, \infty)$.

4.24. Let $X_1, \ldots, X_n$ be i.i.d. with the following pdf or pmf depending on $\theta = (\theta_1, \theta_2)$. Obtain the Fisher information matrix (4.78) in each case.

(i) $X_1 \sim N(\mu, \sigma^2)$, where $\mu \in (-\infty, \infty)$ and $\sigma^2 > 0$, so that $\theta_1 = \mu$ and $\theta_2 = \sigma^2$.

(ii) $X_1 \sim$ Gamma($\alpha, \beta$), whose pdf is given by

$$f(x|\alpha, \beta) = \frac{1}{\Gamma(\alpha)\beta^\alpha} x^{\alpha-1} e^{-x/\beta}, \quad x > 0,$$

where $\alpha > 0$ and $\beta > 0$ are known as the shape and scale parameters, respectively, so that $\theta_1 = \alpha$ and $\theta_2 = \beta$.

(iii) $X_1 \sim$ Beta($\alpha, \beta$), whose pdf is given by

$$f(x|\alpha, \beta) = \frac{\Gamma(\alpha+\beta)}{\Gamma(\alpha)\Gamma(\beta)} x^{\alpha-1} (1-x)^{\beta-1}, \quad 0 < x < 1,$$

where $\alpha > 0$ and $\beta > 0$, so that $\theta_1 = \alpha$ and $\theta_2 = \beta$.

4.25. Show that the Fay–Herriot model of Example 4.18 is a special case of the small-area model (4.85). Specify the matrices $X_i$, $Z_i$, $G_i$, and $R_i$ in this case. Furthermore, show that the BP for $\eta = x_i'\beta + v_i$ is given by $\tilde{\eta} = (1 - B_i)Y_i + B_i x_i'\beta$, where $B_i = D_i/(A + D_i)$.

4.26. Show that the nested-error regression model of Example 4.19 is a special case of the small-area model (4.85). Specify the matrices $X_i$, $Z_i$, $G_i$ and $R_i$ in this case.

4.27. Derive the expression (4.92) for MSPE($\tilde{\eta}$).

4.28. Consider a special case of the Fay–Herriot model (Example 4.18) in which $D_i = D$, $1 \leq i \leq m$. This is known as the balanced case. Without loss of generality, let $D = 1$. Consider the prediction of $\eta_i = x_i'\beta + v_i$. Let $\tilde{\eta}_i$ and

$\hat{\eta}_i$ denote the BLUP and EBLUP, respectively, where the MoM estimator of $A$ is used for the EBLUP [see Example 4.18 (continued) or Exercise 4.25].
(i) Show that

$$\text{MSPE}(\tilde{\eta}_i) = \frac{A}{A+1} + \frac{x_i'(X'X)^{-1}x_i}{A+1},$$

where $X = (x_i')_{1 \le i \le m}$.
(ii) Show that

$$\text{MSPE}(\hat{\eta}_i) = \frac{A}{A+1} + \frac{x_i'(X'X)^{-1}x_i}{A+1} + \frac{2\{1 - x_i'(X'X)^{-1}x_i\}}{(A+1)(m-p)}$$
$$+ \frac{4\{1 - x_i'(X'X)^{-1}x_i\}}{(A+1)(m-p)(m-p-2)}.$$

[Hint: The moment of $(\chi_k^2)^{-1}$ has a closed-form expression, where $\chi_k^2$ denotes a random variable with a $\chi_k^2$-distribution. Find the expression.]
(iii) Let $\eta = (\eta_i)_{1 \le i \le m}$ denote the vector of small-area means and $\hat{\eta} = (\hat{\eta}_i)_{1 \le i \le m}$ denote the vector of EBLUPs. Define the overall MSPE of the EBLUP as $\text{MSPE}(\hat{\eta}) = \text{E}(|\hat{\eta} - \eta|^2) = \text{E}\{\sum_{i=1}^m (\hat{\eta}_i - \eta_i)^2\} = \sum_{i=1}^m \text{E}((\hat{\eta}_i - \eta_i)^2 = \sum_{i=1}^m \text{MSPE}(\hat{\eta}_i)$. Show that

$$\text{MSPE}(\hat{\eta}) = \frac{mA}{A+1} + \frac{p+2}{A+1} + \frac{4}{(A+1)(m-p-2)}.$$

*4.29* [Delta method (continued)]. In Example 4.4 we introduced the delta method for distributional approximations. The method can also be used for moment approximations. Let $T_1, \ldots, T_k$ be random variables whose means and variances exist. Let $g(t_1, \ldots, t_k)$ be a differentiable function. Then, by the Taylor expansion, we can write

$$g(T_1, \ldots, T_k) \approx g(\mu_1, \ldots, \mu_k) + \sum_{i=1}^k \frac{\partial g}{\partial t_i}(T_i - \mu_i),$$

where $\mu_i = \text{E}(T_i)$, $1 \le i \le k$, and $\partial g / \partial t_i$ is evaluated as $(\mu_1, \ldots, \mu_k)$. This leads to the following approximations:

$$\text{E}\{g(T_1, \ldots, T_k)\} \approx g(\mu_1, \ldots, \mu_k),$$
$$\text{var}\{g(T_1, \ldots, T_k)\} \approx \sum_{i=1}^k \left(\frac{\partial g}{\partial t_i}\right)^2 \text{var}(T_i)$$
$$+ 2\sum_{i<j} \left(\frac{\partial g}{\partial t_i}\right)\left(\frac{\partial g}{\partial t_j}\right) \text{cov}(T_i, T_j).$$

(i) Suppose that $T \sim \text{Gamma}(\alpha, \beta)$ with the pdf given in Exercise 4.24(ii). Use the above delta method to approximate the mean and variance of $T^{-1}$.

(ii) Note that the exact mean and variance of $T^{-1}$ can be obtained in this case, given a suitable range of $\alpha$. What is the range of $\alpha$ so that $E(T^{-1})$ exists? What is the range of $\alpha$ so that $E(T^{-2})$ exists?

(iii) Obtain the exact mean and variance of $T^{-1}$ given the suitable range of $\alpha$ and compare the results with the above delta-method approximations. How do the values of $\alpha$ and $\beta$ affect the accuracy of the approximations?

4.30. Let $X_1, \ldots, X_n$ be i.i.d. such that $E(X_1) = 0$, $E(X_1^2) = 1$, and $E(X_1^4) < \infty$. Consider approximation to the mean and variance of

$$Y = \frac{n}{n + \sum_{i=1}^n X_i^2}$$

using the delta method of Exercise 4.29.

(i) Let $g(x_1, \ldots, x_n) = n/(n + \sum_{i=1}^n x_i^2)$. What are the approximations to the mean and variance of $Y = g(X_1, \ldots, X_n)$?

(ii) If we let $g(t_1, \ldots, t_n) = n/(n + \sum_{i=1}^n t_i)$, and $T_i = X_i^2$, $1 \leq i \leq n$, what are the approximations to the mean and variance of $Y = g(T_1, \ldots, T_n)$?

(iii) How does the sample size $n$ affect the approximation to $E(Y)$? In other words, does the accuracy of the approximation improve as $n$ increases? [Hint: First use the dominated convergence theorem (Theorem 2.16) to show that $E(Y)$ converges to a limit as $n \to \infty$.]

(iv) Which approximation [(i) or (ii)] do you think is better? Any general comment(s) on the use of the delta method in moment approximations?

# 5

# Inequalities

We have almost always found, even with the most famous inequalities, that we have a little new to add.

**Hardy, Littlewood, & Pólya (1934)**
*Inequalities*

## 5.1 Introduction

It is said that high school algebra is characterized by equalities, whereas college and more advanced mathematics, inequalities. One may argue that, in a similar way, statistics is characterized by inequalities, too. For example, the words "margin of errors," which nowadays come along routinely with survey results that are published, may be viewed as bounds for typical range of the sampling error. This may be expressed as (i) $P(|\epsilon| \leq b) = 1 - \alpha$ or (ii) $P(|\epsilon| \leq b) \geq 1 - \alpha$, where $\epsilon$ represents the sampling error, $b$ is an upper bound, and $\alpha$ is a small positive number, such as 0.05. In (i), the event inside the probability is characterized by an inequality, whereas in (ii), both the event and the probability itself are characterized by inequalities.

Perhaps the simplest of all is the following basic triangle inequality:

$$|x + y| \leq |x| + |y| \tag{5.1}$$

for all $x$ and $y$. Inequalities such as (5.1) are called numerical inequalities, meaning that they hold for all real numbers. Many of the numerical inequalities can be extended beyond real numbers. For example, extensions of numerical inequalities to matrices have led to many of the matrix inequalities. However, not every numerical inequality has its matrix analogue. For example, if $A$ and $B$ are symmetric matrices such that $A \geq B$, meaning that $A - B$ is nonnegative definite, it is not necessarily true that $A^2 \geq B^2$. See Section 5.3.2 for more counterexamples. Numerical inequalities can be used to establish more sophisticated inequalities, such as moment and probability inequalities,

J. Jiang, *Large Sample Techniques for Statistics*,
DOI 10.1007/978-1-4419-6827-2_5, © Springer Science+Business Media, LLC 2010

but this is not always the case. For example, the covariance inequality states that for any random variables $X$ and $Y$, one has

$$\mathrm{cov}(X, Y) \leq \sqrt{\mathrm{var}(X)\mathrm{var}(Y)}, \tag{5.2}$$

or $\mathrm{E}[\{X - \mathrm{E}(X)\}\{Y - \mathrm{E}(Y)\}] \leq \{\mathrm{var}(X)\}^{1/2}\{\mathrm{var}(Y)\}^{1/2}$, but this is not derived from $\{X - \mathrm{E}(X)\}\{Y - \mathrm{E}(Y)\} \leq \{\mathrm{var}(X)\}^{1/2}\{\mathrm{var}(Y)\}^{1/2}$, which, of course, does not always hold.

Like the Taylor expansion, the value of inequalities to statistics cannot be overstated. There exist a huge number of inequalities: numerical inequalities, matrix inequalities, integral/moment inequalities, and probability inequalities. Instead of trying to come up with a list of all useful inequalities, which is impossible, we focus on developing the basic techniques for making use of existing inequalities and for developing new inequalities. We believe that these methods and techniques are even more important in solving the current and future problems than the inequalities themselves. We should also point out that although this chapter is entitled "Inequalities," it by no means includes all of the inequalities introduced in this book. However, this is the only place that these materials are treated systematically as a single subject.

## 5.2 Numerical inequalities

### 5.2.1 The convex function inequality

The triangle inequality (5.1), of course, can be derived with an elementary argument. Since $x \leq |x|$ and $y \leq |y|$, we have $x + y \leq |x| + |y|$; similarly, $-x - y \leq |x| + |y|$, or $x + y \geq -(|x| + |y|)$, which leads to (5.1). Alternatively, (5.1) is a special case of the convex function inequality. A real-valued function $f(x)$ is convex if for any $x$, $y$, and $\lambda \in [0, 1]$, we have

$$f\{(1 - \lambda)x + \lambda y\} \leq (1 - \lambda)f(x) + \lambda f(y). \tag{5.3}$$

Here, we did not specify the range of $x$, $y$. Typically, it is assumed that $x$, $y \in D$, where $D$ is a convex subset of $R$ in the sense that $x$, $y \in D$ implies $(1 - \lambda)x + \lambda y \in D$ for any $\lambda \in [0, 1]$.

To show that $f(x)$ is convex, one can, of course, verify (5.3) for any $x$, $y \in D$ and $\lambda \in [0, 1]$, but sometimes there are easier ways. For example, if $f'(x)$ exists, a necessary and sufficient condition for $f(x)$ to be a convex function is that $f'(x)$ is nondecreasing; if $f''(x)$ exists, then a necessary and sufficient condition for $f(x)$ to be convex is that $f''(x) \geq 0$.

A concave function may be thought of as a function that has the reversed properties of a convex function; that is, $f(x)$ is concave if and only if (5.3) is satisfied with the reversed inequality. In fact, $f(x)$ is convex if and only if $-f(x)$ is concave. More generally, let $g(x)$ be a linear function of $x$; that is, $g(x) = ax + b$ for some constants $a$ and $b$. Then $f(x)$ is convex if and only if

$g(x) - f(x)$ is concave. Therefore, any convex function inequality (see below) can be reversed for a concave function inequality.

The best know property of a convex function $f(x)$ is the following:

$$f\left(\frac{x_1 + \cdots + x_n}{n}\right) \le \frac{f(x_1) + \cdots + f(x_n)}{n} \tag{5.4}$$

for any $x_1, \ldots, x_n \in D$. To see that (5.1) is a special case of (5.4), note that $f(x) - |x|$ is a convex function; hence, by (5.4), we have

$$\left|\frac{x+y}{2}\right| \le \frac{|x| + |y|}{2},$$

which is the same as (5.1). Note that, in this case, the convex function approach does not really make the derivation simpler if one takes into account that the verification of (5.3) takes about the same as the arguments right above it. However, in many other cases, the convex function approach is very effective. We consider some examples.

*Example 5.1* (Arithmetic, geometric and harmonic means). The harmonic mean is bounded by the geometric mean, which, in turn, is bounded by the arithmetic mean. This string of fundamental inequalities can be expressed as

$$\frac{n}{x_1^{-1} + \cdots + x_n^{-1}} \le \sqrt[n]{x_1 \cdots x_n} \le \frac{x_1 + \cdots + x_n}{n} \tag{5.5}$$

for any positive numbers $x_1, \ldots, x_n$. Both inequalities can be established by the convex function inequality. Let $f(x) = -\log(x)$. Then since $f''(x) = x^{-2} > 0$ for $x > 0$, the function is convex. Therefore, by (5.4), we have

$$-\log\left(\frac{x_1^{-1} + \cdots + x_n^{-1}}{n}\right) \le -\frac{\log(x_1^{-1}) + \cdots + \log(x_n^{-1})}{n},$$

$$-\log\left(\frac{x_1 + \cdots + x_n}{n}\right) \le -\frac{\log(x_1) + \cdots + \log(x_n)}{n}.$$

Inequalities (5.5) then follows by taking the negative and then exponential.

*Example 5.2* (The sample $p$-norm). For any sequence $x_i$, $1 \le i \le n$, and $p > 0$, the sample $p$-norm of the sequence is defined as

$$\|\{x_i\}\|_p = \left(\frac{|x_1|^p + \cdots + |x_n|^p}{n}\right)^{1/p}.$$

The word "sample" corresponds to the case where the $x_1, \ldots, x_n$ are realized values of i.i.d. observations, say, $X_1, \ldots, X_n$, whose $p$-norm is defined as $\|X_1\|_p = \{E(|X_1|^p)\}^{1/p}$. Another look at the sample $p$-norm is to consider the empirical distribution of $x_1, \ldots, x_n$ defined as

$$F_n(x) = \frac{1}{n} \sum_{i=1}^{n} 1_{(x_i \leq x)}.$$

It follows that the sample $p$-norm is simply $\|X\|_p$, where $X$ has the empirical distribution $F_n$ (verify this). A property of the sample $p$-norm is that it is nondecreasing in $p$. In other words, $p \leq q$ implies $\|\{x_i\}\|_p \leq \|\{x_i\}\|_q$. To show this, we may assume, without loss of generality, that the $x_i$'s are positive (why?). Consider $f(x) = x^{q/p}$. Then since $f''(x) = \{q(q-p)/p^2\}x^{q/p-2} \geq 0$ for $x > 0$, $f(x)$ is convex. It follows by (5.4) that

$$\left( \frac{x_1^p + \cdots + x_n^p}{n} \right)^{q/p} \leq \frac{(x_1^p)^{q/p} + \cdots + (x_n^p)^{q/p}}{n}$$

$$= \frac{x_1^q + \cdots + x_n^q}{n}.$$

The claimed property is then verified by taking the $q$th root. Given $x_1, \ldots, x_n$, since the sequence $\|\{x_i\}\|_k$, $k = 1, 2, \ldots$, is nondecreasing, according to §1.5.1.3, the limit $\lim_{k \to \infty} \|\{x_i\}\|_k$ exists if the sequence has an upper bound. In fact, it is easy to show directly that the limit is equal to $\|\{x_i\}\|_\infty \equiv \max_{1 \leq i \leq n} |x_i|$, which is called the $\infty$-norm of the sequence (Exercise 5.1).

An extended property of (5.4) is the following. If $f(x)$ is convex, then for any $x_1, \ldots, x_n \in D$ and $\lambda_1, \ldots, \lambda_n \geq 0$ such that $\lambda_1 + \cdots + \lambda_n = 1$, we have

$$f(\lambda_1 x_1 + \cdots + \lambda_n x_n) \leq \lambda_1 f(x_1) + \cdots + \lambda_n f(x_n). \tag{5.6}$$

Clearly, inequality (5.3), which defines a convex function, is a special case of (5.6) with $n = 2$. We consider a well-known example as an application of (5.6).

*Example 5.3* (Cauchy-Schwarz inequality). For any real numbers $x_1, \ldots, x_n$ and $y_1, \ldots, y_n$, we have

$$(x_1 y_1 + \cdots + x_n y_n)^2 \leq (x_1^2 + \cdots + x_n^2)(y_1^2 + \cdots + y_n^2). \tag{5.7}$$

To show (5.7), assume, without loss of generality, that $\sum_{i=1}^{n} y_i^2 > 0$ [because, otherwise, both sides of (5.7) are zero]. Define $u_i = x_i/y_i$ if $y_i \neq 0$ and $u_i = 0$ if $y_i = 0$. Then it is easy to verify that $u_i y_i^2 = x_i y_i$ and $u_i^2 y_i^2 \leq x_i^2$, $1 \leq i \leq n$ (Exercise 5.2). Now, let $\lambda_i = y_i^2 / \sum_{j=1}^{n} y_j^2$, $1 \leq i \leq n$. Note that the $\lambda_i$'s satisfy the requirements of (5.6). Using the fact that $f(x) = x^2$ is a convex function, we have, by (5.6), $(\sum_{i=1}^{n} y_i^2)^{-2} (\sum_{i=1}^{n} x_i y_i)^2 = (\sum_{i=1}^{n} \lambda_i u_i)^2 \leq \sum_{i=1}^{n} \lambda_i u_i^2 \leq (\sum_{i=1}^{n} y_i^2)^{-1} \sum_{i=1}^{n} x_i^2$, which leads to (5.7).

Hardy, Littlewood, and Pólya (1934) outlined a beautiful argument showing that if $f(x)$ is continuous, the defining inequality (5.3) is actually equivalent to the following seemingly weaker one:

$$f\left(\frac{x+y}{2}\right) \leq \frac{f(x)+f(y)}{2} \tag{5.8}$$

for any $x$ and $y$. Another important result is regarding when the equality holds in (5.6). The same authors showed that if $f(x)$ is continuous, then (5.6) holds with $\leq$ replaced by $<$ unless either (i) all of the $x_i$'s are equal or (ii) $f(x)$ is linear in an interval that contains $x_1, \ldots, x_n$. Based on these results, the authors called $f(x)$ strictly convex if (5.8) holds with $\leq$ replaced by $<$ for any $x$ and $y$ unless $x = y$. In particular, if $f(x)$ is twice differentiable and $f''(x) > 0$, then (5.6) holds with $\leq$ replaced by $<$ unless all of the $x_i$'s are equal.

*Example 5.1 (continued).* Recall in this case the convex function is $f(x) = -\log(x)$ and $f''(x) = x^{-2} > 0$, $x > 0$. Thus, the strict inequalities in (5.5) hold unless all of the $x_i$'s are equal.

*Example 5.2 (continued).* Suppose that at least one of the $x_i$'s is positive. Then, as in Example 5.2, we may focus on the positive $x_i$'s. Recall that, in this case, $f(x) = x^{l/k}$ with $f''(x) = \{l(l-k)/k^2\}x^{l/k-2} > 0$ for $x > 0$ if $k < l$. It follows that $\|\{x_i\}\|_k < \|\{x_i\}\|_l$ if $k < l$, unless all of the positive $x_i$'s are equal.

Although we may use a similar argument to find out when equality occurs in the Cauchy–Schwarz inequality (Example 5.3), we would rather leave this to the next subsection, in which a different method will be used to derive conditions for the equality.

### 5.2.2 Hölder's and related inequalities

The celebrated Hölder's inequality states the following. Let $\alpha, \beta, \ldots, \gamma$ be positive numbers such that $\alpha + \beta + \cdots + \gamma = 1$. Then for any nonnegative numbers $a_i, b_i, \ldots, g_i$, $1 \leq i \leq n$, we have

$$\sum_{i=1}^{n} a_i^\alpha b_i^\beta \cdots g_i^\gamma \leq \left(\sum_{i=1}^{n} a_i\right)^\alpha \left(\sum_{i=1}^{n} b_i\right)^\beta \cdots \left(\sum_{i=1}^{n} g_i\right)^\gamma. \tag{5.9}$$

Moreover, the strict inequality $<$ holds in (5.9) unless either (i) one factor on the right side is zero (e.g., all of the $a_i$'s are zero); or (ii) $a_i, b_i, \ldots, g_i$ are all proportional (i.e., $a_i b_j = a_j b_i, \ldots, a_i g_j = a_j g_i, \ldots$ for all $i$ and $j$).

An alternative expression that is probably more familiar to statisticians is the following. Let $p, q, \ldots, r$ be positive numbers such that

$$\frac{1}{p} + \frac{1}{q} + \cdots + \frac{1}{r} = 1. \tag{5.10}$$

[Note that (5.10) implies that $p, q, \ldots, r$ are all greater than one.] Then for any nonnegative numbers $x_i, y_i, \ldots, z_i$, $1 \leq i \leq n$, we have

$$\sum_{i=1}^{n} x_i y_i \cdots z_i \leq \left( \sum_{i=1}^{n} x_i^p \right)^{1/p} \left( \sum_{i=1}^{n} y_i^q \right)^{1/q} \cdots \left( \sum_{i=1}^{n} z_i^r \right)^{1/r}. \quad (5.11)$$

Moreover, the strict inequality $<$ holds in (5.11) unless either (i) one of the factors on the right side is zero (e.g., all of the $x_i$'s are zero) or (ii) $x_i^p, y_i^q, \ldots, z_i^r$ are all proportional.

A special case of (5.11) is, by far, the most popular (in fact, this is called Hölder's inequality in most books). If $p, q > 0$ and $p^{-1} + q^{-1} = 1$, then for any $x_i, y_i \geq 0$, $1 \leq i \leq n$, we have

$$\sum_{i=1}^{n} x_i y_i \leq \left( \sum_{i=1}^{n} x_i^p \right)^{1/p} \left( \sum_{i=1}^{n} y_i^q \right)^{1/q}. \quad (5.12)$$

The strict inequality holds in (5.12) unless either the $x_i$'s are all zero, or the $y_i$'s are all zero, or $x_i^p y_j^q = x_j^p y_i^q$, $1 \leq i, j \leq n$ (in other words, $x_i^p$ and $y_i^q$ are proportional).

See, for example, Hardy et al (1934, Section 2.7) for two of the various proofs of (5.9). An alternative proof of the special case (5.12) is given in the next subsection.

*Example 5.3 (continued).* The Cauchy–Schwarz inequality is a special case of Hölder's inequality (5.12) with $p = q = 2$. It follows that the equality holds in (5.7) if and only if either the $x_i$'s are all zero, or the $y_i$'s are all zero, or $x_i y_j = x_j y_i$, $1 \leq i, j \leq n$ (i.e., $x_i$ and $y_i$ are proportional). This suggests another proof of the inequality. Consider the difference between the two sides of (5.7). We know the difference is zero if all of the differences $x_i y_j - x_j y_i$, $1 \leq i, j \leq n$, vanish. This means that, perhaps, the difference between the two sides can be expressed as a function of the differences $x_i y_j - x_j y_i$, $1 \leq i, j \leq n$. This conjecture is, indeed, true because

$$\left( \sum_{i=1}^{n} x_i^2 \right) \left( \sum_{i=1}^{n} y_i^2 \right) - \left( \sum_{i=1}^{n} x_i y_i \right)^2 = \frac{1}{2} \sum_{1 \leq i, j \leq n} (x_i y_j - x_j y_i)^2 \quad (5.13)$$

(Exercise 5.6). Thus far we have seen at least three proofs of the Cauchy–Schwarz inequality: by convex function, by Hölder's inequality, and by (5.13).

*Example 5.4.* Let $x_1, \ldots, x_n$ be positive numbers. If we replace $x_i$ and $y_i$ in the Cauchy–Schwarz inequality by $\sqrt{x_i}$ and $1/\sqrt{x_i}$, respectively, we obtain

$$n \leq \left( \sum_{i=1}^{n} x_i \right)^{1/2} \left( \sum_{i=1}^{n} x_i^{-1} \right)^{1/2},$$

which is equivalent to $n(\sum_{i=1}^{n} x_i^{-1})^{-1} \leq n^{-1} \sum_{i=1}^{n} x_i$. This is just the two ends of (5.5) implying that the harmonic mean is bounded by the arithmetic mean. Of course, (5.5) is a stronger result.

So far we have restricted ourselves to nonnegative numbers. If the $x_i$'s, $y_i$'s, and $z_i$'s are not assumed nonnegative, (5.11) and (5.12) continue to hold with $x_i$, $y_i$, and $z_i$ replaced by their absolute values. Then since $|\sum_{i=1}^{n} x_i y_i| \leq \sum_{i=1}^{n} |x_i| \cdot |y_i|$, (5.12) implies that

$$\left| \sum_{i=1}^{n} x_i y_i \right| \leq \left( \sum_{i=1}^{n} |x_i|^p \right)^{1/p} \left( \sum_{i=1}^{n} |y_i|^q \right)^{1/q}. \tag{5.14}$$

There is an interpretation of (5.14) in terms of inner product and norms in a Hilbert space. Consider the space $R^n$ with the inner product $< x, y > = \sum_{i=1}^{n} x_i y_i$ for $x = (x_i)_{1 \leq i \leq n}$ and $y = (y_i)_{1 \leq i \leq n} \in R^n$. If we define the $p$-norm ($p > 1$) of $x$ as $\|x\|_p = (\sum_{i=1}^{n} |x_i|^p)^{1/p}$ (note that this is slightly different from the sample $p$-norm defined in Example 5.2). Then (5.14) simply means that

$$| < x, y > | \leq \|x\|_p \|y\|_q. \tag{5.15}$$

Hölder's inequality can be used to establish another famous inequality: the Minkowski's inequality. The result is better stated in terms of the $p$-norm (see above) as follows. If $p > 1$, then for any $x, y, \ldots, z \in R^n$, we have

$$\|x + y + \cdots + z\|_p \leq \|x\|_p + \|y\|_p + \cdots + \|z\|_p. \tag{5.16}$$

To prove (5.16), it suffices to show

$$\|x + y\|_p \leq \|x\|_p + \|y\|_p \tag{5.17}$$

for any $x$ and $y$ (why?). We have, by (5.12),

$$\sum_i (x_i + y_i)^p = \sum_i (x_i + y_i)^{p-1} x_i + \sum_i (x_i + y_i)^{p-1} y_i$$

$$\leq \left( \sum_i x_i^p \right)^{1/p} \left\{ \sum_i (x_i + y_i)^{(p-1)q} \right\}^{1/q}$$

$$+ \left( \sum_i y_i^p \right)^{1/p} \left\{ \sum_i (x_i + y_i)^{(p-1)q} \right\}^{1/q}$$

$$= (\|x\|_p + \|y\|_p) \left\{ \sum_i (x_i + y_i)^p \right\}^{1/q},$$

which implies (5.17). Note that $p^{-1} + q^{-1} = 1$ implies $(p - 1)q = p$.

Conditions for equality in Minkowski's inequality can be derived from those for equality in Hölder's inequality (Exercise 5.8). Like (5.1), (5.17) is called the triangle inequality, which is one of the basic requirements for $\| \cdot \|_p$ to be (formally) called a norm. A function $\| \cdot \|$ defined on $R^n$ is a norm if (i) $\|x + y\| \leq \|x\| + \|y\|$ for any $x, y \in R^n$, (ii) $\|cx\| = |c| \cdot \|x\|$ for any $x \in R^n$ and

$c \in R$, and (iii) $\|x\| = 0$ implies $x = 0$. The definition can be easily extended beyond $R^n$. It is known that $\| \cdot \|_p$ no longer satisfies (5.16); therefore, it is not a norm if $p < 1$. In fact, the reversed inequality holds in such a case. This is called the reversed Minkowski inequality, which can be derived from the reversed Hölder inequality in the same way as above. See Hardy et al. (1934, Sections 2.8 and 2.11) for more details.

### 5.2.3 Monotone functions and related inequalities

Many useful inequalities can be established by monotone properties of functions. For example, suppose that one wishes to approximate the function $f(x) = \log(1 + x)$ for $x \geq 0$ by something even simpler. An inspection of the Taylor series, $\log(1 + x) = x - x^2/2 + x^3/3 - x^4/4 + \cdots$, suggests

$$x - \frac{x^2}{2} \leq \log(1 + x) \leq x, \quad x \geq 0. \tag{5.18}$$

At this point, (5.18) is only an "educated" guess based on the observation that the terms in the Taylor series have alternate signs when $x \geq 0$. To prove this conjecture, we first consider the function $g(x) = \log(1 + x) - x$. Since $g'(x) = -x/(1 + x) \leq 0$ for $x \geq 0$, $g(x)$ is nonincreasing for $x \geq 0$. Therefore, we have $g(x) \leq g(0) = 0$ for any $x \geq 0$, which is the right-side inequality. The left-side inequality can be proved in a similar way (Exercise 5.9).

In fact, the right-side inequality in (5.18) even holds for $x > -1$, which is the range where the function is well defined. To show this, we once again use $g(x) = \log(1 + x) - x$ and note that $g'(x) > 0$ for $-1 < x < 0$ and $g'(x) \geq 0$ for $x \geq 0$. This means that $g(x)$ is nondecreasing on $(-1, 0)$ and nonincreasing on $[0, \infty)$. Therefore, $g(x)$ has its maximum at $x = 0$. It follows that $g(x) \leq g(0) = 0$, which is the right side of (5.18).

The simple technique illustrated above, which we call the monotone function technique (note that by monotone function it does not mean that the function has to be monotone over the entire range), works quite generally, as long as one can find the "right inequality" to prove. In many cases, such an inequality is hinted at by the Taylor expansion, as in the above example. Sometimes the inequality suggested by the Taylor expansion does hold for all $x$; so, some restriction on the range and modification of the inequality itself are necessary.

*Example 5.5.* Suppose that one is interested in approximating $f(x) = e^x$ for small $x$. Once again, we are looking at the Taylor expansion

$$e^x = 1 + x + \frac{x^2}{2} + \frac{x^3}{6} + \cdots.$$

By (5.18) we know $e^x \geq 1 + x$. The next guess is, perhaps, $e^x \leq 1 + x + x^2/2$, which is false. In other words, the exponential function cannot be bounded by a quadratic function—that is to say, for all $x$. However, if $x$ is small, it is possible to find a constant $a > 0$ such that $e^x \leq 1 + x + ax^2$. To see this, suppose that $|x| \leq b < 2$. Then we have

$$e^x \leq 1 + x + \frac{x^2}{2} + \frac{|x|^3}{3!} + \frac{|x|^4}{4!} + \cdots$$

$$\leq 1 + x + \frac{x^2}{2} + \frac{|x|^3}{2^2} + \frac{|x|^4}{2^3} + \cdots$$

$$= 1 + x + 2 \sum_{k=2}^{\infty} \left(\frac{|x|}{2}\right)^k$$

$$= 1 + x + \frac{x^2}{2 - |x|}$$

$$\leq 1 + x + \frac{x^2}{2 - b};$$

that is, $e^x \leq 1 + x + ax^2$ with $a = (2 - b)^{-1}$ for all $|x| \leq b < 2$.

Alternatively, the last inequality can be proved by the monotone function technique (Exercise 5.12).

We consider an application of the inequalities derived in Example 5.5.

*Example 5.6* (An exponential inequality for bounded independent random variables). Let $X_1, \ldots, X_n$ be independent with $E(X_i) = 0$ and $|X_i| \leq B$ for some constant $B > 0$. According to the WLLN, we have

$$\frac{1}{n} \sum_{i=1}^{n} X_i \overset{P}{\longrightarrow} 0.$$

In other words, for any $\epsilon > 0$, the probability $P(n^{-1}|\sum_{i=1}^{n} X_i| > \epsilon) \to 0$, as $n \to \infty$. The question is how fast does the probability converge to zero. To investigate the convergence rate, let $\lambda$ be an arbitrary positive constant to be determined later. Then we have

$$P\left(\frac{1}{n}\left|\sum_{i=1}^{n} X_i\right| > \epsilon\right) = P\left(\frac{1}{n}\sum_{i=1}^{n} X_i > \epsilon\right) + P\left(\frac{1}{n}\sum_{i=1}^{n} X_i < -\epsilon\right)$$

$$= I_1 + I_2.$$

Furthermore, we have, by Chebyshev's inequality (see Section 5.5),

$$I_1 = P\left(\sum_{i=1}^{n} \lambda X_i > \lambda \epsilon n\right)$$

$$= P\left\{\exp\left(\sum_{i=1}^{n} \lambda X_i\right) > e^{\lambda \epsilon n}\right\}$$

$$\leq e^{-\lambda \epsilon n} \mathrm{E}\left\{\exp\left(\sum_{i=1}^{n} \lambda X_i\right)\right\}$$

$$= e^{-\lambda \epsilon n} \prod_{i=1}^{n} \mathrm{E}\{\exp(\lambda X_i)\}.$$

Since $|\lambda X_i| \leq \lambda B$, according to Example 5.5, as long as $\lambda B < 2$ we have

$$\exp(\lambda X_i) \leq 1 + \lambda X_i + \frac{\lambda^2 X_i^2}{2 - \lambda B}$$

$$\leq 1 + \lambda X_i + \frac{\lambda^2 B^2}{1 - \lambda B};$$

hence, again by Example 5.5, we have

$$\mathrm{E}\{\exp(\lambda X_i)\} \leq 1 + \frac{\lambda^2 B^2}{2 - \lambda B}$$

$$\leq \exp\left(\frac{\lambda^2 B^2}{2 - \lambda B}\right).$$

Thus, continuing, we have

$$I_1 \leq \exp(-\lambda \epsilon n) \exp\left(\frac{\lambda^2 B^2}{2 - \lambda B} n\right)$$

$$= \exp\left\{-\lambda\left(\epsilon - \frac{\lambda B^2}{2 - \lambda B}\right) n\right\}.$$

By similar arguments, one can show that $I_2$ is bounded by the same thing (Exercise 5.10). Thus, we have

$$\mathrm{P}\left(\frac{1}{n}\left|\sum_{i=1}^{n} X_i\right| > \epsilon\right) \leq 2 \exp\left\{-\lambda\left(\epsilon - \frac{\lambda B^2}{2 - \lambda B}\right) n\right\}. \tag{5.19}$$

Note that the $\lambda$ in (5.19) is arbitrary as long as $0 < \lambda < 2B^{-1}$. Consider the function

$$h(\lambda) = \lambda\left(\epsilon - \frac{\lambda B^2}{2 - \lambda B}\right). \tag{5.20}$$

It can be shown that $h(\lambda)$ attains its maxima on $(0, 2B^{-1})$ at

$$\lambda^* = \frac{2}{B}\left(1 - \sqrt{\frac{B}{B + \epsilon}}\right), \tag{5.21}$$

and its maxima is given by

$$h(\lambda^*) = 2\left(\sqrt{1 + \frac{\epsilon}{B}} - 1\right)^2 \tag{5.22}$$

(Exercise 5.10). Thus, by letting $\lambda = \lambda^*$ in (5.19) we obtain

$$P\left(\frac{1}{n}\left|\sum_{i=1}^n X_i\right| > \epsilon\right) \le 2\exp\left\{-2\left(\sqrt{1 + \frac{\epsilon}{B}} - 1\right)^2 n\right\}. \tag{5.23}$$

Such an inequality is often called an exponential inequality because it shows that the convergence rate of the probability on the left side is exponential in $n$. The above arguments are similar to those in Example 2.7 except for the maximization of $h(\lambda)$ [see part (iii) of Exercise 5.10]. Also note that the distributional assumption here is weaker than in Example 2.7 in that the $X_i$'s are not assumed to have the same distribution.

As another application of the monotone function technique, we give another proof of Hölder's inequaliy (5.12) by considering the function

$$g(a) = \frac{a^p}{p} + \frac{b^q}{q} - ab, \quad a > 0,$$

where $b > 0$ and $p$ and $q$ are as in (5.12). (Note that here $b$ is fixed.) Then we have $g'(a) = a^{p-1} - b < 0$ if $a^{p-1} < b$ and $g'(a) \ge 0$ if $a^{p-1} \ge b$. Thus, $g(a)$ has a unique minima at $a_* = b^{1/(p-1)}$, which is zero [note that $p/(p-1) = q$]. It follows that, for any $a, b > 0$, we have

$$ab \le \frac{a^p}{p} + \frac{b^q}{q}. \tag{5.24}$$

Now assume, without loss of generality, that $x_i$ and $y_i, i = 1, \ldots, n$, are positive. Let $a_i = x_i/\|x\|_p$ and $b_i = y_i/\|y\|_q$, $1 \le i \le n$. Then, by (5.24), we have

$$a_i b_i \le \frac{a_i^p}{p} + \frac{b_i^q}{q}, \tag{5.25}$$

$1 \le i \le n$. Taking the sum of (5.25) from 1 to $n$, we get

$$\frac{\sum_{i=1}^n x_i y_i}{\|x\|_p \|y\|_q} = \sum_{i=1}^n a_i b_i$$

$$\le \frac{1}{p}\sum_{i=1}^n a_i^p + \frac{1}{q}\sum_{i=1}^n b_i^q$$

$$= 1,$$

which is (5.12). The argument also shows that the equality holds if and only if $a_i^{p-1} = b_i$, $1 \le i \le n$, which means that $x_i^p$ and $y_i^q$ are proportional.

Our final application involves two monotone functions and a nonnegative function. Suppose that $f(x)$ and $g(x)$ are both nondecreasing, or both nonincreasing, and $h(x) \geq 0$. Then, for any $x_1, \ldots, x_n$, we have

$$\left\{ \sum_{i=1}^{n} f(x_i)h(x_i) \right\} \left\{ \sum_{i=1}^{n} g(x_i)h(x_i) \right\}$$
$$\leq \left\{ \sum_{i=1}^{n} f(x_i)g(x_i)h(x_i) \right\} \left\{ \sum_{i=1}^{n} h(x_i) \right\}. \tag{5.26}$$

If, instead, $f(x)$ is nondecreasing and $g(x)$ is nonincreasing, or $f(x)$ is nonincreasing and $g(x)$ is nondecreasing, the inequality in (5.26) is reversed. To prove (5.26), we use a similar "trick" to (5.13)—namely,

$$\left\{ \sum_{i=1}^{n} f(x_i)g(x_i)h(x_i) \right\} \left\{ \sum_{i=1}^{n} h(x_i) \right\}$$
$$- \left\{ \sum_{i=1}^{n} f(x_i)h(x_i) \right\} \left\{ \sum_{i=1}^{n} g(x_i)h(x_i) \right\}$$
$$= \frac{1}{2} \sum_{1 \leq i \neq j \leq n} h(x_i)h(x_j)\{f(x_i) - f(x_j)\}\{g(x_i) - g(x_j)\}. \tag{5.27}$$

The rest of the proof is left as an exercise (Exercise 5.11).

A special case of (5.26) is when $h(x) = 1$; that is,

$$\left\{ \sum_{i=1}^{n} f(x_i) \right\} \left\{ \sum_{i=1}^{n} g(x_i) \right\} \leq n \sum_{i=1}^{n} f(x_i)g(x_i). \tag{5.28}$$

There is an intuitive explanation of (5.28). If we define $\bar{f} = n^{-1} \sum_{i=1}^{n} f(x_i)$ and $\bar{g} = n^{-1} \sum_{i=1}^{n} g(x_i)$, then (5.28) is equivalent to

$$\frac{1}{n} \sum_{i=1}^{n} \{f(x_i) - \bar{f}\}\{g(x_i) - \bar{g}\} \geq 0.$$

In other words, the sample covariance between the two sets of numbers, $f(x_i)$, $1 \leq i \leq n$ and $g(x_i)$, $1 \leq i \leq n$, is nonnegative if $f$ and $g$ are both nondecreasing or both nonincreasing. A similar interpretation can be given for the case of reversed inequality.

## 5.3 Matrix inequalities

### 5.3.1 Nonnegative definite matrices

In many ways, nonnegative matrices resemble nonnegative numbers. On the other hand, not all results for nonnegative numbers can be extended to nonnegative definite matrices. Some of the basic inequalities involving nonnegative definite matrices and a cautionary tale have already been introduced and told in Section 3.3. We repeat those results for the sake of completeness. We also refer to the notation introduced therein.

(i) $A \geq B \geq 0$ implies $A^{1/2} \geq B^{1/2}$ and $A^{-1} \leq B^{-1}$ if $B$ is nonsingular, but not $A^2 \geq B^2$.

(ii) $A \geq B$ if and only if $C'AC \geq C'BC$ for any matrix $C$ of compatible dimension.

(iii) $A \geq B$ implies $\lambda_{\max}(A) > \lambda_{\max}(B)$, $\lambda_{\min}(A) \geq \lambda_{\min}(B)$, $\text{tr}(A) \geq \text{tr}(B)$, and $\text{tr}(A^2) \geq \text{tr}(B^2)$.

Here are some more results:

(iv) (Another cautionary tale) $A, B \geq 0$ does not imply that $AB + BA \geq 0$ (of course, it does not imply $AB \geq 0$ either, as $AB$ may not be symmetric).

*Example 5.7.* Consider $A = \begin{pmatrix} 2 & 0 \\ 0 & 1 \end{pmatrix}$ and $B = \begin{pmatrix} 1 & 1 \\ 1 & 1 \end{pmatrix}$. Then we have $A \geq 0$ and $B \geq 0$, but $AB + BA = \begin{pmatrix} 4 & 3 \\ 3 & 2 \end{pmatrix}$, which is not $\geq 0$.

The first inequality in (i) can be generalized in several ways. Let $D = \text{diag}(d_1, \ldots, d_k)$ be a diagonal matrix and $f$ a real-valued function; then $f(D)$ is defined as the diagonal matrix $\text{diag}\{f(d_1), \ldots, f(d_k)\}$ as long as $f(d_j)$, $1 \leq j \leq k$, are well defined. For any symmetric matrix $A$ there is an orthogonal matrix $T$ such that $A = TDT'$, where $D = \text{diag}(\lambda_1, \ldots, \lambda_k)$ and the $\lambda$'s are the eigenvalues of $A$. We define $f(A) = Tf(D)T'$ as long as $f(\lambda_j)$, $1 \leq j \leq k$, are well defined. We have the following results (e.g., Zhan 2002, Chapter 1):

(v) (Löwner–Heinz) $A \geq B \geq 0$ implies $A^r \geq B^r$ for any $0 \leq r \leq 1$.

(vi) More generally, $A \geq B \geq 0$ implies

$$(B^p A^r B^p)^{1/q} \geq B^{(2p+r)/q},$$
$$A^{(2p+r)/q} \geq (A^p B^r A^p)^{1/q}$$

for any $p \geq 0$, $q \geq 1$, and $r \geq 0$ such that $(1 + 2p)q \geq 2p + r$.

Clearly, (v) is a special case of (vi) in which $p = 0$, $q = 1$, and $0 \leq r \leq 1$. Another special case is when $p = 1$, $q = 2$, and $r = 2$. Then $A \geq B$ implies $(BA^2B)^{1/2} \geq B^2$ and $A^2 \geq (AB^2A)^{1/2}$.

The next result is regarding a partitioned matrix.

(vii) If $A > 0$, then $\begin{pmatrix} A & B \\ B' & C \end{pmatrix} \geq 0$ if and only if $C \geq B'A^{-1}B$.

As an application of result (vii) we derive the following inequality, which has had important applications in statistics.

**Lemma 5.1.** For any $V, W > 0$ and full rank matrix $X$, we have

$$(X'WX)^{-1}X'WVWX(X'WX)^{-1} \geq (X'V^{-1}X)^{-1}. \tag{5.29}$$

In other words, the left side of (5.29) is minimized when $W = V^{-1}$.

*Proof.* For any vectors $u$ and $v$ of compatible dimensions, we have

$$(u' \ v') \begin{pmatrix} X'V^{-1}X & X'WX \\ X'WX & X'WVWX \end{pmatrix} \begin{pmatrix} u \\ v \end{pmatrix}$$

$$= u'X'V^{-1}Xu + v'X'WXu + u'X'WXv + v'X'WVWXv$$

$$= \left| V^{-1/2}Xu + V^{1/2}WXv \right|^2 \geq 0.$$

Since $X$ is full rank, the matrix $X'V^{-1}X$ is nonsingular and $\geq 0$ by (ii). It follows that $X'V^{-1}X > 0$. Furthermore, the above argument shows that the partitioned matrix is $\geq 0$. Thus, by (vii), we have

$$X'WVWX \geq X'WX(X'V^{-1}X)^{-1}X'WX,$$

which, again by (ii), is equivalent to (5.29). Q.E.D.

The following example shows a specific application of Lemma 5.1.

*Example 5.8* (Weighted least squares). In linear regression it is assumed that $Y = X\beta + \epsilon$, where $Y$ is a vector of responses, $X$ is a matrix of covariates, $\beta$ is a vector of unknown regression coefficients, and $\epsilon$ is the vector errors. It is assumed that $\mathrm{E}(\epsilon) = 0$ and $\mathrm{Var}(\epsilon) = V$, where Var represents covariance matrix. In the classical situation, it is assumed that $V = \sigma^2 I$, where $I$ is the identity matrix and $\sigma^2 > 0$ is an unknown variance. In this case, the best linear unbiased estimator (BLUE) is the least squares (LS) estimator,

$$\hat{\beta} = (X'X)^{-1}X'Y. \tag{5.30}$$

Here, for simplicity, we assume that $X$ is full rank. In general, there may be correlations between the responses; therefore, the assumption $V = \sigma^2 I$ may not be reasonable. In such a case one may instead consider the weighted least squares (WLS) estimator, defined as the vector $\beta$ that minimizes

$$(Y - X\beta)'W(Y - X\beta),$$

where $W$ is a known weighting matrix. In fact, the LS estimator is a special case of the WLS estimator with $W = I$. If $W$ is nonsingular, it can be shown (Exercise 5.14) that the WLS estimator is given by

$$\hat{\beta} = (X'WX)^{-1}X'WY. \tag{5.31}$$

Furthermore, the covariance matrix of the WLS estimator is given by

$$\mathrm{Var}(\hat{\beta}) = (X'WX)^{-1}X'WVWX(X'WX)^{-1}. \tag{5.32}$$

By Lemma 5.1 we know the covariance matrix of the WLS estimator is minimized when $W = V^{-1}$. The corresponding estimator is, again, called BLUE, given by (5.31), with $W = V^{-1}$. In many cases, however, $V$ involves unknown parameters so that the BLUE is not computable. In such cases, it is customary to replace the unknown parameters by their (consistent) estimators. The result is called empirical BLUE or EBLUE. See Chapter 12 for more details.

Our final result of the subsection involves both nonnegative matrices and positive numbers. Let $a_1, \ldots, a_s$ be nonnegative numbers. There are constants $c_i$, $1 \leq i < s$, depending only on $a_1, \ldots, a_s$ such that for any positive numbers $x_1, \ldots, x_s$, we have

$$a_i x_i^2 \leq c_i \left( 1 + \sum_{j=1}^{s} a_j x_j \right)^2 , \quad 1 \leq i \leq s. \tag{5.33}$$

In fact, one may let $c_i = 0$ if $a_i = 0$ and $c_i = a_i^{-1}$ if $a_i > 0$ (Exercise 5.19). An extension of this result to nonnegative definite matrices is the following (Jiang 2000a). We state the result as a lemma for future reference.

**Lemma 5.2.** Let $A_i \geq 0$, $1 \leq i \leq s$. For any $1 \leq i \leq s$ there is a constant $c_i$ depending only on the matrices $A_1, \ldots, A_s$ such that for any $x_1, \ldots, x_s > 0$,

$$x_i^2 A_i \leq c_i \left( I + \sum_{j=1}^{s} x_j A_j \right)^2 , \quad 1 \leq i \leq s, \tag{5.34}$$

where $I$ is the identity matrix.

Some applications of Lemma 5.2 are considered in Section 5.6.

### 5.3.2 Characteristics of matrices

The previous subsection is about inequalities regarding matrices themselves. In this subsection we discuss inequalities regarding characteristics of matrices. These include rank, trace, norm, determinant, and eigenvalues.

We begin with rank. Let $A$ be an $m \times n$ matrix. Then

$$\text{rank}(A) \leq m \wedge n, \tag{5.35}$$

where $a \wedge b = \min(a, b)$. The matrix rank satisfies the triangle inequality, that is,

$$\text{rank}(A + B) \leq \text{rank}(A) + \text{rank}(B). \tag{5.36}$$

The next result is called Sylvester's inequality. For any $m \times n$ matrix $A$ and $n \times s$ matrix $B$, we have

$$\text{rank}(A) + \text{rank}(B) - n \leq \text{rank}(AB) \leq \text{rank}(A) \wedge \text{rank}(B). \tag{5.37}$$

Another result regarding the ranks is known as Fröbenius rank inequality:

$$\text{rank}(AB) + \text{rank}(BC) \leq \text{rank}(B) + \text{rank}(ABC), \tag{5.38}$$

provided that $ABC$ is well defined. We consider an example.

*Example 5.9* (Error contrasts). A general linear model is characterized by the equation $\text{E}(Y) = X\beta$, where $Y$ is a vector of observations (not necessarily independent), $X$ is a matrix of known covariates, and $\beta$ is a vector of unknown parameters. An error contrast of $Y$ is defined as a linear function of $Y$, $a = l'Y$, where $l$ is a (nonrandom) vector of the same dimension as $Y$ such that $l'X = 0$. In other words, $\text{E}(a) = 0$ for any $\beta$. The vector $l$ is called a contrast vector. How many linearly independent error contrasts can one have? If we let $A$ denote a matrix whose columns are contrast vectors, then the question is equivalent to what is the maximum rank of $A$? To answer this question, note that $A'X = 0$. Thus, by the left-side inequality of (5.37), we have $0 = \text{rank}(A'X) \geq \text{rank}(A') + \text{rank}(X) - n = \text{rank}(A) + \text{rank}(X) - n$, which implies $\text{rank}(A) \leq n - \text{rank}(X)$. Therefore, there are, at most, $n - p$ linearly independent error contrasts, where $n$ is the dimension of $Y$ and $p = \text{rank}(X)$.

The matrix trace, norm, and eigenvalues are often connected in inequalities. For example, for any matrices $A$ and $B$, we have

$$|\text{tr}(AB)| \leq \|A\|_2 \|B\|_2, \tag{5.39}$$

provided that $AB$ is well defined. Hereafter, the 2-norm of a matrix $A$ is defined as $\|A\|_2 = \{\text{tr}(A'A)\}^{1/2}$. More generally, for any matrices $A$, $B$, and $C$ such that $B \geq 0$ and $ABC$ is well defined, we have

$$|\text{tr}(ABC)| \leq \lambda_{\max}(B)\|A\|_2\|C\|_2, \tag{5.40}$$

where $\lambda_{\max}$ denotes the largest eigenvalue. Note that (5.39) is a special case of (5.40) with $B = I$ and $C = B$. Another matrix norm, the spectral norm, of a matrix $A$ is defined as $\|A\| = \{\lambda_{\max}(A'A)\}^{1/2}$. Note that $\|A\| = \lambda_{\max}(A)$ if $A \geq 0$. Thus, the right side of (5.40) can be expressed as $\|B\| \cdot \|A\|_2\|C\|_2$. A nice property of the spectral norm is the following. For any vector $x$, we have

$$|Ax| \leq \|A\| \cdot |x|, \tag{5.41}$$

where $|x| = (\sum_i x_i^2)^{1/2}$ is the Euclidean norm of $x = (x_i)$ [the inequality is satisfied with $\|A\|$ replaced by $\|A\|_2$ as well due to the following product inequality (5.45)]. The following triangle inequalities show, in particular, that both $\|\cdot\|$ and $\|\cdot\|_2$ qualify as norms:

$$\|A + B\| \leq \|A\| + \|B\|, \tag{5.42}$$
$$\|A + B\|_2 \leq \|A\|_2 + \|B\|_2. \tag{5.43}$$

It is easy to see that (5.43) is simply Minkowski's inequality (5.17) with $p = 2$ (Exercise 5.20). Another property of matrix norm is the product inequality:

$$\|AB\| \leq \|A\| \cdot \|B\|, \tag{5.44}$$

$$\|AB\|_2 \leq \|A\|_2 \|B\|_2. \tag{5.45}$$

We now take a quick break by considering an example.

*Example 5.8 (continued).* Suppose that the observations $Y$ satisfy a linear mixed model; that is, $Y = X\beta + Z\alpha + \epsilon$, where $Z$ is a known matrix, $\alpha$ is a vector of random effects, and $\epsilon$ is a vector of (additional) errors. It is assumed that $E(\alpha) = 0$, $Var(\alpha) = G$, $E(\epsilon) = 0$, $Var(\epsilon) = R$, and $Cov(\alpha, \epsilon) = 0$. It follows that $V = Var(Y) = ZGZ' + R$. Recall the BLUE is given by (5.31) with $W = V^{-1}$. Here, we assume that $R > 0$, which implies $V > 0$ (why?).

Typically, both $G$ and $R$ depend on some unknown dispersion parameters, or variance components. Let $\theta$ denote the vector of unknown variance components involved in $G$ and $R$; then $V$ depends on $\theta$—that is, $V = V(\theta)$. If we replace $\theta$ by $\hat{\theta}$, a consistent estimator, we obtain the EBLUE as

$$\hat{\beta} = (X'\hat{V}^{-1}X)^{-1}X'\hat{V}^{-1}Y, \tag{5.46}$$

where $\hat{V} = V(\hat{\theta})$. A well-known property of BLUE is its unbiasedness. It is easy to show that any WLS estimator of (5.31) is unbiased [i.e., $E(\hat{\beta}) = \beta$ (verify this)], so, as a special case, the BLUE is unbiased. The EBLUE, on the other hand, is more complicated, as it is no longer linear in $Y$. Nevertheless, Kackar and Harville (1981) showed that the EBLUE remains unbiased if $\hat{\theta}$ satisfies some mild conditions. In deriving their results, the authors avoided an issue about the existence of the expectation of EBLUE (in other words, the authors showed that $E(\hat{\beta}) = \beta$, provided that the expectation exists), which is not obvious. Below we consider a special case in which $G = \sigma^2 I_m$ and $R = \tau^2 I_n$, where $\sigma^2 > 0$ and $\tau^2 > 0$ are unknown variances, and we show the existence of the expectation.

Note that, in this case, the BLUE can be expressed as $\tilde{\beta} = B(\gamma)Y$, where

$$B(\gamma) = \{X'(I + \gamma ZZ')^{-1}X\}^{-1}X'(I + \gamma ZZ')^{-1}.$$

It can be shown that (Exercise 5.22)

$$B(\gamma) = (X'X)^{-1}X'\{I - Z(\delta I + Z'PZ)^{-1}Z'P\},$$

where $P = P_{X^\perp} = I - X(X'X)^{-1}X'$. By (5.44) and (5.42), we have

$$\|B(\gamma)\| \leq \|(X'X)^{-1}X'\| \cdot \|I - Z(\delta I + Z'PZ)^{-1}Z'P\|$$
$$\leq \|(X'X)^{-1}X'\|\{1 + \|Z\| \cdot \|(\delta I + Z'PZ)^{-1}Z'P\|\},$$

where $\delta = \gamma^{-1}$. It can be shown that $\|(X'X)^{-1}X'\| = \lambda_{\min}^{-1/2}(X'X)$ and

$$\|(\delta I + Z'PZ)^{-1}Z'P\| \leq \frac{\|Z\|}{\min \lambda_i > 0 \sqrt{\lambda_i}}, \quad \delta > 0,$$

where $\lambda_1, \ldots, \lambda_m$ are the eigenvalues of $Z'PZ$ (Exercise 5.22). It follows that $\|B(\gamma)\|$ is uniformly bounded for $\gamma \geq 0$. Therefore, by (5.41), $E(\hat{\beta})$ exists for any estimator of $\gamma$ that is nonnegative (which is, of course, reasonable).

Unfortunately, the above arguments do not carry over beyond the special case. In Section 5.6 we use a different method to establish the existence of $E(\hat{\beta})$ in more general situations.

We now continue with some inequalities on traces of nonnegative definite matrices. For any $A, B \geq 0$ and $0 \leq p \leq 1$, we have

$$\text{tr}(A^p B^{1-p}) \leq [\text{tr}\{pA + (1-p)B\}] \wedge [\{\text{tr}(A)\}^p \{\text{tr}(B)\}^{1-p}] \quad (5.47)$$

(see Section 5.3.1 for the definition of $A^p$). The next result is known as Lieb–Thirring's inequality. For any $A, B \geq 0$ and $1 \leq p \leq q$, we have

$$\text{tr}\left[\{A^p B^p\}^q\right] \leq \text{tr}\left[\{A^q B^q\}^p\right]. \quad (5.48)$$

Also, for any matrices $A, B > 0$, we have (Exercise 5.23)

$$\text{tr}\{(A - B)(A^{-1} - B^{-1})\} \leq 0. \quad (5.49)$$

There are, of course, many matrix inequalities. We refer to Section 35.2 of DasGupta (2008) for a collection of matrix inequalities and additional references. Some inequalities were developed purely because of mathematical interest. On the other hand, many inequalities were motivated by practical problems. Quite often one has a conjecture about a matrix inequality due to certain evidences. The next thing is to try to prove the inequality. There are, for the most part, two approaches to proving an inequality. The first is to look for existing inequalities that may help to establish the new inequality (in some rare occasions, one finds in the literature the exact inequality one is trying to prove, so the problem is solved). However, in most cases, this strategy does not work, unless the problem is relatively straightforward. The second approach is to try to establish the inequality oneself using basic knowledge in linear algebra. Sometimes the effort fails after some initial attempts. This might raise doubts about the conjectured inequality, so one instead looks for a counterexample. If, however, a counterexample cannot be found, one has a stronger belief that the conjectured inequality must be true. Such a stronger belief often leads to solving the conjecture. For example, the following inequality, which is Lemma 5.1 of Jiang (1996), was established in exactly the same way as above, using the second approach. We state the result as a lemma for future reference.

**Lemma 5.3.** Let $B = [b_{ij} 1_{(i>j)}]$ be a lower triangular matrix. Then

$$\|B'B\|_2^2 \leq 2\|B' + B\|^2 \|B\|_2^2. \tag{5.50}$$

Lemma 5.3 plays a pivotal role in a case study later in Section 8.1

Another useful inequality in matrix analysis is Weyl's eigenvalue perturbation theorem. Let $A$ and $B$ be $n \times n$ symmetric matrices. Then we have

$$\max_{1 \leq i \leq n} \|\lambda_i^{\downarrow}(A) \quad \lambda_i^{\downarrow}(B)| \leq \|A - B\|, \tag{5.51}$$

where $\lambda_1^{\downarrow}(A) \geq \cdots \geq \lambda_n^{\downarrow}(A)$ are the eigenvalues of $A$ arranged in decreasing order. There are various applications of Weyl's theorem in statistics. For example, in many cases there is a need to estimate the eigenvalues of, say, a covariance matrix $\Sigma$. Suppose that a consistent estimator of $\Sigma$ is obtained, say, $\hat{\Sigma}$. Then by Weyl's theorem we know that eigenvalues of $\hat{\Sigma}$ are consistent estimators of the eigenvalues of $\Sigma$. See Section 12.2 for a more details.

We conclude this subsection with a few inequalities involving determinants. For any matrix $A = (a_{ij})_{1 \leq i,j \leq n}$, let $a_i^r$ denote the $i$th row of $A$; that is, $a_i^r = (a_{ij})_{1 \leq j \leq n}$. Similarly, let $a_j^c$ denote the $j$th column of $A$; that is, $a_j^c = (a_{ij})_{1 \leq i \leq n}$. The well-known Hadamard's inequality states that

$$|A| \leq \left( \prod_{i=1}^{n} |a_i^r| \right) \wedge \left( \prod_{j=1}^{n} |a_j^c| \right). \tag{5.52}$$

Also, for any square matrices $A$ and $B$, we have

$$(|A + B|)^2 \leq |I + AA'| \cdot |I + B'B|. \tag{5.53}$$

Fisher's inequality states that for any $A > 0$ partitioned as $A = \begin{pmatrix} B & C \\ C' & D \end{pmatrix}$, where $B$ and $D$ are square matrices, we have

$$|A| \leq |B| \cdot |D|. \tag{5.54}$$

Finally, Ky Fan's inequality states that for any $A, B \geq 0$ and $0 \leq p \leq 1$,

$$|pA + (1 - p)B| \geq |A|^p |B|^{1-p}. \tag{5.55}$$

## 5.4 Integral/moment inequalities

Integrals and moments are closely related. In fact, a moment is a special integral of a function with respect to a probability measure. Due to this connection, many integral inequalities have their interpretations in terms of the moments and vice versa. On the other hand, some moment inequalities involve random variables with specific properties, such as independence. Such inequalities are better expressed in terms of moments than integrals.

Many numerical inequalities, especially those involving summations, have their integral analogues. For example, we have the following.

*Jensen's inequality.* Let $\varphi$ be a convex function. Then for any random variable $X$, we have

$$\varphi\{E(X)\} \leq E\{\varphi(X)\}, \tag{5.56}$$

provided that the expectations involved exist. There are several forms of (5.56) in terms of integrals. For example, for any measurable functions $f$ and $g$ such that $g \geq 0$, we have

$$\varphi\left\{\frac{\int f(x)g(x)\,dx}{\int g(x)\,dx}\right\} \leq \frac{\int \varphi\{f(x)\}g(x)\,dx}{\int g(x)\,dx}, \tag{5.57}$$

provided that the integrals involved exist and $\int g(x)\,dx > 0$. We consider an application of Jensen's inequality.

*Example 5.10* (A property of the log-likelihood function). Let $X$ be a vector of observations whose pdf with respect to a measure $\mu$ is $f$, where $f \in \mathcal{F}$, a subclass of pdf's with respect to $\mu$. The likelihood function is defined as $L(f) = f(X)$, considered as a function(al) of $f$, where $X$ is the observed data. In a particular case that $f(\cdot) = f(\cdot|\theta)$, where $\theta \in \Theta$, the parameter space [so that $\mathcal{F} = \{f(\cdot|\theta), \theta \in \Theta\}$], this is simply the classical likelihood function $L(\theta) = f(X|\theta)$. Let $f_0$ denote the true pdf of $X$. Then we have

$$E\{\log L(f)\} \leq E\{\log L(f_0)\}, \quad \forall f \in \mathcal{F}. \tag{5.58}$$

In other words, the expected log-likelihood function is maximized at the true pdf of $X$. The result is viewed as one of the fundamental supports for the likelihood principle. In particular, for the parametric likelihood function $L(\theta)$, it shows that the expected log-likelihood is maximized at $\theta = \theta_0$, the true parameter vector. To show (5.58), note that the function $\varphi(x) = -\log(x)$ is convex. Therefore, by (5.56), we have

$$E[\log\{L(f_0)\} - E[\log\{L(f)\}]]$$
$$= \int \log\{f_0(x)\}f_0(x)\,d\mu - \int \log\{f(x)\}f_0(x)\,d\mu$$
$$= \int \left[-\log\left\{\frac{f(x)}{f_0(x)}\right\}\right]f_0(x)\,d\mu$$
$$= E\left[\varphi\left\{\frac{f(X)}{f_0(X)}\right\}\right]$$
$$\geq \varphi\left[E\left\{\frac{f(X)}{f_0(X)}\right\}\right]$$
$$= -\log\left\{\int \frac{f(x)}{f_0(x)}f_0(x)\,d\mu\right\}$$

$$= -\log\left\{\int f(x)\,d\mu\right\}$$
$$= 0.$$

*Hölder's inequality.* Let $(S, \mathcal{F}, \mu)$ be a measure space and $f$ and $g$ be measurable functions on $S$. Then we have

$$\int_S |f(x)g(x)|\,d\mu \leq \left\{\int_S |f(x)|^p\,d\mu\right\}^{1/p}\left\{\int_S |g(x)|^q\,d\mu\right\}^{1/q} \quad (5.59)$$

for any $p, q \geq 1$ such that $p^{-1}+q^{-1} = 1$. A special case is the *Cauchy–Schwarz inequality* with $p = q = 2$,

$$\int_S |f(x)g(x)|\,d\mu \leq \sqrt{\int_S f^2(x)\,d\mu}\sqrt{\int_S g^2(x)\,d\mu}. \quad (5.60)$$

In terms of moments, we have, for any random variables $X$ and $Y$,

$$\mathrm{E}(|XY|) \leq \{\mathrm{E}(|X|^p)\}^{1/p}\{\mathrm{E}(|Y|^q)\}^{1/q}. \quad (5.61)$$

We consider a simple application of Hölder's inequality.

*Example 5.11.* If the $s$th absolute moment of a random variable $X$ exists [i.e., $\mathrm{E}(|X|^s) < \infty$], the $r$th absolute moment of $X$ exists for any $r < s$. This is because, by (5.61) with $p = s/r$ and $q = s/(s - r)$, we have $\mathrm{E}(|X|^r) \leq \{\mathrm{E}(|X|^{rp})\}^{1/p}\{\mathrm{E}(1^q)\}^{1/q} = \{\mathrm{E}(|X|^s)\}^{r/s} < \infty$. Similar to Example 5.2, if we define the $p$-norm of $X$ as $\|X\|_p = \{\mathrm{E}(|X|^p)\}^{1/p}$, then we have $\|X\|_r \leq \|X\|_s$ if $r \leq s$. In other words, $\|X\|_p$ is nondecreasing in $p$.

*Minkowski's inequality.* Using the same notation as in Hölder's inequality and letting $p \geq 1$, we have

$$\left(\int |f(x) + g(x)|^p\,d\mu\right)^{1/p}$$
$$\leq \left(\int |f(x)|^p\,d\mu\right)^{1/p} + \left(\int |g(x)|^p\,d\mu\right)^{1/p}; \quad (5.62)$$

in other words, we have the triangle inequality $\|f+g\|_p \leq \|f\|_p + \|g\|_p$, where $\|f\|_p = (\int |f(x)|^p\,d\mu)^{1/p}$. In terms of the random variables, we have

$$\|X + Y\|_p \leq \|X\|_p + \|Y\|_p. \quad (5.63)$$

*Monotone function inequalities.* Suppose that $f$, $g$, and $h$ are real-valued functions on $R$ such that $f$ and $g$ are both nondecreasing, or both nonincreasing, and $h \geq 0$; then we have

$$\int f(x)g(x)h(x)\,dx \int h(x)\,dx \ge \int f(x)h(x)\,dx \int g(x)h(x)\,dx. \quad (5.64)$$

If, instead, $f$ is nondecreasing and $g$ is nonincreasing, or $f$ is nonincreasing and $g$ is nondecreasing, and $h \ge 0$, the inequality is reversed. If $f$ and $g$ are both strictly increasing, or both strictly decreasing, and $h > 0$, the inequality holds with $\ge$ replaced by $>$. If $f$ is strictly increasing and $g$ is strictly decreasing, or $f$ is strictly decreasing and $g$ is strictly increasing, and $h > 0$, the inequality holds with $\ge$ replaced by $<$. We provide the proof of (5.64).

*Proof.* Since $f$ and $g$ are both nondecreasing and $h \ge 0$, it is easy to see that for any $x, y \in R$,

$$\{f(x) - f(y)\}\{g(x) - g(y)\}h(x)h(y) \ge 0. \quad (5.65)$$

By integrating both sides of (5.65) over $x$ and $y$, we get

$$0 \le \int\int \{f(x) - f(y)\}\{g(x) - g(y)\}h(x)h(y)\,dx\,dy$$

$$= 2\left\{ \int f(x)g(x)h(x)\,dx \int h(x)\,dx - \int f(x)h(x)\,dx \int g(x)h(x)\,dx \right\}.$$

In case $f$ and $g$ are strictly increasing and $h > 0$, (5.65) holds with $\ge$ replaced by $>$ for any $x \ne y$. Therefore, the same argument as above holds with $\le$ replaced by $<$. This completes the proof for one of the cases. The proofs for the other cases are similar (Exercise 5.29). Q.E.D.

Inequality (5.64) is also known as Chebyshev's "other" inequality, in view of the well-known Chebyshev's inequality (see the next section). On the other hand, (5.64) has many applications as well. We consider some examples.

*Example 5.12.* Let $X$ be a random variable that has a pdf $h(x)$ with resepct to the Lebesgue measure. Then (5.64) is equivalent to

$$\mathrm{cov}\{f(X), g(X)\} \ge 0 \quad (5.66)$$

(verify this), where for any random variables $\xi$ and $\eta$,

$$\mathrm{cov}(\xi, \eta) = \mathrm{E}[\{\xi - \mathrm{E}(\xi)\}\{\eta - \mathrm{E}(\eta)\}].$$

This means that if $f$ and $g$ are both nondecreasing, or both nonincreasing, the correlation between $f(X)$ and $g(X)$ is nonnegative, which is, of course, very intuitive. Similarly, if $f$ is nondecreasing and $g$ is nonincreasing, or the other way around, the correlation between $f(X)$ and $g(X)$ is nonpositive.

The next application involves the strict inequality in (5.64).

*Example 5.13* (Jiang 1998a). Suppose that given the random effects $\alpha_i$, $i = 1, \ldots, m$, the binary responses $Y_{ij}$, $i = 1, \ldots, m$, $j = 1, \ldots, n$ are independent such that $\text{logit}\{P(Y_{ij} = 1|\alpha)\} = \mu + \alpha_i$, where $\mu$ is an unknown parameter, $\alpha = (\alpha_i)_{1 \le i \le m}$, and $\text{logit}(p) = \log\{p/(1-p)\}$, $p \in (0,1)$. Furthermore, the random effects are independent and distributed as $N(0, \sigma^2)$, where $\sigma^2$ is an unknown variance. Such a model is called a mixed logistic model, which is a special case of the GLMM (see Chapter 12). In order to estimate the parameters $\mu$ and $\sigma$, one may use the method of moments by solving the following equations:

$$\sum_{i=1}^{m} \sum_{j=1}^{n} Y_{ij} = mn\text{E}\{\psi(\mu + \sigma\xi)\}, \tag{5.67}$$

$$\sum_{i=1}^{m} (Y_{i\cdot}^2 - Y_{i\cdot}) = mn(n-1)\text{E}\{\psi^2(\mu + \sigma\xi)\}, \tag{5.68}$$

where $Y_{i\cdot} = \sum_{j=1}^{n} Y_{ij}$, $\psi(x) = e^x/(1+e^x)$ and $\xi \sim N(0,1)$ (Exercise 5.30). In practice, the expectations on the right sides are approximated by Monte Carlo methods. This is called the method of simulated moments.

A nice property of (5.67) and (5.68) is that the system of equations has a unique solution. To show this, write $M_j(\mu, \sigma) = \text{E}\{\psi^j(\mu + \sigma\xi)\}$, $j = 1, 2$. Since $\psi(\cdot)$ is bounded, continuous, and strictly increasing, it follows that for any given $\sigma$ and $0 < c < 1$, there is a unique solution to

$$M_1(\mu, \sigma) = c \tag{5.69}$$

(Exercise 5.30). Denote this solution by $\mu = \mu_c(\sigma)$. Then the function $\mu_c(\cdot)$ is continuously differentiable (Exercise 5.30). For notation simplicity, write $\mu_c = \mu_c(\sigma)$ and $\mu'_c = \mu'_c(\sigma)$. Then by differentiating both sides of (5.69) (with $\mu$ replaced by $\mu_c$) with respect to $\sigma$, we get

$$\text{E}\left[\frac{\exp(\mu_c + \sigma\xi)}{\{1 + \exp(\mu_c + \sigma\xi)\}^2}(\mu'_c + \xi)\right] = 0. \tag{5.70}$$

Now, consider $M_2(\mu, \sigma)$ along the curve determined by (5.69); that is, $M_c(\sigma) = M_2(\mu_c, \sigma)$. We show that $M_c(\sigma)$ is strictly increasing. It follows that there is a unique solution to $M_c(\sigma) = d$ for any $d$ within the range of $M_c(\sigma)$ (Exercise 5.30). Therefore, there is a unique solution to the system of equations $M_1(\mu, \sigma) = c$ and $M_2(\mu, \sigma) = d$.

It remains to show that $M_c$ is strictly increasing. Note that

$$M'_c(\sigma) = 2\text{E}\left[\frac{\{\exp(\mu_c + \sigma\xi)\}^2}{\{1 + \exp(\mu_c + \sigma\xi)\}^3}(\mu'_c + \xi)\right]$$

$$= 2\int f(x)g(x)h(x)\, dx,$$

where $f(x) = \psi(\mu_c + \sigma x)$, $g(x) = \mu'_c + x$, and $h(x) = f(x)\{1 - f(x)\}\phi(x)$ with $\phi(x) = e^{-x^2/2}/\sqrt{2\pi}$. Since $f$ and $g$ are strictly increasing and $h > 0$, by the monotone function inequality, we have

$$\frac{M_c'(\sigma)}{2} \int h(x)\, dx > \int f(x)h(x)\, dx \int g(x)h(x)\, dx = 0,$$

because $\int g(x)h(x)\, dx = 0$ by (5.70). Thus, $M_c'(\sigma) > 0$, implying that $M_c$ is strictly increasing.

Many of the moment inequalities involve sum of random variables. Historically, inequalities have played important roles in establishing limit theorems for sum of random variables of a certain type. We begin with a classical result.

*Marcinkiewicz–Zygmund inequality.* Let $X_1, \ldots, X_n$ be independent such that $E(X_i) = 0$ and $E(|X_i|^p) < \infty$, $1 \le i \le n$, where $p \ge 1$. Then there are constants $c_1$ and $c_2$ depending only on $p$ such that

$$c_1 E \left( \sum_{i=1}^n X_i^2 \right)^{p/2} \le E \left| \sum_{i=1}^n X_i \right|^p \le c_2 E \left( \sum_{i=1}^n X_i^2 \right)^{p/2}. \tag{5.71}$$

Inequalities (5.71) were first given by Khintchine (1924) for a special case: the sum of independent Bernoulli random variables with equal probability for 1 or 0. Marcinkiewicz and Zygmund (1937a) generalized the result to the above. A further extension to martingale differences was given by Burkholder (1966). Let $X_i$, $1 \le i \le n$, be a sequence of random variables and let $\mathcal{F}_i$, $1 \le i \le n$, be an increasing sequence of $\sigma$-fields (see Appendix A.2); that is, $\mathcal{F}_{i-1} \subset \mathcal{F}_i$, $i \ge 1$, where $\mathcal{F}_0 = \{\emptyset, \Omega\}$. The sequence $X_i$, $\mathcal{F}_i$, $1 \le i \le n$, is called a sequence of martingale differences if $X_i \in \mathcal{F}_i$ (i.e., $X_i$ is $\mathcal{F}_i$ measurable; see Appendix A.2) and $E(X_i | \mathcal{F}_{i-1}) = 0$ a.s., $1 \le i \le n$. An extension of the Marcinkiewicz–Zygmund inequality for the case $p > 1$ is the following.

*Burkholder's inequality.* Let $X_i, \mathcal{F}_i, 1 \le i \le n$ be a sequence of martingale differences and $p > 1$. Then (5.71) holds with $c_1 = (18p^{1/2}q)^{-p}$ and $c_2 = (18pq^{1/2})^p$, where $p^{-1} + q^{-1} = 1$.

*Rosenthal's inequality.* Another well-known result is Rosenthal's inequality, first given for independent random variables (Rosenthal 1970). Hall and Heyde (1980, Section 2.4) gave an extension of the result to martingale differences as follows. Let $X_i, \mathcal{F}_i, 1 \le i \le n$, be a sequence of martingale differences and $p \ge 2$. Then there are constants $c_1$ and $c_2$ depending only on $p$ such that

$$c_1 \left[ E \left\{ \sum_{i=1}^n E(X_i^2 | \mathcal{F}_{i-1}) \right\}^{p/2} + \sum_{i=1}^n E|X_i|^p \right]$$

$$\le E \left| \sum_{i=1}^n X_i \right|^p$$

$$\le c_2 \left[ E \left\{ \sum_{i=1}^n E(X_i^2 | \mathcal{F}_{i-1}) \right\}^{p/2} + \sum_{i=1}^n E|X_i|^p \right]. \tag{5.72}$$

*Example 5.14.* Consider a special case of the Burkholder's inequality with $p = 2$. It can be shown that the martingale differences are orthogonal in the sense that $E(X_i X_j) = 0$ if $i \neq j$ (Exercise 5.31). Thus, we have $E(\sum_{i=1}^{n} X_i)^2 = \sum_{i=1}^{n} E(X_i^2) = E(\sum_{i=1}^{n} X_i^2)$. It follows that, in this case, (5.71) holds with $c_1 = c_2 = 1$. On the other hand, the constants given above for Burkholder's inequality, in general, are $c_1 = (18 \times \sqrt{2} \times 2)^{-2} = 1/2592$ and $c_2 = (18 \times 2 \times \sqrt{2})^2 = 2592$. It is seen that the constants are not very sharp in this case. Of course, $p = 2$ is a very special case that one does not really need an inequality. The constants given above are meant to apply to all cases, not just $p = 2$.

We conclude this section with several inequalities known as maximum inequalities. First, consider a sequence of martingale differences, $X_i, \mathcal{F}_i, 1 \leq i \leq n$. The partial sum $S_m = \sum_{i=1}^{m} X_i$ is called a martingale with respect to the same $\sigma$-fields. A martingale satisfies $S_m \in \mathcal{F}_m$ and $E(S_m | \mathcal{F}_{m-1}) = S_{m-1}$ a.s., $1 \leq m \leq n$ (see Chapter 8). Recall for a random variable $X$, $\|X\|_p = \{E(|X|^p)\}^{1/p}$. The following elegant result is due to Doob (1953).

*Doob's inequality.* For any $p > 1$, we have

$$\|S_n\|_p \leq \left\| \max_{1 \leq m \leq n} |S_m| \right\|_p \leq q \|S_n\|_p, \qquad (5.73)$$

where $p^{-1} + q^{-1} = 1$.

The next inequality is a stronger result than the right side of (5.72) (see Hall and Heyde 1980, Section 2.4). For any $p > 0$, there is a constant $c$ depending only on $p$ such that

$$E \left( \max_{1 \leq m \leq n} |S_i|^p \right)$$
$$\leq c \left[ E \left\{ \sum_{i=1}^{n} E(X_i^2 | \mathcal{F}_{i-1}) \right\}^{p/2} + E \left( \max_{1 \leq i \leq n} |X_i|^p \right) \right]. \qquad (5.74)$$

Finally, a result due to Móricz (1976) regarding a general sequence of random variables $\xi_n$ (not necessarily a partial sum of independent random variables or martingale differences) is useful in many cases for establishing maximum moment inequalities (e.g., Lai and Wei 1984).

*Móricz's inequality.* Let $\xi_n, n = 1, 2, \ldots$, be a sequence of random variables, and $p > 0$ and $\alpha > 1$. If there are nonnegative constants $d_i$ such that

$$E(|\xi_n - \xi_m|^p) \leq \left( \sum_{i=m+1}^{n} d_i \right)^{\alpha}, \quad n > m \geq m_0, \qquad (5.75)$$

where $m_0$ is a positive integer, then there is a constant $c$ depending only on $p$ and $\alpha$ such that

$$E \left( \max_{m \leq k \leq n} |\xi_k - \xi_m|^p \right) \leq c \left( \sum_{i=m+1}^{n} d_i \right)^{\alpha}, \quad n > m \geq m_0. \qquad (5.76)$$

## 5.5 Probability inequalities

Chebyshev's inequality, which has appeared in numerous places so far in the book, is perhaps the simplest probability inequality. This inequality gives an upper bound of a "tail probability" of a random variable in terms of the moment of the random variable. The inequality may be stated as follows.

*Chebyshev's inequality.* For any random variable $\xi$ and $a > 0$, we have

$$P(\xi > a) \leq \frac{E\{\xi 1_{(\xi > a)}\}}{a}. \tag{5.77}$$

*Proof.* The proof is as simple as the inequality itself. Note that $1_{(\xi > a)} \leq a^{-1}\xi 1_{(\xi > a)}$. The result follows by taking expectation on both sides. Q.E.D.

There are many variations of Chebyshev's inequality. For example, if $\xi$ is nonnegative, we get $P(\xi > a) \leq a^{-1}E(\xi)$; thus, for any random variable $\xi$, we have $P(|\xi| > a) \leq a^{-1}E(|\xi|)$. The latter is also known as Markov's inequality. More generally, $P(|\xi| > a) \leq a^{-p}E(|\xi|^p)$ for any $p > 0$; one may also replace the $>$ in (5.77) by $\geq$, and so forth. In a way, Chebyshev's inequality is the weakest because it makes no assumption on specific properties of $\xi$ except perhaps the existence of the expectation. Under further assumptions, much improved inequalities can be obtained. We state a few such results below.

*Bernstein's inequality.* First, assume that $X_1, \ldots, X_n$ are independent with $E(X_i) = 0$ and $|X_i| \leq M$ a.s. Then for any $t > 0$, we have

$$P\left(\sum_{i=1}^{n} X_i > t\right) \leq \exp\left\{-\frac{3t^2}{2Mt + 6\sum_{i=1}^{n} E(X_i^2)}\right\}. \tag{5.78}$$

The proof of (5.78) is an application of Chebyshev's inequality to $\xi = \exp(\lambda \sum_{i=1}^{n} X_i)$ for a suitable choice of $\lambda$ (see below for a more general case). A similar method has been used in the proof of (5.23) (Exercise 5.38).

Several generalizations of Bernstein's inequality are available. For the most part, the generalizations either relax the uniform boundedness of $X_i$ or weaken the assumption that the $X_i$'s are independent. As an example, we derive the following martingale version of Bernstein's inequality. Suppose that $X_i, \mathcal{F}_i, 1 \leq i \leq n$, is a sequence of martingale differences such that

$$E(X_i^k | \mathcal{F}_{i-1}) \leq \frac{k!}{2} B^{k-2} a_i, \quad k \geq 2, \ i \geq 1, \tag{5.79}$$

for some constants $B > 0$ and $a_i \geq 0$. Then for any $t > 0$, we have

$$P\left(\sum_{i=1}^{n} X_i > t\right) \leq \exp\left\{-\frac{A}{2B^2}\left(\sqrt{1 + \frac{2Bt}{A}} - 1\right)^2\right\}, \tag{5.80}$$

where $A = \sum_{i=1}^{n} a_i$.

*Proof.* By Taylor's expansion, we have for any $0 < \lambda < B^{-1}$,

$$e^{\lambda X_i} = 1 + \lambda X_i + \sum_{k=2}^{\infty} \frac{\lambda^k X_i^k}{k!}.$$

By taking conditional expectation with respect to $\mathcal{F}_{i-1}$ on both sides and noting that $E(X_i|\mathcal{F}_{i-1}) = 0$, we have, by (5.79),

$$E(e^{\lambda X_i}|\mathcal{F}_{i-1}) = 1 + \sum_{k=2}^{\infty} \frac{\lambda^k}{k!} E(X_i^k|\mathcal{F}_{i-1})$$

$$\leq 1 + \frac{\lambda^2}{2} a_i \sum_{k=2}^{\infty} (\lambda B)^{k=2}$$

$$= 1 + \frac{\lambda^2 a_i}{2(1 - \lambda B)}$$

$$\leq \exp\left\{\frac{\lambda^2 a_i}{2(1 - \lambda B)}\right\}, \quad i \geq 1,$$

using the inequality $e^x \geq 1 + x$ [see (5.18)]. Note that (5.79) ensures the appropriateness of exchanging the order of conditional expectation and infinite summation (Exercise 5.39). Using the properties of conditional expectation (see Appendix A.2) and the above result, we have

$$E\left\{\exp\left(\lambda \sum_{i=1}^{n} X_i\right)\right\} = E\left[E\left\{\exp\left(\lambda \sum_{i=1}^{n} X_i\right)\Big|\mathcal{F}_{i-1}\right\}\right]$$

$$= E\left\{\exp\left(\lambda \sum_{i=1}^{n-1} X_i\right) E(e^{\lambda X_n}|\mathcal{F}_{i-1})\right\}$$

$$\leq \exp\left\{\frac{\lambda^2 a_n}{2(1 - \lambda B)}\right\} E\left\{\exp\left(\lambda \sum_{i=1}^{n-1} X_i\right)\right\}$$

$$\cdots$$

$$\leq \exp\left\{\frac{\lambda^2 A}{2(1 - \lambda B)}\right\},$$

where $A = \sum_{i=1}^{n} a_i$. It follows that, by Chebyshev's inequality,

$$P\left(\sum_{i=1}^{n} X_i > t\right) = P\left\{\exp\left(\lambda \sum_{i=1}^{n} X_i\right) > e^{\lambda t}\right\}$$

$$\leq e^{-\lambda t} E\left\{\exp\left(\lambda \sum_{i=1}^{n} X_i\right)\right\}$$

$$\leq \exp\left\{\frac{\lambda^2 A}{2(1 - \lambda B)} - \lambda t\right\}. \tag{5.81}$$

Denote the function inside $\{\cdots\}$ on the right side of (5.81) by $g(\lambda)$. It can be shown (Exercise 5.39) that $g(\lambda)$ is minimized for $\lambda \in (0, B^{-1})$ at

$$\lambda = \frac{1}{B}\left(1 - \sqrt{\frac{A}{A + 2Bt}}\right), \tag{5.82}$$

and the minimal value is the one inside $\{\cdots\}$ on the right side of (5.80). This completes the derivation. Q.E.D.

In a way, the right side of (5.80) is a bit complicated and not very easy to interpret. The following inequalities are implied by (5.80), but the bounds are much simplier. First, it can be shown that

$$\frac{A}{2B^2}\left(\sqrt{1 + \frac{2Bt}{A}} - 1\right)^2 \geq \frac{t^2}{2(A + Bt)} \tag{5.83}$$

(Exercise 5.40). Thus, we have, with $A = \sum_{i=1}^{n} a_i$,

$$P\left(\sum_{i=1}^{n} X_i > t\right) \leq \exp\left\{-\frac{t^2}{2(A + Bt)}\right\}. \tag{5.84}$$

Next, if we replace $t$ by $2\sqrt{A}t$ in (5.84), then it follows that

$$P\left(\sum_{i=1}^{n} X_i > 2\sqrt{A}t\right) \leq e^{-t^2}, \quad 0 < t \leq \frac{\sqrt{A}}{2B} \tag{5.85}$$

(Exercise 5.40). In fact, this is the original form of an inequality proved by Bernstein (1937).

Bernstein-type inequalities are useful in evaluating convergence rate in the law of large numbers. We consider some examples.

*Example 5.15.* Suppose that $Y_1, \ldots, Y_n$ are independent and distributed as Bernoulli($p$), where $p \in [0, 1]$. Let $X_i = Y_i - p$. Then, $X_1, \ldots, X_n$ are independent with $E(X_i) = 0$ and $|X_i| \leq 1$. By (5.78), we have

$$P\left(\frac{1}{n}\sum_{i=1}^{n} Y_i > p + \epsilon\right) = P\left(\sum_{i=1}^{n} X_i > n\epsilon\right)$$

$$\leq \exp\left\{-\frac{3n^2\epsilon^2}{2n\epsilon + 6np(1 - p)}\right\}.$$

Using the inequality $p(1 - p) \leq 1/4$ (why?), it is then easy to show that the right side is bounded by $\exp[-\{6\epsilon^2/(4\epsilon + 3)\}n]$. The same inequality is obtained by considering $X_i = p - Y_i$. Thus, in conclusion, we have

$$P\left(\left|\frac{1}{n}\sum_{i=1}^{n}Y_i - p\right| > \epsilon\right) \le 2\,\exp\left(-\frac{6\epsilon^2}{4\epsilon+3}n\right).$$

*Example 5.16.* Suppose that $Y_1,\ldots,Y_n$ are i.i.d. with the Exponential($\lambda$) distribution, where $\lambda > 0$ (see Example 3.8). Then we have $E(Y_i^k) = k!\lambda^k$, $k = 1,2,\ldots$. Let $X_i = Y_i - \lambda$. Then $X_1,\ldots,X_n$ are i.i.d. with $E(X_i) = 0$. Furthermore, using the inequality that for $a,b \ge 0$, $|a - b| \le a \vee b$, we have $E(X_i^k) \le E(|X_i|^k) \le E\{(Y_i \vee \lambda)^k\} \le E(Y_i^k + \lambda^k) = (k! + 1)\lambda^k$. Thus, by letting $\mathcal{F}_i = \sigma(X_1,\ldots,X_i)$ and noting that $E(X_i^k|\mathcal{F}_{i-1}) = E(X_i^k)$, we have $E(X_i^k|\mathcal{F}_{i-1}) \le (k!/2)(1+1/k!)\lambda^{k-1}2\lambda^2 = (k!/2)\lambda^{k-1}4\lambda^2$—that is, (5.79) with $B = \lambda$ and $a_i = 4\lambda^2$. We now apply (5.84) with $t = n\epsilon\lambda$ to get

$$P\left(\frac{1}{n}\sum_{i=1}^{n}X_i > \epsilon\lambda\right) \le \exp\left(-\frac{\epsilon^2}{2\epsilon+8}n\right).$$

The same inequality is obtained by considering $X_i = \lambda - Y_i$. It follows that

$$P\left(\left|\frac{1}{n}\sum_{i=1}^{n}Y_i - \lambda\right| > \epsilon\lambda\right) \le 2\left(-\frac{\epsilon^2}{2\epsilon+8}n\right).$$

See Exercise 5.41 for another application.

As for the moment inequalities, there is a class of probability inequalities known as maximum inequalities. Let us begin with Kolmogorov's well-known inequality. Let $X_1,\ldots,X_n$ be independent with $E(X_i) = 0$. Define $S_m = \sum_{i=1}^{m}X_i$. Then for any $\lambda > 0$, we have

$$P\left(\max_{1\le m\le n}|S_m| \ge \lambda\right) \le \frac{1}{\lambda^2}\sum_{i=1}^{n}E(X_i^2). \tag{5.86}$$

A martingale extension of (5.86) is the following (see Hall and Heyde 1980, Theorem 2.1). Similar to the definition of martigales above (5.73), the sequence $S_m,\mathcal{F}_m,1 \le m \le n$, is called a submartingale if $S_m \in \mathcal{F}_m$ and $E(S_m|\mathcal{F}_{m-1}) \ge S_{m-1}$ a.s., $2 \le m \le n$. Let $S_m,\mathcal{F}_m,1 \le m \le n$, be a submartingale. Then for any $\lambda > 0$, we have

$$P\left(\max_{1\le m\le n}S_m \ge \lambda\right) \le \frac{1}{\lambda}E\left\{S_n 1_{(\max_{1\le m\le n}S_m\ge\lambda)}\right\}. \tag{5.87}$$

If $S_m,\mathcal{F}_m,1 \le m \le n$, is a martingale, then, by Jensen's inequality (5.56), $|S_m|^p,\mathcal{F}_m,1 \le m \le n$, is a submartingale for any $p \ge 1$ (verify this). It follows by (5.87) that, for any $\lambda > 0$,

$$P\left(\max_{1\le m\le n}|S_m| \ge \lambda\right) = P\left(\max_{1\le m\le n}|S_m|^p \ge \lambda^p\right)$$

$$\leq \frac{1}{\lambda^p} \mathrm{E}\left\{|S_n|^p 1_{(\max_{1\leq m\leq n}|S_m|\geq\lambda^p)}\right\}$$

$$\leq \frac{1}{\lambda^p} \mathrm{E}(|S_n|^p). \tag{5.88}$$

By letting $p = 2$ and the fact that $\mathrm{E}(S_n^2) = \sum_{i=1}^n \mathrm{E}(X_i^2)$, where $X_i = S_i - S_{i-1}$ if $\mathrm{E}(X_i^2) < \infty$, $1 \leq i \leq n$, we get (5.86) [note that this inequality is obviously satisfied if any of the $\mathrm{E}(X_i^2)$ is $\infty$]. Another extension of the martingales is called a supermartingale, for which the inequality $\mathrm{E}(S_m|\mathcal{F}_{m-1} \geq S_{m-1}$ is reversed. In other words, $S_m, \mathcal{F}_m, 1 \leq m \leq n$, is a supermartingale if $S_m \in \mathcal{F}_m$ and $\mathrm{E}(S_m|\mathcal{F}_{m-1}) \leq S_{m-1}$ a.s., $2 \leq m \leq n$. Martingale (submartingale, supermartingale) techniques are very useful in establishing maximum inequalities. As an example, we derive the following maximum exponential inequality due to Jiang (1999a). Unlike the previous exponential inequalities such as (5.78) and (5.80), this result does not require the uniform boundedness of $|X_i|$ or moment conditions such as (5.79).

*Example 5.17.* Let $S_m, \mathcal{F}_m, m \geq 0$, be a supermartingale, and $X_i = S_i - S_{i-1}$, $1 \leq i \leq n$. Then, for every $n \geq 1$ and $t > 0$, we have

$$\mathrm{P}\left[\max_{1<m\leq n} \sum_{i=1}^m \left\{X_i - \frac{X_i^2}{6} - \frac{\mathrm{E}(X_i^2|\mathcal{F}_{i-1})}{3}\right\} \geq t\right] \leq e^{-t}. \tag{5.89}$$

The derivation of (5.89) requires the following result.

**Lemma 5.4.** (Stout 1974, p. 299) Let $T_m, \mathcal{F}_m, m \geq 0$ be a nonnegative supermartingale with $T_0 = 1$. Then for any $\lambda > 0$, we have

$$\mathrm{P}\left(\sup_{m\geq 0} T_m \geq \lambda\right) \leq \frac{1}{\lambda}.$$

It is easy to verify the following inequality (Exercise 5.42):

$$\exp\left(x - \frac{x^2}{6}\right) \leq 1 + x + \frac{x^2}{3}, \quad -\infty < x < \infty. \tag{5.90}$$

Now define $T_0 = 1$ and $T_m = \exp[\sum_{i=1}^m \{X_i - (1/6)X_i^2 - (1/3)\mathrm{E}(X_i^2|\mathcal{F}_{i-1})\}]$, $m \geq 1$. We show that $T_m, \mathcal{F}_m, m \geq 0$, satisfies the conditions of Lemma 5.4. It suffices to show that $\mathrm{E}(T_m|\mathcal{F}_{m-1}) \leq T_{m-1}$ a.s., $m \geq 1$. By (5.90) and the inequality $1 + x \leq e^x$, $x \in R$, we have

$\mathrm{E}(T_m|\mathcal{F}_{m-1})$

$$= T_{m-1}\mathrm{E}\left\{\exp\left(X_m - \frac{X_m^2}{6}\right)\middle|\mathcal{F}_{m-1}\right\}\exp\left\{-\frac{\mathrm{E}(X_m^2|\mathcal{F}_{m-1})}{3}\right\}$$

$$\leq T_{m-1}\mathrm{E}\left(1 + X_m + \frac{X_m^2}{3}\middle|\mathcal{F}_{m-1}\right)\exp\left\{-\frac{\mathrm{E}(X_m^2|\mathcal{F}_{m-1})}{3}\right\}$$

$$\leq T_{m-1}\left\{1+\frac{\mathrm{E}(X_m^2|\mathcal{F}_{m-1})}{3}\right\}\exp\left\{-\frac{\mathrm{E}(X_m^2|\mathcal{F}_{m-1})}{3}\right\}$$

$$\leq T_{m-1}.$$

It follows by Lemma 5.4 that the left side of (5.89) equals $\mathrm{P}(\max_{1\leq m\leq n} T_m \geq e^t) \leq \mathrm{P}(\max_{m\geq 0} T_m \geq e^t) \leq e^{-t}$.

Inequality (5.89) can be used to derive an "upper" law of the iterated logarithm for martingales. See Chapter 8 for details.

Inequality (5.88) may be viewed as a strengthening of Chebyshev's inequality by replacing $|S_n|$ on the left side by $\max_{1\leq m\leq n}|S_m|$. The next maximum inequality is another interesting result. It states that, in a way, the tail probability of the maximum of the partial sums is bounded by two times the tail probability of the last partial sum. Let $X_1,\ldots,X_n$ be independent random variables. Then for any $x \in R$, we have

$$\mathrm{P}\left[\max_{1\leq k\leq n}\{S_k - \mathrm{m}(S_k - S_n)\} \geq x\right] \leq 2\mathrm{P}(S_n \geq x), \tag{5.91}$$

where $\mathrm{m}(X)$ denotes the median of $X$. (Here, we use $S_k$ instead of $S_m$ to avoid confusion with the median.) A simple proof of this result can be found in Petrov (1975, pp. 50). In particular, if $X_1,\ldots,X_n$ are independent and symmetrically distributed about zero, then for any $x \in R$,

$$\mathrm{P}\left(\max_{1\leq m\leq n} S_m \geq x\right) \leq 2\mathrm{P}(S_n \geq x). \tag{5.92}$$

Note that (5.91) and (5.92) hold for all $x \in R$, not just $x > 0$.

We conclude this section by presenting an interesting property regarding the multivariate normal distribution. Suppose that $X = (X_1,\ldots,X_n)'$ is multivariate normal with mean vector $\mu$ and covariance matrix $\Sigma = (\sigma_i\sigma_j\rho_{ij})_{1\leq i,j\leq n}$, where $\sigma_i$ is the standard deviation of $X_i$ and $\rho_{ij}$ is the correlation coefficient between $X_i$ and $X_j$. If $\rho_{ij} \geq 0$ for all $i \neq j$. Then

$$\mathrm{P}\left[\bigcap_{i=1}^n\{X_i \leq a_i\}\right] \geq \prod_{i=1}^n \mathrm{P}(X_i \leq a_i), \tag{5.93}$$

$$\mathrm{P}\left[\bigcap_{i=1}^n\{X_i > a_i\}\right] \geq \prod_{i=1}^n \mathrm{P}(X_i > a_i) \tag{5.94}$$

for any $a = (a_1,\ldots,a_n) \in R^n$. More generally, let $\Sigma_d = (\sigma_i\sigma_j\rho_{dij})_{1\leq i,j\leq n}$, $d = 1,2$, be two covariance matrices and let $\mathrm{P}_d(X \in A)$ denote $\mathrm{P}(X \in A)$, where $X \sim N(\mu,\Sigma_d)$, $d = 1,2$. If $\rho_{1ij} \geq \rho_{2ij}$ holds for all $i,j$, then

$$\mathrm{P}_1\left[\bigcap_{i=1}^n\{X_i \leq a_i\}\right] \geq \mathrm{P}_2\left[\bigcap_{i=1}^n\{X_i \leq a_i\}\right] \tag{5.95}$$

for any $a = (a_1, \ldots, a_n) \in R^n$. If, in addition, $\rho_{1ij} > \rho_{2ij}$ holds for at least one pair of $i, j$, then the strict inequality holds in (5.95). The latest results are known as Slepian's inequality (Slepian 1962). A convenient reference for its proof can be found in Tong (1980, p. 11). The result may be interpreted as follows: If $X_1, \ldots, X_n$ are jointly multivariate normal, the more positively correlated these random variables are, the more likely they will lean on the same direction. Note that, although intuitive, the same result may not hold for non-Gaussian random variables. There are many implications of Slepian's inequality, including (5.93) and (5.94) (Exercise 5.43). An application of Slepian's inequality can be found in the sequel (see Example 11.2).

## 5.6 Case study: Some problems on existence of moments

In this section we discuss some applications of Lemma 5.2 in inference about linear mixed models. These models are widely used in practice (e.g., Jiang 2007). See Chapter 12 for more details.

We consider a linear mixed model that can be expressed as

$$Y = X\beta + Z_1\alpha_1 + \cdots + Z_s\alpha_s + \epsilon, \tag{5.96}$$

where $Y = (Y_i)_{1 \leq i \leq n}$ is an $n \times 1$ vector of observations, $X$ is an $n \times p$ matrix of known covariates, $\beta$ is a $p \times 1$ vector of unknown regression coefficients (the fixed effects), $Z_r$, $1 \leq r \leq s$, are known matrices, $\alpha_r$ is a vector of i.i.d. random variables (the random effects) with mean 0 and variance $\sigma_r^2$, $1 \leq r \leq s$, and $\epsilon$ is a vector of errors with mean 0 and variance $\sigma_0^2$. Without loss of generality, we assume that $X$ is of full rank $p < n$, and none of the $Z_r$'s is a zero matrix. Two of the best known methods for estimating $\sigma_r^2$, $0 \leq r \leq s$—known as variance components—are maximum likelihood (ML) and restricted maximum likelihood (REML). See, for example, Jiang (2007). The mean, variance, MSE, and higher moments of REML and ML estimators (REMLE and MLE) were often used in the literature without rigorous justification of the existence of these moments. Note that REMLE and MLE are solutions to systems of nonlinear equations (see below), which have no closed-form expression. Thus, the existence of moments of REMLE and MLE are by no mean obvious.

In addition to variance components estimation, inference about the fixed effects and prediction of the random effects are also of great interest. The best known methods for such inference and prediction are best linear unbiased estimation (BLUE) and best linear unbiased prediction (BLUP), given by

$$\tilde{\beta} = (X'V^{-1}X)^{-1}X'V^{-1}Y, \tag{5.97}$$

$$\tilde{\alpha}_r = \sigma_r Z_r' V^{-1}(Y - X\tilde{\beta}), \quad 1 \leq r \leq s, \tag{5.98}$$

where $V = \mathrm{Var}(Y) = \sum_{r=0}^{s} \sigma_r^2 Z_r Z_r'$, with $Z_0 = I$, the $n$-dimensional identity matrix. Note that the BLUE and BLUP involve the unknown variance components $\sigma_r^2$, $0 \leq r \leq s$. Since the latter are unknown in practice, it is customary to replace them by their REMLE or MLE. The results are usually called empirical BLUE (EBLUE) and BLUP (EBLUP). Note that EBLUE and EBLUP are much more complicated than BLUE and BLUP; in particular, they are no longer linear in $Y$. On the other hand, once again, the mean, variance, and MSE of the EBLUE and EBLUP were frequently used without justification of their existence. For example, Kackar and Harville (1981) showed that the EBLUE and EBLUP remain unbiased if the variance components are estimated by nonnegative, even, and translation invariant estimators. An estimator $\hat{\theta} = \hat{\theta}(Y)$ is even if $\hat{\theta}(-Y) = \hat{\theta}(Y)$ for all $Y$ and translation-invariant if $\hat{\theta}(Y - X\beta) = \hat{\theta}(Y)$. In particular, the REMLE and MLE are both even and translation-invariant. In their arguments showing the unbiasedness property, Kackar and Harville have avoided the issue about the existence of the expectation of EBLUE and EBLUP. Jiang (1998b) proved the existence of the expectations for the special case $s = 1$. The following general results on the existence of moments of REMLE, MLE, EBLUE and EBLUP were given by Jiang (2000a).

Following Jiang (1996), we do not assume that the random effects and errors are normally distributed. In such a case, the REMLE and MLE are understood as the Gaussian REMLE and MLE; that is, they are solutions to the (Gaussian) REML and ML equations, respectively, if such solutions exist and belong to the parameter space $\Theta = \{\sigma^2 = (\sigma_r^2)_{0 \leq r \leq s} : \sigma_0^2 > 0, \sigma_r^2 \geq 0, 1 \leq r \leq s\}$; otherwise, the REMLE and MLE are defined as $\sigma_*^2$, a known point in $\Theta$. The ML equations are equivalent to

$$Y'A(A'VA)^{-1}A'Z_rZ_r'A(A'VA)^{-1}A'Y$$
$$= \mathrm{tr}(Z_r'V^{-1}Z_r), \quad 0 \leq r \leq s, \tag{5.99}$$

where $A$ is any $n \times (n-p)$ full rank matrix such that $A'X = 0$. Similarly, the REML equations are equivalent to

$$Y'A(A'VA)^{-1}A'Z_rZ_r'A(A'VA)^{-1}A'Y$$
$$= \mathrm{tr}(Z_r'A((A'VA)^{-1}A'Z_r)), \quad 0 \leq r \leq s. \tag{5.100}$$

Since these equations do not depend on the choice of $A$ as far as the conditions below (5.99) are satisfied, we assume that $A'A = I$, the $(n-p)$-dimensional identity matrix. We first prove the following result.

**Theorem 5.1.** The $p$th moments $(p > 0)$ of REMLE and MLE are finite, provided that the $2p$th moments of $Y_i$, $1 \leq i \leq n$ are finite.

*Proof.* We provide the proof for MLE only, as the proof for REMLE is very similar (Exercise 5.50). Suppose that $\sigma^2$ satisfies (5.99) and is in $\Theta$. Since

$$A'VA = \sum_{i=0}^{s} \sigma_i^2 A'Z_rZ_r'A, \tag{5.101}$$

by taking the sum of the equations (5.99) over $0 \le r \le s$, we get

$$Y'A(A'VA)^{-1}A'Y$$

$$= \sum_{r=0}^{s} \sigma_r^2 Y'A(A'VA)^{-1}A'Z_r Z_r' A(A'VA)^{-1}A'Y$$

$$= \sum_{r=0}^{s} \sigma_r^2 \operatorname{tr}(V^{-1}Z_r Z_r')$$

$$= \operatorname{tr}\left(V^{-1}\sum_{r=0}^{s} \sigma_r^2 Z_r Z_r'\right)$$

$$= n. \tag{5.102}$$

Define $V(A, \theta) = I + \sum_{r=1}^{s} \theta_r A' Z_r Z_r' A$, where $\theta_r = \sigma_r^2/\sigma_0^2$. Then, by (5.102),

$$\sigma_0^2 = \frac{Y'AV(A, \theta)^{-1}A'Y}{n}$$

$$\le \frac{Y'AA'Y}{n}$$

$$\le \frac{|Y|^2}{n}. \tag{5.103}$$

Note that $V(A, \theta) \ge I$, which implies $V(A, \theta)^{-1} \le I$, and $Y'AA'Y \le \lambda_{\max}(A'A)|Y|^2 = |Y|^2$, using properties (i) and (ii) of Section 5.3.1 and the fact that $x'Bx \le \lambda_{\max}(B)|x|^2$ for any vector $x$ and matrix $B$ (why?).

Suppose $\max_{1 \le r \le s} \sigma_r^2 = \sigma_q^2$. If $\sigma_q^2 < \sigma_0^2$, then $\sigma_q^2 \le |Y|^2/n$ by (5.103). If $\sigma_q^2 \ge \sigma_0^2$, then $\theta_q = \sigma_q^2/\sigma_0^2 \ge 1$. Note that (5.99) (with $r = q$) is equivalent to

$$Y'AV(A, \theta)^{-1}AZ_q Z_q' AV(A, \theta)^{-1}A'Y$$

$$= \sigma_0^2 \operatorname{tr}(Z_q' V_\theta^{-1} Z_q), \tag{5.104}$$

where $V_\theta = I + \sum_{r=1}^{s} \theta_r Z_r Z_r' \le \theta_q I + \sum_{r=1}^{s} \theta_q Z_r Z_r' = \theta_q V_1$, where $1$ is the $(s+1)$-dimensional vector with all components equal to one. It follows that

$$\operatorname{tr}(Z_q' V_\theta^{-1} Z_q) \ge \theta_q^{-1} \operatorname{tr}(Z_q' V_1^{-1} Z_q). \tag{5.105}$$

On the other hand, by Lemma 5.2 and property (ii) of Section 5.3.1,

$$V(A, \theta)^{-1} A' Z_q Z_q' AV(A, \theta)^{-1}$$

$$= \left(1 + \sum_{r=1}^{s} \theta_r A' Z_r Z_r' A\right)^{-1} A' Z_q Z_q' A \left(1 + \sum_{r=1}^{s} \theta_r A' Z_r Z_r' A\right)^{-1}$$

$$\le c_q \theta_q^{-2} I$$

for some constant $c_q > 0$, which implies

$$Y'AV(A,\theta)^{-1}A'Z_qZ_q'AV(A,\theta)^{-1}A'Y$$
$$\leq c_q\theta_q^{-2}Y'AA'Y \leq c_q\theta_q^{-2}|Y|^2, \tag{5.106}$$

again using property (ii) of Section 5.3.1 and an earlier result [below (5.103)]. Combining (5.104)–(5.106), we have

$$\sigma_q^2 \leq \frac{c_q|Y|^2}{\text{tr}(Z_q'V_1^{-1}Z_q)} \leq \frac{c_q}{\|Z_q\|_2^2}\left(1 + \sum_{r=1}^s \|Z_r\|^2\right)|Y|^2 \tag{5.107}$$

(Exercise 5.49). Note that since $Z_r \neq 0$, $\|Z_r\|_2 > 0$ for any $1 \leq r \leq s$.

In conclusion, let $\hat{\sigma}^2 = (\hat{\sigma}_r^2)_{0 \leq r \leq s}$ be the MLE of $\sigma^2$. If the solution to (5.99) does not exist or belong to $\Theta$, we have $\hat{\sigma}^2 = \sigma_*^2$; otherwise, $\max_{0 \leq r \leq s} \hat{\sigma}_r^2$ is bounded either by the right side of (5.103) (when $\max_{1 \leq r \leq s} \hat{\sigma}_r^2 < \hat{\sigma}_0^2$) or by the right side of (5.107) (when $\max_{1 \leq r \leq s} \hat{\sigma}_r^2 \geq \hat{\sigma}_0^2$). In any case, we have

$$\max_{0 \leq r \leq s} \hat{\sigma}_r^2 \leq \sigma_*^2 + \left\{\frac{1}{n} + \frac{c_q}{\|Z_q\|_2^2}\left(1 + \sum_{r=1}^s \|Z_r\|^2\right)\right\}|Y|^2, \tag{5.108}$$

whose $q$th moment is finite (Exercise 5.49). This completes the proof. Q.E.D.

*Note 1.* The moment condition in Theorem 5.1 is seen as minimum. This is because, for example, in the case of balanced mixed ANOVA model (e.g., Jiang 2007, p. 41), the REMLE and MLE are both quadratic functions of the data $Y_i$'s. It follows that the existence of the $2p$th moments of the $Y_i$'s is necessary for the existence of the $p$th moments of REMLE and MLE.

*Note 2.* In particular, if $Y$ is normally distributed, as is often assumed, then Theorem 5.1 implies that any moments of REMLE and MLE are finite. Furthermore, the proof of Theorem 5.1 shows that REMLE and MLE are bounded by quadratic functions of the data.

We now consider the moments of EBLUE and EBLUP. We first state a lemma whose proof is similar to Exercise 5.22 (see Jiang 2000a, p. 141–142). Let $m_r$ be the dimension of $\alpha_r$, $1 \leq r \leq s$, and $D = \text{diag}(\theta_1 I_{m_1}, \cdots, \theta_s I_{m_s})$. Also, write $Z = (Z_1, \cdots, Z_s)$ and denote the projection matrix $P = P_{X^\perp} = I - P_X$ with $P_X = X(X'X)^{-1}X'$.

**Lemma 5.5.** For any $\sigma^2 \in \Theta$, we have

$$(X'V^{-1}X)^{-1}X'V^{-1} = (X'X)^{-1}X'\{I - ZDZ'P(I + PZDZ'P)^{-1}\}.$$

**Theorem 5.2.** The $p$th moments of the EBLUE and EBLUP are finite, provided that the $p$th moments of $Y_i$, $1 \leq i \leq n$, are finite and the estimator of $\sigma^2$ belongs to $\Theta$.

*Proof.* First, consider EBLUE. By Lemma 5.5 and properties of the matrix norm (see Section 5.3.2), we have

$$\|(X'V^{-1}X)^{-1}X'V^{-1}\|$$

$$\leq \|(X'X)^{-1}X'\| \left\{ 1 + \left\| \left( \sum_{r=1}^{s} \theta_r Z_r Z_r' P \right) \left( I + \sum_{r=1}^{s} \theta_r P Z_r Z_r' P \right)^{-1} \right\| \right\}$$

$$\leq \lambda_{\min}^{-1/2}(X'X) \left\{ 1 + \sum_{\theta_r>0} \theta_r \|Z_r\| \left\| Z_r' P \left( I + \sum_{\theta_j>0} \theta_j P Z_j Z_j' P \right)^{-1} \right\| \right\}.$$

Now, apply Lemma 5.2 to obtain

$$\left( I + \sum_{\theta_j>0} \theta_j P Z_j Z_j' P \right)^{-1} P Z_r Z_r' P \left( I + \sum_{\theta_j>0} \theta_j P Z_j Z_j' P \right)^{-1} \leq c_r \theta_r^{-2} I$$

for some constant $c_r > 0$ if $\theta_r > 0$. It follows, by using property (iii) of Section 5.3.1, that

$$\|(X'V^{-1}X)^{-1}X'V^{-1}\| \leq \lambda_{\min}^{-1/2}(X'X) \left( 1 + \sum_{\theta_r>0} \sqrt{c_r}\|Z_r\| \right). \quad (5.109)$$

Note that (5.109) holds for any $\sigma^2 \in \Theta$, including the estimator, say $\hat{\sigma}^2$. That the $p$th moment of EBLUE is finite follows from (5.97) (with $V$ replaced by $\hat{V} = \sum_{r=0}^{s} \hat{\sigma}_r^2 Z_r Z_r'$), (5.41), and (5.109).

Next, we consider EBLUP. Note that the right side of (5.98) can be expressed as $\theta_r Z_r' V_\theta^{-1}\{I - X(X'V^{-1}X)^{-1}X'V^{-1}\}Y$. Suppose that $\theta_r > 0$. Then, by Lemma 5.2, we have

$$\left( I + \sum_{\theta_j>0} \theta_j Z_j Z_j' \right)^{-1} Z_r Z_r' \left( I + \sum_{\theta_j>0} \theta_j Z_j Z_j' \right)^{-1} \leq d_r \theta_r^{-2} I$$

for some constant $d_r > 0$, which implies $\|Z_r' V_\theta^{-1}\|^2 = \lambda_{\max}(V_\theta^{-1} Z_r Z_r' V_\theta^{-1}) \leq d_r \theta_r^{-2}$. Therefore, we have

$$\|\theta_r Z_r V_\theta^{-1}\{I - X(X'V^{-1}X)^{-1}X'V^{-1}\}\|$$

$$\leq \theta_r \|Z_r V_\theta^{-1}\|\{1 + \|X\| \cdot \|(X'V^{-1}X)^{-1}X'V^{-1}\|\}$$

$$\leq \sqrt{d_r} \left\{ 1 + \sqrt{\frac{\lambda_{\max}(X'X)}{\lambda_{\min}(X'X)}} \left( 1 + \sum_{\theta_j>0} \sqrt{c_j}\|Z_j\| \right) \right\}, \quad (5.110)$$

using (5.109). Inequalities (5.110) hold as long as $\theta_r > 0$; they certainly also hold if $\theta_r = 0$. That the $p$th moment of EBLUP is finite follows by (5.98) (with the alternative expression noted above), (5.41), and (5.110). Q.E.D.

*Note 3.* As in Theorem 5.1, the moment condition in Theorem 5.2 is minimum. For example, in some special cases such as the linear regression model, seen as a special case of the linear mixed model, and the balanced random effects model (e.g., Jiang 2007, p. 15), the EBLUE is the same as the BLUE, which is linear in the $Y_i$'s. It follows that the existence of the $p$th moments of the $Y_i$'s is necessary for the existence of the $p$th moments of the EBLUE.

*Note 4.* An observation from the proof of Theorem 5.2 is that the matrix operators $B(\sigma^2) = (X'V^{-1}X)^{-1}X'V^{-1}$ and $B_r(\sigma^2) = \sigma_r^2 Z_r' V^{-1}\{I - B(\sigma^2)\}$, $1 \le r \le s$, are uniformly bounded for $\sigma^2 \in \Theta$. Since the second moments of the data $Y_i$'s exist by the definition of the linear mixed model (why?), Theorem 5.2 implies, in particular, that the mean (expected value) and MSE of the EBLUE and EBLUP exist as long as the estimator of $\sigma^2$ belongs to $\Theta$, which is, of course, a reasonable assumption.

## 5.7 Exercises

5.1. Show that, in Example 5.2, $\|\{x_i\}\|_k \to \|\{x_i\}\|_\infty$ as $k \to \infty$.

5.2. In Example 5.3, define

$$u_i = \begin{cases} x_i/y_i, & y_i \ne 0 \\ 0, & y_i = 0. \end{cases}$$

Show that $u_i y_i^2 = x_i y_i$ and $u_i^2 y_i^2 \le x_i^2$, $1 \le i \le n$.

5.3. Let $x_1, \ldots, x_n$ be positive. Define $\bar{x} = n^{-1}\sum_{i=1}^n x_i$. Show that

$$\bar{x}^{\bar{x}} \le \sqrt[n]{x_1^{x_1} \cdots x_n^{x_n}}.$$

5.4. Show that for any $a_i > 0, b_i > 0$ and $\lambda_i \ge 0$, $1 \le i \le n$ such that $\sum_{i=1}^n \lambda_i = 1$, we have

$$\prod_{i=1}^n a_i^{\lambda_i} + \prod_{i=1}^n b_i^{\lambda_i} \le \prod_{i=1}^n (a_i + b_i)^{\lambda_i}.$$

When does the equality hold?

5.5. A pdf $f(x)$ is called log-concave if $\log\{f(x)\}$ is concave. Show that the following pdf's are log-concave:

(i) the pdf of $N(0,1)$;

(ii) the pdf of $\chi_\nu^2$, where the degrees of freedom $\nu \ge 2$;

(iii) the pdf of the Logistic$(0,1)$ distribution, which is given by $f(x) = e^{-x}(1 + e^{-x})^{-2}$, $-\infty < x < \infty$;

(iv) the pdf of the Dounle Exponential$(0,1)$ distribution, which is given by $f(x) = (1/2)e^{-|x|}$, $-\infty < x < \infty$.

5.6. Verify the identity (5.13).

5.7. Let $x_i, \ldots, x_n$ be real numbers. Define a probability on the space $\mathcal{X} = \{x_1, \ldots, x_n\}$ by

$$P(A) = \frac{\# \text{ of } x_i \in A}{n}$$

for any $A \subset \mathcal{X}$. Show that

$$P(A \cap B) \leq \sqrt{P(A)P(B)}.$$

[Hint: Note that # of $x_i \in A = \sum_{i=1}^{n} 1_{(x_i \in A)}$.]

5.8. Derive the conditions for equality in Minkowski's inequality (5.17).

5.9. Prove the left-side inequality in (5.18); that is

$$\log(1 + x) \geq x - \frac{x^2}{2}, \quad x \geq 0.$$

5.10. This exercise is regarding the latter part of Example 5.6.

(i) By using the same arguments, show that

$$I_2 \leq \exp\left\{-\lambda\left(\epsilon - \frac{\lambda B^2}{2 - \lambda B}\right)n\right\}.$$

(ii) Show that the function $h(\lambda)$ defined by (5.20) attains its maxima on $(0, 2B^{-1})$ at $\lambda^*$ given by (5.21), and the maxima is given by (5.22).

(iii) What is the reason for maximizing $h(\lambda)$?

5.11. This exercise is regarding the inequality (5.26).

(i) Verify the identity (5.27).

(ii) Complete the proof of (5.26).

(iii) Suppose that $f(x)$ and $g(x)$ are both strictly increasing, or both strictly decreasing, and $h(x) > 0$. Find conditions for equality in (5.26).

5.12. This exercise is related to Example 5.5.

(i) Use the monotone function technique to prove the following inequality:

$$e^x \leq 1 + x + \frac{x^2}{2 - b}, \quad |x| \leq b,$$

where $b < 2$. (Hint: Take the logarithm of both sides of the inequality.)

(ii) Suppose that $X_1, \ldots, X_n$ are i.i.d. and distributed as Uniform$[-1, 1]$. Let $\bar{X} = n^{-1} \sum_{i=1}^{n} X_i$. Show that

$$1 \leq E(e^{\bar{X}}) \leq 1 + \frac{1}{3n}.$$

(iii) Prove the following sharper inequality (see below):

$$1 \leq E(e^{\bar{X}}) \leq \exp\left\{\frac{1}{3(2n-1)}\right\}.$$

(iv) Show that the right-side inequality in (iii) is sharper in that

$$\exp\left\{\frac{1}{3(2n-1)}\right\} \leq 1 + \frac{1}{3n}, \quad n = 1, 2, \ldots.$$

*5.13.* Prove the following inequality. For any $x_1, \ldots, x_n$, we have

$$\sum_{1 \leq i \neq j \leq n} x_i^3 x_j^5 \leq \sum_{1 \leq i \neq j \leq n} x_i^6 x_j^2.$$

Can you generalize the result?

*5.14.* Show that in Example 5.8 the WLS estimator is given by (5.31), and its covariance matrix is given by (5.32). (Hint: You may use results in Appendix A.1 on differentiation of matrix expressions.)

*5.15.* Show that for any matrix $A$ of real elements, we have $A'A \geq 0$.

*5.16.* For any matrix $X$ of full rank, the projection matrix onto $\mathcal{L}(X)$, the linear space spanned by the columns of $X$, is defined as $P_X = X(X'X)^{-1}X'$ (the definition can be generalized even if $X$ is not of full rank). The orthogonal projection to $\mathcal{L}(X)$ is defined as $P_{X^\perp} = I - P_X$, where $I$ is the identity matrix. Show that $P_X \geq 0$ and $P_{X^\perp} \geq 0$.

*5.17.* Many of the "cautionary tales" regarding extensions of results for nonnegative numbers to nonnegative definite matrices are due to the fact that matrices are not necessarily commutative. Two matrices $A$ and $B$ are commutative if $AB = BA$. Suppose that $A_1, \ldots, A_s$ are symmetric and pairwise commutative. Then there is an orthogonal matrix $T$ such that $A_i = TD_iT'$, $1 \leq i \leq s$, where $D_i$ is the diagonal matrix whose diagonal elements are the eigenvalues of $A_i$. This is called simultaneous diagonalization (see Appendix A.1). Suppose that $A$ and $B$ are commutative. Prove the following:

(i) $A \geq B$ implies $A^p \geq B^p$ for any $p > 0$ [compare with results (i) and (v) of Section 5.3.1].

(ii) If $A$ and $B$ are both $\geq 0$ or both $\leq 0$, then $AB + BA \geq 0$ [compare with result (iv) of Section 5.3.1].

*5.18.* (Estimating equations) A generalization of the WLS (see Example 5.8) is the following. Let $Y$ denote the vector of observations and $\theta$ a vector of parameters of interest. Consider an estimator of $\theta$, say, $\hat{\theta}$, which is a solution to the equation

$$W(\theta)u(Y, \theta) = 0,$$

where $W(\theta)$ is a matrix depending on $\theta$ and $u(y, \theta)$ is a vector-valued function of $Y$ and $\theta$ satisfying $E\{u(Y, \theta)\} = 0$ if $\theta$ is the true parameter vector (in other words, the estimating equation is unbiased). Write $M(\theta) = W(\theta)u(Y, \theta)$. Then, under some regularity conditions, we have by the Taylor expansion,

$$0 = M(\hat{\theta})$$
$$\approx M(\theta) + \frac{\partial M}{\partial \theta'}(\hat{\theta} - \theta),$$

where $\theta$ represents the true parameter vector. Thus, we have

$$\hat{\theta} - \theta \approx -\left(\frac{\partial M}{\partial \theta'}\right)^{-1} M(\theta)$$

$$\approx - \left\{ \mathrm{E}\left( \frac{\partial M}{\partial \theta'} \right) \right\}^{-1} M(\theta).$$

Here, the approximation means that the neglected term is of lower order in a suitable sense. This leads to the following approximation (whose justification, of course, requires some regularity conditions):

$$\mathrm{Var}(\hat{\theta}) \approx \left\{ \mathrm{E}\left( \frac{\partial M}{\partial \theta'} \right) \right\}^{-1} \mathrm{Var}\{M(\theta)\} \left\{ \mathrm{E}\left( \frac{\partial M'}{\partial \theta} \right) \right\}^{-1}$$

$$\equiv V(\theta).$$

Using a similar argument to that in the proof of Lemma 5.1, show that the best estimator $\hat{\theta}$ corresponds to the estimating equation

$$W_*(\theta)u(Y,\theta) = 0,$$

where $W_*(\theta) = \mathrm{E}(\partial u'/\partial\theta)\{\mathrm{Var}(u)\}^{-1}$, in the sense that for any $W(\theta)$,

$$V(\theta) \geq V_*(\theta)$$

$$= \left\{ \mathrm{E}\left( \frac{\partial M_*}{\partial \theta'} \right) \right\}^{-1} \mathrm{Var}\{M_*(\theta)\} \left\{ \mathrm{E}\left( \frac{\partial M_*'}{\partial \theta} \right) \right\}^{-1}$$

$$= [\mathrm{Var}\{M_*(\theta)\}]^{-1},$$

where $M_*(\theta) = W_*(\theta)u(Y,\theta)$. Here, we assume that $W_*(\theta)$ does not depend on parameters other than $\theta$ (why?). Otherwise, a procedure similar to the EBLUE is necessary (see Example 5.8).

5.19. This exercise is regarding Lemma 5.2.

(i) Show that by letting $c_i = 0$ if $a_i = 0$ and $c_i = a_i^{-1}$ if $a_i > 0$, (5.33) is satisfied for all $x_i > 0$, $1 \leq i \leq s$.

(ii) Prove a special case of Lemma 5.2; that is, (5.34) holds when $A_1, \ldots, A_s$ are pairwise commutative (see Exercise 5.17).

5.20. Derive (5.43) by Minkowski's inequality (5.17).

5.21. Prove the product inequality (5.44). [Hint: For any $A \geq 0$, we have $A \leq \lambda_{\max}(A)I$, where $I$ is the identity matrix; use (iii) of Section 5.3.1]

5.22. This exercise is regarding Example 5.8 (continued) in Section 5.3.2. For parts (i) and (ii) you may use the following matrix identity in Appendix A.1.2: $(D \pm BA^{-1}B')^{-1} = D^{-1} \mp D^{-1}B(A \pm B'D^{-1}B)^{-1}B'D^{-1}$.

(i) Define $H = \delta I + Z'Z$. Show that

$$B(\gamma) = \delta(X'X - X'ZH^{-1}Z'X)^{-1}X'(\delta I + ZZ')^{-1}.$$

(ii) Furthermore, let $Q = \delta I + Z'PZ$. Show by continuing with (i) that

$$B(\gamma) = (X'X)^{-1}X'\{I + ZQ^{-1}Z'(I - P)\}(I - ZH^{-1}Z').$$

(iii) Continuing with (ii), show that $B(\gamma) = (X'X)^{-1}X'(I - ZQ^{-1}Z'P)$.

(iv) Show that $\|(X'X)^{-1}X'\| = \lambda_{\min}^{-1/2}(X'X)$.

(v) Write $A = P'Z$. Show that the positive eigenvalues of $S = A(\delta I + AA')^{-2}A'$ are $\lambda_i(\delta + \lambda_i)^{-2}$, $1 \leq i \leq m$, where $\lambda_i$, $1 \leq i \leq m$, are the positive eigenvalues of $AA'$. [Hint: The positive eigenvalues of $S$ are the same as those of $U = (\delta I + AA')^{-1}A'A(\delta I + AA')^{-1}$.] Use this result to show

$$\|(\delta I + A'A)^{-1}A'\| \leq \frac{\|Z\|}{\min \lambda_i > 0 \sqrt{\lambda_i}}, \quad \delta > 0,$$

where $\lambda_1, \ldots, \lambda_m$ are the eigenvalues of $A'A$.

5.23. Prove inequality (5.49). [Hint: Note that (5.49) is equivalent to $\operatorname{tr}(AB^{-1} + BA^{-1} - 2I) \geq 0$, $\operatorname{tr}(AB^{-1}) = \operatorname{tr}(A^{1/2}B^{-1}A^{1/2})$ and $\operatorname{tr}(BA^{-1}) = \operatorname{tr}(A^{-1/2}BB^{-1/2})$.]

5.24. Prove Schur's inequality: For any square matrices $A$ and $B$, we have

$$\operatorname{tr}\{(A'B)^2\} \leq \|A\|_2^2\|B\|_2^2.$$

[Hint: Let $A = (a_{ij})_{1 \leq i,j \leq n}$ and $B = (b_{ij})_{1 \leq i,j \leq n}$. Express the left side in terms of the elements of $A$ and $B$.]

5.25. (Jiang et al. 2001) let $b > 0$ and $a$, $c_i$, $1 \leq i \leq n$ be real numbers. Define the following matrix

$$A = \begin{pmatrix} 1 & a & 0 & \cdots & 0 \\ a & d & c_1 & \cdots & c_n \\ 0 & c_1 & 1 & \cdots & 0 \\ \vdots & \vdots & \vdots & \ddots & \vdots \\ 0 & c_n & 0 & \cdots & 1 \end{pmatrix},$$

where $d = a^2 + b + \sum_{i=1}^{n} c_i^2$. Show that $\lambda_{\min}(A) \geq b(1 + d)^{-1}$.

5.26. Show that $A \geq B$ implies $|A| \geq |B|$. (Hint: Without loss of generality, let $B > 0$. Then $A \geq B$ iff $B^{-1/2}AB^{-1/2} \geq I$.)

5.27. Use the facts that for any symmetric matrix $A$, we have $\lambda_{\min}(A) = \inf_{|x|=1}(x'Ax/x'x)$ and $\lambda_{\max}(A) = \sup_{|x|=1}(x'Ax/x'x)$ to prove the following string of inequalities. For any symmetric matrices $A$, $B$, we have

$$\lambda_{\min}(A) + \lambda_{\min}(B)$$
$$\leq \lambda_{\min}(A + B)$$
$$\leq \lambda_{\min}(A) + \lambda_{\max}(B)$$
$$\leq \lambda_{\max}(A + B)$$
$$\leq \lambda_{\max}(A) + \lambda_{\max}(B).$$

5.28. Recall that $I_n$ and $1_n$ denote respectively the $n$-dimension identity matrix and vector of 1's, and $J_n = 1_n 1_n'$. You may use the following result (see Appendix A.1) that $|aI_n + bJ_n| = a^{n-1}(a + bn)$ for any $a$, $b \in R$.

Suppose that observations $Y_1, \ldots, Y_n$ satisfy $Y_i = \mu + \alpha + \epsilon_i$, where $\mu$ is an unknown mean, $\alpha$ and $\epsilon_i$'s are independent random variables such that $\mathrm{E}(\alpha) = 0$, $\mathrm{var}(\alpha) = \sigma^2$, $\mathrm{E}(\epsilon_i) = 0$, $\mathrm{var}(\epsilon_i) = \tau^2$, $\mathrm{cov}(\epsilon_i, \epsilon_j) = 0$, $i \neq j$, and $\mathrm{cov}(\alpha, \epsilon_i) = 0$ for any $i$.

(i) Show that the covariance matrix of $Y = (Y_1, \ldots, Y_n)'$ is $V = A + B$, where $A = \tau^2 I_n$ and $B = \sigma^2 J_n$.

(ii) For the matrices $A$ and $B$ in (ii), verify inequality (5.53).

5.29. Prove another case of the monotone function inequality: If $f$ is nondecreasing and $g$ is nonincreasing and $h \geq 0$, then

$$\int f(x)g(x)h(x)\, dx \int h(x)\, dx \leq \int f(x)h(x)\, dx \int g(x)h(x)\, dx.$$

Furthermore, if $f$ is strictly increasing and $g$ is strictly decreasing and $h > 0$, the above inequality holds with $\leq$ replaced by $<$.

5.30. This exercise is associated with Example 5.13.

(i) Show that (5.67) and (5.68) are unbiased in the sense that the expectations of the left sides equal the right sides if $\mu$ and $\sigma$ are the true parameters.

(ii) Show that for any given $\sigma$ and $0 < c < 1$, there is a unique solution to (5.69).

(iii) Show that the function $\mu_c(\cdot)$ is continuously differentiable. (Hint: You may use some well-known results in calculus on differentiability of implicit functions.)

(iv) Given the proved result that $M_c(\cdot)$ is strictly increasing, show that for any $d$ within the range of $M_c$, there is a unique $\sigma$ such that $M_c(\sigma) = d$. (Hint: All you have to show is that $M_c$ is continuous.)

5.31. Show that the martingale differences are orthogonal in the sense that $\mathrm{E}(X_i X_j) = 0$, $i \neq j$ (see Example 5.14).

5.32. Let $X$ be a positive random variable. Show that

$$\mathrm{E}\left\{\frac{X}{X + \mathrm{E}(X)}\right\} \leq \frac{1}{2}.$$

5.33. Let $A$ and $B$ be any events of a probability space $\Omega$. Show the following:

(i) $\mathrm{P}(A \cap B) \leq \sqrt{\mathrm{P}(A)\mathrm{P}(B)}$. The result is an extension of Exercise 5.7.

(ii) $\mathrm{P}(A \triangle B) \leq [\{\mathrm{P}(A)\}^{1/p} + \{\mathrm{P}(B)\}^{1/p}]^p$ for any $p \geq 1$, where $A \triangle B = (A \cap B^c) \cup (B \cap A^c)$.

5.34. Prove Carlson's inequality: If $f \geq 0$ on $[0, \infty)$, then

$$\int_0^\infty f(x)\, dx \leq \sqrt{\pi} \left\{\int_0^\infty f^2(x)\, dx\right\}^{1/4} \left\{\int_0^\infty x^2 f^2(x)\, dx\right\}^{1/4}.$$

[Hint: For any $a, b > 0$, write

$$\int_0^\infty f(x)\, dx = \int_0^\infty \frac{1}{\sqrt{a + bx^2}} \sqrt{a + bx^2}\, f(x)\, dx.$$

Also, note that $\int_0^\infty (a + bx^2)^{-1}\, dx = \pi/2\sqrt{ab}$. ]

5.35. Let $\xi \sim N(0,1)$, and $F(\cdot)$ be any cdf that is strictly increasing on $(-\infty, \infty)$. Show that $E\{\xi F(\xi)\} > 0$. Can you relax the normality assumption?

5.36. Suppose that $X_1, \ldots, X_n$ are i.i.d. random variables such that $E(X_1) = 0$ and $E(|X_1|^p) < \infty$, where $p \geq 2$. Show that

$$E\left(\max_{1 \leq k \leq n} \left|\sum_{i=1}^k X_i\right|^p\right) = O(n^{p/2}).$$

5.37. Let $\xi_1, \xi_2, \ldots$ be a sequence of random variables such that $\xi_i - \xi_j \sim N(0, \sigma^2|i - j|)$ for any $i \neq j$, where $\sigma^2$ is an unknown variance. Show that

$$\sup_{n > m \geq 1} E\left(\frac{\max_{m \leq k \leq n} |\xi_k - \xi_m|}{\sqrt{n - m}}\right)^2 < \infty.$$

5.38. Derive Bernstein's inequality (5.78).

5.39. This exercise is related to the derivation of (5.80).

(i) Let $\xi_n = \sum_{k=2}^n (\lambda^k/k!)X_i^k$, $n = 2, 3, \ldots$, and $\eta = \sum_{k=2}^\infty (\lambda^k/k!)|X_i|^k$. Then we have $\xi_n \to \xi = \sum_{k=2}^\infty (\lambda^k/k!)X_i^k$ and $|\xi_n| \leq \eta$. Show that (5.79) implies that $E(\eta|\mathcal{F}_{i-1}) < \infty$. (Hint: Use the inequality that for any odd $k$, $|X_i|^k \leq 1 + X_i^{k+1}$.)

(ii) Based on the result of (i) and using the dominated convergence theorem (Theorem 2.16), show that $\lim_{n\to\infty} E(\xi_n|\mathcal{F}_{i-1}) = E(\xi|\mathcal{F}_{i-1})$.

(iii) Show that the function $g(\lambda)$ is minimized for $\lambda \in (0, B^{-1})$ at (5.82), and the minimal value is

$$g(\lambda) = -\frac{A}{2B^2}\left(\sqrt{1 + \frac{2Bt}{A}} - 1\right)^2.$$

5.40. Continue on with the martingale extension of Bernstein's inequality.

(i) Prove (5.83).

(ii) Derive (5.85), the original inequality of Bernstein (1937), by (5.84).

5.41. Consider once again Example 5.16. Show that for any $0 \leq t \leq \sqrt{n}$,

$$P\left\{\frac{1}{\sqrt{n\lambda}}\sum_{i=1}^n (Y_i - \lambda) > 2t\right\} \leq e^{-t^2},$$

$$P\left\{\frac{1}{\sqrt{n\lambda}}\sum_{i=1}^n (Y_i - \lambda) < -2t\right\} \leq e^{-t^2}.$$

How would you interpret the results?

5.42. Prove inequality (5.90).

5.43. This exercise is related to Slepian's inequality (5.95), including some of its corollaries.

(i) Show that Slepian's inequality implies (5.93) and (5.94) (Hint: The right sides of these inequalities are the probabilities on the left sides when all of the correlations $\rho_{ij}$ are zero.)

(ii) Show that if $\Sigma_d$, $d = 1, 2$ are positive definite and so is $(1 - \lambda)\Sigma_1 + \lambda\Sigma_2$ for any $\lambda \in [0, 1]$.

(iii) Show that for any fixed $\rho_{kl}$, $(k, l) \neq (i, j)$, the set $R_{ij} = \{\rho_{ij} : \Sigma = (\rho_{kl})_{1 \leq k, l \leq n}$ is positive definite$\}$ is an interval. (Hint: It suffices to show that if $\Sigma$ is positive definite when $\rho_{ij} = \rho'_{ij}$ and $\rho''_{ij}$, it remains so for any $\rho'_{ij} \leq \rho_{ij} \leq \rho''_{ij}$.

(iv) Show that for any fixed $\rho_{kl}$, $(k, l) \neq (i, j)$ and $a = (a_1, \ldots, a_n) \in R^n$, the probability $P[\cap_{i=1}^n \{X_i \leq a_i\}]$ is strictly increasing in $\rho_{ij} \in R_{ij}$.

(v) Suppose that the correlations $\rho_{ij}$ depend on a single parameter $\theta$; that is, $\rho_{ij} = \rho_{ij}(\theta)$, $\theta \in \Theta$, where $\rho_{ij}(\cdot)$ are nondecreasing functions. Show that the probability in (iv) is also a nondecreasing function of $\theta$.

5.44. Suppose that $X_1, \ldots, X_n$ are independent and distributed as $N(0, 1)$.

(i) Determine the constants $B$ and $a_i$ in (5.79), where $\mathcal{F}_i = \sigma(X_1, \ldots, X_i)$. You may use the fact that if $X \sim N(0, 1)$, then $E(X^{2j-1}) = 0$ and $E(X^{2j}) = (2j)!/2^j j!$, $j = 1, 2, \ldots$.

(ii) Determine the right side of inequality (5.84) with $t = \epsilon n$ for any $\epsilon > 0$.

(iii) Can you improve the inequality obtained in (ii) by using the fact that $\sum_{i=1}^n X_i \sim N(0, n)$?

5.45. Let $Y_i, \mathcal{F}_i, 1 \leq i \leq n$, be a sequence of martingale differences. For any $A > 0$, define $X_i = Y_i 1_{\left(\sum_{j<i} Y_j^2 \leq A\right)}$.

(i) Show that $X_i, \mathcal{F}_i, 1 \leq i \leq n$, is also a sequence of martingale differences. Here, the summation $\sum_{j<1} Y_j^2$ is understood as zero.

(ii) Show that $\sum_{i=1}^n X_i^2 \leq A + \max_{1 \leq i \leq n} Y_i^2$. (Hint: Define $i^* = \max\{0 \leq i \leq n : \sum_{j<i} Y_j^2 \leq A\}$ and show that $\sum_{i=1}^n X_i^2 \leq \sum_{i=1}^{i^*} Y_i^2$.)

(iii) Show that $\sum_{i=1}^m Y_i = \sum_{i=1}^m X_i$, $1 \leq m \leq n$, on $\{\sum_{i<n} Y_i^2 \leq A\}$.

(iv) Derive the following inequality. For any $\lambda, A > 0$ and $p > 1$, there is a constant $c$ depending only on $p$ such that

$$P\left(\max_{1 \leq m \leq n} \left|\sum_{i=1}^m Y_i\right| \geq \lambda, \sum_{i<n} Y_i^2 \leq A\right) \leq \frac{c}{\lambda^p}\left\{A^{p/2} + E\left(\max_{1 \leq i \leq n} |Y_i|^p\right)\right\}.$$

[Hint: Use the results of (i)–(iii), (5.88) and Burkholder's inequality.]

5.46. Prove the following extension of (5.80). Let $X_i, \mathcal{F}_i, 1 \leq i \leq n$ be a sequence of martingale differences. Then, for any $t > 0$,

$$P\left\{\sum_{i=1}^n X_i > t, E(X_i^k|\mathcal{F}_{i-1}) \leq \frac{k!}{2}B^{k-2}a_i, k \geq 2, 1 \leq i \leq n\right\}$$

$$\leq \exp\left\{-\frac{A}{2B^2}\left(\sqrt{1 + \frac{2Bt}{A}} - 1\right)^2\right\}.$$

[Hint: Define $Y_i = X_i 1_{(E(X_i^k | \mathcal{F}_{i-1}) \leq 0.5k! B^{k-2} a_i, k \geq 2)}$. Show that $Y_i, \mathcal{F}_i, 1 \leq i \leq n$, is also a sequence of martingale differences and satisfies (5.79) (with $X_i$ replaced by $Y_i$.]

5.47. (i) Show that for any random variable $X$ and $p > 0$, we have

$$E\{X^p 1_{(X \geq 0)}\} = \int_0^\infty p x^{p-1} P(X \geq x) \, dx.$$

[Hint: Note that $X^p 1_{(X \geq 0)} = \int_0^X p x^{p-1} 1_{(X \geq 0)} \, dx = \int_0^\infty p x^{p-1} 1_{(X \geq x)} \, dx$. Use the result in Appendix A.2 to justify the exchange of order of expectation and integration.]

(ii) Show that if $X_1, \ldots, X_n$ are independent and symmetrically distributed about zero, then for any $p > 0$,

$$E\left\{\left(\max_{1 \leq m \leq n} S_m\right)^p 1_{(\max_{1 \leq m \leq n} S_m \geq 0)}\right\} \leq 2E\{S_n^p 1_{(S_n \geq 0)}\},$$

where $S_m = \sum_{i=1}^n X_i$.

5.48. Suppose that $X_1, X_2, \ldots$ are independent Exponential(1) random variables. According to Example 5.16, we have $E(X_i^k) = k!$, $k = 1, 2, \ldots$.

(i) Given $k \geq 2$, define $Y_i = (X_i, X_i^k)'$. Show that

$$\mathrm{Var}(Y_i) = \begin{pmatrix} 1 & c_k \\ c_k & \sigma_k^2 \end{pmatrix} = \mathrm{diag}(1, \sigma_k) \Sigma_k \mathrm{diag}(1, \sigma_k),$$

where $c_k = (k+1)! - k!$, $\sigma_k^2 = (2k!) - (k!)^2$, $\Sigma_k = \begin{pmatrix} 1 & \rho_k \\ \rho_k & 1 \end{pmatrix}$ with $\rho_k = c_k/\sigma_k$.

(ii) Show that

$$\frac{1}{\sqrt{n}} \begin{bmatrix} \sum_{i=1}^n (X_i - 1) \\ \sum_{i=1}^n \sigma_k^{-1}(X_i^k - k!) \end{bmatrix} \xrightarrow{\ d\ } N(0, \Sigma_k).$$

(Hint: Use Theorem 2.14 and the CLT.)

(iii) Show that for any $0 < \alpha < 1$,

$$\lim_{n \to \infty} P\left\{\frac{\sum_{i=1}^n (X_i - 1)}{\sqrt{n}} \leq z_\alpha, \frac{\sum_{i=1}^n (X_i^k - k!)}{\sqrt{n}\sigma_k} \leq z_\alpha\right\} \geq (1-\alpha)^2,$$

where $z_\alpha$ is the $\alpha$-critical value of $N(0,1)$; that is, $P(Z \leq z_\alpha) = 1 - \alpha$ for $Z \sim N(0,1)$.

(iv) Show that the inequality in (iii) is sharp in the sense that for any $\eta > (1-\alpha)^2$, there is $k \geq 1$ such that the limit on the left side is less than $\eta$.

5.49. This exercise is related to the proof of Theorem 5.1.

(i) Verify the inequalities in (5.107).

(ii) Show that the $q$th moment of $|Y|^2$ is finite.

5.50. The proof of Theorem 5.1 for REMLE is very similar to that for MLE. Complete the proof.

# 6

# Sums of Independent Random Variables

## 6.1 Introduction

The classical large-sample theory is about the sum of independent random variables. Even though large-sample techniques have expanded well beyond the classical theory, the foundation set up by the latter remains the best way to understand and further explore elements of large-sample theory. Furthermore, the classical results are often used as examples to illustrate more sophisticated theory, as we have done repeatedly so far in this book, and the "gold standard" for any extensions beyond the classical situation. Here, by gold standard it means that a well-developed, nonclassical large-sample theory should include the classical one as a special case.

The simplest case is the so-called i.i.d case (i.e., the case of independent and identically distributed random variables). In fact, this was the place where the large-sample theory was first developed. In this case, there are three main classical results—namely, the law of large numbers, the central limit theorem, and the law of the iterated logarithm. These results, especially the first two, are well known well beyond the fields of statistics and probability (e.g., James 2006). Let $X_1, X_2, \ldots$ be a sequence of i.i.d. random variables. The weak law of large numbers (WLLN) states that if $E(X_i) = \mu \in (-\infty, \infty)$ (i.e., the expected value is finite), then

$$\bar{X} = \frac{X_1 + \cdots + X_n}{n} \xrightarrow{P} \mu, \qquad (6.1)$$

whereas the strong law of large numbers (SLLN) states that, in fact,

$$\bar{X} = \frac{X_1 + \cdots + X_n}{n} \xrightarrow{\text{a.s.}} \mu. \qquad (6.2)$$

If, in addition, $\text{var}(X_i) = \sigma^2 \in (0, \infty)$ (i.e., the variance is finite and nonzero), the central limit theorem (CLT) states that

$$\frac{\sum_{i=1}^{n}(X_i - \mu)}{\sigma\sqrt{n}} \xrightarrow{\text{d}} N(0, 1), \qquad (6.3)$$

J. Jiang, *Large Sample Techniques for Statistics*,
DOI 10.1007/978-1-4419-6827-2_6, © Springer Science+Business Media, LLC 2010

and the law of the iterated logarithm (LIL) states that

$$\limsup \frac{\sum_{i=1}^{n}(X_i - \mu)}{\sigma\sqrt{2n\log\{\log(n)\}}} = 1 \quad \text{a.s.} \tag{6.4}$$

The WLLN was first discovered by Jacob Bernoulli in 1689 for what is now known as the Bernoulli sequence, and eventually published in 1713, 8 years after his death, in his epic work *Ars Conjectandi*. Later, Siméon-Denis Poisson in 1835 named Bernoulli's theorem "the law of large numbers." The SLLN was first stated by Borel in 1909 for symmetric Bernoulli trials, although a complete proof was not given until Faber (1910). The CLT was first postulated by French mathematician Abraham de Moivre in 1733. The discovery of LIL was much later: Khintchine's 1924 paper was the first.

The WLLN, SLLN, CLT, and LIL for sum of independent, but not necessarily identically distributed random variables are discussed in Sections 6.2–6.5, respectively. Section 6.6 provides further results on invariance principle and probabilities of large deviation. A case study is considered in Section 6.6 regarding the least squares estimator in linear regression. The proofs of most of the theoretical results can be found in Petrov (1975).

## 6.2 The weak law of large numbers

We begin with the i.i.d. case. The following theorem gives necessary and sufficient conditions for WLLN.

**Theorem 6.1.** Let $X_1, X_2, \ldots$ be i.i.d. Then $\bar{X} \xrightarrow{\mathrm{P}} 0$ if and only if $n\mathrm{P}(|X_1| > n) \to 0$ and $\mathrm{E}\{X_1 1_{(|X_1|\leq n)}\} \to 0$, as $n \to \infty$.

Although Theorem 6.1 deals with a special case where the limit of convergence in probability is zero, the result can be easily generalized. For example, $\bar{X} \xrightarrow{\mathrm{P}} \mu$ for some $\mu \in R$ if and only if $n\mathrm{P}(|X_1 - \mu| > n) \to 0$ and $\mathrm{E}\{(X_1 - \mu)1_{(|X_1-\mu|\leq n)}\} \to 0$ as $n \to \infty$. In particular, if $\mathrm{E}(X_1)$ is finite, Theorem 6.1 implies the classical result (6.1). The following example, however, shows a very different situation.

*Example 6.1.* Suppose that $X_1, X_2, \ldots$ are independent Cauchy(0, 1) random variables. Then we have

$$n\mathrm{P}(|X_1| > n) = n\int_{|x|>n} \frac{dx}{\pi(1 + x^2)}$$

$$> \frac{n}{\pi}\int_{|x|>n} \frac{dx}{2x^2}$$

$$= \frac{n}{\pi}\int_{n}^{\infty} \frac{dx}{x^2} = \frac{1}{\pi},$$

which does not go to zero. Therefore, by Theorem 6.1, $\bar{X}$ does not converge to zero in probability.

The result of Example 6.1 is not surprising because, as is well known, the Cauchy distribution does not have the mean or expected value. If the latter exists, we have the classical result noted above (6.1).

We now relax the assumption that the random variables $X_1, X_2, \ldots$ have the same distribution. Furthermore, we allow a general sequence of normalizing constants $0 < a_n \uparrow \infty$; that is, $a_n > 0$, $a_n \leq a_{n+1}$, $n \geq 1$ and $\lim_{n\to\infty} a_n = \infty$. The following theorem gives necessary and sufficient conditions for an extended WLLN defined by (6.5).

**Theorem 6.2.** Let $X_1, X_2, \ldots$ be independent. Then

$$\frac{1}{a_n} \sum_{i=1}^n X_i \xrightarrow{\text{P}} 0 \tag{6.5}$$

if and only if the following three conditions are satisfied:

$$\sum_{i=1}^n \text{P}(|X_i| > a_n) \longrightarrow 0, \tag{6.6}$$

$$\frac{1}{a_n} \sum_{i=1}^n \text{E}\{X_i 1_{(|X_i| \leq a_n)}\} \longrightarrow 0, \tag{6.7}$$

$$\frac{1}{a_n^2} \sum_{i=1}^n \text{var}\{X_i 1_{(|X_i| \leq a_n)}\} \longrightarrow 0. \tag{6.8}$$

For a given sequence $X_1, X_2, \ldots$, it is usually not difficult to find a normalizing sequence $a_n$ such that (6.5) holds. For example, one expects (6.5) to hold if $a_n$ is large enough. However, it is often desirable to choose $a_n$ so that it is "just enough," although such a "cut off" may not exist.

*Example 6.2.* Suppose that $Y_1, Y_2, \ldots$ are independent such that $Y_i \sim$ Poisson($\lambda_i$), where $a \leq \lambda_i \leq b$ for some $a, b > 0$. Consider $X_i = Y_i - \lambda_i$, $i \geq 1$. Since $\text{E}(X_i) = 0$, one would expect (6.5) to hold for some suitable choice of $a_n$. Furthermore, since $\text{E}(\sum_{i=1}^n X_i)^2 = \sum_{i=1}^n \lambda_i$, one may consider $a_n = (\sum_{i=1}^n \lambda_i)^\gamma$ for some positive $\gamma$. First, assume $\gamma \leq 1/2$. We show that in this case, (6.8) is not satisfied; therefore, (6.5) does not hold. Note that

$$\text{var}\{X_i 1_{(|X_i| \leq a_n)}\} = \text{E}\{X_i^2 1_{(|X_i| \leq a_n)}\} - [\text{E}\{X_i 1_{(|X_i| \leq a_n)}\}]^2$$
$$= \lambda_i - \text{E}\{X_i^2 1_{(|X_i| > a_n)}\} - [\text{E}\{X_i 1_{(|X_i| > a_n)}\}]^2,$$

because $\text{E}(X_i) = 0$ and $\text{E}(X_i^2) = \lambda_i$. Furthermore, we have

$$E\{X_i^2 1_{(|X_i|>a_n)}\} \leq E\left\{\frac{X_i^4}{a_n^2} 1_{(|X_i|>a_n)}\right\}$$

$$\leq \frac{E(X_i^4)}{a_n^2} = \frac{\lambda_i + 3\lambda_i^2}{a_n^2}$$

$$|E\{X_i 1_{(|X_i|>a_n)}\}| \leq E\left\{\frac{X_i^2}{a_n} 1_{(|X_i|>a_n)}\right\}$$

$$\leq \frac{E(X_i^2)}{a_n} = \frac{\lambda_i}{a_n}.$$

Here, we used the fact that the fourth central moment of Poisson($\lambda$) is $\lambda+3\lambda^2$. Therefore, we have, for any $i \geq 1$,

$$\mathrm{var}\{X_i 1_{(|X_i|\leq a_n)}\} \geq \lambda_i - \frac{\lambda_i + 4\lambda_i^2}{a_n^2}$$

$$\geq \left(1 - \frac{1+4b}{a_n^2}\right)\lambda_i.$$

It follows that the left side of (6.8) is greater than or equal to

$$\frac{1}{a_n^2}\left(1 - \frac{1+4b}{a_n^2}\right)\sum_{i=1}^{n}\lambda_i = a_n^{1/\gamma-2}\left(1 - \frac{1+4b}{a_n^2}\right).$$

Since $1/\gamma - 2 \geq 0$ and $a_n^2 = (\sum_{i=1}^{n}\lambda_i)^{2\gamma} \geq (an)^{2\gamma} \to \infty$ as $n \to \infty$, we see the left side of (6.8) has a positive lower bound as $n \to \infty$. Next, we assume $\gamma > 1/2$. In this case it is easy to show that (6.6)–(6.8) are satisfied (Exercise 6.1); therefore, (6.5) holds. In conclusion, for $a_n = (\sum_{i=1}^{n}\lambda_i)^{\gamma}$, (6.5) holds if and only if $\gamma > 1/2$. However, there is no smallest $\gamma$ (the so-called cut off) such that (6.5) holds.

A general form of WLLN may be expressed as follows. Let $X_{ni}, i = 1,\ldots,i_n, n = 1,2,\ldots$, be a triangular array of random variables such that for each $n$, the $X_{ni}$'s are independent. We say $X_{ni}$ obeys WLLN if there exists a sequence of constants $b_n$ such that

$$\sum_{i=1}^{i_n} X_{ni} - b_n \xrightarrow{P} 0 \tag{6.9}$$

as $n \to \infty$. Let $m_{ni}$ and $F_{ni}$ denote the median and cdf of $X_{ni}$, respectively. We have the following theorem.

**Theorem 6.3.** $X_{ni}$ obeys WLLN if and only if the following hold:

$$\sum_{i=1}^{i_n} \int_{|x|>1} dF_{ni}(x+m_{ni}) \longrightarrow 0, \tag{6.10}$$

$$\sum_{i=1}^{i_n} \int_{|x|\leq 1} x^2 dF_{ni}(x+m_{ni}) \longrightarrow 0. \tag{6.11}$$

Furthermore, if (6.10) and (6.11) hold, then (6.9) holds, with $b_n$ having the following expression for any $\epsilon > 0$:

$$b_n = \sum_{i=1}^{i_n} \left\{ m_{ni} + \int_{|x| \leq \epsilon} x \, dF_{ni}(x + m_{ni}) \right\} + o(1). \tag{6.12}$$

Note that (6.10) and (6.11) are equivalent to the following (Exercise 6.2):

$$\sum_{i=1}^{i_n} \int \frac{x^2}{1 + x^2} \, dF_{ni}(x + m_{ni}) \longrightarrow 0. \tag{6.13}$$

Thus, another necessary and sufficient condition for $X_{ni}$ to obey WLLN is (6.13). The involvement of the medians can be made "disappear" under the following condition. We say $X_{ni}$ obeys the condition of infinite smallness if

$$\max_{1 \leq i \leq i_n} P(|X_{ni}| > \epsilon) \longrightarrow 0 \tag{6.14}$$

for every $\epsilon > 0$. Combining this with WLLN, we have the following result.

**Theorem 6.4.** Result (6.9) holds with $b_n = 0$ and $X_{ni}$ obeys the condition of infinite smallness if and only if

$$\sum_{i=1}^{i_n} P(|X_{ni}| > \epsilon) \longrightarrow 0, \tag{6.15}$$

$$\sum_{i=1}^{i_n} E\{X_{ni}1_{(|X_{ni}| \leq \tau)}\} \longrightarrow 0, \tag{6.16}$$

$$\sum_{i=1}^{i_n} \text{var}\{X_{ni}1_{(|X_{ni}| \leq \tau)}\} \longrightarrow 0 \tag{6.17}$$

for every $\epsilon > 0$ and some $\tau > 0$.

*Example 6.3.* Suppose that for each $n$, $Y_{ni}$, $1 \leq i \leq i_n$, are independent. Furthermore, there is $B > 0$ such that $|Y_{ni}| \leq B$; in other words, the $Y_{ni}$'s are uniformly bounded. Now, consider $X_{ni} = \{Y_{ni} - E(Y_{ni})\}/a_n$, where $a_n$ is the normalizing constant to be determined. Suppose that $a_n \to \infty$. Then since $|Y_{ni} - E(Y_{ni})| \leq 2B$, we have for every $\epsilon > 0$,

$$P(|X_{ni}| > \epsilon) = P\{|Y_{ni} - E(Y_{ni})| > \epsilon a_n\} = 0$$

for large $n$. For a similar reason, we have

$$E\{X_{ni}1_{(|X_{ni}| \leq 1)}\} = E(X_{ni}) = 0,$$

$$\text{var}\{X_{ni}1_{(|X_{ni}| \leq 1)}\} = \text{var}(X_{ni}) = \frac{\text{var}(Y_{ni})}{a_n^2}.$$

Therefore, for (6.15)–(6.17) to be satisfied, all one needs is

$$\frac{\sum_{i=1}^{i_n} \operatorname{var}(Y_{ni})}{a_n^2} \longrightarrow 0. \tag{6.18}$$

In conclusion, we have $\sum_{i=1}^{i_n} X_{ni} \xrightarrow{P} 0$ or, equivalently,

$$\frac{\sum_{i=1}^{i_n} Y_{ni}}{a_n} - \frac{\sum_{i=1}^{i_n} \operatorname{E}(Y_{ni})}{a_n} \xrightarrow{P} 0,$$

provided that $a_n \to \infty$ and (6.18) holds as $n \to \infty$. In other words, (6.9) holds with $X_{ni} = Y_{ni}/a_{ni}$ and $b_n = a_n^{-1} \sum_{i=1}^{i_n} \operatorname{E}(Y_{ni})$.

There is extensive literature on laws of large numbers for independent random variables. For example, earlier in Section 5.5 we discussed some results involving the convergence rate in WLLN. We conclude this section with another result in this regard.

**Theorem 6.5.** Let $X_1, X_2, \ldots$ be i.i.d. with $\operatorname{E}(X_1) = 0$ and $\operatorname{E}(|X_1|^p) < \infty$ for some $p \geq 1$. Then for every $\epsilon > 0$ we have $\operatorname{P}(|\bar{X}| > \epsilon) = o(n^{1-p})$.

## 6.3 The strong law of large numbers

Following the same strategy, we begin with the i.i.d. case. The theorem below gives a necessary and sufficient condition for SLLN. In particular, it implies the classical result (6.2).

**Theorem 6.6.** Let $X_1, X_2, \ldots$ be i.i.d. Then

$$\bar{X} \xrightarrow{\text{a.s.}} \mu \tag{6.19}$$

for some $\mu \in R$ if and only if

$$\operatorname{E}(|X_1|) < \infty. \tag{6.20}$$

If (6.20) is satisfied, then (6.19) holds with $\mu = \operatorname{E}(X_1)$.

We now relax the assumption that the $X_i$'s are identically distributed. A further extension is that the normalizing constant in SLLN does not have to be $n$, as in the following theorem.

**Theorem 6.7.** Let $X_1, X_2, \ldots,$ be independent, and $0 < a_n \uparrow \infty$. Then

$$\frac{1}{a_n} \sum_{i=1}^n \{X_i - \operatorname{E}(X_i)\} \xrightarrow{\text{a.s.}} 0 \tag{6.21}$$

provided that

$$\sum_{i=1}^{\infty} \frac{\mathrm{var}(X_i)}{a_i^2} < \infty. \tag{6.22}$$

The result of Theorem 6.7 can be generalized. If $X_i$, $i \geq 1$, are independent with mean 0 and $0 < a_n \uparrow \infty$. Then $a_n^{-1} \sum_{i=1}^{n} X_i \xrightarrow{\text{a.s.}} 0$ provided that

$$\sum_{i=1}^{\infty} \frac{\mathrm{E}(|X_i|^p)}{a_i^p} < \infty \tag{6.23}$$

for some $1 \leq p \leq 2$. We consider an example.

*Example 6.2 (continued).* Since $\mathrm{E}(Y_i) = \mathrm{var}(Y_i) = \lambda_i$, by Theorem 6.7,

$$\frac{1}{a_n} \sum_{i=1}^{n} (Y_i - \lambda_i) \xrightarrow{\text{a.s.}} 0, \tag{6.24}$$

provided that

$$\sum_{i=1}^{\infty} \frac{\lambda_i}{a_i^2} < \infty. \tag{6.25}$$

Clearly, there are many choices of $a_n$ that satisfy (6.25). For example, if $a_n = n^p$, then (6.25) holds if and only if $p > 1/2$. Similarly, if $a_n = (\sum_{i=1}^{n} \lambda_i)^\gamma$ as in Example 6.1, then (6.25) holds if and only if $\gamma > 1/2$ (Exercise 6.4). Here, we used the assumption that the $\lambda_i$'s are bounded from above and away from zero. If one only assumes $\lambda_i > 0$ for all $i$, (6.25) still holds with $a_n = (\sum_{i=1}^{n} \lambda_i)^\gamma$ for any $\gamma > 1/2$. To show this, consider the function $f(x) = x^{1-\beta}$, where $\beta = 2\gamma > 1$. Write $\Lambda_i = \sum_{j=1}^{i} \lambda_j$. By Taylor's expansion (see Section 4.2), we have $f(\Lambda_i) - f(\Lambda_{i-1}) = f'(\xi)(\Lambda_i - \Lambda_{i-1})$, where $\Lambda_{i-1} \leq \xi \leq \Lambda_i$, and $f'(x) = (1-\beta)x^{-\beta}$. It follows that

$$\frac{\lambda_i}{a_i^2} = \frac{\Lambda_i - \Lambda_{i-1}}{\Lambda_i^\beta}$$

$$\leq \frac{\Lambda_i - \Lambda_{i-1}}{\xi^\beta}$$

$$= \frac{f(\Lambda_{i-1}) - f(\Lambda_i)}{\beta - 1}$$

$$= \frac{1}{\beta - 1} \left( \frac{1}{\Lambda_{i-1}^{\beta-1}} - \frac{1}{\Lambda_i^{\beta-1}} \right), \quad i \geq 2.$$

Therefore, we have

$$\sum_{i=1}^{\infty} \frac{\lambda_i}{a_i^2} \leq \frac{1}{\lambda_1^{\beta-1}} + \frac{1}{\beta-1} \sum_{i=2}^{\infty} \left( \frac{1}{\Lambda_{i-1}^{\beta-1}} - \frac{1}{\Lambda_i^{\beta-1}} \right)$$

$$\leq \frac{1}{\lambda_1^{\beta-1}} + \frac{1}{\beta-1} \frac{1}{\lambda_1^{\beta-1}}$$

$$= \frac{\beta}{(\beta-1)\lambda_1^{\beta-1}} < \infty.$$

On the other hand, it is easy to give a counterexample that (6.25) is false when $\gamma \leq 1/2$ (e.g., consider $\lambda_i = 1$).

Furthermore, it would also be interesting to know if some version of SLLN still holds when the mean of $X_i$ does not exist. We have the following result.

**Theorem 6.8.** Let $X_1, X_2, \ldots$, be independent, and $0 < a_n \uparrow \infty$. Then

$$\frac{1}{a_n} \sum_{i=1}^{n} [X_i - \mathrm{E}\{X_i 1_{(|X_i| \leq a_i)}\}] \xrightarrow{\text{a.s.}} 0 \tag{6.26}$$

provided that

$$\sum_{i=1}^{\infty} \mathrm{F}_i \left( \frac{X_i^2}{a_i^2 + X_i^2} \right) < \infty. \tag{6.27}$$

We visit another example that was considered earlier.

*Example 6.1 (continued).* We noted that the mean of the Cauchy$(0,1)$ distribution does not exist and, as a result, the WLLN does not hold; that is, $\bar{X}$ does not converge to zero in probability. We now consider a different normalizing constant $a_n$. Note that

$$\mathrm{E}\{X_i 1_{(|X_i| \leq a_i)}\} = \int_{-a_i}^{a_i} \frac{x}{\pi(1+x^2)} \, dx = 0;$$

in other words, the truncated mean of Cauchy$(0,1)$ exists and is equal to zero. Thus, by Theorem 6.8, we have

$$\frac{1}{a_n} \sum_{i=1}^{n} X_i \xrightarrow{\text{a.s.}} 0 \tag{6.28}$$

provided that (6.27) holds. To evaluate the expected value in (6.27), let $b_i$ be a constant to be determined such that $b_i \geq \sqrt{a_i}$ (reason given below). Then

$$\mathrm{E}\left( \frac{X_i^2}{a_i^2 + X_i^2} \right) = \int \frac{x^2}{a_i^2 + x^2} \frac{dx}{\pi(1+x^2)}$$

$$= \frac{2}{\pi} \int_0^{\infty} \frac{x^2}{(a_i^2 + x^2)(1+x^2)} \, dx.$$

Furthermore, we have

$$\int_0^\infty \frac{x^2}{(a_i^2 + x^2)(1 + x^2)}\, dx = \int_0^{b_i} \cdots dx + \int_{b_i}^\infty \cdots dx$$
$$= I_1 + I_2.$$

First, consider the integrand of $I_1$. Write $c = a_i^2$, $d = b_i^2$ and consider the function $\psi(u) = u/(c + u)(1 + u)$ for $0 \leq u \leq d$. It can be shown that $\psi(u)$ attains its maximum over the range at $u = \sqrt{c}$, and the maximum is $(1 + \sqrt{c})^{-2}$ (Exercise 6.5). It follows that the integrand of $I_1$ is bounded by $(1 + a_i)^{-2}$; hence, $I_1 \leq b_i(1 + a_i)^{-2}$, provided that $\sqrt{c} \leq d$ (i.e., $a_i \leq b_i^2$; this explains the range of $b_i$ given above). On the other hand, the integrand of $I_2$ is bounded by $x^{-2}$; hence, $I_2 \leq \int_{b_i}^\infty x^{-2}\, dx = b_i^{-1}$. In conclusion, we have

$$E\left(\frac{X_i^2}{a_i^2 + X_i^2}\right) \leq \frac{2}{\pi}\left\{\frac{b_i}{(1 + a_i)^2} + \frac{1}{b_i}\right\} \tag{6.29}$$

for any $b_i \geq \sqrt{a_i}$. The right side of (6.29) is minimized when $b_i = 1 + a_i \geq 2\sqrt{a_i} > \sqrt{a_i}$ (Exercise 6.5), and the minimum is $4\{\pi(1 + a_i)\}^{-1}$. Therefore, (6.27), hence (6.28), holds if $\sum_{i=1}^\infty (1 + a_i)^{-1} < \infty$. The latter condition is satisfied, for example, by $a_i = i^p$, $i \geq 1$, where $p > 1$.

Historically, the proofs of SLLNs were based on an interesting connection between convergence of an infinite series, say, $\sum_{i=1}^\infty x_i$, and the weighted average $a_n^{-1} \sum_{i=1}^n a_i x_i$ to zero. The connection is built by the following lemma.

**Lemma 6.1** (Kronecker's lemma). If $\sum_{i=1}^\infty x_i$ converges and $a_n \uparrow \infty$, then $a_n^{-1} \sum_{i=1}^n a_i x_i \to 0$.

Here, we are talking about convergence of an infinite series of random variables. The most famous result in this regard is the following.

**Theorem 6.9** (Kolmogorov's three series theorem). Let $X_1, X_2, \ldots$ be a sequence of independent random variables. (i) If $\sum_{i=1}^\infty X_i$ converges a.s., then for every $c > 0$, the following three series converge:

$$\sum_{i=1}^\infty P(|X_i| > c), \tag{6.30}$$

$$\sum_{i=1}^\infty E\{X_i 1_{(|X_i| \leq c)}\}, \tag{6.31}$$

$$\sum_{i=1}^\infty \operatorname{var}\{X_i 1_{(|X_i| \leq c)}\}. \tag{6.32}$$

(ii) Conversely, if the series (6.30)–(6.32) converge for some $c > 0$, then $\sum_{i=1}^\infty X_i$ converges a.s.

As an example, we give a proof of Theorem 6.7 (which was also due to Kolmogorov) using Theorem 6.9.

*Example 6.4* (Proof of Theorem 6.7). Suppose that the condition of Theorem 6.7 are satisfied. Let $Y_i = \{X_i - \mathrm{E}(X_i)\}/a_i$. Then by Chebyshev's inequality, we have $\sum_{i=1}^{\infty} \mathrm{P}(|Y_i| > 1) \le \sum_{i=1}^{\infty} \mathrm{E}(Y_i^2) = \sum_{i=1}^{\infty} \mathrm{var}(X_i)/a_i^2 < \infty$. Next, since $\mathrm{E}(Y_i) = 0$, we have $\mathrm{E}\{Y_i 1_{(|Y_i| \le 1)}\} = -\mathrm{E}\{Y_i 1_{(|Y_i| > 1)}\}$. Thus,

$$
\sum_{i=1}^{\infty} \left| \mathrm{E}\{Y_i 1_{(|Y_i| \le 1)}\} \right| = \sum_{i=1}^{\infty} \left| \mathrm{E}\{Y_i 1_{(|Y_i| > 1)}\} \right|
$$
$$
\le \sum_{i=1}^{\infty} \mathrm{E}\{|Y_i| 1_{(|Y_i| > 1)}\}
$$
$$
\le \sum_{i=1}^{\infty} \mathrm{E}(Y_i^2) < \infty.
$$

Finally, $\sum_{i=1}^{\infty} \mathrm{var}\{Y_i 1_{(|Y_i| \le 1)}\} \le \sum_{i=1}^{\infty} \mathrm{E}\{Y_i^2 1_{(|Y_i| > 1)}\} \le \sum_{i=1}^{\infty} \mathrm{E}(Y_i^2) < \infty$. Therefore, by Theorem 6.9, the series $\sum_{i=1}^{\infty} Y_i$ converges a.s. It then follows by Lemma 6.1 that $a_n^{-1} \sum_{i=1}^{n} \{X_i - \mathrm{E}(X_i)\} = a_n^{-1} \sum_{i=1}^{n} a_i Y_i \xrightarrow{\text{a.s.}} 0$.

Finally, the following theorem uncovers an interesting connection between WLLN and SLLN.

**Theorem 6.10** (Katz 1968). Let $X_1, X_2, \ldots,$ be i.i.d. If $\bar{X}$ converges to zero in probability but not almost surely, then $\limsup \bar{X} = \infty$ a.s. and $\liminf \bar{X} = -\infty$ a.s.

## 6.4 The central limit theorem

We begin with the following landmark theorem due to Lindeberg and Feller.

**Theorem 6.11** (Lindeberg–Feller theorem). Let $X_1, X_2, \ldots$ be a sequence of independent random variables with $\mathrm{E}(X_i) = 0$ $\mathrm{E}(X_i^2) = \sigma_i^2 < \infty$. Define $S_n = \sum_{i=1}^{n} X_i$ and $s_n^2 = \sum_{i=1}^{n} \sigma_i^2$. Then

$$
s_n^{-1} S_n \xrightarrow{\text{d}} N(0, 1) \tag{6.33}
$$

provided that for any $\epsilon > 0$,

$$
\frac{1}{s_n^2} \sum_{i=1}^{n} \mathrm{E}\{X_i^2 1_{(|X_i| > \epsilon s_n)}\} \longrightarrow 0. \tag{6.34}
$$

Condition (6.34) is known as the Lindeberg condition. An easier to verify sufficient condition for the Lindeberg condition is the Liapounov condition:

$$\frac{1}{s_n^{2+\delta}} \sum_{i=1}^{n} \mathrm{E}(|X_i|^{2+\delta}) \longrightarrow 0 \tag{6.35}$$

for some $\delta > 0$ (Exercise 6.13).

In some cases it is more convenient to consider the triangular array introduced in the previous section. Suppose that for each $n$, $X_{ni}, 1 \le i \le i_n$, are independent with $\mathrm{E}(X_{ni}) = 0$ and $\mathrm{E}(X_{ni}^2) = \sigma_{ni}^2 < \infty$. Write $S_n = \sum_{i=1}^{i_n} X_{ni}$ and $s_n^2 = \sum_{i=1}^{i_n} \sigma_{ni}^2$. Then $\sigma_n^{-1} S_n \xrightarrow{d} N(0,1)$ provided that the following (also known as the Lindeberg condition) holds: For any $\epsilon > 0$,

$$\frac{1}{s_n^2} \sum_{i=1}^{i_n} \mathrm{E}\{X_{ni}^2 1_{(|X_{ni}|>\epsilon s_n)}\} \longrightarrow 0. \tag{6.36}$$

Again, a sufficient condition for (6.36) is the following (also known as the Liapounov condition): For some $\delta > 0$,

$$\frac{1}{s_n^{2+\delta}} \sum_{i=1}^{i_n} \mathrm{E}(|X_{ni}|^{2+\delta}) \longrightarrow 0. \tag{6.37}$$

We consider some examples.

*Example 6.5.* Let $Y_1, Y_2, \ldots$ be independent such that $Y_i \sim \mathrm{Bernoulli}(p_i)$, $i \ge 1$. Let $s_n^2 = \sum_{i=1}^{n} \mathrm{var}(Y_i) = \sum_{i=1}^{n} p_i(1 - p_i)$. Then $s_n^{-1} \sum_{i=1}^{n}(Y_i - p_i) \xrightarrow{d} N(0,1)$ provided that $\sum_{i=1}^{\infty} p_i(1 - p_i) = \infty$. To see this, write $X_i = Y_i - p_i$. Then $\mathrm{E}(|X_i|^3) = p_i^3(1 - p_i) + (1 - p_i)^3 p_i \le 2p_i(1 - p_i)$. Thus, we have

$$\frac{1}{s_n^3} \sum_{i=1}^{n} \mathrm{E}(|X_i|^3) \le \frac{2}{s_n^3} \sum_{i=1}^{n} p_i(1 - p_i) = \frac{2}{s_n},$$

which goes to zero as $n \to \infty$. In other words, Liapounov's condition (6.35) is satisfied with $\delta = 1$. The result then follows by Theorem 6.11.

*Example 6.6* (Hájek–Sidak theorem). Suppose that $X_1, X_2, \ldots$ are i.i.d. with mean $\mu$ and variance $\sigma^2 \in (0, \infty)$. Let $c_{ni}, 1 \le i \le n, n = 1, 2, \ldots$, be a triangular array of constants such that as $n \to \infty$,

$$\max_{1 \le i \le n} \frac{c_{ni}^2}{\sum_{j=1}^{n} c_{nj}^2} \longrightarrow 0 \tag{6.38}$$

(note that this also implies that $\sum_{i=1}^{n} c_{ni}^2 > 0$ for large $n$). Then we have

$$\frac{\sum_{i=1}^{n} c_{ni}(X_i - \mu)}{\sigma \sqrt{\sum_{i=1}^{n} c_{ni}^2}} \xrightarrow{d} N(0,1). \tag{6.39}$$

To show this, let $X_{ni} = c_{ni}(X_i - \mu)$. Then $S_n = \sum_{i=1}^{n} c_{ni}(X_i - \mu)$ and $s_n^2 = \sigma^2 \sum_{i=1}^{n} c_{ni}^2$. Thus, $s_n^{-1} S_n$, which is the left side of (6.39), converges in distribution to $N(0,1)$ provided that the Lindeberg condition (6.36) holds. Denote the left side of (6.38) by $\delta_n^2$. Then we have

$$E\{X_{ni}^2 1_{(|X_{ni}|>\epsilon s_n)}\} = c_{ni}^2 E\{(X_1 - \mu)^2 1_{(c_{ni}|X_1-\mu|>\epsilon s_n)}\}$$
$$\le c_{ni}^2 E\{(X_1 - \mu)^2 1_{(\delta_n|X_1-\mu|>\epsilon\sigma)}\}.$$

Thus, the left side of (6.36) is bounded by

$$\frac{1}{s_n^2} \sum_{i=1}^{n} c_{ni}^2 E\{(X_1 - \mu)^2 1_{(\delta_n|X_1-\mu|>\epsilon\sigma)}\}$$
$$= \frac{E\{(X_1 - \mu)^2 1_{(\delta_n|X_1-\mu|>\epsilon\sigma)}\}}{\sigma^2} \longrightarrow 0$$

as $n \to \infty$ by the dominated convergence theorem (Theorem 2.16).

Intuitively, condition (6.38) means that as $n \to \infty$, the contribution of any single term in the summation

$$\sum_{i=1}^{n} \frac{c_{ni}(X_i - \mu)}{\sigma\sqrt{\sum_{j=1}^{n} c_{ni}^2}},$$

which is the left side of (6.39), is negligible. Such a condition is critical for any CLT to hold. For example, consider the following.

*Example 6.7* (A counterexample). Suppose that $X_1, X_2, \ldots$ are i.i.d. with mean $\mu$ and variance $\sigma^2 \in (0, \infty)$, but not normally distributed. Let $c_{n1} = 1$ and $c_{ni} = 0, 2 \le i \le n$. Then the left side of (6.38) is equal to 1 for any $n$. On the other hand, the left side of (6.39) is equal to $(X_1 - \mu)/\sigma$ for any $n$, which is not distributed as $N(0,1)$.

Similar to the LLN, there exist necessary and sufficient conditions for the CLT. We first consider sequences of independent random variables.

**Theorem 6.12.** Let $X_1, X_2, \ldots$ be a sequence of independent random variables, at least one of which has a nondegenerate distribution. Let $\mu_i = E(X_i)$, $\sigma_i^2 = \text{var}(X_i) < \infty$, $i \ge 1$, $s_n^2 = \sum_{i=1}^{n} \sigma_i^2$,

$$F_n(x) = P\left\{\frac{1}{s_n} \sum_{i=1}^{n}(X_i - \mu_i) \le x\right\},$$

and $\Phi(x)$ be the cdf of $N(0,1)$. Then

$$\frac{\max_{1\le i\le n} \sigma_i^2}{s_n^2} \longrightarrow 0, \qquad (6.40)$$

$$\sup_x |F_n(x) - \Phi(X)| \longrightarrow 0 \qquad (6.41)$$

if and only if the following Lindeberg condition is satisfied: For every $\epsilon > 0$,

$$\frac{1}{s_n^2} \sum_{i=1}^{n} \mathrm{E}\{(X_i - \mu_i)^2 1_{(|X_i - \mu_i| > \epsilon s_n)}\} \longrightarrow 0. \tag{6.42}$$

More generally, for triangular arrays of independent random variables, we have the following result.

**Theorem 6.13.** Suppose that for each $n$, $X_{ni}, 1 \leq i \leq i_n$, are independent. Then $X_{ni}$ obeys the condition of infinite smallness [see (6.14)] and $\sum_{i=1}^{i_n} X_{ni} \xrightarrow{\mathrm{d}} N(\mu, \sigma^2)$ if and only if for every $\epsilon > 0$ (6.15) holds and

$$\sum_{i=1}^{i_n} \mathrm{E}\{X_{ni} 1_{(|X_{ni}| \leq \epsilon)}\} \longrightarrow \mu, \tag{6.43}$$

$$\sum_{i=1}^{i_n} \mathrm{var}\{X_{ni} 1_{(|X_{ni}| \leq \epsilon)}\} \longrightarrow \sigma^2. \tag{6.44}$$

*Notes.* Theorem 6.13 does not require the existence of $\mathrm{E}(X_{ni})$ and $\mathrm{var}(X_{ni})$. Similar to (6.41), the convergence to $N(\mu, \sigma^2)$ is uniform, which follows from Pólya's theorem (see Example 1.6). The conditions (6.15), (6.43), and (6.44) for every $\epsilon > 0$ can be replaced by (6.15) for every $\epsilon > 0$ and (6.43) and (6.44) for some $\epsilon > 0$. We consider some examples.

As an application of Theorem 6.13, we prove the following theorem which gives a necessary and sufficient condition for CLT in the i.i.d. case.

**Theorem 6.14.** Let $X_1, X_2, \ldots$ be i.i.d. Then

$$\frac{1}{\sqrt{n}} \sum_{i=1}^{n} (X_i - \mu) \xrightarrow{\mathrm{d}} N(0, \sigma^2) \tag{6.45}$$

for some $\mu \in R$ and $\sigma^2 \in [0, \infty)$ if and only if

$$\mathrm{E}(X_1^2) < \infty. \tag{6.46}$$

If (6.46) is satisfied, then (6.45) holds, with $\mu = \mathrm{E}(X_1)$ and $\sigma^2 = \mathrm{var}(X_1)$.

*Proof.* Suppose that (6.45) holds. Let $X_{ni} = (X_i - \mu)/\sqrt{n}$. Then we have $\sum_{i=1}^{n} X_{ni} \xrightarrow{\mathrm{d}} N(0, \sigma^2)$. Furthermore, for any $\epsilon > 0$, we have $\mathrm{P}(|X_{ni}| > \epsilon) = \mathrm{P}(|X_1 - \mu| > \epsilon\sqrt{n})$, which does not depend on $i$, and goes to zero as $n \to \infty$ (Exercise 6.14). Therefore, the condition of infinite smallness (6.14) is satisfied. It follows by Theorem 6.13 that (6.15), (6.43), and (6.44) hold for any $\epsilon > 0$ and hence in particular for $\epsilon = 1$. In particular, (6.43) implies that $\sqrt{n}\mathrm{E}\{(X_1 - \mu)1_{(|X_1 - \mu| \leq \sqrt{n})}\} \to 0$ (Exercise 6.14); hence, $\mathrm{E}\{(X_1 - \mu)1_{(|X_1 - \mu| \leq \sqrt{n})}\} \to 0$. Furthermore, (6.44) implies that

$$E\{(X_1 - \mu)^2 1_{(|X_1 - \mu| \leq \sqrt{n})}\} - [E\{(X_1 - \mu) 1_{(|X_1 - \mu| \leq \sqrt{n})}\}]^2 \longrightarrow \sigma^2;$$

hence, $E\{(X_1 - \mu)^2 1_{(|X_1 - \mu| \leq \sqrt{n})}\} \longrightarrow \sigma^2$ (Exercise 6.14). It follows by the monotone covergence theorem (see §2.7.7) that $E\{(X_1 - \mu)^2\} = \sigma^2 < \infty$; hence, (6.46) holds.

Conversely, suppose that (6.46) holds. Define

$$X_{ni} = \frac{1}{\sqrt{n}}[X_i 1_{(|X_i| \leq \sqrt{n})} - E\{X_1 1_{(|X_1| \leq \sqrt{n})}\}]. \tag{6.47}$$

It is easy to show that

$$|X_{ni}| \leq 2 \wedge \left\{ \frac{|X_i| + E(|X_1|)}{\sqrt{n}} \right\}.$$

Therefore, for any $\epsilon > 0$, we have, for large $n$,

$$P(|X_{ni}| > \epsilon) \leq P \left\{ \frac{|X_i| + E(|X_1|)}{\sqrt{n}} > \epsilon \right\}$$

$$= P\{|X_1| > \epsilon\sqrt{n} - E(|X_1|)\} = P(|X_1| > \lambda_n),$$

where $\lambda_n = \{\epsilon\sqrt{n} - E(|X_1|)\} \vee 1$. It follows that

$$\sum_{i=1}^{n} P(|X_{ni}| > \epsilon) \leq nP(|X_1| > \lambda_n) \to 0$$

as $n \to \infty$ (Exercise 6.14). In other words, (6.15) holds for any $\epsilon > 0$. Furthermore, we have $E\{X_{ni} 1_{(|X_{ni}| \leq 2)}\} = E(X_{ni}) = 0$; hence, (6.43) is satisfied with $\epsilon = 2$. Finally, we have

$$\sum_{i=1}^{n} \text{var}\{X_{ni} 1_{(|X_{ni}| \leq 2)}\} = \sum_{i=1}^{n} \text{var}(X_{ni})$$

$$= E\{X_1^2 1_{(|X_1| \leq \sqrt{n})}\} - [E\{X_1 1_{(|X_1| \leq \sqrt{n})}\}]^2$$

$$\to E(X_1^2) - \{E(X_1)\}^2 = \sigma^2$$

by the dominated convergence theorem (Theorem 2.16). In other words, (6.44) holds with $\epsilon = 2$. Thus, by Theorem 6.13 (and the note following the theorem),

$$\frac{1}{\sqrt{n}} \sum_{i=1}^{n} [X_i 1_{(|X_i| \leq \sqrt{n})} - E\{X_1 1_{(|X_1| \leq \sqrt{n})}\}] = \sum_{i=1}^{n} X_{ni} \xrightarrow{d} N(0, \sigma^2).$$

On the other hand, we have

$$E \left| \frac{1}{\sqrt{n}} \sum_{i=1}^{n} [X_i 1_{(|X_i| > \sqrt{n})} - E\{X_1 1_{(|X_1| > \sqrt{n})}\}] \right|$$

$$\leq \frac{1}{\sqrt{n}} \sum_{i=1}^{n} [E\{|X_i| 1_{(|X_i| > \sqrt{n})}\} + E\{|X_1| 1_{(|X_1| > \sqrt{n})}\}]$$

$$= 2\sqrt{n} E\{|X_1| 1_{(|X_1| > \sqrt{n})}\}$$

$$\leq 2E\{X_1^2 1_{(|X_1| > \sqrt{n})}\} \longrightarrow 0$$

as $n \to \infty$ by the dominated convergence theorem. Result (6.45) then follows as a result of Theorem 2.15 and Slutsky's theorem (Theorem 2.13). This completes the proof. Q.E.D.

Theorems 6.12 and 6.13 do not apply to all the cases. We conclude this section with an example that shows that in cases where these theorems do not apply, a necessary and sufficient condition may still be found.

*Example 6.5 (continued).* Previously we showed that the condition

$$\sum_{i=1}^{\infty} p_i(1 - p_i) = \infty \qquad (6.48)$$

is sufficient for

$$\frac{1}{s_n} \sum_{i=1}^{n} (Y_i - p_i) \xrightarrow{\text{d}} N(0, 1). \qquad (6.49)$$

In fact, (6.48) is also necessary. However, the result does not follow from Theorem 6.12 or Theorem 6.13. To see this, note that if (6.48) does not hold, then, for example, condition (6.40) fails (Exercise 6.22); therefore, the necessary and sufficient condition of Theorem 6.12 does not apply to this case. Nevertheless, it can be shown by Kolmogorov's three series theorem and a famous result due to Cramér (1936) that (6.49) implies (6.48). We prove this by a contrapositive. Suppose that (6.49) holds but not (6.48). Then we have $\sum_{i=1}^{\infty} p_i(1 - p_i) < \infty$. Let $X_i = Y_i - p_i$. It is easy to show that the three series (6.30)–(6.32) converge for $c = 1$ (Exercise 6.22); hence, by Theorem 6.9 the series $\sum_{i=1}^{\infty} X_i$ converges a.s. to a random variable, say $\xi$. Also, (6.49) implies that at least one of the $p_i$'s is not zero or one. This is because, otherwise, we have $s_n = 0$ for all $n$ and $X_i = 0$ a.s. for all $i$; therefore, the left side of (6.49) is $0/0$, which is not well defined (hence, cannot be convergent to a well-defined distribution). Let $a$ be the first index $i$ ($i \geq 1$) such that $p_i$ is not zero or one. By the same argument, it can be shown that the series $\sum_{i=a+1}^{\infty} X_i$ converges a.s. to a random variable, say $\xi_1$. Also, $s_n^2 \to s^2 = \sum_{i=1}^{\infty} p_i(1 - p_i) \in (0, \infty)$ (Exercise 6.22). Therefore, by taking the limit on both sides of the identity

$$\frac{X_a}{s_n} + \frac{\sum_{i=a+1}^{n} X_i}{s_n} = \frac{\sum_{i=1}^{n} X_i}{s_n}$$

for $n > a$ (note that $X_i = 0$ a.s. for $i < a$), we have, with probability 1,

$$\frac{X_a}{s} + \frac{\xi_1}{s} = \frac{\xi}{s} \sim N(0, 1)$$

by Theorems 2.7 and 2.9 and (6.49). We now apply Cramér's theorem: If $X$ and $Y$ are independent such that $X + Y$ is normally distributed, then both $X$ and $Y$ must be normally distributed. Note that $X_a$ and $\xi_1$ are independent.

It follows that $X_a/s$ is normally distributed and, therefore, $X_a$ is normally distributed, which is, of course, false.

*Note.* Cramér's theorem is a remarkable result. For example, suppose that $X_1, X_2, \ldots$ are independent with mean 0, variance 1 and bounded third absolute moments. Then, by Liapounov's CLT [see (6.35)], the distribution of $n^{-1/2} S_n$ is asymptotically normal as $n$ goes to infinity, where $S_n = \sum_{i=1}^{n} X_i$. On the other hand, suppose that at least one of the $X_i$'s is not normally distributed. Then as long as $n$ is large enough, the distribution $n^{-1/2} S_n$ is never (exactly) normal no matter how large $n$ is (why?).

## 6.5 The law of the iterated logarithm

In a way, the CLT states the convergence rate in LLN. Since the latter implies that, for example, $n^{-1} \sum_{i=1}^{n} (X_i - \mu) \xrightarrow{P} 0$ in the i.i.d. case, where $\mu = E(X_1)$, one would like to know how far one could go by reducing the order of the denominator, say from $n$ to $n^{\gamma}$, where $\gamma < 1$. The CLT states that, in this regard, the best one can do is $1/2 < \gamma < 1$, but not $\gamma = 1/2$ (so $\gamma = 1/2$ is the cut off), because $\xi_n = n^{-1/2} \sum_{i=1}^{n} (X_i - \mu) \xrightarrow{d} N(0, \sigma^2)$ with $\sigma^2 = \mathrm{var}(X_1)$, which implies that for any $\gamma > 1/2$,

$$\frac{1}{n^{\gamma}} \sum_{i=1}^{n} (X_i - \mu) = \frac{\xi_n}{n^{\gamma - 1/2}} \xrightarrow{P} 0$$

by (ii) of Theorem 2.13. On the other hand, even if $\xi_n$ converges in distribution to $N(0, \sigma^2)$, there is still a (small) chance that $\xi_n$ can assume a large value, because a $N(0, \sigma^2)$ random variable is not bounded.

Another way to describe the convergence rate in LLN is the law of the iterated logarithm (LIL). Throughout this section, the value of $\log \log x$ is understood as 1 if $x \leq e$. One of the best known results on LIL is the following theorem due to Kolmogorov (1929).

**Theorem 6.15.** Let $X_1, X_2, \ldots$ be a sequence of independent random variables with mean 0 and finite variance. Suppose that $a_n = \sum_{i=1}^{n} \sigma_i^2 \to \infty$, where $\sigma_i^2 = E(X_i^2)$. If $|X_i| \leq b_i$ a.s., where $b_i$ is a constant such that

$$b_n = o\left( \sqrt{\frac{a_n}{\log \log a_n}} \right).$$
\hfill (6.50)

Then, with $S_n = \sum_{i=1}^{n} X_i$, we have

$$\limsup \frac{S_n}{\sqrt{2 a_n \log \log a_n}} = 1 \quad \text{a.s.}$$
\hfill (6.51)

By replacing $X_i$ with $-X_i$ we obtain "the other half" of the LIL:

$$\liminf \frac{S_n}{\sqrt{2a_n \log\log a_n}} = -1 \quad \text{a.s.} \tag{6.52}$$

Combining (6.51) and (6.52), we get

$$\limsup \frac{|S_n|}{\sqrt{2a_n \log\log a_n}} = 1 \quad \text{a.s.} \tag{6.53}$$

It follows that the a.s. convergence rate of $n^{-1}S_n$ is

$$O\left(\frac{\sqrt{a_n \log\log a_n}}{n}\right). \tag{6.54}$$

We consider some examples.

*Example 6.8.* Suppose that the $X_i$'s are i.i.d. with $E(X_1) = 0$ and $E(X_1^2) = 1$ and bounded (although the latter assumption is unnecessary; see Theorem 6.17 in the sequel). Then we have $a_n = n$; hence, (6.54) becomes

$$O\left(\sqrt{\frac{\log\log n}{n}}\right). \tag{6.55}$$

Although (6.55) appears to be slower than the convergence rate implied by the CLT, which is $O(1/\sqrt{n})$, the meanings of these orders are different. The CLT convergence rate is in the sense of convergence in probability; that is, $n^{-1}S_n = O_P(1/\sqrt{n})$ (see Section 3.4), whereas the LIL convergence rate is in the sense of almost sure convergence, which means that

$$P\left(\limsup \sqrt{\frac{n}{\log\log n}}|n^{-1}S_n| = \sqrt{2}\right) = 1.$$

*Example 6.5 (continued).* Note that in this case we have $|X_i| \le b_i = 1$; hence, (6.50) is satisfied with $a_n = s_n^2$ provided that $\sum_{i=1}^{\infty} p_i(1 - p_i) = \infty$. It follows that (6.51)–(6.53) hold with $a_n = \sum_{i=1}^{n} p_i(1 - p_i)$. For example, suppose that $p_i = i^{-1}$. Then $a_n = \log n + O(1)$, and (6.51) implies that

$$\frac{\sum_{i=1}^{n} Y_i - \log n}{\sqrt{2\log n \log\log\log n}} = 1 \quad \text{a.s.}$$

(see Exercise 6.19), and (6.54) becomes

$$O\left(\frac{\sqrt{\log n \log\log\log n}}{n}\right).$$

It should be pointed out that in this case, the convergence rate of $n^{-1}S_n$ implied by CLT is $O(\sqrt{\log n}/n)$, which is much faster than $O(1/\sqrt{n})$, as in the previous example (see Exercise 6.19).

A key assumption in Theorem 6.15 is that the random variables are bounded. Although this may seem restrictive, as most of the random variables that are commonly in use (such as Poisson or normal) are not bounded, Kolmogorov's theorem was an important step toward LIL for unbounded random variables. The connection (between LIL for bounded random variables and that for unbounded ones) is made by a technique called *truncation*. Suppose that $X_i, i \geq 1$, is a sequence of independent random variables with mean 0 and finite variance. Then one can write

$$X_i = [X_i 1_{(|X_i| \leq b_i)} - \mathrm{E}\{X_i 1_{(|X_i| \leq b_i)}\}] + [X_i 1_{(|X_i| > b_i)} - \mathrm{E}\{X_i 1_{(|X_i| > b_i)}\}]$$
$$= Y_i + Z_i$$

[because $\mathrm{E}(X_i) = 0$], where $b_i$ is a constant satisfying (6.50). Now, Kolmogorov's LIL can be applied to $Y_i$, since the latter is bounded by the "right" constant; hence, all one has to do is to show that

$$\frac{\sum_{i=1}^{n} Z_i}{\sqrt{a_n \log \log a_n}} \xrightarrow{\text{a.s.}} 0$$

as $n \to \infty$, so that the LIL (6.51) will not be affected by the truncation. This idea leads to the proof of the following result.

**Theorem 6.16.** Let $X_1, X_2, \ldots$ be independent with mean 0 and finite variances. Under the notation of Theorem 6.15, the sequence $X_i, i \geq 1$ obeys the LIL; that is, (6.51) holds, provided that $a_n \to \infty$,

$$\frac{1}{a_n} \sum_{i=i_0}^{n} \mathrm{E}\{X_i^2 1_{(|X_i| > \epsilon b_i)}\} \longrightarrow 0, \tag{6.56}$$

$$\sum_{i=i_0}^{\infty} \frac{\mathrm{E}\{X_i^2 1_{(|X_i| > \epsilon b_i)}\}}{a_i \log \log a_i} < \infty \tag{6.57}$$

for every $\epsilon > 0$, where $b_i = (a_i / \log \log a_i)^{1/2}$ and $i_0$ is any index $i$ such that $\log \log a_i > 0$.

We consider an application of Theorem 6.16.

*Example 6.9.* Let $X_i, i \geq 1$, be independent random variables such that $\mathrm{E}(X_i) = 0$, $\mathrm{E}(X_i^2) \geq a$ and

$$\mathrm{E}\{X_i^2 \log |X_i| (\log \log |X_i|)^{\delta}\} \leq b \tag{6.58}$$

for some constants $a, b, \delta > 0$. Then $X_i, i \geq 1$ obeys the LIL. To show this, first note that the assumptions here imply that there is a constant $c > 0$ such that $a \leq \sigma_i^2 \leq c$, $i \geq 1$ (Exercise 6.23); hence, $a_n \propto n$ [definition above (3.6)]. Next, it can be shown that for any $\epsilon > 0$, there is $i_\epsilon \geq i_0$ depending only on $\epsilon$ such that for $i \geq i_\epsilon$, $|X_i| > \epsilon b_i$ implies

$$\log|X_i|(\log\log|X_i|)^\delta > \frac{1}{4}\log a_i(\log\log a_i)^\delta. \tag{6.59}$$

It follows that for $i \geq i_\epsilon$, we have

$$\mathrm{E}\{X_i^2 1_{(|X_i|>\epsilon b_i)}\} \leq \mathrm{E}\left\{X_i^2 \frac{4\log|X_i|(\log\log|X_i|)^\delta}{\log a_i(\log\log a_i)^\delta}\right\}$$

$$\leq \frac{4b}{\log a_i(\log\log a_i)^\delta}$$

by (6.58). Therefore, we have, for $n \geq i_\epsilon$,

$$\frac{1}{a_n}\sum_{i=i_0}^{n}\mathrm{E}\{X_i^2 1_{(|X_i|>\epsilon b_i)}\} \leq \frac{(i_\epsilon - i_0)c}{a_n}$$

$$+\frac{4b}{a_n}\sum_{i=i_\epsilon}^{n}\frac{1}{\log a_i(\log\log a_i)^\delta}, \tag{6.60}$$

which goes to zero as $n \to \infty$ (Exercise 6.23), and

$$\sum_{i=i_0}^{\infty}\frac{\mathrm{E}\{X_i^2 1_{(|X_i|>\epsilon b_i)}\}}{a_i\log\log a_i} \leq c\sum_{i=i_0}^{i_\epsilon-1}\frac{1}{a_i\log\log a_i}$$

$$+4b\sum_{i=i_\epsilon}^{\infty}\frac{1}{a_i\log a_i(\log\log a_i)^{1+\delta}}$$

$$< \infty \tag{6.61}$$

(Exercise 6.23). The result then follows by Theorem 6.16.

The moment condition (6.58) is not minimum, but close to the minimum condition that would be required for LIL. This is because in the i.i.d. case, a finite second moment is both necessary and sufficient for the LIL, as the following theorem states.

**Theorem 6.17.** Let $X_1, X_2, \ldots$ be i.i.d. Then

$$\limsup\frac{\sum_{i=1}^{n}(X_i - \mu)}{\sqrt{2n\log\log n}} = \sigma \quad \text{a.s.} \tag{6.62}$$

for some $\mu \in R$ and $\sigma^2 \in [0, \infty)$ if and only if (6.46) holds. If the latter condition holds, then (6.62) holds, with $\mu = \mathrm{E}(X_1)$ and $\sigma^2 = \mathrm{var}(X_1)$.

The sufficiency part of Theorem 6.17 was first proved by Hartman and Wintner (1941). Therefore, the theorem is often called the Hartman–Wintner LIL. The necessity part of the theorem was due to Strassen (1966).

With Theorem 6.17 we have completed a series of classical results regarding the sum of i.i.d. random variables. We summarize the results as follows.

*Summary.* Let $X_1, X_2, \ldots$ be i.i.d. Then, the following hold:

(i) (WLLN) There exists $\mu \in R$ such that $\bar{X} \xrightarrow{P} \mu$ if and only if $nP(|X_1| > n) \to 0$ and $E\{X_1 1_{(|X_1| \leq n)}\} \to \mu$ (Theorem 6.1 and Exercise 6.24).

(ii) (SLLN) There exists $\mu \in R$ such that $\bar{X} \xrightarrow{a.s.} \mu$ if and only if $E(|X_1|) < \infty$; when the latter condition holds, we have $\mu = E(X_1)$ (Theorem 6.6).

(iii) (CLT) There exist $\mu \in R$ and $\sigma^2 \in [0, \infty)$ such that $n^{-1} \sum_{i=1}^{n} (X_i - \mu) \xrightarrow{d} N(0, \sigma^2)$ if and only if $E(X_1^2) < \infty$; when the latter condition holds, we have $\mu = E(X_1)$ and $\sigma^2 = \text{var}(X_1)$ (Theorem 6.14).

(iv) (LIL) There exist $\mu \in R$ and $\sigma^2 \in [0, \infty)$ such that $\limsup \sum_{i=1}^{n} (X_i - \mu)/\sqrt{2n \log \log n} = \sigma$ a.s. if and only if $E(X_1^2) < \infty$; when the latter condition holds, we have $\mu = E(X_1)$ and $\sigma^2 = \text{var}(X_1)$ (Theorem 6.17).

Note that the condition in (i) for WLLN is weaker than $E(|X_1|) < \infty$ (Exercise 6.24).

## 6.6 Further results

### 6.6.1 Invariance principles in CLT and LIL

Donsker's invariance principle in CLT (Donsker 1951, 1952) is a functional central limit theorem. Roughly speaking, a functional is a function of a function. Here, we consider the space of all continuous functions on $[0, 1]$, denoted by $C$. We can define a distance between two points, $x$ and $y$ in $C$ (note that here $x$ and $y$ denote two continuous functions on $[0, 1]$) by

$$\rho(x, y) = \sup_{t \in [0,1]} |x(t) - y(t)|. \tag{6.63}$$

The space $C$, equipped with the distance $\rho$ is a metric space, which means that $\rho$ satisfies the following basic requirements, held for all $x, y, z \in C$ (to qualify as a distance, or metric):

1. (nonnegativity) $\rho(x, y) \geq 0$;
2. (symmetry) $\rho(x, y) = \rho(y, x)$;
3. (triangle inequality) $\rho(x, z) \leq \rho(x, y) + \rho(y, z)$;
4. (identity of points) $\rho(x, y) = 0$ if and only if $x = y$.

It is easy to show that the distance defined by (6.63) satisfies requirements 1–4 (Exercise 6.31).

As in Section 2.4, we can talk about weak convergence of probability measures on the measurable space $(C, \mathcal{B})$, where $\mathcal{B}$ is the class of Borel sets in $C$, which is a $\sigma$-field (see Appendix A.2). A sequence of probability measures $P_n$ converges weakly to a probability measure $P$, denoted by $P_n \xrightarrow{w} P$, if $P_n(B) \to P(B)$ as $n \to \infty$ for any $P$-continuity set $B$. The latter means that $P(\partial B) = 0$, where $\partial B$ denotes the boundary of $B$ (i.e., the set of points that are limits of sequences of points in $B$ as well as limits of sequences of points outside $B$). Equivalently, $P_n \xrightarrow{w} P$ if $\int_C f \, dP_n \to \int_C f \, dP$ for all bounded, uniformly continous function $f$ on $C$. The space $C$ is, obviously, more complicated than the real line $R$ or any finite-dimensional Euclidean space $R^k$ ($k > 1$). However, there is a connection between the weak convergence in $C$ and that of all finite-dimensional distributions. Let $t_1, \ldots, t_k$ be any set of distinct points in $[0, 1]$. Let $P$ be a probability measure on $(C, \mathcal{B})$. Then the induced probability measure

$$P\pi_{t_1, \ldots, t_k}^{-1}(A) = P\{[x(t_1), \ldots, x(t_k)] \in A\}$$

for any Borel set $A$ in $R^k$ is called a finite-dimensional distribution. Here, $\pi_{t_1, \ldots, t_k}$ denotes the projection that carries to the point $x \in C$ to the point $[x(t_1), \ldots, x(t_k)] \in R^k$. It turns out that weak convergence in $C$ is a little more than weak convergence of all finite-dimensional distributions; that is, $P_n \xrightarrow{w} P$ if and only if $P_n \pi_{t_1, \ldots, t_k}^{-1} \xrightarrow{w} P\pi_{t_1, \ldots, t_k}^{-1}$ for any $k \geq 1$ and any distinct points $t_1, \ldots, t_k \in [0, 1]$ plus that the sequence $P_n$, $n \geq 1$ is *tight*. A family of $\mathcal{P}$ of probability measures on $(C, \mathcal{B})$ is tight if for every $\epsilon > 0$, there is a compact subset of $B$ of $C$ such that $P(B) > 1 - \epsilon$ for all $P \in \mathcal{P}$. A well-known result associated with the latter concept is a probability version of the Arzelá–Ascoli theorem. The sequence $P_n$, $n \geq 1$, is tight in $C$ if and only if the following two conditions hold: (i) For any $\eta > 0$, there exists $M > 0$ such that $P_n(|x(0)| > M) \leq \eta$, $n \geq 1$; and (ii) for any $\epsilon, \eta > 0$, there exist $0 < \delta < 1$ and $N \geq 1$ such that for all $n \geq N$,

$$P_n \left\{ \sup_{|s-t|<\delta} |x(s) - x(t)| \geq \epsilon \right\} \leq \eta.$$

The concept of random variables can now be extended to $C$-valued random variables (i.e., random variables whose values are continuous functions on $[0, 1]$). Such a random variable is often called a stochastic process, denoted by $\xi = (\xi_t, 0 \leq t \leq 1)$, although continuity is not required for the definition. A sequence of $C$-valued random variables $\xi_n$, $n \geq 1$, converges in distribution to a $C$-valued random variable $\xi$, denoted by $\xi_n \xrightarrow{d} \xi$, if $P\xi_n^{-1} \xrightarrow{w} P\xi^{-1}$, where $P\xi_n^{-1}$ is the induced probability measures defined by $P\xi_n^{-1}(B) = P(\xi_n \in B)$ for $B \in \mathcal{B}$, and $P\xi^{-1}$ is defined similarly.

One particular stochastic process is called *Wiener process* or *Brownian motion*. A probability measure $W$ on $(C, \mathcal{B})$ is called a Wiener measure if (i) for each $t \in [0, 1]$, the random variable $x_t \sim N(0, t)$ under $W$–that is,

$$W(x_t \leq \lambda) = \frac{1}{\sqrt{2\pi t}} \int_{-\infty}^{\lambda} e^{-u^2/2t} \, du$$

—and (ii) for any $0 \leq t_0 \leq t_1 \leq \cdots \leq t_k \leq 1$, the random variables

$$x_{t_1} - x_{t_0}, x_{t_2} - x_{t_1}, \ldots, x_{t_k} - x_{t_{k-1}}$$

are independent under $W$. Here, the random variable $x_0$ is understood as equal to zero with probability 1 under $W$ [i.e., $W(x_0 = 0) = 1$]. A $C$-valued random variable, denoted by $W = (W_t, 0 \leq t \leq 1)$, is called a Wiener process if it has the Wiener measure as its distribution [i.e., $P(W \in B) = W(B)$ for any $B \in \mathcal{B}$]. Let $X_1, X_2, \ldots$ be i. i.d. random variables with $E(X_i) = 0$ and $E(X_i^2) = 1$. Let $S_n = \sum_{i=1}^{n} X_i$ denote the partial sum with $S_0 = 0$. Define a sequence of $C$-valued random variables $\xi_n = (\xi_{n,t}, 0 \leq t \leq 1)$ by

$$\xi_{n,t} = \frac{1}{\sqrt{n}} \{ S_{[nt]} + (nt - [nt]) X_{[nt]+1} \}, \tag{6.64}$$

where $[x]$ denotes the integer part of $x$ (i.e., the largest integer less than or equal to $x$). The invariance principle in CLT states the following.

**Theorem 6.18.** Suppose that the $X_i$'s are i.i.d. with mean 0 and variance 1. Then $\xi_n \xrightarrow{d} W$ as $n \to \infty$, where $\xi_n$ is defined by (6.64).

By Theorem 6.18 and the continuous mapping theorem (see Theorem 2.12; note that here we are dealing with $C$-valued random variables, but the same continuous mapping theorem applies), it follows that for any continuous mapping $g$ from $C$ to $R^k$ ($k \geq 1$), we have

$$g(\xi_n) \xrightarrow{d} g(W). \tag{6.65}$$

The name *invariance principle* came from the fact that the distribution of the right side of (6.65) does not depend on the specification of the $X_i$'s other than the first and second moments. On the other hand, it may take some effort to obtain the distribution of $g(W)$. However, because of the invariance principle, one may consider a special, simple sequence of i.i.d. random variables so that one can calculate the limiting distribution of $g(\xi_n)$ for the special sequence. It then follows that the same limiting distribution applies to any sequence of i.i.d. random variables having the same mean and variance. We illustrate this technique with an example.

*Example 6.10.* Suppose that one wishes to obtain the limiting distribution of $n^{-1/2} \max_{1 \leq i \leq n} S_i$, where $S_i$ is defined above. It can be shown that $g(x) = \sup_{0 \leq t \leq 1} x(t)$ is a continous mapping from $C$ to $R$ (Exercise 6.32). Furthermore, we have $g(\xi_n) = n^{-1/2} \max_{1 \leq i \leq n} S_i$ and $g(W) = \sup_{0 \leq t \leq 1} W_t$. Thus, by (6.65), the limiting distribution of $n^{-1/2} \max_{1 \leq i \leq n} S_i$ is the same as the distribution of $\sup_{0 \leq t \leq 1} W_t$. To calculate the latter, we consider one special sequence of i.i.d. random variables $X_i$, known as *random walk*, such that $P(X_i = 1) = P(X_i = -1) = 1/2$. For the random walk, it can be shown (e.g., Billingsley 1968, pp. 71–72) that, for $\lambda \geq 0$,

$$P \left( \frac{1}{\sqrt{n}} \max_{1 \leq i \leq n} S_i \leq \lambda \right) \longrightarrow \sqrt{\frac{2}{\pi}} \int_0^{\lambda} e^{-u^2/2} du. \tag{6.66}$$

Therefore, $P(\sup_{0 \leq t \leq 1} W_t \leq \lambda) = $ the right side of (6.66) for any $\lambda \geq 0$ and $P(\sup_{0 \leq t \leq 1} W_t \leq \lambda) = 0$ for $\lambda < 0$ (Exercise 6.32).

We now consider an invariance principle for LIL. Before we introduce the principle, let us recall the Hartman–Wintner LIL introduced in the previous section (Theorem 6.17), which states that if $X_1, X_2, \ldots$ are i.i.d. random variables with $E(X_1) = 0$ and $E(X_1^2) = 1$, then (6.51) and (6.52) hold, with $a_n = n$. In fact, the Hartman–Wintner law implies a seemingly stronger result. Let $L$ denote the set of limit points of the sequence

$$\eta_{n,1} = \frac{S_n}{\sqrt{2n \log \log n}}, \tag{6.67}$$

where $S_n = \sum_{i=1}^{n} X_i$ and $\log \log n$ is understood as 1 if $n < 3$. Then, under the conditions of the Hartman–Wintner LIL, we have with probability 1 that $L = [-1, 1]$. To see this, note that if $x_n, n \geq 1$, is a sequence of real numbers such that $\liminf x_n = -1$, $\limsup x_n = 1$, and $\lim_{n \to}(x_{n+1} - x_n) = 0$, then the set of limit points of $\{x_n\}$ must coincide with $[-1, 1]$ (Exercise 6.33). Intuitively, since the sequence has to visit $-1$ and 1 infinitely many times with a vanishing move each step, in between it has to visit every neighborhood of every point between $-1$ and 1, infinitely many times. Therefore, it suffices to show that $\Delta_n = \eta_{n+1,1} - \eta_{n,1} \xrightarrow{\text{a.s.}} 0$. Write

$$\Delta_n = \frac{X_{n+1}}{\sqrt{2(n+1) \log \log(n+1)}}$$

$$+ \left\{ \sqrt{\frac{n \log \log n}{(n+1) \log \log(n+1)}} - 1 \right\} \frac{S_n}{\sqrt{2n \log \log n}}$$

$$= \Delta_{1,n+1} + \Delta_{2,n}.$$

To show that $\Delta_{1,n} \xrightarrow{\text{a.s.}} 0$, we use the Borel–Cantelli lemma (Lemma 2.5). Note that there is an alternative statement of this lemma in that $\limsup A_n$ is the event that $A_n$ happens for infinitely many $n$, denoted by $A_n$ i.o. (here i.o. stands for "infinitely often"). Thus, (i) if $\sum_{n=1}^{\infty} P(A_n) < \infty$, then $P(A_n \text{ i.o.}) = 0$ and (ii) if $A_1, A_2, \ldots$ are pairwise independent and $\sum_{n=1}^{\infty} P(A_n) = \infty$, then $P(A_n \text{ i.o.}) = 1$. Here, we only need part (i), and we have

$$\sum_{n=1}^{\infty} P\left( |\Delta_{1,n}| > \frac{1}{\sqrt{2 \log \log n}} \right) = \sum_{n=1}^{\infty} P(|X_n| > \sqrt{n})$$

$$= \sum_{n=1}^{\infty} P(X_1^2 > n)$$

$$= \sum_{n=1}^{\infty} E\{1_{(X_1^2 > n)}\}$$

$$= E\left\{ \sum_{n=1}^{\infty} 1_{(n < X_1^2)} \right\}$$

$$\leq E(X_1^2) < \infty.$$

It follows that, with probability 1, $|\Delta_{1,n}| \leq (2 \log \log n)^{-1/2}$ for large $n$; hence, $\Delta_{1,n} \xrightarrow{\text{a.s.}} 0$. That $\Delta_{2,n} \xrightarrow{\text{a.s.}} 0$ follows from the Hartman–Wintner LIL and the fact that

$$\frac{n \log \log n}{(n+1) \log \log(n+1)} \longrightarrow 1 \quad \text{as } n \to \infty.$$

Strassen (1964) obtained a functional form of the Hartman–Wintner LIL, also known as the invariance principle in LIL, or almost sure invariance principle. Again, let $X_1, X_2, \ldots$ be a sequence of i.i.d. random variables with $E(X_1) = 0$ and $E(X_1^2) = 1$. We extend the definition of $\eta_{n,1}$ in (6.67) to a member $\eta_n = (\eta_{n,t}, 0 \leq t \leq 1)$ in $C$ by

$$\eta_{n,t} = \frac{\xi_{n,t}}{\sqrt{2 \log \log n}} \tag{6.68}$$

for $0 \leq t \leq 1$, where $\xi_{n,t}$ is defined by (6.64) [thus, (6.67) is, indeed, $\eta_{n,t}$ with $t = 1$]. Consider a subset $K$ of $C$ consisting of all absolutely continuous functions $x$ on $[0,1]$ such that $x(0) = 0$ and

$$\int_0^1 \{x'(t)\}^2 \, dt \leq 1. \tag{6.69}$$

Strassen's theorem states the following.

**Theorem 6.19.** Under the assumptions of Theorem 6.18, we have with probability 1 that the set of limit points of $\eta_n, n \geq 1$ with respect to the metric $\rho$ defined by (6.63) coincides with $K$.

Here, again, the invariance principle refers to the fact that the set of limit points $K$ does not depend on the specification of the sequence $X_i$ other than the first two moments. As an application of Theorem 6.19, we derive the "upper half" of Hartman and Wintner's LIL.

*Example 6.11.* Let the $X_i$'s be i.i.d. with mean $E(X_1) = 0$ and $E(X_1^2) = 1$. Consider the mapping (or functional) from $C$ to $R$ defined by $g(x) = x(1)$, $x \in C$. It is easy to show that $g$ is continuous. It follows by Theorem 6.19 that, with probability 1, the set of limit points of

$$g(\eta_n) = \frac{S_n}{\sqrt{2n \log \log n}}$$

is $g(K) = \{g(x) : x \in K\}$ (Exercise 6.34). Therefore, with probability 1, we have $\limsup g(\eta_n) = \sup g(K)$, the supremum of $g(K)$. It remains to show that $\sup g(K) = 1$. For any $x \in K$, since $x(0) = 0$, by the Cauchy–Schwarz inequality [see (5.60)], it can be shown that $x(1) \leq 1$ (Exercise 6.34). It follows that $\sup g(K) \leq 1$. On the other hand, the function $x(t) = t$ belongs to $K$ and it satisfies $g(x) = x(1) = 1$. Thus, we have $\sup g(K) = 1$.

### 6.6.2 Large deviations

The words "large deviations" in the classical framework usually refer to probabilities of the deviation of the sample mean of independent random variables from its expected value. Let $X_1, X_2, \ldots$ be a sequence of independent random variables and $S_n$ be the partial sum $\sum_{i=1}^{n} X_i$. There are two types of results. The first type is associated with the WLLN, which states that, under suitable conditions, the probability $P\{n^{-1}|S_n - E(S_n)| > \epsilon\}$ goes to zero as $n \to \infty$ for any $\epsilon > 0$. A further question is how fast does the probability goes to zero. In other words, we are concerned about the probabilistic convergence rate in WLLN. Such a problem has been encountered (see, for example, Section 5.5), but here we would like to find out a more precise description of the convergence rate. The second type of results is associated with the CLT. Consider, for example, the i.i.d. case so that $E(X_1) = 0$ and $E(X_1^2) = \sigma^2 \in (0, \infty)$. The CLT states that the probability $F_n(x) = P(S_n/\sigma\sqrt{n} \le x) \to \Phi(x)$ for every $x \in R$, where $\Phi(x)$ is the cdf of $N(0, 1)$. Clearly, for any fixed $x$, $F_n(x)$ does not go to 1 or, equivalently, $1 - F_n(x)$ does not go to zero, but what happens when $x \to \infty$ as $n \to \infty$? In other words, we are concerned with the convergence rate (to 1) of the probability $F_n(x_n)$, where $x_n$ is a sequence of nonnegative numbers such that $x_n \to \infty$ as $n \to \infty$. In this subsection we will focus mostly on the i.i.d. case.

*1. Probability of large deviation in WLLN.* This type of large deviation results have been developed following the landmark paper of Varadhan (1966), although the basic idea may be tracked back to Laplace and Cramér. Later, in a series of papers beginning in 1975, Donsker and Varadhan identified three levels of large deviations. Let $X_1, X_2, \ldots$ be a sequence of i.i.d. random variables such that $E(X_1) = \mu$. The level-1 large deviation is regarding the distribution of $n^{-1}S_n$, which we describe below. The level-2 and level-3 large deviations are regarding the empirical distribution and process generated by the i.i.d. sequence, which we will discuss in the next chapter. Let $F$ be the distribution of $X_1$. Let $c_F$ be the logarithm of the mgf of $X_1$; that is,

$$c_F(t) = \log\{E(e^{tX_1})\} = \log\left\{\int e^{tx} F(dx)\right\}, \tag{6.70}$$

which is assumed to exist for all $t \in R$. The funtion $c_F$ is known as the *cumulant generating function*, and it plays an important role in the following theorem of large deviations.

**Theorem 6.20.** Suppose that $c_F(t)$ is finite for all $t \in R$. Then

$$\limsup \frac{1}{n} \log\left\{P\left(\frac{S_n}{n} \in C\right)\right\} \le -\inf_{x \in C} I_F(x) \tag{6.71}$$

for every closed set $C \subset R$, and

$$\liminf \frac{1}{n} \log\left\{P\left(\frac{S_n}{n} \in O\right)\right\} \ge -\inf_{x \in O} I_F(x) \tag{6.72}$$

for every open set $O \subset R$, where

$$I_F(x) = \sup_{t \in R}\{tx - c_F(t)\}. \tag{6.73}$$

The function $I_F$ defined by (6.73) is called *entropy*. To obtain a condition that guarantees equality of the right sides of (6.71) and (6.72), we introduce the following definition. Let $S$ be a subset of an Euclidean space. $x$ is an interior point of $S$ if there is $\epsilon > 0$ such that $S_\epsilon(x) = \{y : |y - x| < \epsilon\} \subset S$; $x$ is a point of closure of $S$ if the distance $d(x, S) = \inf_{s \in S}|x - s| = 0$. The interior of $S$, denoted by $S^o$, is the set of all interior points of $S$; the closure of $S$, denoted by $\bar{S}$, is the set of all points of closure of $S$. An important fact about $S^o$ is that it is the largest open set contained in $S$. Similarly, $\bar{S}$ is the smallest closed set containing $S$. We call a Borel set $A \subset R$ an $I_F$-continuity set if $\inf_{x \in \bar{A}} I_F(x) = \inf_{x \in A^o} I_F(x)$. From Theorem 6.20, it immediately follows that if $A$ is an $I_F$-continuity set, then

$$\lim_{n \to \infty} \frac{1}{n} \log \left\{ P\left(\frac{S_n}{n} \in A\right)\right\} = -\inf_{x \in A} I_F(x). \tag{6.74}$$

We consider some examples.

*Example 6.12.* Suppose that $X_1, X_2, \ldots$ are independent Bernoulli$(1/2)$. It is easy to show (Exercise 6.35) that, in this case, $c_F(t) = \log(1 + e^t) - \log 2$. Furthermore, for any $x \in R$, the function $d_x(t) = xt - c_F(t)$ is strictly concave. For $x \in (0, 1)$, $d_x(t)$ attains its unique maximum at $t = \log\{x/(1 - x)\}$ with $I_F(x) = \log 2 + x \log x + (1 - x) \log(1 - x)$; for $x = 0$ or 1, the supremum of $d_x(t)$ is not attainable, but $I_F(0) = I_F(1) = \log 2$; for $x \notin [0, 1]$, we have $I_F(x) = \infty$. Now, consider the set $A = (-\infty, 1/2 - \epsilon) \cup (1/2 + \epsilon, \infty) \subset R$, where $\epsilon > 0$. Since $A$ is open, we have $A^o = A$. Furthermore, $\bar{A} = (-\infty, 1/2 - \epsilon] \cup [1/2 + \epsilon, \infty]$. If $\epsilon < 1/2$, then $1/2 - \epsilon > 0$ and $1/2 + \epsilon < 1$; therefore, $\inf_{x \in A^o} I_F(x) = \inf_{x \in \bar{A}} I_F(x) = I_F(1/2 - \epsilon) = I_F(1/2 + \epsilon)$ (Exercise 6.35). Hence $A$ is an $I_F$-continuity set. Therefore, by (6.74), we have

$$\lim_{n \to \infty} \frac{1}{n} \log \left\{ P\left(\left|\frac{S_n}{n} - \frac{1}{2}\right| > \epsilon\right)\right\}$$
$$= -\inf_{x \in A} I_F(x)$$
$$= -\left\{\log 2 + \left(\frac{1}{2} - \epsilon\right)\log\left(\frac{1}{2} - \epsilon\right) + \left(\frac{1}{2} + \epsilon\right)\log\left(\frac{1}{2} + \epsilon\right)\right\}$$
$$< 0.$$

On the other hand, if $\epsilon > 1/2$, then $1/2 - \epsilon < 0$ and $1/2 + \epsilon > 1$; hence, $\inf_{x \in A^o} I_F(x) = \inf_{x \in \bar{A}} I_F(x) = \infty$. Hence $A$ is, again, an $I_F$-continuity set. Therefore, by (6.74), we have

$$\lim_{n \to \infty} \frac{1}{n} \log \left\{ P \left( \left| \frac{S_n}{n} - \frac{1}{2} \right| > \epsilon \right) \right\} = - \inf_{x \in A} I_F(x)$$

$$= -\infty. \tag{6.75}$$

Finally, if $\epsilon = 1/2$, then $\inf_{x \in A^o} I_F(x) = \infty$, $\inf_{x \in \bar{A}} I_F(x) = \log 2$; hence, $A$ is not an $I_F$-continuity set. However, since $|S_n/n - 1/2|$ is always bounded by $1/2$, we have $P(|S_n/n - 1/2| > \epsilon) = 0$; hence, (6.75) continues to hold.

*Example 6.13.* Now, consider the case of normal distribution; that is, $X_1, X_2, \ldots$ are independent and distributed as $N(\mu, \sigma^2)$, where $\mu \in R$ and $\sigma^2 > 0$. In this case, we have $c_F(t) = \mu t + \sigma^2 t^2/2$, $t \in R$. Therefore, it is straightforward to show that $I_F(x) = (x - \mu)^2/2\sigma^2$. Now, consider the set $A = (-\infty, \mu - \epsilon) \cup (\mu + \epsilon, \infty)$, where $\epsilon > 0$. Since $I_F(x)$ is a continuous function for all $x$, it is easy to show that $A$ is an $I_F$-continuity set for any $\epsilon > 0$. It then follows by (6.74) that

$$\lim_{n \to \infty} \frac{1}{n} \log \left\{ P \left( \left| \frac{S_n}{n} - \frac{1}{2} \right| > \epsilon \right) \right\} = - \inf_{x \in A} I_F(x)$$

$$= -\frac{\epsilon^2}{2\sigma^2}.$$

An important application of the theory of large deviations is to obtain the convergence rate in WLLN. For example, in Example 6.13 one can write

$$\frac{1}{n} \log \left\{ P \left( \left| \frac{S_n}{n} - \frac{1}{2} \right| > \epsilon \right) \right\} = -\frac{\epsilon^2}{2\sigma^2} + o(1).$$

It follows that

$$P \left( \left| \frac{S_n}{n} - \frac{1}{2} \right| > \epsilon \right) = \exp \left[ \left\{ -\frac{\epsilon^2}{2\sigma^2} + o(1) \right\} n \right].$$

We will use such expressions in subsequent development.

2. *Probability of large deviation in CLT.* First assume that $X_1, X_2, \ldots$ is a sequence of i.i.d. random variables such that the moment generating function $E(e^{tX_1}) < \infty$ for $|t| < \delta$, where $\delta$ is a positive constant. Such a condition is known as Cramér's condition. Without loss of generality, we let $\mu = E(X_1) = 0$ and $\sigma^2 = \text{var}(X_1) > 0$. Again, write $S_n = \sum_{i=1}^{n} X_i$. We are concerned with the convergence of the probability $F_n(x) = P(S_n/\sigma\sqrt{n} \leq x)$ for large $x$. Recall that $\Phi(x)$ denotes the cdf of $N(0,1)$. The *cumulants* of $X_1 \sim F$ are defined as the derivatives of the cumulant generating function $c_F(t)$ at $t = 0$; that is, the $k$th cumulant of $X_1$ is $c_F^{(k)}(0)$, $k = 1, 2, \ldots$.

**Theorem 6.21.** Suppose that Cramér's condition is satisfied. Then for any $x \geq 0$ such that $x = o(\sqrt{n})$, we have

$$\frac{1 - F_n(x)}{1 - \Phi(x)} = \exp\left\{\frac{x^3}{\sqrt{n}}\lambda\left(\frac{x}{\sqrt{n}}\right)\right\}\left\{1 + O\left(\frac{x+1}{\sqrt{n}}\right)\right\}, \tag{6.76}$$

$$\frac{F_n(-x)}{\Phi(-x)} = \exp\left\{-\frac{x^3}{\sqrt{n}}\lambda\left(-\frac{x}{\sqrt{n}}\right)\right\}\left\{1 + O\left(\frac{x+1}{\sqrt{n}}\right)\right\}, \tag{6.77}$$

where $\lambda(t) = \sum_{k=0}^{\infty} a_k t^k$ is a power series with coefficients depending on the cumulants of $X_1$, which converges for sufficiently small $|t|$.

Theorem 6.21 can be extended to sequence of independent random variables not necessarily having the same distribution. Let $X_1, X_2, \ldots$ be independent with mean 0. Let $m_i(z) = \log\{E(e^{zX_i})\}$, where $z$ denotes a complex number. In other words, $m_i(\cdot)$ is the complex cumulant generating function of $X_i$. Here, log denotes the principal value of the logarithm so that $m_i(0) = 0$. We assume that there exists a circle, centered at the point $z = 0$, within which $m_i, i = 1, 2, \ldots$, are analytic. Then, within this circle, $m_i(z)$ can be expanded as a convergent power series

$$m_i(z) = \sum_{k=1}^{\infty} \frac{c_{ik}}{k!} z^k,$$

where $c_{ik}$ is the cumulant of order $k$ of $X_i$. Note that $c_{i1} = E(X_i) = 0$ and $c_{i2} = E(X_i^2) = \sigma_i^2$. Again, let $S_n = \sum_{i=1}^{n} X_i$, and also $s_n^2 - \sum_{i=1}^{n} \sigma_i^2$ and $F_n(x) = P(S_n/s_n \leq x)$. A power series $\sum_{i=1}^{\infty} a_i z^i$ is said to be majorized by another power series $\sum_{i=1}^{\infty} b_i z^i$ if $|a_i| \leq b_i$ for all $i$. The following theorem is an extension of Theorem 6.21.

**Theorem 6.22.** Suppose that there is $\delta > 0$ and constants $c_1, c_2, \ldots$ such that $|m_i(z)| \leq c_i$, $|z| < \delta$ for $i = 1, 2, \ldots$, $\limsup n^{-1}\sum_{i=1}^{n} c_i^{3/2} < \infty$, and $\liminf s_n^2/n > 0$. Then for any $x \geq 0$ such that $x = o(\sqrt{n})$, (6.76) and (6.77) hold with the latest definition of $F_n$ and $\lambda(\cdot)$ replaced by $\lambda_n(\cdot)$, where $\lambda_n(t) = \sum_{k=1}^{\infty} a_{nk} t^k$ is a power series, which, for sufficiently large $n$, is majorized by a power series whose coefficients do not depend on $n$ and is convergent in some circle, so that $\lambda_n(t)$ converges uniformly in $n$ for sufficiently small $|t|$.

The series $\lambda(t)$ in Theorem 6.21 is called the Cramér series, and the series $\lambda_n(t)$ in Theorem 6.22 is called the generalized Cramér series. For example, the coefficients $a_{nk}$ is expressed in terms of the cumulants of $X_1, \ldots, X_n$ of orders up to $k + 3$ for every $k$. In particular, letting $\gamma_{nk} = n^{-1}\sum_{i=1}^{n} c_{ik}$,

$$a_{n0} = \frac{\gamma_{n3}}{6\gamma_{n2}^{3/2}},$$

$$a_{n1} = \frac{\gamma_{n2}\gamma_{n4} - 3\gamma_{n3}^2}{24\gamma_{n2}^3},$$

$$a_{n2} = \frac{\gamma_{n2}^2\gamma_{n5} - 10\gamma_{n2}\gamma_{n3}\gamma_{n4} + 15\gamma_{n3}^3}{120\gamma_{n2}^{9/2}}.$$

Some uesful consequences of Theorem 6.22 are the following.

**Corollary 6.1.** Under the conditions of Theorem 6.22, if $x \geq 0$ and $x = O(n^{1/6})$, then

$$1 - F_n(x) = \{1 - \Phi(x)\} \exp\left(\frac{n\gamma_{n3}}{6s_n^3} x^3\right) + O\left(\frac{e^{-x^2/2}}{\sqrt{n}}\right),$$

$$F_n(-x) = \Phi(-x) \exp\left(-\frac{n\gamma_{n3}}{6s_n^3} x^3\right) + O\left(\frac{e^{-x^2/2}}{\sqrt{n}}\right).$$

One can see some similarity of the above result with the Edgeworth expansion (4.27). However, here the focus is large deviation (i.e., when $x$ is large up to a certain order of $n$). The result implies the following.

**Corollary 6.2.** Under the conditions of Theorem 6.22, if $c_{i3} = 0$, $i = 1, 2, \ldots$ and $|x| = O(n^{1/6})$, then

$$F_n(x) - \Phi(x) = O\left(\frac{e^{-x^2/2}}{\sqrt{n}}\right).$$

If $x$ is not restricted to $O(n^{1/6})$, we have the following results.

**Corollary 6.3.** Suppose that the conditions of Theorem 6.22 are satisfied.
(i) If $c_{i3} = 0$, $i = 1, 2, \ldots$, then for $x \geq 0$ and $x = o(n^{1/4})$, we have

$$\frac{1 - F_n(x)}{1 - \Phi(x)} \longrightarrow 1,$$

$$\frac{F_n(-x)}{\Phi(-x)} \longrightarrow 1$$

as $n \to \infty$.
(ii) If $x \to \infty$ such that $x = o(\sqrt{n})$, then

$$\frac{F_n(x + a/x) - F_n(x)}{1 - F_n(x)} \longrightarrow 1 - e^{-a} \tag{6.78}$$

as $n \to \infty$ for every $a > 0$.

We conclude this section by revisiting two previous examples.

*Example 6.13 (continued).* Consider the case $\mu = 0$ and $\sigma = 1$. Then we have $F_n(x) = \Phi(x)$, the cdf of $N(0, 1)$. It follows, by L'Hospital's rule, that

$$\lim_{x\to\infty} \frac{F_n(x+a/x) - F_n(x)}{1 - F_n(x)} = \lim_{x\to\infty} \frac{\Phi(x+a/x) - \Phi(x)}{1 - \Phi(x)}$$

$$= \lim_{x\to\infty} \frac{\phi(x) - (1 - a/x^2)\phi(x+a/x)}{\phi(x)}$$

$$= 1 - \lim_{x\to\infty} \left(1 - \frac{a}{x^2}\right) \frac{\phi(x+a/x)}{\phi(x)}$$

$$= 1 - \lim_{x\to\infty} \left(1 - \frac{a}{x^2}\right) \exp\left(-a - \frac{a^2}{2x^2}\right)$$

$$= 1 - e^{-a},$$

where $\phi(x) = \Phi'(x) = e^{-x^2/2}/\sqrt{2\pi}$. Note that in this case the limit is derived without using Corollary 6.3, and the result holds without any restriction on how fast $x \to \infty$. Of course, this is a very special in which $F_n$ does not depend $n$. In fact, if $X_1, X_2, \ldots$ are i.i.d. with mean 0 and variance 1, the only possibility that $F_n$ does not depend on $n$ is that $F_n = \Phi$ (why?).

The next example shows somewhat the contrary.

*Example 6.12 (continued).* Let $\tilde{X}_i = X_i - 1/2$. Then we have $E(\tilde{X}_i) = 0$ and $E(\tilde{X}_i)^2 = 1/4$; thus, $\tilde{s}_n^2 = \sum_{i=1}^n E(\tilde{X}_i^2) = n/4$. Let $\tilde{S}_n = \sum_{i=1}^n \tilde{X}_i$. Then $F_n(x) = P(\tilde{S}_n/\tilde{s}_n \le x) = P\{\tilde{S}_n \le (\sqrt{n}/2)x\} = P(\sum_{i=1}^n X_i \le n/2 + (\sqrt{n}/2)x\}$. Now, consider $x = (n-1)/\sqrt{n}$ [which is $O(\sqrt{n})$ instead of $o(\sqrt{n})$] and $a = 1/2$. Then we have $F_n(x) = P(\sum_{i=1}^n X_i \le n-1/2)$ and $F_n(x+a/x) = P\{\sum_{i=1}^n X_i \le n - (n-2)/4(n-1)\}$. If $n > 2$, then $0 < (n-2)/4(n-1) < 1/4$; hence, $F_n(x) = F_n(x+a/x) = P(\sum_{i=1}^n X_i \le n-1) < 1$ (because $\sum_{i=1}^n X_i$ is an integer). It follows that the left side of (6.78) is identical to zero for any $n > 2$, and therefore cannot converge to the right side, which is $1 - e^{-0.5} > 0$. This example shows that the requirement $x = o(\sqrt{n})$ cannot be dropped.

## 6.7 Case study: The least squares estimators

The least squares (LS) method was first introduced by Carl Friedrich Gauss, one of the greatest mathematicians of all time, in 1795, when he was just 18 years old. In 1801, Gauss used his LS method to accurately compute the orbit of the then newly discovered asteroid Ceres. The latter was discovered by Italian astronomer Giuseppe Piazzi, who was able to track its path for 40 days before it got lost in the glare of the sun. Gauss's LS prediction, which was quite different compared to all of the previous solutions that had been proposed, successfully allowed Hungarian astronomer Franz Xaver von Zach to relocate Ceres after it reemerged from behind the sun and therefore confirm Piazzi's assumption that his most famous discovery was, indeed, "better than a comet" (e.g., Federà Serio et al. 2002, p. 19).

The typical situation that the LS method applies is called regression analysis, which has been encountered several times so far in this book (e.g., Example 5.8). Here, we assume that the observation $Y_i$ is associated with a vector of known covariates $x_i$ through the following equation:

$$Y_i = x_i'\beta + \epsilon_i, \quad i = 1, \ldots, n, \tag{6.79}$$

where $\beta$ is a vector of unknown regression coefficients and $\epsilon_i$ represents an error. It is assumed that the errors are i.i.d. with mean 0 and constant variance $\sigma^2 > 0$. Let $Y = (Y_i)_{1 \le i \le n}$ be the vector of observations, $X = (x_i')_{1 \le i \le n}$ be the matrix of covariates, and $\epsilon$ be the vector of errors. Then the linear regression (6.79) can be expressed as

$$Y = X\beta + \epsilon. \tag{6.80}$$

The LS method finds the regression coefficients $\beta$ that minimize $|Y - X\beta|^2 = \sum_{i=1}^{n}(Y_i - x_i'\beta)^2$. For simplicity, assume that $X$ is of full rank $p$. Then the solution, which is called the LS estimator (LSE) of $\beta$, can be expressed as

$$\hat{\beta} = (X'X)^{-1}X'Y. \tag{6.81}$$

The LSE has several nice properties. For example, the Gauss–Markov theorem states that $\hat{\beta}$ is the best linear unbiased estimator (BLUE) of $\beta$; under the normality assumption, $\hat{\beta}$ is the same as the MLE of $\beta$; and $\hat{\beta}$ is consistent and asymptotically normal. The latter are the main subjects of the current section, among other large sample properties of the LSE.

First, consider consistency of LSE. Let $\beta = (\beta_j)_{1 \le j \le p}$. Then we have the following expression (Exercise 6.36):

$$\hat{\beta}_j - \beta_j = \sum_{i=1}^{n} \delta_j'(X'X)^{-1}x_i\epsilon_i, \quad 1 \le j \le p, \tag{6.82}$$

where $\delta_j$ is the $p$-dimensional vector whose $j$th component is one and other components are zero. Fix $1 \le j \le p$. Let $X_{ni} = \delta_j'(X'X)^{-1}x_i\epsilon_i$. Then for each $n$, $X_{ni}$, $1 \le i \le n$, are independent. According to Theorem 6.4, to show $\hat{\beta}_j \xrightarrow{\text{P}} \beta_j$ or, equivalently, $\sum_{i=1}^{n} X_{ni} \xrightarrow{\text{P}} 0$, it suffices to verify conditions (6.15)–(6.17). First look at (6.15). By Chebychev's inequality, we have

$$\sum_{i=1}^{n} \text{P}(|X_{ni}| > \epsilon) \le \frac{1}{\epsilon^2} \sum_{i=1}^{n} \text{E}(X_{ni}^2)$$

$$= \frac{\sigma^2}{\epsilon^2} \sum_{i=1}^{n} \delta_j'(X'X)^{-1}x_ix_i'(X'X)^{-1}\delta_j$$

$$= \frac{\sigma^2}{\epsilon^2} \delta_j'(X'X)^{-1} \left( \sum_{i=1}^{n} x_ix_i' \right) (X'X)^{-1}\delta_j$$

$$= \frac{\sigma^2}{\epsilon^2} \delta_j'(X'X)^{-1}\delta_j$$

because $\sum_{i=1}^{n} x_i x_i' = X'X$. Thus, (6.15) holds provided that

$$\delta_j'(X'X)^{-1}\delta_j \longrightarrow 0. \tag{6.83}$$

In fact, the above arguments show that (6.83) implies $\sum_{i=1}^{n} E(X_{ni}^2) \to 0$, which also implies (6.16) and (6.17) (Exercise 6.36). It is now clear that a sufficient condition for consistency of the LSE is (6.83) for every $1 \leq j \leq p$, which is equivalent to

$$\mathrm{tr}\left\{(X'X)^{-1}\right\} \longrightarrow 0 \tag{6.84}$$

(Exercise 6.36). We consider a simple example.

*Example 6.14.* The case $p = 1$ is called simple linear regression. In this case, (6.79) can be expressed as $Y_i = \beta_0 + \beta_1 x_i + \epsilon_i$, where $x_i$ is the covariate; $\beta_0$ and $\beta_1$ are called the intercept and slope (of the regression), respectively. It follows that $X = (1_n\ x)$, where $x = (x_i)_{1 \leq i \leq n}$. Straightforward calculation shows that $X'X = \begin{pmatrix} n & x. \\ x. & x_.^2 \end{pmatrix}$, where $x. = \sum_{i=1}^{n} x_i$ and $x_.^2 = \sum_{i=1}^{n} x_i^2$; hence,

$$\mathrm{tr}\left\{(X'X)^{-1}\right\} = \frac{1 + \overline{x^2}}{\sum_{i=1}^{n}(x_i - \bar{x})^2},$$

where $\bar{x} = n^{-1}x.$ and $\overline{x^2} = n^{-1}x_.^2$. Therefore, (6.84) holds if and only if $\sum_{i=1}^{n}(x_i - \bar{x})^2 \to \infty$ and $\overline{x^2} = o[\sum_{i=1}^{n}(x_i - \bar{x})^2]$ as $n \to \infty$. For the most part, these assumptions mean that there should be "enough" total variation among the covariates $x_i$ (Exercise 6.36). To see why the assumptions make sense, imagine the extreme opposite where there is no variations among the $x_i$'s (i.e., $x_i = c$, $1 \leq i \leq n$ for some constant $c$). Then the model becomes $Y_i = \beta_0 + \beta_1 c + \epsilon_i$, $1 \leq i \leq n$. Clearly, there is no way one can separate $\beta_0$ and $\beta_1$ from $\beta_0 + \beta_1 c$ if both parameters are unknown. In other words, the parameters $\beta_0$ and $\beta_1$ are not identifiable. Therefore, the LSE (or any other estimators) of these parameters cannot be consistent.

We now consider asymptotic normality of the LSE. By (6.81) we have $\mathrm{Var}(\hat{\beta}) = \sigma^2(X'X)^{-1}$. This suggests that

$$\left(\frac{X'X}{\sigma^2}\right)^{1/2}(\hat{\beta} - \beta) \xrightarrow{d} N(0, I_p), \tag{6.85}$$

where $I_p$ is the $p$-dimensional identity marix (see Appendix A.1 for the definition of $A^{1/2}$ for $A \geq 0$). To show (6.85), we apply Theorem 2.14, and thus to show that for any $\lambda \in R^p$, we have

$$\lambda'\left(\frac{X'X}{\sigma^2}\right)^{1/2}(\hat{\beta} - \beta) \xrightarrow{d} N(0, \lambda'\lambda). \tag{6.86}$$

Without loss of generality, let $\lambda \neq 0$. Then, again, without loss of generality, we may assume that $|\lambda|^2 = \lambda'\lambda = 1$ (why?). (Note how we simplify the arguments step-by-step by using the words "without loss of generality," but make sure that at each step it is, indeed, without loss of generality). Then, similar to (6.82), the left side of (6.86) can be expressed as

$$\frac{\sum_{i=1}^{n} c_{ni}\epsilon_i}{\sigma\sqrt{\sum_{i-1}^{n} c_{ni}^2}} \tag{6.87}$$

with $c_{ni} = \lambda'(X'X)^{-1/2}x_i$ (note that $\sum_{i=1}^{n} c_{ni}^2 = 1$) (Exercise 6.37). According to the Hájek–Sidak theorem (Example 6.6), (6.87) converges in distribution to $N(0,1)$ provided that (6.38) holds, which is equivalent to

$$\max_{1 \leq i \leq n} \lambda'(X'X)^{-1/2}x_i x_i (X'X)^{-1/2}\lambda \longrightarrow 0. \tag{6.88}$$

A sufficient condition for (6.88) to hold for every $\lambda \in R^p$, and hence for the asymptotic normality of the LSE in the sense of (6.85), is thus

$$\max_{1 \leq i \leq n} \{x_i'(X'X)^{-1}x_i\} \longrightarrow 0 \tag{6.89}$$

(Exercise 6.37). We revisit the previous example.

*Example 6.14 (continued).* In this case, it can be shown that the left side of (6.89) is equal to

$$\frac{1}{n} + \frac{\max_{1 \leq i \leq n}(x_i - \bar{x})^2}{\sum_{i=1}^{n}(x_i - \bar{x})^2}. \tag{6.90}$$

Thus, the condition for asymptotic normality of the LSE is that

$$\frac{\max_{1 \leq i \leq n}(x_i - \bar{x})^2}{\sum_{i=1}^{n}(x_i - \bar{x})^2} \longrightarrow 0$$

as $n \to \infty$. Intuitively, this means that the contribution to the total variation by any individual is relatively small compared to the total variation.

Other large-sample properties of the LSE have also been studied. For example, Lai et al. (1979) studied strong consistency property of the LSE. In fact, the authors derived (strong) convergence rate for each component of the LSE. It is assumed that the errors $\epsilon_1, \epsilon_2, \ldots$ in (6.79) is a sequence of random variables such that $\sum_{i=1}^{\infty} c_i\epsilon_i$ converges a.s. for any sequence of real numbers $c_1, c_2, \ldots$ such that $\sum_{i=1}^{\infty} c_i^2 < \infty$. This assumption is weaker than the assumptions we made earlier [below (6.79)]. For example, if $\epsilon_i$, $i \geq 1$, are i.i.d. such that $E(\epsilon_i) = 0$ and $E(\epsilon_i^2) < \infty$, then the above assumption is satisfied (Exercise 6.38). Let $v_{n,jj}$ be the $j$th diagonal element of $(X'X)^{-1}$, $1 \leq j \leq p$. [Note that the matrix $X$ depends on $n$ (i.e., $X = X_n$), but for notation simplicity the subscript $n$ is suppressed; the same note also applies to other notations such as $\hat{\beta}$.] If for any $1 \leq j \leq p$, we have $\lim_{n\to\infty} v_{n,jj} = 0$, then for any $\delta > 0$, we have with probability 1 that

$$\hat{\beta}_j - \beta_j = o\left(\sqrt{v_{n,jj}|\log v_{n,jj}|^{1+\delta}}\right) \tag{6.91}$$

as $n \to \infty$. Thus, if $\lim_{n\to\infty} v_{n,jj} = 0$, $1 \leq j \leq p$, then the LSE is strongly consistent in that $\hat{\beta} \xrightarrow{\text{a.s.}} \beta$, and the (strong) convergence rate for each component of the LSE is given by (6.91).

A more accurate rate of convergence is given by Lai and Wei (1982), who derived a LIL for LSE. Suppose that the $\epsilon_i$'s are independent with $E(\epsilon_i) = 0$, $E(\epsilon_i^2) = \sigma^2$, and $\sup_{i\geq 1} E(|\epsilon_i|^r) < \infty$ for some $r > 2$. Also suppose that $p \geq 2$. Let $X_j$ denote the $j$th column of $X$ and let $X_{-j}$ denote the matrix of $X$ without the $j$th column, $1 \leq j \leq p$. Let $v_{n,j} = P_{X_{-j}^\perp} X_j = (v_{n,j,i})_{1\leq i\leq n}$, where $P_{X_{-j}^\perp} = I - P_{X_{-j}} = I - X_{-j}(X'_{-j}X_{-j})^{-1}X'_{-j}$, and $a_{n,j} = |v_{n,j}|^2$. Fix $1 \leq j \leq p$. If $\lim_{n\to\infty} a_{n,j} = \infty$, $\limsup a_{n+1,j}/a_{n,j} < \infty$, and $\max_{1\leq i\leq n} v_{n,j,i}^2 = o[a_{n,j}(\log a_{n,j})^{-\rho}]$ for all $\rho > 0$, then we have

$$\limsup \left(\frac{a_{n,j}}{2\log\log a_{n,j}}\right)^{1/2} |\hat{\beta}_j - \beta_j| = \sigma \text{ a.s.} \tag{6.92}$$

The normalizing sequence $a_n$ has an intuitive explanation. It is the squared norm of the projection of the vector of covariates corresponding to $\beta_j$ to the space orthogonal to that spanned by the rest of the (vectors of) covariates. Roughly speaking, $a_n$ is a measure of the amount of uncertainty associated with the estimation of $\beta_j$. The amount of uncertainty is closely related to the "effective sample size." To see this, consider an extreme case where there is no uncertainty among the observations. Then all one needs is one sample; that is, the effective sample size is one. On the other hand, the more uncertainty in the sample, the larger the effective sample size has to be in order to achieve the same level of accuracy in estimation.

## 6.8 Exercises

6.1. Show that in Example 6.2, (6.6)–(6.8) are satisfied with $a_n = (\sum_{i=1}^n \lambda_i)^\gamma$ and $\gamma > 1/2$.

6.2. Show that (6.10) and (6.11) together are equivalent to (6.13).

6.3. Use Theorem 6.1 to derive the classical result (6.1).

6.4. This exercise is regarding Example 6.2 and its continuation in Section 6.3.

(i) If $a_n = n^p$, show that (6.25) holds if and only if $p > 1/2$.

(ii) If $a_n = (\sum_{i=1}^n \lambda_i)^\gamma$, show that (6.25) holds if and only if $\gamma > 1/2$.

(iii) Suppose that the assumption that $a \leq \lambda_i \leq b$ for $a, b > 0$ is not made. Instead, the only assumption is that $\lambda_i > 0$ for all $i$. Does the result of (i) necessarily hold?

*6.5.* This exercise is regarding Example 6.1 (continued) in Section 6.3.

(i) Show that the function $\psi(u)$ is maximized at $u = \sqrt{c}$, and the maximum is $(1 + \sqrt{c})^{-2}$.

(ii) Show that the right side of (6.29) is minimized when $b_i = 1 + a_i$, which is greater than $\sqrt{a_i}$, and the minimum is $4\{\pi(1 + a_i)\}^{-1}$.

*6.6.* Suppose that $Y_1, Y_2, \ldots$ are independent random variables. In the following cases, find the conditions for $a_n$ such that

$$\frac{1}{a_n} \sum_{i=1}^{n} \{Y_i - E(Y_i)\} \xrightarrow{P} 0.$$

Give at least one specific example in each case.

(i) $Y_i \sim DE(\mu_i, \sigma_i)$, $i \geq 1$, where $DE(\mu, \sigma)$ is the Double Exponential distribution with pdf $f(x|\mu, \sigma) = (1/2\sigma)e^{-|x-\mu|/\sigma}$, $-\infty < x < \infty$, and $\sigma_i > 0$.

(ii) $Y_i \sim \text{Uniform}[\mu_i - d_i, \mu_i + d_i]$, $i \geq 1$, where $\text{Uniform}[a, b]$ represents the Uniform distribution over $[a, b]$, and $d_i > 0$.

*6.7* (Binomial method of moments). The method of moments (MoM) is widely used to obtained consistent estimators for population parameters. Consider the following special case, in which the observations $X_1, \ldots, X_n$ are i.i.d. with the Binomial$(m, p)$ distribution, where both $m$ and $p$ are unknown. The MoM equates the sample first and second moments of the observations to their expected values. This leads to the following equations:

$$E(X_1) = \bar{X},$$

$$E(X_1^2) = \frac{1}{n} \sum_{i=1}^{n} X_i^2.$$

Note that the left sides of these equations depend on $m$ and $p$. By solving the equations, one obtains the solutions, say, $\hat{m}$ and $\hat{p}$.

(i) Solve the MoM equations to find the solutions $\hat{m}$ and $\hat{p}$.

(ii) Show that $\hat{m}$ and $\hat{p}$ are consistent estimators; that is, $\hat{m} \xrightarrow{P} m$ and $\hat{p} \xrightarrow{P} p$ as $n \to \infty$.

(iii) Is $\hat{m}$ necessarily an integer? Since $m$ needs to be an integer, a modified estimator of $m$ is $\tilde{m}$, defined as the nearest integer to $\hat{m}$. Show that $\tilde{m}$ is also a consistent estimator of $m$ in that $P(\tilde{m} = m) \to 1$ as $n \to \infty$.

*6.8.* Suppose that for each $n$, $X_{ni}$, $1 \leq i \leq n$, are independent with the common cdf $F_n$, and $F_n \xrightarrow{w} F$, where $F$ is a cdf and the weak convergence ($\xrightarrow{w}$) is defined in Chapter 1 above Example 1.6. Define the empirical distribution of $X_{ni}$, $1 \leq i \leq n$, as

$$\hat{F}_n(x) = \frac{1}{n} \sum_{i=1}^{n} 1_{(X_{ni} \leq x)}.$$

$$= \frac{\#\{1 \le i \le n : X_{ni} \le x\}}{n}.$$

Show that $\hat{F}_n(x) \xrightarrow{\text{P}} F(x)$ for every $x$ at which $F$ is continuous.

6.9. Give an example of a sequence of independent random variables $X_1, X_2, \ldots$ such that $\sum_{i=1}^{\infty} \text{var}(X_i)/i^2 = \infty$ and the SLLN is not satisfied.

6.10. A sequence of real numbers $x_i \in [0, 1], i \ge 1$, is said to be uniformly distributed in Weyl's sense on $[0, 1]$ if for any Riemann integrable function $f$ on $[0, 1]$ we have

$$\lim_{n \to \infty} \frac{f(x_1) + \cdots + f(x_n)}{n} = \int_0^1 f(x)\, dx.$$

Let $X_i, i \ge 1$, be independent and Uniform$[0, 1]$ distributed. Show that the sequence $X_i, i \ge 1$, is uniformly distributed in Weyl's sense on $[0, 1]$ almost surely. (Hint: Use §1.5.2.37. Note that, by definition, a Riemann integrable function on $[0, 1]$ is necessarily bounded.)

6.11. Suppose that $X_1, X_2, \ldots$ is a sequence of independent random variables with finite expectation. Show that if

$$\sum_{i=1}^{\infty} \frac{1}{i} \text{E}\{|X_i - \text{E}(X_i)|\} < \infty,$$

then the SLLN holds; that is

$$\frac{1}{n} \sum_{i=1}^{n} \{X_i - \text{E}(X_i)\} \xrightarrow{\text{a.s.}} 0.$$

6.12. Let $Y_1, Y_2, \ldots$ be independent with $Y_i \sim$ Bernoulli$(p_i)$, $i \ge 1$. Show that $\sum_{i=1}^{\infty} (Y_i - p_i)$ converges a.s. if and only if $\sum_{i=1}^{\infty} p_i(1 - p_i) < \infty$.

6.13. Show that the Liapounov condition implies the Lindeberg condition; that is, if (6.35) holds for some $\delta > 0$, then (6.34) holds for every $\epsilon > 0$.

6.14. This exercise is associated with the proof of Theorem 6.14. Parts (i)–(iii) are regarding the necessity part, where $X_{ni} = (X_i - \mu)/\sqrt{n}$; whereas part (iv) is regarding the sufficiency part, where $X_{ni}$ is defined by (6.47).

(i) Show that for any $\epsilon > 0$,

$$\max_{1 \le i \le n} \text{P}(|X_{ni}| > \epsilon) = \text{P}(|X_1 - \mu| > \epsilon \sqrt{n}) \to 0.$$

(Note: You may not use Chebyshev's inequality to show this—why?)

(ii) Show that (6.43) with $\epsilon = 1$ reduces to

$$\sqrt{n}\text{E}\{(X_1 - \mu)1_{(|X_1 - \mu| \le \sqrt{n})}\} \to 0.$$

(iii) Show that

$$\text{E}\{(X_1 - \mu)^2 1_{(|X_1 - \mu| \le \sqrt{n})}\} - [\text{E}\{(X_1 - \mu)1_{(|X_1 - \mu| \le \sqrt{n})}\}]^2 \longrightarrow \sigma^2;$$

hence, $\mathrm{E}\{(X_1 - \mu)^2\} = \lim_{n\to\infty} \mathrm{E}\{(X_1 - \mu)^2 1_{(|X_1-\mu|\le\sqrt{n})}\} = \sigma^2$.

(iv) Show that for any $\epsilon > 0$,

$$\sum_{i=1}^n \mathrm{P}(|X_{ni}| > \epsilon) \le n\mathrm{P}(|X_1| > \lambda_n) \to 0.$$

*6.15.* Show that if $X_1, \ldots, X_n$ are independent Cauchy$(0, 1)$, the sample mean $\bar{X} = n^{-1}(X_1 + \cdots + X_n)$ is also Cauchy$(0, 1)$. Therefore, the CLT does not hold; that is, $\sqrt{n}\bar{X}$ does not converge to a normal distribution.

*6.16* (Sample median). Let $X_1, \ldots, X_n$ be i.i.d. observations with the distribution $\mathrm{P}(X_1 \le x) = F(x - \theta)$, where $F$ is a cdf such that $F(0) = 1/2$; hence, $\theta$ is the median of the distribution of $X_1$. Suppose that $n$ is an odd number: $n = 2m + 1$, say. If $X_{(1)} \le \cdots \le X_{(n)}$ denotes the ordered $X_i$'s, the sample median is defined as $X_{(m)}$. We assume that $F$ has a desity $f$ with respect to the Lebesgue measure such that $f(0) > 0$.

(i) For any $x \in R$, let $S_{n,x}$ be the number of $X_i$'s exceeding $x/\sqrt{n}$. Show that $X_{(m)} \le x/\sqrt{n}$ if and only if $S_{n,x} \le m - 1$.

(ii) Show that $\sqrt{n}\{X_{(m)} - \theta\} \xrightarrow{d} N(0, \sigma^2)$, where $\sigma^2 = \{2f(0)\}^{-2}$.

*6.17.* Suppose that $S_n$ is distributed as Poisson$(\lambda_n)$, where $\lambda_n \to \infty$ as $n \to \infty$. Use two different methods to show that $S_n$ obeys the CLT; that is, $\xi_n = \lambda_n^{-1/2}(S_n - \lambda_n) \xrightarrow{d} N(0, 1)$.

(i) Show that the mgf of $\xi_n$ converges to the mgf of $N(0, 1)$.

(ii) Let $Y_{ni}$, $1 \le i \le n$, be independent and distributed as Poisson$(n^{-1}\lambda_n)$, $n \ge 1$. Show that $\sum_{i=1}^n Y_{ni}$ has the same distribution as $S_n$. Furthermore, show that $X_{ni} = \lambda_n^{-1/2}(Y_{ni} - n^{-1}\lambda_n)$ satisfy Liapounov's condition (6.37) with $\delta = 2$. [Hint: You may use the fact that the fourth central moment of Poisson$(\lambda)$ is $\lambda + 3\lambda^2$.]

*6.18.* Let $Y_1, Y_2, \ldots$ be independent such that $Y_i \sim$ Poisson$(a^i)$, $i \ge 1$, where $a > 1$. Let $X_i = Y_i - a^i$, $i \ge 1$, and $s_n^2 = \sum_{i=1}^n \mathrm{var}(Y_i) = \sum_{i=1}^n a^i = (a - 1)^{-1}(a^{n+1} - 1)$.

(i) Show that Liapounov's condition (6.35) is not satisfied with $\delta = 2$.

(ii) Show that as $n \to \infty$,

$$\left(\frac{a - 1}{a^{n+1} - 1}\right)^{1/2} \sum_{i=1}^n (X_i - a^i) \xrightarrow{d} N(0, 1).$$

(Hint: Use the result of the previous exercise.)

*6.19.* Let the random variables $Y_1, Y_2, \ldots$ be independent and distributed as Bernoulli$(i^{-1})$, $i \ge 1$. Show that

$$\frac{\sum_{i=1}^n Y_i - \log n}{\sqrt{\log n}} \xrightarrow{d} N(0, 1).$$

(Hint: You may recall that $\sum_{i=1}^n i^{-1} - \log n$ converges to a limit known as Euler's constant. The actual value of the constant does not matter.)

6.20 (The delta method). The CLT is often used in conjunction with the delta method introduced in Example 4.4. Here, we continue with Exercise 6.7. Let $\hat{m}$ and $\hat{p}$ be the MoM estimator of $m$ and $p$, respectively, therein.

(i) Show that as $n \to \infty$,

$$\sqrt{n}\left[n^{-1}\sum_{i=1}^{n} X_i^2 - mp\{1 + (m-1)p\}\right] \xrightarrow{\text{d}} N(0, \Sigma),$$

where $\Sigma$ is a covariance matrix. Find $\Sigma$. (Hint: Use Theorem 2.14.)

(ii) Show that the MoM estimators $\hat{m}$ and $\hat{p}$ are jointly asymptotically normal in the sense that

$$\sqrt{n}\begin{pmatrix} \hat{m} - m \\ \hat{p} - p \end{pmatrix} \xrightarrow{\text{d}} N(0, V),$$

where $V$ is another covariance matrix. Find $V$.

(iii) An alternative estimator of $m$ was found (i.e., $\tilde{m}$). Show that $\tilde{m}$ is not asymptotically normal even though it is consistent; that is, $\sqrt{n}(\tilde{m} - m)$ does not converge in distribution to a normal distribution.

6.21. Suppose that for each $n$, $X_{ni}$, $1 \le i \le i_n$, are independent such that $P(X_{ni} = 0) = 1 - p_{ni}$, $P(X_{ni} = -a_{ni}) = P(X_{ni} = a_{ni}) = p_{ni}/2$, where $a_{ni} > 0$ and $0 < p_{ni} < 1$. Suppose that $\max_{1 \le i \le i_n} a_{ni} \to 0$ as $n \to \infty$. Find a necessary and sufficient condition for the triangular array $X_{ni}$ to obey the CLT; that is, $\sum_{i=1}^{i_n} X_{ni} \xrightarrow{\text{d}} N(\mu, \sigma^2)$ as $n \to \infty$ for some $\mu$ and $\sigma^2$.

6.22. This exercise is related to Example 6.5 (continued) at the end of Section 6.4.

(i) Show that for the sequence $Y_i, i \ge 1$, (6.40) fails provided that $s^2 = \sum_{i=1}^{\infty} p_i(1 - p_i) < \infty$. Also show that $s^2 > 0$.

(ii) Show that for $X_i = Y_i - p_i$, the three series (6.30)–(6.32) converge for $c = 1$.

6.23. This exercise is related to Example 6.9.

(i) Show that there is $c > 0$ such that $\sigma_i^2 = \mathrm{E}(X_i^2) \le c$ for all $i$; hence, $a_n \propto n$.

(ii) Show that the right side of (6.60) goes to zero as $n \to \infty$.

(iii) Show (6.61) [Hint: You may use the result of part (i)].

6.24. Let $X$ be a random variable. Show that for any $\mu \in R$, the following two conditions (i) and (ii) are equivalent:

(i) $n\mathrm{P}(|X - \mu| > n) \to 0$ and $\mathrm{E}\{(X - \mu)1_{(|X-\mu|\le n)}\} \to 0$;

(ii) $n\mathrm{P}(|X| > n) \to 0$ and $\mathrm{E}\{X1_{(|X|\le n)}\} \to \mu$.

(iii) Use the equivalence of (i) and (ii) to show the necessary and sufficient condition for WLLN given at the end of Section 6.5.

(iv) Give an example of a random variable $X$ such that $n\mathrm{P}(|X| > n) \to 0$ and $\mathrm{E}\{X1_{(|X|\le n)}\} = 0$ for any $n \ge 1$ and $\mathrm{E}(|X|) = \infty$.

6.25. Let $X_1, X_2, \ldots$ be a sequence of independent random variables with mean 0. Let $p_i = \mathrm{P}(|X_i| > b_i)$, where $b_i$ satisfies (6.50), and

$$a_n = \sum_{i=1}^{n} \mathrm{var}\{X_i 1_{(|X_i| \le b_i)}\}.$$

Suppose that $\sum_{i=1}^{\infty} \mathrm{var}\{X_i 1_{(|X_i| \le b_i)}\} = \infty$, $\sum_{i=1}^{\infty} p_i < \infty$, and

$$\frac{\sum_{i=1}^{n} \mathrm{E}\{X_i 1_{(|X_i| > b_i)}\}}{\sqrt{a_n \log\log a_n}} \longrightarrow 0$$

as $n \to \infty$. Show that $X_i$, $i \ge 1$ obeys the LIL (6.51). [Hint: Use Theorem 6.15 and the Borel–Cantelli lemma (Lemma 2.5).]

*6.26.* Show that if $X_1, X_2, \ldots$ are independent such that $X_i \sim N(0, \sigma_i^2)$, where $a \le \sigma_i^2 \le b$ and $a$ and $b$ are positive constants, then $X_i$, $i \ge 1$ obeys the LIL (6.51).

*6.27.* Suppose that $X_i$, $i \ge 1$, are independent random variables such that $\mathrm{P}(X_i = -i^\alpha) = \mathrm{P}(X_i = i^\alpha) = 0.5 i^{-\beta}$ and $\mathrm{P}(X_i = 0) = 1 - i^{-\beta}$, where $\alpha, \beta > 0$. According to Theorem 6.16, find the condition for $\alpha, \beta$ so that $X_i$, $i \ge 1$, obeys the LIL (6.51).

*6.28.* Let $Y_1, Y_2, \ldots$ be independent such that $Y_i \sim \chi_i^2$. Define $X_i = Y_i - i$. Does the sequence $X_i$, $i \ge 1$, obey the LIL (6.51), where $a_n = \sum_{i=1}^{n} \mathrm{var}(Y_i)$? [Hint: You may use the facts that if $Y \sim \chi_r^2$, then $\mathrm{E}(Y) = r$, $\mathrm{var}(Y) = 2r$, and $\mathrm{E}(Y - r)^4 = 12r(r + 4)$.]

*6.29.* We see that, in the i.i.d. case, the same condition (i.e., a finite second moment) is necessary and sufficient for both CLT and LIL. In other words, a sequence of i.i.d. random variables obeys the CLT if and only if it obeys the LIL. It is a different story, however, if the random variables are independent but not identically distributed. For example, Wittmann (1985) constructed the following example. Let $n_k$ be an infinite sequence of integers such that $n_{k+1} > 2n_k$, $k \ge 1$. Let $X_1, X_2, \ldots$ be independent such that for $n_k + 1 \le i \le 2n_k$, we have $\mathrm{P}(X_i = 1) = \mathrm{P}(X_i = -1) = 1/4$, $\mathrm{P}(X_i = \sqrt{2n_k}) = \mathrm{P}(X_i = -\sqrt{2n_k}) = 1/8n_k$, and $\mathrm{P}(X_i = 0) = 1 - 1/2 - 1/4n_k$; for all other $i \ge 1$, we have $\mathrm{P}(X_i = 1) = \mathrm{P}(X_i = -1) = 1/2$.

(i) Show that $\mathrm{E}(X_i) = 0$ and $\sigma_i^2 = \mathrm{E}(X_i^2) = 1$, therefore $a_n = s_n^2 = \sum_{i=1}^{n} \sigma_i^2 = n$. It follows that (6.40) is satisfied.

(ii) Show that Lindeberg's condition (6.42) does not hold for $\epsilon = 1$.

(iii) Show by Theorem 6.12 that $X_i$, $i \ge 1$, does not obey the CLT. (Hint: You may use the result of Example 1.6.)

Wittmann (1985) further showed that the sequence obeys the LIL. On the other hand, Marcinkiewicz and Zygmund (1937b) constructed a sequence of independent random variables that obeys the CLT but not the LIL.

*6.30.* Show that if $X_1, X_2, \ldots$ are i.i.d. with mean 0 and variance 1, then $\xi_n = S_n / \sqrt{2n \log\log n} \xrightarrow{P} 0$ as $n \to \infty$, where $S_n = \sum_{i=1}^{n} X_i$. However, $\xi_n$ does not converge to zero almost surely. This gives another example that convergence in probability does not necessarily imply almost sure convergence.

*6.31.* Show that the distance $\rho$ defined by (6.63) is, indeed, a distance or metric by verifying requirements 1–4 below (6.63).

*6.32.* This exercise is related to Example 6.10.

(i) Show that the mapping $g(x) = \sup_{0 \le t \le 1} x(t)$ is a continuous mapping from $C$ to $R$.

(ii) Show that $g(X_n) = \sup_{0 \le t \le 1} X_{n,t} = n^{-1/2} \max_{1 \le i \le n} S_i$ and $g(W) = \sup_{0 \le t \le 1} W_t$.

(iii) Show that $P(\sup_{0 \le t \le 1} W_t \le \lambda) = 0$ for $\lambda < 0$.

*6.33.* Let $x_n, n \ge 1$, be a sequence of real numbers such that $\liminf x_n = a$, $\limsup x_n = b$, where $a < b$, and $\lim_{n \to \infty} (x_{n+1} - x_n) = 0$. Show that the set of limit points of $\{x_n\}$ coincide with $[a, b]$.

*6.34.* This exercise is related to Example 6.11.

(i) Show that the functional $g(x) = x(1)$ defines a continuous mapping from $C$ to $R$.

(ii) Show that with probability 1, the set of limit points of $g(\eta_n)$ is $g(K)$.

(iii) Show that $x(1) \le 1$ for any $x \in K$.

*6.35.* Consider Example 6.12.

(i) Show that in this case we have $c_F(t) = \log(1 + e^t) - \log 2$.

(ii) Show that for any $x \in R$, the function $d_x(t) = xt - c_F(t)$ is strictly concave.

(iii) Show that

$$
I_F(x) = \begin{cases}
\log 2 + x \log x + (1 - x) \log(1 - x), & x \in (0, 1) \\
\log 2, & x = 0 \text{ or } 1 \\
\infty, & \text{otherwise.}
\end{cases}
$$

(iv) Show that for $A = (-\infty, 1/2 - \epsilon) \cup (1/2 + \epsilon, \infty)$ with $0 < \epsilon < 1/2$, we have $\inf_{x \in A^\circ} I_F(x) = \inf_{x \in \bar{A}} I_F(x) = I_F(1/2 - \epsilon) = I_F(1/2 + \epsilon)$, which is

$$
\log 2 + \left( \frac{1}{2} - \epsilon \right) \log \left( \frac{1}{2} - \epsilon \right) + \left( \frac{1}{2} + \epsilon \right) \log \left( \frac{1}{2} + \epsilon \right) > 0.
$$

*6.36.* This exercise is associated with the proof of consistency of the LSE in Section 6.7, where $X_{ni}$ is defined below (6.82).

(i) Verify expression (6.82).

(ii) Show that (6.83) implies $\sum_{i=1}^n E(X_{ni}^2) \to 0$, which, in turn, implies (6.16) and (6.17); that is, in fact,

$$
\sum_{i=1}^n E\{X_{ni} 1_{(|X_{ni}| \le \tau)}\} \longrightarrow 0,
$$

$$
\sum_{i=1}^n \mathrm{var}\{X_{ni} 1_{(|X_{ni}| \le \tau)}\} \longrightarrow 0
$$

for any $\tau > 0$.

(iii) Show that (6.83) for every $1 \le j \le p$ is equivalent to (6.84).

(iv) Interpret the quantity $\sum_{i=1}^n (x_i - \bar{x})^2$ in Example 6.14.

*6.37.* This exercise is associated with the proof of asymptotic normality of the LSE in Section 6.7.

(i) Show that the left side of (6.86) is equal to (6.87) and that $\sum_{i=1}^{n} c_{ni}^2 = 1$.

(ii) Show that (6.88) holds for every $\lambda \in R^p$ provided that (6.89) holds.

(iii) Show that in Example 6.14, the left side of (6.89) reduces to (6.90).

6.38. Suppose that $\epsilon_i$, $i \geq 1$, are i.i.d. such that $E(\epsilon_i) = 0$ and $E(\epsilon_i^2) < \infty$. Show that $\sum_{i=1}^{\infty} c_i \epsilon_i$ converges a.s. for any sequence of constants $c_i$, $i \geq 1$, such that $\sum_{i=1}^{\infty} c_i^2 < \infty$.

# 7

# Empirical Processes

## 7.1 Introduction

In Section 6.6.1 we discussed a topic that was somehow different from the rest of Chapter 6. There, the subject being dealt with was a random function, instead of a random variable. A closer look reveals that the random function was constructed based on sum of i.i.d. random variables and equal to the latter at particular values of its variable. Since, in practice, random variables often represent observations, we call a function constructed from observed random variables a *statistical function*.

As it turns out, these statistical functions are of great practical interest and therefore deserve some more extensive discussion. For the most part, we will focus on one particular class of statistical functions, called empirical processes. Let $X_1, X_2, \ldots$ be a sequence of i.i.d. random variables with the common distribution function $F$. The empirical distribution function (empirical d.f.) is defined as

$$F_n(x) = \frac{1}{n} \sum_{i=1}^{n} 1_{(X_i \leq x)}, \quad -\infty < x < \infty. \tag{7.1}$$

Although it might look simple, (7.1) is not the easiest thing in the world to understand. Here, the $X_i$'s are observations and $x$ is the variable of the function. For each realization of the $X_i$'s (i.e., realized values of $X_1, \ldots, X_n$), (7.1) defines a function of $x$, which is a step function with jumps at the realized values $X_1, \ldots, X_n$ (Exercise 7.1). Note that the indicator $1_{(X_i \leq x)} = 1$ if $X_i \leq x$, and 0 otherwise; or, in terms of a function of $x$, $1_{(X_i \leq x)} = 1$ if $x \geq X_i$, and 0 otherwise. After all, since the $X_i$'s are random, the function (7.1) is also random. In other words, for different realized values $X_1, \ldots, X_n$, (7.1) defines a different function.

According to the SLLN (see Section 6.3), for each $x$ the empirical d.f. converges a.s. to $E\{1_{(X_1 \leq x)}\} = P(X_1 \leq x) = F(x)$ as $n \to \infty$. In fact, a stronger result holds: The a.s. convergence is uniform in that

J. Jiang, *Large Sample Techniques for Statistics*,
DOI 10.1007/978-1-4419-6827-2_7, © Springer Science+Business Media, LLC 2010

$$\sup_{x} |F_n(x) - F(x)| \xrightarrow{\text{a.s.}} 0 \qquad (7.2)$$

as $n \to \infty$ (see below). We then consider a centralized and normalized version of the empirical d.f. defined by

$$\sqrt{n}\{F_n(x) - F(x)\}, \quad -\infty < x < \infty. \qquad (7.3)$$

The (random) function (7.3) is called an empirical process.

Dehling and Philipp (2002) noted that, to the surprise of many statisticians and probabilists, the study of empirical processes can be traced back to a paper by German mathematician Hermann Weyl in 1916. In this seminal paper, Weyl streamlined the theory of uniform distribution mod 1, and here is what it is. Let $n_i$, $i = 1, 2, \ldots$, be an increasing sequence of integers. For any $\omega \in [0, 1)$, define $X_i(\omega) = \{n_i \omega\}$, where $\{x\}$ denotes the fractional part of $x$. The sequence $X_1, X_2, \ldots$ can be viewed as random variables defined on the probability space $([0, 1), \mathcal{B}, P)$, where $\mathcal{B}$ denotes the Borel sets on $[0, 1)$ and $P$ denotes the Lebesgue measure. Each $X_i$ has a uniform distribution in that $P(X_i \le x) = x$ for $0 \le x \le 1$; however, the $X_i$'s are dependent (Exercise 7.2). Let $F_n(x)$ denote the empirical d.f. of $X_i$, $1 \le i \le n$, defined by (7.1). Weyl proved that $\sup_{x \in [0,1)} |F_n(x) - x| \to 0$ for all $\omega \in [0, 1)$, except possibly on a set of Lebesgue measure 0.

The restriction to uniform distribution as Weyl did is, actually, without loss of generality. In fact, the Uniform distribution on $(0, 1)$ plays a particular and very important role in the study of empirical processes due to the following theorem called inverse transformation.

**Theorem 7.1** (The inverse transformation). Let $\xi \sim \text{Uniform}(0, 1)$ and $F$ be a cdf. Define

$$F^{-1}(t) = \inf\{x : F(x) \ge t\}, \quad 0 < t < 1. \qquad (7.4)$$

Then $X = F^{-1}(\xi) \sim F$. In fact, $X \le x$ if and only if $\xi \le F(x)$.

*Proof.* By the definition it can be shown that $X \le x$ if and only if $\xi \le F(x)$. Therefore, $P(X \le x) = P\{\xi \le F(x)\} = F(x)$ and this completes the proof. Q.E.D.

The function $F^{-1}$ corresponds to the quantiles of the distribution $F$.

Theorem 7.1 allows us to simplify the study of empirical processes to that of one particular empirical process, the one of Uniform random variables. More precisely, let $\xi_1, \xi_2, \ldots$ be a sequence of independent Uniform$(0, 1)$ random variables. Denote the empirical d.f. and empirical process of the $\xi_i'$ by $G_n(t)$ and $U_n(t)$, respectively; that is, $G_n(t) = n^{-1} \sum_{i=1}^{n} 1_{(\xi_i \le t)}$ and $U_n(t) = \sqrt{n}\{G_n(t) - t\}$ for $t \in [0, 1]$. Then the empirical d.f. $F_n$ defined by (7.1) and $G_n(F)$ have identical distribution in the sense that for any $k \ge 1$ and $x_1 < \cdots < x_k$, the random vectors

$$[F_n(x_1), \ldots, F_n(x_k)] \quad \text{and} \quad [G_n\{F(x_1)\}, \ldots, G_n\{F(x_k)\}] \qquad (7.5)$$

have an identical joint distribution (Exercise 7.3). Similarly, the empirical processes $\sqrt{n}(F_n - F)$ and $U_n(F)$ have an identical distribution. Conversely, if we begin with the sequence $\{\xi_i\}$ and define $X_i = F^{-1}(\xi_i)$, $i \geq 1$, then for the sequence $\{X_i\}$, we have

$$F_n = G_n(F) \quad \text{and} \quad \sqrt{n}(F_n - F) = U_n(F). \qquad (7.6)$$

For these reasons we often focus on the empirical process $U_n$ in the sequel, which we call the uniform empirical process, with the understanding that similar results may be easily derived for $F_n$ using the connection. It should be noted that although the chapter is entitled "Empirical Processes" following the tradition of the literature in this field, the discussions in the sequel involve both the empirical d.f. and the empirical process. Most of the proofs of the results can be found in Shorack and Wellner (1986); otherwise, references will be given at the specific places.

A more convenient notation for the uniform empirical process is $U_n = \sqrt{n}(G_n - I)$, where $I$ represents the identical function, $I(t) = t$ for $t \in [0, 1]$. Similarly, we call $G_n$ the uniform empirical d.f. For any functions $x$ and $y$ on $[0, 1]$, define the uniform or supremum metric

$$\|x - y\| = \sup_{0 \leq t \leq 1} |x(t) - y(t)|. \qquad (7.7)$$

Note that this is the same metric introduced earlier by (6.63) for functions in the space $C$ of continuous functions on $[0, 1]$ (see Section 6.6.1). It is easy to verify that (7.7) remains as a metric for all functions on $[0, 1]$. Another subspace of functions on $[0, 1]$ is all functions on $[0, 1]$ that are right-continuous and possess left-limit at each point. This subspace is denoted by $D$.

## 7.2 Glivenko–Cantelli theorem and statistical functionals

We begin with the following celebrated result due to Glivenko and Cantelli.

**Theorem 7.2** (Glivenko-Cantelli theorem). $\|G_n - I\| \xrightarrow{\text{a.s.}} 0$ as $n \to \infty$.

The Glivenko–Cantelli theorem may be regarded as a uniform SLLN for the empirical d.f. It might appear that the result follows directly from Pólya's theorem (Example 1.6), because $G_n(t) \xrightarrow{\text{a.s.}} t$ for each $t$ by SLLN (Section 6.3), and the function $F(t) = t$ is continuous. However, the convergence here is a.s., which means that for each $t \in [0, 1]$, there is a set of probability 0 for which the convergence does not hold, and this set may be different for different $t$. On the other hand, to derive the Glivenko–Cantelli theorem from Pólya's theorem one needs to verify that $G_n \xrightarrow{\text{w}} I$ a.s.; that is,

$$P\left\{\lim_{n\to\infty} G_n(t) = t, \forall t \in [0,1]\right\} = 1,$$

which may not be so obvious. However, a similar $\epsilon$-$\delta$ argument to Example 1.6 leads to the proof of Theorem 7.2 (Exercise 7.4). We consider some applications of the Glivenko–Cantelli theorem.

*Example 7.1.* The previous result (7.2) is now seen as a consequence of, and therefore equivalent to, the Glivenko–Cantelli theorem. This is because, by Theorem 7.1 and independence, the sequence $\tilde{X}_i = F^{-1}(\xi_i)$, $i \geq 1$, has the same (joint) distribution as $X_i$, $i \geq 1$. Therefore, the empirical d.f. $F_n$ of the $X_i$'s has the same probabilistic behavior as the empirical d.f. of the $\tilde{X}_i$'s, denoted by $\tilde{F}_n$. On the other hand, we have, by (7.6), $\sup_x |\tilde{F}_n(x) - F(x)| = \sup_x |G_n\{F(x)\} - F(x)| \leq \sup_t |G_n(t) - t| = \|G_n - I\| \xrightarrow{\text{a.s.}} 0$ as $n \to \infty$, which implies (7.2). Of course, (7.2) implies the Glivenko–Cantelli theorem as a special case. This example shows, once again, how effective the strategy is to simply focus on the uniform empirical d.f.

*Example 7.2.* The inverse uniform empirical d.f. $G_n^{-1}$ is, by (7.4), the function $G_n^{-1}(t) = \inf\{x : G_n(x) \geq t\}$. There is a more explicit expression of $G_n^{-1}$. Let $\xi_{n,i}$ denote the $i$th order statistic of $\xi_1, \ldots, \xi_n$; that is, $\xi_{n,1} \leq \cdots \leq \xi_{n,n}$ is $\xi_1, \ldots, \xi_n$ arranged in an increasing order. Then we have

$$G_n^{-1}(t) = \xi_{n,i} \quad \text{if } \frac{i-1}{n} < t \leq \frac{i}{n} \quad \text{for } 1 \leq i \leq n \tag{7.8}$$

(Exercise 7.5). Furthermore, it can be shown that $\|G_n^{-1} - I\| = \|G_n - I\|$ (Exercise 7.5). Thus, by Theorem 7.2, we have $\|G_n^{-1} - I\| \xrightarrow{\text{a.s.}} 0$ as $n \to \infty$. By (7.8), this implies that

$$\max_{1 \leq i \leq n} \sup_{(i-1)/n < t \leq i/n} |\xi_{n,i} - t| \xrightarrow{\text{a.s.}} 0, \quad \text{as } n \to \infty.$$

In particular, the result implies that

$$\max_{1 \leq i \leq n} \left|\xi_{n,i} - \frac{i}{n}\right| \xrightarrow{\text{a.s.}} 0$$

as $n \to \infty$. The latter result may be interpreted as that, asymptotically, the $i$th order statistic of $\xi_1, \ldots, \xi_n$ converges to $i/n$ uniformly in $i$.

Some more applications can be brought about by considering statistical functionals. The concept of functionals was introduced in Section 6.6.1. Let $h$ be a functional defined on the space $\mathcal{D}$ of cdf's. In other words, $h(F)$ is a map from $F \in \mathcal{D}$ to $h(F) \in R$. Below are some examples.

*Example 7.3.* Let $a$ be a fixed point and consider $h(F) = F(a)$.

*Example 7.4* (The quantile). For any fixed $0 < t < 1$, consider $h(F) = F^{-1}(t)$ defined by (7.4). In particular, if $t = 0.95$, then $h(F)$ is the 95th quantile of the distribution $F$.

*Example 7.5.* The mean or expectation functional is defined as $h(F) = \mu_F = \int x \, dF(x)$. More generally, for any positive integer $p$, the $p$th moment of $F$ is the functional $h(F) = \mathrm{E}_F(X^p) = \int x^p \, dF(x)$; the $p$th central moment of $F$ is the functional $h(F) = \int (x - \mu_F)^p \, dF(x)$.

A functional $h$ is continuous at $F$ if for any sequence $H_n$ of cdf's, $\|H_n - F\| \to 0$ implies $h(H_n) \to h(F)$. As an immediate consequence of Theorem 7.2 (and Example 7.1), we have the following theorem.

**Theorem 7.3.** Let $X_1, X_2, \ldots$ be i.i.d. with distribution $F$ and let $F_n$ be the empirical d.f. of (7.1). If $h$ is continuous at $F$, the $h(F_n) \xrightarrow{\text{a.s.}} h(F)$ as $n \to \infty$.

We consider some applications of Theorem 7.3.

*Example 7.3 (continued).* It is easy to verify that $h$ is continuous (Exercise 7.6); hence, by Theorem 7.3, we have $F_n(a) = h(F_n) \xrightarrow{\text{a.s.}} h(F) = F(a)$. On the other hand, the latter follows directly from the SLLN.

*Example 7.4 (continued).* It can be shown that, for any fixed $t$, the quantile functional $h$ is continuous provided that $F^{-1}$ is continuous in a neighborhood of $t$ (Exercise 7.7). It follows by Theorem 7.3 that $F_n^{-1}(t) \xrightarrow{\text{a.s.}} F^{-1}(t)$ as $n \to \infty$. It should be pointed out that the continuity of $F^{-1}$ (in a neighborhood of $t$) cannot be dropped (Exercise 7.7).

*Example 7.5 (continued).* The expectation functional $h$ is not continuous at $F$ even if $F$ has a finite expectation. To see this, let $\epsilon_n$ be a sequence of positive numbers such that $\epsilon_n \to 0$ as $n \to \infty$. Let $\Delta_n$ be a distribution such that $h(\Delta_n) = \int x \, d\Delta_n(x) = \epsilon_n^{-1}$. Consider $H_n = (1 - \epsilon_n)F + \epsilon_n \Delta_n$. Then we have $\|H_n - F\| = \epsilon_n \|\Delta_n - F\| \to 0$ as $n \to \infty$ (why?). On the other hand, we have $h(H_n) = (1 - \epsilon_n)h(F) + \epsilon_n h(\Delta_n) = (1 - \epsilon_n)h(F) + 1 \to h(F) + 1$ as $n \to \infty$; hence, $h$ is not continuous at $F$. By a similar argument, it can be shown that the functionals of higher moments are not continuous at $F$ even if the corresponding moments of $F$ exist. On the other hand, the result

$$h(F_n) = \frac{1}{n} \sum_{i=1}^{n} X_i \xrightarrow{\text{a.s.}} \mathrm{E}_F(X_1) = h(F)$$

as $n \to \infty$ follows directly from the SLLN.

The above examples show that whereas being a useful result conceptually, Theorem 7.3 may not be an efficient way of establishing almost sure convergence. It is most useful, in practice, if the functional $h$ is known to be continuous (so that one does not need to verify its continuity); otherwise, checking the continuity of $h$ may take as much effort as directly showing the a.s. convergence (see Exercises 7.6 and 7.7). Furthermore, there are situations where $h$ is not continuous, such as Example 7.5, so that Theorem 7.3 does not apply; nevertheless, a.s. convergence of $h(F_n)$ to $h(F)$ may still be easily established. The reason for this is that the sequence of empirical d.f. $F_n$ is not an arbitrary class of cdf's. In the next section we further explore the asymptotic behavior of this special class of distributions.

## 7.3 Weak convergence of empirical processes

Consider the uniform empirical process $U_n$ defined in Section 7.1. For any fixed $t \in [0,1]$, by the CLT we have, as $n \to \infty$,

$$
\begin{aligned}
U_n(t) &= \sqrt{n}\{G_n(t) - t\} \\
&= \frac{1}{\sqrt{n}} \sum_{i=1}^{n} \{1_{(\xi_i \leq t)} - t\} \\
&\xrightarrow{d} N\{0, t(1-t)\}.
\end{aligned}
$$

However, this result is considered "the easier part" compared to a much stronger result to follow. More generally, for any distinct $t_1, \ldots, t_k \in [0,1]$, the joint distribution of $U_n(t_1), \ldots, U_n(t_k)$ is asymptotically multivariate normal with mean 0 and covariance matrix $\Sigma = (\sigma_{uv})_{1 \leq u,v \leq k}$, where $\sigma_{uv} = \mathrm{cov}\{U_n(t_u), U_n(t_v)\} = t_u \wedge t_v - t_u t_v$. This follows from a multivariate version of the CLT or can be derived from the (univariate) CLT and Theorem 2.14 (Exercise 7.8). It also uses the fact that $E\{U_n(t)\} = 0$ and $\mathrm{cov}\{U_n(t), U_n(s)\} = s \wedge t - st$, $s, t \in [0,1]$. Still, the latest result is considered "the easier part."

The "harder part" is to establish convergence in distribution of the empirical process in a functional space. To state this much stronger result, we need to first introduce an important process. Recall the definition of Wiener process, or Brownian motion, in Section 6.6.1. Here, for notation convenience we denote the Wiener process by $W(t)$, $0 \leq t \leq 1$. The stochastic process $U(t) = W(t) - tW(1)$ is called a *Brownian bridge*. A stochastic process $\{x(t)\}$ is called a Gaussian process if for any $t_1 < \cdots < t_k$, the joint distribution of $x(t_1), \ldots, x(t_k)$ is (multivariate) normal. Note that the Wiener process is a Gaussian process such that for any $0 \leq t_1 < \cdots < t_k \leq 1$, the random variables $W(t_2) - W(t_1), \ldots, W(t_k) - W(t_{k-1})$ are independent and distributed as $N(0, t_2 - t_1), \ldots, N(0, t_k - t_{k-1})$, respectively. It follows that a Brownian bridge is also a Gaussian process such that $E\{U(t)\} = 0$ and $\mathrm{cov}\{U(s), U(t)\} = s \wedge t - st$, $s, t \in [0,1]$ (Exercise 7.9). Earlier we introduced the space $D$ of right-continuous functions on $[0,1]$ that possess left-limit at each point. Let $\mathcal{D}$ denote the $\sigma$-field generated by the finite-dimensional subsets of $D$ (see Appendix A.2). Also, recall the uniform metric $\| \cdot \|$ defined by (7.7). In Section 6.6.1 we extended the concept of weak convergence to a metric space. Let $\xi_n$ be a sequence of $D$-valued random variables on a common probability space $(\Omega, \mathcal{A}, P)$. We say $\xi_n$ converges in distribution to $\xi$, a $D$-valued random variable on $(\Omega, \mathcal{A}, P)$, and denote this by $\xi_n \xrightarrow{d} \xi$ on $(D, \mathcal{D}, \| \cdot \|)$ if $P\xi_n^{-1} \xrightarrow{w} P\xi^{-1}$ as $n \to \infty$, where $P\zeta_n^{-1}$ is the induced probability measure by $\zeta_n$; that is, $P\xi_n^{-1}(B) - P(\xi_n \subset B)$ for $B \in \mathcal{D}$, and $P\xi^{-1}$ is the induced probability measure by $\xi$ in a similar way. An important and useful result is that $\xi_n \xrightarrow{d} \xi$ if and only if $E\{g(\xi_n)\} \to E\{g(\xi)\}$ for any bounded continuous function $g$ on the metric space $(D, \| \cdot \|)$. It then follows another very useful result called continuous mapping theorem, which is an extension of Theorem 2.12 to a metric space: If $\xi_n \xrightarrow{d} \xi$, then $g(\xi_n) \xrightarrow{d} g(\xi)$ for any continuous function $g$ on $(D, \| \cdot \|)$. (Note that the boundedness of $g$ is not required for the continuous mapping theorem.) In 1949, Doob conjectured the following result in a landmark paper, which was later proved by Donsker (1952).

**Theorem 7.4** (Doob–Donsker). $U_n \xrightarrow{d} U$ on $(D, \mathcal{D}, \| \cdot \|)$ as $n \to \infty$, where $U$ is the Brownian bridge.

It should be noted that a stronger result such as Theorem 7.4 is not motivated by, or developed for, mathematical interest. There are situations of applications where the weaker results given at the beginning of this section are simply not enough. As an example, we consider some applications of Theorem 7.4 to the well-known Kolmogorov–Smirnov statistics.

Let $X_1, \ldots, X_n$ be independent observations with an unknown common cdf $F$. The problem of interest is to test the hypothesis

$$H_0 : F(x) = F_0(x), \quad -\infty < x < \infty \qquad (7.9)$$

against one of the following alternatives:

$$H_1 : F(x) \neq F_0(x) \quad \text{for some } x$$
$$H_1^+ : F(x) \geq F_0(x) \quad \text{and } > \text{ hold for some } x$$
$$H_1^- : F(x) \leq F_0(x) \quad \text{and } < \text{ hold for some } x,$$

where $F_0$ is the hypothesized cdf. The Kolmogorov–Smirnov test statistics for $H_0$ against the three alternatives are respectively

$$D_n = \sup_x |F_n(x) - F_0(x)|, \qquad (7.10)$$

$$D_n^+ = \sup_x \{F_n(x) - F_0(x)\}, \qquad (7.11)$$

$$D_n^- = \sup_x \{F_0(x) - F_n(x)\}, \qquad (7.12)$$

where $F_n$ is the empirical d.f. defined by (7.1).

These statistics are, of course, very intuitive, but their null distributions which are used to determine the critical values of the tests are not easy to obtain, especially for large $n$. Fortunately, Theorem 7.4 allows us to derive the asymptotic null distributions of the Kolmogorov–Smirnov statistics, which can be used as approximations when $n$ is large. To see this, note that under (7.9) and by (7.6) we have, for any $\lambda$,

$$P(\sqrt{n}D_n^+ \leq \lambda) = P\left[\sup_x \sqrt{n}\{F_n(x) - F(x)\} \leq \lambda\right]$$

$$= P\left[\sup_x U_n\{F(x)\} \leq \lambda\right]. \tag{7.13}$$

Suppose that $F_0$ is continuous. Then, under the null hypothesis, the range of $F$, $R(F) = \{F(x), -\infty < x < \infty\} = [0, 1]$ (why?). Therefore, we have

$$\sup_x U_n\{F(x)\} = \sup_{t \in R(F)} U_n(t) = \sup_{0 \leq t \leq 1} U_n(t).$$

It can be shown (Exercise 7.10) that the function $g(x) = \sup_{0 \leq t \leq 1} x(t)$ is continuous on $(D, \|\cdot\|)$. Thus, by Theorem 7.4 and the continuous mapping theorem (above Theorem 7.4), we have $\sup_{0 \leq t \leq 1} U_n(t) = g(U_n) \xrightarrow{d} g(U) = \sup_{0 \leq t \leq 1} U(t)$; hence, the right side of (7.13) converges to $P\{\sup_{0 \leq t \leq 1} U(t) \leq \lambda\} = 1 - P(\|U^+\| > \lambda)$ as $n \to \infty$, where $u^+ = u \vee 0$. Note that $\sup_{0 \leq t \leq 1} U(t) = \sup_{0 \leq t \leq 1} U^+(t) = \|U^+\|$ with probability 1. It can be shown (e.g., Shorack and Wellner 1986, pp. 34–37) that for any $\lambda > 0$,

$$P(\|U^+\| > \lambda) = \exp(-2\lambda^2).$$

Therefore, we conclude that, under (7.9),

$$\lim_{n \to \infty} P(\sqrt{n}D_n^+ \leq \lambda) = 1 - \exp(-2\lambda^2) \tag{7.14}$$

for $\lambda > 0$, and 0 otherwise. Similar arguments show that

$$\lim_{n \to \infty} P(\sqrt{n}D_n^- \leq \lambda) = 1 - \exp(-2\lambda^2), \tag{7.15}$$

$$\lim_{n \to \infty} P(\sqrt{n}D_n \leq \lambda) = 1 - 2\sum_{j=1}^{\infty}(-1)^{j-1}\exp(-2j^2\lambda^2) \tag{7.16}$$

for $\lambda > 0$, and 0 otherwise under the null hypothesis (7.9).

The limits (7.14)–(7.16) are derived under the assumption that $F_0$ is continuous and may not hold if the latter assumption fails. In fact, Wood and Altavela (1978) considered Kolmogorov–Smirnov tests for discrete hypothesized distribution. By virtually the same arguments as above, the authors showed that when $F_0$ is discrete with set $J$ of discontinuity points, the right sides of (7.14) and (7.15) should be replaced by

$$P\left[\max_{x\in J} U\{F_0(x)\} \leq \lambda\right] \tag{7.17}$$

and the right side of (7.16) should be replaced by

$$P\left[\max_{x\in J} |U\{F_0(x)\}| \leq \lambda\right]. \tag{7.18}$$

Unlike (7.14)–(7.16), there are no closed form expressions for (7.17) and (7.18), in general. Nevertheless, these expressions may be evaluated by Monte–Carlo methods (Exercise 7.11).

## 7.4 LIL and strong approximation

Let us begin with the following theorem due to Smirnov (1944).

**Theorem 7.5.** $\limsup \|U_n\|/\sqrt{2\log\log n} = 1/2$ a.s. The same result holds with $U_n$ replaced by $U_n^+$ or $U_n^-$.

Chung (1949) strengthened Smirnov's result by showing that for any non-decreasing sequence of positive numbers $\lambda_n$, the probability $P(\|U_n\| \geq \lambda_n \text{ i.o.})$ is 0 or 1 depending on whether or not the infinite series

$$\sum_{n=1}^{\infty} \frac{\lambda_n^2}{n} \exp(-2\lambda_n^2)$$

converges [recall the definition of i.o. in Section 6.6.1, halfway between (6.67) and (6.68)]. That Chung's result implies Smirnov's is left to the reader as an exercise (Exercise 7.12).

Another interesting result called the "other LIL" is the following.

**Theorem 7.6** (Mogulskii). $\liminf \sqrt{2\log\log n}\|U_n\| = \pi/2$ a.s.

Note that there is no "contradiction" between Theorem 7.5 and Theorem 7.6. Theorem 7.5 is regarding the upper limit of $\|U_n\|$ divided by $\sqrt{2\log\log n}$, whereas Theorem 7.6 is about the lower limit of $\|U_n\|$ multiplied by $\sqrt{2\log\log n}$, and $\|U_n\|/\sqrt{2\log\log n} \leq \sqrt{2\log\log n}\|U_n\|$ for large $n$. In fact, Theorem 7.6 is similar to another classical result due to Chung (1948): If $X_1, X_2, \ldots$ are i.i.d. with mean 0 and variance 1, then

$$\liminf \sqrt{\frac{2\log\log n}{n}} \max_{1\leq k\leq n} \left|\sum_{i=1}^{k} X_i\right| = \frac{\pi}{2} \quad \text{a.s.}$$

Since the sample path of $U_n$ belongs to $D$, it it natural to consider a functional LIL similar to what we considered in Section 6.6.1. Let $(M, \rho)$

be a metric space and $S \subset M$. Let $\xi_n, n \geq 1$, be a sequence of $M$-valued random variables on a probability space $(\Omega, \mathcal{A}, P)$. We say the sequence is a.s. relatively compact with respect to $\rho$ on $M$ with limit set $S$, denoted by $\xi_n$ r.c. $S$ w.r.t. $\rho$ on $M$ a.s., if there exists $A \in \mathcal{A}$ with $P(A) = 1$ such that the following conditions (i)–(iii) hold for each $\omega \in A$:

(i) Every subsequence $n'$ has a further subsequence $n''$ for which $\xi_{n''}(\omega)$ converges with respect to $\rho$ [in other words, $\xi_n(\omega), n \geq 1$ is a Cauchy sequence with respect to $\rho$].

(ii) All of the $\rho$-limit points of $\xi_n(\omega)$ belong to $S$.

(iii) For any $s \in S$, there is a subsequence $n'$ (which may depend on $s$ and $\omega$) such that $\rho\{\xi_{n'}(\omega), s\} \to 0$.

Recall the subset $K$ of $C$ defined in Section 6.6.1 [see (6.69)]. Since $C \subset D$, $K$ is also a subset of $D$. Finkelstein (1971) proved the following result.

**Theorem 7.7.** $U_n/\sqrt{2 \log \log n}$ r.c. $K$ w.r.t. $\|\cdot\|$ on $D$ a.s.

We consider an example as an application of Theorem 7.7.

*Example 7.6.* Finkelstein (1971) showed that

$$\sup \left\{ \int_0^1 x^2(t) \, dt : x \in K \right\} = \frac{1}{\pi^2}.$$

Also, it can be shown that the functional $g(x) = \int_0^1 x^2(t) \, dt$ is continuous with respect to $\|\cdot\|$ on $D$ (Exercise 7.13). It follows from Theorem 7.7 that

$$\limsup \frac{\int_0^1 U_n^2(t) \, dt}{2 \log \log n} = \frac{1}{\pi^2} \quad \text{a.s.} \tag{7.19}$$

A few words about LIL for a general empirical process (7.3). An extension of Theorem 7.5 states that

$$\limsup \frac{\|\sqrt{n}(F_n - F)\|}{\sqrt{2 \log \log n}} \leq \frac{1}{2} \quad \text{a.s.} \tag{7.20}$$

with equality if $1/2$ is in the range of $F$. For example, if $F$ is continuous, then the latter certainly holds; hence, the equality holds in (7.20). On the other hand, Theorem 7.6 extends without any modification; that is,

$$\liminf \sqrt{2 \log \log n} \|\sqrt{n}(F_n - F)\| = \frac{\pi}{2} \quad \text{a.s.} \tag{7.21}$$

In a way, the LIL describes the precise a.s. (or strong) convergence rate of the empirical process. There are similar results on the a.s. convergence rate of the empirical process, which may not be as precise as the LIL in terms of the rate but more useful in some other regard. For example, sometimes a "second-order" approximation is needed in applications. To illustrate this, note that Theorem 7.4 states that the weak convergence limit of $U_n$ is $U$, the Brownian bridge. Although, in general, weak convergence does not necessarily imply a.s. convergence (see Chapter 2), the Skorokhod representation theorem (Theorem 2.18) states that there is a version of $U_n$ and $U$ defined on a common probability space such that $U_n \xrightarrow{\text{a.s.}} U$. Here, a version of $U_n$ and $U$ means a sequence of random variables having the same distributions as $U_n$ and $U$, respectively. See Theorem 2.18 for the precise definitions. So, in a certain sense, the Brownian bridge is also the a.s. limit of $U_n$. The question then is what is the a.s. convergence rate of $U_n - U$ in the same sense? Such a problem is often referred to as the strong approximation of empirical process.

The Skorokhod representation is useful in establishing results of weak convergence, or convergence in probability; however, it does not help in deriving results of a.s. convergence. The reason is that Skorokhod representation tells nothing about the joint distribution of $U_1, U_2, \ldots$, which is something involved in the a.s. convergence. An improvement of the Skorokhod representation is called the Hungarian construction, which began with the pioneering work of Csörgő and Révész (1975). The following is one of the fundamental results.

**Theorem 7.8** (The Hungarian construction). There exists a sequence of independent Uniform$(0, 1)$ random variables $\xi_1, \xi_2, \ldots$ and a sequence of Brownian bridges $U^{(n)}$, $n \geq 1$, such that

$$\limsup \frac{\sqrt{n}}{(\log n)^2} \left\| U_n - U^{(n)} \right\| < c \quad \text{a.s.},$$

where $U_n$ is the empirical process of $\xi_1, \ldots, \xi_n$ and $c$ is a finite constant.

See Chapter 12 of Shorack and Wellner (1986) for further results on the Hungarian construction. An application is considered later in Section 7.8.

## 7.5 Bounds and large deviations

There is a rich class of probability inequalities for empirical processes (e.g., Shorack and Wellner 1986). These inequalities play important roles not only in establishing the limit laws, such as SLLN and LIL, but also for obtaining bounds for deviations of the empirical processes. Many of these inequalities are maximum inequalities. For example, those regarding $\|U_n\|$ are maximum inequalities because, by definition, $\|U_n\|$ is the supremum, or maximum, of $U_n(t)$ for $t \in [0, 1]$. We begin with the following well-known James inequality.

**Theorem 7.9.** For any $0 < p \leq 1/2$ and $\lambda > 0$, we have

$$P \left( \left\| \frac{U_n^+}{1 - I} \right\|_0^p \geq \frac{\lambda}{q} \right) \leq \exp \left\{ -\frac{\lambda^2}{2pq} \psi \left( \frac{\lambda}{p\sqrt{n}} \right) \right\},$$

where $\|x\|_a^b = \sup_{a \le t \le b} |x(t)|$, $U_n^+/(1-I)$ denotes the process $U_n^+(t)/(1-t), t \in [0,1)$ (recall $a^+ = a \vee 0$), $q = 1 - p$, and

$$\psi(u) = \frac{2}{u^2}[(1+u)\{\log(1+u) - 1\} + 1]. \tag{7.22}$$

Some properties of $\psi$ are left as an exercise (Exercise 7.14). Regarding $U_n^-$, we have the following result.

**Theorem 7.10** (Shorack). For any $0 < p \le 1/2$ and $0 < \lambda \le \sqrt{np}$, we have

$$P\left(\left\|\frac{U_n^-}{1-I}\right\|_0^p \ge \frac{\lambda}{q}\right) \le \exp\left\{-\frac{\lambda^2}{2p}\psi\left(-\frac{\lambda}{p\sqrt{n}}\right)\right\} \wedge \exp\left(-\frac{\lambda^2}{2pq}\right),$$

where $U_n^-/(1-I)$ denotes the process $U_n^-(t)/(1-t), t \in [0,1)$ (recall $a^- = -a \wedge 0$), and $\psi$ is the same function defined by (7.22).

*Example 7.7.* Consider the special case of $p = q = 1/2$. Let $\lambda = \epsilon p \sqrt{n}$, where $0 < \epsilon < 1$. Then Theorem 7.9 implies

$$P\left(\left\|\frac{U_n^+}{1-I}\right\|_0^{1/2} \ge \epsilon\sqrt{n}\right) \le \exp\left\{-\frac{\epsilon^2}{2}\psi(\epsilon)n\right\}, \tag{7.23}$$

whereas Theorem 7.10 implies

$$P\left(\left\|\frac{U_n^-}{1-I}\right\|_0^{1/2} \ge \epsilon\sqrt{n}\right) \le \exp\left\{-\frac{\epsilon^2}{4}\psi(-\epsilon)n\right\} \wedge \exp\left(-\frac{\epsilon^2}{2}n\right). \tag{7.24}$$

Note that $1 \le \psi(-\epsilon) < 2$ [part (e) of Exercise 7.14]. Thus, the first term on the right side of (7.24) is greater than the second term. It follows that

$$P\left(\left\|\frac{U_n^-}{1-I}\right\|_0^{1/2} \ge \epsilon\sqrt{n}\right) \le \exp\left(-\frac{\epsilon^2}{2}n\right). \tag{7.25}$$

Also note that $\psi(\epsilon) \le 1$ and $\psi(\epsilon) \to 1$ as $\epsilon \to 0$ [parts (d) and (a) of Exercise 7.14]. It follows that the bound on the right side of (7.23) is greater than that on the right side of (7.25), but, as $\epsilon \to 0$, the bounds are approximately equal.

Another interesting result is a maximum inequality regarding a uniform empirical process indexed by subintervals of $[0,1]$. For any $C = (s,t]$, where $0 \le s \le t \le 1$, define $U_n(C) = U_n(t) - U_n(s)$ and $|C| = t - s$. Mason, Shorack and Wellner (1983) proved the following.

**Theorem 7.11.** For any $0 < a \le b \le 1/2$ and $\lambda > 0$, we have

$$P\left\{\sup_{|C|\leq a} |U_n(C)| \geq \lambda\sqrt{a}\right\} \leq \frac{20}{ab^3} \exp\left\{-(1-b)^4 \frac{\lambda^2}{2}\psi\left(\frac{\lambda}{\sqrt{an}}\right)\right\},$$

where $\psi$ is the function defined by (7.22).

A celebrated inequality for empirical processes is known as DKW inequality, named after Dvoretzky, Kiefer, and Wolfowitz.

**Theorem 7.12** (DKW inequality). There exists a constant $c$ such that

$$\frac{1}{2}P(\|U_n\| \geq \lambda) \leq P(\|U_n^-\| \geq \lambda) \leq ce^{-2\lambda^2}, \quad \lambda \geq 0. \tag{7.26}$$

In the original paper of Dvoretzky et al. (1956), the authors did not specify the value of the constant $c$. Birnbaum and McCarty (1958) conjectured that $c$ can be chosen as 1. By tracking down the original proof of Dvoretzky et al. (1956), Shorack and Wellner (1986) showed that $c = 29$ is good enough while acknowledging that this is not the minimum possible value. Hu (1985) showed that the constant can be improved to $c = 2\sqrt{2}$. Massart (1990) finally proved Birnbaum and McCarty's conjecture by showing that $c$ can be chosen as 1 as long as $e^{-2\lambda^2} \leq 1/2$ [of course, it has to be because the left side of (7.26) is bounded by 1/2], and the value cannot be further improved.

As a demonstration of the DKW inequality (with the best constant $c$), consider the following example.

*Example 7.8.* Let $X_1, \ldots, X_n$ be i.i.d. observations with an unknown continuous distribution $F$. Suppose that one wishes to determine the sample size $n$ so that the probability is at least 95% that the maximum difference between the empirical d.f. $F_n$ and $F$ is less than 0.1. This means that one needs to determine $n$ such that

$$P\left\{\sup_x |F_n(x) - F(x)| < 0.1\right\} \geq 0.95. \tag{7.27}$$

By (7.6) and continuity of $F$, we see the left side of (7.27) is equal to

$$P\left[\sup_x |G_n\{F(x)\} - F(x)| < 0.1\right]$$
$$= P\left\{\sup_{0\leq t\leq 1} |G_n(t) - t| < 0.1\right\}$$
$$= P(\|U_n\| < 0.1\sqrt{n})$$
$$= 1 - P(\|U_n\| \geq 0.1\sqrt{n}),$$

which is $\geq 1 - 2e^{-0.02n}$ by the DKW inequality with $c = 1$. Thus, it suffices to let $1 - 2e^{-0.02n} \geq 0.95$, or $n \geq 185$.

We conclude this section with some results on large deviations of the empirical d.f. The results are similar to those discussed in Section 6.6.2. For simplicity, we will focus on the uniform empirical d.f. $G_n$. First, note that the latter can be expressed as $n^{-1}S_n$, where $S_n = \sum_{i=1}^{n} Y_i$ with $Y_i = 1_{(X_i \leq x)}$ and $Y_1, Y_2, \ldots$ is a sequence of i.i.d. random variables. Using the general result of Section 6.6.2, it can be shown that for each $t \in [0, 1]$ and $\delta \geq 0$, we have

$$\lim_{n \to \infty} \frac{1}{n} \log[P\{G_n(t) \geq t + \delta\}] = -f(\delta, t), \tag{7.28}$$

where $f(\delta, t) = (t + \delta) \log\{(t + \delta)/t\} + (1 - t - \delta) \log\{(1 - t - \delta)/(1 - t)\}$ if $0 \leq \delta \leq 1 - t$ and $f(\delta, t) = \infty$ if $\delta > 1 - t$ (Exercise 7.16).

To derive a result of large maximum deviation regarding $G_n$, we consider $D_{n,h} = \|(G_n - I)h\|$, $D_{n,h}^+ = \|(G_n - I)^+h\|$, and $D_{n,h}^- = \|(G_n - I)^-h\|$, where $h$ is any function on $(0, 1)$ satisfying the following conditions:

(i) $h$ is positive and continuous on $(0, 1)$;

(ii) $h$ is symmetric about $1/2$ and $\lim_{t \to 0} h(t)$ exists or is $\infty$.

An obvious example of $h$ is $h = 1$ (or any positive constant). Another example is $h(t) = -\log\{t(1 - t)\}$. For any such function $h$, we define

$$I_h(\lambda) = \inf_{t \in (0,1)} f\{\lambda/h(t), t\}. \tag{7.29}$$

Some properties of $I_h$ are explored in an exercise (Exercise 7.17). Shorack and Wellner (1986) proved the following result.

**Theorem 7.13.** For any $h$ satisfying conditions (i) and (ii) above, we have, for each $\lambda \geq 0$,

$$\lim_{n \to \infty} \frac{1}{n} \log\{P(D_{n,h} \geq \lambda)\} = -I_h(\lambda),$$

where $I_h(\lambda)$ is defined by (7.29). The same result holds with $D_{n,h}$ replaced by $D_{n,h}^+$ or $D_{n,h}^-$.

## 7.6 Non-i.i.d. observations

There have been a number of extensions of the Glivenko–Cantelli theorem. One extension considers the so-called triangular arrays $X_{n1}, \ldots, X_{nn}$ so that for each $n$, the $X_{ni}$'s are independent with $X_{ni} \sim F_{ni}$. We then define

$$\bar{F}_n(x) = \frac{1}{n} \sum_{i=1}^{n} F_{ni}(x), \quad -\infty < x < \infty. \tag{7.30}$$

The empirical d.f. of the $X_{ni}$'s is defined as

$$F_n(x) = \frac{1}{n} \sum_{i=1}^{n} 1_{(X_{ni} \leq x)}, \quad -\infty < x < \infty. \tag{7.31}$$

The following theorem extends the Glivenko–Cantelli theorem to triangular arrays, where the supremum norm $\|\cdot\|$ is defined similarly as (7.7); that is,

$$\|F - G\| = \sup_{-\infty < x < \infty} |F(x) - G(x)|. \tag{7.32}$$

**Theorem 7.14.** For the $F_n$ and $\bar{F}_n$ defined by (7.31) and (7.30), respectively, we have $\|F_n - \bar{F}_n\| \xrightarrow{\text{a.s.}} 0$, as $n \to \infty$.

Another extension of the Glivenko–Cantelli theorem is to stationary ergodic sequences. A sequence of random variables $X_i$, $i \geq 0$, is said to be (strictly) stationary if for any $k \geq 0$, the joint distribution of $(X_{k+1}, X_{k+2}, \ldots)$ is the same as that of $(X_0, X_1, \ldots)$. If $(\Omega, \mathcal{A}, P)$ is a probability space, a measurable map $T: \Omega \to \Omega$ is said to be measure-preserving if $P(T^{-1}A) = P(A)$ for all $A \in \mathcal{A}$, where $T^{-1}A = \{\omega \in \Omega, T(\omega) \in A\}$. Any stationary sequence $\{X_i\}$ may be thought of as being generated by a measure-preserving transformation $T$ in the sense that there exists a random variable $X$ defined on a probability space $(\Omega, \mathcal{A}, P)$ and a map $T: \Omega \to \Omega$ such that the sequence $XT^i, i \geq 0$, has the same joint distribution as $X_i, i \geq 0$, where $XT^i(\omega) = X\{T^i(\omega)\}$, $\omega \in \Omega$, and $XT^0 = X$. The sequence $X_i, i \geq 0$ is said to be ergodic if the transformation $T$ satisfies the following: For any $A \in \mathcal{A}$, $T^{-1}A = A$ implies $P(A) = 0$ or 1. An extension of the SLLN is the following.

**Ergodic Theorem.** If $T$ is measure-preserving and $\mathrm{E}(|X|) < \infty$, then

$$\frac{1}{n} \sum_{i=0}^{n-1} XT^i \xrightarrow{\text{a.s.}} \mathrm{E}(X|\mathcal{I}),$$

where $\mathcal{I} = \{A \in \mathcal{A} : T^{-1}A = A\}$, which is a $\sigma$-field called the invariant $\sigma$-field (with respect to $T$).

The ergodic theorem can be used to establish the following extension of the Glivenko–Cantelli theorem [see Dehling and Philipp (2002) for the proof].

**Theorem 7.15.** Let $X_i, i \geq 0$ be a stationary ergodic sequence with common cdf $F$, and $F_n(x) = n^{-1} \sum_{i=0}^{n-1} 1_{(X_i \leq x)}$. Then $\|F_n - F\| \xrightarrow{\text{a.s.}} 0$ as $n \to \infty$.

The behavior of the empirical d.f. (7.31) is closely related to that of the so-called generalized binomial distribution. Let $\xi_1, \ldots, \xi_n$ be independent Bernoulli random variables with probabilities of success $p_1, \ldots, p_n$, respectively. The distribution of $\xi = \xi_1 + \cdots + \xi_n$ is called generalized binomial. It is clear that the summation in (7.31) has a generalized binomial distribution, in which $\xi_i = 1_{(X_{ni} \leq x)}$ and $p_i = F_{ni}(x)$, $1 \leq i \leq n$. On the other hand, let $\bar{p} = (p_1 + \cdots + p_n)/n$, and let $\eta$ denote a random variable that has a

Binomial$(n, \bar{p})$ distribution. Hoeffding's (1956) inequalities show that the generalized binomial random variable $\xi$ is more dispersed than its counterpart $\eta$ in the sense described by the following lemma.

**Lemma 7.1.** (i) For any $a$ and $b$ such that $0 \leq a \leq n\bar{p} \leq b \leq n$ we have $P(a \leq \xi \leq b) \leq P(a \leq \eta \leq b)$ with equality holds if and only if $p_1 = \cdots = p_n$, unless $a = 0$ and $b = n$. (ii) For any function $g$ satisfying

$$g(k) + g(k+2) \geq 2g(k+1), \quad 0 \leq k \leq n - 2, \tag{7.33}$$

we have $E\{g(\xi)\} \geq E\{g(\eta)\}$.

Note that condition (7.33) is satisfied by all convex functions. Hoeffding originally required $g(k) + g(k+2) > 2g(k+1), 0 \leq k \leq n-2$ instead of (7.33) (also see Shorack and Wellner 1986, p. 805). With a simple argument, it can be shown that this requirement can be relaxed to (7.33) (Exercise 7.18).

The result of weak convergence of empirical processes discussed in Section 7.3 also has extensions to non-i.i.d. cases. Let $F_{ni}$, $1 \leq i \leq n, n \geq 1$, be an array of arbitrary distributions on $[0, 1]$ and let $X_{ni}$, $1 \leq i \leq n, n \geq 1$, be an array of random variables such that for each $n \geq 1$, $X_{ni}, \ldots, X_{nn}$ are independent with distributions $F_{n1}, \ldots, F_{nn}$, respectively. Let $w_n = (w_{ni})_{1 \leq i \leq n}, n \geq 1$, be a sequence of nonzero constant vectors. Consider the following weighted empirical process:

$$Z_n(t) = \frac{1}{|w_n|} \sum_{i=1}^{n} w_{ni}\{1_{(X_{ni} \leq t)} - F_{ni}(t)\}, \quad 0 \leq t \leq 1. \tag{7.34}$$

It is easy to show (Exercise 7.19) that

$$\text{cov}\{Z_n(s), Z_n(t)\} = \frac{1}{|w_n|^2} \sum_{i=1}^{n} w_{ni}^2 \{F_{ni}(s \wedge t) - F_{ni}(s)F_{ni}(t)\}. \tag{7.35}$$

Consider a function closely related to (7.35): $v_n(t) = |w_n|^{-2} \sum_{i=1}^{n} w_{ni}^2 F_{ni}(t)$. The following theorem, which is a special case of Theorem 3.3.1 of Shorack and Wellner (1986), extends the Doob–Donsker theorem (Theorem 7.4).

**Theorem 7.16.** If $|w_n|^{-2} \max_{1 \leq i \leq n} w_{ni}^2 \rightarrow 0$ and $\max_{1 \leq i \leq n} \|F_{ni} - I\| \rightarrow 0$ as $n \rightarrow \infty$, then $Z_n \xrightarrow{d} U$, the Brownian bridge, on $(D, \mathcal{D}, \|\cdot\|)$ as $n \rightarrow \infty$.

As for sequence of dependent random variables, Billingsley (1968, Section 22) proved weak convergence of empirical process of stationary $\varphi$-mixing sequence. Let $\ldots, X_{-1}, X_0, X_1, \ldots$ be a stationary sequence of random variables. For any $-\infty < k < \infty$, let $\mathcal{F}_{-\infty}^{k} = \sigma(X_i, i \leq k)$, where $\sigma(X_i, i \in I)$ represents the $\sigma$-field generated by $X_i$, $i \in I$, and $\mathcal{F}_{k}^{\infty} = \sigma(X_i, i \geq k)$. The sequence is said to be $\varphi$-mixing if for any $-\infty < k < \infty$ and $n \geq 1$, we have

$$|P(E_1 \cap E_2) - P(E_1)P(E_2)| \le \varphi(n)P(E_1) \qquad (7.36)$$

for any $E_1 \in \mathcal{F}_{-\infty}^k$ and $E_2 \in \mathcal{F}_{k+n}^\infty$. Note that if $P(E_1) > 0$, (7.36) is equivalent to $|P(E_2|E_1) - P(E_2)| \le \varphi(n)$. Roughly speaking, the mixing condition states that there is a decay in dependence as the random variables in the sequence are further apart and the rate of decay is controlled by $\varphi$. It is required that

$$\lim_{n \to \infty} \varphi(n) = 0. \qquad (7.37)$$

For example, in the following theorem due to Billingsley, the rate of decay $\varphi(n)$ is further specified.

**Theorem 7.17.** Let $\{X_i\}$ be stationary $\varphi$-mixing and $X_i \in [0, 1]$. Let $F$ be the cdf of $X_i$ and let $F_n$ be the empirical cdf defined by (7.1). If $F$ is continuous and $\sum_{n=1}^\infty n^2 \sqrt{\varphi(n)} < \infty$, then $\sqrt{n}(F_n - F) \xrightarrow{d} Z$ on $(D, \mathcal{D}, \|\cdot\|)$ as $n \to \infty$, where $Z$ is a Gaussian process satisfying $E\{Z(t)\} = 0$ and

$$\text{cov}\{Z(s), Z(t)\} = E\{g_s(X_0)g_t(X_0)\}$$
$$+ \sum_{i=1}^\infty [E\{g_s(X_0)g_t(X_i)\} + E\{g_s(X_i)g_t(X_0)\}],$$

with $g_t(x) = 1_{(0 \le x \le t)} - F(x)$ and $P(Z \in C) = 1$ (i.e., with probability 1 the sample path of $Z$ is continuous).

There is vast literature on extensions of the results of empirical processes to various non-i.i.d. cases. See, for example, Dehling et al. (2002).

## 7.7 Empirical processes indexed by functions

Another way of extending the results is to think of the empirical processes as a statistical functional (see Section 7.2). Note that (7.1) can be written as

$$P_n(f) = \frac{1}{n} \sum_{i=1}^n f(X_i), \qquad (7.38)$$

where $f(y) = 1_{(y \le x)}$. Alternatively, one may define the empirical measure as $P_n = n^{-1} \sum_{i=1}^n \delta_{X_i}$, where $\delta_y$ represents a point mass at $y$. Then the functional (7.38) can be expressed as $P_n(f) = \int f dP_n$. The empirical process, with the original definition of (7.1), may be viewed as the image of a special class of functions under $P_n$; that is, $\{P_n(1_{(-\infty,x]}), x \in R\}$, where $1_A(y) = 1$ if $y \in A$ and 0 otherwise. More generally, one may consider the process $\{P_n(f), f \in \mathcal{F}\}$ for an arbitrary class of functions $\mathcal{F}$ and call it an empirical process (indexed by functions). Note that for each $f \in \mathcal{F}$, (7.38) is a random variable [which is why $\{P_n(f), f \in \mathcal{F}\}$ is a process].

Suppose that $X_1, \ldots, X_n$ are i.i.d. with cdf $F$. Note that $F(x)$, too, can be expressed as a functional; that is, $F(x) = P(f) = \int f dP$, where $f = 1_{(-\infty, x]}$ and $P(A) = P(X_1 \in A)$. The Glivenko–Cantelli theorem (Theorem 7.2) can be expressed as

$$\sup_{f \in \mathcal{F}} |P_n(f) - P(f)| \xrightarrow{\text{a.s.}} 0 \quad \text{as } n \to \infty \tag{7.39}$$

for the special class $\mathcal{F} = \mathcal{F}_1 = \{1_{(-\infty, x]}, x \in R\}$. More generally, one may question whether or not (7.39) holds for a given class $\mathcal{F}$; if it does, $\mathcal{F}$ is said to be a $P$-Glivenko–Cantelli class. Here, $P$ refers to the fact that the supremum in (7.31) depends on the underlying distribution $F$ or $P$.

To extend the Glivenko–Cantelli theorem in this direction, some regularity conditions need to be imposed on $\mathcal{F}$. For the most part, these conditions attempt to control the complexity of $\mathcal{F}$, which is necessary. Note that the classical Glivenko–Cantelli theorem states that (7.39) holds for $\mathcal{F} = \mathcal{F}_1$ without any restriction on $F$. However, the following example shows that without restrictions on $\mathcal{F}$, (7.39) may not hold for some $F$.

*Example 7.9* (A counterexample). Let $F$ be a continuous distribution; therefore, $P$ is nonatomic in the sense that $P(\{x\}) = 0$ for every $x$. Let $\mathcal{A}$ be the class of all finite subsets of $R$ and $\mathcal{F} = \{1_A, A \in \mathcal{A}\}$. Now, let $\hat{A} = \{X_1, \ldots, X_n\}$. Clearly, we have $\hat{A} \in \mathcal{A}$; hence, $\hat{f} = 1_{\hat{A}} \in \mathcal{F}$ (for any realization of the random variables). However, we have $P_n(\hat{f}) = P_n(\hat{A}) = 1$ and $P(\hat{f}) = P(\hat{A}) = 0$. Thus, the left side of (7.39) is equal to 1 for every $n$; hence, cannot converge to zero almost surely.

The complexity of $\mathcal{F}$ is measured by a quantity called entropy. For $1 \leq r < \infty$, let $\mathcal{L}_r(P)$ be the collection of functions $f$ such that

$$\|f\|_{r, P} = \left( \int |f|^r dP \right)^{1/2} < \infty.$$

An $\epsilon$-bracket in $\mathcal{L}_r(P)$ is a pair of functions $g, h \in \mathcal{L}_r(P)$ such that

$$P\{g(X) \leq h(X)\} = 1 \quad \text{and} \quad \|h - g\|_{r, P} \leq \epsilon.$$

A function $f$ lies in the $\epsilon$-bracket $g, h$ if $P\{g(X) \leq f(X) \leq h(X)\} = 1$. The bracketing number, denoted by, $N\{\epsilon, \mathcal{F}, \mathcal{L}_r(P)\}$, is the minimum number of $\epsilon$-brackets in $\mathcal{L}_r(P)$ needed to cover $\mathcal{F}$ so that every $f \in \mathcal{F}$ lies in at least one $\epsilon$-bracket. In most cases, one does not need to know the exact bracketing number—only an estimate of its order is sufficient. We consider an example.

*Example 7.10.* Let $X_1, \ldots, X_n$ be i.i.d. random variables with distribution $F$ (which corresponds to $P$) on $R$. Let $\mathcal{F} = \mathcal{F}_1$ [defined below (7.39)]. It can be shown that $N\{\epsilon, \mathcal{F}, \mathcal{L}_r(P)\} < \infty$ for every $\epsilon > 0$ (Exercise 7.22).

With the definition of the bracketing number, we can state an extension of the Glivenko–Cantelli theorem.

**Theorem 7.18.** If $N\{\epsilon, \mathcal{F}, \mathcal{L}_r(P)\} < \infty$ for every $\epsilon > 0$, then (7.39) holds; that is, $\mathcal{F}$ is a $P$-Glivenko–Cantelli class.

In a similar way, we can extend the result in Section 7.3 on weak convergence of the empirical process. Define the entropy with bracketing as the logarithm of the bracketing number, denoted by $E\{\epsilon, \mathcal{F}, \mathcal{L}_r(P)\} = \log[N\{\epsilon, \mathcal{F}, \mathcal{L}_r(P)\}]$. Let $l^\infty(\mathcal{F})$ denote the collection of all bounded functionals $P\colon \mathcal{F} \mapsto R$. We say $\mathcal{F}$ is $P$-Donsker if $\sqrt{n}(P_n - P) \xrightarrow{d} G$ in $l^\infty(\mathcal{F})$ as $n \to \infty$, where $G$ is a Gaussian process indexed by $f \in \mathcal{F}$ with mean 0 and covariance $\mathrm{cov}\{G(f_1), G(f_2)\} = \mathrm{cov}\{f_1(X), f_2(X)\}$, $f_1, f_2 \in \mathcal{F}$, and $X \sim F$ (or $P$). Here, $\xrightarrow{d}$ is defined the same way as in Section 7.3 (above Theorem 7.4) with $D$ replaced by $l^\infty(\mathcal{F})$, $\mathcal{D}$ replaced by the Borel $\sigma$-field generated by the open balls in $l^\infty(\mathcal{F})$ [i.e., sets of the form $\{Q \in l^\infty(\mathcal{F}) : \rho(Q, P) < \epsilon\}$ for some $P \in l^\infty(\mathcal{F})$ and $\epsilon > 0$; see below] and $\|\cdot\|$ is replaced by the metric $\rho(P, Q) = \sup_{f \in \mathcal{F}} |P(f) - Q(f)|$. The following theorem extends Theorem 7.4.

**Theorem 7.19.** If $\int_0^\infty \sqrt{E\{\epsilon, \mathcal{F}, \mathcal{L}_2(P)\}}\, d\epsilon < \infty$, then $\mathcal{F}$ is $P$-Donsker.

The proofs of Theorems 7.18 and 7.19 and much more on empirical processes indexed by functions can be found in Kosorok (2008).

# 7.8 Case study: Estimation of ROC curve and ODC

The receiver operating characteristic (ROC) curve is a measure of the accuracy of a continuous diagnostic test. Typically, the patients are classified as "healthy" or "diseased" according to a cutoff point, $c$, so that the patients whose test scores are higher than $c$ are classified as "diseased"; otherwise they are classified as"healthy" or "normal." Let $X$ denote the test score of a randomly selected healthy patient and let $Y$ denote that of a randomly selected diseased patient. We assume that both $X$ and $Y$ are continuous random variables with cdf (pdf) $F$ ($f$) and $G$ ($g$), respectively, and that $X$ and $Y$ are independent. The sensitivity of the test is defined as $\mathrm{SE}(c) = \mathrm{P}(Y > c) = 1 - G(c)$. In other words, the sensitivity is the probability that a diseased individual is (correctly) classified as diseased when the cutoff $c$ is used. On the other hand, the specificity of the test is $\mathrm{SP}(c) = \mathrm{P}(X \leq c) = F(c)$, which is the probability of correctly classifying a healthy individual. These concepts are similar to the complements of type II and type I errors in statistical hypothesis testing (e.g., Lehmann 1986). The ROC curve is then defined as a plot of the fraction of "true positive," $\mathrm{SE}(c)$ (on the vertical axis), versus that of "false positive," $1 - \mathrm{SP}(c)$ (on the horizontal axis), for $-\infty < c < \infty$. Equivalently, the ROC curve can be viewed as a plot of

$$\text{ROC}(t) = 1 - G\{F^{-1}(1-t)\} \quad \text{versus } t \text{ for } t \in [0,1], \qquad (7.40)$$

where $F^{-1}$ is defined by (7.4). Another closely related plot is the ordinal dominance curve (ODC; Bamber 1975), which is obtained by reversing the axes; that is,

$$\text{ODC}(t) = F\{G^{-1}(t)\} \quad \text{versus } t \text{ for } t \in [0,1]. \qquad (7.41)$$

It is easy to verify that both the ROC curve and ODC have the following properties (Exercise 7.23):

(i) Invariance under monotonically increasing transformations of the measurement scale.

(ii) If $X$ is stochastically smaller than $Y$—that is, $F(x) \geq G(x)$ for all $x$—then the curve lies above the diagonal of the unit square.

(iii) The curve is concave if $f$ and $g$ have a monotone likelihood ratio in the sense that $f(x)/g(x)$ is nondecreasing in $x$.

(iv) The area under the curve is the probability $P(X < Y)$.

Swets and Pickett (1982) listed a variety of areas where ROC curves are used. The areas range from signal detection, psychology, to nutrition and medicine. A more recent example was given in Peng and Zhou (2004), in which the authors considered estimation of the ROC curve of a carbohydrate antigenic determinant (CA 19-9) in distinguishing between case and control patients. The data were originally used by Wieand et al. (1989) to demonstrate the superiority of CA 19-9 in detecting pancreatic cancer. The control and case groups consisted respectively of 51 patients with pancreatitis and 90 patients with pancreatic cancer. Concentrations of CA 19-9 in sera from all the patients were studied at the Mayo Clinic in Rochester, Minnesota, USA.

Typically, two datasets are collected: $X_1, \ldots, X_m$ from the healthy population and $Y_1, \ldots, Y_n$ from the diseased population. We assume that these observations are independent. An empirical ROC curve is then obtained by replacing $F$ and $G$ in (7.40) by $F_m$ and $G_n$, the empirical d.f.'s defined by (7.1) (with, of course, some changes in notation), respectively. Similarly, an empirical ODC is obtained by replacing $F$ and $G$ in (7.41) by $F_m$ and $G_n$, respectively. Hsieh and Turnbull (1996) described asymptotic properties of the empirical ODC. Similar results can also be derived for the empirical ROC curve. The authors assumed that $m = m(n)$ such that $n/m \to \lambda \in (0, \infty)$ as $n \to \infty$. It is also assumed that the slope of the ODC—that is,

$$\text{ODC}'(t) = \frac{f\{G^{-1}(t)\}}{g\{G^{-1}(t)\}}$$

—is bounded on any subinterval $(a, b)$ of $(0, 1)$, where $0 < a < b < 1$. Then by applying the Glivenko–Cantelli theorem (see Theorem 7.2) and the DKW inequality (see Theorem 7.12), the authors showed that

$$\|F_m G_n^{-1} - FG^{-1}\| = \sup_{0 \leq t \leq 1} |F_m\{G_n^{-1}(t)\} - \text{ODC}(t)| \xrightarrow{\text{a.s.}} 0$$

as $n \to \infty$. Furthermore, by using results of strong approximation of the empirical process (see Section 7.4), the authors showed that there exists a probability space on which one can define the empirical processes $F_m$ and $G_n$ and two independent Brownian bridges $U_1^{(n)}$ and $U_2^{(n)}$ such that

$$\sqrt{n}[F_m\{G_n^{-1}(t)\} - \mathrm{ODC}(t)]$$

$$= \sqrt{\lambda}U_1^{(n)}\{\mathrm{ODC}(t)\} + \frac{f\{G^{-1}(t)\}}{g\{G^{-1}(t)\}}U_2^{(n)}(t) + o\left\{\frac{(\log n)^2}{\sqrt{n}}\right\}$$

a.s. uniformly on $[a, b]$ $\hspace{6cm}$ (7.42)

for any $0 < a < b < 1$.

Another quantity of interest is the area under either the ROC curve or ODC, which is the probability $P(X < Y)$ according to property (iv) above. This area is known as a measure of accuracy on how well the test separates the subjects being tested into those with and without the disease in question. The traditional academic point system assigns letter grades to a diagnostic test according to its area under the ROC curve or ODC as follows: above 0.9, excellent (A); 0.8–0.9, good (B); 0.7–0.8, fair (C); 0.6–0.7, poor (D); below 0.5, fail (F). A natural estimate of the area under the ODC is the area under the empirical ODC; that is,

$$\hat{P}(X < Y) = \int_0^1 F_m\{G_n^{-1}(t)\}\, dt = \frac{1}{mn} \sum_{1 \le i \le m, 1 \le j \le n} 1_{(X_i < Y_j)} \quad (7.43)$$

(Exercise 7.24). Using the result of (7.42), Hsieh and Turnbull showed that

$$\sqrt{n}\{\hat{P}(X < Y) - P(X < Y)\} \xrightarrow{\mathrm{d}} N(0, \sigma^2)$$

as $n \to \infty$, where

$$\sigma^2 = \mathrm{var}\left[\sqrt{\lambda}\int_0^1 U_1\{\mathrm{ODC}(t)\}\, dt + \int_0^1 \frac{f\{G^{-1}(t)\}}{g\{G^{-1}(t)\}}U_2(t)\, dt\right]$$

$$= \lambda\, \mathrm{var}\left(\int_0^1 U_1[F\{G^{-1}(t)\}]\, dt\right) + \mathrm{var}\left(\int_0^1 U_2[G\{F^{-1}(t)\}]\, dt\right)$$

$$= \lambda\|FG^{-1}\|_*^2 + \|GF^{-1}\|_*^2,$$

where $U_1$ and $U_2$ are two independent Brownian bridges and

$$\|h\|_*^2 = \int_0^1 h^2(t)dt - \left\{\int_0^1 h(t)dt\right\}^2.$$

## 7.9 Exercises

7.1. Use a computer to draw two realizations of $X_1, \ldots, X_{10}$ from the standard normal distribution and then plot the empirical d.f. (7.1) based on each realization of $X_1, \ldots, X_{10}$. Compare the two plots by making them (i) side-by-side and (ii) one on top of the other (same axes).

*7.2.* Show that Weyl's sequence $X_i$, $i = 1, 2, \ldots$, defined in Section 7.1 [below (7.3)] has identical uniform distribution in that $P(X_i \leq x) = x$, $0 \leq x \leq 1$ for all $i$, where $P$ denotes Lebesgue measure; however, the $X_i$'s are not independent.

*7.3.* Show that for any $k \geq 1$ and $x_1 < \cdots < x_k$, the two random vectors in (7.5) have identical joint distribution.

*7.4.* Prove the Glivenko–Cantelli theorem (Theorem 7.2) by an $\epsilon$-$\delta$ argument. (Hint: See Example 1.6; consider the points $j/k$, $0 \leq j \leq k$, $k = 1, 2, \ldots$.)

*7.5.* This exercise is regarding Example 7.2.
(i) Verify expression (7.8).
(ii) Show that $\|G_n^{-1} - I\| = \|G_n - I\|$.

*7.6.* Show that the statistical functional in Example 7.3 is continuous.

*7.7.* Show that the quantile functional of Example 7.4 is continuous provided that $F^{-1}$ is continuous in a neighborhood of $t$. Give a counterexample to show that if $F^{-1}$ is not continuous in a neighborhood of $t$, the result may not be true.

*7.8.* Use Theorem 2.14 and the CLT to show that for any distinct $t_1, \ldots, t_k \in [0, 1]$, the joint distribution of $U_n(t_1), \ldots, U_n(t_k)$ is asymptotically multivariate normal with mean 0 and covariance matrix $\Sigma = (\sigma_{uv})_{1 \leq u, v \leq k}$, where $\sigma_{uv} = \mathrm{cov}\{U_n(t_u), U_n(t_v)\} = t_u \wedge t_v - t_u t_v$.

*7.9.* Show that the Brownian bridge $U(t)$ defined in Section 7.3 satisfies $E\{U(t)\} = 0$ and that $\mathrm{cov}\{U(s), U(t)\} = s \wedge t - st$ for all $0 \leq s, t \leq 1$.

*7.10.* Show that the following functions $g$ are continuous on $(D, \|\cdot\|)$:
(i) $g(x) = \sup_{0 \leq t \leq 1} x(t)$;
(ii) $g(x) = \sup_{0 \leq t \leq 1} |x(t)|$.

*7.11.* Consider a one-sided Kolmogorov–Smirnov test for the null hypothesis (7.9), where $F_0$ is a discrete distribution with the following jumps:

| $x$ | 1 | 2 | 3 | 4 | 5 | 6 |
|---|---|---|---|---|---|---|
| $F_0(x)$ | 0.033 | 0.600 | 0.833 | 0.933 | 0.961 | 1.000 |

[the exercise is based on an example given in Wood and Altavela (1978)]. The alternative is $H_1^-$ given below (7.9), so the statistic $D_n^-$ of (7.12) is considered.
(i) Show that for any $\lambda > 0$,

$$\lim_{n \to \infty} P(\sqrt{n} D_n^- > \lambda) = 1 - P(Z_1 \leq \lambda, \ldots, Z_5 \leq \lambda), \qquad (7.44)$$

where $(Z_1, \ldots, Z_5)$ has a multivariate normal distribution with means 0 and covariances given by

$$\mathrm{cov}(Z_i, Z_j) = F_0(i) \wedge F_0(j) - F_0(i) F_0(j), \quad 1 \leq i, j \leq 5.$$

(ii) The observed value of $\sqrt{n} D_n^-$ in Wood and Alravela (1978) was 1.095. For each sample size $n$, where $n = 30$, 100, and 200, generate 10,000 random vectors $(Z_1, \ldots, Z_5)'$ as above and evaluate the right side of (7.44) with $\lambda = 1.095$ by Monte-Carlo method.

*7.12.* Show that Chung's result (given at the beginning of Section 7.4) implies Smirnov's LIL (Theorem 7.5).

*7.13.* This exercise is regarding Example 7.6 in Section 7.4.

(i) Show that the functional $g(x) = \int_0^1 x^2(t)\, dt$, $x \in D$, is continuous with respect to $\|\cdot\|$.

(ii) Derive (7.19).

*7.14.* Verify the following properties for the function $\psi$ defined by (7.22):

(a) $\psi(u)$ is nonincreasing for $u \geq -1$ with $\psi(0) = 1$;

(b) $u\psi(u)$ is nondecreasing for $u \geq -1$;

(c) $\psi(u) \sim (2\log u)/u$ as $u \to \infty$;

(d) $0 \leq 1 - \psi(u) \leq u/3$ for $0 \leq u \leq 3$;

(e) $0 \leq \psi(u) - 1 \leq |u|$ for $-1 \leq u \leq 0$;

(f) $\psi'(0) = -1/3$, $\psi(-1) = 2$ and $\psi'(-1) = -\infty$;

(g) $u\psi(u)$ equals 0 and $-2$ respectively for $u = 0$ and $-1$ and has derivative 1 for $u = 0$;

(h) for $|u| < 1$, we have the Taylor expansion

$$\psi(u) = 1 - \frac{u}{3} + \frac{u^2}{6} - \frac{u^3}{10} + \cdots + \frac{(-1)^k 2u^k}{(k+1)(k+2)} + \cdots.$$

*7.15.* Show that for any $0 < a \leq 1/2$ and $\lambda > 0$, we have

$$P\left\{ \sup_{0 \leq h \leq a} \sup_{0 \leq t \leq 1-h} |U_n(t+h) - U_n(t)| \geq \lambda\sqrt{a} \right\}$$
$$\leq \frac{160}{a} \exp\left\{ -\frac{\lambda^2}{32} \psi\left( \frac{\lambda}{\sqrt{an}} \right) \right\},$$

where $\pi$ is the function defined by (7.22).

*7.16.* Derive (7.28) using the general result of Section 6.6.2.

*7.17.* This exercise explores some properties of the function $I_h$ defined by (7.29).

(i) Take $h = 1$. Show that $I_1$ is nondecreasing on $(0, 1)$, $I_1(\lambda) = 2\lambda^2 + O(\lambda^3)$ as $\lambda \to 0$, and $I_1(\lambda) \to \infty$ as $\lambda \to 1$.

(ii) Take $h(t) = -\log\{t(1-t)\}$. Show that $I_h(\lambda) \sim (e\lambda)^2/8$ as $\lambda \to 0$ and $I_h(\lambda) \to \infty$ as $\lambda \to 1$.

(iii) Continue part (ii). Let $t_\lambda$ be the value of $t$ at which the infimum in (7.29) is attained. Find the limit of $t_\lambda$ as $\lambda \to 0$.

*7.18.* This exercise is regarding (7.33), which is a key condition in Lemma 7.1. Hoeffding (1956) originally required

$$g(k) + g(k+2) > 2g(k+1), \quad 0 \leq k \leq n - 2, \tag{7.45}$$

instead of (7.33) (also see Shorack and Wellner 1986, p. 805). Show by a simple argument that this requirement can be relaxed to (7.33). [Hint: Suppose that

$g$ satisfies (7.33). Let $h(x) = g(x) + \epsilon x^2$, where $\epsilon$ is an arbitrary positive constant. Show that $h$ satisfies (7.45) (with $g$ replaced by $h$, of course).]

*7.19.* For the weighted empirical process defined by (7.34), verify the covariance function (7.35).

*7.20.* Show that Billingsley's theorem on weak convergence of the empirical process of the stationary $\varphi$-mixing sequence (Theorem 7.17) implies the Doob–Donsker theorem (Theorem 7.4); so the former is an extension of the latter.

*7.21.* Give a specific example of a stantionary $\varphi$-mixing (but not i.i.d.) sequence that satisfies the conditions of Theorem 7.17.

*7.22.* Show that in Example 7.10 we have $N\{\epsilon, \mathcal{F}, \mathcal{L}_r(P)\} < \infty, \forall \epsilon > 0$.

*7.23.* Verify properties (i)–(iv) for the ROC curve and ODC defined in Section 7.8.

*7.24.* Verify the second identity in (7.43).

# 8

# Martingales

The mathematical modeling of physical reality and the inherent non-determinism of many systems provide an expanding domain of rich pickings in which martingale limit results are demonstrably of great usefulness.

**Hall and Heyde (1980)**
*Martingale Limit Theory and Its Application*

## 8.1 Introduction

The term *martingale* originally referred to a betting strategy. Imagine a gambler playing a blackjack game (also known as twenty-one) in a casino (if you have not been in a casino or have never heard about the blackjack, there is nothing to worry, as far as this book is concerned). He begins with an initial bet of $5, which is the minimal according to the rule of the casino table. Every time he loses, he doubles the bet; otherwise he returns to the minimal bet. For example, a sequence of bettings may be $5 (lose), $10 (lose), $20 (lose), $40 (lose), $80 (win), $5 (lose), $10 (lose), .... It is easy to see that with this strategy, as long as the gambler does not keep losing, whenever he wins he recovers all his previous losses, plus an additional $5, which is equal to his initial bet (Exercise 8.1). However, $5 is as much as he can win at the end of any losing sequence, and he is risking more and more in order to win the $5 as the sequence extends longer and longer. On the fourth bet of the sequence, the gambler is risking $40 to win $5; on the eighth bet, he is risking $640; on the 17th bet, he would be risking $327,680, still for a chance to win $5. So, why would anyone (ever) want to play this game? Well, there are at least two reasons. First, when someone loses, there is a tendency or desire to "get it back" (in other words, once the gambler starts lossing, it is difficult for him to stop). Second and perhaps more importantly, the gambler figures that sooner or later he has to win; however, is he right?

J. Jiang, *Large Sample Techniques for Statistics*,
DOI 10.1007/978-1-4419-6827-2_8, © Springer Science+Business Media, LLC 2010

There are a few places in real life where the theory and practice do not seem to work together. Unfortunately for the gambler, this is one of those places. The problem is that the condition of the theory is never met in practice. To keep playing with this betting strategy, it takes not only a lot of courage (to keep playing despite heavy losses) but also unlimited resources (i.e., money), which no gambler has in real life. There is another "untold secret," so far, of the casino, which turns out to be a "killer." Just like the gambler, the casino knows well about this betting strategy, so it has a way to stop the gambler from playing with it. On each gambling table there is a maximum bet, say, $500. This makes it impossible for the gambler to keep playing the martingale strategy, because the maximum number of consecutive bets he can make with this strategy is seven (Exercise 8.1). There are other tiny little "tricks" that give the casinos small edges (which is why they stay in business).

The probabilistic definition of a (discrete-time) martingale is the following. Let $S_1, S_2, \ldots$ be a sequence of random variables satisfying

$$E(S_{n+1}|S_1, \ldots, S_n) = S_n \quad \text{a.s.;} \tag{8.1}$$

that is, the conditional expectation (see Appendix A.2) of the next observation, given all the past observations, is (almost surely) equal to the last observation. Then the sequence is called a *martingale*. More generally, let $(\Omega, \mathcal{F}, P)$ be a probability space. Let $I$ represent an index set of integers. For example, $I = \{1, 2, \ldots\}$ or $I = \{\ldots, -1, 0, 1, \ldots\}$. Let $\mathcal{F}_n$, $n \in I$, be a non-decreasing sequence of $\sigma$-fields of $\mathcal{F}$ sets. This means that $\mathcal{F}_n \subset \mathcal{F}_m \subset \mathcal{F}$ for any $m, n \in I$ such that $n < m$, and we will keep this notation/assumption throughout this chapter. A sequence of random variables $S_n$, $n \in I$, is called a (discrete-time) martingale with respect to $\mathcal{F}_n$, $n \in I$, or $S_n$, $\mathcal{F}_n$, $n \in I$, is a martingale if it satisifies the following condition (or two conditions):

$$S_n \in \mathcal{F}_n, \quad E(S_m|\mathcal{F}_n) = S_n \quad \text{a.s.} \quad \forall m, n \in I, m > n. \tag{8.2}$$

Here, $\xi \in \mathcal{G}$ means that the random variable $\xi$ is measurable with respect to the $\sigma$-field $\mathcal{G}$. Note that the condition also implies the existence of the expectation of $S_n$ (although the expectation is not necessarily finite) for all $n \in I$. If $S_n$, $\mathcal{F}_n$, $n \geq 1$ is a martingale according to the (8.2), then $S_n$, $n \geq 1$, is also a martingale according to (8.1); but the converse is not necessarily true (Exercise 8.2). In this chapter we only consider the discrete-time situations. Extension to continuous time will be considered in the next chapter.

Similarly, a sequence $S_n$, $\mathcal{F}_n$, $n \in I$, is a *submartingale (supermartingale)* if the equality in (8.2) is replaced by $\geq$ ($\leq$). In terms of (8.1), a submartingale (supermartingale) means $E(S_{n+1}|S_1, \ldots, S_n) \geq S_n$ a.s. $[E(S_{n+1}|S_1, \ldots, S_n) \leq S_n$ a.s.]. Returning to the gambling problem, if the gambler's fortune over time is a supermartingale, his expected future given the current decreases; if it is a submartingale, his expected future given the current increases; if it is a martingale, then his expected future fortune will be the same as his current fortune. So, as a gamble, he wishes that his fortune over time is a submartingale, or at least a martingale. We consider a more specific example.

*Example 8.1.* Suppose that the gambler has probability $p$ of winning each blackjack game. After the initial bet, the gamble starts a sequence of plays (here we assume that there is no maximum bet set on the table and that the gambler has unlimited resources, so that he can continuously play the game) such that his win/loss total after the $n$th game is $S_n$, $n \geq 1$. Furthermore, let $X_n$ be the result of his $n$th play [so the value of $X_n$ is either $-a$ (loss) or $a$ (win) for some positive integer $a$].

First, suppose that $p < 1/2$. If $X_n$ is a loss, say, $X_n = -a$ for some $a > 0$, his next bet is $2a$; thus, his expectation for the $(n+1)$st play is

$$(2a) \times p + (-2a) \times (1-p) = 2a(2p-1) \leq 0.$$

If $X_n$ is a win, his next bet is 5; thus, similarly, the expectation is

$$5 \times p + (-5) \times (1-p) = 5(2p-1) \leq 0.$$

In conclusion, no matter what the value of $X_n$, the gambler's conditional expectation for his $(n+1)$st play given the results of his previous plays is less than or equal to zero; that is, $E(X_{n+1}|X_1,\ldots,X_n) \leq 0$. [In fact, it can be seen that $X_{n+1}$ depends on $X_n$ but not on $X_1,\ldots,X_{n-1}$, so that $E(X_{n+1}|X_1,\ldots,X_n) = E(X_{n+1}|X_n)$.] Note that $S_n = X_1 + \cdots + X_n$, so that $S_{n+1} = S_n + X_{n+1}$. If we define $\mathcal{F}_n = \sigma(X_1,\ldots,X_n)$, then we have $S_n \in \mathcal{F}_n$ and $E(S_{n+1}|\mathcal{F}_n) = S_n + E(X_{n+1}|\mathcal{F}_n) \leq S_n$, $n \geq 1$. Thus, by Lemma 8.1 in the sequel, $S_n$, $\mathcal{F}_n$, $n \geq 1$, is a supermartingale.

By similar arguments, it can be shown that if $p \geq 1/2$, $S_n$, $\mathcal{F}_n$, $n \geq 1$, is a submartingale; if $p = 1/2$, $S_n$, $\mathcal{F}_n$, $n \geq 1$, is a martingale.

The name martingale was first introduced to the modern probabilistic literature by French mathematician Jean Ville in 1939. Early developments of martingale theory were influenced by S. Bernstein and P. Lévy, who considered the martingale as a generalization of sums of independent random variables (see Example 8.2 below). It was J. L. Doob in his landmark book, *Stochastic Processes* (1953), that brought a complete new look to the subject. Among Doob's most celebrated work was his discovery of the martingale convergence theorem, which we introduce in Section 8.3.

## 8.2 Examples and simple properties

First, let us derive a simpler, equivalent definition of martingale (submartingale, supermartingale).

**Lemma 8.1.** Let $I$ be a set of all integers between $a$ and $b$, where $a$ can be $-\infty$ and $b$ can be $\infty$. Then $S_n$, $\mathcal{F}_n$, $n \in I$, is a martingale (submartingale, supermartingale) if and only if $S_n \in \mathcal{F}_n$ and $E(S_{n+1}|\mathcal{F}_n) = (\geq, \leq)S_n$ a.s. for all $n$ such that $n, n+1 \in I$.

The proof is left as an exercise (Exercise 8.3). We consider some examples below, and in between the examples we introduce more concepts and properties of martingales, submartingales, and supermartingales.

*Example 8.2.* A classical example of a martingale is sums of independent random variables. Let $X_1, X_2, \ldots$ be a sequence of independent random variables such that $E(X_i) = 0$ for all $i$. Let $S_n = \sum_{i=1}^{n} X_i$ and $\mathcal{F}_n = \sigma(X_1, \ldots, X_n)$, $n \geq 1$. Then, $S_n$, $\mathcal{F}_n$, $n \geq 1$ is a martingale (Exercise 8.4). See Exercise 8.11 for an extension of this example.

In the above example, $S_n$ is the sum of $X_1, \ldots, X_n$ and, conversely, $X_n$ is the difference of $S_n$ and $S_{n-1}$. We can extend this notion to the martingales, which in many cases is more convenient. Let $X_n$, $n \in I$, be a sequence of random variables, where $I$ is as in Lemma 8.1. We say $X_n$, $\mathcal{F}_n$, $n \in I$, is a sequence of martingale differences if

$$X_n \in \mathcal{F}_n, \quad E(X_{n+1}|\mathcal{F}_n) = 0 \quad \text{a.s.} \tag{8.3}$$

for all $n$ such that $n, n+1 \in I$. The connection between martingale and martingale differences is illustrated by the following lemma.

**Lemma 8.2.** If $X_n$, $\mathcal{F}_n$, $n > 1$, is a sequence of martingale differences, then $S_n = \sum_{i=1}^{n} X_i$, $\mathcal{F}_n$, $n \geq 1$, is a martingale. Conversely, if $S_n$, $\mathcal{F}_n$, $n \geq 1$, is a martingale, define $X_1 = S_1$ and $X_n = S_n - S_{n-1}$, $n \geq 2$, then $X_n$, $\mathcal{F}_n$, $n \geq 1$, is a sequence of martingale differences.

The martingale differences provide a convenient way of contructing a martingale: One may first construct a sequence of martingale differences and then take the sums. In particular, the techniques used in the next lemma are sometimes very useful. A sequence of random variables $\xi_n$, $n \in I$, is said to be *adapted* to $\mathcal{F}_n$, $n \in I$, if $\xi_n \in \mathcal{F}_n$, $n \in I$. A sequence $\eta_n$, $n \in I$, is said to be *predictable* with respect to $\mathcal{F}_n$, $n \in I$, if $\eta_n \in \mathcal{F}_{n-1}$, $n-1, n \in I$.

**Lemma 8.3.** (i) If $\xi_n$, $n \geq 1$ is adapted to $\mathcal{F}_n$, $n \geq 1$, let $X_1 = \xi_1$, $X_n = \xi_n - E(\xi_n|\mathcal{F}_{n-1})$, $n \geq 2$. Then $X_n$, $\mathcal{F}_n$, $n \geq 1$, is a sequence of martingale differences; hence, $S_n = \sum_{i=1}^{n} X_i$, $\mathcal{F}_n$, $n \geq 1$, is a martingale.
(ii) If $X_n$, $\mathcal{F}_n$, $n \in I$, is a sequence of martingale differences, where $I$ is as in Lemma 8.1, and $\eta_n$, $n \in I$, is predictable with respect to $\mathcal{F}_n$, $n \in I$, then $\eta_n X_n$, $\mathcal{F}_n$, $n \in I$, is a sequence of martingale differences.

The proof is left as an exercise (Exercise 8.5). We consider an example.

*Example 8.3 (Quadratic form).* Let $X_1, \ldots, X_n$ be independent random variables such that $E(X_i) = 0$ and $E(X_i^2) < \infty$, $1 \leq i \leq n$, and $A = (a_{ij})_{1 \leq i,j \leq n}$ is a symmetric, constant matrix. The random variable $Q = X'AX = \sum_{i,j=1}^{n} a_{ij} X_i X_j$, where $X = (X_1, \ldots, X_n)'$, is called a quadratic form in $X_1, \ldots, X_n$. There is an interesting, and useful, decomposition of the quadratic form as a sum of martingale differences after subtracting its mean. To see this, note that $E(Q) = \sum_{i=1}^{n} a_{ii} E(X_i^2)$. Thus,

$$Q - E(Q) = \sum_{i,j=1}^{n} a_{ij} X_i X_j - \sum_{i=1}^{n} a_{ii} E(X_i^2)$$

$$- \sum_{i=1}^{n} a_{ii}\{X_i^2 - E(X_i^2)\} + \sum_{i \neq j} a_{ij} X_i X_j$$

$$= \sum_{i=1}^{n} a_{ii}\{X_i^2 - E(X_i^2)\} + 2 \sum_{i>j} a_{ij} X_i X_j$$

$$= \sum_{i=1}^{n} a_{ii}\{X_i^2 - E(X_i^2)\} + 2 \sum_{i=1}^{n} \left( \sum_{j<i} a_{ij} X_j \right) X_i$$

$$= \sum_{i=1}^{n} Y_i,$$

where $Y_i = a_{ii}\{X_i^2 - E(X_i^2)\} + 2\left(\sum_{j<i} a_{ij} X_j\right) X_i$, $1 \le i \le n$ [here the summation $\sum_{j<1}(\cdots)$ is understood as zero]. Let $\mathcal{F}_i = \sigma(X_1, \ldots, X_i)$, $1 \le i \le n$. By Lemma 8.3(ii) and Lemma 8.4 below, it is easy to show that $Y_i$, $\mathcal{F}_i$, $1 \le i \le n$, is a sequence of martingale differences (Exercise 8.6).

The above decomposition was used by Jiang (1996) to derive a CLT for quadratic forms (see Section 8.8).

Some simple properties of martingale differences, martingales, submartingales, and supermartingales are summarized in the next three lemmas. The proofs are left as exercises (Exercises 8.7–8.9).

**Lemma 8.4.** (i) If $X_n^{(j)}$, $\mathcal{F}_n$, $n \in I$, $j = 1, 2$, are two sequences of martingale differences, where $I$ is as in Lemma 8.1, then $X_n^{(1)} + X_n^{(2)}$, $\mathcal{F}_n$, $n \in I$, is a sequence of martingale differences.

(ii) If $S_n^{(j)}$, $\mathcal{F}_n$, $n \in I$, $j = 1, 2$, are two martingales (submartingales, supermartingales), then $S_n^{(1)} + S_n^{(2)}$, $\mathcal{F}_n$, $n \in I$, is a martingale (submartingale, supermartingale).

(iii) If $S_n^{(j)}$, $\mathcal{F}_n$, $n \in I$, $j = 1, 2$, are two submartingales, then $S_n^{(1)} \vee S_n^{(2)}$, $\mathcal{F}_n$, $n \in I$, is a submartingale.

(iv) If $S_n^{(j)}$, $\mathcal{F}_n$, $n \in I$, $j = 1, 2$, are two supermartingales, then $S_n^{(1)} \wedge S_n^{(2)}$, $\mathcal{F}_n$, $n \in I$, is a supermartingale.

Note that it is important that the $\sigma$-fields $\mathcal{F}_n$ for the two martingales differences (martingales, submartingales, supermartingales) are the *same*.

**Lemma 8.5.** (i) If $S_n$, $\mathcal{F}_n$, $n \in I$ is a martingale, then for any convex (concave) function $\psi$, $\psi(S_n)$, $\mathcal{F}_n$, $n \in I$, is a submartingale (supermartingale).

(ii) If $S_n$, $\mathcal{F}_n$, $n \in I$, is a submartingale (supermartingale), then for any nondecreasing convex (concave) function $\psi$, $\psi(S_n)$, $\mathcal{F}_n$, $n \in I$, is a submartingale (supermartingale).

(iii) If $S_n$, $\mathcal{F}_n$, $n \in I$, is a supermartingale (submartingale), then for any nonincreasing convex (concave) function $\psi$, $\psi(S_n)$, $\mathcal{F}_n$, $n \in I$, is a submartingale (supermartingale).

**Lemma 8.6.** Suppose that $X_i$, $\mathcal{F}_i$, $1 \leq i \leq n$, is a sequence of martingale differences. Then the following hold.

(i) $\mathrm{E}(X_i) = 0$, $2 \leq i \leq n$.

(ii) If $\mathrm{E}(X_i^2) < \infty$, $1 \leq i \leq n$, then the sequence $X_i$, $1 \leq i \leq n$, is orthogonal; that is, $\mathrm{E}(X_i X_j) = 0$, if $i \neq j$. It follows that

$$\mathrm{E}\left(\sum_{i=1}^{n} X_i\right)^2 = \sum_{i=1}^{n} \mathrm{E}(X_i^2).$$

Time for some more examples.

*Example 8.4.* For any real number $x$, $x^+ = x$ if $x > 0$ and $0$ if $x \leq 0$; similarly, $x^- = -x$ if $x < 0$ and $0$ if $x \geq 0$. If $S_n$, $\mathcal{F}_n$, $n \in I$, is a submartingale, then $S_n^+$, $\mathcal{F}_n$, $n \in I$, is also a submartingale. There are two ways to show this. First note that $x^+$ is a nondecreasing convex function of $x$. The conclusion then follows by (ii) of Lemma 8.5. Second, note that $x^+ = x \vee 0$. Since $S_n$, $\mathcal{F}_n$, $n \in I$, is a submartingale and, obviously, $0$, $\mathcal{F}_n$, $n \in I$, is a submartingale. Thus, the conclusion follows from (iii) of Lemma 8.4.

Similarly, if $S_n$, $\mathcal{F}_n$, $n \in I$, is a superbmartingale, there are two ways to show that $-S_n^-$, $\mathcal{F}_n$, $n \in I$, is also a supermartingale (try it!).

Since a martingale is both a submartingale and a supermartingale, and $x = x^+ - x^-$, it follows that any martingale can be decomposed as a sum of a submartingale and a supermartingale, $S_n = S_n^+ - S_n^-$, both of which are not (necessarily) martingales (this rules out a trivial decomposition such as $S_n = S_n + 0$).

*Example 8.5 (Likelihood ratio).* Let $P$ and $Q$ be two probability measures on $(\Omega, \mathcal{F})$ and let $X_i$, $i \geq 1$, be a sequence of random variables (not necessarily independent). Suppose that the joint pdf of $X_1, \ldots, X_n$ under $P$ and $Q$ are $f_n$ and $g_n$, respectively. Define

$$S_n = \frac{g_n(X_1, \ldots, X_n)}{f_n(X_1, \ldots, X_n)}$$

if $f_n(X_1, \ldots, X_n) > 0$ and $S_n = 0$ if $f_n(X_1, \ldots, X_n) = 0$. When $X_1, \ldots, X_n$ are observations, $S_n$ is called the *likelihood ratio* (of $Q$ with respect to $P$). For example, in a classical hypothesis testing problem, $P$ represents the distribution under the null hypothesis and $Q$ represents that under a given alternative. In this case, the likelihood ratio measures how likely the data are from the alternative compared to from the null hypothesis. An important, and interesting, property of the likelihood ratio is that $S_n$, $n \geq 1$, is a supermartingale with respect to $\mathcal{F}_n = \sigma(X_1, \ldots, X_n)$, $n \geq 1$, under the probability distribution $P$. To show this, note that, obviously, $S_n \in \mathcal{F}_n$. Furthermore, let $\mathcal{B}$ be the Borel sets in $R$; we have, for any $B \in \mathcal{B}^n$,

$$
\int_{(X_1,\ldots,X_n)\in B} S_{n+1}\, dP
$$

$$
= \int_{(X_1,\ldots,X_n)\in B, f_{n+1}(X_1,\ldots,X_{n+1})>0} \frac{g_{n+1}(X_1,\ldots,X_{n+1})}{f_{n+1}(X_1,\ldots,X_{n+1})}\, dP
$$

$$
= \int \cdots \int_{(B\times R)\cap\{f_{n+1}(x_1,\ldots,x_{n+1})>0\}} g_{n+1}(x_1,\ldots,x_{n+1})\, dx_1 \cdots dx_{n+1}
$$

$$
= \int \cdots \int_{(B\times R)\cap\{f_n(x_1,\ldots,x_n)>0, f_{n+1}(x_1,\ldots,x_{n+1})>0\}} \cdots
$$

$$
+ \int \cdots \int_{(B\times R)\cap\{f_n(x_1,\ldots,x_n)=0, f_{n+1}(x_1,\ldots,x_{n+1})>0\}} \cdots
$$

$$
\leq \int \cdots \int_{B\cap\{f_n(x_1,\ldots,x_n)>0\}} \left\{ \int g_{n+1}(x_1,\ldots,x_{n+1}) dx_{n+1} \right\} dx_1 \cdots dx_n
$$

$$
+ \int \cdots \int_{f_n(x_1,\ldots,x_n)=0} \left\{ \int_{f_{n+1}(x_1,\ldots,x_{n+1})>0} g_{n+1}(x_1,\ldots,x_{n+1})\, dx_{n+1} \right\}
$$

$$
dx_1 \cdots dx_n
$$

$$
= \int \cdots \int_{B\cap\{f_n(x_1,\ldots,x_n)>0\}} g_n(x_1,\ldots,x_n)\, dx_1 \cdots dx_n
$$

$$
= \int_{(X_1,\ldots,X_n)\in B} S_n\, dP.
$$

The second to last equation used the facts that

$$
\int g_{n+1}(x_1,\ldots,x_{n+1})\, dx_{n+1} = g_n(x_1,\ldots,x_n),
$$

and that because

$$
\int f_{n+1}(x_1,\ldots,x_{n+1})\, dx_{n+1} = f_n(x_1,\ldots,x_n),
$$

$f_n(x_1,\ldots,x_n) = 0$ implies that the set $\{x_{n+1} : f_{n+1}(x_1,\ldots,x_{n+1}) > 0\}$ has Lebesgue measure 0; hence, the integral of $g_{n+}(x_1,\ldots,x_{n+1})$ over this set is zero. In conclusion, we have shown that for any $B \in \mathcal{B}^n$,

$$
\int_{(X_1,\ldots,X_n)\in B} S_{n+1}\, dP \leq \int_{(X_1,\ldots,X_n)\in B} S_n\, dP;
$$

thus, $E_P(S_{n+1}|\mathcal{F}_n) \leq S_n$ a.s. $P$ (see Appendix A.2).

The implication of this example to hypothesis testing is that suppose that the data actually come from the null hypothesis. Then the more data one collects, the less likely the data would look like they were coming from the alternative.

*Example 8.6 (Branching process).* Let $X_{n,i}, i \geq 1, n \geq 1$, be an array of i.i.d. random variables taking values in nonnegative integers. We assume that $E(X_{n,i}) = \mu > 0$. Let $T_0 = 1$ and

$$T_n = \sum_{i=1}^{T_{n-1}} X_{n,i}, \quad n \geq 1. \tag{8.4}$$

The sequence $T_n$, $n \geq 1$, is called a *branching process*. The name comes from a process in which bacterials reproduce themselves. Starting with one bacterial, suppose that from time $n-1$ to time $n$, the $i$th bacterial becomes $X_{n,i}$ bacterials. Let the total number of bacterials at time $n$ be $T_n$. It is easy to see that $T_n$ can be expressed as (8.4). Consider the normalized branching process $S_n = T_n/\mu^n$. We show that $S_n$, $n \geq 1$, is a martingale with respect to the $\sigma$-fields $\mathcal{F}_n = \sigma(T_1, \ldots, T_n)$, $n \geq 1$. It is obvious that $S_n \in \mathcal{F}_n$. Also, we have

$$E(T_{n+1}|\mathcal{F}_n) = E\left(\sum_{i=1}^{T_n} X_{n+1,i} \,\middle|\, \mathcal{F}_n\right)$$

$$= E\left\{\sum_{k=1}^{\infty} 1_{(T_n=k)} \sum_{i=1}^{k} X_{n+1,i} \,\middle|\, \mathcal{F}_n\right\}$$

$$= \sum_{k=1}^{\infty} 1_{(T_n=k)} E\left(\sum_{i=1}^{k} X_{n+1,i} \,\middle|\, \mathcal{F}_n\right)$$

$$= \sum_{k=1}^{\infty} 1_{(T_n=k)} k\mu$$

$$= \mu T_n \quad \text{a.s.;}$$

hence, $E(S_{n+1}|\mathcal{F}_n) = S_n$ a.s.

We conclude this section with the intruduction of an important concept in martingale theory. A measurable function $\tau$ taking values in $\{1, 2, \ldots, \infty\}$ is called a stopping time with respect to $\mathcal{F}_n$, $n \geq 1$, if $\{\tau = n\} \in \mathcal{F}_n$, $n \geq 1$. For each stopping time $\tau$ we can define a corresponding $\sigma$-field

$$\mathcal{F}_\tau = \{A \in \mathcal{F}_\infty : A \cap \{\tau = n\} \in \mathcal{F}_n, n \geq 1\}, \tag{8.5}$$

where $\mathcal{F}_\infty = \sigma(\cup_{n=1}^{\infty} \mathcal{F}_n)$. We consider an example.

*Example 8.7.* Let $S_n$, $n \geq 1$, be a sequence of random variables and let $\mathcal{F}_n = \sigma(S_1, \ldots, S_n)$, $n \geq 1$. For any Borel set $B$, define $\tau = \inf\{n \geq 1, S_n \in B\}$, where $\inf\{\emptyset\} \equiv \infty$. Intuitively, $\tau$ is the first time that the sequence $S_n$ enters $B$. For any $n \geq 1$, we have $\{\tau = n\} = \{S_k \notin B, 1 \leq k < n$ and $S_n \in B\} \in \mathcal{F}_n$. Therefore, $\tau$ is a stopping time with respect to $\mathcal{F}_n$, $n \geq 1$.

Some basic properties of stopping times are summarized below. The proof is left as an exercise (Exercise 8.10).

**Lemma 8.7.** Suppose that $\tau$ is a stopping time with respect to $\mathcal{F}_n$, $n \geq 1$.
(i) $\mathcal{F}_\tau$ is a $\sigma$-field and $\tau \in \mathcal{F}_\tau$.
(ii) If $S_n$, $n \geq 1$, is adapted to $\mathcal{F}_n$, $n \geq 1$, and $S_\infty$ is defined as $\limsup S_n$, then $S_\tau \in \mathcal{F}_\tau$.
Now suppose that $\tau_1$ and $\tau_2$ are stopping times with respect to $\mathcal{F}_n$, $n \geq 1$.
(iii) $\tau_1 \vee \tau_2$ and $\tau_1 \wedge \tau_2$ are both stopping times with respect to $\mathcal{F}_n$, $n \geq 1$.
(iv) If $\tau_1 \leq \tau_2$, then $\mathcal{F}_{\tau_1} \subset \mathcal{F}_{\tau_2}$.

## 8.3 Two important theorems of martingales

### 8.3.1 The optional stopping theorem

Property (ii) of Lemma 8.7 suggests that the first part of the defining property of a martingale (submartingale, supermartingale) (i.e., $S_n \in \mathcal{F}_n$) is preserved, if the fixed time $n$ is replaced by a stopping time $\tau$. Now, suppose that we have a nondecreasing sequence of stopping times, $\tau_1 \leq \tau_2 \leq \cdots$. Property (iv) of Lemma 8.7 then implies that $\mathcal{F}_{\tau_k}$, $k \geq 1$, is a nondecreasing sequence of $\sigma$-fields. One may thus conjecture that $S_{\tau_k}$, $\mathcal{F}_{\tau_k}$, $k \geq 1$, remains a martingale (submartingale, supermartingale). The following theorem, known as Doob's optional stopping theorem (or optional sampling theorem; Doob 1953), implies that the conjecture is true under certain regularity conditions.

**Theorem 8.1.** Let $S_n$, $\mathcal{F}_n$, $n \geq 1$, be a submartingale and let $\tau_2$ be a stopping time with respect to $\mathcal{F}_n$, $n \geq 1$, such that $P(\tau_2 < \infty) = 1$ and $E(S_{\tau_2})$ exists. If

$$\liminf E\{S_n^+ 1_{(\tau_2 > n)}\} = 0, \tag{8.6}$$

then for any stopping time $\tau_1$ with respect to $\mathcal{F}_n$, $n \geq 1$, as long as $E\{S_{\tau_1} 1_{(\tau_1 \leq \tau_2)}\}$ exists, we have

$$E(S_{\tau_2} | \mathcal{F}_{\tau_1}) \geq S_{\tau_1} \quad \text{a.s. } \{\tau_1 \leq \tau_2\}. \tag{8.7}$$

Because the negative of a supermartingale is a submartingale, and a martingale is both a submartingale and a supermartingale (and also note that $(-x)^+ = x^-$, $|x| = x^+ + x^-$), Theorem 8.1 immediately implies the following.

**Corollary 8.1.** (i) If the word submartingale in Theorem 8.1 is replaced by supermartingale and (8.6) is replaced by

$$\liminf \mathrm{E}\{S_n^- 1_{(\tau_2 > n)}\} = 0, \tag{8.8}$$

then (8.7) is replaced by

$$\mathrm{E}(S_{\tau_2}|\mathcal{F}_{\tau_1}) \leq S_{\tau_1} \quad \text{a.s.} \ \{\tau_1 \leq \tau_2\}. \tag{8.9}$$

(ii) If the word submartingale in Theorem 8.1 is replaced by martingale, and (8.6) is replaced by

$$\liminf \mathrm{E}\{|S_n| 1_{(\tau_2 > n)}\} = 0, \tag{8.10}$$

then (8.7) is replaced by

$$\mathrm{E}(S_{\tau_2}|\mathcal{F}_{\tau_1}) = S_{\tau_1} \quad \text{a.s.} \ \{\tau_1 \leq \tau_2\}. \tag{8.11}$$

We consider some applications of the optional stopping theorem.

*Example 8.8.* If $S_n$, $\mathcal{F}_n$, $n \geq 1$, is a martingale (submartingale, supermartingale) and $\tau$ is a stopping time with respect to $\mathcal{F}_n$, $n \geq 1$, then $S_{\tau \wedge k}$, $\mathcal{F}_{\tau \wedge k}$, $k \geq 1$, is a martingale (submartingale, supermartingale). To see this, note that by (iii) of Lemma 8.7 it is easy to see that $\tau \wedge k$ is a stopping time for any $k \geq 1$. Furthermore, since

$$S_{\tau \wedge k} = \sum_{l=1}^{k} S_l 1_{(\tau = l)} + S_k 1_{(\tau > k)},$$

$\mathrm{E}(S_{\tau \wedge k})$ exists for all $k \geq 1$. Also, note that $|S_n| 1_{(\tau \wedge k > n)} = 0$ when $n \geq k$; therefore, (8.10) is satisfied (with $\tau_2$ replaced by $\tau \wedge k$). Finally, for any $k_1 \leq k_2$, we have $\tau \wedge k_1 \leq \tau_2 \wedge k_2$; hence, $S_{\tau \wedge k_1} 1_{(\tau \wedge k_1 \leq \tau \wedge k_2)} = S_{\tau \wedge k_1}$, whose expectation exists as shown. It follows by (8.11) that $\mathrm{E}(S_{\tau \wedge k_2}|\mathcal{F}_{\tau \wedge k_1}) = S_{\tau \wedge k_1}$ a.s. The arguments for submartingale and supermartingale are similar (Exercise 8.17).

In particular, if $S_n$, $\mathcal{F}_n$, $n \geq 1$, is a nonnegative supermartingale, then since $\tau \wedge k \geq \tau \wedge 1 = 1$ for any $k \geq 1$, we have $\mathrm{E}(S_{\tau \wedge k}|\mathcal{F}_1) \leq S_1$ a.s., which implies $\mathrm{E}(S_{\tau \wedge k}) \leq \mathrm{E}(S_1)$. Also, since $\lim_{k \to \infty} S_{\tau \wedge k} 1_{(\tau < \infty)} = S_\tau 1_{(\tau < \infty)}$, we have, by Fatou's lemma (Lemma 2.4),

$$\begin{aligned}
\mathrm{E}\{S_\tau 1_{(\tau < \infty)}\} &= \mathrm{E}\left\{\lim_{k \to \infty} S_{\tau \wedge k} 1_{(\tau < \infty)}\right\} \\
&\leq \liminf \mathrm{E}\{S_{\tau \wedge k} 1_{(\tau < \infty)}\} \\
&\leq \liminf \mathrm{E}(S_{\tau \wedge k}) \\
&\leq \mathrm{E}(S_1).
\end{aligned} \tag{8.12}$$

*Example 8.9.* Earlier in Lemma 5.4 we introduced an inequality due to Stout (1974) without giving the proof. Now, with the result of (8.12) we can give a simple proof as follows. For any $0 < \alpha < \lambda$, let $\tau$ be the smallest integer $m \geq 0$ such that $T_m > \alpha$, if such an $m$ exist; otherwise, let $\tau = \infty$. By Example 8.7 we know that $\tau$ is a stopping time. It follows, by (8.12), that $\alpha P(\tau < \infty) \leq E\{T_\tau 1_{(\tau < \infty)}\} \leq E(T_0) = 1$; thus, by Chebyshev's inequality,

$$P\left(\sup_{m \geq 0} T_m > \alpha\right) = P(\tau < \infty) \leq \frac{1}{\alpha}. \tag{8.13}$$

Now, forget about the middle step of (8.13). The bottom line is that this inequality holds for any $\alpha < \lambda$; therefore, we must have $P(\sup_{m \geq 0} T_m \geq \lambda) \leq \lambda^{-1}$ (why?).

*Example 8.10.* A submartingale analogue of (8.12) is the following. Suppose that $S_n$, $\mathcal{F}_n$, $n \geq 1$, is a submartingale and $\tau$ is a stopping time with respect to $\mathcal{F}_n$, $n \geq 1$, such that $E(\tau) < \infty$ and there is a constant $c > 0$ such that

$$E(|X_i||\mathcal{F}_{i-1}) \leq c \quad \text{a.s.} \ \{\tau \geq i\}, \ i \geq 1,$$

where $X_1 = S_1$ and $X_i = S_i - S_{i-1}$, $i \geq 2$. Then we have $E(|S_\tau|) < \infty$ and $E(S_\tau) \geq E(S_1)$. To show this, note that, similar to Example 8.8, we have $E(S_{\tau \wedge k}|\mathcal{F}_1) \geq S_1$ a.s., which implies $E(S_{\tau \wedge k}) \geq E(S_1)$. The question is whether one can exchange the order of limit (as $n \to \infty$) and expectation (because here we cannot use Fatou's lemma). Note that

$$|S_{\tau \wedge k}| = \left|\sum_{i=1}^{\infty} X_i 1_{(i \leq \tau \wedge k)}\right|$$

$$\leq \sum_{i=1}^{\infty} |X_i| 1_{(i \leq \tau)} \equiv \eta.$$

If we can show

$$E(\eta) < \infty, \tag{8.14}$$

then by the dominated convergence theorem (Theorem 2.16), we have

$$E(S_\tau) = E\left(\lim_{k \to \infty} S_{\tau \wedge k}\right)$$

$$= \lim_{n \to \infty} E(S_{\tau \wedge k})$$

$$\geq E(S_1). \tag{8.15}$$

The first equation in (8.15) is because $E(\tau) < \infty$ implies that $\tau < \infty$ a.s. It remains to show (8.14). This follows because

$$E(\eta) = \sum_{i=1}^{\infty} E\{|X_i|1_{(\tau \geq i)}\}$$

$$= \sum_{i=1}^{\infty} E\{1_{(\tau \geq i)}E(|X_i||\mathcal{F}_{i-1})\}$$

$$[\text{because } \{\tau \geq i\} = \Omega \setminus \{\tau \leq i-1\} \in \mathcal{F}_{i-1}]$$

$$\leq c \sum_{i=1}^{\infty} E\{1_{(i \leq \tau)}\}$$

$$= c\, E\left(\sum_{i=1}^{\infty} 1_{(i \leq \tau)}\right)$$

$$= c\, E(\tau) < \infty.$$

Similarly, if the word submartingale is changed to martingale (and all other conditions remain), the conclusion is that $E(S_\tau) = E(S_1)$.

### 8.3.2 The martingale convergence theorem

The martingale convergence theorem may be motivated from convergence of monotonic sequences of real numbers. Recall that (see §1.5.1.3) an increasing sequence of real numbers $a_n$, $n = 1, 2, \ldots$, converges if it has an upper bound. According to the definition, a submartingale satisfies

$$E(S_{n+1}|\mathcal{F}_n) \geq S_n \quad \text{a.s.,} \tag{8.16}$$

which looks almost like an increasing sequence (except that there is a conditional expectation on the left side). Here is another way to look at it. What (8.16) means is that for any $A \in \mathcal{F}_n$, we have

$$\int_A S_{n+1}\, dP \geq \int_A S_n\, dP;$$

in other words, $S_{n+1}$ is greater than or equal to $S_n$ on any set of nonzero probability that belongs to $\mathcal{F}_n$. These observations lead one to conjecture that a submartingale would converge in some sense if it has an upper bound. The question is: What kind of upper bound? Since $S_n$ is a random variable, it may not be realistic to assume that it is uniformly bounded, so the bound would be better in some other sense. Doob (1953) found that $L^1$ boundedness is sufficient for the almost sure convergence of a submartingale. This result is known as the martingale convergence theorem. Note that in the following theorem we extend the definition of a submartingale by allowing $-\infty$ and $\infty$ to be its possible values.

**Theorem 8.2.** Suppose that $S_n$, $\mathcal{F}_n$, $n \geq 1$, is a submartingale such that

$$\sup_{n \geq 1} E(S_n^+) < \infty. \tag{8.17}$$

Then $S_\infty = \lim_{n \to \infty} S_n$ exists almost surely, and it has the following properties: (i) $E(S_\infty^+) < \infty$; (ii) $|S_\infty| < \infty$ a.s. $\{S_1 > -\infty\}$; and (iii) if

$$E(|S_n|) < \infty, \quad n \geq 1, \tag{8.18}$$

then $E(|S_\infty|) < \infty$.

Note that although (8.17) is sufficient for the a.s. existence of $S_\infty$, it does not imply that the latter is a.s. finite. (For a trivial example, consider $S_n = -\infty$, $n \geq 1$, which satisfies the condition, but $S_\infty = -\infty$.) However, if (8.17) is strengthened to

$$\sup_{n \geq 1} E(|S_n|) < \infty, \tag{8.19}$$

then according to (iii) of Theorem 8.1, $S_\infty \in L^1$ and therefore is a.s. finite. Also note that, under (8.17), (8.19) is equivalent to (8.18) (Exercise 8.18). In other words, the $L^1$ boundedness of a submartingale is all that is needed for the a.s. convergence of the submartingale in the usual sense of, say, §1.5.1 (i.e., the limit is a.s. a finite number).

The martingale convergence theorem is a milestone in martingale theory not only because of its applications in various fields, some of which are discussed in the sequel, but also because of a proof of the theorem using Doob's upcrossing inequality (e.g., Hall and Heyde 1980, p. 17) that inspired a generation of methodology based on stopping times. The version presented below is in a slightly stronger form than the original one given in Doob (1953, p. 314). For any $a < b$, let $U_n(a,b)$ denote the number of times that $S_1, \ldots, S_n$ crosses from a value $\leq a$ to one $\geq b$ (known as upcrossing).

**Lemma 8.8** (Doob 1960). Suppose that $S_k$, $\mathcal{F}_k$, $1 \leq k \leq n$, is a submartingale. Then for any $a < b$, we have

$$E\{U_n(a,b)\} \leq \frac{E\{(S_n - a)^+\}}{b - a}.$$

See, for example, Hall and Heyde (1980, p. 15–16) for a proof of Lemma 8.8. The proof makes use of a sequence of stopping times that are the times that the sequence $S_1, \ldots, S_n$ upcrosses the interval $(a,b)$ and the optional stopping theorem (Theorem 8.1).

Of course, Theorem 8.2 also holds if the word submartingale is replaced by martingale. We consider an example.

*Example 8.11.* For any random variable $\xi \in L^1$, the sequence $E(\xi|\mathcal{F}_n)$, $n \geq 1$, converges almost surely. To see this let $S_n = E(\xi|\mathcal{F}_n)$, $n \geq 1$. Then $S_n$,

$\mathcal{F}_n$, $n \geq 1$, is a martingale. Furthermore, by Jensen's inequality [see (5.56); note that here we consider the conditional expectation], we have $\mathrm{E}(|S_n|) = \mathrm{E}\{|\mathrm{E}(\xi|\mathcal{F}_n)|\} \leq \mathrm{E}\{\mathrm{E}(|\xi||\mathcal{F}_n)\} = \mathrm{E}(|\xi|) < \infty$; hence, (8.19) is satisfied. It follows that $S_n \xrightarrow{\text{a.s.}} S$ for some random variable $S$. In fact, it can be shown that $S_n$ also converges in $L^1$ and $S = \mathrm{E}(\xi|\mathcal{F}_\infty)$ a.s., where $\mathcal{F}_\infty$ is defined below (8.5) (e.g., Chow and Teicher 1988, Section 11.1).

As for a supermartingale, we have the following result.

**Corollary 8.2.** Let $S_n$, $\mathcal{F}_n$, $n \geq 1$ be a nonnegative supermartingale. Then $S_n$ converges almost surely to a limit $S_\infty$. Furthermore, if $\mathrm{E}(S_1) < \infty$, then $\mathrm{E}(|S_\infty|) < \infty$.

This is because $-S_n$, $\mathcal{F}_n$, $n \geq 1$, is a submartingale and $(-S_n)^+ = S_n^- = 0$. Thus, by Theorem 8.2, $S_n$ converges a.s. to a limit $S_\infty$. If $\mathrm{E}(S_1) < \infty$, then $\mathrm{E}(|S_n|) = \mathrm{E}(S_n) \leq \mathrm{E}(S_1) < \infty$. Thus, by (iii) of Theorem 8.2 we have $\mathrm{E}(|S_\infty|) < \infty$. Again, we consider an example.

*Example 8.5 (continued).* Earlier we showed that the likelihood ratio $S_n$ is a supermartingale with respect to the $\sigma$-fields $\mathcal{F}_n = \sigma(X_1, \ldots, X_n)$ and the probability measure $P$. This supermartingale is certainly nonnegative. Furthermore, we have

$$\mathrm{E}_P(S_1) = \mathrm{E}_P\left\{\frac{g_1(X_1)}{f_1(X_1)}\right\}$$
$$= \int \frac{g_1(x_1)}{f_1(x_1)} f_1(x_1) \, dx_1$$
$$= \int g_1(x_1) \, dx_1 = 1.$$

Thus, by Corollary 8.2, the likelihood ratio $S_n$ converges a.s. $P$ to $S_\infty \in L^1(P)$.

## 8.4 Martingale laws of large numbers

### 8.4.1 A weak law of large numbers

In the following we often use the convenient notation $S_n = \sum_{i=1}^n X_i$ for a martingale $S_n$, meaning that $X_i$ is the corresponding martingale difference defined in Lemma 8.2. Hall and Heyde (1980) gave the following extension of Theorem 6.2, where the sequence of normalizing constants, $a_n$, satisfy the condition above (6.5). The proof is left as an exercise (Exercise 8.20).

**Theorem 8.3.** Let $S_n = \sum_{i=1}^n X_i$, $\mathcal{F}_n$, $n \geq 1$, be a martingale. Then $a_n^{-1} S_n \xrightarrow{P} 0$ as $n \to \infty$ if the following conditions hold:

(i) $\sum_{i=1}^{n} P(|X_i| > a_n) \longrightarrow 0$;

(ii) $a_n^{-1} \sum_{i=1}^{n} E\{X_i 1_{(|X_i| \le a_n)} | \mathcal{F}_{i-1}\} \xrightarrow{P} 0$;

(iii) $a_n^{-2} \sum_{i=1}^{n} E[\text{var}\{X_i 1_{(|X_i| \le a_n)} | \mathcal{F}_{i-1}\}] \longrightarrow 0$.

Note that in the case of independent variables, conditions (i)–(iii) of Theorem 8.3 are also necessary. See Theorem 6.2. The following example given by Hall and Heyde (1980, pp. 29–30) shows that these conditions are not necessary in the martingale case.

*Example 8.12.* Let $Y_i$, $i \ge 1$, be a sequence of independent random variables such that $Y_1 = 1$ and, for $i > 1$, $P(Y_i = 0) = i^{-1}$, $P(Y_i = -2) = P\{Y_i = 2(2i-1)/(i-1)\} = (i-1)/2i$. Note that $E(Y_i) = 1$ for all $i$. Now, let

$$S_n = \left( \prod_{i=1}^{n} Y_i \right) - 1, \quad n \ge 1,$$

and $\mathcal{F}_n = \sigma(Y_1, \ldots, Y_n)$. It is easy to show that $S_n, \mathcal{F}_n, n \ge 1$, is a martingale with $E(S_n) = 0$ (Exercise 8.21). Furthermore, we have

$$P(S_n \ne -1) = P\left( \prod_{i=1}^{n} Y_i \ne 0 \right)$$

$$= P(Y_2 \ne 0, \ldots, Y_n \ne 0)$$

$$= \prod_{i=2}^{n} \left( \frac{i-1}{i} \right)$$

$$= \frac{1}{n} \longrightarrow 0$$

as $n \to \infty$. Thus, $S_n \xrightarrow{P} -1$. It follows that $a_n^{-1} S_n \xrightarrow{P} 0$ for any sequence $a_n$ satisfying the conditions above (6.5). Now, consider one special such sequences, $a_n = n$, $n \ge 1$. Note that we have $X_i = S_i - S_{i-1} = Y_1 \cdots Y_{i-1}(Y_i - 1)$, $i \ge 2$. For any $n \ge 1$, consider any $i \ge 1$ such that $3 \times 2^{i-2} > n$. Then if $Y_2, \ldots, Y_i$ are nonzero, we have $|X_i| \ge 3 \times 2^{i-2} > n$ (why?). Thus,

$$P(|X_i| > n) \ge P(Y_2 \ne 0, \ldots, Y_i \ne 0)$$

$$= \prod_{j=2}^{i} \left( \frac{j-1}{j} \right) = \frac{1}{i};$$

hence, $\sum_{i=1}^{n} P(|X_i| > n) \ge \sum_{l_n < i \le n} i^{-1} \to \infty$ as $n \to \infty$, where

$$l_n = 2 + \frac{\log n - \log 3}{\log 2}$$

(Exercise 8.21). Therefore, condition (i) of Theorem 8.3 is not satisfied despite the convergence to zero of $a_n^{-1} S_n$.

## 8.4.2 Some strong laws of large numbers

As in Section 6.3, Kronecker's lemma (Lemma 6.1) is a useful tool for establishing the SLLN for martingales. Recall the idea is that the convergence of an infinite series, say, $\sum_{i=1}^{\infty} x_i$, implies $a_n^{-1} \sum_{i=1}^{n} a_i x_i \to 0$ for any nondecreasing sequence of positive numbers $a_n$ such that $a_n \to \infty$. Now, suppose that $S_n = \sum_{i=1}^{n} X_i$, $\mathcal{F}_n$, $n \geq 1$, is a martingale. According to Kronecker's lemma, whenever the series

$$\sum_{i=1}^{\infty} \frac{X_i}{a_i} \tag{8.20}$$

converges, we have

$$\begin{aligned}
\frac{S_n}{a_n} &= \frac{1}{a_n} \sum_{i=1}^{n} X_i \\
&= \frac{1}{a_n} \sum_{i=1}^{n} a_i \left( \frac{X_i}{a_i} \right) \\
&\to 0. \tag{8.21}
\end{aligned}$$

The problem then is to find out when the series (8.20) converges. The following theorem can be derived from a result due to Chow (1965).

**Theorem 8.4.** For any $1 \leq p \leq 2$, the series (8.20) converges and (8.21) holds a.s. on the set $\{\sum_{i=1}^{\infty} a_i^{-p} E(|X_i|^p | \mathcal{F}_{i-1}) < \infty\}$, where $\mathcal{F}_0 = \{\emptyset, \Omega\}$.

*Example 8.13.* As another example of the stopping time techniques, we give a proof of a special case of Theorem 8.4: the case $p = 2$. The proof is essentially the same as the proof of Theorem 2.15 of Hall and Heyde (1980). For any $B > 0$, let $\tau$ be the smallest integer $n \geq 1$ such that

$$\sum_{i=1}^{n+1} a_i^{-2} E(X_i^2 | \mathcal{F}_{i-1}) > B$$

if such an $n$ exists; otherwise, let $\tau = \infty$. It can be shown by Example 8.7 that $\tau$ is a stopping time with respect to $\mathcal{F}_n$, $n \geq 1$ (Exercise 8.22). Note that $1_{(\tau \geq i)}$, $i \geq 1$, is predictable with respect to $\mathcal{F}_i$, $i \geq 1$ (why?). It follows, by (ii) of Lemma 8.3, that $1_{(\tau \geq i)} X_i$, $\mathcal{F}_i$, $i \geq 1$, is a sequence of martingale differences; therefore, $S_{\tau \wedge n} = \sum_{i=1}^{n} 1_{(\tau \geq i)} X_i$, $\mathcal{F}_n$, $n \geq 1$, is a martingale. We now show that $S_{\tau \wedge n}$ is $L^2$ bounded. This is because, by property (ii) of Lemma 8.6,

$$\begin{aligned}
E(S_{\tau \wedge n}^2) &= \sum_{i=1}^{n} E\{1_{(\tau \geq i)} X_i^2\} \\
&= \sum_{i=1}^{n} E\{1_{(\tau \geq i)} E(X_i^2 | \mathcal{F}_{i-1})\} \quad [\text{because } 1_{(\tau \geq i)} \in \mathcal{F}_{i-1}]
\end{aligned}$$

$$= E\left\{\sum_{i=1}^{n} 1_{(\tau \geq i)} E(X_i^2 | \mathcal{F}_{i-1})\right\}$$

$$= E\left\{\sum_{i=1}^{\tau \wedge n} E(X_i^2 | \mathcal{F}_{i-1})\right\} \leq B,$$

using the definition of $\tau$ for the last inequality. It follows from the martingale convergence theorem (Theorem 8.2) (note that $L^2$ boundedness implies $L^1$ boundedness) that $S_{\tau \wedge n}$ converges a.s. as $n \to \infty$. Therefore, $S_n$ converges a.s. on $\{\tau = \infty\}$. In other words, $S_n$ converges a.s. on $\{\sum_{i=1}^{\infty} E(X_i^2 | \mathcal{F}_{i-1}) \leq B\}$. The result then follows by the arbitrariness of $B$ [Exercise 8.23, part (iv)].

When the range of $p$ is not $[1, 2]$, we have the following result. Note that in part (i), the sequence is not required to be martingale differences. Again, part (ii) is due to Chow (1965).

**Theorem 8.5.** (i) Let $X_i$, $i \geq 1$, be any sequence of random variables. Then, the conclusion of Theorem 8.4 holds for any $p \in (0, 1)$.

(ii) Let $X_i, \mathcal{F}_i, i \geq 1$, be a sequence of martingale differences. For any $p > 2$ and any sequence $b_i > 0$, $i \geq 1$, such that $\sum_{i=1}^{\infty} b_i < \infty$, (8.20) converges and (8.21) holds a.s. on $\{\sum_{i=1}^{\infty} a_i^{-p} b_i^{1-p/2} E(|X_i|^p | \mathcal{F}_{i-1}) < \infty\}$.

The proof is left as exercises (Exercises 8.23 and 8.24).

*Note.* Although we have assumed that $a_n$ is a sequence of normalizing constants, Theorem 8.4 and Theorem 8.5 continue to hold if $a_n$ is a sequence of predictable random variables with respect to $\mathcal{F}_n$, $n \geq 1$ (i.e., $a_n \in \mathcal{F}_{n-1}$, $n \geq 1$), provided that $a_n > 0$ and $a_n \uparrow \infty$ a.s.

A special case of interest is $a_n = n$, $n \geq 1$. In this case we obtain the following SLLN for martingales.

**Corollary 8.3.** $n^{-1} S_n \xrightarrow{\text{a.s.}} 0$ as $n \to \infty$ provided either

$$\sum_{i=1}^{\infty} \frac{E(|X_i|^p)}{i^p} < \infty \tag{8.22}$$

for some $1 \leq p \leq 2$ or

$$\sum_{i=1}^{\infty} \frac{E(|X_i|^p)}{i^p b_i^{p/2-1}} < \infty \tag{8.23}$$

for some $p > 2$ and $b_i > 0$, $i \geq 1$, such that $\sum_{i=1}^{\infty} b_i < \infty$.

To see this, note that, for example, (8.22) implies that

$$E\left\{\sum_{i=1}^{\infty} \frac{E(|X_i|^p | \mathcal{F}_{i-1})}{i^p}\right\} = \sum_{i=1}^{\infty} \frac{E(|X_i|^p)}{i^p} < \infty,$$

which implies that $\sum_{i=1}^{\infty} i^{-p} \mathrm{E}(|X_i|^p | \mathcal{F}_{i-1}) < \infty$ a.s. The desired result then follows from Theorem 8.4.

It should be noted that although the martingale convergence theorem and Kronecker's lemma are often used together to establish the SLLN for martingales, other methods have also been used for similar purposes. For example, the following result, which is not a consequence of Theorem 8.4 or 8.5, can be derived by using Doob's maximum inequality [see (5.87)] and Burkholder's inequality for martingales [see (5.71) and the subsequent discussion]. In a way, this approach is more similar to the traditional methods for establishing SLLN for sums of independent random variables (see Chapter 6).

**Theorem 8.6.** Let $S_n = \sum_{i=1}^{n} X_i$, $\mathcal{F}_n$, $n \geq 1$ be a martingale. If for some $p \geq 1$, we have

$$\sum_{i=1}^{\infty} \frac{\mathrm{E}(|X_i|^{2p})}{i^{p+1}} < \infty, \tag{8.24}$$

then $n^{-1} S_n \xrightarrow{\text{a.s.}} 0$ as $n \to \infty$.

The theorem can be derived from Theorem 2 of Chow (1960). Note that, for example, (8.22) and (8.24) do not imply each other (Exercise 8.25). As an application of Theorem 8.6 (or Corollary 8.3), consider the following.

*Example 8.14.* Let $X_i$, $\mathcal{F}_i$, $i \geq 1$, be adapted and there are a constant $c > 0$ and a random variable $X$ with $\mathrm{E}(|X|) < \infty$ such that

$$P(|X_i| > x) \leq cP(|X| > x), \quad x \geq 0, \; i \geq 1. \tag{8.25}$$

Then we have

$$\frac{1}{n} \sum_{i=1}^{n} \{X_i - \mathrm{E}(X_i | \mathcal{F}_{i-1})\} \xrightarrow{\text{P}} 0. \tag{8.26}$$

If the moment condition for $X$ is strengthened to $\mathrm{E}(|X| \log^+ |X|) < \infty$, then (8.26) can be strengthened to a.s. convergence. To show (8.26), we write

$$\begin{aligned} X_i - \mathrm{E}(X_i | \mathcal{F}_{i-1}) &= [X_i 1_{(|X_i| \leq i)} - \mathrm{E}\{X_i 1_{(|X_i| \leq i)} | \mathcal{F}_{i-1}\}] \\ &\quad + X_i 1_{(|X_i| > i)} - \mathrm{E}\{X_i 1_{(|X_i| > i)} | \mathcal{F}_{i-1}\}\} \\ &= Y_i + Z_i + W_i. \end{aligned}$$

It can be shown that (8.25) implies

$$\mathrm{E}\{|X_i| 1_{(|X_i| > i)}\} \leq c\mathrm{E}\{|X| 1_{(|X| > i)}\}, \tag{8.27}$$

$$\mathrm{E}\{X_i^2 1_{(|X_i| \leq i)}\} \leq 2c \int_0^i x P(|X| > x) \, dx \tag{8.28}$$

for any $i \geq 1$, and that $\mathrm{E}(|X|) < \infty$ implies

$$\sum_{i=1}^{\infty} \frac{1}{i^2} \int_0^i x \mathrm{P}(|X| > x) \, dx < \infty \tag{8.29}$$

(Exercise 8.26). It follows that

$$\begin{aligned}
\frac{1}{n} \mathrm{E} \left| \sum_{i=1}^n Z_i \right| &\leq \frac{1}{n} \sum_{i=1}^n \mathrm{E}\{|X_i| 1_{(|X_i| > i)}\} \\
&\leq \frac{c}{n} \sum_{i=1}^n \mathrm{E}(|X| 1_{(|X| > i)}\} \\
&\to 0
\end{aligned}$$

as $n \to \infty$ (why?). Thus, we have $n^{-1} \sum_{i=1}^n Z_i \to 0$ in $L^1$, hence in probability (Theorem 2.15). By similar arguments, it can be shown that $n^{-1} \sum_{i=1}^n W_i \xrightarrow{\mathrm{P}} 0$. Furthermore, using the fact that the variance of a random variable is bounded by its second moment, we have

$$\begin{aligned}
\mathrm{E}(Y_i^2) &= \mathrm{E}[\mathrm{var}\{X_i 1_{(|X_i| \leq i)} | \mathcal{F}_{i-1}\}] \\
&\leq \mathrm{E}[\mathrm{E}\{X_i^2 1_{(|X_i| \leq i)} | \mathcal{F}_{i-1}\}] \\
&= \mathrm{E}\{X_i^2 1_{(|X_i| \leq i)}\}.
\end{aligned}$$

Thus, we have, by (8.28) and (8.29), $\sum_{i=1}^{\infty} i^{-2} \mathrm{E}(Y_i^2) < \infty$. It follows by Theorem 8.6 (or Corollary 8.3) that $n^{-1} \sum_{i=1}^n Y_i \to 0$ a.s., hence in probability (Theorem 2.7). This shows (8.26). The a.s. convergence under the stronger moment condition is left as an exercise (Exercise 8.26).

## 8.5 A martingale central limit theorem and related topic

It is often more convenient to consider an array, instead of a single sequence, of martingales, as far as the CLT is concerned (see below for further explanation), and the results may be presented more explicitly if we consider an array of martingale differences. This means that $S_{ni} = \sum_{j=1}^i X_{nj}$, $\mathcal{F}_{ni}$, $1 \leq i \leq k_n$, $n \geq 1$, is an array such that for each $n$, $S_{ni}$, $\mathcal{F}_{ni}$, $1 \leq i \leq k_n$, is a martingale, where $k_n$ is a nondecreasing sequence of positive integers such that $k_n \to \infty$ as $n \to \infty$ (e.g., $k_n = n$). Throughout this section, we assume that $S_{ni}$ has mean 0 and a finite second moment for all $n$ and $i$. It follows that $X_{n1} = S_{n1}$ and $X_{ni} = S_{ni} - S_{ni-1}$, $2 \leq i \leq k_n$. Here, for convenience we define $S_{n0} = 0$ and $\mathcal{F}_{n0} = \{\emptyset, \Omega\}$. Then for each $n$, $X_{ni}$, $\mathcal{F}_{ni}$, $1 \leq i \leq k_n$, is a sequence of martingale differences with $\mathrm{E}(X_{ni}) = 0$ and $\mathrm{E}(X_{ni}^2) < \infty$, $1 \leq i \leq k_n$.

We begin with the following well-known martingale CLT (Hall and Heyde 1980, p. 58). Let $Y_n$, $n \geq 1$, be a sequence of random variables on the probability space $(\Omega, \mathcal{F}, P)$ converging in distribution to a random variable $Y$. We say the convergence is *stable*, denoted by $Y_n \xrightarrow{P} Y$ (stably), if for all continuity points $y$ of $Y$ and all $A \in \mathcal{F}$, the limit $\lim_{n\to\infty} P(\{Y_n \leq y\} \cap A) = P_y(A)$ exists and $P_y(A) \to P(A)$ as $y \to \infty$.

**Theorem 8.7.** Let $X_{ni}$, $\mathcal{F}_{ni}$, $1 \leq i \leq k_n$, $n \geq 1$, be an array of martingale differences as above. Suppose that

$$\max_{1\leq i\leq k_n} |X_{ni}| \xrightarrow{P} 0, \tag{8.30}$$

$$\sum_{i=1}^{k_n} X_{ni}^2 \xrightarrow{P} \eta^2, \tag{8.31}$$

where $\eta^2$ is a random variable, and

$$E\left(\max_{1\leq i\leq k_n} X_{ni}^2\right) \text{ is bounded in } n. \tag{8.32}$$

In addition, assume that the $\sigma$-fields satisfy

$$\mathcal{F}_{ni} \subset \mathcal{F}_{n+1i}, \quad 1 \leq i \leq k_n, \ n \geq 1. \tag{8.33}$$

Then we have, as $n \to \infty$,

$$S_{nk_n} = \sum_{i=1}^{k_n} X_{ni} \xrightarrow{d} Z \text{ (stably)}, \tag{8.34}$$

where the random variable $Z$ has characteristic function

$$c_Z(t) = E\{\exp(-\eta^2 t^2/2)\}. \tag{8.35}$$

Note that the $\eta^2$ in (8.31) is allowed to be a random variable. In particular, if $\eta^2$ is a constant, say, $\eta^2 = 1$, then, by (8.35), we have $Z \sim N(0,1)$, which is the form of the classical CLT (see Section 6.4). Hall and Heyde (1980, pp. 59) noted that the restriction (8.33) on the $\sigma$-fields can be dropped if $\eta^2$ is a constant, provided that the word stably is removed from (8.34). This note turns out to be useful in many applications (see, for example, Section 8.8). On the other hand, Hall and Heyde (1980, p. 59–60) gave an example of an array of martingale differences for which all the conditions of Theorem 8.7 are satisfied, and yet $\eta^2$ is not a constant. We consider another example.

*Example 8.15* (Conditional logistic model). Suppose that given a random variable, $\alpha$, $X_1, X_2, \ldots$ are independent Bernoulli observations such that

$$\text{logit}\{P(X_i = 1|\alpha)\} = \mu + \alpha, \tag{8.36}$$

where $\mu$ is an unknown parameter, and $\text{logit}(p) = \log\{p/(1-p)\}$ for $p \in (0,1)$. Furthermore, suppose that $\alpha$ is distributed as $N(0, \sigma^2)$, where $\sigma^2$ is an unknown variance. Equation (8.36) suggests that the sum $S_n = \sum_{i=1}^n X_i$ is an important statistic in estimating the parameter $\mu$ (why?). Therefore, the asymptotic behavior of $S_n$ is of interest. Let $h(x) = e^x/(1+e^x)$, $-\infty < x < \infty$, which is the inverse function of logit. Define $X_{ni} = \{X_i - h(\mu + \alpha)\}/\sqrt{n}$ and $\mathcal{F}_{ni} = \mathcal{F}_i = \sigma(\alpha, X_1, \ldots, X_i)$, $1 \le i \le n$. We show that $X_{ni}$, $\mathcal{F}_{ni}$, $1 \le i \le n$, is an array of martingale differences.

Clearly, we have $X_{ni} \in \mathcal{F}_{ni}$, $1 \le i \le n$. Next, we show that $E(X_i|\mathcal{F}_{ni-1}) = h(\mu + \alpha)$ a.s. It suffices to show (see Appendix A.2) that for any Borel measurable function $f(x)$ and $g(x_1, \ldots, x_{i-1})$, we have

$$E\{f(\alpha)g(X_1, \ldots, X_{i-1})X_i\}$$
$$= E\{f(\alpha)g(X_1, \ldots, X_{i-1})h(\mu + \alpha)\}. \tag{8.37}$$

The proof of (8.37) is left as an exercise (Exercise 8.29). It follows that $E(X_{ni}|\mathcal{F}_{ni-1}) = 0$ a.s., $1 \le i \le n$.

Condition (8.30) is clearly satisfied because $|X_{ni}| \le 1/\sqrt{n}$ and so is (8.32). It is also clear that (8.33) is satisfied. It remains to verify condition (8.31) (which is usually the more challenging part compared to the other conditions). For this, we write

$$\sum_{i=1}^n X_{ni}^2 = \frac{1}{n} \sum_{i=1}^n \{X_i - h(\mu + \alpha)\}^2$$

$$= \frac{1}{n} \sum_{i=1}^n X_i^2 - 2h(\mu + \alpha)\left(\frac{1}{n}\sum_{i=1}^n X_i\right) + h^2(\mu + \alpha)$$

$$= \{1 - 2h(\mu + \alpha)\}\left(\frac{1}{n}\sum_{i=1}^n X_i\right) + h^2(\mu + \alpha),$$

because $X_i$ is 0 or 1; hence, $X_i^2 = X_i$, $1 \le i \le n$. Furthermore, it can be shown by the result derived above and Example 8.14 that $n^{-1}\sum_{i=1}^n X_i \overset{P}{\longrightarrow} h(\mu + \alpha)$ (Exercise 8.29). Therefore, we have $\sum_{i=1}^n X_{ni}^2 \overset{P}{\longrightarrow} h(\mu + \alpha)\{1 - h(\mu + \alpha)\} = \eta^2$.

It follows by Theorem 8.7 that $\sqrt{n}\{n^{-1}S_n - h(\mu + \alpha)\} = \sum_{i=1}^n X_{ni} \overset{d}{\longrightarrow} Z$ (stably) as $n \to \infty$, where $Z$ is a random variable having the cf (8.35).

A situation of this kind of observations may occur in practice when the population has clusters or subpopulations. Suppose that the probability of an individual having a certain disease within a certain cluster depends on a "latent" variable, $\alpha$, that depends on the cluster. In other words, there is a conditional probability of disease given $\alpha$, which is modeled by (8.36) in this example. This result shows that if one only samples from a given cluster, the asymptotic distribution of the sample proportion of disease, $n^{-1}S_n$, depends on the cluster-specific random variable $\alpha$, which, of course, makes sense. However, quite often in practice, people would collect samples from different clusters. In such a case, the asymptotic distribution of estimators of population parameters, such as $\mu$ and $\sigma^2$, will be unconditional [i.e., not dependent on $\alpha$ (see Chapter 12)].

Another application of Theorem 8.7 is considered later in Section 8.8.

A related topic to the martingale CLT is its convergence rate. Here, we consider two types of results. The first is the uniform convergence rate over all $x \in R$; the second is the nonuniform convergence rate, in which the bounds depends on $x$. The following theorem on the uniform convergence rate is the same as Theorem 3.7 of Hall and Heyde (1980), but presented in the form of a martingale array. We believe the latter form is more convenient to use in practice. One reason is that in many applications the observations are not "nested" as the same size $n$ increases; in other words, the observations under a smaller sample size are not necessarily a subset of those under a larger one. For example, a larger-scale survey run by one organization may not include samples from a smaller-scale survey run by a different organization. See Chapters 12 and 13 for many applications involving this type of data. Another reason is that normalization (or standardization) of the sequence is made more explicit under martingale array than under a single sequence of martingale. See Example 8.15 and another example in the sequel.

**Theorem 8.8.** Let $S_{ni} = \sum_{j=1}^{i} X_{nj}$, $\mathcal{F}_{ni}$, $1 \leq i \leq n$, be an array of martingales, where $\mathcal{F}_{ni} = \sigma(X_{n1}, \ldots, X_{ni})$, $1 \leq i \leq n$. Let $V_{ni}^2 = \sum_{j=1}^{i} \mathrm{E}(X_{nj}^2 | \mathcal{F}_{n\underline{j-1}})$, $1 \leq i \leq n$. Write $S_n = S_{nn}$ and $V_n^2 = V_{nn}^2$. If

$$\max_{1 \leq i \leq n} |X_{ni}| \leq \frac{M}{\sqrt{n}}, \tag{8.38}$$

$$P\left\{|V_n^2 - 1| > 9M^2 D \frac{(\log n)^2}{\sqrt{n}}\right\} \leq B \frac{\log n}{n^{1/4}} \tag{8.39}$$

for some constants $M$, $B$, and $D$ with $D \geq e$, then for $n \geq 2$, we have

$$\sup_{-\infty < x < \infty} |P(S_n \leq x) - \Phi(x)| \leq c \frac{\log n}{n^{1/4}}, \tag{8.40}$$

where $\Phi$ is the cdf of $N(0,1)$ and $c = 2 + B + 7M\sqrt{D}$.

Comparing (8.38) with the well-known Berry–Esseen bound (4.26), it is seen that the convergence rate is considerably slower for martingales than for sums of independent random variables. In fact, it can be shown that $n^{-1/4} \log n$ is the best possible rate for martingales (e.g., Hall and Heyde 1980, p. 84). The reason for the slower convergence rate is that the martingale differences are not independent. What the dependence does is reduce the effective sample size. [Think about an extreme case: If the same story is repeated twice, the effective same size is 1 (story), not 2 (stories).] On the other hand, the $n$ in the Berry–Essen bound represents the effective sample size, which equals the sample size in the independence case. If, however, there were a dependence among the sequence, by replacing $n$ with the effective sample size one would have ended up with a slower rate.

Also, note that conditions (8.38) and (8.39) imply that $\eta^2 = 1$ in (8.31) (with $k_n = n$). To see this, note that (8.39) implies that $V_n^2 \xrightarrow{P} 1$ as $n \to \infty$. Furthermore, write $U_n^2 = \sum_{i=1}^n X_{ni}^2$ and $Y_{ni} = X_{ni}^2 - \mathrm{E}(X_{ni}^2|\mathcal{F}_{ni-1})$, $1 \le i \le n$. Note that (8.38) implies $|Y_{ni}| \le M^2/n$ a.s. Thus, by (ii) of Lemma 8.6,

$$
\mathrm{E}(U_n^2 - V_n^2)^2 = \mathrm{E}\left(\sum_{i=1}^n Y_{ni}\right)^2
$$

$$
= \sum_{i=1}^n \mathrm{E}(Y_{ni}^2) \le \frac{M^4}{n}.
$$

It follows that $U_n^2 - V_n^2 \to 0$ in $L^2$, and hence in probability as $n \to \infty$. We consider a specific example.

*Example 8.16.* Suppose that $\xi_1, \xi_2, \ldots$ are independent such that $\mathrm{P}(\xi_i = -1) = \mathrm{P}(\xi_i = 1) = 1/2$, $i \ge 1$. Define $X_i = \xi_1 \cdots \xi_i$, $i \ge 1$, and we would like to obtain the convergence rate of (8.40) for $\sum_{i=1}^n X_i$ after a suitable normalization. Let $X_{ni} = X_i/\sqrt{n}$ and $\mathcal{F}_{ni} = \sigma(X_{n1}, \ldots, X_{ni}) = \sigma(X_1, \ldots, X_i) = \sigma(\xi_1, \ldots, \xi_i)$ (why?), $1 \le i \le n$. Then we have $X_{ni} \in \mathcal{F}_{ni}$ and

$$
\mathrm{E}(X_{ni}|\mathcal{F}_{ni-1}) = \frac{\xi_1 \cdots \xi_{i-1}}{\sqrt{n}} \mathrm{E}(\xi_i|\xi_1, \ldots, \xi_{i-1})
$$

$$
= \frac{\xi_1 \cdots \xi_{i-1}}{\sqrt{n}} \mathrm{E}(\xi_i) = 0.
$$

Thus, $X_{ni}$, $\mathcal{F}_{ni}$, $1 \le i \le n$, is an array of martingale differences or, equivalently, $S_{ni} = \sum_{j=1}^i X_{nj}$, $\mathcal{F}_{ni}$, $1 \le i \le n$, is an array of martingales. Also note that $X_i^2 = 1$ for all $i$. It follows that $X_{ni}^2 = 1/n$, $1 \le i \le n$, and $V_n^2 = 1$. Thus, (8.38) and (8.39) are satisfied with $M = 1$ and $B = 0$. Therefore, if we let $D = e$ (the smallest value for $D$), we have (8.40) for $n \ge 2$ with $S_n = \sum_{i=1}^n X_{ni} = n^{-1/2} \sum_{i=1}^n X_i$ and $c = 2 + 7\sqrt{e} \approx 13.6$.

Now, consider the nonuniform convergence rate in CLT. The following theorem, again in the form of martingale array, can be derived Theorem 3.9 of Hall and Heyde (1980).

**Theorem 8.9.** With the same notation of Theorem 8.8 and $U_n^2 = \sum_{i=1}^n X_{ni}^2$, define, for any $0 < \delta \le 1$,

$$
p_n = \sum_{i=1}^n \mathrm{E}\left\{|X_{ni}|^{2(1+\delta)}\right\} + \mathrm{E}\left(|U_n^2 - 1|^{1+\delta}\right),
$$

$$
q_n = \sum_{i=1}^n \mathrm{E}\left\{|X_{ni}|^{2(1+\delta)}\right\} + \mathrm{E}\left(|V_n^2 - 1|^{1+\delta}\right).
$$

There is a constant $c_\delta$ depending only on $\delta$ such that for all $x$,

$$|P(S_n \leq x) - \Phi(x)| \leq c_\delta \frac{(p_n \wedge q_n)^{1/(3+2\delta)}}{1 + |x|^{4(1+\delta)^2/(3+2\delta)}}. \tag{8.41}$$

Again, we consider a simple example.

*Example 8.16 (continued).* Since $X_{ni}^2 = 1$ for all $i$, if we let $\delta = 1$, we have $p_n = q_n = n$. Thus, the left side of (8.41) is bounded by $c_1 n^{1/5} (1 + |x|^{16/5})^{-1}$. It might seem that by letting $\delta \to 0$, one might be able to obtain a nonuniform convergence rate of $n^{1/3}(1 + |x|^{4/3})^{-1}$. However, this is not going to happen, as far as Theorem 8.9 is concerned, because the constant $c_\delta \to \infty$ as $\delta \to 0$.

## 8.6 Convergence rate in SLLN and LIL

After a discussion on the convergence rate in the martingale CLT, we now turn our attention to convergence rate in the martingale SLLN. Let $S_n = \sum_{i=1}^n X_i$, $\mathcal{F}_n$, $n \geq 0$, be a martingale with $S_0 = 0$. We assume that $E(X_i^2) < \infty$ for all $i$, and let $\sigma_n^2 = \sum_{i=1}^n E(X_i^2)$. Lagodowski and Rychlik (1986) proved the following result, where $\lim_{\epsilon \to 0+}$ means that $\epsilon \to 0$ while $\epsilon > 0$.

**Theorem 8.10.** Suppose that there are constants $b_i > 0$ such that

$$E(X_i^2 | \mathcal{F}_{i-1}) \leq b_i \quad \text{a.s.} \tag{8.42}$$

and constants $0 < c_1 < c_2$ such that

$$c_1 \sigma_n^2 \leq \sum_{i=1}^n b_i \leq c_2 \sigma_n^2. \tag{8.43}$$

Furthermore, suppose that

$$\sup_{-\infty < x < \infty} |P(S_n < x\sigma_n) - \Phi(x)| \longrightarrow 0 \tag{8.44}$$

as $n \to \infty$. For any $q \geq 2$ and $q/2 < p \leq q$, if

$$\lim_{A \to \infty} \lim_{\epsilon \to 0+} \epsilon^{r(p-1)} \sum_{n > A/\epsilon^r} n^{p-2} \sum_{i=1}^n P(|X_i| \geq \epsilon \sigma_n n^{1/r}) = 0, \tag{8.45}$$

where $r = 2q/(2p - q)$, then we have

$$\lim_{\epsilon \to 0+} \epsilon^{r(p-1)} \sum_{n=1}^\infty n^{p-2} P(|S_n| \geq \epsilon \sigma_n n^{1/r})$$
$$= \frac{2^{r(p-1)/2} \Gamma[\{1 + r(p-1)\}/2]}{(p-1)\Gamma(1/2)}, \tag{8.46}$$

where $\Gamma$ is the gamma function.

Note that condition (8.44) requires uniform convergence in the CLT. Also note that if $\xi_n$, $n \geq 1$, is a sequence of random variables and $F$ is a cdf that is continuous on $(-\infty, \infty)$, then $\sup_{-\infty < x < \infty} |P(\xi_n < x) - F(x)| \to 0$ if and only if $\sup_{-\infty < x < \infty} |P(\xi_n \leq x) - F(x)| \to 0$ (Exercise 8.32). Therefore, one can replace the $<$ in (8.44) by $\leq$. We consider an example.

*Example 8.16 (continued).* It is obvious that (8.42) and (8.43) are satisfied in this case and $\sigma_n^2 = n$. Furthermore, by Theorem 8.8 (with $X_{ni} = X_i/\sqrt{n}$) and the above note, (8.44) is satisfied. Now, let $p = q = 2$; hence, $r = 2$. Since $P(|X_i| \geq \epsilon n) = 0$ if $n > \epsilon^{-1}$ and $1$ if $n \leq \epsilon^{-1}$, we have, for any $A > 0$,

$$\sum_{n > A/\epsilon^2} \sum_{i=1}^{n} P(|X_i| \geq \epsilon n) = \sum_{A\epsilon^{-2} < n \leq \epsilon^{-1}} n = 0$$

if $\epsilon \leq A$; hence, the inside limit of (8.45) is zero for every $A > 0$. Therefore (8.45) is satisfied. It follows by (8.46) that

$$\lim_{\epsilon \to 0^+} \epsilon^2 \sum_{n=1}^{\infty} P(|S_n| \geq \epsilon n) = 1$$

using the equation $\Gamma(x+1) = x\Gamma(x)$, $x > 0$.

Theorem 8.10 describes the convergence rate in the SLLN in terms of the decay of the probability $P(|\sigma_n^{-1} S_n| \geq \epsilon n^{1/r})$ as $n \to \infty$. Note that $r \geq 2$; hence, $0 < 1/r \leq 1/2$. Another way to describe the convergence rate in the SLLN is the LIL (see Section 6.5). For example, Hall and Heyde (1980) gave the following result. Let $W_n, n \geq 1$, be a nondecreasing sequence of positive random variables (i.e., $0 < W_1 \leq W_2 \leq \cdots$) and $Z_n$, $n \geq 1$, be a sequence of nonnegative random variables. Suppose that both sequences are predictable with respect to $\mathcal{F}_n$, $n \geq 1$. Define $\phi(t) = \sqrt{2t \log \log t}$ if $t > e$ and $\phi(t) = 1$ otherwise.

**Theorem 8.11.** Suppose that $W_n \xrightarrow{\text{a.s.}} \infty$ and $W_n/W_{n+1} \xrightarrow{\text{a.s.}} 1$ as $n \to \infty$ and that the following conditions are satisfied:

$$\frac{1}{\phi(W_n^2)} \sum_{i=1}^{n} [X_i 1_{(|X_i| > Z_i)} - E\{X_i 1_{(|X_i| > Z_i)} | \mathcal{F}_{i-1}\}] \xrightarrow{\text{a.s.}} 0, \qquad (8.47)$$

$$\frac{1}{W_n^2} \sum_{i=1}^{n} \text{var}\{X_i 1_{(|X_i| \leq Z_i)} | \mathcal{F}_{i-1}\} \xrightarrow{\text{a.s.}} 1, \qquad (8.48)$$

$$\sum_{i=1}^{\infty} \frac{1}{W_i^4} E\{X_i^4 1_{(|X_i| \leq Z_i)} | \mathcal{F}_{i-1}\} < \infty \text{ a.s.} \qquad (8.49)$$

Then we have $\limsup S_n/\phi(W_n^2) = 1$ a.s. and $\liminf S_n/\phi(W_n^2) = -1$ a.s.

We consider some examples.

*Example 8.16 (continued).* Let $W_n = \sqrt{n}$ and $Z_i = 1$. Then the left side of (8.47) is identical to zero. Also, we have

$$\text{var}\{X_i 1_{(|X_i| \le Z_i)}|\mathcal{F}_{i-1}\} = \text{var}(X_i|\mathcal{F}_{i-1})$$
$$= \text{E}(X_i^2|\mathcal{F}_{i-1}) = 1,$$

and similarly $\text{E}(X_i^4|\mathcal{F}_{i-1}) = 1$. Therefore, the left side of (8.48) is equal to 1, and the left side of (8.49) is equal to $\sum_{i=1}^{\infty} i^{-2} < \infty$. Therefore, all of the conditions of Theorem 8.11 are satisfied. It follows that $\limsup S_n/\sqrt{2n \log \log n} = 1$ a.s. and $\liminf S_n/\sqrt{2n \log \log n} = -1$ a.s.

On the other hand, the conditions of Theorem 8.11 are not necessary in the sense that for given sequences $W_i$ and $Z_i$ satisfying the conditions of the theorem, there exists a martingale $S_n = \sum_{i=1}^{n} X_i$, $\mathcal{F}_n$, $n \ge 1$, that does not satisfy conditions (8.47)–(8.49), and yet the conclusion of the theorem still holds for the martingale. To see an example, consider the following.

*Example 8.17.* Let $W_n = \sqrt{n}$ and $Z_i = i$. Let $X_1, X_i, \ldots$ be a sequence of i.i.d. random variables such that $\text{E}(X_i) = 0$, $\text{E}(X_i^2) = 1$, and $\text{E}(|X_i|^3) = \infty$ (e.g., let $X_i = \xi_i/\sqrt{3}$, where $\xi_i \sim t_3$). Then $S_n = \sum_{i=1}^{n} X_i$, $\mathcal{F}_n = \sigma(X_1, \ldots, X_n)$, $n \ge 1$, is a martingale, and the conclusion of Theorem 8.11 holds by Hartman and Wintner's LIL (Theorem 6.17). On the other hand, we show that the sequence $X_i$, $i \ge 1$, does not satisfy (8.49). To see this, note that for any $a \ge 1$, we have

$$\sum_{i \ge a} \frac{1}{i^2} \ge \sum_{i=[a]}^{\infty} \int_i^{i+1} \frac{dx}{x^2} = \int_{[a]}^{\infty} \frac{dx}{x^2} = \frac{1}{[a]},$$

where $[a]$ represents the largest integer $\le a$. It follows that

$$\sum_{i=1}^{\infty} \frac{1}{W_i^4} \text{E}\{X_i^4 1_{(|X_i| \le Z_i)}\}|\mathcal{F}_{i-1}\}$$

$$= \sum_{i=1}^{\infty} \frac{1}{i^2} \text{E}\{X_1^4 1_{(|X_1| \le i)}\}$$

$$= \text{E}\left\{X_1^4 \sum_{i=1}^{\infty} \frac{1}{i^2} 1_{(|X_1| \le i)}\right\}$$

$$= \text{E}\left(X_1^4 \sum_{i \ge |X_1| \vee 1} \frac{1}{i^2}\right)$$

$$\geq \mathrm{E}\left(\frac{X_1^4}{[|X_1| \vee 1]}\right)$$

$$\geq \mathrm{E}\left\{\frac{X_1^4}{[|X_1| \vee 1]}1_{(|X_1|\geq 1)}\right\}$$

$$= \mathrm{E}\{|X_1|^3 1_{(|X_1|\geq 1)}\}$$

$$= \infty.$$

## 8.7 Invariance principles for martingales

This section deals with a similar topic as Section 6.6.1. The first invariance principle for martingales was derived by Billingsley (1968), who considered stationary and ergodic (see the definition following Theorem 7.14) martingale differences. In the following we assume that $S_n = \sum_{i=1}^n X_i$, $\mathcal{F}_n$, $n \geq 1$, is a martingale with mean 0 and a finite second moment. Let $S_0 = 0$ and $\mathcal{F}_0 = \{\emptyset, \Omega\}$ for convenience. In the case of stationary martingale differences, a straightforward extension of the results of Section 6.6.1 for sums of i.i.d. random variables would be to consider (6.64) with $\sqrt{n}$ replaced by $\sigma\sqrt{n}$, where $\sigma^2 = \mathrm{E}(X_i^2)$ for $t \in [0,1]$. However, without the stationarity assumption, such an extension may not be meaningful. Hall and Heyde (1980) considered the following variation of (6.64):

$$\xi_n(t) = \frac{1}{U_n}\left(S_i + \frac{tU_n^2 - U_i^2}{X_{i+1}}\right) \quad \text{for} \quad \frac{U_i^2}{U_n^2} \leq t < \frac{U_{i+1}^2}{U_n^2}, \qquad (8.50)$$

$0 \leq i \leq n-1$, and $\xi_n(1) = U_n^{-1}S_n$, where $U_0^2 = 0$ and $U_i^2 = \sum_{j=1}^i X_j^2$, $i \geq 1$. Intuitively, $\xi_n$ is a function on $[0,1]$ obtained by linear interpolating between the (two-dimensional) points $(U_i^2/U_n^2, S_i/U_n)$, $i = 0, \ldots, n$ (Exercise 8.33). Since $\xi_n$ is continuous, it is a member of the space $\mathcal{C}$ of continous functions on $[0,1]$ equipped with the uniform distance $\rho$ of (6.63). Then we have the following result.

**Theorem 8.12.** Suppose that the following Lindeberg condition holds:

$$\frac{1}{s_n^2}\sum_{i=1}^n \mathrm{E}\{X_i^2 1_{(|X_i|>\epsilon s_n)}\} \longrightarrow 0 \qquad (8.51)$$

as $n \to \infty$ for every $\epsilon > 0$, where $s_n^2 = \mathrm{E}(S_n^2)$, and that

$$\frac{U_n^2}{s_n^2} \overset{\mathrm{P}}{\longrightarrow} \eta^2, \qquad (8.52)$$

where the random variable $\eta^2$ is a.s. positive. Then $\xi_n \overset{\mathrm{d}}{\longrightarrow} W$, where $W$ is the Brownian motion on $[0,1]$.

Note that here the convergence in distribution is in $(\mathcal{C}, \rho)$, the same as in Theorem 6.18. We consider some examples.

*Example 8.18.* Note that (8.50) does not reduce to (6.64), even in the special case of i.i.d. observations. However, in the latter case, it is trivial to verify the conditions (of Theorem 8.12). To see this, note that if $X_1, X_2, \ldots$ are i.i.d. with $E(X_i) = 0$, $E(X_i^2) = \sigma^2 \in (0, \infty)$, then we have $s_n^2 = \sigma^2 n$. It follows that

$$\frac{1}{s_n^2} \sum_{i=1}^{n} E\{X_i^2 1_{(|X_i| > \epsilon s_n)}\} = \frac{1}{\sigma^2} E\{X_1^2 1_{(|X_1| > \epsilon \sigma \sqrt{n})}\},$$

which goes to zero as $n \to \infty$ for every $\epsilon > 0$ (why?). Furthermore, we have

$$\frac{U_n^2}{s_n^2} = \frac{1}{\sigma^2 n} \sum_{i=1}^{n} X_i^2 \xrightarrow{P} 1$$

by the WLLN. Therefore the conditions of Theorem 8.12 are satisfied.

*Example 8.16 (continued).* Note that in this case we have $X_i^2 = 1$, $i \geq 1$; hence, $s_n^2 = n$. It follows that $E\{X_i^2 1_{(|X_i| > \epsilon s_n)}\} = P(1 > \epsilon\sqrt{n}) = 0$ if $\epsilon\sqrt{n} \geq 1$, and $s_n^{-2} U_n^2 = 1$. Thus, once again, the conditions of Theorem 8.12 are obvious.

*Example 8.19.* As in Sections 6.6.1 (also see Section 7.3), a result like Theorem 8.12 may have many applications. This is because $\xi_n \xrightarrow{d} W$ implies that $h(\xi_n) \xrightarrow{d} h(W)$ for any continuous function $h$ on $\mathcal{C}$. In particular, if one considers $h(x) = x(1)$ for $x \in \mathcal{C}$, then the result implies $U_n^{-1} S_n \xrightarrow{d} W(1) \sim N(0, 1)$. In other words, we have a CLT for a martingale $S_n$ normalized by $U_n$. If one considers $h(x) = \sup_{t \in [0,1]} x(t)$ and notes that $\sup_{t \in [0,1]} \xi_n(t) = U_n^{-1} \max_{0 \leq i \leq n} S_i$, then we have $U_n^{-1} \max_{0 \leq i \leq n} S_i \xrightarrow{d} \sup_{t \in [0,1]} W(t)$.

Hall and Heyde proved their result by using the following Skorokhod representation and limit theorem for Brownian motion. If $S_n = \sum_{i=1}^{n} X_i, \mathcal{F}_n, n \geq 1$, is a zero-mean, square-integrable martingale, then there exists a standard Brownian motion $W$ defined on a probability space and a sequence of non-negative random variables $\tau_n$, $n \geq 1$, with the following properties, where $T_n = \sum_{i=1}^{n} \tau_i$, $\tilde{S}_n = W(T_n)$, $\tilde{X}_1 = \tilde{S}_1$, $\tilde{X}_n = \tilde{S}_n - \tilde{S}_{n-1}$, $n \geq 2$, and $\mathcal{G}_n$ is the $\sigma$-field generated by $\tilde{S}_i, 1 \leq i \leq n$, and $W(t), 0 \leq t \leq T_n$: (i) $S_n$, $n \geq 1$, has the same joint distribution as $\tilde{S}_n$, $n \geq 1$; (ii) $T_n \in \mathcal{G}_n$, $n \geq 1$; and (iii) $E(\tau_n | \mathcal{G}_{n-1}) = E(\tilde{X}_n^2 | \mathcal{G}_{n-1})$ a.s. The limit theorem for Brownian motion states that if $W(t)$, $t \geq 0$, is the standard Brownian motion and $T_n$, $n \geq 1$, is a sequence of positive random variables, then $\tilde{\xi}_n \xrightarrow{d} W_1$ in $(\mathcal{C}, \rho)$, where $\tilde{\xi}_n(t) = W(tT_n)/\sqrt{T_n}$, $t \in [0, 1]$, and $W_1$ is the restriction of $W$ to $[0, 1]$, provided that there is a sequence of constants $c_n$ such that $T_n/c_n \xrightarrow{P} \eta^2$, where $\eta^2$ is a.s. positive. See Section 10.5 for more details.

We now consider the invariance principle in the LIL. Heyde and Scott (1973) extended Strassen's (1964) invariance principle in the LIL to martingales. Recall the space $K$ of absolutely continuous functions $x$ on $[0, 1]$ with $x(0) = 0$ and satisfying (6.69), and the function $\phi(t)$ defined above Theorem 8.11. Hall and Heyde (1980) considered normalizing the martingale $S_n$ based on a general sequence of random variables $W_i$, $i \geq 1$, satisfying $0 < W_1 \leq W_2 \leq \cdots$, and

$$\zeta_n(t) = \frac{1}{\phi(W_n^2)} \left( S_i + \frac{tW_n^2 - W_i^2}{W_{i+1}^2 - W_i^2} X_{i+1} \right)$$

$$\text{for } \frac{W_i^2}{W_n^2} \leq t < \frac{W_{i+1}^2}{W_n^2}, \tag{8.53}$$

$0 \leq i \leq n-1$, and $\zeta_n(1) = \phi^{-1}(W_n^2)S_n$. It is clear that, except for the different denominators, (8.50) is a special case of (8.53) with $W_i = U_i$, provided that $X_1^2 > 0$ a.s. Theorem 8.11 in the previous section can now be extended to an invariance principle for $\zeta_n$. Recall the definition of an a.s. relative compact sequence in Section 7.4 (above Theorem 7.7).

**Theorem 8.13.** Under the conditions of Theorem 8.11 we have $\zeta_n$ r.c. $K$ w.r.t. $\rho$ of (6.63) on $\mathcal{C}$ a.s.

In words, we have with probability 1 that the sequence $\zeta_n$ is relative compact in $\mathcal{C}$ and its set of $\rho$ limit points coincides with $K$, where $\rho$ is the uniform distance of (6.63). Note that the assumption that $W_n$ is predictable does not exclude $U_n$ from application. This is because one may replace $U_n$ by $U_{n-1}$, which is predictable. On the other hand, the assumption that $U_n/U_{n+1} \xrightarrow{\text{a.s.}} 1$ ensures that normalizing by $U_n$ is asymptotically equivalent to normalizing by $U_{n-1}$. We consider a simple example.

*Example 8.16 (continued).* Earlier in Section 8.6 we showed that the sequence $X_i$ in this example satisfies all the conditions of Theorem 8.11 with $W_n = \sqrt{n}$ and $Z_n = 1$. It follows that $\zeta_n$ r.c. $K$ w.r.t. $\rho$ on $\mathcal{C}$ a.s., where

$$\zeta_n(t) = \frac{S_i + (tn - i)X_{i+1}}{\sqrt{2n \log \log n}}, \quad \frac{i}{n} \leq t < \frac{i+1}{n},$$

$0 \leq i \leq n - 1$, and $\zeta_n(1) = S_n/\sqrt{2n \log \log n}$.

## 8.8 Case study: CLTs for quadratic forms

There is a great deal of statistical inference based on quadratic functions of random variables. For example, the log-likelihood function under a Gaussian model depends quadratically on the data; many of the goodness-of-fit (or lack-of-fit) measures involve the data in squared Euclidean distance; of course, estimators of variances and covariances are usually quadratic functions of the data. Let $\xi_{ni}$, $1 \leq i \leq k_n$, $n \geq 1$, be an array of random variables such that for each $n$, $\xi_{ni}$, $1 \leq i \leq k_n$, are independent with mean 0, and let $A_n = (a_{nij})_{1 \leq i,j \leq k_n}$ be a sequence of (nonrandom) real symmetric matrices. Write $\xi_n = (\xi_{ni})_{1 \leq i \leq k_n}$. We are interested in the limiting behavior of

$$Q_n = \xi'_n A_n \xi_n. \tag{8.54}$$

Such a problem is of direct interest in statistical inference, even if the observations themselves are not independent. For example, in Chapter 12 we discuss application of large-sample techniques in linear mixed models. The latter is defined as observations $y_1, \ldots, y_n$ satisfying

$$y = X\beta + Z_1\alpha_1 + \cdots + Z_s\alpha_s + \epsilon, \tag{8.55}$$

where $y = (y_i)_{1 \leq i \leq n}$, $X$ is matrix of known covariates, $\beta$ is a vector of unknown fixed effects, $Z_r$, $1 \leq r \leq s$, are known matrices, $\alpha_r$, $1 \leq r \leq s$, are vectors of (unobservable) random effects, and $\epsilon$ is a vector of errors. (As will be seen in later chapters, it is more customary in statistical literature to use lowercase letters, such as $y$, to represent observed data and uppercase letters, such as $X$, for known covariate or design matrices, and we will gradually adopt such changes in notation.) It is further assumed that the components of $\alpha_r$ are independent with mean 0 and unknown variance $\sigma_r^2$, $1 \leq r \leq s$; the components of $\epsilon$ are independent with mean 0 and unknown variance $\sigma_0^2$; and $\alpha_1, \ldots, \alpha_s, \epsilon$ are independent. It is easy to see that, even if the random effects and errors are independent, the observations $y_1, \ldots, y_n$ are typically correlated. This is because the same random effects may be "shared" by many observations. For example, consider the following.

*Example 8.20.* Suppose that the observations $y_{ij}$, $1 \leq i \leq m_1$, $1 \leq j \leq m_2$, satisfy $y_{ij} = \mu + u_i + v_j + e_{ij}$, where $\mu$ is an unknown mean, the $u_i$'s and $v_j$'s are independent random effects such that $u_i \sim N(0, \sigma_1^2)$ and $v_j \sim N(0, \sigma_2^2)$, the $\epsilon_{ij}$'s are independent errors such that $\epsilon_{ij} \sim N(0, \sigma_0^2)$, and the random effects and errors are independent. It can be shown that the model is a special case of (8.55) (Exercise 8.34). Under the assumed model, there are multiple observations sharing the same random effects. For example, the random effect $u_i$ is shared by all of the observations $y_{ij}$, $1 \leq j \leq m_2$; similarly, the random effect $v_j$ is shared by all of the observations $y_{ij}$, $1 \leq i \leq m_1$. As a result, there are correlations among the observations. Such a model is often called a variance components model. For example, in animal and dairy science, variance components models are used to model different sources of variations, such as the sire (i.e., male animals) and environmental effects.

According to our earlier discussion in Section 5.6—in particular, (5.99) and (5.100)—the ML or REML estimators of the variance components $\sigma_r^2$, $0 \leq r \leq s$, depend on $y$ through the quadratic forms

$$Q = y'PZ_rZ_r'Py, \quad 0 \le r \le s, \tag{8.56}$$

where $Z_0 = I$, the identity matrix, and $P = A(A'VA)^{-1}A'$. (Again, it is customary in statistical literature to suppress the subscript $n$ representing the sample size, so, for example, we write $Q$ instead of $Q_n$; but keep in mind that the objects we are dealing with depend on the sample size if asymptotics are under consideration.) Recall that $A$ is a full rank matrix such that $A'X = 0$. Thus, if we let $\xi$ represent the combined vector of random effects and errors [i.e., $\zeta - (\epsilon', \alpha_1', \ldots, \alpha_s')'$], then (8.56) is equal to

$$\begin{aligned} Q &= (y - X\beta)'PZ_rZ_r'P(y - X\beta) \\ &= \xi'Z'PZ_rZ_r'PZ\xi, \quad 0 \le r \le s, \end{aligned}$$

where $Z = (I \; Z_1 \; \cdots \; Z_s)$. It follows that the ML and REML estimators depend on quadratic forms in independent random variables.

In fact, such problems as asymptotic behavior of REML estimators have led Jiang (1996) to consider CLTs for quadratic forms in independent random variables expressed in the general form of (8.54). There had been results on similar topics prior to Jiang's study. Some of these applied only to a special kind of random variables (e.g., Guttorp and Lockhart 1988) or to $A_n$ with a special structure (e.g., Fox and Taqqu 1985). Rao and Kleffe (1988) derived a more general form of CLT for quadratic forms in independent random variables, extending an earlier result of Schmidt and Thrum (1981). However, as noted by Rao and Kleffe (1988, p. 51), "the applications (of the theorem) might be limited as it is essentially based on the assumption that the off diagonal blocks of $A_n$ tend to zero." Such restrictions were removed by Jiang (1996), whose approach is a classical application of the martingale CLT introduced in Section 8.5. Note that $E(Q_n) = \sum_{i=1}^{k_n} a_{nii}E(\xi_{ni}^2)$. Thus, we have

$$\begin{aligned} Q_n - E(Q_n) &= \sum_{1 \le i,j \le k_n} a_{nij}\xi_{ni}\xi_{nj} - \sum_{i=1}^{k_n} a_{nii}E(\xi_{ni}^2) \\ &= \sum_{i=1}^{k_n} a_{nii}\{\xi_{ni}^2 - E(\xi_{ni}^2)\} + \sum_{i \ne j} a_{nij}\xi_{ni}\xi_{nj} \\ &= \sum_{i=1}^{k_n} a_{nii}\{\xi_{ni}^2 - E(\xi_{ni}^2)\} + 2\sum_{i=1}^{k_n}\left(\sum_{j<i} a_{nij}\xi_{nj}\right)\xi_{ni} \\ &= \sum_{i=1}^{k_n} X_{ni}, \end{aligned} \tag{8.57}$$

where $X_{ni} = a_{nii}\{\xi_{ni}^2 - E(\xi_{ni}^2)\} + 2(\sum_{j<i} a_{nij}\xi_{nj})\xi_{ni}$. Let $\mathcal{F}_{ni} = \sigma(\xi_{nj}, 1 \le j \le i)$, $1 \le i \le k_n$. It is easy to verify that $X_{ni}, \mathcal{F}_{ni}, 1 \le i \le k_n$, is an array of martingale differences (see Example 8.3). Due to this important observation, the martingale CLT (Theorem 8.7) becomes a natural tool to derive the CLT

for $Q_n$. Before we explore Jiang's results in further detail, let us first consider some examples to see what to expect.

*Example 8.21.* If $\xi_{ni}$, $1 \leq i \leq k_n$, are distributed as $N(0,1)$, then

$$\frac{Q_n - \mathrm{E}(Q_n)}{\{\mathrm{var}(Q_n)\}^{1/2}} \xrightarrow{\mathrm{d}} N(0,1) \tag{8.58}$$

if and only if

$$\frac{\|A_n\|}{\|A_n\|_2} \longrightarrow 0, \tag{8.59}$$

where $\|A\|$ and $\|A\|_2$ are the spectral norm and 2-norm of a matrix $A$ defined above (5.40) and (5.41), respectively. The proof of this result is left as an exercise (Exercise 8.35).

However, such a nice result may not hold for a general array of independent random variables, as the following example shows.

*Example 8.22.* Let $A_n$ be the $n$-dimensional identity matrix and $\xi_{ni}$, $1 \leq i \leq n$, be independent such that $\mathrm{P}(\xi_{ni} = -1) = \mathrm{P}(\xi_{ni} = 1) = (n-2)/2n$, $\mathrm{P}(\xi_{ni} = -\sqrt{2}) = \mathrm{P}(\xi_{ni} = \sqrt{2}) = 1/2n$, and $\mathrm{P}(\xi_{ni} = 0) = 1/n$, $1 \leq i \leq n$, $n \geq 2$. By Theorem 6.13, it can be shown that (8.58) fails, despite the fact that (8.59) holds (Exercise 8.36).

The situation in Example 8.22 is somehow extreme because the (squares of the) random variables are asymptotically degenerated. Such cases must be excluded if one attempts to generalize the result of Example 8.21. Jiang (1996) proved the following theorems. Let $A_n^{\circ} = A_n - \mathrm{diag}(a_{nii}, 1 \leq i \leq k_n)$ (here the superscript o refers to "off-diagonal") and $\mathcal{A}_n = \{1 \leq i \leq k_n, a_{nii} \neq 0\}$.

**Theorem 8.14.** If

$$\inf_{n \geq 1} \left\{ \min_{1 \leq i \leq k_n} \mathrm{var}(\xi_{ni}) \right\} \wedge \left\{ \min_{i \in \mathcal{A}_n} \mathrm{var}(\xi_{ni}^2) \right\} > 0,$$

$$\sup_{n \geq 1} \left[ \max_{1 \leq i \leq k_n} \mathrm{E}\{\xi_{ni}^2 1_{(|\xi_{ni}| > x)}\} \right] \vee \left[ \max_{i \in \mathcal{A}_n} \mathrm{E}\{\xi_{ni}^4 1_{(|\xi_{ni}| > x)}\} \right] \longrightarrow 0$$

as $x \to \infty$, then (8.59) implies (8.58).

To state the next result we first introduce some notation. Let $b_{ni}$, $1 \leq i \leq k_n$, $n \geq 1$, be an array of nonnegative constants. Define $\gamma_{ni}^{(1)} = \mathrm{E}\{\xi_{ni}^4 1_{(|\xi_{ni}| \leq b_{ni})}\}$, $\gamma_{ni}^{(2)} = \mathrm{E}\{(\xi_{ni}^2 - 1)^4 1_{(|\xi_{ni}| \leq b_{ni})}\}$, $\delta_{ni}^{(1)} = \mathrm{E}\{X_{ni}^2 1_{(|\xi_{ni}| > b_{ni})}\}$, and $\delta_{ni}^{(2)} = \mathrm{E}\{(\xi_{ni}^2 - 1)^2 1_{(|\xi_{ni}| > b_{ni})}\}$. Then define

$$\gamma_{nij} = \begin{cases} \gamma_{ni}^{(1)}\gamma_{nj}^{(1)} & \text{if } i \neq j \\ \gamma_{ni}^{(2)} & \text{if } i = j; \end{cases}$$

$$\delta_{nij} = \begin{cases} \{\delta_{ni}^{(1)} + \delta_{nj}^{(1)}\}/2 & \text{if } i \neq j, \\ \delta_{ni}^{(2)} & \text{if } i = j \in \mathcal{A}_n, \\ 0 & \text{otherwise.} \end{cases}$$

**Theorem 8.15.** Suppose that $\mathrm{E}(\xi_{ni}^2) = 1$ for all $i$ and $n$ and there are $b_{ni}$ as above such that

$$\frac{1}{\sigma_n^2} \sum_{i,j=1}^{k_n} a_{nij}^2 \delta_{nij} \longrightarrow 0,$$

$$\frac{1}{\sigma_n^4} \left\{ \sum_{i,j=1}^{k_n} a_{nij}^4 \gamma_{nij} + \sum_{i=1}^{k_n} \left( \sum_{j \neq i} a_{nij}^2 \right)^2 \gamma_{ni}^{(1)} \right\} \longrightarrow 0,$$

where $\sigma_n^2 = \mathrm{var}(Q_n)$. Then (8.58) holds provided that $\|A_n^\circ\|/\sigma_n \to 0$.

It might seem that Theorem 8.15 is more restrictive than Theorem 8.14 because of the assumption $\mathrm{E}(\xi_{ni}^2) = 1$. It is, in fact, the opposite. Jiang (1996) showed that Theorem 8.14 can be derived from Theorem 8.15 with a special choice of $b_{ni}$ and a simple transformation.

As for the proof of Theorem 8.15, the key steps are to verify the conditions of Theorem 8.7—namely, (8.30)–(8.32). [As noted following Theorem 8.7, (8.33) is not needed if $\eta^2$ is a constant and the word stably is removed from (8.34).] However, sometimes these conditions are not easy to verify directly, such as in this case. A technique that is often used in such situations is called *truncation*. Let $u_{ni} = \xi_{ni} 1_{(|\xi_{ni}| \leq b_{ni})} - \mathrm{E}\{\xi_{ni} 1_{(|\xi_{ni}| \leq b_{ni})}\}$, and $U_{ni} = (\xi_{ni}^2 - 1)1_{(|\xi_{ni}| \leq b_{ni})} - \mathrm{E}\{(\xi_{ni}^2 - 1)1_{(|\xi_{ni}| \leq b_{ni})}\}$, and

$$Y_{ni} = \frac{1}{\sigma_n} \left\{ a_{nii} U_{ni} + 2 \left( \sum_{j<i} a_{nij} u_{nj} \right) u_{ni} \right\}.$$

It is easy to verify that $Y_{ni}, \mathcal{F}_{ni}, 1 \leq i \leq k_n$, is an array of martingale differences. Furthermore, it can be shown that

$$\frac{Q_n - \mathrm{E}(Q_n)}{\sigma_n} = \sum_{i=1}^{k_n} Y_{ni} + \Delta_n,$$

where $\Delta_n \to 0$ in $L^2$, and hence in probability (Theorem 2.15). Conditions (8.30) and (8.32) are then verified for $Y_{ni}, 1 \leq i \leq k_n, n \geq 1$. As for condition (8.31), it can be shown that

$$\sum_{i=1}^{k_n} Y_{ni}^2 = \sum_{t=1}^{3} V_t + o_P(1),$$

where $V_1 = \sigma_n^{-2} \sum_{i \in A_n} a_{nii}^2 \mathrm{var}(\xi_{ni}^2)$,

$$V_2 = \frac{4}{\sigma_n^2} \sum_{i=1}^{k_n} a_{nii} \mathrm{E}(U_{ni} u_{ni}) \left( \sum_{j<i} a_{nij} u_{nj} \right),$$

and $V_3 = u_n' C_n u_n$, with $u_n = (u_{ni})_{1 \le i \le k_n}$, $C_n = (4/\sigma_n^2) B_n' B_n$, and $B_n$ being the lower triangular matrix of $A_n$—that is, $B_n = [a_{nij} 1_{(i>j)}]_{1 \le i,j \le k_n}$.

The next thing is to show that $V_2 \to 0$ in $L^2$. Let $l_n$ be the $k_n$-dimensional vector whose $i$th component is $a_{nii} \mathrm{E}(U_{ni} u_{ni})$. Also, write $c_n = 16/\sigma_n^4$. By exchanging the order of summations, we have

$$\mathrm{E}(V_2^2) = c_n \mathrm{E} \left[ \sum_{j=1}^{k_n} \left\{ \sum_{i>j} a_{nij} a_{nii} \mathrm{E}(U_{ni} u_{ni}) \right\} u_{nj} \right]^2$$

$$= c_n \sum_{j=1}^{k_n} \left\{ \sum_{i>j} a_{nij} a_{nii} \mathrm{E}(U_{ni} u_{ni}) \right\}^2 \mathrm{E}(u_{nj}^2). \qquad (8.60)$$

Since $\mathrm{E}(u_{nj}^2) \le \mathrm{E}(\xi_{nj}^2) = 1$, the summation of the right side of (8.60) is bounded by $|B_n' l_n|^2 \le \lambda_{\max}(B_n' B_n) |l_n|^2 \le \|B_n' B_n\|_2 |l_n|^2$, using the fact that the spectral norm of a matrix is bounded by its 2-norm. We now apply Lemma 5.3 to get $\mathrm{E}(V_2^2) \le c_n \sqrt{2} \|A_n^o\| \cdot \|B_n\|_2 |l_n|^2$. Finally, note that

$$\sigma_n^2 = \sum_{i \in A_n} a_{nii}^2 \mathrm{var}(\xi_{ni}^2) + 2 \sum_{i \ne j} a_{nij}^2, \qquad (8.61)$$

which implies $\|B_n\|_2 \le \sigma_n/2$ and $|l_n|^2 \le \sigma_n^2$ (why?). It follows that $\mathrm{E}(V_2^2) \le (16/\sqrt{2})(\|A_n^o\|/\sigma_n) \to 0$ according to the assumption.

The last thing is to show that $V_3 - \mathrm{E}(V_3) \to 0$ in $L^2$. Here, we use the following result, whose proof is left as an exercise (Exercise 8.37).

**Lemma 8.9.** Let $u_{ni}$, $1 \le i \le k_n$, be independent such that $\mathrm{E}(u_{ni}) = 0$, $\mathrm{E}(u_{ni}^2) = \sigma_{ni}^2$, and $\mathrm{E}(u_{ni}^4) < \infty$, and let $C_n = (c_{nij})_{1 \le i,j \le k_n}$ be symmetric. Then $u_n' C_n u_n - \mathrm{E}(u_n' C_n u_n) \to 0$ in $L^2$ provided that $\sum_{i=1}^{k_n} c_{nii}^2 \mathrm{var}(u_{ni}^2) \to 0$ and $\sum_{i>j} c_{nij}^2 \sigma_{ni}^2 \sigma_{nj}^2 \to 0$.

We verify the conditions of Lemma 8.9 for the current $u_n$ and $C_n$. The assumption of Theorem 8.15 implies that

$$\sum_{i=1}^{k_n} c_{nii}^2 \mathrm{var}(u_{ni}^2) = \frac{16}{\sigma_n^4} \sum_{i=1}^{k_n} \left( \sum_{j<i} a_{nij}^2 \right)^2 \mathrm{var}(u_{ni}^2) \longrightarrow 0.$$

Once again, we apply Lemma 5.3 to argue that

$$\sum_{i>j} c_{nij}^2 \mathrm{E}(u_{ni}^2)\mathrm{E}(u_{nj}^2) \le \frac{1}{2}\sum_{i\ne j} c_{nij}^2$$

$$\le \frac{8}{\sigma_n^4}\|B_n'B_n\|_2^2$$

$$\le \frac{4\|A_n^\circ\|^2}{\sigma_n^2} \longrightarrow \mathsf{U}$$

by the assumption of the theorem.

In conclusion, we have shown that

$$\sum_{i=1}^{k_n} Y_{ni}^2 = \sigma_n^{-2}\sum_{i\in\mathcal{A}_n} a_{nii}^2\,\mathrm{var}(\xi_{ni}^2) + \mathrm{E}(u_n'C_nu_n) + o_{\mathrm{P}}(1).$$

However, $\mathrm{E}(u_n'C_nu_n) = 2\sigma_n^{-2}\sum_{i\ne j}a_{nij}^2 + o(1)$ (Exercise 8.37). Thus, in view of (8.61), we have $\sum_{i=1}^{k_n}Y_{ni}^2 = 1 + o_{\mathrm{P}}(1)$, which implies (8.31) with $\eta^2 = 1$.

## 8.9 Case study: Martingale approximation

Martingale limit theory is useful in deriving limit theorems for random processes that may not be martingales themselves. A technique that often makes these derivations possible is called martingale approximation. The idea is to obtain an (a.s.) error bound for the difference between the random process and the approximating martingale that is good enough so that the desired limit theorem for the random process follows as a result of the corresponding limit theorem for the approximating martingale. As an example, we consider a recent work by Wu (2007), who used the martingale approximation to derive strong limit theorems for sums of dependent random variables associated with a Markov chain (see Section 10.2).

Suppose that $\xi_i, i \in Z$, is a stationary and ergodic Markov chain, where $Z$ is the set of all integers and the stationary Markovian property implies that

$$\mathrm{P}(\xi_{n+1} = y|\xi_n = x, \xi_{n-1} = x_{n-1}, \ldots) = \mathrm{P}(\xi_1 = y|\xi_0 = x) \qquad (8.62)$$

for all $n \in Z$ and $y, x, x_{n-1}, \ldots$. Let $X_i = g(\xi_i)$, where $g$ is a measurable function, and $S_n = \sum_{i=1}^n X_i$. The interest is to obtain strong (i.e., a.s.) limit theorems for $S_n, n \ge 1$. Note that such topics were discussed in Chapter 6, where the $X_i$'s are assumed to be independent random variables.

Wu (2007) considered the following approximating martingale. Let $\mathcal{F}_k = \sigma(\xi_j, j \le k)$. For any random variable $Z$ with finite first moment, define the projection $\mathcal{P}_k Z = \mathrm{E}(Z|\mathcal{F}_k) - \mathrm{E}(Z|\mathcal{F}_{k-1})$. Let $D_k = \sum_{i=k}^\infty \mathcal{P}_k g(\xi_i)$, provided that the infinite series converges almost surely. Then $D_k, \mathcal{F}_k, k \in Z$, is a sequence of martingale differences that is stationary and ergodic. It follows that $M_n = \sum_{k=1}^{n} D_k, \mathcal{F}_n, n \geq 1$, is a martingale. Furthermore, the following error bound for $S_n - M_n$ is obtained. Let $\delta_{i,q} = \|\mathcal{P}_0 g(\xi_i)\|_q$, where for any random variables $Z$ and $q > 0$, $\|Z\|_q = \{E(|Z|^q)\}^{1/q}$, and $\Delta_{j,q} = \sum_{i=j}^{\infty} \delta_{i,q}$.

**Theorem 8.16.** Let $E\{g(\xi_0)\} = 0$ and $g(\xi_0) \in L^q$ for some $q > 1$. Then

$$\|S_n - M_n\|_q^r \leq 3B_q^r \sum_{j=1}^{n} \Delta_{j,q}^r,$$

where $r = q \wedge 2$ and $B_q = 18q^{3/2}(q-1)^{-1/2}$ if $q \in (1,2) \cup (2,\infty)$ and $1$ if $q = 2$.

By Theorem 8.16 and Borel-Cantelli lemma (Lemma 2.5), an a.s. bound for $S_n - M_n$ can be obtained, as follows.

**Corollary 8.4.** Under the assumptions of Theorem 8.16, we have $S_n - M_n = o(n^{1/q})$ a.s., provided that $\Delta_{0,q} < \infty$ and

$$\sum_{j=1}^{\infty} j^{-a} \Delta_{j,q}^b < \infty, \tag{8.63}$$

where $a = \{(q+4)/2(q+1)\} \wedge 1$ and $b = q/(q+1)$.

Here, $S_n - M_n = o(n^{1/q})$ a.s. means that $(S_n - M_n)/n^{1/q} \xrightarrow{\text{a.s.}} 0$ as $n \to \infty$. Based on the martingale approximation, a number of strong limit results were obtained for $S_n$. The first theorem below gives some SLLNs. We say a function $h$ is slowly varying if for any $\lambda > 0$, $\lim_{x \to \infty} h(\lambda x)/h(x) = 1$.

**Theorem 8.17.** Under the assumption of Theorem 8.16, let $h$ be a positive, nondecreasing slowly varying function.
   (i) If $q > 2$, $\Delta_{n,q} = O[(\log n)^{-\alpha}]$ for some $0 \leq \alpha \leq 1/q$, and

$$\sum_{j=1}^{\infty} \{j^{\alpha} h(2^j)\}^{-q} < \infty,$$

then $S_n/\sqrt{n} h(n) \xrightarrow{\text{a.s.}} 0$, as $n \to \infty$.
   (ii) If $1 < q \leq 2$, $\Delta_{0,q} < \infty$, and

$$\sum_{j=1}^{\infty} \{h(2^j)\}^{-q} < \infty,$$

then $S_n/n^{1/q} h(n) \xrightarrow{\text{a.s.}} 0$ as $n \to \infty$.
   (iii) If $1 < q < 2$ and (8.63) holds, then $S_n/n^{1/q} \xrightarrow{\text{a.s.}} 0$ as $n \to \infty$.

The next result is a LIL. Define $\sigma = \| \sum_{i=0}^{\infty} \mathcal{P}_0 g(\xi_i) \|_2$.

**Theorem 8.18.** (i) Suppose that $\sigma < \infty$ and that, for some $q > 2$, we have $E\{g(\xi_0)\} = 0$, $g(\xi_0) \in L^q$ and

$$\sum_{k=1}^{\infty} \{(\log k)^{-1/2} \Delta_{2^k,q}\}^q < \infty.$$

Then we have, for either choice of sign,

$$\limsup_{n \to \infty} \pm \frac{S_n}{\sqrt{2n \log \log n}} = \sigma \quad \text{a.s.}$$

The final result is a strong invariance principle. We have considered a.s. invariance principles for sums of independent random variables in Section 6.6.1 and for martingales in Section 8.7, but here it is in the sense of an a.s. approximation of $S_n$ by a Brownian motion (see Section 10.5). For such a result to hold, it is often necessary to enlarge the underlying probability space and redefine the stationary process without changing its distribution. This is what we mean below by a richer probability space. Define

$$\chi_q(n) = \begin{cases} n^{1/q} (\log n)^{1/2}, & 2 < q < 4 \\ n^{1/4} (\log n)^{1/2} (\log \log n)^{1/4}, & q \geq 4, \end{cases}$$

and $\tau_q(n) = n^{1/q} (\log n)^{1/2+1/q} (\log \log n)^{2/q}$. Recall for a sequence of nonnegative random variables $\eta_n$ and a sequence of normalizing constants $a_n > 0$, $\eta_n = O(a_n)$ a.s. means that $\limsup_{n \to \infty} \eta_n / a_n < \infty$ a.s.

**Theorem 8.19.** Under the assumption of Theorem 8.18, let $s = q \wedge 4$.
(i) If $\Delta_{n,s} = O[n^{1/s-1/2} (\log n)^{-1}]$ and

$$\sum_{k=1}^{\infty} \left\| E\left(D_k^2 | \mathcal{F}_0\right) - \sigma^2 \right\|_{s/2} < \infty, \tag{8.64}$$

where $\sigma$ is the same as in Theorem 8.18, then, on a richer probability space, there exists a standard Brownian motion $B$ such that

$$\left| S_n - B(\sigma^2 n) \right| = O[\chi_q(n)] \quad \text{a.s.}$$

(ii) If $\Delta_{n,s} = O(n^{1/s-1/2})$ and

$$\sum_{k=1}^{\infty} \left\| \mathcal{P}_0 \left(D_k^2\right) \right\|_{s/2} < \infty, \tag{8.65}$$

then, on a richer probability space, there exists a standard Brownian motion $B$ such that

$$\left|S_n - B(\sigma^2 n)\right| = O[\tau_s(n)] \quad \text{a.s.}$$

As an application of the strong limit theorems, Wu (2007) considered the prediction problem in an input/output system. The system may be expressed as $X_n = g(\xi_n)$, where $\xi_n = (\ldots, \epsilon_{n-1}, \epsilon_n)$ and $\epsilon_i, i \in Z$, are i.i.d. random variables. Here, the inputs are $\epsilon_i, i \leq n$, $X_n$ is the output of the system, and $g$ is a filter. It is easy to show that the sequence $\xi_n, n \in Z$, satisfies the Markovian property (8.62) (Exercise 8.38). The author showed that, in this case, simple and easy-to-use bounds for the norms involved in (8.64) and (8.65) can be obtained.

## 8.10 Exercises

*8.1.* This exercise is in connection with the opening problem on casino gambling (Section 8.1).

(i) Show that whenever the gambler wins, he recovers all his previous losses plus an additional $5.

(ii) Suppose that the maximum bet on the casino table is $500 and your initial bet is $5. How many consecutive times can you bet with the martingale strategy?

(iii) Use a computer to simulate 100 sequences of plays. Each play consists of a betting and flipping a fair coin. You win if the coin lands head, and you lose otherwise. Start with $5, then follow the martingale betting strategy until winning (but don't forget the $500 limit). How many times do you hit the limit?

(iv) Now, suppose the coin is biased so that the probability of landing head is 0.4 instead of 0.5. What happends this time when you play the games in (iii)?

*8.2.* Show that if $S_n$, $\mathcal{F}_n$, $n \geq 1$, is a martingale according to the extended definition (8.2), then $S_n$, $n \geq 1$, is a martingale according to (8.1). Give an example to show that the converse is not necessarily true.

*8.3.* Prove Lemma 8.1 (note that the "only if" part is obvious).

*8.4.* Show that $S_n$, $\mathcal{F}_n$, $n \geq 1$, in Example 8.2 is a martingale.

*8.5.* Prove Lemma 8.3.

*8.6.* Show that $Y_i$, $\mathcal{F}_i$, $1 \leq i \leq n$, in Example 8.3 is a sequence of martingale differences.

*8.7.* Verify properties (i)–(iv) of Lemma 8.4.

*8.8.* Verify properties (i)–(iii) of Lemma 8.5.

*8.9.* Prove properties (i) and (ii) of Lemma 8.6.

*8.10.* Prove properties (i)–(iv) of Lemma 8.7.

*8.11.* A sequence of random variables $X_n$, $n \geq 1$, is said to be $m$-dependent if for any $n \geq 1$, $\sigma(X_1, \ldots, X_n)$ and $\sigma(X_{n+m+1}, \ldots)$ are independent. Suppose, in addition, that $E(X_n) = 0$, $n \geq 1$. Define $\mathcal{F}_n = \sigma(X_1, \ldots, X_n)$ and

$$S_n = E\left(\sum_{i=1}^{n+m} X_i \,\middle|\, \mathcal{F}_n\right),$$

$n \geq 1$. Show that $S_n$, $\mathcal{F}_n$, $n \geq 1$, is a martingale. Note that Example 8.2 is a special case of this exercise with $m = 0$.

8.12. Suppose that $X_1, \ldots, X_n$ are i.i.d. with finite expectation. Define $S_k = \sum_{i=1}^{k} X_i$, $M_k = (n - k + 1)^{-1} S_{n-k+1}$, and $\mathcal{F}_k = \sigma(S_n, \ldots, S_{n-k+1})$, $1 \leq k \leq n$. Show that $M_k$, $\mathcal{F}_k$, $1 \leq k \leq n$ is a martingale. [Hint: Note that $M_k = E(X_1|\mathcal{F}_k)$, $1 \leq k \leq n$.]

8.13 (U-statistics). A sequence of random variables $X_n$, $n \geq 1$, is said to be exchangeable if for any $n > 1$ and any permutation $i_1, \ldots, i_n$ of $1, \ldots, n$, $(X_{i_1}, \ldots, X_{i_n})$ has the same distribution as $(X_1, \ldots, X_n)$. For a fixed $m \geq 1$, a Borel-measurable function $\psi$ on $R^m$ is called symmetric if for any permutation $j_1, \ldots, j_m$ of $1, \ldots, m$, we have $\psi(x_{j_1}, \ldots, x_{j_m}) = \psi(x_1, \ldots, x_m)$ for all $(x_1, \ldots, x_m) \in R^m$. Let $\psi$ be symmetric and $E\{|\psi(X_1, \ldots, X_m)|\} < \infty$. Define

$$U_{m,n} = \binom{n}{m}^{-1} \sum_{1 \leq j_1 < \cdots < j_m \leq n} \psi(X_{j_1}, \ldots, X_{j_m}), \quad n \geq m,$$

and $\mathcal{G}_n = \sigma(U_{m,k}, k \geq n)$. Let $N > m$ be a fixed integer. Then define $U_n^* = U_{m,N-n}$ and $\mathcal{F}_n = \mathcal{G}_{N-n}$. Show that $U_n^*$, $\mathcal{F}_n$, $n \leq N - m$ is a martingale. [Hint: First show that $E(U_{m,n}|\mathcal{G}_{n+1}) = U_{m,n+1}$ a.s.]

8.14 (Record-breaking time). Let $X_n$, $n \geq 1$ be a sequence of random variables. Define $\tau_1 = 1$ and

$$\tau_{k+1} = \begin{cases} \inf\{n > \tau_k : X_n > X_{\tau_k}\} & \text{if } \tau_k < \infty \text{ and } \{n \geq 1 : X_n > X_{\tau_k}\} \neq \emptyset \\ \infty & \text{otherwise,} \end{cases}$$

$k \geq 1$. The sequence $\tau_k$, $k = 1, 2, \ldots$, may be interpreted as record-breaking times. Show by induction that $\tau_k$, $k \geq 1$, is a sequence of stopping times with respect to $\mathcal{F}_n = \sigma(X_1, \ldots, X_n)$, $n \geq 1$.

8.15. Let $X_i$, $i \geq 1$, be i.i.d. with cdf $F$. Define $\tau_k$ as in Exercise 8.14. Also, let $\omega_F = \sup\{x : F(x) < 1\}$. Show that (i)–(iii) are equivalent:
  (i) $\tau_k < \infty$ a.s. for every $k \geq 1$.
  (ii) $\tau_k < \infty$ a.s. for some $k > 1$.
  (iii) $\omega_F = \infty$ or $\omega_F < \infty$ and $F$ is continuous at $\omega_F$.

8.16. Show that if $\tau_1$ and $\tau_2$ are both stopping times with respect to $\mathcal{F}_n$, $n \geq 1$, then $\{\tau_1 \leq \tau_2\} \in \mathcal{F}_{\tau_2}$.

8.17. Complete the arguments for submartingale and supermartingale in Example 8.8.

8.18. Suppose that $S_n$, $\mathcal{F}_n$, $n \geq 1$, is a submartingale. Show that conditions (i) and (ii) are equivalent:
  (i) condition (8.17) and $E(|S_1|) < \infty$;
  (ii) condition (8.19).

*8.19.* Suppose that $X_n$, $n \geq 1$, are $m$-dependent as in Exercise 8.11 and $E(X_n) = \mu \in (-\infty, \infty)$, $n \geq 1$. Let $\tau$ be a stopping time with respect to $\mathcal{F}_n = \sigma(X_1, \ldots, X_n)$ such that $E(\tau) < \infty$. Show that $E(S_{\tau+m}) = \{E(\tau) + m\}\mu$.

*8.20.* The proof of Theorem 8.3 is fairly straightforward. Try it.

*8.21.* This exercise is associated with Example 8.12.

(i) Show that the sequence $S_n$, $\mathcal{F}_n$, $n \geq 1$, is a martingale with $E(S_n) = 0$, $n \geq 1$.

(ii) Show that $3 \times 2^{i-2} > n$ if and only if $i > l_n$.

(iii) Show that $\sum_{l_n < i \leq n} i^{-1} \to \infty$ as $n \to \infty$.

*8.22.* Show, by Example 8.7, that the $\tau$ defined in Example 8.13 is a stopping time with respect to the $\sigma$-fields $\mathcal{F}_n$, $n \geq 1$.

*8.23.* In this exercise you have an opportunity to practice the stopping time technique that we used in Example 8.13 by giving a proof for part (i) of Theorem 8.5.

(i) Show that for any $p \in (0, 1)$ and $c_i \geq 0$, $1 \leq i \leq n$, we have

$$\left( \sum_{i=1}^n c_i \right)^p \leq \sum_{i=1}^n c_i^p.$$

(Hint: For any $0 \leq \alpha_i \leq 1$, we have $\alpha_i^p \geq \alpha_i$; consider $\alpha_i = c_i / \sum_{j=1}^n c_j$, $1 \leq i \leq n$.)

(ii) Use a similar stopping time technique as in Example 8.13 to show that

$$E \left( \sum_{i=1}^{\tau \wedge n} \frac{|X_i|}{a_i} \right) \leq B.$$

(iii) Use Fatou's lemma to show that

$$\lim_{n \to \infty} \sum_{i=1}^{\tau \wedge n} \frac{|X_i|}{a_i} < \infty \quad \text{a.s.}$$

and hence $\sum_{i=1}^{\infty} |X_i|/a_i < \infty$ a.s. on $\{\tau = \infty\} = \{\sum_{i=1}^{\infty} E(|X_i|^p|\mathcal{F}_{i-1})/a_i^p \leq B\}$ for any $B > 0$.

(iv) Conclude that $\sum_{i=1}^{\infty} X_i/a_i$ converges a.s. on

$$\left\{ \sum_{i=1}^{\infty} E(|X_i|^p|\mathcal{F}_{i-1})/a_i^p < \infty \right\}.$$

Note that here it is not required that the $X_i$'s are martingale differences.

*8.24.* In this exercise you are asked to provide a proof for part (ii) of Theorem 8.5.

(i) Let $Y_i = X_i/a_i$, $i \geq 1$. Show that

$$E(Y_i^2|\mathcal{F}_{i-1}) \leq \begin{cases} b_i & \text{if } E(|X_i|^p|\mathcal{F}_{i-1}) \leq a_i^p b_i^{p/2} \\ a_i^{-p} b_i^{1-p/2} E(|X_i|^p|\mathcal{F}_{i-1}) & \text{otherwise.} \end{cases}$$

[Hint: First show that $E(Y_i^2|\mathcal{F}_{i-1}) \leq a_i^{-2}\{E(|X_i|^p|\mathcal{F}_{i-1})\}^{2/p}$. In the case that $E(|X_i|^p|\mathcal{F}_{i-1}) > a_i^p b_i^{p/2}$, write $\{E(|X_i|^p|\mathcal{F}_{i-1})\}^{2/p}$ as

$$E(|X_i|^p|\mathcal{F}_{i-1})\{E(|X_i|^p|\mathcal{F}_{i-1})\}^{2/p-1}$$

and note that $2/p - 1 < 0$.]

(ii) Use a special case of Theorem 8.4 with $p = 2$ to complete the proof of part (ii) of Theorem 8.5.

8.25. Give two examples to show that (8.22) and (8.24) do not imply each other. In other words, construct two sequences of martingale differences so that the first sequence satisfies (8.22) but not (8.24) and the second sequence satisfies (8.24) but not (8.22).

8.26. This exercise is related to Example 8.14.

(i) Show that condition (8.25) implies (8.27) and (8.28) for every $i \geq 1$.

(ii) Show that $E(|X|) < \infty$ implies (8.29).

(iii) Show that (8.27) and $E(|X|\log^+|X|) < \infty$ implies

$$\sum_{i=1}^{\infty} i^{-1} E\{|X_i|1_{(|X_i|>i)}\} < \infty.$$

(iv) Use the result of (iii) and Kronecker's lemma to show that

$$n^{-1} \sum_{i=1}^{n} Z_i \xrightarrow{\text{a.s.}} 0,$$
$$n^{-1} \sum_{i=1}^{n} W_i \xrightarrow{\text{a.s.}} 0;$$

hence, (8.26) can be strengthened to a.s. convergence under the stronger moment condition.

8.27. Suppose that $\xi_1, \xi_2, \ldots$ are independent such that $\xi_i \sim \text{Bernoulli}(p_i)$, where $p_i \in (0, 1)$, $i \geq 1$. Show that as $n \to \infty$,

$$\frac{1}{n} \sum_{i=1}^{n} \xi_1 \cdots \xi_{i-1}(\xi_i - p_i) \xrightarrow{\text{a.s.}} 0.$$

8.28. Derive the classical CLT from the martingale CLT; that is, show by Theorem 8.7 that if $X_1, X_2, \ldots$ are i.i.d. with $E(X_i) = 0$ and $E(X_i^2) = \sigma^2 \in (0, \infty)$, then $n^{-1/2} \sum_{i=1}^{n} X_i \xrightarrow{d} N(0, \sigma^2)$.

8.29. This exercise is related to Example 8.15.

(i) Verify (8.37).

(ii) Show that $n^{-1} \sum_{i=1}^{n} X_i \xrightarrow{P} h(\mu + \alpha)$. [Hint: Use a result derived in the example on $E(X_i|\mathcal{F}_{i-1})$ and Example 8.14.]

8.30. Let $Z_0, Z_1, \ldots$ be independent $N(0,1)$ random variables. Find a suitable sequence of normalizing constants, $a_n$, such that

$$\frac{1}{a_n} \sum_{i=1}^n Z_{i-1} Z_i \xrightarrow{\text{d}} N(0, 1)$$

and justify your answer. For the justification part, note that Example 8.14 is often useful in establishing (8.31). Also, note the following facts (and you do need to verify them):

(i) For any $M > 0$, we have

$$\max_{1 \le i \le n} |Z_{i-1} Z_i| \le 2 \left\{ M^2 + \sum_{i=0}^n Z_i^2 1_{(|Z_i| > M)} \right\}.$$

(ii) $\max_{1 \le i \le n} (Z_{i-1} Z_i)^2 \le \sum_{i=0}^n Z_i^4$.

8.31 [MA(1) process]. A time series $X_t$, $t \in T = \{\ldots, -1, 0, 1, \ldots\}$, is said to be a moving-average process of order 1, denoted by MA(1), if it satisfies

$$X_t = \epsilon_t + \theta \epsilon_{t-1}$$

for all $t$, where $\theta$ is a parameter and $\epsilon_t, t \in T$, is a sequence of i.i.d. random variables with mean 0 and variance $\sigma^2 \in (0, \infty)$ (the i.i.d. assumption can be relaxed; see the next chapter). Given $t_0 \in T$, find a suitable sequence of normalizing constants $a_n$ such that

$$\frac{1}{a_n} \sum_{t=t_0+1}^{t_0+n} X_t \xrightarrow{\text{d}} N(0, 1)$$

and justify your answer using similar methods as outlined in the previous exercise. Does the sequence $a_n$ depend on $t_0$?

8.32. Show that if $\xi_n$, $n \ge 1$ is a sequence of random variables and $F$ is a continuous cdf, then

$$\sup_{-\infty < x < \infty} |P(\xi_n \le x) - F(x)| \longrightarrow 0$$

if and only if

$$\sup_{-\infty < x < \infty} |P(\xi_n < x) - F(x)| \longrightarrow 0.$$

8.33. Show that the function $\xi_n$ defined by (8.50) is simply linear interpolations between the points

$$(0, 0), \left( \frac{U_1^2}{U_n^2}, \frac{S_1}{U_n} \right), \ldots, \left( 1, \frac{S_n}{U_n} \right).$$

8.34. Show that the model in Example 8.20 can be expressed as (8.55) (this includes determination of the number $s$ and matrices $X, Z_1, \ldots, Z_s$) with all the subsequent assumptions satisfied.

*8.35.* Show that if $\xi_{ni}$, $1 \leq i \leq k_n$, are independent $N(0,1)$ random variables, then (8.58) holds if and only if (8.59) holds.

*8.36.* Show that the array of random variables defined in Example 8.22 satisfies (8.59) but not (8.58).

*8.37.* This exercise is related to the proof of Theorem 8.15.

(i) Prove Lemma 8.9.

(ii) Verify (8.61).

*8.38.* Verify that the sequence $\xi_n, n \in \mathbb{Z}$, in the input/output system considered at the end of Section 8.9 satisfies the Markovian property (8.62).

# 9

# Time and Spatial Series

## 9.1 Introduction

Time series occur naturally in a wide range of practices. For example, the opening price of a certain stock at the New York Stock Exchange, the monthly rainfall total of a certain region, and the CD4+ cell count over time of an individual infected with the HIV virus may all be viewed as time series. A time series is usually denoted by $X_t$, $t \in T$, where $T$ is a set of times, or $X(t)$, $t \in T$, although in this book the latter notation is reserved for (continuous-time) stochastic processes (see the next chapter). An observed time series is a sequence of numbers, one at each observational time. For example, Table 9.1 shows the seasonal energy consumption (coal in the unit of ton) of a certain city from 1991 to 1996. The numbers may be viewed as an observed time series $X_t$, $t = 1, \ldots, 24$, where the times $t = 1, 2, \ldots$ correspond to the first season of 1991, the second season of 1991, and so on. Figure 9.1 shows a plot of $X_t$ against $t$.

Table 9.1. Seasonal energy consumption

| Year | Jan.–March | April–June | July–Sept. | Oct.–Dec. |
|------|-----------|-----------|-----------|-----------|
| 1991 | 6878.4 | 5343.7 | 4847.9 | 6421.9 |
| 1992 | 6815.4 | 5532.6 | 4745.6 | 6406.2 |
| 1993 | 6634.4 | 5658.5 | 4674.8 | 6645.5 |
| 1994 | 7130.2 | 5532.6 | 4989.6 | 6642.3 |
| 1995 | 7413.5 | 5863.1 | 4997.4 | 6776.1 |
| 1996 | 7476.5 | 5965.5 | 5202.1 | 6894.1 |

A statistical model is often used to describe a time series. In fact, there are many such models. The following are some examples.

J. Jiang, *Large Sample Techniques for Statistics*,
DOI 10.1007/978-1-4419-6827-2_9, © Springer Science+Business Media, LLC 2010

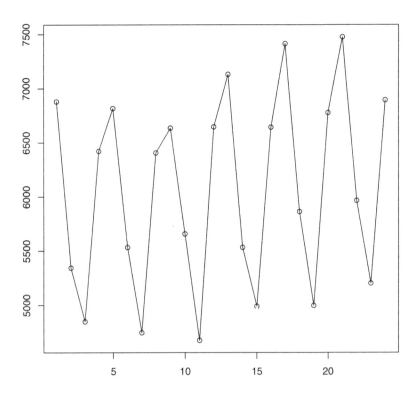

**Fig. 9.1.** *Energy consumption against time*

*Example 9.1* (White noise). A white noise process, denoted by $WN(0, \sigma^2)$, is defined as $X_t$, $t \in T$, such that $E(X_t) = 0$, $var(X_t) = \sigma^2$, and $cov(X_s, X_t) = 0$ for $s \neq t$.

Throughout this chapter, $Z$ denotes the set of integers; that is, $Z = \{\ldots, -1, 0, 1, \ldots\}$.

*Example 9.2* (AR, MA and ARMA processes). A time series $X_t$, $t \in Z$, is called an autoregressive process of order $p$, or $AR(p)$ process, if it satisfies

$$X_t = b_1 X_{t-1} + \cdots + b_p X_{t-p} + \epsilon_t, \tag{9.1}$$

where $\epsilon_t$, $t \in Z$, is a $WN(0, \sigma^2)$ process and the $b$'s are unknown parameters. $X_t$ is a moving-average process of order $q$, or $MA(q)$ process, if it satisfies

$$X_t = \epsilon_t + a_1\epsilon_{t-1} + \cdots + a_q\epsilon_{t-q}, \tag{9.2}$$

where $\epsilon_t$ is the same as in (9.1) and the $a$'s are unknown parameters. A special case of the MA($q$) process was considered in Example 8.31. Finally, the time series is called an autoregressive moving-average process of order $(p, q)$, or ARMA($p, q$) process, if it satisfies

$$X_t = b_1X_{t-1} + \cdots + b_pX_{t-p} + \epsilon_t + a_1\epsilon_{t-1} + \cdots + a_q\epsilon_{t-q}, \tag{9.3}$$

where $\epsilon_t$ is the same as in (9.1) and the $a$'s and $b$'s are unknown parameters.

A time series $X_t$, $t \in Z$, is said to be strictly stationary if for any $t_1, \ldots, t_N, h \in Z$, the joint distribution of $X_{t_1+h}, \ldots, X_{t_N+h}$ is the same as that of $X_{t_1}, \ldots, X_{t_N}$. The time series is said to be second-order stationary if it satisfies $\mathrm{E}(X_t) = \mu$, a constant, $\mathrm{E}(X_t^2) < \infty$, and $\mathrm{cov}(X_t, X_{t+h})$ does not depend on $t$. For a second-order stationary time series $X_t$, its autocovariance function is defined by (4.48)—namely,

$$\gamma(h) = \mathrm{cov}(X_t, X_{t+h}), \quad t, h \in Z. \tag{9.4}$$

It follows that for any $s, t \in Z$, $\mathrm{cov}(X_s, X_t) = \gamma(t - s)$. It is easy to see that an autocovariance function has the following basic properties:

(i) (symmetry) $\gamma(-h) = \gamma(h)$, $h \in Z$.

(ii) (nonnegative definitness) For any $n \geq 1$, the matrix of autocovariances $\Gamma_n = [\gamma(t - s)]_{1 \leq t, s \leq n}$ is nonnegative definite.

(iii) (boundedness) $|\gamma(h)| \leq \gamma(0)$, $h \in Z$.

The autocorrelation function of the time series is then defined as

$$\rho(h) = \gamma(h)/\gamma(0), \quad h \in Z. \tag{9.5}$$

By property (iii) above (or the Cauchy–Schwarz inequality), the value of $\rho(h)$ is always between $-1$ and $1$.

A well-known result in time series analysis is the spectral representation theorem, as follows (e.g., Hannan 1970, p. 46).

**Theorem 9.1.** For any autocovariance function $\gamma$ of a second-order stationary time series we have the representation

$$\gamma(h) = \int_{-\pi}^{\pi} e^{ih\lambda} F(d\lambda) \tag{9.6}$$

for all $h \in Z$, where $F$ is a nondecreasing function, which is uniquely defined if we require in addition that (i) $F(-\pi) = 0$ and (ii) $F$ is right-continuous.

The function $F$ is called the spectral distribution function. In the case that $F$ is absolutely continuous, there exists a function $f$, called the spectral density function, such that

$$\gamma(h) = \int_{-\pi}^{\pi} e^{ih\lambda} f(\lambda) \, d\lambda \tag{9.7}$$

for all $h \in Z$. A sufficient condition for the existence of the spectral density is that the autocovariance function is absolutely summable; that is, (4.49) holds. In addition, we have the following result regarding a linear time series that can be expressed as

$$X_t = \sum_{j=-\infty}^{\infty} a_j \epsilon_{t-j}, \quad t \in Z, \tag{9.8}$$

where $\epsilon_t$ is as in (9.1).

**Theorem 9.2** (Herglotz theorem). If $\sum_{j=-\infty}^{\infty} a_j^2 < \infty$, then $X_t$ has the spectral density $f(\lambda) = (\sigma^2/2\pi)|\sum_{j=-\infty}^{\infty} a_j e^{ij\lambda}|^2$.

Another celebrated result in time series analysis is the Wold decomposition (Wold 1938), discussed earlier in Section 4.4.3. It states that any second-order stationary time series $X_t$ can be expressed as

$$X_t = v_t + \sum_{j=0}^{\infty} a_j \epsilon_{t-j}, \tag{9.9}$$

where $\epsilon_t$ is as in (9.1), $v_t$ is purely deterministic in the sense that it belongs to the linear space spanned by $v_{t-1}, v_{t-2}, \ldots$ in the $L^2$ sense, $\epsilon_t$ and $v_t$ are orthogonal in that $E(\epsilon_s v_t) = 0$ for all $s, t \in Z$, and the coefficients $a_j$ satisfy $\sum_{j=0}^{\infty} a_j^2 < \infty$. Note that the infinite series in (9.9) is different from that in (9.8) in that the summation is restricted to nonnegative integers. This difference is important in time series prediction theory, in which $\epsilon_t$ represents innovations, or forecast errors. Wold decomposition ensures the ability to linearly forecast any second-order stationary time series by means of a process purely determined by its past values plus a moving average of current and past innovations. [Note that such an implication does not prevail from (9.8) because the summation also involve future innovations.]

An extension of time series is a spatial series. A spatial series is also denoted by $X_t$ except that $t \in Z^k$, where $k > 1$; in other words, $t = (t_1, \ldots, t_k)$, where $t_j \in Z$, $1 \le j \le k$. For example, $k = 2$ gives rise to a collection of random variables in the plane. The analysis of spatial series is of interest in a number of fields such as geography, econometrics, geology, and ecology. As it turns out, the development for spatial series analysis is essentially the same for $k = 2$ and for $k > 2$, but there are major differences between $k = 1$ (i.e., time series) and $k = 2$. For such a reason, we mainly focus on $k = 2$ for spatial series.

Some of the time series models can be extended to spatial series. For example, a spatial WN spatial series is defined the same way as Example 9.1, with the understanding that $s, t \in Z^2$. Similarly, a spatial ARMA model is defined as $X_t$, $t \in Z^2$, satisfying

$$X_t = \sum_{s \in \langle 0,p]} b_s X_{t-s} + \sum_{s \in [0,q]} a_s W_{t-s}, \quad t \in Z^2, \tag{9.10}$$

where $p = (p_1, p_2)$, $q = (q_1, q_2)$, $\langle 0,p] = \{s = (s_1, s_2) \in Z^2 : 0 \le s_j \le p_j, j = 1,2, s \ne (0,0)\}$, $[0,q] = \langle 0,q] \cup \{(0,0)\}$, $W_t$ is a spatial WN$(0, \sigma^2)$ series; and the $a$'s and $b$'s are unknown parameters. A spatial series $X_t$, $t \in Z^2$, is said to be second-order stationary if $E(X_t)$ is a constant and $E(X_{s+h} X_{t+h}) = E(X_s X_t)$ for all $s, t, h \in Z^2$. It is easy to show that this condition is equivalent to that $E(X_t)$ and $E(X_t X_{t+h})$ does not depend on $t$ for any $t, h \in Z^2$ (Exercise 9.2). For a second-order stationary spatial series, we can define its autocovariance function in the same way as (9.4), except that $t, h \in Z^2$. A spectral representation theorem similar to (9.6) for second-order stationary spatial series was given by Yaglom (1957). Furthermore, Tjøstheim (1978) extended the Wold decomposition to a second-order stationary purely nondeterministic spatial series. The decomposition has

$$X_t = \sum_{s \ge 0} a_s W_{t-s}, \tag{9.11}$$

where $W_t$ is as in (9.10), and, for $s = (s_1, s_2) \in Z^2$, $s \ge 0$ means that $s_1 \ge 0$ and $s_2 \ge 0$. This corresponds to (9.9) with $v_t = 0$, which is what the term "purely nondeterministic" means intuitively.

## 9.2 Autocovariances and autocorrelations

Suppose that a second-order stationary time series $X_t$ with $E(X_t) = 0$ is observed for $t = 1, \ldots, n$. Its sample autocovariance function is defined as

$$\hat{\gamma}(h) = \begin{cases} n^{-1} \sum_{s=h+1}^{n} X_s X_{s-h}, & 0 \le h \le n-1 \\ 0, & h \ge n. \end{cases} \tag{9.12}$$

The definition naturally extends to negative integers by $\hat{\gamma}(-h) = \hat{\gamma}(h)$. The sample autocorrelation function is defined as $\hat{\rho}(h) = \hat{\gamma}(h)/\hat{\gamma}(0)$. We will focus on a second-order stationary, purely nondeterministic time series, which, by Wold decomposition, can be expressed as (9.9) without $v_t$; that is,

$$X_t = \sum_{j=0}^{\infty} a_j \epsilon_{t-j} \tag{9.13}$$

with $\sum_{j=0}^{\infty} a_j^2 < \infty$ [in some cases, a constant $\mu$ is added to the right side of (9.13); here, for simplicity, we let $\mu = 0$]. We also assume, without loss of generality, that $a_0 = 1$. There have been strong interests in the asymptotic properties of the sample autocaviance and autocorrelation functions. For example, the identification of ARMA models (see Example 9.2 and subsequent sections of this chapter) relies on the a.s. convergence rate of the sample autocaviances and autocorrelations. The classical theory of inference about the time series (9.13) assumes that the $\epsilon_t$'s are i.i.d. However, as several authors have noted (e.g., Hall and Heyde 1980, p. 194), this assumption is somewhat unrealistic. A more natural assumption is

$$E(\epsilon_t | \mathcal{F}_{t-1}) = 0 \quad \text{a.s.} \tag{9.14}$$

for all $t$, where $\mathcal{F}_t = \sigma(\epsilon_s, s \leq t)$. In other words, $\epsilon_t$, $\mathcal{F}_t$, $t \in Z$, is a sequence of martingale differences. As observed in Hannan and Heyde (1972, p. 2059), in the case where $X_t$ is strictly stationary and $\epsilon_t$ are prediction errors [i.e., $\epsilon_t = X_t - E(X_t | \mathcal{G}_{t-1})$, where $\mathcal{G}_t = \sigma(X_s, s \leq t)$], (9.14) is equivalent to the assertion that the best linear predictor is the best predictor, both in the least squares sense. By using the martingale laws of large numbers (see Section 8.4.1), these latter authors further obtained weak and strongs law as well as a CLT for the sample autocovariances. We state the strong law result below and leave the CLT for later discussion. Note that the first part of the theorem does not require that $\epsilon_t$ be strictly stationary. Instead, a weaker distributional assumption is imposed for this part.

**Theorem 9.3.** Suppose that (9.14) holds and there is a constant $c$ and a nonnegative random variable $\xi$ with $E(\xi^2) < \infty$ such that

$$P(|\epsilon_t| > x) \leq cP(\xi > x) \tag{9.15}$$

for all $x > 0$ and $t \in Z$. Furthermore, suppose that

$$\sum_{j=0}^{\infty} |a_j| < \infty. \tag{9.16}$$

(i) If, in addition, we have

$$\frac{1}{n} \sum_{t=1}^{n} E(\epsilon_t^2 | \mathcal{F}_{t-1}) \xrightarrow{\text{a.s.}} \sigma^2 > 0, \tag{9.17}$$

then $\hat{\gamma}(h) \xrightarrow{P} \gamma(h)$ as $n \to \infty$ for all $h \in Z$. (ii) If (9.15) is strengthened so that $\epsilon_t$ is strictly stationary and (9.17) is strengthened so that

$$E(\epsilon_t^2 | \mathcal{F}_{t-1}) = \sigma^2 > 0 \quad \text{a.s.} \tag{9.18}$$

for all $t \in Z$, then $\hat{\gamma}(h) \xrightarrow{\text{a.s.}} \gamma(h)$ as $n \to \infty$, for all $h \in Z$.

We now consider deeper asymptotic results for $X_t$. For the rest of this section we assume that $X_t$ is strictly stationary and ergodic (see the definition following Theorem 7.14). As discussed, (9.14) is a reasonable assumption for the innovations $\epsilon_t$, but not (9.18). In fact, the only reasonable conditions that would make (9.18) hold is, perhaps, that the innovations are Gaussian (Exercise 9.8). Furthermore, some inference of time series requires a.s. convergence rates of the sample autocovariances and autocorrelations (see the next section). For example, An et al. (1982) proved the following results. Here, a sequence of random variables $\xi_n$ is a.s. $o(a_n)$ $(O(a_n))$ for some sequence of positive constants $a_n$ if $\limsup_{n \to \infty} |\xi_n|/a_n = 0$ $(< \infty)$ a.s.

**Theorem 9.4.** Suppose that $\sum_{j=0}^{\infty} |a_j| < \infty$ and (0.14) and (9.18) hold. Furthermore, there is $r \geq 4$ such that $E(|\epsilon_r|^r) < \infty$. Then for any $\delta > 0$ and $P_n \leq n^a$, where $a = r/\{2(r-2)\}$, we have

$$\max_{0 \leq h \leq P_n} |\hat{\gamma}(h) - \gamma(h)|$$

$$= o\left\{ n^{-1/2} (P_n \log n)^{2/r} (\log \log n)^{2(1+\delta)/r} \right\} \quad \text{a.s.} \quad (9.19)$$

In particular, if $r = 4$ and $X_t$ is an ARMA process and, furthermore, $P_n = O\{(\log n)^b\}$ for some $b < \infty$, then we have

$$\sup_{0 \leq h \leq P_n} |\hat{\gamma}(h) - \gamma(h)| = O\left( \sqrt{\frac{\log \log n}{n}} \right) \quad \text{a.s.} \quad (9.20)$$

Note that the convergence rate on the right side of (9.20) is the best possible. The proof of (9.19) is an application of two well-known inequalities: Doob's maximum inequality [see (5.73)] and Burkholder's inequality [below (5.71)]. Note that the two inequalities are often used together in an argument involving martingales. The proof also used an argument due to Móricz (1976, p. 309) dealing with maximum moment inequalities that was briefly discussed at the end of Section 5.4. The proof of (9.20) is more tedious.

Once again, condition (9.18) is assumed in the Theorem 9.4. In An et al. (1982), the authors discussed possibilities of weakening this condition. Here, the authors focused on the sample autocovariances, with the understanding that similar results for the sample autocorrelations can be obtained as a consequence of those for the sample autocovariances. However, in some applications, it is the sample autocorrelations that are of direct interest. Huang (1988a) showed that for the convergence rate of sample autocorrelations, condition (9.18) can be completely removed. We state Huang's results as follows.

**Theorem 9.5.** Let $X_t$ be an ARMA process, (9.14) holds, and $E(\epsilon_t^4) < \infty$. Then (9.20) holds with $\gamma$ replaced by $\rho$.

**Theorem 9.6.** Suppose that $X_t$ satisfies (9.13) and (9.14).

(i) (CLT) If $\sum_{j=1}^{\infty} \sqrt{j} a_j^2 < \infty$, then for any given positive integer $K$, the joint distribution of $\sqrt{n}\{\hat{\rho}(h) - \rho(h)\}$, $h = 1, \ldots, K$, converges to $N(0, V_K)$, where the $(s, t)$ element of $V_K$ is $E(\eta_s \eta_t)$, $1 \leq s, t \leq K$, with

$$\eta_t = \frac{\epsilon_0}{\sigma^2} \sum_{u=1}^{\infty} \{\rho(u+t) + \rho(u-t) - 2\rho(u)\rho(t)\}\epsilon_{-u}$$

and $\sigma^2 = E(\epsilon_0^2)$.

(ii) (LIL) If $\sum_{j=1}^{\infty} j a_j^2 < \infty$, then for any given positive integer $K$ and constants $c_1, \ldots, c_K$, we have with probability 1 that the set of limit points of the sequence

$$\left(\frac{n}{\log \log n}\right)^{1/2} \sum_{h=1}^{K} c_h\{\hat{\rho}(h) - \rho(h)\}, \quad n \geq 3,$$

coincides with $[-\sqrt{2}\tau, \sqrt{2}\tau]$, where $\tau^2 = \sum_{s,t=1}^{K} c_s c_t E(\eta_s \eta_t)$ and $\eta_t$ is given above.

(iii) (Uniform convergence rate) If $\sum_{j=3}^{\infty} j(\log \log j)^{1+\delta} a_j^2 < \infty$ for some $\delta > 0$, and $\sum_{j=1}^{\infty} |a_j| < \infty$, then we have

$$\sup_{h \geq 0} |\hat{\rho}(h) - \rho(h)| = O\left(\sqrt{\frac{\log n}{n}}\right) \quad \text{a.s.}$$

We omit the proofs of Theorem 9.5 and Theorem 9.6, which are highly technical, but remark that martingale techniques play important roles in these proofs. Huang (1988a) also obtained results of CLT, LIL and the uniform convergence rate for the sample autocovariances under conditions weaker that (9.18) (but without having it completely removed).

## 9.3 The information criteria

On the morning of March 16, 1971, Hirotugu Akaike, as he was taking a seat on a commuter train, came up with the idea of a connection between the relative Kullback–Leibler discrepancy and the empirical log-likelihood function, a procedure that was later named Akaike's information criterion, or AIC (Akaike 1973, 1974; see Bozdogan 1994 for the historical note). The idea has allowed major advances in model selection and related fields (e.g., de Leeuw 1992), including model identifications in time series (see the next section) .

The problem of model selection arises naturally in time series analysis. For example, in an ARMA model (see Example 9.2), the orders $p$ and $q$ are unknown and therefore need to be identified from the information provided by the data. Practically speaking, there may not be an ARMA model for the true data-generating process—and this is true not only for time series models but for all models that are practically used. George Box, one of the most influential statisticians of the 20th century, once said, and has since been quoted, that "all models are wrong; some are useful." What it means is that, even though there may not exist, say, a "true" ARMA model, a suitable choice of one may still provide a good (or perhaps the best) approximation from a practical point of view.

The idea of AIC may be described as follows. Suppose that one wishes to approximate an unknown pdf, $g$, by a given pdf, $f$. The Kullback–Leibler discrepancy, or information, defined as

$$I(g; f) = \int g(x) \log g(x)\, dx - \int g(x) \log f(x)\, dx, \qquad (9.21)$$

provides a measure of lack of approximation. It can be shown, by Jensen's inequality, that the Kullback–Leibler information is always nonnegative and it equals zero if and only if $f = g$ a.e. [i.e., $f(x) = g(x)$ for all $x$ except on a set of Lebesgue measure zero]. However, it is not a distance (Exercise 9.9). Note that the first term on the right side of (9.21) does not depend on $f$. Therefore, to best approximate $g$, one needs to find an $f$ that minimizes

$$-\int g(x) \log f(x)\, dx = -\mathrm{E}_g\{\log f(X)\},$$

where $\mathrm{E}_g$ means that the expectation is taken with $X \sim g$. Since we do not know $g$, the expectation is not computable. However, suppose that we have independent observations $X_1, \ldots, X_n$ from $g$. Then we may replace the expectation by the sample mean, $n^{-1}\sum_{i=1}^{n} \log f(X_i)$, which is an unbiased estimator for the expectation. In particular, under a parametric model, denoted by $M$, the pdf $f$ depends on a vector $\theta_M$ of parameters, denoted by $f = f_M(\cdot|\theta_M)$. For example, under an ARMA$(p, q)$ model, we have $M = (p, q)$ and $\theta_M = (b_1, \ldots, b_p, a_1, \ldots, a_q)'$. Then the AIC is a two-step procedure. The first step is to find the $\theta_M$ that minimizes

$$-\frac{1}{n}\sum_{i=1}^{n} \log f_M(X_i|\theta_M) \qquad (9.22)$$

for any given $M$. Note that (9.22) is simply the negative log-likelihood function under $M$. Therefore, the $\theta_M$ that minimizes (9.22) is the MLE, denoted by $\hat{\theta}_M$. Then, the second step of AIC is to find the model $M$ that minimizes

$$-\frac{1}{n}\sum_{i=1}^{n} \log f_M(X_i|\hat{\theta}_M). \qquad (9.23)$$

However, there is a serious drawback in this approach: Expression (9.23) is no longer an unbiased estimator for $-\mathrm{E}_g\{\log f_M(X|\theta_M)\}$ due to overfitting. The latter is caused by double-use of the same data—for estimating the expected log-likelihood and for estimating the parameter vector $\theta_M$. Akaike (1973) proposed to retify this problem by correcting the bias, which is

$$\frac{1}{n}\sum_{i=1}^{n} \mathrm{E}_g\{\log f_M(X_i|\hat{\theta}_M)\} - \mathrm{E}_g\{\log f_M(X|\theta_M)\}.$$

He showed that, asymptotically, the bias can be approximated by $|M|/n$, where $|M|$ denotes the dimension of $M$ defined as the number of estimated parameters under $M$. For example, if $M$ is an ARMA$(p, q)$ model, then $|M| = p + q + 1$ (the 1 corresponds to the unknown variance of the WN). Thus, a term $|M|/n$ is added to (9.23), leading to

$$-\frac{1}{n} \sum_{i=1}^{n} \log f_M(X_i|\hat{\theta}_M) + \frac{|M|}{n}.$$

The expression is then multiplied by the factor $2n$, which does not depend on $M$ and therefore does not affect the choice of $M$, to come up with the AIC:

$$\text{AIC}(M) = -2 \sum_{i=1}^{n} \log f_M(X_i|\hat{\theta}_M) + 2|M|. \qquad (9.24)$$

In words, the AIC is minus two times the maximized log-likelihood plus two times the number of estimated parameters.

A number of similar criteria have proposed since the AIC. These criteria may be expressed in a general form as

$$\hat{D}_M + \lambda_n|M|, \qquad (9.25)$$

where $\hat{D}_M$ is a measure of lack-of-fit by the model $M$ and $\lambda_n$ is a penalty for complexity of the model. The measure of lack-of-fit is such that a model of greater complexity fits better, therefore it has a smaller $\hat{D}_M$; on the other hand, such a model receives more penalty for having a larger $|M|$. Therefore, criterion (9.25), known as the generalized information criterion, or GIC (Nishii 1984; Shibata 1984), is a trade-off between model fit and model complexity. Note that AIC corresponds to (9.25) with $\hat{D}_M$ being $-2$ times the maximized log-likelihood and $\lambda_n = 2$. We consider some other special cases below. In all of these cases, the measure of lack-of-fit is the same as in AIC.

*Example 9.3.* Hurvich and Tsai (1989) argued that in the case of the ARMA$(p, q)$ model, a better bias correction could be obtained if one replaces $p + q + 1$ by an asymptotically equivalent quantity,

$$\frac{n(p + q + 1)}{n - p - q - 2}.$$

This leads to a modified criterion known as AICC. The AICC corresponds to (9.25) with $\lambda_n = 2n/(n - p - q - 2)$. So, if $n \to \infty$ while the ranges of $p$ and $q$ are bounded, AICC is asymptotically equivalent to AIC.

One concern about AIC is that it does not lead to consistent model selection if the dimension of the optimal model is finite. Here, an optimal model means a true model with minimum dimension. For example, suppose that the true underlying model is AR(2); then AR(3) is also a true model (by letting the additional coefficient, $b_3$, equal to zero [see (9.1)], but not an optimal model. On the other hand, AR(1) is an incorrect model (or wrong model). So, if one consider all AR models as candidates, the only optimal model is AR(2). Furthermore, consistency of model selection is defined as that the probability of selecting an optimal model goes to 1 as $n \to \infty$.

*Example 9.4* (BIC). The Bayesian information criterion, or BIC (Schwarz 1978), corresponds to (9.25) with $\lambda_n = \log n$. Unlike the AIC, the BIC is a consistent model selection procedure (e.g., Hannan 1980).

*Example 9.5* (The HQ criterion). Hannan and Quinn (1979) proposed a criterion for determine the order $p$ of an AR model based on a LIL for the partial autocorrelations (e.g., Hanna 1970, pp. 21–23). Their criterion corresponds to (9.25) with $\lambda_n = c \log\{\log(n)\}$, where $c > 2$ is a constant.

The idea of choosing $\lambda_n$ so that (9.25) leads to a consistent model selection strategy is, actually, quite simple. The AIC is not consistent because it does not put enough penalty for complex models. For example, suppose that the true underlying model is AR($p$). Then AIC tends to choose an order higher than $p$ in selecting the order for the AR model. This problem is called overfitting. It can be shown that AIC does not have the other kind of problem—underfitting, meaning that the procedure tends to select an order less than $p$, in this case. This means that, asymptotically, AIC is expected to select, at least, a true model; but the selected model may not be optimal in that it can be further simplified. For a procedure to be consistent, one needs to control both overfitting and underfitting. On the one hand, one needs to increase the penalty $\lambda_n$ in order to reduce overfitting; on the other hand, one cannot overdo this because otherwise, the underfitting will again make the procedure inconsistent. The question is: What it the "right" amount of penalty?

The way to find out the answer is to evaluate the asymptotic order (see Chapter 3) of the first term in (9.25) (i.e., the measure of lack-of-fit). As it turns out, in typical situations there is a difference in the order of $\hat{D}_M$ depending on whether $M$ is a true (but not necessarily optimal) model or wrong model. Roughly speaking, let $\hat{D}_M = O(a_n)$ when $M$ is true and $\hat{D}_M = O(b_n)$ when $M$ is wrong, where $a_n = o(b_n)$. Then if we choose $\lambda_n$ such that $a_n = o(\lambda_n)$ and $\lambda_n = o(b_n)$, we have a consistent model selection criterion. To see this, let $M_0$ be an optimal model and denote (9.25) by $c(M)$. If $M$ is a wrong model, then, asymptotically, we have $c(M) = O(b_n) + o(b_n)|M| = O(b_n)$ while $c(M_0) = O(a_n) + o(b_n)|M| = o(b_n)$. So, asymptotically, one expects $c(M) > c(M_0)$. On the other hand, if $M$ is a true but nonoptimal model, meaning $|M| > |M_0|$, we have $c(M) = O(a_n) + \lambda_n|M| = o(\lambda_n) + \lambda_n|M|$ while $c(M_0) = O(a_n) + \lambda_n|M_0| = o(\lambda_n) + \lambda_n|M_0|$. Therefore,

$$c(M) - c(M_0) = \lambda_n(|M| - |M_0|) + o(\lambda_n),$$

which is expected to be positive if $\lambda_n \to \infty$. It follows that, asymptotically, neither a wrong model nor a true but nonoptimal model cannot possibly be the minimizer of (9.25), so the minimizer has to be $M_0$. Of course, to make a rigorous argument, one needs to clearly define what is meant by $o(a_n)$, and so on., because $\hat{D}_M$ is a random quantity. Usually, this is in the probability sense (see Section 3.4).

Also, it is clear that the choice of $\lambda_n$ for consistent model selection, if it exists, is not unique. In fact, there may be many choices of $\lambda_n$ that all lead to consistent model selection criteria, but their finite sample performance can be quite different. This issue was recently addressed by Jiang et al. (2008), where the authors proposed a new strategy for model selection, called a *fence*.

## 9.4 ARMA model identification

We first write the ARMA model (9.3) in a more convenient form:

$$\sum_{j=0}^{p} \alpha_j X_{t-j} = \sum_{j=0}^{q} \beta_j \epsilon_{t-j}, \tag{9.26}$$

where $\alpha_0 = \beta_0 = 1$. It is assumed that

$$A(z) = \sum_{j=0}^{p} \alpha_j z^j \neq 0,$$

$$B(z) = \sum_{j=0}^{q} \beta_j z^j \neq 0 \tag{9.27}$$

for $|z| \leq 1$. Here, $z$ denotes a complex number. It is also assumed that the polynomials $A(z)$ and $B(z)$ are coprime, meaning that they do not have a common (polynomial) factor. Let $\mathcal{L}$ denote the (backward) lagoperator; that is, $\mathcal{L}\xi_t = \xi_{t-1}$ for any time series $\xi_t$, $t \in Z$. Then (9.26) can be expressed as

$$A(\mathcal{L})X_t = B(\mathcal{L})\epsilon_t, \quad t \in Z. \tag{9.28}$$

Condition (9.27) implies that there is $\rho > 1$ such that the function $\Phi(z) = A^{-1}(z)B(z)$ is analytic on $\{z : |z| \leq \rho\}$, and therefore has the Taylor expansion

$$\Phi(z) = \sum_{j=0}^{\infty} \phi_j z^j, \quad |z| \leq \rho. \tag{9.29}$$

It follows that

$$X_t = \Phi(\mathcal{L})\epsilon_t = \sum_{j=0}^{\infty} \phi_j \epsilon_{t-j}, \quad t \in Z. \tag{9.30}$$

The $\phi_j$'s are called the Wold coefficients of $X_t$. It can be shown that $\phi_0 = 1$ and $\phi_j \rho^j \to 0$ as $j \to \infty$ (Exercise 9.10). Furthermore, the autocovariance function of $X_t$ can be expressed in terms of its Wold coefficients; that is,

$$\gamma(h) = \sigma^2 \sum_{j=0}^{\infty} \phi_j \phi_{j+h}, \quad h \geq 0, \tag{9.31}$$

and $\gamma(-h) = \gamma(h)$. It can be shown that there is a constant $c > 0$ such that $|\gamma(h)| \leq c\rho^{-h}$, $h \geq 0$, where $\rho > 1$ is the number in (9.29) (Exercise 9.11). In other words, the autocorrelations of an ARMA process decay at an exponential rate. If the coefficients $\alpha_j$, $1 \leq j \leq p$, and $\beta_j$, $1 \leq j \leq q$, are known, the Wold coefficients can be computed by the following recursive method:

$$\beta_j = \sum_{k=0}^{p} \alpha_k \phi_{j-k}, \quad j = 0, 1, \ldots, \tag{9.32}$$

where we define $\beta_j = 0$ if $j > q$ and $\phi_j = 0$ if $j < 0$. Thus, in view of (9.31), the autocovariance function is uniquely determined by the ARMA coefficients $\alpha_j$'s and $\beta_j$'s. In practice, however, the reverse problem is of interest: Given the autocovariances, how do we estimate the ARMA coefficients? This problem is of interest because the autocovariances can be estimated from the observed data (see Section 9.2).

A traditional method of estimation for ARMA models is called the Yule–Walker (Y-W) estimation. For simplicity, let us consider a special case, the AR($p$) model defined by (9.1). It can be shown that the autocovariances and AR coefficients jointly satisfy the following the Yule–Walker equation:

$$\begin{bmatrix} \gamma(1) \\ \gamma(2) \\ \vdots \\ \gamma(p) \end{bmatrix} = \begin{bmatrix} \gamma(0) & \gamma(1) & \cdots & \gamma(p-1) \\ \gamma(1) & \gamma(0) & \cdots & \gamma(p-2) \\ \vdots & \vdots & & \vdots \\ \gamma(p-1) & \gamma(p-2) & \cdots & \gamma(0) \end{bmatrix} \begin{bmatrix} b_1 \\ b_2 \\ \vdots \\ b_p \end{bmatrix} \tag{9.33}$$

(Exercise 9.12). Furthermore, we have

$$\sigma^2 = \gamma(0) - \sum_{j=1}^{p} b_j \gamma(j) \tag{9.34}$$

(Exercise 9.12). Thus, one may estimate the AR coefficients by solving (9.33) with $\gamma(j)$ replaced by $\hat{\gamma}(j)$, the sample autocovariance, $0 \leq j \leq p$. Let the estimate of $b_j$ be $\hat{b}_j$, $1 \leq j \leq p$. Then $\sigma^2$ is estimated by the right side of (9.34) with $\gamma(j)$ replaced by $\hat{\gamma}(j)$, $0 \leq j \leq p$, and $b_j$ by $\hat{b}_j$, $1 \leq j \leq p$. Two alternative methods of estimation are the least squares (LS) estimation and maximum likelihood (ML) estimation, the latter under the normality assumption.

When the orders $p$ and $q$ are known, large-sample properties of the Y-W, LS, and ML estimators in ARMA models, including consistency, asymptotic normality, and strong convergence rate, are well known (e.g., Brockwell and Davis 1991). Note that all of these estimators are functions of the sample auto-covariances (autocorrelations); hence, asymptotic properties of the estimators can be derived from the corresponding asymptotic theory of the sample auto-covariances and autocorrelations discussed in Section 2. In practice, however, not only the parameters of the ARMA model are unknown, but the orders $p$ and $q$ also need to be determined. Naturally, the orders would need to be determined before the parameters are estimated. Hannan (1980) showed that if the orders $p$ and $q$ are determined either by the BIC (Example 9.4) or by the HQ criterion, the latter being an extension of Example 9.5 to ARMA models, the resulting estimators of the orders, $\hat{p}$ and $\hat{q}$, are strongly consistent. Here, strong consistency means that, with probability 1, one has $\hat{p} = p$ and $\hat{q} = q$ for large $n$, where $p$ and $q$ are the true orders of the ARMA model. The author also showed that AIC is not consistent (even in the weak sense, as defined earlier) for determining the orders and obtained the asymptotic distribution for the limits of $\hat{p}$ and $\hat{q}$. Note that if $\hat{p}$ and $\hat{q}$ are consistent, we must have $\lim_{n\to\infty} P(\hat{p} = p, \hat{q} = q) = 1$, where $p$ and $q$ are the true orders. Instead, for AIC the limit is not 1, but has a distribution over the range of overfitted $p$ and $q$ (i.e., orders higher than the true orders). The result thus confirms an earlier statement that AIC does not have the underfitting problem asymptotically. It should be pointed out that AIC is designed for a situation quite different from this, in which the underlying time series $X_t$ is *not* generated from an ARMA model. In other words, an ARMA model is only used as an approximation. Therefore, it would be unfair to judge AIC solely based on the consistency property. Also, consistency is a large-sample property, which is not always an indication of finite sample performance. Hannan (1980) also studied the (weak) consistency property of the criterion (9.25) in general. The main result (i.e., Theorem 3 of Hannan 1980) states that the criterion is consistent as long as $\lambda_n \to \infty$, but this result is clearly in error. To see this, suppose that the true orders, $p$ and $q$, are greater than zero (i.e., we have a nondegenerate ARMA model). If $\lambda_n$ approaches infinity at such a fast rate that the second term in (9.25) almost surely dominates the first term whenever $|M| = p + q > 0$, the procedure almost surely will not select any orders other than zeros.

Nevertheless, the main interest here is strong consistency of the order esti-mation. The key assumptions of Hannan (1980) are that the innovations $\epsilon_t$ are stationary satisfying (9.14) and (9.18) plus finiteness of the fourth moment. As discussed earlier (see Section 9.2), all of the assumptions are reasonable except (9.18). The author did offer some discussion on the possibility of weak-ening this assumption. Huang (1988b, 1989) was able to completely remove this assumption for ARMA model identification. As noted near the end of Section 9.3, to derive a (strongly) consistent criterion for the order determination, one needs to evaluate the asymptotic order of the first term in (9.25). As it turns out, for ARMA models this term is asymptotically equivalent to a function of the sample autocorrelations. Therefore, it is not surprising that the a.s. convergence rate of the sample autocorrelations plays an important role in deriving a strongly consistent model selection criterion. Since Huang was able to remove (9.18) in obtaining the a.s. convergence rate for the sample autocorrelations (see Section 9.2), as a consequence he was able to remove (9.18) as a requirement for ARMA model identification.

We briefly describe Huang's first method of ARMA model identification (Huang 1988b). The idea was motivated by the Wold decomposition (9.30). It is seen that under the basic assumptions for ARMA models made at the beginning of this section, the roles of $X_t$ and $\epsilon_t$ are exchangeable. Therefore, there is a reversed Wold decomposition,

$$\epsilon_t = \sum_{j=0}^{\infty} \psi_j X_{t-j}, \quad t \in Z, \tag{9.35}$$

with $\psi_0 = 1$. From this, a method was suggested by Durbin (1960) to fit the ARMA model. The motivation is that it would be much easier to identify the ARMA parameters if the innovations $\epsilon_t$ were observable. Of course, the $\epsilon_t$ are not observed, but we have expression (9.35). Therefore, Durbin suggested to first fit a long autoregression to the data to get the estimated $\epsilon_t$'s and then to solve a LS problem to find the estimates of the $\alpha_j$'s and $\beta_j$'s. This approach has been used by several authors. See, for example, Hannan and Rissanen (1982). Huang (1988b) combined this approach with a new idea. If one defines a stationary times series by (9.30) with $\phi_j = \psi_j$—that is,

$$Y_t = \sum_{j=0}^{\infty} \psi_j \epsilon_{t-j}, \tag{9.36}$$

—then $Y_t$ satisfies the following reversed ARMA model:

$$\sum_{j=0}^{q} \beta_j Y_{t-j} = \sum_{j=0}^{p} \alpha_j \epsilon_{t-j}. \tag{9.37}$$

Similar to (9.32), we have the following connection between the coefficients $\psi_j$ and the ARMA parameters:

$$\alpha_j = \sum_{k=0}^{q} \beta_k \psi_{j-k}, \quad j = 0, 1, \dots, \tag{9.38}$$

where $\alpha_j = 0$ if $j > p$ and $\psi_j = 0$ if $j < 0$ (Exercise 9.13).

Here is another way to look at (9.36)—it is simply (9.35) with $X_t$ replaced by $\epsilon_t$. From a theoretical point of view, there is an advantage dealing with (9.36). The reason is that, in expansion (9.36), the innovations $\epsilon_t$ are independent, whereas in expansion (9.35), the $X_t$'s are correlated. In fact, the innovations correspond to the orthogonal elements in the Hilbert space defined below. Let $\mathcal{L}(\xi_t, t \in T\}$ denote the Hilbert space spanned by the random variables $\xi_t$, $t \in T$, with the inner product and norm defined by $\langle \xi, \eta \rangle = \mathrm{E}(\xi\eta)$ and $\|\xi\| = \{\mathrm{E}(\xi^2)\}^{1/2}$ for $\xi$ and $\eta$ in the Hilbert space. It follows that $\epsilon_t$, $t \in Z$, is an orthogonal sequence of the Hilbert space. More specifically, let $H(s) = \mathcal{L}(\epsilon_u, u \le s)$ and $P_H$ denote the projective operator to the subspace $H$ of the Hilbert space. Define the random variables $v_s(t) = P_{H(s)} Y_t$ for all $s$ and $t$. For any $p, q$, Let $\mu_{p,q}^2$ denote the normalized squared prediction error (SPE) of $v_{-p-1}(0)$ by $v_{-p-1}(-k)$, $k = 1, \ldots, q$, and let $\sigma_{p,q}^2$ denote the SPE of $Y_0$ by $Y_{-k}$, $k = 1, \ldots, q$, and $\epsilon_{-j}$, $j = 1, \ldots, p$; that is,

$$\mu_{p,q}^2 = \left\| v_{-p-q}(0) - P_{\mathcal{L}\{v_{-p-1}(-k), k=1,\ldots,q\}} v_{-p-1}(0) \right\|^2 / \sigma^2,$$

$$\sigma_{p,q}^2 = \left\| Y_0 - P_{\mathcal{L}\{Y_{-k}, k=1,\ldots,q; \epsilon_{-j}, j=1,\ldots,p\}} Y_0 \right\|^2.$$

Huang (1988b) showed that $\mu_{p,q}^2 = \sigma_{p,q}^2 / \sigma^2 - 1$. From this he realized that $\mu_{p,q}^2$ can be used as a tool to determine the orders. Because $\sigma^2$ is the minimum variance of a linear predictor of $Y_0$ using all the past, $Y_s$, $s < 0$, and $\sigma_{p,q}^2$ is the minimum variance of a linear predictor of $Y_0$ using $Y_{-k}$, $k = 1, \ldots, q$, and $\epsilon_{-j}$, $j = 1, \ldots, p$, $\mu_{p,q}^2$ may be viewed as a measure of deficiency of using the information provided by $Y_s$, $s < 0$, when we linearly predict $Y_0$ from $Y_{-k}$, $1 \le k \le q$, and $\epsilon_{-j}$, $1 \le j \le p$. The higher the $\mu_{p,q}^2$, the higher the deficiency in using the information; and the information in the past has been completely used if and only if $\mu_{p,q}^2 = 0$. Based on this idea, Huang proposed a method of ARMA model identification.

First, he obtained an estimated $\mu_{p,q}^2$, $\hat{\mu}_{p,q}^2$. To do so, he first found an expression of $\mu_{p,q}^2$ as a function of the $\psi_j$'s in (9.36). He then obtained estimators of the $\psi_j$'s using the LS method and thus the estimator $\hat{\mu}_{p,q}^2$ using the plug-in method (i.e., by replacing the $\psi_j$'s by their estimators in the function). He then determined the orders $p$ and $q$ as follows: Let $K_n = [\log n)^\alpha]$ for some $\alpha > 1$ (here, $[x]$ represents the largest integer $\le x$). Define $T_n = \max\{0 \le k \le K_n : \hat{\psi}_k^2 > (\log n/n) \sum_{j=0}^{k-1} \hat{\psi}_j^2\}$, where $\hat{\psi}_j$ is the estimator of $\psi_j$ mentioned above and $P_n = [(1 + \delta)T_n]$, where $\delta$ is a small number (e.g., $\delta = 0.1$). Let $G$ be the set of all $(p, q)$ such that $0 \le p, q \le P_n$ and $\hat{\mu}_{p,q}^2 \le P_n \log n/n$. Define $(\hat{p}, \hat{q})$ as the element in $G$ such that $\hat{p} + \hat{q} = \min\{p + q : (p, q) \in G\}$. Huang showed that $\hat{p}$ and $\hat{q}$ are strongly consistent estimators of the true orders $p$ and $q$, respectively.

On the other hand, giving the orders, the parameters $\beta_j$'s and $\alpha_j$'s can be expressed as functions of the $\psi_j$'s. Thus, by plugging in the estimators $\hat{\psi}_j$'s, we obtain estimators of the $\beta_j$'s and $\alpha_j$'s. Huang then applied this method of estimation with the estimated orders as above to obtain estimators of the $\beta_j$'s and $\alpha_j$'s without assuming that the orders are known. He showed that the estimators are asymptotically normal in the sense that for any sequence of constants $\lambda_j$, $j \ge 1$, the distribution of $\sqrt{n} \sum_{j=1}^{\hat{q}} \lambda_j \{\hat{\beta}_j(\hat{p}, \hat{q}) - \beta_j\}$ converges to $N(0, \tau^2)$, where $\hat{\beta}_j(p, q)$ is the estimator of $\beta_j$ with the given $p$ and $q$ as mentioned above, $\tau^2 = \lambda' W \lambda$ with $\lambda = (\lambda_1, \ldots, \lambda_q)'$, $q$ being the true order, and $W$ being a covariance matrix depending on the parameters; a similar result holds for the estimators of the $\alpha_j$'s. Furthermore, the estimators obey the LIL in the sense that, with probability 1, the set of limit points of $\sqrt{n/2 \log \log n} \sum_{j=1}^{\hat{q}} \lambda_j \{\hat{\beta}_j(\hat{p}, \hat{q}) - \beta_j\}$ is $[-\tau, \tau]$; a similar result holds for the estimators of the $\alpha_j$'s.

Note that because of the strong consistency of $\hat{p}$ and $\hat{q}$, we have with probability 1 that $\hat{p} = p$ and $\hat{q} = q$ for large $n$, where $p$ and $q$ are the true orders (why?). Therefore, to establish the CLT and LIL for $\hat{\beta}_j(\hat{p}, \hat{q})'$, $1 \leq j \leq q$, and so forth. all one has to do is to prove the same results for $\hat{\beta}_j(p, q)$, $1 \leq j \leq q$, and so forth, where $p$ and $q$ are the true orders (again, why?). Also note that although Huang's procedure for the order determination is different from the information criterion (9.25), it is not the reason why he was able to remove (9.18) as a critical condition for ARMA model identification, as noted earlier. The reason is, once again, that he dropped such a condition in obtaining the uniform convergence rate for the sample autocorrelations (Section 9.2).

## 9.5 Strong limit theorems for i.i.d. spatial series

Let us now switch our attention to spatial series. The classical limit theorems, as discussed in Chapter 6, are regarding sums of i.i.d. random variables. Similarly, there are "classical" limit theorems for sums of i.i.d. spatial series. To the surprise of some people (including the author himself when he first discovered these results), some of these are not-so-straightforward generalizations of the classical results. For example, suppose that $X_t$, $t \in N$, is an i.i.d. time series, where $N = \{1, 2, \ldots\}$. Then, according to the SLLN, we have $n^{-1} \sum_{t=1}^{n} X_t \xrightarrow{\text{a.s.}} E(X_1)$ as $n \to \infty$, provided that $E(|X_1|) < \infty$. Now, suppose that $X_t$, $t \in N^2$, is an i.i.d. spatial series. One would expect a similar result to hold—that is,

$$\frac{1}{n_1 n_2} \sum_{t_1=1}^{n_1} \sum_{t_2=1}^{n_2} X_{(t_1, t_2)} \xrightarrow{\text{a.s.}} E\{X_{(1,1)}\} \tag{9.39}$$

as $n_1, n_2 \to \infty$, provided that $E\{|X_{(1,1)}|\} < \infty$—but this result is false! At first, the surprise might seem a bit counterintuitive, as one can, perhaps, rearrange the spatial series as a time series and then apply the classical SLLN, so why would not (9.39) hold?

The problem is that there are so many ways by which $n_1$ and $n_2$ can (independently) go to infinity. In fact, if $n_1$ and $n_2$ are restricted to increase in a certain manner—for example, $n_1 = n_2 \to \infty$, —then (9.39) holds as long as $E\{|X_{(1,1)}|\} < \infty$ (Exercise 9.14). As for the rearrangement, note that the order of the terms in the rearranged time series that appear in the summation in (9.39) depends on how $n_1, n_2 \to \infty$—and this is true no matter how one rearranges the spatial series. For example, one method of rearrangement is called the *diagonal method*, namely, $Y_1 = X_{(1,1)}, Y_2 = X_{(1,2)}, Y_3 = X_{(2,1)}, \ldots$, and so on. Consider $n_1 = n_2 = k$. When $k = 1$, the summation in (9.39) only involves $Y_1$; when $k = 2$, it involves $Y_1, Y_2, Y_3$, and $Y_5$ in that order. Now, consider $n_1 = 2k$ and $n_2 = k$. When $k = 1$, the summation involves $Y_1$ and $Y_3$; when $k = 2$, it involves $Y_1, Y_2, Y_3, Y_5, Y_6, Y_9, Y_{10}$, and $Y_{14}$, in that order (Exercise 9.15). All of the strong (a.s.) classical limit theorems, including the SLLN and LIL, may not hold if the order of terms in the summation is allowed to change during the limiting process. On the other hand, all of the weak classical limit theorems, including the WLLN and CLT, are not affected by the change of order in the summation (why?). For example, we have

$$\frac{1}{\sqrt{n_1 n_2}} \sum_{t_1=1}^{n_1} \sum_{t_2=1}^{n_2} X_{(t_1,t_2)} \xrightarrow{d} N(0,\sigma^2) \tag{9.40}$$

as $n_1, n_2 \to \infty$, provided that $X_t$, $t \in N^2$, are i.i.d. with $\mathrm{E}\{X_{(1,1)}\} = 0$ and $\sigma^2 = \mathrm{E}\{X_{(1,1)}^2\} \in (0,\infty)$. Therefore, the focus of the current section is strong limit theorems for i.i.d. spatial series.

Smythe (1973) showed that for (9.39) to hold, all one has to do is to strengthen the moment condition, by a little. More specifically, define $\log^+(x) = \log(x)$ if $x > 1$ and $\log^+(x) = 0$ otherwise. The moment condition for the classical SLLN is that $\mathrm{E}(|X_1|) < \infty$. For (9.39) to hold for an i.i.d. spatial series $X_t$, $t \in N^2$, one needs

$$\mathrm{E}\left[|X_{(1,1)}| \log^+\{|X_{(1,1)}|\}\right] < \infty, \tag{9.41}$$

and this condition is also necessary. More generally, consider an i.i.d. spatial series $X_t$, $t \in N^d$, where $d$ is a positive integer. We use the notation $|n| = n_1 \cdots n_d$ for $n = (n_1, \ldots, n_d)$, $1 \le t \le n$, for $1 \le t_j \le n_j$, $1 \le j \le d$, $1 = (1, \ldots, 1)$ ($d$-dimensional), and $n \to \infty$ for $n_j \to \infty$, $1 \le j \le d$. Then

$$\frac{1}{|n|} \sum_{1 \le t \le n} X_t \xrightarrow{a.s.} \mathrm{E}(X_1) \tag{9.42}$$

as $n \to \infty$ if and only if

$$\mathrm{E}\left\{|X_1|\{\log^+(|X_1|)\}^{d-1}\right\} < \infty. \tag{9.43}$$

So, in particular, when $d = 1$, (9.43) is equivalent to $\mathrm{E}(|X_1|) < \infty$, which is the classical condition; for $d = 2$, (9.43) reduces to (9.41).

Wichura (1973) considered the LIL for the independent spatial series $X_t$, $t \in N^d$. A special case of his results is the i.i.d. case, as follows. Define $\log(x) = 1$ if $x < e$, and $\log\log(x) = \log\{\log(x)\} = 1$ if $x < e^e$; maintain other notation as above. If $X_t, t \in N^d$, are i.i.d. with $\mathrm{E}(X_1) = 0$ and $\mathrm{E}(X_1^2) = 1$, where $d > 1$, then with probability 1 as $n \to \infty$, the set of limit points of

$$\zeta_n = \frac{\sum_{1 \le t \le n} X_t}{\sqrt{2d|n| \log \log(|n|)}}, \quad n \in N^d, \tag{9.44}$$

is $[-1, 1]$ if and only if

$$E\left[\frac{X_1^2 \{\log(|X_1|)\}^{d-1}}{\log \log(|X_1|)}\right] < \infty. \tag{9.45}$$

Recall in the classical situation (see a summary of the classical results at the end of Section 6.5), the necessary and sufficient condition for the LIL is $E(X_1^2) < \infty$. Comparing this condition with (9.45), it seems that there is a discontinuity between $d = 1$ and $d > 1$. Wichura (1973, p. 280) gave the following interpretation for this difference: It "is in precisely the latter case that one can deduce the finiteness of" $E(X_1^2)$ from (9.45) (Exercise 9.16). Note that there is no such discontinuity in $d$ in Smythe's SLLN result, as above. In particular, when $d = 2$, we have with probability 1 that the set of limit points of $\zeta_n$, $n \in N^2$, is $[-1, 1]$ if and only if

$$E\left\{\frac{X_1^2 \log(|X_1|)}{\log \log(|X_1|)}\right\} < \infty. \tag{9.46}$$

## 9.6 Two-parameter martingale differences

Given the roles that martingale differences haveplayed in time series analysis, it is not surprising that similar tools have been used in the analysis of spatial series. The major difference, as noted by Tjøstheim (1978, p. 131), is that "a time series is unidirectional following the natural distinction made between past and present. A similar obvious ordering does not seem to exist for a general spatial series, and this fact reflects itself in the available methods of analysis." To define a two-parameter martingale, which is termed for the lattice analogy of martingales (see Chapter 8), one needs first to be clear what is the past, as the present is usually quite easy to define. Suppose that $t = (t_1, t_2)$ is the present. Tjøstheim (1983) defined the past as $P(t) = \{s = (s_1, s_2) : s_1 < t_1 \text{ or } s_2 < t_2\} \equiv P_1(t)$. Also see Jiang (1989). Jiang (1991a) considered a different definition, in which the past is defined according to a single direction, $P(t) = \{s = (s_1, s_2) : s_1 < t_1\} \equiv P_3(t)$. Jiang (1999a) considered $P(t) = \{s = (s_1, s_2) : s_1 < t_1 \text{ or } s_1 = t_1, s_2 < t_2\} \equiv P_2(t)$. Another possible definition of the past is $P(t) = \{s = (s_1, s_2) : s_1 < t_1 \text{ and } s_2 < t_2\} \equiv P_4(t)$. It is easy to see $P_1(t) \supset P_2(t) \supset P_3(t) \supset P_4(t)$ (Exercise 9.17). Some other types of past will be considered later. A spatial series $\epsilon_t$, $t \in Z^2$, is called a two-parameter martingale differences (TMD) if it satisfies for all $t \in Z^2$,

$$E\{\epsilon_t | \epsilon_s, s \in P(t)\} = 0 \quad \text{a.s.} \tag{9.47}$$

Here, the conditional expectation is with respect to the $\sigma$-field generated by $\epsilon_s$, $s \in P(t)$, and $P(t)$ is a defined past of $t$. We consider some examples.

*Example 9.6.* Let $W_t, t \in Z^2$, be an independent spatial series with $E(W_t) = 0$ and $E(W_t^2) < \infty$. Consider

$$\epsilon_t = W_{(t_1, t_2 - 1)} W_{(t_1, t_2)}$$

for $t = (t_1, t_2)$. Then, $\epsilon_t, t \in Z^2$, is a TMD with respect to $P(t) = P_j(t)$, $j = 1, 2, 3, 4$. To see this, note that

$$\sigma\{\epsilon_s, s \in P_1(t)\} \subset \sigma\{W_s, s \in P_1(t)\}$$

[every $\epsilon_s$, where $s \in P_1(t)$, is a function of $W_s$, $s \in P_1(t)$]. It follows that

$$\begin{aligned}
&E\{\epsilon_t | \epsilon_s, s \in P_1(t)\} \\
&= E\left[E\{W_{(t_1, t_2 - 1)} W_t | W_s, s \in P_1(t)\} | \epsilon_s, s \in P_1(t)\right] \\
&= E\left[W_{(t_1, t_2 - 1)} E\{W_t | W_s, s \in P_1(t)\} | \epsilon_s, s \in P_1(t)\right] \\
&= 0
\end{aligned}$$

because $E\{W_t | W_s, s \in P_1(t)\} = E(W_t) = 0$. This verifies that $\epsilon_t$ is a TMD with $P(t) = P_1(t)$. The rest are left as an exercise (Exercise 9.18).

*Example 9.7.* Let $W_t, t \in Z^2$, be any spatial series. Define

$$\epsilon_t = W_t - E\{W_t | W_s, s \in P_2(t)\}. \tag{9.48}$$

Then $\epsilon_t, t \in Z^2$, is a TMD with respect to $P(t) = P_2(t)$. This is because for any $s \in P_2(t)$, $\sigma\{W_r, r \in P_2(s)\} \subset \sigma\{W_r, r \in P_2(t)\}$; hence,

$$\epsilon_s = W_s - E\{W_s | W_r, r \in P_2(s)\} \in \sigma\{W_r, r \in P_2(t)\}.$$

Therefore, we have $\sigma\{\epsilon_s, s \in P_2(t)\} \subset \sigma\{W_s, s \in P_2(t)\}$. It follows that

$$\begin{aligned}
&E\{\epsilon_t | \epsilon_s, s \in P_2(t)\} \\
&= E\{W_t | \epsilon_s, s \in P_2(t)\} - E\left[E\{W_t | W_s, s \in P_2(t)\} | \epsilon_s, s \in P_2(t)\right] \\
&= E\{W_t | \epsilon_s, s \in P_2(t)\} - E\{W_t | \epsilon_s, s \in P_2(t)\} \\
&= 0.
\end{aligned}$$

On the other hand, $\epsilon_t, t \in Z^2$, is not necessarily a TMD with respect to $P(t) = P_1(t)$. To see this, note that $\epsilon_s, s \in P_1(t)$, involve all of the $W_s, s \in Z^2$ (why?). Therefore, we can write

$$\begin{aligned}
&E\{\epsilon_t | \epsilon_s, s \in P_1(t)\} \\
&= E\{W_t | \epsilon_s, s \in P_1(t)\} - E\left[E\{W_t | W_s, s \in P_2(t)\} | \epsilon_s, s \in P_1(t)\right],
\end{aligned}$$

but this is as far as we can go (Exercise 9.20).

Tjøstheim (1983) considered an extension of the martingale CLT (Theorem 8.7) to TMD satisfying (9.47) with $P(t) = P_1(t)$, but the proof appears to involve some flaws. Jiang (1991b) proved a CLT for triangular arrays of spatial series satisfying a weaker TMD condition than that assumed by Tjøstheim (1983). A similar result was also obtained by Huang (1992). We state Jiang's result below, where the following notation will be used through the rest of the chapter: $s = (s_1, s_2)$, $t = (t_1, t_2)$, $n = (n_1, n_2)$, $|n| = n_1 n_2$, $0 = (0, 0)$, $1 = (1, 1)$, and $s \leq t$ if and only if $s_j \leq t_j$, $j = 1, 2$.

**Theorem 9.7.** Let $\epsilon_{n,t}$, $n \geq 1$, $1 < t < n$, be a triangular array of spatial series. Suppose that there exists a family of $\sigma$-fields $\mathcal{F}_t$, $t \geq 0$, satisfying $\mathcal{F}_s \subset \mathcal{F}_t$ if $s \leq t$. Let $\mathcal{F}_j(t-)$ denote the smallest $\sigma$-field containing $\mathcal{F}_s$, $s_j < t_j$ or $s_j = t_j$, $s_{3-j} < t_{3-j}$, $j = 1, 2$. If

$$\epsilon_{n,t} \in \mathcal{F}_t, \quad \mathrm{E}\{\epsilon_{n,t} | \mathcal{F}_j(t-)\} = 0 \text{ a.s.}, \quad j = 1, 2, \tag{9.49}$$

and furthermore, as $|n| \to \infty$,

$$\max_{1 \leq t \leq n} |\epsilon_{n,t}| \xrightarrow{\mathrm{P}} 0, \tag{9.50}$$

$$\sum_{1 \leq t \leq n} \epsilon_{n,t}^2 \xrightarrow{\mathrm{P}} \eta^2, \tag{9.51}$$

where $\eta$ is a bounded random variable, and

$$\mathrm{E}\left(\max_{1 \leq t \leq n} \epsilon_{n,t}^2\right) \text{ is bounded in } n, \tag{9.52}$$

then as $|n| \to \infty$, we have

$$\sum_{1 \leq t \leq n} \epsilon_{n,t} \xrightarrow{\mathrm{d}} Z \text{ (stably)}, \tag{9.53}$$

where the random variable $Z$ has characteristic function

$$c_Z(\lambda) = \mathrm{E}\{\exp(-\eta^2 \lambda^2 / 2)\}. \tag{9.54}$$

*Note 1.* The limiting process here is $|n| \to \infty$, which is weaker than $n \to \infty$ (i.e., $n_1, n_2 \to \infty$).

*Note 2.* An analogue to condition (8.33) of Theorem 8.7 is not needed because here the $\sigma$-fields do not depend on $n$ (in other words, such a condition is automatically satisfied).

*Note 3.* In the special case where $\epsilon_{n,t} = \epsilon_t / a_n$, $a_n$ being a normalizing constant depending on $n$, one may let $\mathcal{F}_t = \sigma(\epsilon_s, s \leq t)$. Then the first condition of (9.49) (i.e., $\epsilon_{n,t} \in \mathcal{F}_t$) is obviously satisfied; the second condition is equivalent to

$$\mathrm{E}(\epsilon_t | \epsilon_s, s_1 < t_1 \text{ or } s_1 = t_1, s_2 < t_2) = 0 \text{ a.s.}, \tag{9.55}$$

$$\mathrm{E}(\epsilon_t | \epsilon_s, s_2 < t_2 \text{ or } s_2 = t_2, s_1 < t_1) = 0 \text{ a.s.} \tag{9.56}$$

Condition (9.55) is the same as (9.47) with $P(t) = P_2(t)$, whereas (9.56) is the condition with the coordinates switched. Note that (9.47) with $P(t) = P_1(t)$ implies (9.55) and (9.56) (Exercise 9.21).

*Note 4.* Later Jiang (1993) was able to weaken (9.49) to that with $j = 1$ only and (9.52) to that with $\epsilon_{n,t}^2$ replaced by $|\epsilon_{n,t}|^p$ for some $p > 1$.

Furthermore, Jiang (1999a) obtained a LIL for a strictly stationary spatial series $\epsilon_t$, $t \in Z^2$, satisfying

$$\mathrm{E}\{\epsilon_0 | \epsilon_s, s_1 < 0 \text{ or } s_2 < 0\} = 0 \quad \text{a.s.} \tag{9.57}$$

The definition of a strictly stationary spatial series is similar to that for a strictly stationary time series; that is, for any $k \geq 1$ and $t_1, \ldots, t_k, s \in Z^2$, the joint distribution of $\epsilon_{t_j+s}$, $j = 1, \ldots, k$, does not depend on $s$. Note that because of the strict stationarity, (9.57) is equivalent to (9.47) for all $t$ (Exercise 9.22). Let $(\Omega, \mathcal{F}, \mathrm{P})$ be the probability space Define the measure-preserving transformations, $U$ and $V$, on the induced probability space $\left(R^{Z^2}, \mathcal{B}^{Z^2}, \mathrm{P}\epsilon^{-1}\right)$, where $\mathcal{B}$ represents the Borel $\sigma$-field and $\epsilon = (\epsilon_t)_{t \in Z^2}$, by $(Ux)_t = x_{t+u}$, $(Vx)_t = v_{t+v}$, $t \in Z^2$, for $x = (x_t)_{t \in Z^2} \in R^{Z^2}$, where $u = (1, 0)$ and $v = (0, 1)$. In other words, $U$ is the shift by 1 in the first coordinate and $V$ is that in the second coordinate. Denote the a.s. invariant $\sigma$-field (see Appendix A.2) corresponding to $U$ and $V$ by $\bar{\tau}_U$ and $\bar{\tau}_V$, respectively. Let $\bar{\tau} = \bar{\tau}_U \cap \bar{\tau}_V$. The spatial series $\epsilon_t$ is said to be ergodic if $\epsilon^{-1}(\bar{\tau}) = \overline{\{\emptyset, \Omega\}}$, the $\sigma$-field whose elements have probabilities either 0 or 1, and strongly ergodic if $\epsilon^{-1}(\bar{\tau}_U) = \epsilon^{-1}(\bar{\tau}_V) = \overline{\{\emptyset, \Omega\}}$.

**Theorem 9.8.** Let $\epsilon_t$, $t \in Z^2$, be strictly stationary with $\mathrm{E}(X_1^2) = 1$ and $\mathrm{E}(|X_1|^q) < \infty$ for some $q > 2$ and (9.57) holds. Define $\zeta_n$ as (9.44) with $d = 2$.

(i) If $\epsilon_t$ is ergodic, then with probability 1 as $n \to \infty$, the set of limit points of $\zeta_n$ is $[-1, 1]$.

(ii) If $\epsilon_t$ is strongly ergodic, then with probability 1 as $|n| \to \infty$, the set of limit points of $\zeta_n$ is $[-1, 1]$.

In fact, Jiang proved the theorem under a TMD condition slightly weaker than (9.57); that is, for every $m \geq 0$,

$$\mathrm{E}\{\epsilon_0 | \epsilon_s, s_1 < t_1 \text{ or } t_1 \leq s_1 \leq t_1 + m, s_2 < t_2\} = 0 \quad \text{a.s.} \tag{9.58}$$

(Exercise 9.21). The author also discussed a situation where the moment condition, $\mathrm{E}(|X_1|^q) < \infty$ for some $q > 2$, can be reduced to (9.46), which is the minimum moment condition for the i.i.d. case (Section 9.5).

## 9.7 Sample ACV and ACR for spatial series

The subjects of this section are similar to Section 9.2 but for spatial series. Let $X_t$, $t \in Z^2$, be a spatial series (not necessarily stationary). The sample autocovariance (ACV) function for $X_t$ is defined as

$$\hat{\gamma}(u,v) = \frac{1}{|n|} \sum_{1 \le t \le n} X_{t-u} X_{t-v}, \quad u, v \in Z^2. \tag{9.59}$$

Practically speaking, the range of the summation needs to be adjusted according to $u$ and $v$ (and the range of the observed $X_t$'s), as in Section 9.2, but we ignore such an adjustment for the sake of simplicity. The sample autocorrelation (ACR) function is $\hat{\rho}(u,v) = \hat{\gamma}(u,v)/\hat{\gamma}(0,0)$. Note that here we do not assume that $X_t$ is (second-order) stationary; otherwise the notation would be simpler. Nevertheless, for the rest of the section we focus on a linear spatial series that can be expressed as

$$X_t = \sum_{s \in Z^2} a_s W_{t-s}, \tag{9.60}$$

$t \in Z^2$, where $W_t$, $t \in Z^2$, is a spatial $WN(0, \sigma^2)$ series with $\sigma^2 > 0$. Furthermore, we assume that the (constant) coefficients $a_s$ satisfy

$$|a_s| \le c\phi^{|s_1|+|s_2|} \tag{9.61}$$

for some constants $c > 0$ and $0 < \phi < 1$ and all $s = (s_1, s_2) \in Z^2$. It follows that $E(X_t) = 0$ [and this is why there is no need to subtract the sample mean from $X_t$ in the definition of sample ACV; i.e., (9.59)] A special case for which (9.60) and (9.61) are satisfied is considered in the next section.

Jiang (1989) obtained the uniform convergence rate of the iterated logarithm for ACV and ACR under (9.60), (9.61), and the following TMD condition:

$$W_t \in \mathcal{F}_t, \quad E\{W_t|\mathcal{F}_j(t-)\} = 0 \text{ a.s.}, \quad j = 1, 2, \tag{9.62}$$

for all $t$, where $\mathcal{F}_t$, $t \in Z^2$, is a family of $\sigma$-fields satisfying $\mathcal{F}_s \subset \mathcal{F}_t$ if $s \le t$ and $\mathcal{F}_j(t-)$ is the smallest $\sigma$-field containing all the $\mathcal{F}_s$, $s_j < t_j$ or $s_j = t_j$, $s_{3-j} < t_{3-j}$, $j = 1, 2$. Jiang (1991a) obtained the same results under a weaker TMD condition:

$$W_t \in \mathcal{F}_1(t), \quad E\{W_t|\mathcal{F}_1(t-)\} = 0 \quad \text{a.s.}, \tag{9.63}$$

where $\mathcal{F}_1(t)$ is the smallest $\sigma$-field containing all the $\mathcal{F}_s$, $s_1 < t_1$ or $s_1 = t_1$, $s_2 \le t_2$ (Exercise 9.23). Let $D_n = D([(\log n_1)^a], [(\log n_2)^a])$, where $a$ is a positive constant and $D$ is a positive (constant) integer. For the uniform convergence rate of the sample ACV, it is also assumed that

$$E\{W_t^2|\mathcal{F}_1(t-)\} = 1 \quad \text{a.s.} \tag{9.64}$$

for all $t$ and that

$$\limsup_{|n| \to \infty} \frac{1}{|n|} \sum_{t \le n} \left( |W_t|^{4p} + [E\{W_t^4|\mathcal{F}_1(t-)\}]^p \right) < \infty \quad \text{a.s.} \tag{9.65}$$

for some $p > 1$, where the notation $\bar{t}$ denotes $(|t_1|, |t_2|)$. Then we have

$$\max_{\bar{u},\bar{v} \leq D_n} |\hat{\gamma}(u,v) - \gamma(u,v)| = O\left(\sqrt{\frac{\log \log |n|}{|n|}}\right) \quad \text{a.s.} \quad (9.66)$$

If, in addition, we have $\sum_{s \in Z^2} a_s^2 > 0$, then (9.66) holds for the (sample) ACR (i.e., with $\gamma$ replaced by $\rho$). The condition (9.64) can be weakened to some extent, depending on whether ACV or ACR is considered.

Note that if $W_t$ is strictly stationary and strongly ergodic (see Section 9.6) and $E(|W_t|^q) < \infty$ for some $q > 4$, then (9.65) is a consequence of the ergodic theorem (e.g., Zygmund 1951; Jiang 1991b). In fact, the strong ergodicity condition can be weakened for (9.65) to hold (Exercise 9.24).

An exponential inequality, which may be regarded as a two-parameter analogue (and extension) of (5.89), plays an important role in establishing (9.66). For $u = (u_1, u_2)$, $v = (v_1, v_2) \in Z^2$, the notation $u \overset{1}{<} v$ means that $u_1 < v_1$ or $u_1 = v_1$ and $u_2 < v_2$. Let $\xi_t$ be a TMD satisfying (9.63) with $W$ replaced by $\xi$, and let $\eta_t$ be another spatial series satisfying $\eta_t \in \mathcal{F}_1(t)$. Then for any $u \overset{1}{<} v$, $1 \leq m = (m_1, m_2) \leq n = (n_1, n_2)$, and $\lambda > 0$, we have

$$P\left(\max_{m_1 \leq k_1 \leq n_1} \sum_{t_1=m_1}^{k_1} \sum_{t_1=m_2}^{n_2} \left[\xi_{t-u}\eta_{t-v} - \frac{1}{6}\xi_{t-u}^2\eta_{t-v}^2\right.\right.$$
$$\left.\left. -\frac{1}{3}E\{\xi_{t-u}^2 | \mathcal{F}_1(t-u-)\}\eta_{t-v}^2\right] \geq \lambda\right) \leq e^{-\lambda}. \quad (9.67)$$

To see how (9.67) works, suppose that one wishes to show that

$$\frac{1}{\sqrt{|n| \log \log |n|}} \sum_{1 \leq t \leq n} W_{t-u}W_{t-v} \quad (9.68)$$

is bounded a.s. [see Exercise 9.25 for a connection between (9.66) and (9.68)]. In view of the ergodic theorem, this is equivalent to that

$$\frac{1}{\sqrt{|n| \log \log |n|}} \sum_{1 \leq t \leq n} W_{t-u}W_{t-v}$$
$$-\frac{1}{6|n|} \sum_{1 \leq t \leq n} [W_{t-u}^2 W_{t-v}^2 + 2E\{W_{t-u}^2 | \mathcal{F}_1(t-u-)\}W_{t-v}^2] \quad (9.69)$$

is bounded a.s. (because of the subtracted average converges a.s.). Now, consider the probability that (9.69) is $\geq a$ for some $a > 0$. If we multiply both sides of the inequality by $\log \log |n|$, we come up with an inequality like (9.67) with $\xi_t = (\log \log |n|/|n|)^{1/2}W_t$ and $\lambda = a \log \log |n|$, and the right side of the inequality is $e^{-\lambda} = (\log |n|)^{-a}$ (verify this). This upper bound of the probability goes to zero as $|n| \to \infty$, but apparently not fast enough to directly imply that (9.69) is a.s. bounded. However, the upper bound is "good enough" to allow the following subsequence method, which is often used in obtaining strong limit results. Consider $n_i = ([e^{i_1}], [e^{i_2}])$ ($[x]$ is the largest integer $\leq x$), which is a subsequence of $n$ indexed by $i = (i_1, i_1) \geq 1 = (1, 1)$. It can be shown (Exercise 9.26) that $\sum_{i \geq 1} (\log |n_i|)^{-a} < \infty$ for sufficiently large $a$. It then follows by a lattice version of the Borel–Cantelli lemma (Lemma 2.5) that (9.69) is bounded a.s., at least for the subsequence $n_i$. The question then is how to extend the a.s. convergence to the entire sequence $n$, and this is where one needs a maximum inequality, such as (9.67). Note that inside the probability sign is the maximum of sums rather than a single sum. This covers the "gaps" between the subsequence $n_i$ and therefore the entire sequence. There are, of course, technical details, but this is the main idea.

It is worthy to mention another inequality, which may be regarded as a two-parameter extension of Burkholder's inequality (see Section 5.4). This inequality was used in obtaining the uniform convergence rate of sample ACR for, say, strictly stationary spatial series. We omit the details of the latter result (see Jiang 1991a), but the inequality is nevertheless useful in perhaps other problems as well (See Exercise 9.27). Let $\xi_t$ be as in (9.67). Define $S_{k,n} = \sum_{k+1 \leq t \leq k+n} \xi_t$. For every $p > 1$, there is a constant $B_p$ depending only on $p$ such that for any $k \in Z^2$ and $N \geq 1$, we have

$$E\left(\max_{1 \leq n \leq N} |S_{k,n}|^p\right) \leq B_p (1 + \log_2 N_2)^p \left\{ \sum_{k+1 \leq t \leq k+N} E(|\xi_t|^p) \right\} \quad (9.70)$$

if $1 < p \leq 2$, where $\log_2(\cdot)$ is the logarithmic function with base 2, and

$$E\left(\max_{1 \leq n \leq N} |S_{k,n}|^p\right) \leq B_p \left( \sum_{k+1 \leq t \leq k+N} \|\xi_t\|_p^2 \right)^{p/2} \quad (9.71)$$

if $p > 2$, where $\|\xi_t\|_p = \{E(|\xi_t|^p)\}^{1/p}$.

## 9.8 Case study: Spatial AR models

There is extensive literature on spatial AR models, introduced in Section 9.1, as well as their applications in fields such as ecology and economics. For example, Lichstein et al. (2002) used Gaussian spatial AR models to examine breeding habitat relationships for three common neotropical migrant songbirds in the southern Appalachian Mountains of North Carolina and Tennessee (USA). Langyintuo and Mekuria (2008) discussed an application of spatial AR models in assessing the influence of neighborhood effects on the adoption of improved agricultural technologies in developing countries.

A spatial AR model is defined by (9.10) with $q = 0$ or, equivalently,

$$X_t - \sum_{s \in \langle 0, p]} b_s X_{t-s} = W_t, \quad t \in Z^2, \tag{9.72}$$

where $p = (p_1, p_2)$ is the order of the spatial AR model and $W_t$, $t \in Z^2$, is a spatial $WN(0, \sigma^2)$ series. It is required that the corresponding polynomial of two complex variables, $z_1$ and $z_2$, satisfy

$$1 - \sum_{s \in \langle 0, p]} b_s z_1^{s_1} z_2^{s_2} \neq 0, \quad |z_1| \leq 1, |z_2| \leq 1. \tag{9.73}$$

In engineering literature, (9.73) is known as the minimum phase property. Chiang (1987) noted that (9.73) corresponds a special kind of Markov property in random fields, called the quadrant Markov property. Also see Chiang (1991). The term "random fields" is often used interchangeably with "spatial series," although the former also includes continuous multiparameter processes (in other words, the term "random fields" to spatial series is like the term "stochastic processes" to time series). The Markovian property is defined by dependence of the present on the past through the immediate past (see the next chapter). As noted earlier, in the lattice case the past is not uniquely defined, and, depending on the definition of the past (and immediate past), there are several different types of Markovian properties of random fields (e.g., Chiang 1991). The minimum phase property also implies the Wold decomposition (9.11), where the coefficients $a_s$ satisfy (9.61) for $s \geq 0$.

Tjøstheim (1978) considered a similar Y-W equation to that in the time series [see (9.33)] for estimating the AR coefficients in (9.72), namely,

$$\sum_{v \in \langle 0, p]} b_v \gamma(u, v) = \gamma(u, 0), \quad u \in \langle 0, p]. \tag{9.74}$$

The Y-W estimator of $b = (b_v)_{v \in \langle 0, p]}$ is defined as the solution to (9.74) with $\gamma$ replaced by $\hat{\gamma}$, the sample ACV. The Y-W equation can be expressed in a compact form as $Gb = g$, where $G = [\gamma(u, v)]_{u,v \in \langle 0, p]}$ and $g = [\gamma(u, 0)]_{u \in \langle 0, p]}$. The estimator $\hat{b} = (\hat{b})_{v \in \langle 0, p]}$ satisfies $\hat{G}\hat{b} = \hat{g}$, where $\hat{G}$ and $\hat{g}$ are $G$ and $g$, respectively, with $\gamma$ replaced by $\hat{\gamma}$. Under the assumption that the innovations $W_t$ are i.i.d., the author showed that $\hat{b}$ is consistent; that is, $\hat{b} \xrightarrow{P} b$ as $|n| \to \infty$. Furthermore, the estimator is asymptotically normal in that, as $n \to \infty$, $\sqrt{|n|}(\hat{b}-b)$ converges in distribution to a multivariate normal distribution with mean vector 0 and a certain covariance matrix. Note that here the limiting process for the consistency is $n_1 n_2 \to \infty$, whereas that for the asymptotic normality is $n_1, n_2 \to \infty$.

Tjøstheim (1983) considered the strong consistency of the Y-W estimator as well as the LS estimator of $b$. Let $L(n, p) = \{t : 1 \leq t, t - s \leq n, \forall s \in \langle 0, p]\}$. The LS estimator is defined by the vector $b$ that minimizes

$$\sum_{t \in L(n,p)} \left| X_t - \sum_{s \in \langle 0, p]} b_s X_{t-s} \right|^2. \tag{9.75}$$

Under the assumption that the innovations are i.i.d. and that $n_1$ and $n_2$ go to infinity at the same rate; that is, $n_j = h_j k$, where $h_j$ is a fixed integer, $j = 1, 2$, and $k \to \infty$. The author showed that both the Y-W and LS estimators converge a.s. to $b$. The author also obtained asymptotic normality of these estimators under the same limiting process. So far, the asymptotic results are based on the assumption that $p$ is known. Tjøstheim (1983) also considered the situation where $p$ is unknown and therefore has to be determined from the data. He considered a GIC-type criterion (see Section 9.3) in the form of

$$C(p) = \log \hat{\sigma}^2(p) + \frac{l(|n|)}{|n|} d(p), \tag{9.76}$$

where $d(p)$ is the number of AR coefficients involved in (9.72) [i.e., $d(p) = (p_1 + 1)(p_2 + 1) - 1$] and $\hat{\sigma}^2(p)$ is the residual sum of squares (RSS) after fitting the LS problem; that is,

$$\hat{\sigma}^2(p) = \frac{1}{|n|} \sum_{t \in L(n,p)} \left| X_t - \sum_{x \in (0,p]} \hat{b}_s^{(p)} X_{t-s} \right|^2,$$

where $\hat{b}^{(p)}$ is the minimizer of (9.75) for the given $p$. The function $l(\cdot)$ depends on the criterion. For the AIC, BIC, and HQ (see Section 9.3), the corresponding $l(|n|)$ are 2, $\log(|n|)$, and $2 \log \log(|n|)$, respectively. Tjøstheim defined the estimator of $p$, $\hat{p}$, as the minimizer of (9.76) over $0 \leq p \leq P$, where $P$ is known. He showed that $\hat{p}$ is consistent in the sense that $P(\hat{p} \neq p) \to 0$ under the limiting process $n_1 = n_2 = k \to \infty$, provided that $l(|n|) \to \infty$ and $l(|n|)/|n| \to 0$ as $k \to \infty$. Thus, in particular, the BIC and HQ are consistent, whereas the AIC is not. These results are similar to those of Section 9.3. Furthermore, the author considered the extension of his results by replacing the i.i.d. assumption on the innovations $W_t$ by the following TMD assumption:

$$W_t \in \mathcal{F}_t, \quad \mathrm{E}\{W_t | \mathcal{F}(t-)\} = 0 \quad \text{a.s.}, \tag{9.77}$$

where $\mathcal{F}(t-)$ is the smallest $\sigma$-field containing all $\mathcal{F}_s$, $s_1 < t_1$ or $s_2 < t_2$, but the proofs appear to be flawed. Another limitation of Tjøstheim's results is that $P$, the upper bound of the range of $p$ over which (9.76) is minimized, is assumed known, whereas in practice, such an upper bound may not be known.

   Jiang (1991b) proved that if $X_t$ is a spatial AR($p$) series satisfying the minimum phase property (9.73), where the WN innovation series $W_t$ is strictly stationary and strongly ergodic (see Section 9.6) with $\mathrm{E}\{W_0^2 \log^+(|W_0|)\} < \infty$, then the Y-W estimator, $\hat{b}$, is strongly consistent as $|n| = n_1 n_2 \to \infty$; that is, $\hat{b} \xrightarrow{\text{a.s.}} b$ as $|n| \to \infty$. The author argued that the same result holds under the following alternative to the strong ergodicity condition: $W_t$ is a TMD satisfying (9.62) with $\mathcal{F}_t = \sigma(W_s, s \leq t)$ [note that in such a case, the first condition in (9.62) automatically holds], and the series $\sum_{i=0}^{\infty} \mathrm{E}|\mathrm{E}\{W_{(i,0)}^2 | \mathcal{F}_{(-1,0)}\} - 1|$ and $\sum_{i=0}^{\infty} \mathrm{E}|\mathrm{E}\{W_{(0,i)}^2 | \mathcal{F}_{(0,-1)}\} - 1|$ are both finite. Furthermore, the author obtained asymptotic normality of the Y-W estimator. For example, suppose that $W_t$ is strictly stationary, strongly ergodic, $E\{W_0^4 \log^+(|W_0|)\} < \infty$, and the TMD condition (9.62) is satisfied for some family of $\sigma$-fields $\mathcal{F}_t$ [it is not necessarily to have $\mathcal{F}_t = \sigma(W_s, s \leq t)$]. Then we have, as $|n| \to \infty$,

$$\sqrt{|n|}(\hat{b} - b) \xrightarrow{d} N(0, \Gamma^{-1}\Sigma\Gamma^{-1}), \tag{9.78}$$

where $\Gamma = [\gamma(u, v)]_{u,v \in \langle 0,p]}$ and $\Sigma = [\sigma(u, v)]_{u,v \in \langle 0,p]}$, with

$$\sigma(u, v) = E(W_0^2 X_{-u} X_{-v}).$$

Note that the limiting process here is, again, $|n| = n_1 n_2 \to \infty$, which is more general than $n_1, n_2 \to \infty$ at the same rate (Tjøstheim 1983), or even $n_1$ and $n_2$ independently go to infinity. Once again, the author proposed an alternative to the strong ergodicity condition. The alternative is that for any $x, y \in \langle 0, \infty) = \{s = (s_1, s_2), s_1, s_2 \geq 0, (s_1, s_2) \neq (0,0)\}$, $j = 1, 2$, and $t_{3-j} \in Z$, both

$$\frac{1}{n_j} \sum_{t_j=1}^{n_j} \{E(W_t^2|\mathcal{F}_{t-}) - 1\},$$

where $\mathcal{F}_{t-}$ is the smallest $\sigma$-field containing all $\mathcal{F}_s$, $s \leq t$ and $s \neq t$, and

$$\frac{1}{n_j} \sum_{t_j=1}^{n_j} W_{t-x} W_{t-y} \{E(W_t^2|\mathcal{F}_{t-}) - 1\}$$

converges to zero in probability as $n_j \to \infty$.

An underlying assumption of Jiang (1991b) is that $p$, the order of the spatial AR model, is known. Jiang (1993) considered the more practical situation where $p$ is unknown and therefore has to be determined from the observed data. He considered a criterion of the form (9.76) except with $l(|n|)$ replaced by $l(n)$ [the difference is that the former depends only on $|n| = n_1 n_2$, whereas the latter depends on $n = (n_1, n_2)$]. He showed that if $l(n) \to \infty$ and $l(n)/|n| \to 0$ as $|n| \to \infty$, then the minimizer of the criterion function over $0 \leq p \leq P$, $\hat{p}$, is a consistent estimator of $p$; that is, $P(\hat{p} \neq p) \to 0$, as $|n| \to \infty$. Note that a similar result was obtained by Tjøstheim (1983) under the restricted limiting process $n_1 = n_2 \to \infty$. Once again, the result assumed a known upper bound $P$, which may not be practical. One idea of relaxing this assumption is to let $P$ increase with the sample size; that is, $P = P_n \to \infty$, as $n \to \infty$. Another challenging task is to obtain strong consistency of $\hat{p}$. The following result was proven in Jiang (1993). Suppose that $X_t$ is a spatial AR($p$) series satisfying the minimum phase property, where $W_t$ is a TMD satisfying (9.63). Furthermore, suppose that

$$\liminf_{n \to \infty} \frac{1}{|n|} \sum_{1 \leq t \leq n} E\{W_t^2|\mathcal{F}_1(t-)\} > 0 \quad \text{a.s.},$$

$\sup_t \mathrm{E}\{W_t^2|\mathcal{F}_1(t-)\} < \infty$ a.s., and $\mathrm{E}(|W_t|^q) < \infty$ for some $q > 4$. Let $\hat{p}$ be the minimizer of (9.76), with $l(|n|)$ replaced by $l(n)$, over $0 \leq p \leq P_n$, where $P_n = ([(\log n_1)^\alpha], [(\log n_2)^\alpha])$ for some $\alpha > 0$. If $l(n)$ satisfies

$$\frac{l(n)}{\log\log(|n|)} \to \infty, \quad \frac{l(n)}{|n|}\{\log(|n|)\}^{2\alpha} \to 0$$

as $n \to \infty$, then $\hat{p} \xrightarrow{\text{a.s.}} p$ as $n \to \infty$. Note that the latter result implies that, with probability 1, we have $\hat{p} = p$ for large $n$. This property ensures that if the Y-W estimator of $b$ is obtained using $\hat{p}$ instead of $p$, the resulting estimator of $b$ has the same asymptotic properties, such as strong consistency and asymptotic normality, as that obtained using the true $p$ (Exercise 6.30). Also, note that the limiting process for the strong consistency of $\hat{p}$ is $n \to \infty$ instead of $|n| \to \infty$. This makes sense because as the sample size increases, $P_n$ needs to increase in both directions corresponding to $n_1$ and $n_2$ to make sure that the range of minimization eventually covers the true $p$. Therefore, $n_1$ and $n_2$ both have to increase.

The strong consistency of $\hat{b}$ is a consequence of the ergodic theorem (Jiang 1991b). A main tool for establishing the asymptotic normality of $\hat{b}$ is the CLT for triangular arrays of the TMD (Theorem 9.7). In fact, the TMD condition assumed in Jiang (1993) is weaker than that of Jiang (1991b), namely, (9.63) instead of (9.62), and an extension of Theorem 9.7 under the weaker condition was given in Jiang (1993). The uniform convergence rate of the sample ACV (ACR) discussed in Section 9.7 played a key role in obtaining the strongly consistent order determination for the spatial AR model.

## 9.9 Exercises

*9.1.* Verify the basic properties (i)–(iii) of an autocovariance function [below (9.4) in Section 9.1].

*9.2.* Let $X_t$, $t \in Z^2$, be a spatial series such that $\mathrm{E}(X_t^2) < \infty$ for any $t$. Show that the following two statements are equivalent:

(i) $\mathrm{E}(X_t)$ is a constant and $\mathrm{E}(X_{s+h}X_{t+h}) = \mathrm{E}(X_s X_t)$ for all $s, t, h \in Z^2$;

(ii) $\mathrm{E}(X_t)$ and $\mathrm{E}(X_t X_{t+h})$ does not depend on $t$ for any $t, h \in Z^2$.

*9.3* (Poisson process and WN). A stochastic process $P(t)$, $t \geq 0$ is called a Poisson process if it satisfies the following: (i) $P(0) = 0$; (ii) for any $0 \leq s < t$ and nonnegative integer $k$,

$$\mathrm{P}\{P(t) - P(s) = k\} = \frac{\{\lambda(t-s)\}^k}{k!}e^{-\lambda(t-s)},$$

where $\lambda$ is a positive constant; and (iii) the process has independent increments; that is, for any $n > 1$ and $0 \leq t_0 < t_1 < \cdots < t_n$, the random variables $P(t_j) - P(t_{j-1})$, $j = 1, \ldots, n$, are independent. The constant $\lambda$ is called the strength of the Poisson process. Derive the mean and variance of $P(t)$ and show that $\epsilon_n = P(n+1) - P(n) - \lambda$, $n = 1, 2, \ldots$, is a WN$(0, \lambda)$ process.

9.4 (Brownian motion and WN). Recall a stochastic process $B(t)$, $t \geq 0$, a Brownian motion if it satisfies (i) $B(0) = 0$, (ii) for any $0 \leq s < t$, $B(t) - B(s) \sim N(0, t-s)$, and (iii) the process has independent increments. Show that $\epsilon_n = B(n+1) - B(n)$, $n = 1, 2, \ldots$, is a standard normal WN process.

9.5. The time series $X_t$, $t \in Z$, satisfies (i) $E(X_t^2) < \infty$, (ii) $E(X_t) = \mu$, a constant, and (iii) $E(X_s X_t) = \psi(t - s)$ for some function $\psi$, for $s, t \in Z$. Show that $X_t$, $t \in Z$, is second-order stationary and find its autocovariance function.

9.6. Suppose that $X_t$, $t \in Z$, is second-order stationary. Show that if the $n$th-order covariance matrix of $X_t$, $\Gamma_n = [\gamma(i-j)]_{1 \leq i, j \leq n}$, is singular, then there are constants $a_j$, $0 \leq j \leq n-1$, such that for any $t > s$ we have

$$X_t = a_0 + \sum_{j=1}^{n-1} a_j X_{s-j} \quad \text{a.s.}$$

9.7. Suppose that $X_t$ and $Y(t)$ are both second-order stationary and the two time series are independent with the same mean and autocovariance function. Define a "coded" time series as

$$Z_t = \begin{cases} X_t \text{ if } t \text{ is odd} \\ Y_t \text{ if } t \text{ is even.} \end{cases}$$

Is $Z_t$ a second-order stationary time series? Justify your answer.

9.8. Show that if the innovations $\epsilon_t$ is a Gaussian WN$(0, \sigma^2)$ process with $\sigma^2 > 0$, then (9.18) holds.

9.9. Show that the Kullback–Leibler information defined by (9.21) is $\geq 0$ with equality holding if and only if $f = g$ a.e.; that is, $f(x) = g(x)$ for all $x \notin A$, where $A$ has Lebesgue measure zero. However, it is not a distance [as defined below (6.63)].

9.10. Show that the Wold coefficients of an ARMA process $X_t$ in (9.30) satisfy the following:
(i) $\phi_0 = 1$.
(ii) $\phi_j \rho^j \to 0$ as $j \to \infty$, where $\rho > 1$ is the number in (9.29).

9.11. Show that the autocovariance function of an ARMA$(p, q)$ process can be expressed as (9.31). Furthermore, there is a constant $c > 0$ such that $|\gamma(h)| \leq c\rho^{-h}$, $h \geq 0$, where $\rho > 1$ is the number in (9.29).

9.12. Verify the Yule–Walker equation (9.33) as well as (9.34).

9.13. Verify the reversed ARMA model (9.37) as well as (9.38).

9.14. Suppose that $X_t$, $t \in N^2$, is an i.i.d. spatial series. Show that (9.39) holds when $n_1 = n_2 \to \infty$, provided that $E\{|X_{(1,1)}|\} < \infty$.

9.15. Regarding the diagonal method of rearranging a spatial series as a time series (see the second paragraph of Section 9.6), write the order of terms in the summation in (9.39) for the following cases:

(i) $n_1 = n_2 = k$, $k = 3$;
(ii) $n_2 = n_2 = k$, $k = 4$;
(iii) $n_1 = 2k$, $n_2 = k$, $k = 3$;
(iv) $n_2 = 2k$, $n_2 = k$, $k = 4$.

*9.16.* (i) Show that for any random variable $X$,

$$\mathrm{E}\left[\frac{X^2\{\log(|X|)\}^{d-1}}{\log\log(|X|)}\right] < \infty,$$

where $d > 1$, implies $\mathrm{E}(X^2) < \infty$.

(ii) Give an example of a random variable $X$ such that

$$\mathrm{E}\left\{\frac{X^2}{\log\log(|X|)}\right\} < \infty$$

but $\mathrm{E}(X^2) = \infty$.

*9.17.* Draw diagrams of the different pasts, $\mathrm{P}_j(t)$, $j = 1, 2, 3, 4$, defined in Section 9.6 and show $\mathrm{P}_1(t) \supset \mathrm{P}_2(t) \supset \mathrm{P}_3(t) \supset \mathrm{P}_4(t)$.

*9.18.* Verify that the spatial series $\epsilon_t$ defined in Example 9.6 is a TMD with respect to $P(t) = P_j(t)$, $j = 2, 3, 4$ (the case $j = 1$ was already verified).

*9.19.* Let $W_t$ be as in Example 9.6. Define $\epsilon_t = W_{(t_1+1, t_2-1)}$. Show that $\epsilon_t$, $t \in Z^2$, is a TMD with respect to $P(t) = P_j(t)$, $j = 1, 2, 3, 4$.

*9.20.* This exercise is related to Example 9.7.

(i) It was shown that $\epsilon_t$, $t \in Z^2$, is a TMD with respect to $P(t) = P_2(t)$, but not necessarily a TMD with respect to $P(t) = P_1(t)$. Is $\epsilon_t$, $t \in Z^2$, a TMD with respect to $P(t) = P_3(t)$, or $P(t) = P_4(t)$?

(ii) If we switch the roles of $P_1(t)$ and $P_2(t)$ [i.e., define, instead of (9.48), $\epsilon_t = W_t - \mathrm{E}\{W_s, s \in P_1(t)\}$], is $\epsilon_t$, $t \in Z^2$, a TMD with respect to $P(t) = P_2(t)$?

(iii) Is there a general rule that you can draw from Example 9.7 and this exercise?

*9.21.* (i) Show that (9.47) with $P(t) = P_1(t)$ implies (9.55) and (9.56).

(ii) Give an example of a spatial series $\epsilon_t$ satisfying (9.55) but not (9.56).

(iii) Show that (9.57) implies (9.58) for all $m \geq 0$.

*9.22.* Show that if $\epsilon_t$, $t \in Z^2$, is strictly stationary, then (9.57) holds if and only if (9.47) holds for all $t$.

*9.23.* Show that (9.63) is a weaker TMD condition than (9.62).

*9.24.* Let $X_t$, $t \in Z^2$ be a strictly stationary spatial series. Define the invariant $\sigma$-fields $X^{-1}(\bar{\tau}_U)$, $X^{-1}(\bar{\tau}_V)$, and $X^{-1}(\bar{\tau})$ as in Section 9.6 (above Theorem 9.8) with $\epsilon$ replaced by $X$. For this exercise, however, all you need to know is that these are some $\sigma$-fields depending on the stationary spatial series. Then, according to the ergodic theorem (see Jiang 1991b), we have

$$\frac{1}{|n|} \sum_{1 \leq t \leq n} X_t \xrightarrow{\text{a.s.}} \mathrm{E}\{X_1 | X^{-1}(\bar{\tau})\}$$

as $|n| \to \infty$, provided that (9.41) holds and $X^{-1}(\bar{\tau}_U) = X^{-1}(\bar{\tau}_V)$. You may assume that this latter condition is satisfied for whatever spatial series we are dealing with in this exercise. Suppose that $W_t$, $t \in Z^2$, is strictly stationary.

(i) Consider $X_t = W_{(t_1, -t_2)}$. Show that $X_t$, $t \in Z^2$, is strictly stationary. Similar results hold for $X_t = W_{(-t_1, t_2)}$, and $X_t = W_{-t}$.

(ii) Suppose that $E(|W_0|^q) < \infty$ for some $q > 4$. Use the above ergodic theorem and the facts in (i) to argue that

$$\limsup_{|n| \to \infty} \frac{1}{|n|} \sum_{t \le n} |W_t|^{4p} < \infty \quad \text{a.s.}$$

for some $p > 1$.

(iii) Show that $X_t = E\{W_t^4 | \mathcal{F}_1(t-)\}$, $t \in Z^2$, is strictly stationary. Hint: Suppose that

$$E\{W_0^4 | \mathcal{F}_1(0-)\} = g\left(W_s, s \overset{1}{<} 0\right) \quad \text{a.s.}$$

for some function $g$, where $s \overset{1}{<} 0$ if and only if $s_1 < 0$ or $s_1 = 0$ and $s_2 < 0$. Then

$$E\{W_t^4 | \mathcal{F}_1(t-)\} - g\left(W_s, s \overset{1}{<} t\right) \quad \text{a.s.}$$

(iv) Using similar arguments, show that

$$\limsup_{|n| \to \infty} \frac{1}{|n|} \sum_{t \le n} [E\{W_t^4 | \mathcal{F}_1(t-)\}]^p < \infty \quad \text{a.s.}$$

for some $p > 1$, provided that $E(|W_0|^q) < \infty$ for some $q > 4$.

9.25. Let $W_t$, $t \in Z^2$, be a $WN(0, \sigma^2)$ spatial series and $u \ne v$. Show that (9.68) is bounded a.s. if and only if

$$\hat{\gamma}(u, v) - \gamma(u, v) = O\left(\sqrt{\frac{\log \log |n|}{|n|}}\right) \quad \text{a.s.,}$$

where $\gamma(u, v)$ and $\hat{\gamma}(u, v)$ are the ACV and sample ACV of $W_t$ at $u$ and $v$, respectively.

9.26. Let $n_i = ([e^{i_1}], [e^{i_2}])$ for $i = (i_1, i_2)$, where $[x]$ represents the largest integer $\le x$. Show that $\sum_{i \ge 1} (\log |n_i|)^{-a} < \infty$ for sufficiently large $a$, where $|n| = n_1 n_2$ for any $n = (n_1, n_2)$.

9.27. Suppose that $W_t$, $t \in Z^2$, satisfy (9.63) and $E(|W_t|^p) < \infty$ for some $p > 1$. Use the TMD extension of Burkholder's inequality [i.e., (9.70) and (9.71)] to establish the following SLLN: $|n|^{-1} \sum_{1 \le t \le n} W_t \overset{\text{a.s.}}{\longrightarrow} 0$, as $|n| \to \infty$. [Hint: You may use a similar subsequence method as described in the second to last paragraph of Section 9.7, with $n_i = (2^{i_1}, 2^{i_2})$ for $i = (i_1, i_2)$.]

*9.28.* Suppose that $X_t$ is a spatial AR series satisfying

$$X_{(t_1,t_2)} - 0.5X_{(t_1-1,t_2)} - 0.5X_{(t_1,t_2-1)} + 0.25X_{(t_1-1,t_2-1)} = W_{(t_1,t_2)},$$

$t \in Z^2$, where $W_t$ is a spatial $WN(0, \sigma^2)$ series.

(i) What is the order $p$ of the spatial AR model? What are the coefficients?

(ii) Verify that the minimum phase property (9.73) is satisfied.

(iii) Write out the Y-W equation (9.72) for the current model.

(iv) Suppose that $X_t$, $1 \le t \le n$, are observed, where $n = (n_1, n_2)$. Without using the consistency results discussed in Section 9.8, show that the Y-W estimators of the AR coefficients converge in probability to the true values of those coefficients as $n_1 n_2 \to \infty$. You may assume that $\hat{\gamma}(u, v) \overset{P}{\longrightarrow} \gamma(u, v)$ as $n_1 n_2 \to \infty$ for any $u, v$.

*9.29.* Continue with the previous exercise.

(i) Find an expression for the LS estimator of $b$, the vector of the AR coefficients, that minimizes (9.75). Is the LS estimator different from the Y-W estimator in the previous exercise?

(ii) Show that the LS estimator of $b$ converges in probability to $b$ as $n_1 n_2 \to \infty$. Again, you may assume that the sample ACV converges to the ACV in probability as $n_1 n_2 \to \infty$.

*9.30.* Show that $\hat{p} \overset{\text{a.s.}}{\longrightarrow} p$ as $n \to \infty$ if and only if $P(\hat{p} = p$ for large $n) = 1$. Here, $n = (n_1, n_2)$ is large if and only if both $n_1$ and $n_2$ are large. Let $\tilde{b}$ and $\hat{b}$ denote the Y-W estimator of $b$, the vector of spatial AR coefficients, obtained using $p$ and $\hat{p}$, respectively. Show the following:

(i) If $\tilde{b} \overset{\text{a.s.}}{\longrightarrow} b$ as $n \to \infty$, then $\hat{b} \overset{\text{a.s.}}{\longrightarrow} b$ as $n \to \infty$.

(ii) If $\sqrt{|n|}(\tilde{b} - b) \overset{d}{\longrightarrow} N(0, R)$ as $n \to \infty$, where $R$ is a covariance matrix, then $\sqrt{|n|}(\hat{b} - b) \overset{d}{\longrightarrow} N(0, R)$ as $n \to \infty$ for the same $R$.

# 10

# Stochastic Processes

## 10.1 Introduction

A stochastic process may be understood as a continuous-time series or as an extension of the time series that includes both discrete-time and continuous-time series. In this chapter we discuss a few well-known stochastic processes, which in a certain sense define the term stochastic processes. These include both discrete-time and continuous-time processes that have not been previously discussed in details. Again, our focus is the limiting behaviors of these processes.

During the author's time as a graduate student, one of the classroom examples that struck him the most was given by Professor David Aldous in his lectures on Probability Theory. The example was taken from Durrett (1991, p. 275). A modified (and expanded) version is given below.

*Example 10.1.* Professor E. B. Dynkin used to entertain the students in his probability class with the following counting trick. A professor asks a student to write 100 random digits from 0 to 9 on the blackboard. Table 10.1 shows 100 such digits generated by a computer. The professor then asks another

**Table 10.1. Random digits and the student's sequence**

9 6 3 2 2 8 <u>7</u> 1 1 0 1 7 8 <u>7</u> 0 9 4 6 7 6 <u>3</u> 9 7 <u>9</u> 6
7 9 5 4 4 9 7 <u>8</u> 6 7 9 9 4 3 5 <u>1</u> <u>9</u> 1 1 6 7 5 4 0 5
<u>8</u> 7 9 4 0 5 5 2 <u>8</u> 4 0 9 9 3 7 6 <u>3</u> 1 3 0 <u>7</u> 7 7 9 0
5 1 0 <u>7</u> <u>7</u> 4 0 4 2 2 3 <u>3</u> 7 9 <u>5</u> 6 5 3 0 <u>9</u> 3 4 2 8 7

student to choose one of the first 10 digits without telling him. Here, we use the computer to generate a random number from 1 to 10. The generated number is 7, and the 7th number of the first 10 digits in the table is also 7. Suppose that this is the number that the second student picks. She then

J. Jiang, *Large Sample Techniques for Statistics,*
DOI 10.1007/978-1-4419-6827-2_10, © Springer Science+Business Media, LLC 2010

counts 7 places along the list, starting from the number next to 7. The count stops at (another) 7. She then counts 7 places along the list, again. This time the count stops at 3. She then counts 3 places along the list, and so on. In the case that the count stops at 0, the student then counts 10 places on the list. The student's counts are underlined in Table 10.1. The trick is that these are all secretly done behind the professor, who then turns around and points out where the student's counts finally ends, which is the last 9 in the table.

Table 10.2 shows how the professor does the trick. Regardless of what the student is doing, he simply picks a first digit of his own, say, the very first digit, which is 9. He then forms his own sequence according to the same rules as the student. The professor's sequence are overlined in Table 10.2. It is seen

**Table 10.2. The professor's trick**

$$\overline{9\,6\,3\,2\,2\,8\,7}\,\underline{1}\,1\,\overline{0}\,1\,7\,8\,\underline{7}\,0\,9\,4\,6\,7\,\overline{6}\,\underline{3}\,\overline{9}\,7\,\underline{9}\,6$$
$$7\,9\,5\,4\,4\,9\,7\,\overline{8}\,6\,7\,9\,9\,4\,3\,5\,\underline{1}\,\overline{9}\,1\,1\,6\,7\,5\,4\,0\,5$$
$$\overline{8}\,7\,9\,4\,0\,5\,5\,2\,\overline{8}\,4\,0\,9\,9\,3\,7\,6\,\overline{3}\,1\,3\,\overline{0}\,7\,7\,7\,9\,0$$
$$5\,1\,0\,7\,\overline{7}\,4\,0\,4\,2\,2\,3\,\overline{3}\,7\,9\,\overline{5}\,6\,5\,3\,0\,\overline{9}\,3\,4\,2\,8\,7$$

that, at some point (first 8 in the second row), the two sequences hit and then move together. Therefore, the professor's sequence will end exactly where the student's does.

Now we know the professor's trick, but what is the "trick"? The random digits written on the blackboard may be thought of as realizations of the first 100 of a sequence of independent random variables $\xi_1, \xi_2, \ldots$ having the same distribution $P(\xi_i = j) = 1/10$, $j = 1, \ldots, 10$ (here 0 is treated the same as 10). Starting with an initial location $X_1 = I$ $(1 \leq I \leq 10)$, the locations of the sequence of digits formed either by the professor or by the student satisfy

$$X_{n+1} = X_n + \xi_{X_n}, \quad n = 1, 2, \ldots \tag{10.1}$$

(Exercise 10.1). An important property of the sequence $X_n$ is the following:

$$P(X_{n+1} = j | X_1 = i_1, \ldots, X_{n-1} = i_{n-1}, X_n = i)$$
$$= P(X_{n+1} = j | X_n = i), \tag{10.2}$$

$n = 1, 2, \ldots$, for any $i_1, \ldots, i_{n-1}, i, j$ such that $i_1 = i$, $i_{s-1}+1 \leq i_s \leq i_{s-1}+10$, $2 \leq s \leq n - 1$, $i_{n-1} + 1 \leq i \leq i_{n-1} + 10$, and $i + 1 \leq j \leq i + 10$. To see this, note that $X_n$ is strictly increasing and is a function of $\xi_1, \ldots, \xi_k$, where $k = X_{n-1}$ (Exercise 10.1). Therefore, the left side of (10.2) is equal to

$$P(i + \xi_i = j | X_1 = i_1, \ldots, X_{n-1} = i_{n-1}, X_n = i)$$
$$= P(\xi_i = j - i | \text{something about } \xi_1, \ldots, \xi_k, \text{ where } k = i_{n-1} < i)$$
$$= P(\xi_i = j - i) = 0.1,$$

and the same result is obtained for the right side of (10.2) (Exercise 10.1). A process that satisfies (10.2) is call a *Markov chain*. Furthermore, the chains of the professor and student may be considered as being independent. Using the Markovian property and independence of the two chains, it can be shown (see the next section) that, sooner or later, the chains will hit. More precisely, let $X_n$ and $Y_n$ denote the chains of the professor and student, respectively. Then for some $m$ and $n$ we have $X_m = Y_n$. On the other hand, by the way that these chains are constructed [i.e., (10.1)], once $X_m = Y_n$, we have $X_{m+1} = Y_{n+1}$, $X_{m+2} = Y_{n+2}$, and so on. In other words, once the chains hit, they will never be apart. The most striking part of this story is, perhaps, that the chains will hit wherever they start. For example, in Table 10.1, one may start at any of the first 10 digits and then follow the rules. The chains will hit each other at some point and then follow the same path.

"Sooner or later" or "at some point" turn out to be the key words as our story unfolds. The implication is that the chains do not have to hit within the first 100 digits. In fact, numerical computations done by one of Professor Dynkin's graduate students suggested that there is an approximate .026 chance that the two chains will not hit within the first 100 digits. Table 10.3 gives an example of such an "accident." Once again, the chains of professor and student are overlined and underlined, respectively. (Ironically, this was the very first example that the author tried, and it did not work! The example in Table 10.1 and Table 10.2 was the author's second attempt.)

**Table 10.3. An example of "accident"**

$$\overline{5}\ 3\ 8\ 7\ 8\ \overline{3}\ 8\ 5\ \overline{4}\ \underline{4}\ 2\ 4\ \overline{5}\ \underline{0}\ 3\ 6\ 0\ \overline{2}\ 7\ \underline{5}\ 2\ 7\ 9\ \underline{5}\ 7$$
$$8\ 4\ 3\ \underline{2}\ 4\ \underline{9}\ \overline{7}\ 2\ 1\ 9\ 2\ 2\ 3\ \overline{9}\ 2\ \underline{8}\ \underline{8}\ 0\ 1\ 5\ 3\ 5\ \overline{7}\ 1\ \underline{7}$$
$$8\ 2\ 4\ 8\ \overline{9}\ \underline{2}\ 4\ 9\ 4\ 3\ \underline{0}\ 0\ 4\ \overline{4}\ 3\ 8\ 2\ \overline{0}\ 1\ 5\ \underline{4}\ 1\ 9\ 7\ \underline{9}$$
$$6\ 8\ \overline{8}\ 0\ 7\ 6\ 1\ 9\ 9\ \underline{8}\ \overline{5}\ 4\ 6\ 5\ 2\ \overline{0}\ 0\ \underline{7}\ 1\ 9\ 5\ 5\ 5\ 1\ \underline{9}$$

The example has led to a natural topic for the next section.

## 10.2 Markov chains

The defining feature of a Markov chain is (10.2). We now express it under a more general framework. Consider a stochastic process $X_n$, $n = 0, 1, 2, \ldots$, that takes on a finite or countable number of possible states, where each state is a possible value of the process. The notation $X_0$ usually represents the initial state of the process. Without loss of generality, we assume that the set of states is a subset of $\{0, 1, 2, \ldots\}$, denoted by $S$. The process is said to be a homogeneous Markov chain, or simply Markov chain, if it satisfies

$$P(X_{n+1} = j | X_n = i, X_{n-1} = i_{n-1}, \ldots, X_0 = i_0) = p(i, j) \qquad (10.3)$$

for all $i_0, \ldots, i_{n-1}, i, j \in S$ and some function $0 \leq p(i,j) \leq 1$ such that $\sum_{j \in S} p(i,j) = 1$. From (10.3) it immediately implies that

$$
\begin{aligned}
&P(X_{n+1} = j | X_n = i, X_{n-1} = i_{n-1}, \ldots, X_0 = i_0) \\
&= P(X_{n+1} = j | X_n = i) \quad = \quad p(i,j)
\end{aligned}
\tag{10.4}
$$

(Exercise 10.2). Equation (10.4) is known as the Markov property. Intuitively, it may be interpreted as that the future depends on the present but not on the past or the present depends on the past only through the immediate past. The function $p(i,j)$ is known as the transition probability of the Markov chain. Note that the distribution of the Markov chain is determined by its transition probability and the distribution of the initial state; that is, $p_0(j) = P(X_0 = j)$ (Exercise 10.3). Another implication of (10.3) is that the conditional probability on the left side does not depend on $n$. More generally, we have

$$
\begin{aligned}
&P(X_{n+m} = j | X_n = i, X_{n-1} = i_{n-1}, \ldots, X_0 = i_0) \\
&= P(X_{n+m} = j | X_n = i)
\end{aligned}
\tag{10.5}
$$

and it does not depend on $n$. To see this, note that the left side of (10.5) can be written as

$$
\begin{aligned}
&\sum_{k \in S} P(X_{n+m} = j, X_{n+m-1} = k | X_n = i, X_{n-1} = i_{n-1}, \ldots, X_0 = i_0) \\
&= \sum_{k \in S} P(X_{n+m} = j | X_{n+m-1} = k, X_n = i, X_{n-1} = i_{n-1}, \ldots, X_0 = i_0) \\
&\quad \times P(X_{n+m-1} = k | X_n = i, X_{n-1} = i_{n-1}, \ldots, X_0 = i_0) \\
&= \sum_{k \in S} P(X_{n+m-1} = k | X_n = i, X_{n-1} = i_{n-1}, \ldots, X_0 = i_0) p(k,j),
\end{aligned}
\tag{10.6}
$$

and a similar argument also carries for the right side of (10.5). The claimed results thus follow by induction. Equation (10.5) is known as the $m$-step transition probability, denoted by $p^{(m)}(i,j)$. Clearly, we have $p^{(1)}(i,j) = p(i,j)$. The transition probabilities satisfies the Chapman–Kolmogorov identity:

$$
p^{(m+l)}(i,j) = \sum_{k \in S} p^{(m)}(i,k) p^{(l)}(k,j),
\tag{10.7}
$$

which can be established using a similar argument as (10.6) (Exercise 10.4). Equation (10.7) resembles the rule for matrix products. In fact, if we denote by $P$ the (possibly infinite-dimensional) matrix of transition probabilities, $P = [p(i,j)]_{i,j \in S}$, and, similarly, by $P^{(m)}$ the matrix of $m$-step transition probabilities, then (10.7) simply states that

$$
P^{(m+l)} = P^{(m)} P^{(l)},
\tag{10.8}
$$

where the right side is the matrix product. In particular, we have $P^{(m)} = P^{(m-1)} P = P^{(m-2)} P^2 = \cdots = P^m$. We consider some examples.

*Example 10.2* (Random walk). Let $\xi_i$, $i \geq 1$, be i.i.d. with $P(\xi_i = j) = a_j$, $j = 0, \pm 1, \ldots$. Define $X_0 = 0$ and $X_n = \sum_{i=1}^{n} \xi_i$, $n \geq 1$. It is easy to see that $X_n$, $n \geq 1$ is a Markov chain with states $S = \{0, \pm 1, \ldots\}$ and transition probability $p(i, j) = a_{j-i}$ (Exercise 10.5). A special case is called a simple random walk, for which $a_j = p$ if $j = 1$, $1 - p$ if $j = -1$, and 0 otherwise. It follows that for the simple random walk, we have $p(i, i + 1) = p$, $p(i, i - 1) = 1 - p$, and $p(i, j) = 0$ otherwise. In this case, the process may be thought of as the wanderings of a drunken man. Each time he takes a random step either to the left $(-1)$ with probability $1 - p$ or to the right $(+1)$ with probability $p$.

*Example 10.3* (Branching process). Consider the branching process of Example 8.6 (with the notation $T_n$ replaced by $X_n$). We show that the process is a Markov chain with $S = \{0, 1, \ldots\}$ and derive its transition probability. To see this, write $i_0 = 1$ and define $\sum_{k=1}^{0} X_{n,k} = 0$. Then we have

$$P(X_{n+1} = j | X_n = i, X_{n-1} = i_{n-1}, \ldots, X_0 = i_0)$$

$$= P\left(\sum_{k=1}^{i} X_{n+1,k} = j \,\middle|\, \text{something about } X_{m,k}, \right.$$

$$\left. 1 \leq m \leq n, 1 \leq k \leq \max_{0 \leq u \leq n-1} i_u\right)$$

$$= P\left(\sum_{k=1}^{i} X_{n+1,k} = j\right)$$

$$= p(i, j).$$

*Example 10.1 (continued).* Here, the Markov chain has the states $S = \{1, 2, \ldots\}$, and the transition probability is given by

$$p(i, j) = \frac{1}{10} 1_{(i+1 \leq j \leq i+10)}, \quad i, j \in S. \tag{10.9}$$

Furthermore, the two-step transition probability is given by

$$p^{(2)}(i, j) = \frac{10 - |j - i - 11|}{100}, \quad i, j \in S. \tag{10.10}$$

It is easy to verify that the transition probabilities satisfy $\sum_{j \in S} p(i, j) = 1$ and $\sum_{j \in S} p^{(2)}(i, j) = 1$ for any $i \in S$ (Exercise 10.6). Earlier it was claimed that the chain of the professor and that of the student will eventually hit. We now outline a proof of this claim, referring the details to Exercise 10.6.

For notation convenience, let $X = \{X(n), n \geq 0\}$ and $Y = \{Y(n), n \geq 0\}$ denote the chains of the professor and student, respectively. Suppose that the $X$ chain starts at $X(0) = a$ and the $Y$ chain starts at $Y(0) = b$. Without loss of generality, let $a < b$. First, note that by (10.9), it can be shown that

$$P\{X(n+1) \neq j | X(n), \ldots, X(1)\}$$
$$= \frac{9}{10} \quad \text{if } j - 10 \leq X(n) \leq j - 1 \tag{10.11}$$

(Exercise 10.6). Let $i_0 = b, i_1, \ldots, i_s$, be any sequence of positive integers such that $i_r - i_{r-1} \geq 10$, $1 \leq r \leq s$. Define the stopping time $T_r$ as the first time that the $X$ chain is within "striking distance" of $i_r$—that is, the smallest $n$ such that $X(n) \geq i_r - 10$, $1 \leq r \leq s$. The key idea of the proof is to show that at the time immediately after $T_r$, the $X$ chain will eventually hit $i_r$ for some $r$; that is, $X(T_r + 1) = i_r$ for some $r \geq 1$. This makes sense because at the time $T_r$, the chain is within the striking distance of $i_r$; in other words, $i_r$ is among the next 10 integers after $X(T_r)$, so why does $X(T_r + 1)$ always have to miss $i_r$ if it has an equal chance of hitting any of the 10 integers? To make this idea a rigorous argument, note that $T_r = n$ if and only if $X(n-1) < i_r - 10$ and $X(n) \geq i_r - 10$. Also, it is easy to see that $T_1 < \cdots < T_s$. Furthermore, let $j_1, \ldots, j_s$ be any possible values for $T_1, \ldots, T_s$. It can be shown that

$$\mathcal{A} = \{X(j_1 + 1) \neq i_1, \ldots, X(j_{s-1} + 1) \neq i_{s-1}, T_1 = j_1, \ldots, T_s = j_s\}$$
$$\in \sigma\{X(1), \ldots, X(j_s)\} \tag{10.12}$$

(Exercise 10.6). By (10.11) and (10.12), it follows that

$$P\{X(T_1 + 1) \neq i_1, \ldots, X(T_s + 1) \neq i_s, T_1 = j_1, \ldots, T_s = j_s\}$$
$$= P[\mathcal{A} \cap \{X(j_s + 1) \neq i_s\}]$$
$$= E[1_{\mathcal{A}} P\{X(j_s + 1) \neq i_s | X(j_s), \ldots, X(1)\}]$$
$$= \frac{9}{10} P\{X(T_1 + 1) \neq i_1, \ldots, X(T_{s-1} + 1) \neq i_{s-1}, T_1 = j_1, \ldots, T_s = j_s\}$$

because $i_s - 10 \leq X(j_s) \leq i_s - 1$ on $\mathcal{A}$ (Exercise 10.6). It follows that

$$P\{X(T_1 + 1) \neq i_1, \ldots, X(T_s + 1) \neq i_s\}$$
$$= \frac{9}{10} P\{X(T_1 + 1) \neq i_1, \ldots, X(T_{s-1} + 1) \neq i_{s-1}\}.$$

Continue with this argument; we arrive at the conclusion that

$$P\{X(T_1 + 1) \neq i_1, \ldots, X(T_s + 1) \neq i_s\} = \left(\frac{9}{10}\right)^s. \tag{10.13}$$

Now, let $U_0 = 0$ and $U_r$ be the first $n \geq U_{r-1}$ such that $Y(n) \geq Y(U_{r-1}) + 10$, $r = 1, 2, \ldots$. By (10.13) and independence of $X$ and $Y$, it can be shown that

$$P(X, Y \text{ do not hit})$$
$$\leq P\{X(T_1 + 1) \neq Y(U_1), \ldots, X(T_s + 1) \neq Y(U_s)\}$$
$$= \left(\frac{9}{10}\right)^s \tag{10.14}$$

(Exercise 10.6). Since $s$ is arbitrary, the left side of (10.14) must be zero.

We now introduce three important concepts of Markov chain. They are irreducibility, aperiodicity, and recurrency.

A state $j$ is said to be accessible from a state $i$ if $p^{(m)}(i,j) > 0$ for some $m \geq 0$. Two states that are accessible to each other are called communicate, denoted by $i \leftrightarrow j$. Communication is an equivalence relation in that (i) $i \leftrightarrow i$, (ii) $i \leftrightarrow j$ implies $j \leftrightarrow i$, and (iii) $i \leftrightarrow j$ and $j \leftrightarrow k$ imply $i \leftrightarrow k$ Two states that communicate are said to be in the same class. It is clear that any two classes are either disjoint or identical (Exercise 10.7). We say the Markov chain is *irreducible* if there is only one class; that is, all states communicate with each other .

The period of a state $i$ is defined as $d(i) = \sup\{k : p^{(m)}(i,i) = 0$ whenever $m/k$ is not an integer$\}$. It is clear that the latter set is not empty (as it includes at least $k = 1$), so $d(i)$ is well defined if we let $d(i) = \infty$ if $p^{(m)}(i,i) = 0$ for all $m \geq 1$. A state $i$ with $d(i) = 1$ is said to be *aperiodic*. It can be shown that if $i \leftrightarrow j$, then $d(i) = d(j)$. In other words, the states in the same class have the same period (Exercise 10.8).

For any states $i$ and $j$, define $q(i,j)$ as the probability that, starting in $i$, the chain ever makes a transition into $j$; that is,

$$q(i,j) = P(X_n = j \text{ for some } n \geq 1 | X_0 = i)$$
$$= \sum_{n=1}^{\infty} q^{(n)}(i,j),$$

where $q^{(n)}(i,j) = P(X_n = j, X_k \neq j, 1 \leq k \leq n-1 | X_0 = i)$. It is clear that $q(i,j) > 0$ if and only if $i \leftrightarrow j$. A state $i$ is said to be *recurrent* if $q(i,i) = 1$; that is, starting in $i$, the chain will return with probability 1. A state that is not recurrent is called *transient*. The following result is useful in checking the recurrency of a given state: State $i$ is recurrent if and only if.

$$\sum_{n=1}^{\infty} p^{(n)}(i,i) = \infty. \tag{10.15}$$

Furthermore, if state $i$ is recurrent and $i \leftrightarrow j$, then state $j$ is also recurrent (Exercise 10.9). Thus, the states in the same class are either all recurrent or all transient. We consider an example.

*Example 10.2 (continued).* First, we show that $i \leftrightarrow j$ for any $i, j \in S$, provided that $a_j > 0$ for $j = -1, 1$. Without loss of generality, let $i < j$. Let $m = j - i$. Then

$$p^{(m)}(i,j) = P(X_{n+m} = j | X_n = i)$$
$$\geq P(X_{n+m} = j, X_{n+m-1} = j-1, \ldots, X_{n+1} = i+1 | X_n = i)$$
$$= P(\xi_{n+1} = 1, \ldots, \xi_{n+m} = 1)$$
$$= a_1^m > 0.$$

Thus, the Markov chain is irreducible if $a_{-1} > 0$ and $a_1 > 0$. In the special case of a simple random walk, this means $0 < p < 1$. Furthermore, the Markov chain is aperiodic if and only if $a_0 \neq 0$. In particular, for the special case of a simple random walk with $0 < p < 1$, we have $d(i) = 2$ for all $i \in S$ (Exercise 10.11). Finally, we consider recurrency for the case of a simple random walk with $0 < p < 1$. Since in this case the chain is irreducible, we only need to check one of its states, say, $i = 0$ (why?). Clearly, we have $p^{(n)}(0,0) = 0$ if $n$ is odd (it takes an even number of steps to return). Now, suppose that $n$ is even, say, $n = 2k$. Then, starting at $i = 0$, the chain will return in $n$ steps if and only if it takes $k$ steps to the right and $k$ steps to the left. In other words, there are exactly $k$ ones and $k$ minus ones among $\xi_1, \ldots, \xi_n$. It follows from the binomial distribution that $p^{(n)}(0,0) = C_k^n p^k (1-p)^k$, $n = 1, 2, \ldots$. By using Stirling's approximation (see Example 3.4), it can be shown that

$$p^{(n)}(0,0) \sim \frac{\{4p(1-p)\}^k}{\sqrt{\pi k}} \tag{10.16}$$

(Exercise 10.12), where $a_n \sim b_n$ if $\lim_{n \to \infty}(a_n/b_n) = 1$. Therefore, (10.15) holds (with $i = 0$) if and only if $p = 1/2$. In other words, the chain is recurrent if $p = 1/2$, and transient if $p \neq 1/2$.

Some important properties of Markov chains are associated with their asymptotic behavior. To describe these properties, we first introduce the concept of a stationary distribution. A probability measure $\pi(\cdot)$ on $S$ is called a stationary distribution with respect to a Markov chain with states $S$ and transition probability $p(\cdot, \cdot)$ if it satisfies

$$\sum_{i \in S} \pi(i) p(i,j) = \pi(j), \quad j \in S. \tag{10.17}$$

Consider the limiting behavior of $p^{(n)}(i,j)$. If $j$ is transient, then we have

$$\sum_{n=1}^{\infty} p^{(n)}(i,j) < \infty \tag{10.18}$$

for all $i \in S$ (Exercise 10.13). To see what this means, define $N_j = \sum_{n=1}^{\infty} 1_{(X_n = j)}$, which is the total number of visits to $j$ by the chain $X_n$. Then we have $E(N_j | X_0 = i) = \sum_{n=1}^{\infty} P(X_n = j | X_0 = i) = \sum_{n=1}^{\infty} p^{(n)}(i,j)$. Thus, the left side of (10.18) is the expected number of visits to $j$ when the chain starts in $i$. This means that if $j$ is transient, then starting in $i$, the expected number of transitions into $j$ is finite, and this is true for all $i$. It follows that $p^{(n)}(i,j) \to 0$ as $n \to \infty$ for all $i$ if $j$ is transient.

To further explore the asymptotic behavior of $p^{(n)}(i,j)$ we define, for a given Markov chain $X_n$, $T_j$ as the first time that the chain visits $j$ after the initial state—that is, $T_j = \inf\{n \geq 1 : X_n = j\}$ if such a time exists (i.e., finite); otherwise, define $T_j = \infty$. Let $\mu_j = E(T_j | X_0 = j)$ (i.e., the expected number of transitions needed to return to state $j$). A state $j$ is called positive recurrent if $\mu_j < \infty$ and null recurrent if $\mu_j = \infty$. It is clear that a state that is positive recurrent must be recurrent; hence, a transient state $j$ must have $\mu_j = \infty$ (Exercise 10.14). Like recurrency, positive (null) recurrency is a class property; that is, the states in the same class are either all positive recurrent or all null recurrent (Exercise 10.14). Next, define $N_n(j) = \sum_{m=1}^{n} 1_{(X_m=j)}$, which is the total number of visit to $j$ by time $n$.

**Theorem 10.1.** If $i$ and $j$ communicate, then the following hold:
(i) $\mathrm{P}\{\lim_{n\to\infty} n^{-1} N_n(j) = \mu_j^{-1} | X_0 = i\} = 1$;
(ii) $\lim_{n\to\infty} n^{-1} \sum_{k=1}^{n} p^{(k)}(i,j) = \mu_j^{-1}$;
(iii) $\lim_{n\to\infty} p^{(n)}(i,j) = \mu_j^{-1}$ if $j$ is aperiodic.

Theorem 10.1(i) should help explain the terms "positive" and "null recurrent." Starting from any state $i$ that communicates with $j$, the asymptotic fraction of times that the chain spends at $j$ is equal to a positive constant if $j$ is positive recurrent; otherwise, the asymptotic fraction of times spent at $j$ is zero. If we go one step further by considering irreducibility, we come up with the following theorem.

**Theorem 10.2** (Markov-chain convergence theorem). An irreducible, aperiodic Markov chain belongs to one of the following two classes:
  (i) All states are null recurrent (which include those that are transient), in which case we have $\lim_{n\to\infty} p^{(n)}(i,j) = 0$ for all $i,j$, and there exists no stationary distribution.
  (ii) All states are positive recurrent, in which case we have

$$\pi(j) = \lim_{n\to\infty} p^{(n)}(i,j) = \frac{1}{\mu_j} \tag{10.19}$$

for all $j$, and $\pi(\cdot)$ is the unique stationary distribution for the Markov chain.

We illustrate Theorem 10.2 with an example.

*Example 10.4* (Birth and death process). A birth and death process is a Markov chain with states $S = \{0, 1, 2, \ldots\}$ and transition probabilities given by $p(i, i+1) = p_i$, $p(i, i-1) = q_i$ and $p(i,i) = r_i$, $i \in S$, where $q_0 = 0$ and $p_i, q_i$, and $r_i$ are nonnegative numbers such that $p_i + q_i + r_i = 1$. Such a process is also called a birth and death chain with reflecting barrier 0, or simply *birth and death chain*. Intuitively, if the chain $X_n$ represents the total number of a biological population (e.g., bacteria), then $+1$ ($-1$) correspond to a birth (death) in the population, or no birth or death occurs, at a given time $n$. A birth and death chain is irreducible if $p_i > 0$ and $q_{i+1} > 0$, $i \in S$, and aperiodic if $r_i > 0$ for some $i$ (Exercise 10.16).

Now, focusing on the irreducible and aperiodic case, assume $p_i > 0, i \in S$, $q_i > 0, i \geq 1$, and $r_i > 0$ for some $i$. By Theorem 10.2, there is a limiting distribution of $p^{(n)}(i,j)$ that is independent of the initial state $i$. To determine the limiting distribution, we consider $\pi(i) = \prod_{k=1}^{i}(p_{k-1}/q_k)$. It can be shown (Exercise 10.16) that $\pi(\cdot)$ satisfies

$$\pi(i)p(i,j) = \pi(j)p(j,i), \quad i,j \in S. \tag{10.20}$$

A Markov chain that satisfies (10.20) is called *(time) reversible*. Intuitively, this means that the rate at which the chain goes from $i$ to $j$ is the same as that from $j$ to $i$. Any distribution $\pi$ that satisfies the reversal condition (10.20) is necessarily stationary. To see this, simple sum over $i$ on both sides of (10.20) and we get (10.17). It follows by Theorem 10.2 that $\lim_{n\to\infty} p^{(n)}(i,j) = \pi(j)$.

It is seen that the "trick" is to find the unique stationary distribution $\pi$ using whatever method. One method that is often used is to solve (10.17), or its matrix form $P'\pi = \pi$, where $P'$ is the transpose of the matrix $P$ of transition probabilities and $\pi = [\pi(i)]_{i \in S}$. In some cases, a solution can be guessed that satisfies the stronger condition (10.20).

A variation of the birth and death chain that has a finite state space is considered in Exercise 10.17. Some important applications of the Markov-chain convergence theorem are discussed in Chapter 15.

## 10.3 Poisson processes

The Poisson process is a special case of what is called a counting process. The latter means a process $N(t), t \geq 0$, that represents the number of events that have occurred up to time $t$. Obviously, a counting process must satisfy the following: (a) the values of $N(t)$ are nonnegative integers; (b) $N(s) \leq N(t)$ if $s < t$; and (c) for $s < t$, $N(t) - N(s)$ equals the number of events that have occurred in the interval $(s, t]$. There are, at least, three equivalent definitions of a Poisson process. The first is the most straightforward and anticipated. Suppose that (i) the counting process satisfies $N(0) = 0$; (ii) the process has independent increments—that is, the numbers of events that occur in disjoint time intervals are independent; and (iii) the number of events in any interval of length $t$ follows a Poisson distribution with mean $\lambda t$—that is,

$$P\{N(s+t) - N(s) = x\} = e^{-\lambda t}\frac{(\lambda t)^x}{x!}, \quad x = 0, 1, \ldots, \tag{10.21}$$

where $\lambda > 0$ is called the rate of the Poisson process. From a practical point of view, (10.21) is not something that may be easily checked. This makes the second definition somewhat more appealing. A counting process $N(t)$ is said to have stationary increments if for any $t_1 < t_2$, the distribution of $N(s+t_2) - N(s+t_1)$ does not depend on $s$. A counting process $N(t)$ is a Poisson process if (i) $N(0) = 0$, (ii) the process has independent and stationary increments, (iii) $P\{N(u) = 1\} = \lambda u + o(u)$ as $u \to 0$, and (iv) $P\{N(u) \geq 2\} =$

$o(u)$ as $u \to 0$. The third definition of a Poisson process is built on a connection between a Poisson process and the sum of independent exponential random variables. Let $T_1, T_2, \ldots$ be a sequence of independent exponential random variables with mean $1/\lambda$. Then we can define a Poisson process as

$$N(t) = \sup\{n : S_n \le t\}, \tag{10.22}$$

where $S_0 = 0$ and $S_n = \sum_{i=1}^{n} T_i$, $n \ge 1$. The Poisson process (10.22) has an intuitive explanation. Imagine $T_i$ as the interarrival time of the $i$th event; that is, $T_1$ is the time of the first event and $T_i$ is the time between the $(i-1)$st and $i$th event, $i \ge 2$. Then $S_n$ is the arrival time, or "waiting time," for the $n$th event. A counting process is Poisson if its interarrival times are i.i.d. exponential or, equivalently, its arrival times can be expressed as $S_n$ (Exercise 10.18). It can be shown that the three definitions of a Poisson process are equivalent (e.g., Ross 1983, Section 2.1).

As mentioned, the second equivalent definition is especially useful in justifying the assumptions of a Poisson process. It is related to a fundamental asymptotic theory of Poisson distribution, known as Poisson approximation to binomial. To see this, suppose that someone is unaware of the mathematical equivalence of these definitions but, nevertheless, wants to justify a Poisson process based on properties (i)–(iv) of the second definition. Divide the interval $[0, t]$ into $n$ subintervals so that each has length $t/n$, where $n$ is large. Then (iv) implies that

P(2 or more events in some subinterval)

$$\le \sum_{j=1}^{n} P(2 \text{ or more events in subinterval } j)$$

$$= n \left(\frac{t}{n}\right) o(1) = to(1) \to 0$$

as $n \to \infty$. Thus, with probability tending to 1, $N(t)$ is the sum of $n$ independent random variables (which are the numbers of events in those subintervals) taking the values of 0 or 1 (i.e., Bernoulli random variables). It follows that the distribution of $N(t)$ is asymptotically Binomial$(n, p)$, where $p = P\{N(t/n) = 1\} = \lambda(t/n) + o(t/n)$, according to (iii); that is,

$$P\{N(t) = x\}$$

$$\approx \binom{n}{x} p^x (1-p)^{n-x}$$

$$= \frac{n!}{x!(n-x)!} \left\{\frac{\lambda t}{n} + o\left(\frac{t}{n}\right)\right\}^x \left\{1 - \frac{\lambda t}{n} - o\left(\frac{t}{n}\right)\right\}^{n-x}. \tag{10.23}$$

It is now a simple exercise of calculus to show that the right side of (10.23) converges to $e^{-\lambda t}(\lambda t)^x / x!$ for every $x$ (Exercise 10.19). In general, if $X \sim$ Binomial$(n, p)$, where $n$ is large and $p$ is small, such that $np \approx \lambda$, then the distribution of $X$ is approximately Poisson($\lambda$), and this is called *Poisson approximation to binomial*. We consider some applications of this approximation.

*Example 10.5* (The Prussian horse-kick data). This famous example was given by von Bortkiewicz in his book entitled *The Law of Small Numbers* published in 1898. The number of fatalities resulting from being kicked by horses was recorded for 10 corps of Prussian cavalry over a period of 20 years, giving a total of 200 observations. The numbers in Table 10.4 are the observed relative frequencies as well the corresponding probabilities computed under a Poisson distribution with mean $\lambda = .61$ (see below). The approximations are

**Table 10.4. Prussian horse-kick data and Poisson approximation**

| # of Deaths per Year | # of Cases Recorded | Relative Frequency | Poisson Probability |
|---|---|---|---|
| 0 | 109 | .545 | .543 |
| 1 | 65 | .325 | .331 |
| 2 | 22 | .110 | .101 |
| 3 | 3 | .015 | .021 |
| 4 | 1 | .005 | .003 |

amazingly close, especially for lower numbers of deaths. This can be justified by the Poisson approximation to binomial. Consider the event that a given soldier is kicked to death by a horse in a given corps-year. Obviously, this event can only happen once during the 20 years if we trace down the same corp over the 20 years. However, the point is to consider the total number of deaths for each of the 200 corps-years as a realization of a random variable $X$. If we assume that these events are independent over the soliders, then $X$ is the sum of $n$ independent Bernoulli random variables that are the event indicators (1 for death and 0 otherwise), where $n$ is the total number of soldiers in a corps. Notice that a corps is a very large army unit (in the United States Army, a corps consists of two to five divisions, each with 10,000 to 15,000 soldiers; depending on the country at the different times of history, the actual number of soldiers in a corps varied), so $n$ is expected to be very large. On the other hand, the probability $p$ that a cavalry soldier is kicked to death is expected to be very small (if the soldier is careful about his horse). So, we are in a situation of a Binomial($n, p$) distribution, where $n$ is large and $p$ is small. It follows that the distribution of $X$ can be approximated by Poisson($\lambda$), where $\lambda$ is approximately equal to $np$. The value of $\lambda$ can be estimated by the maximum likelihood method. Let $X_1, \ldots, X_{200}$ denote the total numbers of deaths for the 200 corps-years. Then the MLE for $\lambda$ is given by $\hat{\lambda} = 200^{-1} \sum_{i=1}^{200} X_i = (0 \times 109 + 1 \times 65 + 2 \times 22 + 3 \times 3 + 4 \times 1)/200 = .61$.

*Example 10.6* (Fisher's dilution assay). Another well-known example in the statistics literature is Fisher's dilution assay problem (Fisher 1922b). A solution containing an infective organism is progressively diluted. At each dilution, a number of agar plates are streaked. From the number of sterile plates observed at each dilution, an estimate of the concentration of infective organisms in the original solution is obtained. For simplicity, suppose that the dilution is doubled each time so that after $k$ dilutions, the expected number of infective organisms per unit volume is given by $\rho_k = \rho_0/2^k$, $k = 0, 1, \ldots$, where $\rho_0$, which is the density of infective organisms in the original solution, is what we wish to estimate. The idea is that if $k$ is sufficiently large, one can actually count the number of organisms on each plate and therefore obtain an estimate of $\rho_0$. A critical assumption made here is that at the $k$th dilution, the actual number of organisms, $N_k$, follows a Poisson distribution with mean $\rho_k v$, where $v$ is the volumn of solution for each agar plate. Under this assumption, the unknown density $\rho_0$ can be estimated using the maximum likelihood. Again, we can justify this assumption using Poisson approximation to binomial. Imagine that the plate is divided into many small parts of equal volume, say, $v_0$, so that within each small part there is at most one organism. Then the number of organism in each small part is a Bernoulli random variable with probability $p_k$ of having an organism. If we further assume that the number of organisms in different small parts are independent, then $N_k$ is the sum of $n$ independent Bernoulli random variables, where $n = v/v_0$ is the total number of small parts, and therefore has a Binomial$(n, p_k)$ distribution. If $n$ is sufficiently large, $p_k$ must be sufficiently small so that $np_k$ is approximately equal to a constant, which is $\rho_k v$. It follows that $N_k$ has an approximate Poisson distribution with mean $\rho_k v$.

Durret (1991, p. 125) gives the following extension of Poisson approximation to binomial to "nearly binomial." The similarity to the second definition of Poisson process is evident.

**Theorem 10.3.** Suppose that for each $n$, $X_{n,i}, 1 \leq i \leq n$, are independent nonnegative integer-valued random variables such that
  (i) $P(X_{n,i} = 1) = p_{n,i}$, $P(X_{n,i} \geq 2) = \epsilon_{n,i}$;
  (ii) $\lim_{n \to \infty} \sum_{i=1}^{n} p_{n,i} = \lambda \in (0, \infty)$;
  (iii) $\lim_{n \to \infty} \max_{1 \leq i \leq n} p_{n,i} = 0$;
  (iv) $\lim_{n \to \infty} \sum_{i=1}^{n} \epsilon_{n,i} = 0$.
Then we have $S_n = \sum_{i=1}^{n} X_{n,i} \xrightarrow{\text{d}} \xi \sim \text{Poisson}(\lambda)$.

It should be pointed out that Poisson approximation to binomial works in a different way than the well-known normal approximation to binomial. The latter assumes that $p$ is fixed and lies strictly between 0 and 1 and then $n \to \infty$ in the Binomial$(n, p)$ distribution; whereas the former is under the limiting process that $n \to \infty$, $p \to 0$, and $np \to \lambda \in (0, \infty)$. Nevertheless, the Poisson and normal distributions are asymptotically connected in that a Poisson distribution with large mean can be approximated by a normal distribution. More precisely, we have the following result regarding the Poisson process.

**Theorem 10.4** (CLT for Poisson process). Let $N(t), t \geq 0$, be a Poisson process with rate $\lambda$. Then

$$\lim_{\lambda \to \infty} P\left\{\frac{N(t) - \lambda t}{\sqrt{\lambda t}} \leq x\right\} = \Phi(x)$$

for all $x$, where $\Phi(\cdot)$ is the cdf of $N(0, 1)$.

An extension of Theorem 10.4 is given in the next section.

Finally, we consider limiting behavior of the arrival times of a Poisson process. The following theorem points out an interesting connection between the conditional arrival times and order statistics of independent uniformly distributed random variables. The proof is left as an exercise (Exercise 10.20).

**Theorem 10.5.** Let $N(t), t \geq 0$ be a Poisson process. Given that $N(t) = n$, the consecutive arrival times $S_1, \ldots, S_n$ have the same joint distribution as the order statistics of $n$ independent Uniform$(0, t)$ random variables.

Theorem 10.5 allows us to study, after a suitable normalization, asymptotic behavior of $S_1, \ldots, S_n$ through $U_{(1)}, \ldots, U_{(n)}$, where $U_{(1)}, \ldots, U_{(n)}$ are the order statistics of $U_1, \ldots, U_n$ which are independent Uniform$(0, 1)$ random variables. For example, we have (Weiss 1955)

$$\frac{1}{n} \sum_{i=1}^{n} 1_{\{U_{(i)} - U_{(i-1)} > x/n\}} \xrightarrow{P} e^{-x}, \tag{10.24}$$

$x > 0$, as $n \to \infty$. Also, we have (Exercise 10.21)

$$\frac{n}{\log n} \max_{1 \leq i \leq n+1} \{U_{(i)} - U_{(i-1)}\} \xrightarrow{P} 1, \tag{10.25}$$

$$P\left\{n^2 \min_{1 \leq i \leq n+1} \{U_{(i)} - U_{(i-1)}\} > x\right\} \longrightarrow e^{-x}. \tag{10.26}$$

The corresponding results regarding the arrival times of a Poisson process are therefore obtained via Theorem 10.5 (Exercise 10.21).

Some further asymptotic theory will be introduced under a more general framework in the next section.

## 10.4 Renewal theory

The interarrival times of a Poisson process can be generalized in a way called renewal process. Suppose that $X_1, X_2, \ldots$ is a sequence of independent non-negative random variables with a common distribution $F$ such that $F(0) < 1$.

The $X_i$'s may be interpreted the same way as the Poisson interarrival times $T_i$'s following (10.22). Similarly, we define the arrival times $S_n$ as $S_0 = 0$ and $S_n = \sum_{i=1}^{n} X_i$, $n \geq 1$, and a counting process $N(t)$ by (10.22). The process is then called a *renewal process*. The term "renewal" refers to the fact that the process "starts afresh" after each arrival; that is, $S_{n+k} - S_n$, $k = 1, 2, \ldots$, have the same (joint) distribution regardless of $n$. In some books, $N(t)$ is defined as $\inf\{n : S_n > t\}$ instead of (10.22) (so the difference is 1), but the basic asymptotic theory, which we outline below, is the same.

Let $\mu = \mathrm{E}(X_i)$. For simplicity we assume that $\mu$ is finite, although many of the asymptotic results extend to the case $\mu = \infty$. The first result states that $N(t) \to \infty$ as $t \to \infty$.

**Theorem 10.6.** $N(t) \xrightarrow{\text{a.s.}} \infty$ as $t \to \infty$.

This is because $N(t)$ is nondecreasing with $t$; so $N(\infty) \equiv \lim_{t \to \infty} N(t)$ exists (see §1.5.1.3). Furthermore, we have $\mathrm{P}\{N(\infty) < \infty\} = \mathrm{P}(X_i = \infty$ for some $i) = 0$, because $X_i$ is a.s. finite for every $i$. Theorem 10.6 is used to derive the next asymptotic result.

**Theorem 10.7** (SLLN for renewal processes). $N(t)/t \xrightarrow{\text{a.s.}} 1/\mu$ as $t \to \infty$.

The proof is left as an exercise (Exercise 10.24). Theorem 10.7 states that, asymptotically, the rate at which renewals occur is equal to the reciprocal of the mean of the interarrival time, which, of course, makes sense. A related result says that not only does the convergence hold almost surely, it also holds in expectation. This is known as the *elementary renewal theorem*. Note that, in general, a.s. convergence does not necessarily imply convergence in expectation (see Chapter 2; also Exercise 10.25). Define $m(t) = \mathrm{E}\{N(t)\}$, known as the *renewal function* (Exercise 10.26).

**Theorem 10.8.** $m(t)/t \to 1/\mu$ as $t \to \infty$.

The proof is not as "elementary" as the name might suggest. For example, one might attempt to prove the theorem by Theorem 10.7 and the dominated convergence theorem (Theorem 2.16). This would not work, however. The standard proof involves the well-known Wald equation, as follows.

**Theorem 10.9.** Let $\tau$ be a stopping time such that $\mathrm{E}(\tau) < \infty$. Then

$$\mathrm{E}\left(\sum_{i=1}^{\tau} X_i\right) = \mathrm{E}(\tau)\mu. \tag{10.27}$$

Equation (10.27) can be derived as a simple consequence of Doob's optional stopping theorem—namely, Corollary 8.1. To see this, define $\xi_i = X_i - \mu$

and $M_n = \sum_{i=1}^{n} \xi_i$. Then $M_n, \mathcal{F}_n = \sigma(X_1, \ldots, X_n)$, $n \geq 1$, is a martingale. Therefore, by (8.11), we have

$$E(M_\tau | \mathcal{F}_\tau) = M_1 \quad \text{a.s.,}$$

which implies $E(M_\tau) = E(M_1) = E(\xi_1) = 0$. On the other hand, we have $S_n = n\mu + M_n$; hence $E(S_\tau) = E(\tau\mu + M_\tau) = E(\tau)\mu$, which is (10.27). To apply the Wald equation to the renewal process, note that $N(t) + 1$ is a stopping time (Exercise 10.27). Therefore, by (10.27), we obtain

$$E\left\{S_{N(t)+1}\right\} = \mu\{m(t) + 1\}. \tag{10.28}$$

Identity (10.28) plays important roles not only in the proof of the elementary renewal theorem but also in other aspects of the renewal theory.

The next result may be viewed as an extension of Theorem 10.4.

**Theorem 10.10.** Suppose that the variance of the interarrival time $\sigma^2$ is finite and positive. Then

$$\lim_{t \to \infty} P\left\{\frac{N(t) - t/\mu}{\sigma\sqrt{t/\mu^3}} \leq x\right\} = \Phi(x)$$

for all $x$, where $\Phi(\cdot)$ is the cdf of $N(0, 1)$.

We outline a proof below and let the reader complete the details (Exercise 10.28). First, note the following fact:

$$N(t) < n \quad \text{if and only if} \quad S_n > t \tag{10.29}$$

for any positive integer $n$. For any given $x$, we have

$$P\left\{\frac{N(t) - t/\mu}{\sigma\sqrt{t/\mu^3}} \leq x\right\} = P\{N(t) \leq x_t\}$$

$$\leq P\{N(t) < [x_t] + 1\}$$

$$= P(S_{[x_t]+1} > t) \quad \text{[by (10.29)]}$$

$$= P\left\{\frac{S_{[x_t]+1} - ([x_t] + 1)\mu}{\sigma\sqrt{[x_t] + 1}} > u_t\right\},$$

where $x_t = t/\mu + \sigma x\sqrt{t/\mu^3}$, $[x_t]$ is the largest integer $\leq x_t$, and

$$u_t = \frac{t - ([x_t] + 1)\mu}{\sigma\sqrt{[x_t] + 1}}.$$

It is easy to show that as $t \to \infty$, $x_t \to \infty$ while $u_t \to -x$. Therefore, for any $\epsilon > 0$, we have $u_t > -x - \epsilon$ for large $t$; hence,

$$P\left\{\frac{N(t) - t/\mu}{\sigma\sqrt{t/\mu^3}} \leq x\right\} \leq P\left\{\frac{S_{[x_t]+1} - ([x_t] + 1)\mu}{\sigma\sqrt{[x_t] + 1}} > -x - \epsilon\right\}$$

for large $t$. It follows by the CLT that

$$\limsup_{t\to\infty} P\left\{\frac{N(t) - t/\mu}{\sigma\sqrt{t/\mu^3}} \leq x\right\} \leq 1 - \Phi(-x - \epsilon)$$

$$= \Phi(x + \epsilon). \tag{10.30}$$

By a similar argument, it can be shown that

$$\liminf_{t\to\infty} P\left\{\frac{N(t) - t/\mu}{\sigma\sqrt{t/\mu^3}} \leq x\right\} \geq \Phi(x - \epsilon). \tag{10.31}$$

The conclusion then follows from (10.30), (10.31), and the arbitrariness of $\epsilon$.

*Example 10.7.* In the case of Poisson process discussed in the previous section, we have $X_i \sim$ Exponential($1/\lambda$), so $\mu = \sigma = 1/\lambda$. Thus, in this case, Theorem 10.10 reduces to Theorem 10.4. More generally, if $X_i$ has a Gamma($\alpha, \beta$) distribution (note that the Exponential distribution is a special case with $\alpha = 1$), then we have $\mu = \alpha\beta$ and $\sigma = \sqrt{\alpha}\beta$. It follows by Theorem 10.10 that $\{N(t) - t/\alpha\beta\}/\sqrt{t/\alpha^2\beta} \xrightarrow{d} N(0, 1)$ as $t \to \infty$.

*Example 10.8.* Now, suppose that $X_i$ has a Bernoulli($p$) distribution, where $0 < p < 1$. This is a case of a discrete interarrival time, where $X_i = 0$ means that the arrival of the $i$th event takes no time (i.e., arriving at the same time as the previous event). In this case, we have $\mu = p$ and $\sigma = \sqrt{p(1-p)}$; so by Theorem 10.10 we have $\{N(t) - t/p\}/\sqrt{t(1-p)/p^2} \xrightarrow{d} N(0, 1)$, as $t \to \infty$.

We are now ready to introduce some deeper asymptotic theory. The following famous theorem is due to David Blackwell. A nonnegative random variable $X$ is said to be *lattice* if there exists $d \geq 0$ such that $\sum_{k=0}^{\infty} P(X = kd) = 1$. The largest $d$ that has this property is called the period of $X$. If $X$ is lattice and $X \sim F$, then $F$ is said to be lattice, and the period of $X$ is also called the period of $F$. For example, in Example 10.8, we have $P(X_1 = 0) + P(X_1 = 1) = 1$ while $P(X_1 = 0) < 1$; so $F$ is lattice and has period 1. On the other hand, the $F$ in Example 10.7 is clearly not lattice.

**Theorem 10.11.** If $F$ is not lattice, then $m(t+a) - m(t) \to a/\mu$ as $t \to \infty$, for all $a \geq 0$.

*Example 10.7 (continued).* In the case of the Poisson process, we have $\mu = 1/\lambda$; hence, by Blackwell's theorem, $m(t + a) - m(t) \to \lambda a$. In particular, if $a = 1$, we have $E\{N(t + 1)\} - E\{N(t)\} \to \lambda$ as $t \to \infty$. This means that, in the longrun, the mean number of arrivals between time $t$ and time $t + 1$ is approximately equal to the reciprocal of $\mu$, the mean of the interarrival time. This, of course, makes sense. For example, if $\mu = 0.2$, meaning that it takes, on average, 0.2 second for a new event to arrive, then there are, on average, approximately five arrivals within a second, in the longrun.

Our last asymptotic result is known as the key renewal theorem. Let $h$ be a function that satisfies (i) $h(t) \geq 0$ for all $t \geq 0$; (ii) $h(t)$ is nonincreasing, and (iii) $\int_0^\infty h(t)\, dt < \infty$.

**Theorem 10.12.** If $F$ is not lattice and $h$ is as above, then

$$\lim_{t \to \infty} \int_0^t h(t - x)\, dm(x) = \frac{1}{\mu} \int_0^\infty h(t)\, dt. \tag{10.32}$$

Note the following alternative expression of $\mu = E(X_i)$:

$$\mu = \int_0^\infty x\, dF(x) = \int_0^\infty \bar{F}(t)\, dt, \tag{10.33}$$

where $\bar{F}(t) = 1 - F(t)$, which can be derived by Fubini's theorem (Exercise 10.29). Therefore, the key renewal theorem may be written as

$$\int_0^t h(t - x)\, dm(x) \longrightarrow \frac{\int_0^\infty h(t)\, dt}{\int_0^\infty \bar{F}(t)\, dt} \quad \text{as } t \to \infty.$$

In particular, if $h(t) = 1_{[0,a]}(t)$, then $\int_0^t h(t - x)\, dm(x) = \int_{t-a}^t dm(x) = m(t) - m(t - a)$ and $\int_0^\infty h(t)\, dt = \int_0^a dt = a$; thus, (10.32) reduces to $m(t) - m(t - a) \to a/\mu$ as $t \to \infty$, which is simply Blackwell's theorem.

## 10.5 Brownian motion

The term *Brownian motion* has appeared in various places so far in this book. It originated as a physics phenomenon, discovered by English botanist Robert Brown in 1827. While studying pollen particles floating in water, Brown observed minute particles in the pollen grains executing the jittery motion. After repeating the experiment with particles of dust, he concluded that the motion was due to pollen being "alive," but the origin of the motion remained unclear. Later in 1900, French mathematician Louis Bachelier wrote a historical Ph.D. thesis, The Theory of Speculation, in which he developed the first theory about Brownian motion. His work, however, was somewhat overshadowed by Albert Einstein, who in 1905 used a probabilistic model to explain Brownian motion. According to Einstein's theory, if the kinetic energy of fluids is "right," the molecules of water move at random. Thus, a small particle would receive a random number of impacts of random strength and from random directions in any short period of time. This random bombardment by the molecules of the fluid would cause a sufficiently small particle, such as that in a pollen grain, to move in the way that Brown described.

In a series of papers originating in 1918, Norbert Wiener, who received his Ph.D. at the age of 18, defined Brownian motion as a stochastic process $B(t), t \geq 0$ satisfying the following conditions:

(i) $B(t)$ has independent and stationary increments
(ii) $B(t) \sim N(0, \sigma^2 t)$ for every $t > 0$, where $\sigma^2$ is a constant.
(iii) With probability 1, $B(t)$ is a continuous function of $t$.

Thinking of Brown's experiment of pollen grains, (iii) is certainly reasonable, assuming that the particles could not "jump" from one location to another; the assumption of independent increments of (i) can be justified by Einstein's theory of "random bombardments." As for (ii), it may be argued that this is implied by (i) and the central limit theorem (Exercise 10.30). In particular, Wiener (1923) proved the existence of a Brownian motion according to the above definition. For these reasons, Brownian motion is also called a Wiener process in honor of Wiener's significant contributions [and the notation $W(t)$ is also often used for Brownian motion].

*Note.* By condition (iii), the sample paths of Brownian motion are almost surely continuous. On the other hand, these paths are never smooth, as one would expect, in that, with probability 1, $B(t)$ is nowhere differentiable as a function of $t$. This remarkable feature of Brownian motion was first discovered by Paley, Wiener, and Zygmund (1933). See Dvoretzky, Erdös, and Kakutani (1961) for a "short" proof of this result.

A simple consequence of the definition is that $B(0) = 0$ with probability 1. To see this, note that by (ii), (iii), and Fatou's lemma (Lemma 2.4), we have $E\{B^2(0)\} = E\{\lim_{t \to 0} B^2(t)\} \leq \liminf_{t \to 0} E\{B^2(t)\} = \liminf_{t \to 0} \sigma^2 t = 0$; hence, $B(0) = 0$ a.s. Therefore, without loss of generality, we assume that $B(0) = 0$. Also, as any Brownian motion can be converted to one with $\sigma = 1$, known as standard Brownian motion [by considering $B(t)/\sigma$], we will focus on the latter case only.

Brownian motion is a Gaussian process defined in Section 7.3. It is also a special case of what is called a continuous-time Markov process, which is an extension of the Markov chains discussed in Section 10.2, in that the conditional distribution of $B(s + t)$ given $B(u), 0 < u \leq s$, depends only on $B(s)$. To see this, note that by independent increments, we have

$$P\{B(s + t) \leq y | B(s) = x, B(u), 0 < u < s\}$$
$$= P\{B(s + t) - B(s) \leq y - x | B(s) = x, B(u), 0 < u < s\}$$
$$= P\{B(s + t) - B(s) \leq y - x\}.$$

On the other hand, by a similar argument, we have $P\{B(s + t) \leq y | B(s) = x\} = P\{B(s + t) - B(s) \leq y - x\}$, verifying the Markovian property. In fact, a *strong Markov property* holds, as follows. For each $t \geq 0$, let

$\mathcal{F}_0(t) = \sigma\{B(u), u \leq t\}$ and $\mathcal{F}_+(t) = \cap_{s>t}\mathcal{F}_0(s)$. The latter is known as the right-continuous filtration. It is known that $\mathcal{F}_0(t)$ and $\mathcal{F}_+(t)$ have the same completion (e.g., Durrett 1991, p. 345). We call $\mathcal{F}(t), t \geq 0$, a Brownian filtration if (i) $\mathcal{F}(t) \supset \mathcal{F}_0(t)$, and (ii) for all $t \geq 0$ the process $B(t+s) - B(t), s \geq 0$, is independent of $\mathcal{F}(t)$. A random variable $\tau$ is a stopping time for a Brownian filtration $\mathcal{F}(t), t \geq 0$, if $\{\tau \leq a\} \in \mathcal{F}(t)$ for all $t$. The strong Markov property states that if $\tau$ is a stopping time for the Brownian filtration $\mathcal{F}(t), t \geq 0$, then the process $B(\tau + t) - B(\tau), t \geq 0$, is a Brownian motion that is independent of $\mathcal{F}(\tau)$, where $\mathcal{F}(\tau) = \{A : A \cap \{\tau \leq t\} \in \mathcal{F}(t), \forall t\}$. This result was proved independently by Hunt (1956) and Dynkin (1957) (the latter author being the same professor who gave the counting-trick example discussed in our openning section; see Example 10.1).

The strong Markov property is used to establish the following theorem called the *reflection principle* of Brownian motion.

**Theorem 10.13.** Suppose that $\tau$ is a stopping time for the Brownian filtration $\mathcal{F}(t), t \geq 0$. Define

$$B^*(t) = \begin{cases} B(t), & t \leq \tau \\ 2B(\tau) - B(t), & t > \tau \end{cases}$$

(known as Brownian motion reflected at time $\tau$). Then $B^*(t), t \geq 0$, is a standard Brownian motion.

The proof is left as an exercise (with a hint; see Exercise 10.31). The reflection is one of many Brownian motions "constructed" from Brownian motion. To mention a couple more, let $B(t), t \geq 0$, be a standard Brownian motion, then (1) (scaling relation) for any $a \neq 0$, $a^{-1}B(a^2t), t \geq 0$, is a standard Brownian motion and (2) (time inversion) $W(t) = 0, t = 0$, and $tB(1/t), t > 0$, is a Brownian motion (Exercise 10.32).

Furthermore, Brownian motion is a *continuous martingale*, which extends the martingales discussed in Chapter 8 to continuous-time processes. This means that for any $s < t$, we have $E\{B(t)|\mathcal{F}(s)\} = B(s)$, where $\mathcal{F}(t), t \geq 0$, is the Brownian filtration. To see the property more clearly, note that

$$\begin{aligned} E\{B(t)|B(u), u \leq s\} &= E\{B(s) + B(t) - B(s)|B(u), u \leq s\} \\ &= B(s) + E\{B(t) - B(s)|B(u), u \leq s\} \\ &= B(s) + E\{B(t) - B(s)\} \\ &= B(s). \end{aligned}$$

By a similar argument, it can be shown that $B^2(t) - t, t \geq 0$ is a continuous martingale (Exercise 10.33).

Another well-known result for Brownian motion is regarding its hitting time, or maximum over an interval. For any $a > 0$, let $T_a$ be the first time the Brownian motion hits $a$. It follows that $T_a \leq t$ if and only if $\max_{0 \leq s \leq t} B(s) \geq a$. On the other hand, we have

$$P\{B(t) \geq a\} = P\{B(t) \geq a | T_a \leq t\}P(T_a \leq t)$$

because $P\{B(t) \geq a | T_a > t\} = 0$. Furthermore, if $T_a \leq t$, the process hits $a$ somewhere on $[0,t]$; so, by symmetry, at the point $t$, the process could go either way, either above or below $a$, with equal probability. Therefore, we must have $P\{B(t) \geq a | T_a \leq t\} = 1/2$. This implies

$$P\left\{\max_{0 \leq s \leq t} B(s) \geq a\right\} = P(T_a \leq t)$$

$$= 2P\{B(t) \geq a\}$$

$$= \sqrt{\frac{2}{\pi}} \int_{a/\sqrt{t}}^{\infty} e^{-x^2/2} \, dx. \qquad (10.34)$$

We consider an example.

*Example 10.9.* Suppose that one has the option of purchasing one unit of a stock at a fixed price $K$ at time $t \geq 0$. The value of the stock at time 0 is \$1 and its price varies over time according to the *geometric Brownian motion*; that is, $P(t) = e^{B(t)}$, where $B(t), t \geq 0$, is Brownian motion. What is the expected maximum worth of owning the option up to a future time $T$? As the option will be exercised at time $t$ if the stock price at the time is $K$ or higher, the expected value is $E[\max_{0 \leq t \leq T}\{P(t) - K\}^+]$, where $x^+ = \max(x, 0)$. To obtain a further expression, note that for any $u > 0$, $\max_{0 \leq t \leq T}\{P(t) - K\}^+ \geq u$ if and only if $\max_{0 \leq t \leq T}\{P(t) - K\}^+ \geq u$ (why?). Also, for any nonnegative random variable, $X$ we have

$$E(X) = E\left\{\int_0^\infty 1_{(u \leq X)} \, du\right\} = \int_0^\infty P(X \geq u) \, du.$$

Thus, we have, by (10.34),

$$E\left[\max_{0 \leq t \leq T}\{P(t) - K\}^+\right] = \int_0^\infty P\left[\max_{0 \leq t \leq T}\{P(t) - K\}^+ \geq u\right] du$$

$$= \int_0^\infty P\left[\max_{0 \leq t \leq T}\{P(t) - K\} \geq u\right] du$$

$$= \int_0^\infty P\left\{\max_{0 \leq t \leq T} P(t) \geq K + u\right\} du$$

$$= \int_0^\infty P\left\{\max_{0 \leq t \leq T} B(t) \geq \log(K + u)\right\} du$$

$$= \sqrt{\frac{2}{\pi}} \int_0^\infty \int_{\log(K+u)/\sqrt{T}}^{\infty} e^{-x^2/2} dx \, du$$

$$= 2 \int_0^\infty \left[1 - \Phi\left\{\frac{\log(K + u)}{\sqrt{T}}\right\}\right] du,$$

where $\Phi(\cdot)$ is the cdf of $N(0,1)$.

Brownian motion obeys the following strong law of large numbers, whose proof is left as an exercise (Exercise 10.34).

**Theorem 10.14** (SLLN for Brownian motion). $B(t)/t \xrightarrow{\text{a.s.}} 0$ as $t \to \infty$.

A deeper result is the law of the iterated logarithm. Let $\psi(t) = \sqrt{2t \log \log t}$.

**Theorem 10.15** (LIL for Brownian motion). Lim $\sup_{t \to \infty} B(t)/\psi(t) = 1$ a.s. and, by symmetry, $\lim \inf_{t \to \infty} B(t)/\psi(t) = -1$ a.s.

Furthermore, Brownian motion is often associated with the limiting process of a sequence of stochastic processes, just like Gaussian (or normal) distribution, which often emerges as the limiting distribution of a sequence of random variables. One of the fundamental results in this regard is the convergence of empirical process (see Section 7.3). Let $B(t), t \geq 0$, be a Brownian motion. The conditional stochastic process $B(t), 0 \leq t \leq 1$, given $B(1) = 0$ is called the *Brownian bridge* (or tied-down Brownian motion). Another way of defining the Brownian bridge is by the process $U(t) = B(t) - tB(1), 0 \leq t \leq 1$. It is easy to show that the Brownian bridge is a Gaussian process with mean 0 and co-variances $\text{cov}\{U(s), U(t)\} = s(1-t)$, $s \leq t$ (Exercise 10.35). Let $X_1, X_2, \ldots$ be a sequence of independent random variables with the common distribution $F$. The empirical process is defined by (7.3); that is, $\sqrt{n}\{F_n(x) - F(x)\}$, where $F_n$ is the empirical d.f. defined by (7.1). As noted in Section 7.1, we may assume, without loss of generality, that $F$ is the Uniform$(0, 1)$ distribution and hence consider $U_n(t) = \sqrt{n}\{G_n(t) - t\}, 0 \leq t \leq 1$, where $G_n(t) = n^{-1} \sum_{i=1}^{n} 1_{(\xi_i \leq t)}$ and $\xi_1, \xi_2, \ldots$ are independent Uniform$(0, 1)$ random variables. It follows by Theorem 7.4 that

$$\sup_{0 \leq t \leq 1} |U_n(t)| \xrightarrow{\text{d}} \sup_{0 \leq t \leq 1} |U(t)| \tag{10.35}$$

as $n \to \infty$, where $U(t)$ is the Brownian bridge. We consider a well-known application of (10.35).

*Example 10.10.* One of the Kolmogorov–Smirnov statistics for testing goodness-of-fit is defined as $D_n = \sup_x |F_n(x) - F_0(x)|$, where $F_0$ is the hypothesized distribution under (7.9). Suppose that $F_0$ is continuous. Then we have [see (7.13) and the subsequent arguments]

$$
\begin{aligned}
P(\sqrt{n}D_n \leq \lambda) &= P\{ \sup_{0 \leq t \leq 1} |U_n(t)| \leq \lambda \} \\
&\longrightarrow P\{ \sup_{0 \leq t \leq 1} |U(t)| \leq \lambda \} \\
&= 1 - 2 \sum_{j=1}^{\infty} (-1)^{j-1} \exp(-2j^2\lambda^2).
\end{aligned}
$$

The derivation of the last equation [i.e., (7.16)] can be found, for example, in Durrett (1991, pp. 388–391).

Another well-known asymptotic theory in connection with the Brownian motion is Donsker's invariance principle (see Section 6.6.1). It states that if $X_1, X_2, \ldots$ are i.i.d. random variables with mean 0 and variance 1, $S_n = \sum_{i=1}^{n} X_i$ with $S_0 = 0$, and

$$\xi_n(t) = \frac{1}{\sqrt{n}} \{ S_{[nt]} + (nt - [nt]) X_{[nt]+1} \}$$

([$x$] is the integer part of $x$), then $\xi_n \overset{d}{\longrightarrow} W$ as $n \to \infty$, where $W(t) = B(t), 0 \le t \le 1$, and $B(t), t \ge 0$, is Brownian motion. An application of the invariance principle was considered in Example 6.10. Below is another one.

*Example 10.11.* Let the $X_i$'s be as above. Consider the functional $\psi(x) = \int_0^1 x(t)\,dt$, which is continuous on $\mathcal{C}$, the space of continuous functions on $[0,1]$ equipped with the uniform distance $\rho$ of (6.63). It follows that $\psi(\xi_n) \overset{d}{\longrightarrow} \psi(W)$ as $n \to \infty$. Furthermore, it can be shown that

$$\psi(\xi_n) = n^{-3/2} \sum_{k=1}^{n} S_k = n^{-3/2} \sum_{i=1}^{n} (n + 1 - i) X_i, \tag{10.36}$$

$$\psi(W) = \int_0^1 B(t)\,dt \sim N(0, 1/3) \tag{10.37}$$

(see the next section). Thus, the left side of (10.36) converges in distribution to $N(0, 1/3)$. This result can also be established directly using the Lindeberg–Feller theorem (the extended version following Theorem 6.11; Exercise 10.36).

Many applications of Brownian motion are made possible by the following result known as Skorokhod's representation theorem. Suppose that $X$ is a random variable with mean 0. We wish to find a stopping time $\tau$ at which the Brownian motion has the same distribution as $X$; that is, $B(\tau) \overset{d}{=} X$. Let us first consider a simple case.

*Example 10.12.* If $X$ has a two-point distribution on $a$ and $b$, where $a < 0 < b$, $B(\tau)$ can be constructed by using a continuous-time version of Wald's equation (see Theorem 10.9). The latter states that if $\tau$ is a bounded stopping time for the Brownian filtration, then $E\{B(\tau)\} = 0$ and $E\{B^2(\tau)\} = E(\tau)$ (e.g., Durrett 1991, p. 357). Define $\tau = \inf\{t : B(t) = a$ or $b\}$. It can be shown that $\tau$ is a stopping time for the Brownian filtration and $\tau < \infty$ a.s. (Exercise 10.37). By Wald's equation, we have $E\{B(\tau \wedge n)\} = 0$ for any fixed $n \ge 1$. On the other hand, we have $B(\tau \wedge n) = B(\tau) = a$ or $b$ if $\tau \le n$ and $B(\tau \wedge n) = B(n) \in (a, b)$ if $\tau > n$; so in any case, we have $|B(\tau \wedge n)| \le |a| \vee |b|$ and $B(\tau \wedge n) \to B(\tau)$ as $n \to \infty$. It follows by the dominated convergence theorem (Theorem 2.16) that

$$0 = E\{B(\tau)\} = aP\{B(\tau) = a\} + bP\{B(\tau) = b\}$$

(here we use the fact that $\tau < \infty$ a.s.). Also, we have

$$1 = P\{B(\tau) = a\} + P\{B(\tau) = b\}.$$

On the other hand, the same equations are satisfied with $B(\tau)$ replaced by $X$; therefore, we must have $P\{B(\tau) = a\} = P(X = a)$ and $P\{B(\tau) = b\} = P(X = b)$. It also follows that $E(\tau) = E\{B^2(\tau)\} = |a|b$ (verify this).

In general, we have the following.

**Theorem 10.16** (Skorokhod's representation). Let $B(t), t \geq 0$, be the standard Brownian motion. (i) For any random variable $X$, there exists a stopping time $\tau$ for the Brownian filtration, which is a.s. finite, such that $B(\tau) \overset{d}{=} X$. (ii) If $E(X) = 0$ and $E(X^2) < \infty$, then $\tau$ can be chosen such that $E(\tau) < \infty$.

Consider a sequence of independent random variables $X_1, X_2, \ldots$ with mean 0 and finite variance. Let $\tau_1$ be a stopping time with $E(\tau_1) = E(X_1^2)$ such that $B(\tau_1) \overset{d}{=} X_1$. By the strong Markov property (above Theorem 10.13), $B(\tau_1 + t) - B(\tau_1), t \geq 0$, is again a Brownian motion that is independent of $\mathcal{F}(\tau_1)$. We then find another stopping time $\tau_2$, independent of $\mathcal{F}(\tau_1)$, such that $E(\tau_2) = E(X_2^2)$ and $B(\tau_1 + \tau_2) - B(\tau_1) \overset{d}{=} X_2$, and so on. In this way we construct a sequence of stopping times $\tau_i, i \geq 1$, and let $T_n = \sum_{i=1}^{n} \tau_i$ so that $B(T_n + \tau_{n+1}) - B(T_n) \overset{d}{=} X_{n+1}$ and is independent of $\mathcal{F}(T_n)$, $n \geq 1$. It follows that $B(T_n) \overset{d}{=} S_n = \sum_{i=1}^{n} X_i$ and $E(T_n) = \sum_{i=1}^{n} E(\tau_i) = \sum_{i=1}^{n} E(X_i^2)$. This is a very useful representation. For example, suppose that $X_i, i \geq 1$, are i.i.d. with $E(X_1) = 0$ and $E(X_1^2) = 1$. Then we have $S_n = \sum_{i=1}^{n} X_i = B(T_n)$. By the LIL for Brownian motion (Theorem 10.15), we have

$$\limsup_{n \to \infty} \frac{S_n}{\sqrt{2n \log \log n}} = 1 \quad \text{a.s.}$$

This result was first proved by Strassen (1964; see Theorem 6.17).

## 10.6 Stochastic integrals and diffusions

The diffusion process is closely related to stochastic integral and differential equations. In fact, we already have encountered one such integral in Example 10.11 of the previous section, where we considered the integral of Brownian motion over the interval $[0, 1]$ [see (10.37)]. This is understood as the integral of the sample path of Brownian motion, which is continuous and therefore integrable over any finite interval almost surely. Furthermore, the integral can be computed as the limit

$$\int_0^1 B(t)\, dt = \lim_{n\to\infty} \frac{1}{n} \sum_{k=1}^n B\left(\frac{k-1}{n}\right). \tag{10.38}$$

Equation (10.38) is an example of what we call *stochastic integrals*. Here, the integration is with respect to $t$ (i.e., the Lebesgue measure). There is another kind of stochastic integrals, with respect to Brownian motion. To see this, note that by integration by parts we can write the left side of (10.38) as

$$\int_0^1 B(t)\, dt = B(1) - \int_0^1 t\, dB(t). \tag{10.39}$$

The integral on the right side, which is with respect to $B(t)$, has to be well defined because the one on the left side is. This is defined similarly as the Riemann–Stieltjes integral (see §1.5.4.36), namely,

$$\int_0^1 t\, dB(t) = \lim_{n\to\infty} \sum_{k=1}^n \frac{k-1}{n} \left\{ B\left(\frac{k}{n}\right) - B\left(\frac{k-1}{n}\right) \right\}.$$

More generally, consider a stochastic process $X(t) = X(t,\omega)$, where $t \in [0,\infty)$ and $\omega \in \Omega$, $(\Omega, \mathcal{F}, P)$ being the probability space. This means that $X(t)$ is measurable in the sense that $\{(\omega, t) : X(t,\omega) \in B\} \in \mathcal{F} \times \mathcal{B}_{[0,\infty)}$ for every $B \in \mathcal{B}_R$, where $\mathcal{B}_I$ represents all of the Borel subsets of $I$, a finite or infinite interval of the real line $R$. As in the previous section, we define $\mathcal{F}(t), t \geq 0$ as a nondecreasing family of $\sigma$-fields, called filtration, in the sense that $\mathcal{F}(s) \subset \mathcal{F}(t) \subset \mathcal{F}$ for any $0 \leq s \leq t$. We say the process $X(t)$ is $\mathcal{F}(t)$-adapted if $X(t) \in \mathcal{F}(t)$ [i.e., $X(t)$ is $\mathcal{F}(t)$ measurable] for every $t \geq 0$. Furthermore, an $\mathcal{F}(t)$-adapted process $X(t)$ is *progressively measurable* if

$$\{(\omega, s) : s < t, X(s,\omega) \in B\} \in \mathcal{F}(t) \times \mathcal{B}_{[0,\infty)}$$

for any $t \geq 0$ and $B \in \mathcal{B}_R$ (see Appendix A.2). In the following we will assume that $\mathcal{F}(t)$ is the Brownian filtration (see the previous section) and say $X(t)$, or simply $X$, is adapted, or progressively measurable, without having to mention the filtration. The stochastic Itô integral, named after the Japanese mathematician Kiyoshi Itô, with respect to Brownian motion over an interval $[0,T]$ is defined as follows. If $X$ is an elementary process in the sense that there is a partition of $[0,T]$, $0 = t_{n,0} < t_{n,1} < \cdots < t_{n,K_n} = T$, such that $X(t)$ does not change with $t$ over each subinterval $[t_{n,k-1}, t_{n,k})$, $1 \leq k \leq K_n$, then

$$\int_0^T X(t)\, dB(t) = \sum_{k=1}^{K_n} X(t_{n,k-1})\{B(t_{n,k}) - B(t_{n,k-1})\}. \tag{10.40}$$

In general, let $\mathcal{M}_T$ be the class of progressively measurable processes $X$ such that $P\left\{\int_0^T X^2(t)\, dt < \infty\right\} = 1$. Then any $X \in \mathcal{M}_T$ can be approximated by a sequence of elementary processes $X_n, n \geq 1$, such that

$$\int_0^T \{X(t) - X_n(t)\}^2 \, dt \xrightarrow{\text{P}} 0$$

as $n \to \infty$. We therefore define the Itô intergral

$$\mathcal{I}_T(X) = \int_0^T X(t) \, dB(t) \tag{10.41}$$

as the limit of convergence in probability of $\mathcal{I}_T(X_n)$, defined by (10.40), as $n \to \infty$. The Itô integral has the following nice properties. Let $\mathcal{M}_T^2$ be the class of progressively measurable processes $X$ such that $\mathrm{E}\left\{\int_0^T X^2(t) \, dt\right\} < \infty$. It is clear that $\mathcal{M}_T^2$ is a subset of $\mathcal{M}_T$.

**Lemma 10.1.** (i) For $X \in \mathcal{M}_T^2$, we have $\mathrm{E}\{\mathcal{I}_T(X)\} = 0$,

$$\mathrm{E}\{\mathcal{I}_T^2(X)\} = \mathrm{E}\left\{\int_0^T X^2(t) \, dt\right\},$$

and $\mathrm{E}\{\mathcal{I}_T(X)|\mathcal{F}(t)\} = \mathcal{I}_t(X),\ 0 \le t \le T$. (ii) For $X, Y \in \mathcal{M}_T^2$, we have

$$\mathrm{E}\{\mathcal{I}_T(X)\mathcal{I}_T(Y)\} = \mathrm{E}\left\{\int_0^T X(t)Y(t) \, dt\right\}.$$

Lemma 10.1 provides us a convenient way of computing the variances and covariances of Itô integrals. As a simple example, we verify that the variance of the integral in (10.37) is, indeed, equal to $1/3$ [by the way, the value was mistakenly stated as $1/2$ in Durrett (1991, p. 367)].

*Example 10.13.* Note that a Borel measurable function, $h(t), t \ge 0$, is a special case of a stochastic process that is deterministic at each $t$. Therefore, we can write (10.39) as

$$\int_0^1 B(t) \, dt = \int_0^1 dB(t) - \int_0^1 t \, dB(t)$$
$$= \mathcal{I}_1(1) - \mathcal{I}_1(t).$$

It follows that $\mathrm{E}\{\int_0^1 B(t) \, dt\} = \mathrm{E}\{\mathcal{I}_1(1)\} - \mathrm{E}\{\mathcal{I}_1(t)\} = 0$ and

$$\mathrm{E}\left\{\int_0^1 B(t) \, dt\right\}^2 = \mathrm{E}\{\mathcal{I}_1^2(1)\} - 2\mathrm{E}\{\mathcal{I}_1(1)\mathcal{I}_1(t)\} + \mathrm{E}\{\mathcal{I}_1^2(t)\}$$
$$= \int_0^1 dt - 2\int_0^1 t \, dt + \int_0^1 t^2 \, dt = \frac{1}{3}.$$

In fact, the exact distribution of the stochastic integral (10.41) can be obtained, not only for a fixed $T$ but also for a stopping time $\tau$ in the sense that $\{\tau < t\} \in \mathcal{F}(t)$ for every $t$, as follows.

**Lemma 10.2.** Let $X \in \mathcal{M}_T$. If for some $a > 0$ we have

$$P\left\{\int_0^T X^2(t)\, dt \geq a\right\} = 1,$$

then the stopping time

$$\tau_a = \inf\left\{t : \int_0^t X^2(s)\, ds \geq a\right\} \tag{10.42}$$

is well defined (Exercise 10.38) and

$$\mathcal{I}_{\tau_a}(X) = \int_0^{\tau_a} X(t)\, dB(t) \sim N(0, a).$$

In particular, if $X(t) \in \mathcal{M}_T$ and $X(t) \neq 0$ a.e. $t \in [0, T]$, then, with $a = \int_0^T X^2(t)\, dt$, we have $\tau_a = T$ a.s. (why?). It follows by Lemma 10.2 that

$$\int_0^T X(t)\, dB(t) \sim N(0, a). \tag{10.43}$$

This is a very useful result, for example, in determining the limiting distribution of the result of Donsker's invariance principle. We consider an example.

*Example 10.13 (continued).* We now verify that the limiting distribution in (10.37) is, indeed, $N(0, 1/3)$. This follows by writting

$$\int_0^1 B(t)\, dt = \int_0^1 (1 - t)\, dB(t)$$

and (10.43) with $X(t) = 1 - t$ and $T = 1$. Here, $a = \int_0^1 (1 - t)^2\, dt = 1/3$.

A stochastic process $X(t), 0 \leq t \leq T$, that is defined as the solution to the stochastic integral equation

$$X(t) = X(0) + \int_0^t \mu\{X(s)\}\, ds$$

$$+ \int_0^t \sigma\{X(s)\}\, dB(s), \quad 0 \leq t \leq T, \tag{10.44}$$

is called a (homogeneous) *diffusion process*, or *diffusion*, where $\mu(x), \sigma^2(x), x \in R$, are nonrandom functions called the trend and diffusion coefficients, respectively. Equivalently, a diffusion $X(t)$ is defined as the solution to the following stochastic differential equation (SDE):

$$dX(t) = \mu\{X(t)\}\,dt + \sigma\{X(t)\}\,dB(t), \quad X(0), \ \ 0 \le t \le T. \quad (10.45)$$

Here, $X(0)$ specifies the initial state of the process. The diffusion is a special case of the Itô process, defined as

$$dX(t) = \mu(t)\,dt + \sigma(t)\,dB(t), \quad (10.46)$$

where $\mu(t)$ and $\sigma(t)$ are adapted processes to the Brownian filtration $\mathcal{F}(t)$. The following theorem is a special case of what is known as Itô's formula.

**Theorem 10.17.** Let $X$ be an Itô process satisfying (10.46). For any twice continuously differentiable function $f$, the process $f(X)$ satisfies

$$\begin{aligned}
df\{X(t)\} &= f'\{X(t)\}\,dX(t) + \frac{1}{2}f''\{X(t)\}\sigma^2(t)\,dt \\
&= \left[ f'\{X(t)\}\mu(t) + \frac{1}{2}f''\{X(t)\}\sigma^2(t) \right]dt \\
&\quad + f'\{X(t)\}\sigma(t)\,dB(t). \quad (10.47)
\end{aligned}$$

Itô's formula may be regarded as the chain rule for change of variables in *stochastic calculus*. It differs from the standard chain rule due to the additional term involving the second derivative. For a derivation of Itô's formula (in a more general form), see, for example, Arnold (1974, Section 5.5). In particular, for the diffusion (10.45), which is a special case of (10.46) with $\mu(t) = \mu\{X(t)\}$ and $\sigma(t) = \sigma\{X(t)\}$, it follows that $f\{X(t)\}$ is also a diffusion satisfying (10.45) with trend and diffusion coefficients given by $f'(x)\mu(x) + (1/2)f''(x)\sigma^2(x)$ and $\{f'(x)\sigma(x)\}^2$, respectively. We consider some examples.

*Example 10.14* (Brownian motion). The standard Brownian motion restricted to $[0, T]$ is a diffusion with $\mu(x) = 0$ and $\sigma^2(x) = 1$.

The next example gives an explanation why the process is called diffusion.

*Example 10.15* (The heat equation). Let $u(t, x)$ denote the temperature in a rod at position $x$ and time $t$. Then $u(t, x)$ satisfies the *heat equation*

$$\frac{\partial u}{\partial t} = \frac{1}{2}\frac{\partial^2 u}{\partial x^2}, \quad t > 0. \quad (10.48)$$

It can be verified that for any continuous function $f$, the function

$$\begin{aligned}
u(t, x) &= \mathrm{E}[f\{x + B(t)\}] \\
&= \frac{1}{\sqrt{2\pi t}}\int_{-\infty}^{\infty} f(y)\exp\left\{ -\frac{(y - x)^2}{2t} \right\}dy \quad (10.49)
\end{aligned}$$

solves the heat equation (Exercise 10.39). Note that $x + B(t)$ has the $N(x, t)$ distribution, whose pdf actually satisfies the heat equation (see Exercise 10.39). Furthermore, $x + B(t), t \geq 0$, is the Brownian motion with initial state $x$; so, intuitively speaking, the expected functional value of the Brownian motion initiated at $x$ satisfies the heat equation. Now, according to the previous example, the standard Brownian motion is a diffusion with $\mu(x) = 0$ and $\sigma^2(x) = 1$. If we assume that $f$ is twice continuously differentiable, then by Itô's formula with $X(t) = x + B(t)$, we have

$$df\{x + B(t)\} = f'\{x + B(t)\}dB(t) + \frac{1}{2}f''\{x + B(t)\} \, dt$$

or, in its integral form,

$$f\{x + B(t)\} = f(x) + \int_0^t f'\{x + B(t)\} \, dB(t)$$
$$+ \frac{1}{2}\int_0^t f''\{x + B(t)\} \, dt. \tag{10.50}$$

By taking expectations on both sides of (10.50) and applying Lemma 10.1(i), we get (Exercise 10.39)

$$u(t, x) = \mathrm{E}[f\{x + B(t)\}]$$
$$= f(x) + \frac{1}{2}\int_0^t \mathrm{E}[f''\{x + B(t)\}] \, dt$$
$$= f(x) + \frac{1}{2}\int_0^t \frac{\partial^2 u}{\partial x^2} \, dt. \tag{10.51}$$

Differentiating both sides of (10.51) with respect to $t$ leads to the heat equation (10.48). Of course, (10.48) can be verified directly (see Exercise 10.39), but the point is to show a connection between diffusion processes and the heat equation, which holds not only for this special case but in more general forms.

Going back to the diffusion SDE (10.45). We must show the existence and uniqueness of a solution to the SDE. For this we need the following definition. We say the SDE (10.45) has a weak solution if there exists a probability space $(\Omega, \mathcal{F}, P)$, a family of nondecreasing $\sigma$-fields $\mathcal{F}(t) \subset \mathcal{F}, 0 \leq t \leq T$, a Brownian motion $B(t), 0 \leq t \leq T$, and a continuous-path process $X(t), 0 \leq t \leq T$, both adapted to $\mathcal{F}(t), 0 \leq t \leq T$, such that

$$P\left(\int_0^T [|\mu\{X(t)\}| + \sigma^2\{X(t)\}] \, dt < \infty\right) = 1$$

and (10.44) holds. The following result is proven in Durrett (1996, p. 210).

**Theorem 10.18.** Suppose that the function $\mu$ is locally bounded, the function $\sigma^2$ is continuous and positive, and there is $A > 0$ such that

$$x\mu(x) + \sigma^2(x) \le A(1 + x^2), \quad x \in R. \tag{10.52}$$

Then the SDE (10.45) has a unique weak solution.

We now consider some limit theorems for stochastic integrals. For simplicity, we consider the Itô integral

$$Y(t) = \mathcal{I}_t(X) = \int_0^t X(s)\, dB(s), \quad t \ge 0, \tag{10.53}$$

where the process $X(t)$ has continuous and nonvanishing path. It follows that

$$\tau(t) = \int_0^t X^2(s)\, ds \tag{10.54}$$

is finite and positive for every $t > 0$, which is called the *intrinsic time* of $Y(t)$. Let $\tau_a = \inf\{t : \tau(t) \ge a\}$ [see (10.42)]. It can be shown that $W(a) = 0$ if $a = 0$ and $W(a) = Y(\tau_a), a \ge 0$, is a standard Brownian motion (Exercise 10.40). Therefore, by the SLLN of Brownian motion (Theorem 10.14), we have $W(a)/a \xrightarrow{\text{a.s.}} 0$ as $a \to \infty$, which implies

$$\lim_{t \to \infty} \frac{Y(t)}{\tau(t)} = 0 \text{ a.s.} \tag{10.55}$$

Furthermore, by the LIL of Brownian motion (Theorem 10.15), we have

$$\limsup_{t \to \infty} \frac{Y(t)}{\sqrt{2\tau(t) \log \log \tau(t)}} = 1 \text{ a.s.} \tag{10.56}$$

Finally, we consider some limit theorems for diffusion processes. Let $\tau(a) = \inf\{t \ge 0 : X(t) = a\}$ and $\tau(a, b) = \inf\{t \ge \tau(a) : X(t) = b\}$. The process $X(t), t \ge 0$, is said to be *recurrent* if $P\{\tau(a, b) < \infty\} = 1$ for all $a, b \in R$; it is called *positive recurrent* if $E\{\tau(a, b)\} < \infty$ for all $a, b \in R$. The following results are proven in Kutoyants (2004, Section 1.2.1). We say the process $X$ has ergodic properties if there exists a distribution $F$ such that for any measurable function $h$ with finite first moment with respect to $F$, we have

$$P\left\{ \frac{1}{T} \int_0^T h\{X(t)\}\, dt \to \int h(x)\, dF(x) \right\} = 1.$$

**Theorem 10.19** (SLLN and CLT for diffusion process). Let $X$ be a process satisfying $dX(t) = \mu\{X(t)\}\, dt + \sigma\{X(t)\}\, dB(t)$, $X(0) = x_0$, $t \ge 0$, and suppose that it is positive recurrent. Then $X$ has ergodic properties with the density function of $F$ given by

$$f(x) = \frac{\sigma^{-2}(x) \exp[2 \int_0^x \{\mu(u)/\sigma^2(u)\}\, du]}{\int_{-\infty}^\infty \sigma^{-2}(y) \exp[2 \int_0^y \{\mu(u)/\sigma^2(u)\}\, du]\, dy}. \tag{10.57}$$

Furthermore, for any measurable function $g$ such that $\rho^2 = \int g^2(x)\, dF(x) < \infty$, we have as $T \to \infty$

$$\frac{1}{\sqrt{T}} \int_0^T g\{X(t)\}\, dB(t) \xrightarrow{\;\mathrm{d}\;} N(0, \rho^2).$$

We conclude this section with a simple example.

*Example 10.16.* Consider the diffusion process $dX(t) = -\mathrm{sign}\{X(t) - \mu\}\, dt + \sigma\, dB(t)$, $X_0$, $t \geq 0$, where $\mathrm{sign}(x) = -1$ if $x < 0$ and $1$ if $x \geq 0$ and $\mu$, $\sigma > 0$ are unknown parameters. Then the process has ergodic properties with $f(x|\mu) = e^{-2|x-\mu|}$, which does not depend on $\sigma$ (Exercise 10.41).

## 10.7 Case study: GARCH models and financial SDE

Historically, Brownian motion and diffusion were used to model physical processes, such as the random motion of molecules from a region of higher concentration to one of lower concentration (see Example 10.15), but later they had found applications in many other areas. One such areas that was developed relatively recently is theoretical finance. In 1973, Black and Scholes derived the price of a call option under the assumption that the underlying stock obeys a geometric Brownian motion (i.e., the logarithm of the price follows a Brownian motion; Black and Scholes 1973). Since then, continuous-time models characterized by diffusion and SDE have taken the center stage of modern financial theory. A continuous-time financial model typically assmes that a security price $X(t)$ obeys the following SDE:

$$dX(t) = \mu_t X(t)\, dt + \sigma_t X(t)\, dB(t), \quad 0 \leq t \leq T, \tag{10.58}$$

where $B(t)$ is a standard Brownian motion; $\mu_t$ and $\sigma_t^2$ are called the mean return and conditional volatility in finance. In particular, the Black–Scholes model corresponds to (10.58) with $\mu_t = \mu$ and $\sigma_t = \sigma$, which are unknown parameters. It is clear that the latter is a special case of the diffusion process defined in the previous section; that is, (10.45), where the trend and diffusion coefficients are given by $\mu(x) = \mu x$ and $\sigma^2(x) = \sigma^2 x^2$. In general, (10.58) may be regarded as a more general form of diffusion, where $\mu_t$ and $\sigma_t^2$ are also called the drift in probability and diffusion variance in probability, respectively. Considering (10.58), we can write this as

$$X^{-1}(t)\, dX(t) = \mu_t\, dt + \sigma_t\, dB(t), \quad 0 \leq t \leq T.$$

What this means is that the relative change of the price over time is due to two factors and is expressed as the sum of them. The first is a mean chance over time; the second is a random change over time that follows the rule of

Brownian motion. It is important to note that we must talk about relative change, not actual change, because the latter is likely to depend on how high, or low, the price is at time $t$.

On the other hand, in reality, virtually all economic time series data are recorded only at discrete intervals. For such a reason, empiricists have favored discrete-time models. The most widely used discrete-time models are of autoregressive conditionally heteroscedastic (ARCH) type, first introduced by Engle (1982). These models may be expressed, in general, as

$$x_k = \mu_k + y_k, \quad y_k = \sigma_k \epsilon_k, \tag{10.59}$$
$$\sigma_k^2 = \sigma^2(y_{k-1}, y_{k-2}, \ldots, k, a_k, \alpha), \tag{10.60}$$

$k = 1, 2, \ldots$, where $\epsilon_k$ is a sequence of independent $N(0, 1)$ random variables, $\sigma_k^2$ is the conditional variance of $x_k$ given the information at time $k$, $\mu_k$ corresponds to a drift which may depend on $k$, $\sigma_k^2$, and $x_{k-1}, x_{k-2}, \ldots$, $a_k$ is a vector of exogenous and lagged endogenous variables, and $\alpha$ is a vector of parameters. In reality, $x_k$ represents the observation at the frequency $(k/n)T$, where $[0, T]$ is the observed time interval, $n$ is the total number of observations, and $h = T/n$ is the length of the basic (or unit) time interval. In particular, Engle's (1982) model corresponds to (10.59) and (10.60) with

$$\sigma_k^2 = \alpha_0 + \sum_{j=1}^{p} \alpha_j y_{k-j}^2, \tag{10.61}$$

where the $\alpha$'s are nonnegative parameters. This is known as the ARCH($p$) model; the generalized ARCH, or GARCH model, of Bollerslev (1986) can be expressed as (10.59) and (10.60) with

$$\sigma_k^2 = \alpha_0 + \sum_{i=1}^{p} \alpha_i \sigma_{k-i}^2 + \sum_{j=1}^{q} \alpha_{p+j} y_{k-j}^2. \tag{10.62}$$

The latter model is denoted by GARCH($p, q$). It is clear that the ARCH model is similar to the MA model, and the GARCH model similar to the ARMA model in time series (see Section 9.1). However, unlike in MA or ARMA models, here the random quantities involved are nonnegative. The motivation for GARCH is that the documented econometric studies show that financial time series tend to be highly heteroskedastic. Such a heteroskedasticity is characterized by the conditional variance, modeled as a function of conditional variances and residuals in the past.

Until the early 1990s these two types of models—the contiuous-time models defined by the SDE and discrete-time GARCH models—had developed very much independently with little attempt to reconcile each other. However, these models are used to describe and analyze the same financial data; therefore, it would be reasonable to expect some kind of connection between the two. More specifically, consider two processes, the first being the GARCH process and the second being the continuous-time process observed at the same discrete-time points. Suppose that the time points are equally spaced. What happens when the length of the time interval goes to zero or, equivalently, the total number of observations, $n$, goes to infinity? Nelson (1990) bridged a partial connection between the two processes by giving conditions under which the GARCH process converges weakly to the diffusion process that govern the discrete-time observations of the continuous-time process as the length of the discrete time intervals goes to zero. Nelson derived the result by utilizing a general theory developed by Stroock and Varadhan (1979). Suppose that for each $h > 0$, $X_{h,k}, k \geq 0$, are $d$-dimensional random variables with the probability distribution $P_h$ and the Markovian property

$$P_h(X_{h,k+1} \in B | X_{h,0}, \ldots, X_{h,k}) = \pi_{h,k}(X_{h,k}, B) \text{ a.s. } P_h$$

for all Borel sets $B$ of $R^d$ and $k \geq 0$. Define a continuous-time process $X_h(t), 0 \leq t \leq T$, as a step-function such that

$$X_h(t) = X_{h,k}, \quad kh \leq t < (k+1)h, \ 0 \leq k \leq T/h.$$

Then, under suitable regularity conditions, the process $X_h(t)$ converges weakly as $h \to 0$ to the process $X(t)$ defined by the stochastic integral equation

$$X(t) = X(0) + \int_0^t \mu\{X(s), s\} \, ds$$
$$+ \int_0^t \sigma\{X(s), s\} \, dB^{(d)}(s), \ 0 \leq t \leq T, \tag{10.63}$$

where $B^{(d)}(t), 0 \leq t \leq T$, is a standard $d$-dimensional Brownian motion that is independent of $X(0)$. Here, a $d$-dimensional Brownian motion is defined by modifying assumption (iii) of the one-dimension Brownian motion (see Section 10.5) by $B^{(d)}(t) \sim N(0, \sigma^2 t I_d)$, where $I_d$ is the $d$-dimensional identity matrix, and the standard $d$-dimensional Brownian motion has $\sigma = 1$. The equivalent SDE to (10.63) is

$$dX(t) = \mu\{X(t), t\}dt + \sigma\{X(t), t\}dB^{(d)}(t), \ X(0), \ 0 \leq t \leq T, \tag{10.64}$$

which defines a more general form of diffusion process than (10.44) (why?). The functions $\mu(x, t)$ and $\sigma(x, t)$ and initial state $X(0)$ are determined by the following limits, whose existence is part of the regularity conditions:

$$\lim_{h \to 0} \sup_{|x| \leq R, 0 \leq t \leq T} \|\mu_h(x, t) - \mu(x, t)\| = 0,$$

$$\lim_{h \to 0} \sup_{|x| \leq R, 0 \leq t \leq T} \|a_h(x, t) - a(x, t)\| = 0,$$

and $a(x, t) = \sigma(x, t)\sigma(x, t)'$, where $\|A\| = \{\text{tr}(A'A)\}^{1/2}$,

$$\mu_h(x, t) = \frac{1}{h} \int_{\|y-x\| \leq 1} (y-x) \pi_{h, [t/h]}(x, dy),$$

$$a_h(x, t) = \frac{1}{h} \int_{\|y-x\| \leq 1} (y-x)(y-x)' \pi_{h, [t/h]}(x, dy)$$

($[t/h]$ denotes the largest integer $\leq t/h$), and $X_{h,0} \xrightarrow{d} X(0)$ as $h \to 0$. It should be noted that here the weak convergence is not merely in the sense of convergence in distribution of $X_h(t)$ for each given $t$: The distribution of the entire sample path of $X_h(t), 0 \leq t \leq T$ converges to the distribution of the sample path of $X(t), 0 \leq t \leq T$, as $h \to 0$. This is very similar to the weak convergence in Donsker's invariance principle (see Section 6.6.1).

Consider, for example, the MGARCH$(1, 1)$ model, defined by

$$x_k = \sigma_k \epsilon_k,$$
$$\log \sigma_k^2 = \alpha_0 + \alpha_1 \log \sigma_{k-1}^2 + \alpha_2 \log \epsilon_{k-1}^2. \tag{10.65}$$

Suppressing in the notation the dependence on $n$, we can rewrite (10.65) as

$$z_k = \frac{\sigma_k \epsilon_k}{\sqrt{n}},$$
$$\log \sigma_k^2 = \frac{\beta_0}{n} + \left(1 + \frac{\beta_1}{n}\right) \log \sigma_{k-1}^2 + \frac{\beta_2}{\sqrt{n}} \xi_k,$$

where $\xi_k = (\log \epsilon_k^2 - c_0)/c_1$, $c_0 = E(\log \epsilon_k^2)$, $c_1 = \sqrt{\text{var}(\log \epsilon_k^2)}$, and the $\beta$'s are new parameters. Define the bivariate process $[Z_n(t), \sigma_n^2(t)], t \in [0, 1]$, as

$$Z_n(t) = z_k, \quad \sigma_n^2(t) = \sigma_{k+1}^2, \quad t \in \left[\frac{k}{n}, \frac{k+1}{n}\right).$$

Then, as $n \to \infty$ (which is equivalent to $h \to 0$), the bivariate process converges in distribution to the bivariate diffusion process $[X(t), \sigma^2(t)]$ satisfying

$$dX(t) = \sigma(t) \, dB_1(t),$$
$$d \log \sigma^2(t) = \{\beta_0 + \beta_1 \log \sigma^2(t)\} \, dt + \beta_2 \, dB_2(t),$$

$t \in [0, 1]$, for the same parameters $\beta_j$, $j = 0, 1, 2$, where $B_1(t)$ and $B_2(t)$ are two independent standard Brownian motions.

Weak convergence is one way to study the connection between the discrete and continuous-time models. On the other hand, Wang (2002) showed that GARCH models and its diffusion limit are not asymptotically equivalent in the sense of La Cam deficiency, unless the volatility $\sigma_t^2$ is deterministic, which is considered a trivial case. Le Cam's deficiency measure is for comparison of two statistical experiments (Le Cam 1986; Le Cam and Yang 2000). Here, a statistical experiment consists of a sample space $\Omega$, a $\sigma$-field $\mathcal{F}$, and a family of distributions indexed by $\theta$, a vector of parameters, say $\{P_\theta, \theta \in \Theta\}$, where $\Theta$ is the parameter space. Consider two statistical experiments

$$\mathcal{E}_i = (\Omega_i, \mathcal{F}_i, \{P_{i,\theta}, \theta \in \Theta\}), \quad i = 1, 2.$$

Let $\mathcal{A}$ denote an action space and $L : \Theta \times \mathcal{A} \to [0, \infty)$ be a loss function. Define $\|L\| = \sup\{L(\theta, a) : \theta \in \Theta, a \in \mathcal{A}\}$. Let $d_i$ be a decision procedure for the $i$th experiment and $R_i(d_i, L, \theta)$ be the risk from using $d_i$ when $L$ is the loss function and $\theta$ is the true parameter vector, $i = 1, 2$. Le Cam's deficiency measure is defined as $\Delta(\mathcal{E}_1, \mathcal{E}_2) = \delta(\mathcal{E}_1, \mathcal{E}_2) \vee \delta(\mathcal{E}_2, \mathcal{E}_1)$, where

$$\delta(\mathcal{E}_1, \mathcal{E}_2) = \inf_{d_1} \sup_{d_2} \sup_{\theta \in \Theta} \sup_{\|L\|=1} |R_1(d_1, L, \theta) - R_2(d_2, L, \theta)|$$

and $\delta(\mathcal{E}_2, \mathcal{E}_1)$ is defined likewise. Two sequences of experiments, $\mathcal{E}_{n,1}$ and $\mathcal{E}_{n,2}$, where $n$ denotes the sample size, are said to be asymptotically equivalent if $\Delta(\mathcal{E}_{n,1}, \mathcal{E}_{n,2}) \to 0$ as $n \to \infty$. Wang (2002) considered the GARCH(1, 1) model (which is found to be adequate in most applications) for the sake of simplicity. He showed that the GARCH and diffusion experiments are not asymptotically equivalent according to the above definition.

The problem was further investigated by Brown et al. (2003), who considered the MGARCH(1, 1) process observed at "lower frequencies" (although the authors believe their results can be extended to GARCH models in general). Suppose that the diffusion process is observed at the time points $t_k = (k/n)T, k = 1, \ldots, n$. Thus, $T/n$ is the length of the basic time interval and $\phi_1 = n/T$ is the corresponding basic frequency. Let $u_k$ be the observed diffusion process at time $t_k$, $k = 1, \ldots, n$. Consider the MGARCH process observed at time $t_{ls}, l = 1, \ldots, [n/s]$, where $s$ is some positive integer. Then $\phi_s = n/(sT)$ is called a lower frequency if $s > 1$. Let $x_{ls}$ be the MGARCH process observed at time $t_{ls}, l = 1, \ldots, [n/s]$. Brown et al. (2003) considered the MGARCH experiment with observations $x_{ls}, 1 \leq l \leq [n/s]$, and the diffusion experiment with observations $u_{ls}, 1 \leq l \leq [n/s]$. They showed that the two experiments are asymptotically equivalent if $n^{1/2}/s \to 0$ as $n \to \infty$. For example, $s = n^{2/3}$ works; on the other hand, the result of Wang (2002) shows that this is not the case for $s = 1$.

## 10.8 Exercises

*10.1.* This exercise is related to Example 10.1.

(i) Verify that the locations of the sequence of digits formed either by the professor or by the student, as shown in Table 10.2, satisfy (10.1).

(ii) Show by induction that $X_n$ is a function of $\xi_1, \ldots, \xi_k$, where $k = X_{n-1}$.

(iii) Show that the right side of (10.2) is (also) equal to 0.1.

(iv) In Table 10.3, if, instead, the student starts at a digit (among the first 10 digits) other than the 10th (which is a 4), will her chain end at the same spot as the professor's (which is the second to last 0)?

*10.2.* Show that (10.3) implies the Markov property (10.4).

*10.3.* Show that the finite-dimensional distributions of a Markov chain $X_n$, $n = 0, 1, 2, \ldots$, are determined by its transition probability $p(\cdot, \cdot)$ and initial distribution $p_0(\cdot)$.

*10.4.* Derive the Chapman–Kolmogorov identity (10.7).

*10.5.* Show that the process $X_n$, $n \geq 1$, in Example 10.2 is a Markov chain with transition probability $p(i, j) = a_{j-i}$, $i, j \in S = \{0, \pm 1, \ldots\}$.

*10.6.* This exercise is related to Example 10.1 (continued in Section 10.2).

(i) Show that the one- and two-step transition probabilities of the Markov chain are given by (10.9) and (10.10), respectively. Also verify $\sum_{j \in S} p(i, j) = 1$ and $\sum_{j \in S} p^{(2)}(i, j) = 1$ for all $i \in S$

(ii) Derive (10.11).

(iii) Show that (10.12) holds for any possible values $j_1, \ldots, j_s$ of $T_1, \ldots, T_s$, respectively.

(iv) Show that $i_s - 10 \leq X(j_s) \leq i_s - 1$ on $\mathcal{A}$, where $\mathcal{A}$ is defined in (10.12).

(v) Derive (10.14) using (10.13) and independence of $X$ and $Y$. [Note that the first inequality in (10.14) is obvious.]

*10.7.* Show that any two classes of states (see Section 10.2.1) are either disjoint or identical.

*10.8.* Show that $i \leftrightarrow j$ implies $d(i) = d(j)$.

*10.9.* Show that if $i$ is recurrent and $i \leftrightarrow j$, then $j$ is recurrent.

*10.10.* In Example 10.2, if $a_{-1} = a_1 = 0$ but $a_{-2}$ and $a_2$ are nonzero, what states communicate? What if $a_1 = 0$ but $a_{-1} \neq 0$?

*10.11.* Show that in Example 10.2, the Markov chain is aperiodic if and only if $a_0 \neq 0$. Also show that in the special case of simple random walk with $0 < p < 1$, we have $d(i) = 2$ for all $i \in S$.

*10.12.* Derive the approximation (10.16) using Stirling's approximation (see Example 3.4).

*10.13.* Show that if state $j$ is transient, then (10.18) holds for all $i$. [Hint: By the note following (10.18), the left side of (10.18) is equal to the expected number of visits to $j$ when the chain starts in $i$. If $j$ is not accessible from $i$, then the expected number of visits to $j$ is zero when the chain starts in $i$; otherwise, the chains makes $k$ visits to $j$ ($k \geq 1$) if and only if it makes its first visit to $j$ and then returns $k - 1$ times to $j$.]

*10.14.* Show that positive recurrency implies recurrency. Also show that positive (null) recurrency is a class property.

*10.15.* Consider a Markov chain with states $0, 1, 2, \ldots$ such that $p(i, i+1) = p_i$ and $p(i, i - 1) = 1 - p_i$, where $p_0 = 1$. Find the necessary and sufficient condition on the $p_i$'s for the chain to be positive recurrent and determine its limiting probabilities in the latter case.

*10.16.* This exercise is related to the birth and death chain of Example 10.4.

(i) Show that the chain is irreducible if $p_i > 0$, $i \geq 0$, and $q_i > 0$, $i \geq 1$.

(ii) Show that the chain is aperiodic if $r_i > 0$ for some $i$.

(iii) Show that the chain has period 2 if $r_i = 0$ for all $i$.

(iv) Show that the simple random walk (see Example 10.2) with $0 < p < 1$ is a special case of the birth and death chain that is irreducible but periodic with period 2.

(v) Verify (10.20).

*10.17.* Consider a birth and death chain with two reflecting barriers (i.e., the state space is $\{0, 1, \ldots, l\}$); the transition probabilities are given as in Example 10.4 for $1 \leq i \leq l - 1$; $q_0 = r_0 = 0$, $p_0 = 1$; $r_l = p_l = 0$, $q_l = 1$; and $p(i, j) = 0$ otherwise.

(i) Show that the chain is irreducible if $p_i > 0$ and $q_i > 0$ for all $1 \leq i \leq l - 1$.

(ii) Show that the chain is aperiodic if $r_i > 0$ for some $1 \leq i \leq l - 1$.

(iii) Determine the stationary distribution for the chain.

*10.18.* For the third definition of a Poisson process, derive the pdf of $S_n$, the waiting time until the $n$th event. To what family of distribution does the pdf belong?

*10.19.* Show that the right side of (10.23) converges to $e^{-\lambda t}(\lambda t)^x/x!$ for $x = 0, 1, \ldots$.

*10.20.* Prove Theorem 10.5. [Hint: First derive an expression for $P\{t_i \leq S_i \leq t_i + h_i, 1 \leq i \leq n | N(t) = n\}$; then let $h_i \to 0$, $1 \leq i \leq n$.]

*10.21.* Derive (10.25) and (10.26). Also obtain the corresponding results for a Poisson process using Theorem 10.5.

*10.22.* Two balanced dice are rolled 36 times. Each time the probability of "double six" (i.e., six on each die) is $1/36$. Consider this as a situation of the Poisson approximation to binomial. The binomial distribution is Binomial$(36, 1/36)$; so the mean of the approximating Poisson distribution is $36 * (1/36) = 1$. Compare the two probability distributions for $k = 0, 1, 2, 3$, where $k$ is the total number of double sixes out of the 36 times.

*10.23.* Compare the distribution of a Poisson process $N(t)$ with rate $\lambda = 1$ with the approximating normal distribution. According to Theorem 10.4, we have $\{N(t) - t\}/\sqrt{t} \xrightarrow{\text{d}} N(0, 1)$ as $t \to \infty$. Compare (the histogram of) the distribution of $\{N(t) - t\}/\sqrt{t}$ with the standard normal distribution for $t = 1, 10, 50$. What do you conclude?

*10.24.* Give a proof of Theorem 10.7. [Hint: Note that $S_{N(t)} \leq t < S_{N(t)+1}$; then use the result of Theorem 10.6.]

*10.25.* Let $U$ be a random variable that has the Uniform$(0, 1)$ distribution. Define $\xi_n = n1_{(U \leq n^{-1})}$, $n \geq 1$.

(i) Show that $\xi_n \xrightarrow{\text{a.s.}} 0$ as $n \to \infty$.

(ii) Show that $E(\xi_n) = 1$ for every $n$, and therefore does not converge to $E(0) = 0$ as $n \to \infty$.

*10.26.* Show that the renewal function has the following expression:

$$m(t) = \sum_{n=1}^{\infty} F_n(t),$$

where $F_n(\cdot)$ is the cdf of $S_n$.

*10.27.* Let $N(t)$ be a renewal process. Show that $N(t) + 1$ is a stopping time with respect to the $\sigma$-fields $\mathcal{F}_n = \sigma(X_1, \ldots, X_n)$, $n \geq 1$ [see Section 8.2, above (8.5), for the definition of a stopping time].

*10.28.* This exercise is related to the proof of Theorem 10.10.
(i) Verify (10.29).
(ii) Show that $u_t \to -x$ as $t \to \infty$.
(iii) Derive (10.31).

*10.29.* Derive (10.33) by Fubini's theorem (see Appendix A.2).

*10.30.* This exercise shows how to justify assumption (ii) given assumption (i) of Brownian motion (see Section 10.5). Suppose that assumption (i) holds such that $E\{B(t) - B(s)\} = 0$, $E\{B(t) - B(s)\}^2 < \infty$, and

$$n P\{|B(t + 1/n) - B(t)| > \epsilon\} \longrightarrow 0$$

as $n \to \infty$. Use an appropriate CLT in Section 6.4 to argue that $B(t) - B(s)$ has a normal distribution with mean 0 and variance $\sigma^2(t - s)$.

*10.31.* In this exercise you are encouraged to give a proof of the reflection principle of Brownian motion (Theorem 10.13).
(i) Show that if $X$, $Y$, and $Z$ are random vectors such that (a) $X$ and $Y$ are independent, (b) $X$ and $Z$ are independent, and (c) $Y$ and $Z$ have the same distribution, then $(X, Y)$ and $(X, Z)$ have the same distribution.
(ii) Let $X = \{B(t)\}_{0 \leq t \leq \tau}$, $Y = \{B(t + \tau) - B(\tau)\}_{t \geq 0}$, and $Z = \{B(\tau) - B(t + \tau)\}_{t \geq 0}$. Use the strong Markov property of Brownian motion (see Section 10.5) and the result of (i) to show that $(X, Y)$ and $(X, Z)$ have the same distribution. The reflection principle then follows.

*10.32.* Let $B(t), t \geq 0$, be a standard Brownian motion. Show that each of the following is a standard Brownian motion:
(1) (Scaling relation) $a^{-1} B(a^2 t), t \geq 0$, where $a \neq 0$ is fixed.
(2) (Time inversion) $W(t) = 0$, $t = 0$ and $t B(1/t)$, $t > 0$.

*10.33.* Let $B(t), t \geq 0$, be a Brownian motion. Show that $B^2(t) - t, t \geq 0$, is a continuous martingale in that for any $s < t$,

$$E\{B^2(t) - t | B(u), u \leq s\} = B^2(s) - s.$$

*10.34.* Prove the SLLN for Brownian motion (Theorem 10.14). [Hint: First, show the sequence $B(n)/n, n = 1, 2, \ldots$, converges to zero almost surely as $n \to \infty$; then show that $B(t)$ does not oscillate too much between $n$ and $n + 1$.]

*10.35.* Show that the Brownian bridge $U(t), 0 \leq t \leq 1$ (defined below Theorem 10.15), is a Gaussian process with mean 0 and covariances $\text{cov}\{U(s), U(t)\} = s(1 - t)$, $s \leq t$.

*10.36.* This exercise is associated with Example 10.11.
(i) Verify (10.36) and (10.37).
(ii) Show, by using Lindeberg–Feller's theorem (Theorem 6.11; use the extended version following that theorem), that the right side of (10.36) converges in distribution to $N(0, 1/3)$.

*10.37.* Let $B(t), t \geq 0$, be Brownian motion and $a < 0 < b$. Define $\tau = \inf\{t : B(t) = a \text{ or } b\}$.

(i) Show that $\tau = \inf\{t : B(t) \notin (a, b)\}$.

(ii) Show that $\tau$ is a stopping time for the Brownian filtration.

(iii) Show that $\tau < \infty$ a.s. [Hint: Use (10.34).]

*10.38.* Verify that $\tau_a$ defined by (10.42) is a stopping time (whose definition is given above Lemma 10.2).

*10.39.* This exercise is related to the heat equation (Example 10.15).

(i) Verify that the pdf of $N(x, t)$,

$$f(y, t, x) = \frac{1}{\sqrt{2\pi t}} \exp\left\{-\frac{(y - x)^2}{2t}\right\}, \quad -\infty < y < \infty,$$

satisfies the heat equation (10.48).

(ii) Verify that the function $u(t, x)$ defined by (10.49) satisfies the heat equation.

(iii) Show, by taking expectations under the integral signs, that (10.50) implies (10.51); then obtain the heat equation by taking the partial derivatives with respect to $t$ on both sides of (10.51).

*10.40.* Show that the process $W(a), a \geq 0$, defined below (10.54) is a Brownian motion (Hint: Use Lemmas 10.1 and 10.2).

*10.41.* Verify that for the diffusion process in Example 10.16, the density function (10.57) reduces to $e^{-2|x-\mu|}$.

# 11

# Nonparametric Statistics

## 11.1 Introduction

This is the first of a series of five chapters on applications of large-sample techniques in specific areas of statistics. Nonparametric statistics are becoming increasingly popular in research and applications. Some of the earlier topics include statistics based on ranks and orders, as discussed in Lehmann's classical text *Nonparametrics* (Lehmann 1975). The area is expanding quickly to include some modern topics such nonparametric curve estimation and functional data analysis. A classical parametric model assumes that the observations $X_1, \ldots, X_n$ are realizations of i.i.d. samples from a population distribution $F_\theta$, where $\theta$ is a vector of unknown parameters. For example, the normal distribution $N(\mu, \sigma^2)$ has $\theta = (\mu, \sigma^2)'$ and the binomial distribution Binomial$(n, p)$ has $\theta = p$. In contrast, a nonparametric model would not specify the form of the distribution, up to a number of unknown parameters such as the above. Thus, the population distribution will be denoted by $F$ instead of $F_\theta$.

It can be said that a nonparametric model is not making much of a model assumption, if at all. For this reason, many nonparametric methods are based on "common sense" instead of model assumptions (of course, it may be argued that common sense is an assumption). For example, consider the following.

*Example 11.1* (Permutation test). Suppose that $m + n$ subjects are randomly assigned to control and treatment groups so that there are $m$ subjects in the control group and $n$ subjects in the treatment group. The treatment group receives a treatment (e.g., a drag); the control group receives a placebo (a placebo is a dummy or pretend treatment that is often used in controlled experiments). Because the subjects are randomly assigned to the groups, it may be assumed that the only population difference between the two groups is the treatment. Let $X_1, \ldots, X_m$ and $Y_1, \ldots, Y_n$ represent the observations from the control and treatment groups, respectively. A parametric model for assessing the treatment effect may be that the observations are independent

J. Jiang, *Large Sample Techniques for Statistics,*
DOI 10.1007/978-1-4419-6827-2_11, © Springer Science+Business Media, LLC 2010

such that $X_i \sim N(\mu_1, \sigma^2)$, $1 \le i \le m$, and $Y_j \sim N(\mu_2, \sigma^2)$, $1 \le j \le n$. Under this model, evidence of the treatment effect may be obtained by testing

$$H_0 : \quad \mu_1 = \mu_2. \tag{11.1}$$

A standard test statistic for the hypothesis (11.1) is the two-sample $t$-statistic with pooled sample variance,

$$t = \frac{\bar{Y} - \bar{X}}{s_p \sqrt{m^{-1} + n^{-1}}}, \tag{11.2}$$

where $\bar{X}$ and $\bar{Y}$ are the sample means defined by $\bar{X} = m^{-1} \sum_{i=1}^{m} X_i$ and $\bar{Y} = n^{-1} \sum_{j=1}^{n} Y_j$, respectively, $s_p^2$ is the pooled sample variance defined by

$$s_p = \frac{m-1}{m+n-2} s_X^2 + \frac{n-1}{m+n-2} s_Y^2,$$

and $s_X^2$ and $s_Y^2$ are the sample variances defined by $s_X^2 = (m-1)^{-2} \sum_{i=1}^{m} (X_i - \bar{X})^2$ and $s_Y^2 = (n-1)^{-1} \sum_{j=1}^{n} (Y_j - \bar{Y})^2$, respectively. The idea is that under (11.1), the $t$-statistic (11.2) has a $t$-distribution with $m + n - 2$ degrees of freedom; therefore, the $p$-value for testing (11.1) agaist the alternative

$$H_1 : \quad \mu_1 < \mu_2 \tag{11.3}$$

is the probability $P(t_{m+n-2} \ge t)$, where $t_\nu$ represents a random variable with the $t$-distribution with $\nu$ degrees of freedom, and $t$ is the observed $t$ of (11.2). Clearly, this procedure makes (heavy) use of the parametric model assumption (i.e., normality with equal population variance).

Now, consider a different strategy based on common sense: If the treatment really makes no difference, then the same thing is expected to happen with any assignment of $n$ out the $m + n$ subjects to the treatment group (and the rest to the control group). Therefore, the observed difference $\bar{Y} - \bar{X}$ is equally likely to be equaled or exceeded for any such assigment. Suppose that there are a total of $k$ different assignments of $n$ subjects to the treatment that result in the difference in sample means (i.e., $\bar{Y} - \bar{X}$ recomputed for each reassignment of the observations to the control and treatment groups) greater than or equal to the observed $\bar{Y} - \bar{X}$. Then the $p$-value for testing the null hypothesis that there is no treatment effect against the alternative that there is a positive treatment effect is $k / \binom{m+n}{n}$. If $m + n$ is is small, the exact $p$-value can be obtained; otherwise, the following Monte Carlo method is often used in practice. Combine the observations as $Z_i, i = 1, \ldots, m + n$, where $Z_i = X_i, 1 \le i \le m$, and $Z_i = Y_{i-m}, m + 1 \le i \le m + n$. Draw a large number, say $N$, of random permutations of the labels $1, \ldots, m + n$. For each permutation, assign the first $m$ labels as control and last $n$ labels as treatment, and compute the difference between the sample means of the treatment and control groups. More specifically, let the permutation be $\pi(1), \ldots, \pi(m + n)$, which is a rearrangement of $1, \ldots, m + n$. Then we compute

$$\Delta_\pi = \frac{1}{n} \sum_{i=m+1}^{m+n} Z_{\pi(i)} - \frac{1}{m} \sum_{i=1}^{m} Z_{\pi(i)}.$$

Suppose that out of the $N$ permutations, $l$ have the value of $\Delta_\pi$ greater than or equal to $\Delta = \bar{Y} - \bar{X}$. Then the $p$-value of the permutation test is approximately equal to $l/N$.

The idea behind the Monte Carlo method is the law of large numbers. Consider the space $\Pi$ of all different permutations of $1, \ldots, m + n$. On the one hand, we have

$$\begin{aligned}
p-\text{value} &= \frac{k}{\binom{m+n}{n}} \\
&= \frac{k \times m! n!}{\binom{m+n}{n} \times m! n!} \\
&= \frac{\# \text{ of permutations with } \Delta_\pi \geq \Delta}{\text{total } \# \text{ of permutations}} \\
&= \mathrm{P}(\Delta_\pi \geq \Delta),
\end{aligned}$$

where the probability is with respect to the random permutation $\pi \in \Pi$. On the other hand, let $\pi^{(1)}, \ldots, \pi^{(N)}$ denote the random sample of permutations drawn; we have, by the SLLN (see Section 6.3),

$$\begin{aligned}
\frac{l}{N} &= \frac{1}{N} \sum_{i=1}^{N} 1_{(\Delta_{\pi^{(i)}} \geq \Delta)} \\
&\xrightarrow{\text{a.s.}} \mathrm{P}(\Delta_\pi \geq \Delta)
\end{aligned}$$

as $N \to \infty$. Thus, the Monte Carlo method gives an approximate $p$-value.

The Monte Carlo method, or the law of large numbers, is one way to obtain the approximate $p$-value. Another method that is often used is the CLT—or more generally, the invariance principle—to obtain the asymptotic distribution of the test statistics. This will be discussed in the sequel.

An apparent advantage of nonparametric methods is robustness. Intuitively, the more specific assumptions are made regarding a (parametric) model, the more likely some of these assumptions are not going to hold in practice. Therefore, by making less assumptions, one potentially makes the method more robust against violations of (the parametric) model assumptions. However, there is a price that one is expected to pay. This happens when the parametric assumptions actually hold. For example, in Example 11.1, what if the normality assumption is indeed valid? (Statisticians refer to such a situation as that "sometimes, life is good.") If the parametric assumption is valid, but nevertheless not used, one has not fully utilized the available information (which may come from both the data and the knowledge about the distribution of the data). This may result in a loss of efficiency, which is the price we pay. In Section 11.3 we study this problem in the case of

Wilcoxon and other nonparametric testing procedures compared to the $t$-test based on the normality assumption. Although it was believed at first that a heavy price in loss of efficiency would have to be paid for the robustness, it turns out, rather surprisingly, that the efficiencies of Wilcoxon tests, as well as some other nonparametric tests, hold up quite well, even under the normality assumption. On the other hand, these nonparametric tests have considerable advantages in situations when the normality assumption fails.

## 11.2 Some classical nonparametric tests

Let us begin with (still) another proposal for the testing problem of Example 1.1. This time we rank the combined observations $X_1, \ldots, X_m, Y_1, \ldots, Y_n$ in increasing order (so the smallet observation receives the rank of 1, the second smallest the rank of 2, and so on). For simplicity, assume that there are no ties (if the underlying distributions are continuous, the probability of having ties is zero). If $S_1 < \cdots < S_n$ denote the ranks of the $Y$'s (among all the $m+n$ observations), define

$$W_s = S_1 + \cdots + S_n, \tag{11.4}$$

called the rank-sum. The idea is that if the null hypothesis of no treatment effect holds, the distribution of the rank-sum is something we can expect (i.e., determine); otherwise, if the rank-sum is much larger than what we expect, the null hypothesis should be rejected. This procedure is called the two-sample Wilcoxon test. The question then is: What do we expect? If $m$ and $n$ are relatively small, the exact distribution of the rank-sum can be determined. An alternative method, which is attractive when $m$ and $n$ are large, is based on the following CLT. To make a formal statement, let $X_1, \ldots, X_m$ be i.i.d. with distribution $F$, $Y_1, \ldots, Y_n$ be i.i.d. with distribution $G$, and the $X$'s and $Y$'s be independent. Suppose that both $F$ and $G$ are continuous but otherwise unknown. We are concerned with the hypothesis

$$\mathrm{H}_0: \quad F = G \tag{11.5}$$

against a suitable alternative $\mathrm{H}_1$. It can be shown (Exercise 11.1) that under the null hypothesis, we have

$$E(W_s) = \frac{1}{2}n(m+n+1), \tag{11.6}$$

$$\mathrm{var}(W_s) = \frac{1}{12}nm(m+n+1). \tag{11.7}$$

Furthermore, as $m, n \to \infty$,

$$\frac{W_s - n(m+n+1)/2}{\sqrt{mn(m+n+1)/12}} \xrightarrow{\mathrm{d}} N(0,1). \tag{11.8}$$

Therefore, in a large sample, we have the following approximation:

$$P(W_s \leq x) \approx \Phi \left\{ \frac{x - n(m+n+1)/2}{\sqrt{mn(m+n+1)/12}} \right\}, \tag{11.9}$$

where $\Phi(\cdot)$ is the cdf of $N(0, 1)$. It is found that for a moderate sample size, the following finite-sample correction improves the approximation. The idea is based on the fact that $W_s$ is an integer. It follows that for any integer $x$, $W_s \leq x$ if and only if $W_s \leq x + \delta$ for any $\delta \in (0, 1)$. Therefore, to be fair, $\delta$ is chosen as $1/2$. This leads to

$$P(W_s \leq x) \approx \Phi \left\{ \frac{x + 1/2 - n(m+n+1)/2}{\sqrt{mn(m+n+1)/12}} \right\}. \tag{11.10}$$

Table 11.1, taken from part of Table 1.1 of Lehmann (1975), shows the accuracy of the normal approximation for $m = 3$ and $n = 6$.

**Table 11.1. Normal approximation to $P(W_s \leq x)$**

| $x$ | 6 | 7 | 8 | 9 | 10 |
|---|---|---|---|---|---|
| Exact | .012 | .024 | .048 | .083 | .131 |
| (11.9) | .010 | .019 | .035 | .061 | .098 |
| (11.10) | .014 | .026 | .047 | .078 | .123 |

In connection with the two-sample Wilcoxon test, there is a Wilcoxon one-sample test. Suppose that $X_1, \ldots, X_n$ are i.i.d. observations from a continuous distribution that is symmetric about $\zeta$. We are interested in testing $H_0: \zeta = 0$ against $H_1: \zeta > 0$. The standard parametric test is the $t$-test, assuming that $F$ is normal. The test statistic is given by

$$t = \frac{\bar{X}}{s/\sqrt{n}}, \tag{11.11}$$

where $s^2 = (n-1)^{-1} \sum_{i=1}^{n} (X_i - \bar{X})^2$ is the sample variance. Alternatively, we may consider the ranks of the absolute values of the observations, $|X_1|, \ldots, |X_n|$. Let $R_1 < \cdots < R_a$ and $S_1 < \cdots < S_b$ denote the ranks of the absolute values of the negative and positive observations, respectively. For example, if $X_1 = -5$, $X_2 = 1$, and $X_3 = 4$, we have $a = 1$, $b = 2$, $R_1 = 3$, $S_1 = 1$, and $S_2 = 2$. The one-sample Wilcoxon test, also known as the Wilcoxon signed-rank test, rejects $H_0$ if

$$V_s = S_1 + \cdots + S_b > c, \tag{11.12}$$

where $c$ is a critical value depending on the level of significance. Similar to (11.6)–(11.8), it can be shown that

$$E(V_s) = \frac{1}{4}n(n+1), \tag{11.13}$$

$$\mathrm{var}(V_s) = \frac{1}{24}n(n+1)(2n+1), \tag{11.14}$$

$$\frac{V_s - n(n+1)/4}{\sqrt{n(n+1)(2n+1)/24}} \xrightarrow{\;d\;} N(0,1) \ \text{as } n \to \infty \tag{11.15}$$

(Exercise 11.2). Still another alternative is the sign test. Let $N_+$ denote the total number of positive observations. Then the null hypothesis is rejected if $N_+$ exceeds some critical value. Note that under $H_0$, $N_+$ has a Binomial$(n, 1/2)$ distribution (why?), so the critical value can be determined exactly. Alternatively, a large-sample critical value may be obtained via the CLT—namely,

$$\frac{2}{\sqrt{n}}\left(N_+ - \frac{n}{2}\right) \xrightarrow{\;d\;} N(0,1) \ \text{as } n \to \infty \tag{11.16}$$

(Exercise 11.3). The following example addresses an issue regarding some (undesirable) practices of using these tests.

*Example 11.2.* It is unfortunately not uncommon for researchers to apply two or more tests to their data, each at level $\alpha$, but to report only the outcome of the most significant one, thus claiming more significance for their results than is justified. A statistician following this practice could be accused of a lack of honesty but could rejoin the community of trustworth statisticians by stating the true significance level of this procedure. Consider, for example, the following small dataset extracted from Table I of Forrester and Ury (1969): $-16, -87, -5, 0, 8, -90, 0, 0, -31, -12$. The numbers are differences in tensile strength between tape-closed and sutured wounds (tape minus suture) on 10 experimental rats measured after 10 days of healing. If one applies the $t$-test to the data, it gives a $t$-statistic of $-2.04$, which corresponds to a (two-sided) $p$-value of .072. If one uses the Wilcoxon signed-rank test, it leads to a sum of ranks for the negative differences, 44, plus half of the sum of ranks for the zero differences, 3 [this is an extended version of (11.12) to deal with the cases with ties]. This gives a total of 47 and a $p$-value of .048. Finally, if the sign-test is used, one has 6 negative signs out of a total of 7 after eliminating the ties (again, this is an extended procedure of the sign-test when there are ties) and thus a $p$-value of .125. Suppose that all three tests have been performed. An investigator eager to get the result published might simply report the result of the signed-rank test, which is (barely) significant at 5% level, while ignoring those of the other tests. However, this may be misleading.

Jiang (1997b) derived sharp upper and lower bounds for the asymptotic significance level of a testing procedure that rejects when the largest of several standardized test statistics exceeds $z_\alpha$, the $\alpha$-critical value of $N(0,1)$. To state Jiang's results, first note that when considering asymptotic significance levels, one may replace $s$ in the denominator of (11.1) by $\sigma$, the population standard deviation (why?). Thus, we consider, without loss of generality,

$$S_j^* = \frac{S_j - \mathrm{E}(S_j)}{\sqrt{\mathrm{var}(S_j)}}, \quad j = 1, 2, 3, \tag{11.17}$$

which are (11.11) with $s$ replaced by $\sigma$, the left side of (11.15), and the left side of (11.16), respectively, where the expectations and variances are computed under the null hypothesis. The $S_j$'s are special cases of the $U$-statistics—to be discussed in Section 11.5 and therefore have a joint asymptotic normal distribution; that is,

$$\begin{pmatrix} S_1 \\ S_2 \\ S_3 \end{pmatrix} \xrightarrow{d} N \left\{ \begin{pmatrix} 0 \\ 0 \\ 0 \end{pmatrix}, \begin{pmatrix} 1 & \rho_w & \rho_s \\ \rho_w & 1 & \sqrt{3}/2 \\ \rho_s & \sqrt{3}/2 & 1 \end{pmatrix} \right\} \tag{11.18}$$

as $n \to \infty$ (Exercise 11.4). It follows that the asymptotic significance level of rejecting $H_0$ when $\max(S_1, S_2, S_3) > z_\alpha$ is $p_\alpha = \mathrm{P}\{\max(\xi_1, \xi_2, \xi_3) > z_\alpha\}$, where $(\xi_1, \xi_2, \xi_3)'$ has the distribution on the right side of (11.18). According to Slepian's inequality (see the end of Section 5.5), for fixed $\alpha$, the probability $p_\alpha$ is a decreasing function of $\rho_w$ and $\rho_s$, respectively. Thus, $p_\alpha$ is bounded by the probability when both $\rho_w$ and $\rho_s$ are zero, which is

$$\mathrm{P}\{\max(\xi_1, \xi_2, \xi_3) > z_\alpha, \xi_1 > z_\alpha\} + \mathrm{P}\{\max(\xi_1, \xi_2, \xi_3) > z_\alpha, \xi_1 \leq z_\alpha\}$$
$$= \mathrm{P}(\xi_1 > z_\alpha) + \mathrm{P}(\xi_1 \leq z_\alpha)\mathrm{P}\{\max(\xi_2, \xi_3) > z_\alpha\}$$
$$= \alpha + (1 - \alpha)p_\alpha^*$$

with $p_\alpha^* = \mathrm{P}\{\max(\xi_2, \xi_3) > z_\alpha\}$, where $\xi_2$ and $\xi_3$ are jointly bivariate normal with means 0, variances 1, and correlation coefficient $\sqrt{3}/2$. On the other hand, obviously, we have $p_\alpha \geq p_\alpha^*$. Therefore, we have

$$p_\alpha^* \leq p_\alpha \leq p_\alpha^* + (1 - p_\alpha^*)\alpha. \tag{11.19}$$

It can be shown that both sides of the inequalities (11.19) are sharp in the sense that there are distributions $F$ that are continuous and symmetric about zero for which the left- or right-side equalities are either attained or approached with arbitrary closeness (Exercise 11.5).

Note that the probabilities $p_\alpha$ and $p_\alpha^*$ depend on the underlying distribution $F$ (hence, the bounds are for all the distributions $F$ that are continuous and asymmetric about 0), but the dependence is only through $\rho_w$ and $\rho_s$. Jiang (2001) computed the analytic or numerical values of these correlation coefficients for a number of distributions that are symmetric about 0, as shown in Table 11.2, where DE represents the Double Exponential distribution, $\mathrm{NM}(\epsilon, \tau)$ denotes a normal mixture distribution with the cdf $F(x) = (1 - \epsilon)\Phi(x) + \epsilon\Phi(x/\tau)$, and $\Phi$ is the cdf of $N(0, 1)$. Given the values of $\rho_w$ and $\rho_s$, the corresponding actual asymptotic significance levels $p_\alpha$ can be calculated approximately. Again, see Table 11.2, which combines Table II and Table IV of Jiang (2001).

**Table 11.2. Exact or approximate values of $\rho_w$ and $\rho_s$ and corresponding approximate asymptotic significance levels**

| $F$ | $\rho_w$ | $\rho_s$ | $\alpha = 0.05$ | $\alpha = 0.025$ | $\alpha = 0.01$ |
|---|---|---|---|---|---|
| Normal | $\sqrt{3/\pi}$ | $\sqrt{2/\pi}$ | 0.079 | 0.041 | 0.017 |
| DE | $(3\sqrt{3})/(4\sqrt{2})$ | $1/\sqrt{2}$ | 0.086 | 0.045 | 0.019 |
| Rectangular | 1 | $\sqrt{3/2}$ | 0.071 | 0.037 | 0.015 |
| $t_3$ | 0.825 | 0.637 | 0.093 | 0.049 | 0.021 |
| $t_{10}$ | 0.961 | 0.774 | 0.081 | 0.043 | 0.018 |
| $NM(0.5, 2)$ | 0.953 | 0.757 | 0.082 | 0.043 | 0.018 |
| $NM(0.1, 4)$ | 0.850 | 0.656 | 0.091 | 0.048 | 0.021 |
| $NM(0.1, 10)$ | 0.648 | 0.459 | 0.102 | 0.054 | 0.023 |

## 11.3 Asymptotic relative efficiency

This section is concerned with asymptotic comparisons of tests that include, in particular, the comparison between a nonparametric and a parametric test. We begin with a heuristic derivation of the asymptotic power of a test. Suppose that we are interested in testing the hypothesis

$$\text{II}_0 : \quad \theta = \theta_0, \tag{11.20}$$

where $\theta$ is a (vector-valued) parameter associated with $F$, the underlying distribution of $X_1, \ldots, X_n$. Consider a statistic, $T_n$, that has the asymptotic property

$$\frac{\sqrt{n}\{T_n - \mu(\theta)\}}{\tau(\theta)} \xrightarrow{\text{d}} N(0, 1) \tag{11.21}$$

as $n \to \infty$ if $\theta$ is the true parameter, where $\mu(\cdot)$ and $\tau(\cdot)$ are some functions and the latter is assumed to be positive and may depend on some additional parameters. For example, in the problem of testing for the center of symmetry discussed in the previous section, let the cdf under $\theta$ be $F(x - \theta)$, where $F$ is continuous and symmetric about 0, and $E_\theta$ and $P_\theta$ denote the expectation and probability under $\theta$. The $t$-test is associated with $T_n = \bar{X}$ and we have $E_\theta(T_n) = \theta$. Furthermore, we have $\sqrt{n}(\bar{X} - \theta) \xrightarrow{\text{d}} N(0, \sigma^2)$, where $\sigma^2 = \text{var}(X_i)$. Thus, (11.21) holds with $\mu(\theta) = \theta$ and $\tau(\theta) = \sigma$. The latter depends on an additional unknown parameter $\sigma$ but not on $\theta$. For the sign test, let $T_n = n^{-1}N_+$. Then we have $E_\theta(T_n) = P_\theta(X_1 > 0) = P(X_1 - \theta > -\theta) = 1 - F(-\theta) = F(\theta)$. Similar to (11.16), we have, by the CLT,

$$\sqrt{n}\{T_n - F(\theta)\} \xrightarrow{\text{d}} N[0, F(\theta)\{1 - F(\theta)\}] \tag{11.22}$$

if $\theta$ is the true parameter (Exercise 11.6). Thus, (11.21) holds with $\mu(\theta) = F(\theta)$ and $\tau(\theta) = \sqrt{F(\theta)\{1 - F(\theta)\}}$. Finally, for the Wilcoxon signed-rank test, we consider $T_n = V_s/\binom{N}{2}$. It is shown in Section 11.5 that (11.21) holds with

$$\mu(\theta) = \mathrm{E}\{F(Z_1 + 2\theta)\}, \tag{11.23}$$

$$\tau^2(\theta) = 4\left(\mathrm{E}\{F^2(Z_1 + 2\theta)\} - [\mathrm{E}\{F(Z_1 + 2\theta)\}]^2\right), \tag{11.24}$$

where the expectations are taken with respect to $Z_1 \sim F$. In particular, when $\theta = 0$, (11.23) and (11.24) reduce to $1/2$ and $1/3$, respectively. In this case, it is easy to show that (11.21) is equivalent to (11.15) (Exercise 11.7).

Now, consider a class of large-sample tests that reject $H_0$ when

$$\frac{\sqrt{n}\{T_n - \mu(\theta_0)\}}{\tau(\theta_0)} \geq z_\alpha, \tag{11.25}$$

where $z_\alpha$ is the $\alpha$-critical value of $N(0, 1)$. Note that, in (11.25), $T_n$, $\mu$, and $\tau$ depend on the test. If we restrict the class to those satisfying (11.21), then all of the tests have asymptotic significance level $\alpha$ and therefore are considered equally good as far as the level of significance is concerned. The comparison of these tests is then focused on the power of the tests, defined as the probability of rejecting the null hypothesis when it is false. Suppose that we wish this probability to be $\beta$ when $\theta \neq \theta_0$ is the true parameter. Then we have

$$\beta = \mathrm{P}\left[\frac{\sqrt{n}\{T_n - \mu(\theta_0)\}}{\tau(\theta_0)} \geq z_\alpha\right]$$

$$= \mathrm{P}\left(\frac{\sqrt{n}\{T_n - \mu(\theta)\}}{\tau(\theta)} \geq \frac{\tau(\theta_0)}{\tau(\theta)}\left[z_\alpha + \frac{\sqrt{n}\{\mu(\theta_0) - \mu(\theta)\}}{\tau(\theta_0)}\right]\right).$$

Thus, in view of (11.21), we would expect

$$\frac{\tau(\theta_0)}{\tau(\theta)}\left[z_\alpha + \frac{\sqrt{n}\{\mu(\theta_0) - \mu(\theta)\}}{\tau(\theta_0)}\right] \approx z_\beta. \tag{11.26}$$

The point is, for any $\theta \neq \theta_0$ such that $\mu(\theta) > \mu(\theta_0)$, as long as $n$ is large enough, one is expected to have the power of at least $\beta$ of rejecting the null hypothesis. This can be seen from (11.26): As $n \to \infty$, the left side of (11.26) goes to $-\infty$ and therefore is $\leq z_\beta$ for large $n$, implying that the probability of rejection is $\geq \beta$. On the other hand, a test is more efficient than another test if it can achieve the same power with a smaller sample size. From (11.26), we can solve for the required sample size, $n$, for achieving power of $\beta$; that is,

$$n \approx \left\{\frac{\tau(\theta)}{\mu(\theta) - \mu(\theta_0)}\right\}^2 \left\{z_\beta - \frac{\tau(\theta_0)}{\tau(\theta)}z_\alpha\right\}^2. \tag{11.27}$$

Suppose that test 1 and test 2 are being compared with corresponding sample sizes $n_1$ and $n_2$; we thus have

$$n_j \approx \left\{\frac{\tau_j(\theta)}{\mu_j(\theta) - \mu_j(\theta_0)}\right\}^2 \left\{z_\beta - \frac{\tau_j(\theta_0)}{\tau_j(\theta)}z_\alpha\right\}^2, \quad j = 1, 2.$$

By taking the ratio, we have

$$\frac{n_1}{n_2} \approx \left\{\frac{\mu_2(\theta) - \mu_2(\theta_0)}{\mu_1(\theta) - \mu_1(\theta_0)}\right\}^2 \left\{\frac{\tau_1(\theta)}{\tau_2(\theta)}\right\}^2$$

$$\times \left\{z_\beta - \frac{\tau_1(\theta_0)}{\tau_1(\theta)}z_\alpha\right\}^2 \left\{z_\beta - \frac{\tau_2(\theta_0)}{\tau_2(\theta)}z_\alpha\right\}^{-2}. \tag{11.28}$$

The expression depends on $\theta$, the alternative. To come up with something independent of the alternative, we let $\theta \to \theta_0$. This means that we are focusing on the ability of a test in detecting a small difference from $\theta_0$. It follows, by L'Hôspital's rule, that the right side of (11.28) converges to

$$e_{2,1} = \left(\frac{c_2}{c_1}\right)^2, \tag{11.29}$$

where $c_j = \mu_j'(\theta_0)/\tau_j(\theta_0)$, $j = 1, 2$. The quantity $|c| = |\mu'(\theta_0)/\tau(\theta_0)|$ is called the efficacy of the test $T_n$ and $e_{2,1}$ is the asymptotic relative efficiency (ARE) of test 2 with respect to test 1 for the reason given above.

Now, consider, once again, the problem of testing the center of symmetry discussed in the previous section. Suppose that $F$ has a pdf, $f$. Then for the $t$-test we have $c = 1/\sigma$; for the sign test, we have $c = 2f(0)$; and for the Wilcoxon test, we have $c = 2\sqrt{3} \int f^2(z)\, dz$ (Exercise 11.8). It follows that the AREs for the comparison of each pair of these tests are given by

$$e_{S,t} = 4\sigma^2 f^2(0), \tag{11.30}$$

$$e_{W,t} = 12\sigma^2 \left\{\int f^2(z)\, dz\right\}^2, \tag{11.31}$$

$$e_{S,W} = \frac{f^2(0)}{3\{\int f^2(z)\, dz\}^2}. \tag{11.32}$$

The values of the AREs depend on the underlying distribution $F$. For example, when $F$ is the $N(0, \sigma^2)$ distribution, we have $e_{S,t} = 2/\pi \approx 0.637$ and $e_{W,t} = 3/\pi \approx 0.955$. It is remarkable that even in this case, where the $t$-test is supposed to be the standard and preferred strategy, the Wilcoxon test is a serious competitor. On the other hand, when $F$ is (very) different from the normal distribution, the nonparametric tests may have substantial advantages. We consider some examples.

*Example 11.3.* Suppose that $F$ has the pdf

$$f(x) = \frac{1}{2}\phi(x) + \frac{1}{4}\{\phi(x - \mu) + \phi(x + \mu)\}, \quad -\infty < x < \infty,$$

where $\phi(\cdot)$ is the pdf of $N(0, 1)$. This is a mixture of $N(0, 1)$ $N(\mu, 1)$ and $N(-\mu, 1)$ with probabilities $1/2$, $1/4$, and $1/4$, respectively, where $\mu > 0$. It can be shown that, in this case, we have

$$e_{S,t} = \frac{1}{2\pi}\left(1 + \frac{\mu^2}{2}\right)\left(1 + e^{-\mu^2/2}\right)^2, \tag{11.33}$$

which goes to $\infty$ as $\mu \to \infty$ (Exercise 11.10).

*Example 11.4.* Let $F$ be the NM$(\epsilon, \tau)$ distribution considered near the end of the previous section. It can be shown that, in this case,

$$e_{\mathrm{W,t}} = \frac{3}{\pi}(1 - \epsilon + \epsilon\tau^2)$$

$$\times \left\{ (1 - \epsilon)^2 + 2\sqrt{2}\frac{\epsilon(1-\epsilon)\tau}{\sqrt{1+\tau^2}} + \frac{\epsilon^2}{\tau} \right\}^2 \qquad (11.34)$$

(Exercise 11.11). Thus, $e_{\mathrm{W,t}} \to \infty$ as $\tau \to \infty$ for any fixed $\epsilon > 0$.

A more rigorous treatment of ARE can be given by considering a sequence of alternatives $\theta_n$ that can be expressed as

$$\theta_n = \theta_0 + \frac{\Delta}{\sqrt{n}} + o\left(\frac{1}{\sqrt{n}}\right), \qquad (11.35)$$

where $\Delta$ is a constant. Suppose that we have

$$\frac{\sqrt{n}\{T_n - \mu(\theta_n)\}}{\tau(\theta_0)} \overset{\mathrm{d}}{\longrightarrow} N(0,1), \qquad (11.36)$$

where the underlying distribution for $T_n$ has the parameter $\theta_n$. More precisely, what (11.36) means is the following. Consider a sequence of tests $T_n, n \geq 1$, where $T_n$ is based on independent samples $X_{n,1}, \ldots, X_{n,n}$ from the distribution that has $\theta_n$ as the true parameter. Then (11.36) holds as $n \to \infty$. Note that there is no need to change the denominator to $\tau(\theta_n)$ if we assume that $\tau(\cdot)$ is continuous (why?). By the Taylor expansion, we have

$$\mathrm{P}\left[ \frac{\sqrt{n}\{T_n - \mu(\theta_0)\}}{\tau(\theta_0)} \geq z_\alpha \right]$$

$$= \mathrm{P}\left[ \frac{\sqrt{n}\{T_n - \mu(\theta_n)\}}{\tau(\theta_0)} \geq z_\alpha - c\Delta + o(1) \right]$$

where $c = \mu'(\theta_0)/\tau(\theta_0)$ (verify this). Thus, in view of (11.36), we have

$$\lim_{n \to \infty} \mathrm{P}\left[ \frac{\sqrt{n}\{T_n - \mu(\theta_0)\}}{\tau(\theta_0)} \geq z_\alpha \right] = 1 - \Phi(z_\alpha - c\Delta).$$

It follows that the asymptotic power is an increasing linear function of $\Delta$ if $\mu'(\theta_0) > 0$. The slope of the linear function depends on the test through $c$, but the intercept does not depend on the test. This naturally leads to the comparison of $c$ and hence the ARE. A remaining question is how to verify (11.36). In some cases, this can be shown by applying the CLT for triangular arrays of independent random variables (see Section 6.4). For example, in the case of testing for the center of symmetry, let $X_{ni}, 1 \leq i \leq n$, be independent observations with the cdf $F(x - \theta_n)$. Then for the $t$-test, we have $T_n = \bar{X}_n = n^{-1} \sum_{i=1}^{n} X_{ni}$ and

$$\frac{\sqrt{n}(T_n - \theta_n)}{\sigma} = \sum_{i=1}^{n} Y_{ni},$$

where $Y_{ni} = (X_{ni} - \theta_n)/\sigma\sqrt{n}$. It is easy to verify that the $Y_{ni}, 1 \leq i \leq n$, $n \geq 1$, satisfy the conditions given below (6.35)—namely, that for each $n$, $Y_{ni}, 1 \leq i \leq n$, are independent, with $\mathrm{E}(Y_{ni}) = 0$, $\sigma_{ni}^2 = \mathrm{E}(Y_{ni}^2) = 1/n$, and $s_n^2 = \sum_{i=1}^{n} \sigma_{ni}^2 = 1$, and that the Lindeberg condition (6.36) holds for every $\epsilon > 0$. It follows that (11.36) holds (Exercise 11.12). In fact, in this case, expression (11.35) is not needed and the result holds for any sequence $\theta_n$. For the sign test, we have $T_n = n^{-1} \sum_{i=1}^{n} 1_{(X_{ni}>0)}$ and

$$2\sqrt{n}\{T_n - F(\theta_n)\} = \sum_{i=1}^{n} Y_{ni},$$

where $Y_{ni} = (2/\sqrt{n})\{1_{(X_{ni}>0)} - F(\theta_n)\}$. Once again, the conditions below (6.35) can be verified (Exercise 11.13); hence, $s_n^{-1} \sum_{i=1}^{n} Y_{ni} \xrightarrow{d} N(0,1)$. This time, we do need (11.35) (or a weaker condition that $\theta_n \to \theta_0$ as $n \to \infty$) in order to derive (11.36) because then $s_n = 2\sqrt{F(\theta_n)\{1 - F(\theta_n)\}} \to 1$. For the Wilcoxon signed-rank test, the verification of (11.36) is postponed to Section 11.5. We conclude this section with another example.

*Example 11.5* (Two-sample Wilcoxon vs. $t$). Consider the two-sample tests discussed at the beginning of Section 11.2. More specifically, we assume that $G(y) = F(y - \theta)$, where $F$ has a pdf, $f$. The null hypothesis (11.5) is then equivalent to (11.20) with $\theta_0 = 0$. Consider the two-sample $t$-test based on

$$t = \left\{ \left( \frac{1}{m} + \frac{1}{n} \right) S_p^2 \right\}^{-1/2} (\bar{Y} - \bar{X}),$$

where $S_p^2 = (m+n-2)^{-1}\{\sum_{i=1}^{m}(X_i - \bar{X})^2 + \sum_{j=1}^{n}(Y_j - \bar{Y})^2\}$. For simplicity, we assume that both $m, n \to \infty$ such that $m/N \to \rho$ and $n/N \to 1 - \rho$, where $N = m + n$ is the total sample size and $\rho \in (0,1)$. This restriction can be eliminated by using an argument of subsequences (see §1.5.1.6; also see Exercise 11.14). It is easy to show that $S_p^2 \xrightarrow{P} \sigma^2$, where $\sigma^2$ is the variance of $F$ (Exercise 11.14). Thus, without loss of generality, we consider a large-sample version of $t$ by replacing $S_p^2$ by $\sigma^2$. Let $T_N = \bar{Y} - \bar{X}$ and consider $\theta_N = \Delta/\sqrt{N} + o(1/\sqrt{N})$. Then we have

$$\sqrt{N}(T_N - \theta_N)$$
$$= \sqrt{N} \left\{ \frac{1}{n} \sum_{j=1}^{n}(Y_j - \theta_N - \mu) - \frac{1}{m} \sum_{i=1}^{m}(X_i - \mu) \right\}$$

$$= \left(\frac{N}{n}\right)^{1/2} \frac{1}{\sqrt{n}} \sum_{j=1}^{n} (Y_j - \theta_N - \mu) - \left(\frac{N}{m}\right)^{1/2} \frac{1}{\sqrt{m}} \sum_{i=1}^{m} (X_i - \mu)$$

$$= \xi_N - \eta_N,$$

where $\mu = \int x f(x) \, dx$, the mean of $F$. Now, $\xi_N$ and $\eta_N$ are two sequences of random variables such that $\xi_N \xrightarrow{d} N(0, \sigma^2/\rho)$, $\eta_N \xrightarrow{d} N\{0, \sigma^2/(1-\rho)\}$, and $\xi_N$ is independent of $\eta_N$. It follows that

$$\sqrt{N}(T_N - \theta_N) \xrightarrow{d} N\left\{0, \frac{\sigma^2}{\rho(1-\rho)}\right\} \tag{11.37}$$

as $N \to \infty$, provided that $\theta_N$ is the true $\theta$ for $T_N$ (Exercise 11.15). It follows that (11.36) holds with $n$ replaced by $N$, $\mu(\theta) = \theta$, and $\tau(0) = \sigma/\sqrt{\rho(1-\rho)}$. Following the same arguments, it can be shown that the asymptotic power of the two-sample $t$-test is $1 - \Phi(z_\alpha - c_t \Delta)$, where $c_t = \sqrt{\rho(1-\rho)}/\sigma$.

Next, we consider the two-sample Wilcoxon test. There is an alternative expression that associate the statistic $W_s$ of (11.4) to a $U$-statistic, to be discussed in Section 11.5; that is,

$$W_s = W_{XY} + \frac{1}{2}n(n+1), \tag{11.38}$$

where $W_{XY}$ = the number of pairs $(i,j)$ for which $X_i < Y_j$. In other words, $W_{XY} = \sum_{i,j} 1_{(X_i < Y_j)}$ (Exercise 11.16). By applying the asymptotic theory of $U$-statistics, it can be shown that (11.36) holds with $T_N = W_{XY}/mn$, $\mu(\theta) = P_\theta(X < Y) = \int \{1 - F(x - \theta)\} f(x) \, dx$, and $\tau(0) = \lim\{N(N+1)/12mn\}^{1/2} = 1/\sqrt{12\rho(1-\rho)}$. It follows that the asymptotic power of the two-sample Wilcoxon test is $1 - \Phi(z_\alpha - c_W \Delta)$, where $c_W = \sqrt{12\rho(1-\rho)} \int f^2(z) \, dz$ (Exercise 11.16). It turns out that the ARE for Wilcoxon versus $t$ is, once again,

$$e_{W,t} = 12\sigma^2 \left\{\int f^2(z) \, dz\right\}^2$$

[see (11.31)]. As shown, the ARE is approximately 0.955 when $F$ is the standard normal distribution, which is supposed to be the ideal case for the $t$-test; on the other hand, the ARE may be much in favor of the Wilcoxon test when the underlying distribution is different from normal. A remaining question then is: How do we know if $F$ is normal or not? This problem will be discussed in the next section.

## 11.4 Goodness-of-fit tests

There are at least two reasons why the topic of this section should be part of a chapter called *Nonparametric Statistics*. First, as discussed in the previous section, a parametric procedure such as the $t$-test is more powerful than a nonparametric procedure if the parametric distributional assumption, such as normality, holds. On the other hand, a nonparametric procedure is more robust in that it performs well under a wide range of distributions. Therefore, if one is using a parametric procedure, it is important to confirm that the distributional assumption holds; if it does, the more powerful parametric procedure can be used; otherwise, a nonparametric or robust procedure may be required. Such a confirmation may be carried out by a goodness-of-fit test. Second, the tests considered in this section are based on the empirical distribution of the data, which are closely related to the order statistics, one of the traditional topics of nonparametric statistics [see, e.g., Chapter 7 of David and Nagaraja (2003)]. Furthermore, it is seen below that, under suitable conditions, not only the asymptotic but the exact null distributions of these test statistics do not depend on the underlying distribution. In other words, the null distributions are "distribution free."

The null hypothesis that we wish to test is

$$H_0: \quad F = F_0, \tag{11.39}$$

where $F$ is the (unknown) underlying distribution and $F_0$ is a distribution that is either completely specified or specified up to some unknown parameters. We will assume that both $F$ and $F_0$ are continuous distributions and $F_0$ has pdf $f_0$. Earlier in Section 2.6 we considered one class of goodness-of-fit tests—namely, the $\chi^2$-test that was initially proposed by Pearson (1900). Here, we consider a different class of goodness-of-fit tests. Recall from Chapter 7 that one can estimate the distribution $F$ by its empirical d.f., $F_n(x) = n^{-1} \sum_{i=1}^n 1_{(X_i \leq x)}$. Therefore, it is natural to consider the difference between $F_n$ and $F_0$ and to use it as a springboard for goodness-of-fit tests. The first of such tests is the Kolmogorov–Smirnov test that has been discussed earlier (e.g., Section 7.3). The test is based on the Kolmogorov–Smirnov statistic

$$D_n = \sup_x |F_n(x) - F_0(x)|,$$

where $F_n$ is based on independent observations $X_1, \ldots, X_n$ from $F$. Another test is the Cramér–von Mises test, based on the statistic

$$W_n = \int \{F_n(x) - F_0(x)\}^2 f_0(x) \, dx.$$

A third test is the Anderson–Darling test based on the statistic

$$A_n = \int \frac{\{F_n(x) - F_0(x)\}^2}{F_0(x)\{1 - F_0(x)\}} f_0(x) \, dx.$$

First, note that by a similar argument to the one below (7.13), it is seen that the (exact) null distribution of the Kolmogorov–Smirnov test does not depend on $F_0$. This distribution-free property makes the testing procedure convenient because all one needs is a single table that applies to any $F_0$ (e.g., Owen 1962). The Cramér–von Mises and Anderson–Darling tests share the same distributional-free property as the Kolmogorov–Smirnov test (see below). Furthermore, asymptotic null distributions of these tests can be derived, which are more convenient to use in large-sample situations. Earlier in Section 7.3, it was shown that the asymptotic null distribution of $\sqrt{n}D_n$ is given by the right side of (7.16). A similar technique can be used to derive the asymptotic null distributions of the other two tests. Note that both $W_n$ and $A_n$ are special cases of a statistical functional of the form

$$\Psi(F_n) = \int \psi\{F_0(x)\}\{F_n(x) - F_0(x)\}^2 \, dF_0(x) \tag{11.40}$$

for some function $\psi$ on $[0, 1]$. By Theorem 7.1, $\Psi(F_n)$ has the same distribution as that with $X_i = F_0^{-1}(\xi_i)$, $1 \le i \le n$, where $\xi_1, \ldots, \xi_n$ are independent Uniform$(0, 1)$ random variables. Since $F_0^{-1}(\xi_i) \le x$ if and only if $\xi_i \le F_0(x)$ (see Theorem 7.1), we have, with $X_i = F_0^{-1}(\xi_i)$, $1 \le i \le n$, that $F_n(x) = G_n\{F_0(x)\}$, where $G_n(t) = n^{-1}\sum_{i=1}^{n} 1_{(\xi_i \le t)}$. Thus, by making a change of variables, $t = F_0(x)$, we obtain another expression for $\Psi(F_n)$:

$$\Psi(F_n) = \int \psi\{F_0(x)\}[G_n\{F_0(x)\} - F_0(x)]^2 \, dF_0(x)$$

$$= \int_0^1 \psi(t)\{G_n(t) - t\}^2 \, dt. \tag{11.41}$$

The latter expression shows, in particular, that both Cramér–von Mises and Anderson–Darling test statistics are distributional free in that their null distributions do not depend on $F_0$. Further expressions can be obtained under the null hypothesis. Let $\xi_{(1)} < \cdots < \xi_{(n)}$ be the order statistics of $\xi_1, \ldots, \xi_n$. It can be shown (Exercise 11.17) that, under $H_0$, we have

$$\Psi(F_n) = \frac{2}{n}\sum_{i=1}^{n}\left[\phi_1\{\xi_{(i)}\} - \frac{2i-1}{2n}\phi_0\{\xi_{(i)}\}\right]$$

$$+ \int_0^1 (1-t)^2\psi(t) \, dt, \tag{11.42}$$

where $\phi_0(t) = \int_0^t \psi(s) \, ds$ and $\phi_1(t) = \int_0^t s\psi(s) \, ds$. In particular, we have

$$W_n = \frac{1}{n}\sum_{i=1}^{n}\left\{\xi_{(i)} - \frac{2i-1}{2n}\right\}^2 + \frac{1}{12n^2}, \tag{11.43}$$

$$A_n = -1 - \frac{1}{n^2}\sum_{i=1}^{n}(2i-1)[\log\{\xi_{(i)}\} + \log\{1 - \xi_{(n-i+1)}\}]. \tag{11.44}$$

Equations (11.43) and (11.44) suggest a way to evaluate the critical values of the Cramér–von Mises and Anderson–Darling tests by a Monte Carlo method. Simply generate independent random variables $\xi_1, \ldots, \xi_n$ from the Uniform$(0,1)$ distribution and compute (11.43) and (11.44) for each Monte Carlo sample. Over a large number of Monte Carlo samples, the $100(1-\alpha)$th percentile of the computed values for (11.43) and (11.44) then approximate the critical values of the corresponding tests at the significance level $\alpha$ for any $\alpha \in (0,1)$ according to the law of large numbers.

The Monte Carlo method can provide approximations to the critical values at any accuracy, but it can be computationally intensive if $n$ is large. On the other hand, when $n$ is large, asymptotic distributions can be used to obtain the critical values. By using large-sample techniques of the empirical distribution discussed in Section 7.3, we can derive the asymptotic null distribution of $n\Psi(F_n)$ from the Doob–Donsker theorem (Theorem 7.4). By (11.41), we have

$$n\Psi(F_n) = \int_0^1 \psi(t)U_n^2(t)\, dt \;=\; h(U_n), \tag{11.45}$$

where $U_n(t) = \sqrt{n}\{G_n(t) - t\}$ and $h$ is the functional defined by $h(G) = \int_0^t \psi(t)G^2(t)\, dt$ for any $G \in D$, the space of all functions on $[0,1]$ that are right-continuous and possess left-limit at each point. Recall that $D$ is equipped with the uniform metric $\| \cdot \|$ [see (7.7)]. By Theorem 7.4, we have $U_n \xrightarrow{d} U$ as $n \to \infty$, where $U$ is the Brownian bridge. Thus, we have, as $n \to \infty$,

$$n\Psi(F_n) \xrightarrow{d} h(U) \;=\; \int_0^1 \psi(t)U^2(t)\, dt, \tag{11.46}$$

provided that one can show the continuity of the functional $h$ on $(D, \|\cdot\|)$. For the Cramér–von Mises test, the verification of continuity is left as an exercise (Exercise 11.18). However, this approach encounters some difficulties for the Anderson–Darling test due to the fact that the function $1/t(1-t)$ is not continuous at $t = 0$ or $1$. Nevertheless, it can be shown that (11.46) remains valid (e.g., Rosenblatt 1952). For the Cramér–von Mises test, the right side of (11.46) is $\omega^2 = \int_0^1 U^2(t)dt$. Smirnov (1936) showed that

$$P(\omega^2 \leq x)$$
$$= 1 - \frac{1}{\pi}\sum_{j=1}^{\infty}(-1)^{j-1}\int_{(2j-1)^2\pi^2}^{(2j)^2\pi^2} u^{-1}\left(\frac{-\sqrt{u}}{\sin\sqrt{u}}\right)^{1/2} e^{-xu/2}\, du \tag{11.47}$$

for $x > 0$. Smirnov obtained (11.47) by inverting the characteristic function (cf) of $\omega^2$, which is given by

$$c_W(t) = \left(\frac{\sqrt{2it}}{\sin\sqrt{2it}}\right)^{1/2}, \tag{11.48}$$

where $i = \sqrt{-1}$. For the Anderson–Darling test, the corresponding cf is

$$c_A(t) = \prod_{j=1}^{\infty} \left\{ 1 - \frac{2it}{j(j+1)} \right\}^{-1/2}, \tag{11.49}$$

where $i = \sqrt{-1}$. The expression for the cdf is more complicated and therefore omitted (see Anderson and Darling 1954).

## 11.5 *U* statistics

We mentioned a few times that the asymptotic null distributions of the Wilcoxon one- and two-sample tests are normal. These results are not directly implied by the CLT for sum of independent random variables as discussed in Chapter 6. To see this, note that, for example, the statistic $V_s$ for the Wilcoxon signed-rank test has the following alternative expression:

$$\begin{aligned}
V_s &= S + W \\
&= \sum_{i=1}^{n} 1_{(X_i > 0)} + \sum_{1 \le i < j \le n} 1_{(X_i + X_j > 0)}
\end{aligned} \tag{11.50}$$

(Exercise 11.19). The first summation, $S$, is a sum of i.i.d. random variables. However, the second double summation, $W$, which is actually the dominant factor for $V_s$ in terms of the order, is not a sum of independent random variables. Such a statistic can be characterized more generally as follows.

Let $X_1, \ldots, X_n$ be independent observations from the same distribution with cdf $F$. A statistic of the following type is called a $U$-statistic:

$$U = \binom{n}{m}^{-1} \sum_{1 \le i_1 < \ldots < i_m \le n} \phi(X_{i_1}, \cdots, X_{i_m}), \tag{11.51}$$

where $\phi: R^m \to R$ is a symmetric function of $m$ variables, known as the kernel of the $U$-statistic, and the summation is over all possible indexes $i_1, \ldots, i_m$ such that $1 \le i_1 < \cdots < i_m \le n$. Note that the $U$-statistic depends on the sample size $n$ and therefore may be denoted by $U_n$. However, in the statistics literatute, such a dependence on the sample size is often suppressed in the notation. This signifies a difference between probability theory and statistics, a transition that we have already been making since the beginning of this chapter. We hope the reader will get used to this kind of changes. A complete notation can make the concept clear, such as in probability theory where the subscript $n$ is always used, but could also limit understanding of the concept, in that a good reader should be able to see beyond the notation. It is easy to see that the $U$-statistic (11.51) is an unbiased estimator of the following parameter, viewed as a statistical functional (see Section 7.2):

$$\theta(F) = \int \cdots \int \phi(x_1, \ldots, x_m) \, dF(x_1) \cdots dF(x_m), \tag{11.52}$$

provided that the (multidimensional) integral is finite. Many of the well-known statistics are $U$-statistics. Below are some examples.

*Example 11.6* (Sample mean). $U = n^{-1} \sum_{i=1}^{n} X_i$, $m = 1$, $\phi(x) = x$, and $\theta(F) = \mu = \mathrm{E}(X_1) = \int x \, dF(x)$.

*Example 11.7* (Sample variance). It can be shown that

$$\frac{1}{n-1} \sum_{i=1}^{n} (X_i - \bar{X})^2 = \binom{n}{2}^{-1} \sum_{1 \leq i < j \leq n} \frac{(X_i - X_j)^2}{2}; \qquad (11.53)$$

so it is a $U$-statistic with $m = 2$, $\phi(x_1, x_2) = (x_1 - x_2)^2/2$, and $\theta(F) = \mathrm{var}(X_1) = \int (x - \mu)^2 \, dF(x)$, where $\mu$ is given above (Exercise 11.20).

*Example 11.8* (One-sample Wilcoxon statistic). Consider

$$U = \binom{n}{2}^{-1} \sum_{1 \leq i < j \leq n} 1_{(X_i + X_j > 0)},$$

which corresponds to the second term on the right side of (11.50). Then we have $m = 2$, $\phi(x_1, x_2) = 1_{(x_1 + x_2 > 0)}$, and $\theta(F) = \mathrm{P}(X_1 + X_2 > 0)$.

*Example 11.9* (Gini's mean difference). This is defined by

$$U = \binom{n}{2}^{-1} \sum_{1 \leq i < j \leq n} |X_i - X_j|;$$

so $m = 2$, $\phi(x_1, x_2) = |x_1 - x_2|$, and $\theta(F) = \mathrm{E}(|X_1 - X_2|)$.

*Example 11.10.* Consider the estimation of $\mu^2$, where $\mu$ is as in Example 11.6. An obvious estimator is $\bar{X}^2$, although this is not an unbiased estimator (why and what is the bias?). An unbiased estimator is given by the $U$-statistic

$$U = \binom{n}{2}^{-1} \sum_{1 \leq i < j \leq n} X_i X_j$$

with $m = 2$, $\phi(x_1, x_2) = x_1 x_2$, and $\theta(F) = \mathrm{E}(X_1 X_2) = \mu^2$.

Our main focus is the asymototic distribution of $U$-statistics. In this regard, a nice representation, discovered by Hoeffding (1961), is very useful. To introduce the representation, let us first define the following, known as the canonical functions of $U$-statistics. Let

$$\phi_c(x_1, \ldots, x_c) = \mathrm{E}\{\phi(x_1, \ldots, x_c, X_{c+1}, \ldots, X_m)\}$$
$$= \mathrm{E}\{\phi(X_1, \ldots, X_m) | X_1 = x_1, \ldots, X_c = x_c\}, \quad (11.54)$$

$c = 0, 1, \ldots, m$, where the expectation is taken with respect to $X_{c+1}, \ldots, X_m$ and when $c = 0$, this means $E\{\phi(X_1, \ldots, X_m)\} = \theta(F)$. Write $\theta = \theta(F)$ for notation simplicity. Note that $E\{\phi_c(X_1, \ldots, X_c)\} = \theta$ for every $0 \le c \le m$. We then centralize the $\phi$'s by letting $\tilde{\phi} = \phi - \theta$ and $\tilde{\phi}_c = \phi_c - \theta$, $1 \le c \le m$. The canonical functions are defined recursively by

$$
\begin{aligned}
g_1(x_1) &= \tilde{\phi}_1(x_1), \\
g_2(x_1, x_2) &= \tilde{\phi}_2(x_1, x_2) - \{g_1(x_1) + g_1(x_2)\} \\
g_3(x_1, x_2, x_3) &= \tilde{\phi}_3(x_1, x_2, x_3) - \sum_{i=1}^{3} g_1(x_i) - \sum_{1 \le i < j \le 3} g_2(x_i, x_j), \\
&\cdots \\
g_m(x_1, \ldots, x_m) &= \tilde{\phi}_m(x_1, \ldots, x_m) - \sum_{i_1=1}^{m} g_1(x_{i_1}) - \cdots
\end{aligned}
$$

$$
- \sum_{1 \le i_1 < \cdots < i_{m-1} \le m} g_{m-1}(x_{i_1}, \ldots, x_{i_{m-1}}). \tag{11.55}
$$

The canonical functions are clearly symmetric in their arguments and satisfy the following property known as complete degeneracy (Exercise 11.21):

$$
E\{g_c(x_1, \ldots, x_{c-1}, X_c)\} = 0, \quad 1 \le c \le m. \tag{11.56}
$$

We now express the $U$-statistics in terms of their canonical functions.

**Theorem 11.1** (Hoeffding representation). The $U$-statistic (11.51) can be expressed as

$$
U = \theta + \sum_{c=1}^{m} \binom{m}{c} \binom{n}{c}^{-1} S_{nc}, \tag{11.57}
$$

where $S_{nc} = \sum_{1 \le i_1 < \cdots < i_c \le n} g_c(X_{i_1}, \ldots, X_{i_c})$.

The first integer $r$ such that $g_r \ne 0$ is called the rank of the $U$-statistic. Thus, a $U$-statistic can be expressed as

$$
U = \theta + \sum_{c=r}^{m} \binom{m}{c} U_{nc}, \tag{11.58}
$$

where $r$ is the rank of the $U$-statistic and

$$
U_{nc} = \binom{n}{c}^{-1} \sum_{1 \le i_1 < \cdots < i_c \le n} g_c(X_{i_1}, \ldots, X_{i_c})
$$

is the U-statistic with kernal $g_c$, $r \le c \le m$. Note that $S_{nc}$, $1 \le c \le m$, satisfy $E(S_{nc}) = 0$ and the following nice orthogonality property:

$$E(S_{nc}S_{nd}) = \binom{n}{c}\delta_c 1_{(c=d)}, \tag{11.59}$$

where $\delta_c = E\{g_c^2(X_1,\ldots,X_c)\}$, provided that

$$E\{\phi^2(X_1,\ldots,X_m)\} < \infty \tag{11.60}$$

(Exercise 11.22). Furthermore, (11.57) leads to a martingale representation of $U$-statistics. Let $\mathcal{F}_k = \sigma(X_1,\ldots,X_k)$. Then we have $S_{nc} \in \mathcal{F}_n$ and

$$E(S_{nc}|\mathcal{F}_k) = S_{kc}, \quad c \le k \le n \tag{11.61}$$

(Exercise 11.23). Equation (11.61) shows that $S_{nc}, \mathcal{F}_n, n \ge c$, is a martingale for every $1 \le c \le m$. Therefore, $U = U_n$, $\mathcal{F}_n$, $n \ge m$ is a martingale. An alternative expression is in terms of martingale differences; that is,

$$U = \theta + \sum_{k=1}^{n}\xi_{nk}, \tag{11.62}$$

where $\xi_{nk} = \sum_{c=1}^{m}\binom{m}{c}\binom{n}{c}^{-1}(S_{kc} - S_{k-1c})$, $\mathcal{F}_k$, $k \ge 1$, is a sequence of martingale differences (Exercise 11.23). With expression (11.62), it is certainly feasible to establish a full array of limit theorems for $U$-statistics using the martingale limit theory (see Chapter 8). However, here we are concerned with asymptotic distribution of $U$-statistics for which (11.60) holds and

$$\sigma_1^2 = \text{var}\{\phi_1(X_1)\} > 0. \tag{11.63}$$

Under these conditions, the asymptotic distribution can be derived using a much simpler argument of CLT for sums of independent random variables (see Chapter 6). To see this, note that by (11.57), we can write

$$U - \theta = \frac{m}{n}\sum_{i=1}^{n}g_1(X_i) + R_n, \tag{11.64}$$

where $R_n$ is the remaining term. By (11.59), we have

$$E(R_n^2) = \sum_{c=2}^{m}\binom{m}{c}^2\binom{n}{c}^{-1}\delta_c,$$

where $\delta_c < \infty$, $2 \le c \le m$. It follows that $nE(R_n^2) \to 0$; hence, $\sqrt{n}R_n \xrightarrow{P} 0$ as $n \to \infty$. We now apply the CLT for sum of i.i.d. random variables and Slutsky's theorem (Theorem 2.13) to (11.64) to conclude that

$$\sqrt{n}(U - \theta) = \frac{m}{\sqrt{n}}\sum_{i=1}^{n}g_1(X_i) + o_P(1)$$
$$\xrightarrow{d} N(0, m^2\sigma_1^2). \tag{11.65}$$

The same argument can be used to derive the asymptotic joint distribution of several $U$-statistics. Consider two $U$-statistics:

$$U^{(1)} = \binom{n}{a}^{-1} \sum_{1 \le i_1 < \cdots < i_a \le n} \phi^{(1)}(X_{i_1}, \ldots, X_{i_a}),$$

$$U^{(2)} = \binom{n}{b}^{-1} \sum_{1 \le i_1 < \cdots < i_b \le n} \phi^{(2)}(X_{i_1}, \ldots, X_{i_b}).$$

Recall notation (11.54), so that we have

$$\phi_c^{(1)}(x_1, \ldots, x_c) = \mathrm{E}\{\phi^{(1)}(x_1, \ldots, x_c, X_{c+1}, \ldots, X_a)\},$$
$$\phi_d^{(2)}(x_1, \ldots, x_d) = \mathrm{E}\{\phi^{(2)}(x_1, \ldots, x_d, X_{d+1}, \ldots, X_b)\}.$$

Define $\sigma_{cd} = \mathrm{cov}\{\phi_c^{(1)}(X_1, \ldots, X_c), \phi_d^{(2)}(X_1, \ldots, X_d)\}$. Then we have the following formula for the covariance between $U^{(1)}$ and $U^{(2)}$.

**Theorem 11.2.** For any $a \le b$ we have

$$\mathrm{cov}\{U^{(1)}, U^{(2)}\} = \binom{n}{a}^{-1} \sum_{c=1}^{a} \binom{b}{c}\binom{n-b}{a-c} \sigma_{cc}. \tag{11.66}$$

See, for example, Lee (1990) for a derivation of (11.66). Two immediate consequences of Theorem 11.2 are

$$\mathrm{var}(U) = \binom{n}{m}^{-1} \sum_{c=1}^{m} \binom{m}{c}\binom{n-m}{m-c} \sigma_c^2, \tag{11.67}$$

where $\sigma_c^2 = \sigma_{cc} = \mathrm{var}\{\phi_c(X_1, \ldots, X_c)\}$, and

$$n\,\mathrm{cov}\{U^{(1)}, U^{(2)}\} \longrightarrow ab\sigma_{11} \tag{11.68}$$

as $n \to \infty$ (Exercise 11.24).

In the derivations below we allow the distribution of $X_1, \ldots, X_n$ to be dependent on $n$ (i.e., $F = F_n$). Consider the $U$-statistics

$$U^{(j)} = \binom{n}{m[j]}^{-1} \sum_{1 \le i_1 < \cdots < i_{m[j]} \le n} \phi^{(j)}(X_{i_1}, \ldots, X_{i_{m[j]}}), \tag{11.69}$$

$1 \le j \le s$. We assume that (i) $\max_{1 \le j \le s} \mathrm{var}\{\phi^{(j)}(X_1, \ldots, X_{m[j]})\}$ are bounded (note that the variances now depend on $n$; so merely finiteness of the variances is not sufficient); (ii) as $n \to \infty$,

$$\mathrm{cov}\{\phi_1^{(j)}(X_1), \phi_1^{(k)}(X_1)\} \longrightarrow \sigma(j, k), \quad 1 \le j, k \le s, \tag{11.70}$$

and $\Sigma = [m[j]m[k]\sigma(j,k)]_{1 \leq j,k \leq s}$ is positive definite; and (iii)

$$E[\{g_1^{(j)}(X_1)\}^2 1_{(|g_1^{(j)}(X_1)| > \epsilon\sqrt{n})}] \longrightarrow 0, \quad 1 \leq j \leq s, \tag{11.71}$$

as $n \to \infty$ for every $\epsilon > 0$, where $g_1^{(j)}$ is $g_1$ with $m = m[j]$ and $\phi = \phi^{(j)}$.

**Theorem 11.3.** Under assumptions (i)–(iii), we have that as $n \to \infty$,

$$\sqrt{n} \begin{bmatrix} U^{(1)} - \theta_1 \\ \vdots \\ U^{(s)} - \theta_s \end{bmatrix} \xrightarrow{\text{d}} N(0, \Sigma), \tag{11.72}$$

where $\theta_j = E\{\phi^{(j)}(X_1, \ldots, X_{m[j]})\}, 1 \leq j \leq s$.

To show (11.72), note that by Theorem 2.14, this is equivalent to that, for every $\lambda = (\lambda_j)_{1 \leq j \leq s} \in R^s$, we have

$$\lambda' \sqrt{n} \begin{bmatrix} U^{(1)} - \theta_1 \\ \vdots \\ U^{(s)} - \theta_s \end{bmatrix} \xrightarrow{\text{d}} N(0, \lambda'\Sigma\lambda). \tag{11.73}$$

To show (11.73), note that by (11.64), we can write the left side as

$$\sqrt{n} \sum_{j=1}^{s} \lambda_j \{U^{(j)} - \theta_j\} = \sqrt{n} \sum_{j=1}^{s} \left\{ \lambda_j \frac{m[j]}{n} \sum_{i=1}^{n} g_1^{(j)}(X_i) + \lambda_j R_{n,j} \right\}$$

$$= \sum_{i=1}^{n} \frac{1}{\sqrt{n}} \sum_{j=1}^{s} \lambda_j m[j] g_1^{(j)}(X_i) + \sum_{j=1}^{s} \lambda_j \sqrt{n} R_{n,j}$$

$$= I_1 + I_2.$$

By assumption (i) and the argument following (11.64), it can be shown that $E(I_2^2) \to 0$ [make sure that assumption (i) is sufficient for this argument]. Thus, it remains to show that $I_1$ converges in distribution to the right side of (11.73) (and then apply Slutsky's theorem). To this end, we write

$$\xi_{ni} = \frac{1}{\sqrt{n}} \sum_{j=1}^{s} \lambda_j m[j] g_1^{(j)}(X_i).$$

Then for each $n \geq 1$, $\xi_{ni}, 1 \leq i \leq n$, are independent with $E(\xi_{ni}) = 0$, and by (11.70), $E(\xi_{ni}^2) = n^{-1}\{\lambda'\Sigma\lambda + o(1)\}$ (verify this); hence, $s_n^2 = \sum_{i=1}^{n} E(\xi_{ni}^2) = \lambda'\Sigma\lambda + o(1)$. By the CLT for triangular arrays of independent random variables (see Section 6.4), to show that $s_n^{-1} \sum_{i=1}^{n} \xi_{ni} \xrightarrow{\text{d}} N(0,1)$ or, equivalently, $(\lambda'\Sigma\lambda)^{-1/2} I_1 \xrightarrow{\text{d}} N(0,1)$, we need to verify the Lindeberg condition (6.36) (with $i_n = n$ and $X$ replaced by $\xi$) or, equivalently,

$$\frac{1}{n}\sum_{i=1}^{n}\mathrm{E}\left[\left\{\sum_{j=1}^{s}\lambda_j m[j]g_1^{(j)}(X_i)\right\}^2 1_{(|\xi_{ni}|>\epsilon)}\right]\longrightarrow 0 \qquad (11.74)$$

for every $\epsilon>0$. The left side of (11.74) is equal to

$$\mathrm{E}\left[\left\{\sum_{j=1}^{s}\lambda_j m[j]q_1^{(j)}(X_1)\right\}^2 1_{(|\xi_{n1}|>\epsilon)}\right]\le s\sum_{j=1}^{s}\mathrm{F}\{\eta_j^2 1_{(|\xi_{n1}|>\epsilon)}\},$$

where $\eta_j=\lambda_j m[j]g_1^{(j)}(X_1)$, using, for example, the Cauchy–Schwarz inequality. Furthermore, it is easy to see that $|\xi_{n1}|>\epsilon$ implies $|\eta_k|>\epsilon\sqrt{n}/s$ for some $1\le k\le s$; hence, $1_{(|\xi_{n1}|>\epsilon)}\le\sum_{k=1}^{s}1_{(|\eta_k|>\epsilon\sqrt{n}/s)}$. It follows that

$$\text{the left side of (11.74)} \le s\sum_{j,k=1}^{s}\mathrm{E}\{\eta_j^2 1_{(|\eta_k|>\epsilon\sqrt{n}/s)}\}.$$

Now, use the inequality

$$x^2 1_{(y>a)}\le x^2 1_{(x>a)}+y^2 1_{(y>a)}, \qquad (11.75)$$

which holds for any (nonnegative) numbers $x$, $y$, and $a$ (Exercise 11.25), to conclude that the left side of (11.74) is bounded by

$$s\sum_{j,k=1}^{s}\left[\mathrm{E}\{\eta_j^2 1_{(|\eta_j|>\epsilon\sqrt{n}/s)}\}+\mathrm{E}\{\eta_k^2 1_{(|\eta_k|>\epsilon\sqrt{n}/s)}\}\right]$$

$$=2s^2\sum_{j=1}^{s}\mathrm{E}\{\eta_j^2 1_{(|\eta_j|>\epsilon\sqrt{n}/s)}\}$$

$$=2s^2\sum_{\lambda_j\ne 0}\lambda_j^2 m^2[j]\mathrm{E}\left[\{g_1^{(j)}(X_1)\}^2 1_{(|g_1^{(j)}(X_1)|>\epsilon\sqrt{n}/s|\lambda_j|m[j])}\right],$$

which converges to zero by assumption (iii).

Now, let us revisit the problem of testing for the center of symmetry discussed in Sections 11.2 and 11.3. For simplicity, suppose that the underlying distribution of $X_1,\ldots,X_n$ is $F(x-\theta_n)$, where $F$ is a continuous cdf with a finite first moment, and $\theta_n$ has the expression (11.35) with $\theta_0=0$. Let $\phi^{(1)}(x)=x$, $\phi^{(2)}(x,y)=1_{(x+y>0)}$, and $\phi^{(3)}(x)=1_{(x>0)}$. Then we have

$$U^{(1)}=n^{-1}\sum_{i=1}^{n}\phi^{(1)}(X_i)=\bar{X},$$

$$U^{(2)}=\binom{n}{2}^{-1}\sum_{1\le i<j\le n}\phi^{(2)}(X_i,X_j)$$

$$= \binom{n}{2}^{-1} \sum_{1 \le i < j \le n} 1_{(X_i + X_j > 0)},$$

$$U^{(3)} = n^{-1} \sum_{i=1}^{n} \phi^{(3)}(X_i) = n^{-1} \sum_{i=1}^{n} 1_{(X_i > 0)}.$$

Furthermore, we have $\theta_1 = \mathrm{E}\{\phi^{(1)}(X_1)\} = \theta_n$ (excuse us for a light abuse of the notation), $\theta_2 = \mathrm{E}\{\phi^{(2)}(X_1, X_2)\} = \mathrm{E}\{F(Z_1 + 2\theta_n)\}$, where $Z_1$ has the cdf $F$, and $\theta_3 = F(\theta_n)$. Let us verify (11.70) for $j = 1$ and $k = 2$ and leave the rest of the verifications to an exercise (Exercise 11.26). We have $\mathrm{cov}\{\phi^{(1)}(X_1), \phi^{(2)}(X_1)\} = \mathrm{cov}\{X_1, F(X_1 + \theta_n)\} = \mathrm{E}\{Z_1 F(Z_1 + 2\theta_n)\}$. Since $F$ is continuous, we see by dominated convergence theorem (Theorem 2.16) that the covariance converges to $\mathrm{E}(Z_1 F(Z_1))\}$ as $n \to \infty$. Thus, assumption (ii) holds with $\Sigma$ being the covariance matrix with $\theta_n = 0$. Next, we verify (11.71) for $j = 2$ and leave the rest to an exercise (Exercise 11.26). Note that $g_1^{(2)}(X_1) = F(X_1 + \theta_n) - \mathrm{E}\{F(Z_1 + 2\theta_n)\}$, which is bounded in absolute value by 1. Thus, the left side of (11.71) (with $j = 2$) is zero for sufficiently large $n$. It follows that assumption (iii) holds. Therefore, our earlier claims (in Sections 11.2 and 11.3) regarding the (joint) asymptotic distributions of these statistics are justified.

We conclude this section with a brief discussion on two-sample $U$-statistics. Let $X_1, \ldots, X_m$ and $Y_1, \ldots, Y_n$ be independent samples from $F$ and $G$, respectively. A two-sample U-statistic with kernel $\phi$ is defined as

$$U = \binom{m}{a}^{-1} \binom{n}{b}^{-1} \sum \phi(X_{i_1}, \ldots, X_{i_a}, Y_{j_1}, \ldots, Y_{j_b}), \tag{11.76}$$

where the summation is over all indexes $1 \le i_1 < \cdots < i_a \le m$ and $1 \le j_1 < \cdots < j_b \le n$. Similar to (11.67) and (11.68), we have

$$\mathrm{var}(U) = \binom{m}{a}^{-1} \binom{n}{b}^{-1} \sum_{c=1}^{a} \sum_{d=1}^{b} \binom{a}{c} \binom{m-a}{a-c} \binom{b}{d} \binom{n-b}{b-d} \sigma_{cd}, \tag{11.77}$$

where $\sigma_{cd}$ is the covariance between $\phi(X_1, \ldots, X_a, Y_1, \ldots, Y_b)$ and

$$\phi(X_1, \ldots, X_c, X'_{c+1}, \ldots, X'_a, Y_1, \ldots, Y_d, Y'_{d+1}, \ldots, Y'_b),$$

in which the $X$'s, $X'$s and $Y$'s, $Y'$s are independently distributed as $F$ and $G$, respectively. Here, we assume that all of the $\sigma_{cd}$ are finite, which is equivalent to $\sigma_{ab} < \infty$ (why?). In particular, if $m, n \to \infty$ such that $m/N \to \rho$ and $n/N \to 1 - \rho$ for some $\rho \in [0, 1]$, where $N = m + n$, then

$$N \, \mathrm{var}(U) \longrightarrow \frac{a^2}{\rho} \sigma_{10} + \frac{b^2}{1-\rho} \sigma_{01}. \tag{11.78}$$

For simplicity, we now focus on the special case $a = b = 1$. For a general treatment of the subject, see, for example, Koroljuk and Borovskich (1994).

The following theorem states the asymptotic distribution of the two-sample $U$-statistic when $\sigma_{10}$ and $\sigma_{01}$ are positive. Again, for simplicity, we assume that $F$ and $G$ do not depend on $m$ and $n$; extension to the case where the distributions depend on the sample sizes can be made along the same lines as for the one-sample case (see Theorem 11.3).

**Theorem 11.4.** If $\sigma_{11} < \infty$ and $\sigma_{01}, \sigma_{10} > 0$, then

$$\frac{U - \theta}{\sqrt{\mathrm{var}(U)}} \xrightarrow{\ \mathrm{d}\ } N(0, 1), \tag{11.79}$$

where $\theta = \mathrm{E}\{\phi(X_1, Y_1)\}$.

We give an outline of the proof and leave the details to an exercise (Exercise 11.27; also see Lehmann 1999, pp. 378–380). The basic idea is similar to the one-sample case. First, consider a special case in which the limits above (11.78) hold for some $\rho \in [0, 1]$. We find a first-order approximation as follows. Write $\zeta_N = \sqrt{N}(U - \theta)$ and

$$\zeta_N^* = \sqrt{\frac{N}{m}} \frac{1}{\sqrt{m}} \sum_{i=1}^{m} \{\phi_{10}(X_i) - \theta\} + \sqrt{\frac{N}{n}} \frac{1}{\sqrt{n}} \sum_{j=1}^{n} \{\phi_{01}(Y_j) - \theta\}$$

$$= \sqrt{\frac{N}{m}} \eta_{N,1} + \sqrt{\frac{N}{n}} \eta_{N,2},$$

where $\phi_{10}(x) = \mathrm{E}\{\phi(x, Y)\}$ and $\phi_{01}(y) = \mathrm{E}\{\phi(X, y)\}$. It can be shown that both $\mathrm{var}(\zeta_N^*)$ and $\mathrm{cov}(\zeta_N, \zeta_N^*)$ converge to the right side of (11.78) (with $a = b = 1$), so that $\mathrm{E}(\zeta_N - \zeta_N^*)^2 = \mathrm{var}(\zeta_N) - 2\mathrm{cov}(\zeta_N, \zeta_N^*) + \mathrm{var}(\zeta_N^*) \to 0$. On the other hand, by CLT for the sum of independent random variables, we have $\eta_{N,1} \xrightarrow{\mathrm{d}} N(0, \sigma_{10})$, $\eta_{N,2} \xrightarrow{\mathrm{d}} N(0, \sigma_{01})$, and $\zeta_{N,1}$ and $\zeta_{N,2}$ are independent. It then follows, by (11.78) (with $a = b = 1$), that (11.79) holds under the limiting process $m/N \to \rho \in [0, 1]$. However, note that the limiting distribution does not depend on $\rho$. That (11.79) holds without this restriction follows by the argument of subsequences (see §1.5.1.6).

As a special case, consider the two-sample Wilcoxon test, which is closely related to the statistic $W_{XY}$ in (11.38). Here, we have

$$U = \frac{1}{mn} \sum_{i=1}^{m} \sum_{j=1}^{n} \phi(X_i, Y_j),$$

where $\phi(x, y) = 1_{(x < y)}$. It is easy to verify that

$$\sigma_{10} = \mathrm{E}\{G^2(X_1)\} - [\mathrm{E}\{G(X_1)\}]^2,$$
$$\sigma_{01} = \mathrm{E}\{F^2(Y_1)\} - [\mathrm{E}\{F(Y_1)\}]^2, \tag{11.80}$$

where $X_1$ and $Y_1$ are independent and distributed as $F$ and $G$, respectively. Since both $F$ and $G$ are continuous, $\sigma_{10}$ and $\sigma_{01}$ are positive. Furthermore, it is obvious that $\sigma_{11} < \infty$. Thus, the assumptions of Theorem 11.4 are satisfied. In particular, under the null hypothesis $F = G$, we have $\sigma_{10} = \sigma_{01} = 1/12$ (Exercise 11.28).

## 11.6 Density estimation

In a way, nonparametric estimation problems are extensions of parametric estimation problems, but the nature of the former is quite different from the latter. Consider, for example, the situation of i.i.d. observations, say, $X_1, \ldots, X_n$. In a parametric problem we assume that the distribution of $X_i$ is $F_\theta$, which is fully specified up to the parameter (vector) $\theta$; so the problem is essentially the estimation of $\theta$. In a nonparametric problem, the distribution is entirely unknown (with, perhaps, some restrictions on general properties; see below) and therefore is denoted by $F$. In Chapter 7 we considered estimation of $F$ in terms of its cdf. In this section, we consider the estimation of $F$ in terms of its pdf, $f$. The pdf has the advantage of providing a visually more informative representation of the underlying distribution. For example, the histogram often gives a rough idea about the shape of the distribution. In fact, according to Scott (1992, p. 125), the latter "stood as the only nonparametric density estimator until the 1950s." For such a reason, our discussion will begin with the histograms.

Although the histograms are extensively used, it is not that often that a mathematical definition is needed. One way to define it is via the empirical d.f. Note that $f$ is the derivative of $F$; so it can be expressed as $f(x) = \lim_{h \to 0} h^{-1}\{F(x+h) - F(x)\}$ or

$$f(x) = \lim_{h \to 0} \frac{F(x+h) - F(x-h)}{2h}. \tag{11.81}$$

The latter expression has the advantage of faster convergence. In fact, if $F$ is twice continuously differentiable at $x$, then we have

$$\frac{F(x+h) - F(x)}{h} - f(x) = O(h),$$

$$\frac{F(x+h) - F(x-h)}{2h} - f(x) = o(h) \tag{11.82}$$

(Exercise 11.29). Expression (11.81) also appears to be more "fair," or "balanced" than the previous expression. Because the empirical d.f., $\hat{F}$, is an estimator of $F$, it is natural to consider (11.81) with $F$ replaced by $\hat{F}$. However, one cannot do so because then this limit is either zero or $\infty$ (Exercise 11.29). So at some point one has to stop; in other words, $h$ cannot get too close to zero. If the latter is fixed, it is called the *bin width*, or *bandwidth*. We can then write the estimator of $f$ as

$$\hat{f}(x) = \frac{\hat{F}(x+h) - \hat{F}(x-h)}{2h}$$

$$= \frac{1}{2nh} \sum_{i=1}^{n} 1_{(x-h < X_i < x+h)} \text{ (with probability 1)} \qquad (11.83)$$

(Exercise 11.29). Note that the summation in (11.83) has a Binomial$(n, p)$ distribution with $p = F(x+h) - F(x-h)$. Thus, the (asymptotic) behavior of the histogram can be derived from that of the Binomial distribution. For example, we have

$$\mathrm{E}\{\hat{f}(x)\} = \frac{F(x+h) - F(x-h)}{2h},$$

$$\mathrm{var}\{\hat{f}(x)\} = \frac{p(1-p)}{4nh^2}.$$

It follows that $\hat{f}(x)$ is a pointwise consistent estimator of $f(x)$ if

$$h \to 0 \quad \text{and} \quad nh \to \infty \qquad (11.84)$$

(not $nh^2 \to \infty$; see Exercise 11.29). Hereafter the limiting process is understood as $h = h_n$ such that $h_n \to 0$ and $nh_n \to \infty$, but for notation simplicity, the subscript of $h_n$ is often suppressed. The condition may be interpreted as that $h$ needs to go to zero, but not too fast. This is exactly what we have speculated [below (11.82)] except that now we have the exact rate of convergence, which can be written as $h^{-1} = o(n)$.

Although the histogram is consistent under (11.84), it turns out that one can do better. The improvement is also motivated by a practical concern that the histogram is not smooth, a property that one may expect the true density function to have. This leads to the kernel estimator, defined by

$$\hat{f}(x) = \frac{1}{nh} \sum_{i=1}^{n} K\left(\frac{x - X_i}{h}\right), \qquad (11.85)$$

where $K(\cdot)$ is a function known as the *kernel*. It is typically assumed that $K$ is nonnegative, symmetric about zero, and satisfies $\int K(u)\, du = 1$. It is clear that the histogram is a special case of the kernel estimator if $K$ is chosen as the pdf of Uniform$(-1, 1)$. The latter is not a smooth function, and this is why the histogram is not smooth; but by choosing $K$ as a smooth function, one has an estimator of $f(x)$ that is smooth. For example, the pdf of $N(0, 1)$ is often used (known as the *Gaussian* kernel) and so is the symmetric Beta pdf,

$$K(u) = \frac{\Gamma(\nu + 3/2)}{\Gamma(1/2)\Gamma(\nu + 1)}(1 - u^2)^\nu, \quad -1 < u < 1,$$

and $K(u) = 0$ elsewhere. The special cases $\nu = 0, 1, 2, 3$ correspond to the *uniform*, *Epanechnikov*, *biweight*, and *triweight* kernels, respectively. An important practical problem in kernel density estimation is how to choose the bandwidth $h$. Note that given conditions such as (11.84), there are still plenty of choices for $h$; so, in a way, the order of convergence (or divergence) does not solve the problem. A solution to this problem is known as the bias–variance trade-off. Before we go into the details let us first state a result regarding the asymptotic bias of the kernel estimator. Here, the bias is defined as bias$\{\hat{f}(x)\} = E\{\hat{f}(x)\} - f(x)$ for a given $x$.

**Theorem 11.5.** Suppose that $f$ is continuous and bounded. Then the bias of the kernel estimator goes to zero as $h \to 0$ for every $x$.

The argument that leads to the conclusion of Theorem 11.5 is simple. First, write (verify this)

$$E\{\hat{f}(x)\} = \frac{1}{n}\sum_{i=1}^{n}\frac{1}{h}\int K\left(\frac{x-y}{h}\right)f(y)\,dy$$

$$= \int K(u)f(x-hu)\,du$$

$$= f(x) + \int K(u)\{f(x-hu) - f(x)\}\,du.$$

Then use the dominated convergence theorem (Theorem 2.16) to complete the argument. A further investigation into the bias as well as the variance leads to the following theorem, drawn from Lehmann (1999, p. 410).

**Theorem 11.6.** Suppose that $f$ is three times differentiable with a bounded third derivative in a neighborhood of $x$ and that $K$ satisfies

$$\int K^2(u)\,du < \infty \text{ and } \int |u|^3 K(u)\,du < \infty.$$

(i) If $h \to 0$ as $n \to \infty$, then we have

$$\text{bias}\{\hat{f}(x)\} = \frac{h^2}{2}f''(x)\int u^2 K(u)\,du + o(h^2). \tag{11.86}$$

(ii) If, in addition, $nh \to \infty$ as $n \to \infty$, then we have

$$\text{var}\{\hat{f}(x)\} = \frac{f(x)}{nh}\int K^2(u)\,du + o\{(nh)^{-1}\}. \tag{11.87}$$

The proof is based on the Taylor expansion—namely,

$$f(x-hu) = f(x) - huf'(x) + \frac{h^2 u^2}{2}f''(x) - \frac{h^3 u^3}{6}f'''(\xi),$$

where $\xi$ lies between $x - hu$ and $x$. The details are left as an exercise (Exercise 11.30). A measure of accuracy in estimation is the mean squared error (MSE):

$$\text{MSE}\{\hat{f}(x)\} = \text{E}\{\hat{f}(x) - f(x)\}^2.$$

It is easy to show (Exercise 11.31) that the MSE combines the bias and variance in such a way that

$$\text{MSE}\{\hat{f}(x)\} = [\text{bias}\{\hat{f}(x)\}]^2 + \text{var}\{\hat{f}(x)\}. \tag{11.88}$$

By (11.88), (11.86), and (11.87), we see that, under the condition (11.84) and if we ignore the lower-order terms, we have

$$\text{MSE}\{\hat{f}(x)\} \approx \frac{h^4}{4}\{f''(x)\}^2\tau^4 + \frac{f(x)}{nh}\gamma^2, \tag{11.89}$$

where $\tau^2 = \int u^2 K(u)\, du$ and $\gamma^2 = \int K^2(u)\, du$. The right side of (11.89) is minimized when

$$h = \left[\frac{\gamma^2 f(x)}{\tau^4\{f''(x)\}^2}\right]^{1/5} n^{-1/5} \tag{11.90}$$

(Exercise 11.31). Note that (11.90) is not yet the optimal solution because $f$ is unknown in practice. However, it gives us at least some idea about the optimal rate at which $h \to 0$. The optimal rate is $O(n^{-1/5})$, which is much more specific than (11.84).

When $f$ is unknown, a natural approach would be to replace it by an estimator and hence obtain an estimated optimal bandwidth. One complication is that the optimal bandwidth depends on $x$, but, ideally, one would like to use a bandwidth that works for different $x$'s within a certain interval, if not all of the $x$'s. To obtain an optimal bandwidth that does not depend on $x$, we integrate both sides of (1.89) with respect to $x$. This leads to

$$\int \text{MSE}\{\hat{f}(x)\}\, dx \approx \frac{\tau^4 h^4}{4}\int\{f''(x)\}^2\, dx + \frac{\gamma^2}{nh}\int f(x)\, dx$$

$$= \frac{\tau^4\theta^2 h^4}{4} + \frac{\gamma^2}{nh} \tag{11.91}$$

with $\theta^2 = \int\{f''(x)\}^2\, dx$. By the same argument, the right side of (11.91) is minimized when

$$h = \left(\frac{\gamma^2}{\tau^4\theta^2}\right)^{1/5} n^{-1/5}. \tag{11.92}$$

This time, the optimal $h$ does not depend on $x$. Furthermore, the minimum integrated MSE (IMSE) is given by (verify this)

$$\text{IMSE} = \frac{5}{4}\left(\tau\gamma^2\right)^{4/5}\theta^{2/5}n^{-4/5}. \tag{11.93}$$

An implication of (11.92) and (11.93) is the following. Note that the IMSE depends on the kernel $K$ through $c_K = (\tau\gamma^2)^{4/5}$. It has been shown (e.g., Fan and Yao 2003, Table 5.1) that for the commonly used kernels such as those listed below (11.85), the performance of the corresponding kernel estimators are nearly the same in terms of the values of $c_K$. On the other hand, the optimal bandwidths, $h_1$ and $h_2$, corresponding to two different kernels, $K_1$ and $K_2$, satisfy $h_1/h_2 = \kappa_1/\kappa_2$, where $\kappa = (\gamma^2/\tau^4)^{1/5}$ and the subscript $j$ corresponds to $K_j$, $j = 1, 2$. This means that one can adjust the (optimal) bandwidth $h_2$ such that $h_2 = (\kappa_2/\kappa_1)h_1$ so that the kernel $K_2$ using the bandwidth $h_2$ performs nearly the same as the kernel $K_1$ using the bandwidth $h_1$. This is the idea behind the *canonical kernel* (Marron and Nolan 1988).

Going back to the problem on the estimation of the optimal bandwidth, from (11.92), we see that all we need is to find an (consistent) estimator of $\theta^2$. If $f$ is the pdf of a normal distribution with standard deviation $\sigma$, then it can be shown that $\theta^2 = 3/8\sqrt{\pi}\sigma^5$ (Exercise 11.32). Of course, if one knows $f$ is normal, then nonparametric density estimation would not be necessary (because a parametric method would probably do better). In general, one may expand $f$ around the Gaussian density using the Edgeworth expansion (see Section 4.3). Using this approach, Hjort and Jones (1996) obtained the following estimator of the optimal bandwidth:

$$\hat{h} = \hat{h}_0 \left(1 + \frac{35}{48}\hat{\gamma}_4 + \frac{35}{32}\hat{\gamma}_3^2 + \frac{385}{1024}\hat{\gamma}_4^2\right)^{-1/5}, \tag{11.94}$$

where $\hat{h}_0$ is the estimated optimal bandwidth assuming that $f$ is normal—that is, (11.92) with $\theta^2$ replaced by $3/8\sqrt{\pi}\hat{\sigma}^5$, or, more explicitly,

$$\hat{h}_0 = 1.06 \left(\frac{\hat{\sigma}}{n^{1/5}}\right) \tag{11.95}$$

(we call $\hat{h}_0$ the baseline bandwidth), and $\hat{\sigma}^2$ is the sample variance given by

$$\hat{\sigma}^2 = \frac{1}{n-1}\sum_{i=1}^{n}(X_i - \bar{X})^2.$$

Furthermore, $\hat{\gamma}_3$ and $\hat{\gamma}_4$ are the sample skewness and kurtosis given by

$$\hat{\gamma}_3 = \frac{1}{(n-1)\hat{\sigma}^3}\sum_{i=1}^{n}(X_i - \bar{X})^3,$$

$$\hat{\gamma}_4 = \frac{1}{(n-1)\hat{\sigma}^4}\sum_{i=1}^{n}(X_i - \bar{X})^4 - 3,$$

respectively. There have been other approaches for bandwidth selection, including the *cross-validation* method and plug-in method. The latter estimates $\theta^2$ as a functional. See, for example, Jones et al. (1996) for an overview. We conclude this section with a numerical example.

*Example 11.11* (A numerical example). We generate $X_1, \ldots, X_n$, where $n = 30$, from a double exponential distribution with mean 0 and standard deviation 1 [i.e., $DE(0, 1)$]. The generated data are given by Table 11.3 (up to the third digit). Now, suppose that the true density function is unknown

**Table 11.3. Data generated by a computer**

| | | | | | |
|---|---|---|---|---|---|
| 1.814 | 5.056 | 2.434 | 0.113 | −0.822 | 0.531 |
| −0.784 | 0.098 | −3.063 | 1.558 | 0.665 | 2.235 |
| 1.612 | −0.426 | 0.092 | −1.661 | −0.925 | 0.744 |
| 0.714 | −2.864 | −0.829 | −1.309 | −0.408 | 0.558 |
| −1.228 | 0.381 | 0.241 | 1.030 | 0.417 | 1.366 |

and one has to estimate it nonparametrically using the kernel method. We use the Gaussian kernel with $k(u) = e^{-u^2/2}/\sqrt{2\pi}$. The sample variance is $\hat{\sigma}^2 = 2.59$; the baseline bandwidth is computed by (11.95) as $\hat{h}_0 = 0.86$; and the sample skewness and kurtosis are computed as $\hat{\gamma}_3 = 0.45$ and $\hat{\gamma}_4 = 1.27$, respectively. These lead to the estimated optimal bandwidth, computed by (11.94), as $\hat{h} = 0.71$. A plot of the kernel estimate of the density is shown in Figure 11.1. The true density is also plotted in dash line for comparison. It appears that the kernel estimate is missing some of the height in the middle. However, the sample size is $n = 30$, which is not very large. What happens when $n$ increases? The reader is encouraged to explore this in an exercise (Exercise 11.33).

## 11.7 Exercises

*11.1.* Verify (11.6)–(11.8).

*11.2.* Verify (11.13)–(11.15).

*11.3.* Verify (11.6).

*11.4.* Show that the asymptotic correlation coefficient between $S_2$ and $S_3$ in (11.18), which correspond to the test statistics of the signed-rank and sign tests, is equal to $\sqrt{3}/2$.

*11.5.* This exercise is to show that both sides of inequality (11.19) are sharp in that there are distributions $F$ that are continuous and symmetric about zero for which the left- or right-side equalities are either attained or approached with arbitrary closeness.

(i) Let $F$ has a rectangular distribution that is symmetric about zero. Show that $\rho_w = 1$; hence, the left side equalilty holds.

(ii) Let $F$ have the pdf

$$f_p(x) = \begin{cases} 0, & |x| \leq 1 \\ \{(p-1)/2\}|x|^{-p}, & |x| > 1, \end{cases}$$

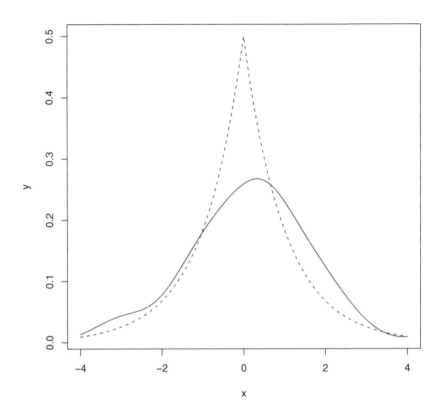

**Fig. 11.1.** *Numerical example: solid line: kernel estimate; dash line: true density*

where $p > 3$. Verify that $f_p$ is a pdf. Then show that $\rho_\mathrm{w}, \rho_\mathrm{s} \to 0$ as $p \to 3$. Therefore, the right side of (11.19) can be approached with arbitrary closeness by choosing $p$ sufficiently close to 3.

*11.6.* Show that (11.22) holds as $n \to \infty$ if $\theta$ is the true center of symmetry.

*11.7.* This exercise has several parts.

(i) Suppose that $X$ has a continuous distribution $F$. Show that $F(X)$ has the Uniform[0, 1] distribution [Hint: Use (7.4) and the facts that $F(x) \geq u$ if and only if $x \geq F^{-1}(u)$ and that $F\{F^{-1}(u)\} = u$.]

(ii) Show that when $\theta = 0$, (11.23) and (11.24) reduce to 1/2 and 1/3, respectively.

(iii) Show that when $\theta = 0$, (11.15) is equivalent to (11.21) with $\mu(\theta)$ and $\tau(\theta)$ given by (11.23) and (11.24).

*11.8.* Show that for the problem of testing for the center of symmetry discussed in Sections 11.2 and 11.3, the efficacies of the $t$, sign, and Wilcoxon signed-rank tests are given by $1/\sigma$, $2f(0)$, and $2\sqrt{3}\int f^2(z)\,dz$, respectively, where $\sigma$ and $f$ are the standard deviation and pdf of $F$, respectively.

*11.9.* Evaluate the AREs (11.30)–(11.32) when $F$ is the following distribution:

(i) Double Exponential with pdf $f(z) = (1/2\sigma)e^{-|x|/\sigma}$, $-\infty < x < \infty$, where $\sigma > 0$.

(ii) Logistic with pdf $f(z) = (1/\beta)e^{-z/\beta}/(1 + e^{-z/\beta})^2$, $-\infty < x < \infty$, where $\beta > 0$.

(iii) Uniform$[-a, a]$, where $a > 0$.

*11.10.* Verify that the ARE $e_{S,t}$ of (11.30) is given by (11.33) in the case of Example 11.3.

*11.11.* Verify that the ARE $e_{W,t}$ of (11.31) is given by (11.34) in the case of Example 11.4.

*11.12.* In the case of testing for the center of symmetry, suppose that $X_{ni}, 1 \le i \le n$, are independent observations with the cdf $F(x - \theta_n)$. Then for the $t$-test, we have $T_n = \bar{X}_n = n^{-1}\sum_{i=1}^n X_{ni}$ and

$$\frac{\sqrt{n}(T_n - \theta_n)}{\sigma} = \sum_{i=1}^n Y_{ni},$$

where $Y_{ni} = (X_{ni} - \theta_n)/\sigma\sqrt{n}$. Show that $Y_{ni}, 1 \le i \le n, n \ge 1$, satisfy the Lindeberg condition (6.36) with $X_{ni}$ replaced by $Y_{ni}$, $i_n = n$ and $s_n^2 = 1$.

*11.13.* Continuing with the previous exercise. For the sign test, we have $T_n = n^{-1}\sum_{i=1}^n 1_{(X_{ni}>0)}$ and

$$2\sqrt{n}\{T_n - F(\theta_n)\} = \sum_{i=1}^n Y_{ni},$$

where $Y_{ni} = (2/\sqrt{n})\{1_{(X_{ni}>0)} - F(\theta_n)\}$. Once again, verify the conditions below (6.35) with $X_{ni}$ replaced by $Y_{ni}$.

*11.14.* Consider the pooled sample variance, $S_p^2$, of Example 11.5.

(i) Show that $S_p^2 \xrightarrow{P} \sigma^2$ as $m, n \to \infty$ such that $m/N \to \rho \in (0,1)$, where $N = m + n$ and $\sigma^2$ is the variance of $F$.

(ii) Show that the assumption of (i) remains valid even if $\rho = 0$ or 1.

(iii) Show that the conclusion of (i) remains valid as $m, n \to$ without any restriction [Hint: Suppose otherwise. Then there is an $\epsilon > 0$ and a sequence $(m_k, n_k), k = 1, 2, \ldots$, such that $|S_p^{-1} - \sigma^2| \ge \epsilon$ for $(m, n) = (m_k, n_k), k = 1, 2, \ldots$. Without loss of generality, one may assume that $m_k/N_k \to \rho \in [0,1]$ (otherwise choose a subsequence that has this property (using §1.5.1.4).]

*11.15.* (i) Show that (11.37) holds under the limiting process of (i) of the previous exercise, provided that $\theta_N$ is the true $\theta$ for $T_N$. You may use a similar argument as in Example 11.4 and the following fact: If $\xi_N$ and $\zeta_N$ are two sequences of random variables such that $\xi_N \xrightarrow{d} \xi$, $\eta_N \xrightarrow{d} \eta$, and $\xi_N$ and $\eta_N$ are independent for each $N$, then $\xi_N \pm \eta_N \xrightarrow{d} \xi \pm \eta$.

(ii) Based on (11.37), derive the asymptotic power of the two-sample $t$-test and show that it is equal to $1 - \Phi(z_\alpha - c_t\Delta)$, where $c_t = \sqrt{\rho(1 - \rho)}/\sigma$ and $\Phi$ is the cdf of $N(0, 1)$.

*11.16.* Continue with the previous exercise.

(i) Verify the identity (11.38).

(ii) Given that (11.36) holds with $T_N = W_{XY}/mn$, $\mu(\theta) = P_\theta(X < Y) = \int\{1 - F(x - \theta)\}f(x)\,dx$, and $\tau(0) = 1/\sqrt{12\rho(1 - \rho)}$, derive the asymptotic power of the two-sample Wilcoxon test and show that it is equal to $1 - \Phi(z_\alpha - c_W\Delta)$, where $c_W = \sqrt{12\rho(1 - \rho)}\int f^2(z)\,dz$.

*11.17.* Verify (11.42) and thus, in particular, (11.43) and (11.44) under the null hypothesis (11.39).

*11.18.* Show that the functional $h$ defined below (11.45) is continuous on $(D, \|\cdot\|)$.

*11.19.* Verify the identity (11.50).

*11.20.* Verify (11.53) and also show that $\theta(F) = \text{var}(X_1)$ for the sample variance.

*11.21.* Verify the property of complete degeneracy (11.56).

*11.22.* This exercise is concerned with moment properties of $S_{nc}$, $1 \le c \le m$, that are involved in the Hoeffding representation (11.57).

(i) Show that $E(S_{nc}) = 0$, $1 \le c \le m$.

(ii) Show that

$$E\{g_c(X_{i_1}, \ldots, X_{i_c})g_d(X_{j_1}, \ldots, X_{j_d})\} = 0$$

except that $c = d$ and $\{i_1, \ldots, i_c\} = \{j_1, \ldots, j_d\}$.

(iii) Verify the orthogonality property (11.59).

*11.23.* Verify the following.

(i) The martingale property (11.61).

(ii) The expression (11.57), considered as a sequence of random variables, $U_n$, $\mathcal{F}_n = \sigma(X_1, \ldots, X_n)$, $n \ge m$, is a martingale.

(iii) The expression (11.62) and that $\xi_{nk}, \mathcal{F}_k, k \ge 1$, is a sequence of martingale differences.

(iv) An alternative expression for $\xi_{nk}$,

$$\xi_{nk} = E(U|X_1, \ldots, X_k) - E(U|X_1, \ldots, X_{k-1}).$$

*11.24.* Show that (11.68) holds as $n \to \infty$.

*11.25.* Verify the numerical inequality (11.75) for any $x, y, a \ge 0$.

*11.26.* Consider once again the problem of testing for the center of symmetry. More specifically, refer to the continuing discussion near the end of Section 11.5.

(i) Verify (11.70) for $1 \le j \le k \le 3$ except for $j = 1$ and $k = 2$, which has been verified.

(ii) Verify (11.71) for $j = 1$ and $j = 3$.

*11.27.* This exercise involves some details regarding the proof of Theorem 11.4 at the end of Section 11.5.

(i) Show that both $\text{var}(\zeta_N^*)$ and $\text{cov}(\zeta_N, \zeta_N^*)$ converge to the right side of (11.78) with $a = b = 1$, provided that $m/N \to \rho \in [0, 1]$.

(ii) Show that $\eta_{N,1} \xrightarrow{d} N(0, \sigma_{10})$, $\eta_{N,2} \xrightarrow{d} N(0, \sigma_{01})$.

(iii) Combine the results of (i) and (ii) to show that (11.79) holds under the limiting process in (i).

(iv) Using the argument of subsequences (see §1.5.1.6), show that (11.79) holds without the restriction on the limiting process.

*11.28.* Consider the $U$-statistic associated with Wilcoxon two-sample test (see the discussion at the end of Section 11.5).

(i) Verify (11.80).

(ii) Show that under the null hypothesis $F = G$, we have $\sigma_{10} = \sigma_{01} = 1/12$.

*11.29.* This exercise is related to the expression of the histogram (see Section 11.6).

(i) Show that (11.82) holds provided that $F$ is twice continuously differentiable.

(ii) Show that the limit (11.81) is either zero or $\infty$ if $F$ is replaced by $\hat{F}$, the empirical d.f.

(iii) Show that the histogram is equal to the right side of (11.83) with probability 1.

(iv) Show that the histogram is pointwise consistent under the limiting process (11.84).

*11.30.* Give a proof of Theorem 11.6. As mentioned, the proof is based on the Taylor expansion. The details can be found in Lehmann's book but you are encouraged to explore without looking at the book (or check with it after you have done it independently).

*11.31.* (i) Verify (11.88).

(ii) Show that the right side of (11.89) is minimized when $h$ is given by (11.90).

*11.32.* Regarding the parameter $\theta^2$ defined below (11.91), show that $\theta^2 = 3/8\sqrt{\pi}\sigma^5$ if $f$ is the pdf of $N(\mu, \sigma^2)$.

*11.33.* This exercise is related to Example 11.11 at the end of the chapter.

(i) Verify the calculations of $\hat{\sigma}^2$, $\hat{h}_0$, $\hat{\gamma}_3$, $\hat{\gamma}_4$, and $\hat{h}$ in the example.

(ii) Simulate a larger data set, say, with $n = 100$, and repeat the calculations and plots in the example. Does the estimated density better approximate the true density? Note that a $DE(0, 1)$ random variable, $X$, can be generated by first generating $X_1$ and $X_2$ independently from the Exponential(1) distribution and then letting $X = X_1 - X_2$.

# 12

# Mixed Effects Models

## 12.1 Introduction

Mixed effects models, or simply mixed models, are widely used in practice. These models are characterized by the involvement of the so-called random effects. To understand the basic elements of a mixed model, let us first recall a linear regression model, which can be expressed as $y = X\beta + \epsilon$, where $y$ is a vector of observations, $X$ is a matrix of known covariates, $\beta$ is a vector of unknown regression coefficients, and $\epsilon$ is a vector of (unobservable random) errors. In this model, the regression coefficients are considered fixed. However, there are cases in which it makes sense to assume that some of these coefficients are random. These cases typically occur when the observations are correlated. For example, in medical studies, observations are often collected from the same individuals over time. It may be reasonable to assume that correlations exist among the observations from the same individual, especially if the times at which the observations are collected are relatively close. In animal breeding, lactation yields of dairy cows associated with the same sire may be correlated. In educational research, test scores of the same student may be related. Now, let us see how a linear mixed model may be useful for modeling the correlations among the observations.

*Example 12.1.* Consider, for example, the above example of medical studies. Assume that each individual is associated with a random effect whose value is unobservable. Let $y_{ij}$ denote the observation from the $i$ individual collected at time $t_j$ and let $\alpha_i$ be the random effect associated with the $i$th individual. Assume that there are $m$ individuals. For simplicity, let us assume that the observations from all individuals are collected at a common set of times, say, $t_1, \ldots, t_k$. Then, a linear mixed model may be expressed as $y_{ij} = x_{ij}'\beta + \alpha_i + \epsilon_{ij}$, $i = 1, \ldots, m$, $j = 1, \ldots, k$, where $x_{ij}$ is a vector of known covariates; $\beta$ is a vector of unknown regression coefficients; the random effects $\alpha_1, \ldots, \alpha_m$ are assumed to be i.i.d. with mean 0 and variance $\sigma^2$; the $\epsilon_{ij}$'s are errors which are i.i.d. with mean 0 and variance $\tau^2$; and the

J. Jiang, *Large Sample Techniques for Statistics,*
DOI 10.1007/978-1-4419-6827-2_12, © Springer Science+Business Media, LLC 2010

random effects and errors are independent. It is easy to show (Exercise 12.1) that the correlation between any two observations from the same individual is $\sigma^2/(\sigma^2 + \tau^2)$, whereas observations from different individuals are uncorrelated. This model is a special case of the linear mixed models for analysis of longitudinal data (e.g., Diggle et al. 1996). There are different types of linear or nonlinear mixed models that can be used to model the correlations among the observations.

There is no general consensus among mixed model users on the roles that the random effects play. For many users, the main purpose of introducing the random effects is to model the correlations among observations, such as in the analysis of longitudinal data. On the other hand, in many cases the random effects represent unobserved variables of practical interest, which for good reasons should be considered random. This is the case, for example, in small-area estimation (e.g., Rao 2003). Robinson (1991) gave a wide-ranging account of the estimation (or prediction) of random effects in linear mixed models with examples and applications. Jiang and Lahiri (2006) provided an overview of the prediction theory for random effects and its applications in small-area estimation.

A general linear mixed model may be expressed as

$$y = X\beta + Z\alpha + \epsilon, \tag{12.1}$$

where $y$ is a vector of observations, $X$ is a matrix of known covariates, $\beta$ is a vector of unknown regression coefficients, which are often called the fixed effects, $Z$ is known matrix, $\alpha$ is a vector of random effects, and $\epsilon$ is a vector of errors. Both $\alpha$ and $\epsilon$ are unobservable. Compared with the linear regression model, it is clear that the difference is $Z\alpha$, which may take many different forms, and thus creates a rich class of models, as we will see. The basic assumptions for (12.1) are that the random effects and errors have mean 0 and finite variances. Typically, the covariance matrices $G = \text{Var}(\alpha)$ and $R = \text{Var}(\epsilon)$ involve some unknown dispersion parameters, or variance components. It is also assumed that $\alpha$ and $\epsilon$ are uncorrelated.

If the normality assumption is made, as is often the case, the linear mixed model is called a Gaussian linear mixed model, or *Gaussian mixed model*. This means that both $\alpha$ and $\epsilon$ are normal, in addtion to the basic assumptions above. Otherwise, if normality is not assumed, the model is called a *non-Gaussian linear mixed model* (Jiang 2007). Another way of classifying the linear mixed models is in terms of the $Z$ matrix, or the expression of $Z\alpha$. The model is called a (Gaussian) *mixed ANOVA model* if

$$Z\alpha = Z_1\alpha_1 + \cdots + Z_s\alpha_s, \tag{12.2}$$

where $Z_1, \ldots, Z_s$ are known matrices and $\alpha_1, \ldots, \alpha_s$ are vectors of random effects such that for each $1 \leq i \leq s$, the components of $\alpha_i$ are independent and distributed as $N(0, \sigma_i^2)$. It is also assumed that the components of $\epsilon$ are independent and distributed as $N(0, \tau^2)$, and $\alpha_1, \ldots, \alpha_s$, and $\epsilon$ are independent. If the normality assumption is not made, but instead the components of $\alpha_i, 1 \leq i \leq s$, and $\epsilon$ are assumed i.i.d., we have a non-Gaussian mixed ANOVA model. It is clear that a mixed ANOVA model is a special case of (12.1) with $Z = (Z_1 \; \cdots \; Z_s)$ and $\alpha = (\alpha_1' \; \cdots \; \alpha_s')'$. For mixed ANOVA models (Gaussian or non-Gaussian), a natural set of *variance components* are $\tau^2, \sigma_1^2, \ldots, \sigma_s^2$. Alternatively, the Hartley–Rao form of variance components (Hartley and Rao 1967) are $\lambda = \tau^2, \gamma_1 = \sigma_1^2/\tau^2, \ldots, \gamma_s = \sigma_s^2/\tau^2$. We consider an example.

*Example 12.2* (One-way random effects model). A model is called a random effects model if the only fixed effect is an unknown mean. Suppose that the observations $y_{ij}$, $i = 1, \ldots, m$, $j = 1, \ldots, n_i$, satisfy $y_{ij} = \mu + \alpha_i + \epsilon_{ij}$ for all $i$ and $j$, where $\mu$ is an unknown mean, $\alpha_i$, $i = 1, \ldots, m$, are random effects which are distributed independently as $N(0, \sigma^2)$, $\epsilon_{ij}$'s are errors that are distributed independently as $N(0, \tau^2)$, and the random effects are independent of the errors. It is easy to see that the one-way randon effects model is a special case of the mixed ANOVA model with $X = 1_n$, where $n = \sum_{i=1}^{m} n_i$ is the total number of observations, and $Z = \mathrm{diag}(1_{n_i}, 1 \leq i \leq m)$ (recall that $1_k$ denotes the $k$-dimensional vector of 1's).

A different type of linear mixed model is called the *longitudinal model*. Following Datta and Lahiri (2000), a longitudinal model can be expressed as

$$y_i = X_i\beta + Z_i\alpha_i + \epsilon_i, \quad i = 1, \ldots, m, \tag{12.3}$$

where $y_i$ represents the vector of observations from the $i$th individual, $X_i$ and $Z_i$ are known matrices, $\beta$ is an unknown vector of regression coefficients, $\alpha_i$ is a vector of random effects, and $\epsilon_i$ is a vector of errors. It is assumed that $\alpha_i, \epsilon_i$, $i = 1, \ldots, m$, are independent with $\alpha_i \sim N(0, G_i)$ and $\epsilon_i \sim N(0, R_i)$, where the covariance matrices $G_i$ and $R_i$ are known up to a vector $\theta$ of variance components. Example 12.1 is a special case of the longitudinal model, in which $X_i = (x_{ij}')_{1 \leq j \leq k}$, $Z_i = 1_k$, $G_i = \sigma^2$, and $R_i = \tau^2 I_k$ ($I_k$ denotes the $k$-dimensional identity matrix), and so $\theta = (\sigma^2, \tau^2)'$. Note that the one-way random effects model of Example 12.2 is a special case of both the mixed ANOVA model and longitudinal model. However, in general, these two types of linear mixed models are different (Exercises 12.2 and 12.3).

For the most part, linear mixed models have been used in situations where the observations are continuous. However, discrete, or categorical, observations are often encountered in practice. For example, the number of heart attacks of a patient during the past year takes the values $0, 1, 2, \ldots$; the blood pressure is often measured in the categories low, median, and high; and many survey results are binary such as yes (1) and no (0). McCullagh and Nelder (1989) introduced an extension of linear models, known as generalized linear models, or GLM, that applies to discrete of categorical observations. They noted that the key elements of a classical linear model (i.e., a linear regression model) are (i) the observations are independent, (ii) the mean of the observation is a linear function of covariates, and (iii) the variance of the observation is a constant. The extension to GLM consists of modification of (ii) and (iii) above by (ii)$'$ the mean of the observation is associated with a linear function of covariates through a link function; and (iii)$'$ the variance of the observation is a function of the mean. It is clear that independence of the observations is still a basic requirement for GLM. To come up with a broader class of models that apply to correlated discrete or categorical observations, we take a similar approach as above in extending the classical linear model to linear mixed models by introducing random effects to the GLM.

To motivate the extension, let us first consider an alternative expression of the Gaussian mixed model. Suppose that, given a vector of random effects, $\alpha$, the observations $y_1, \ldots, y_n$ are (conditionally) independent such that $y_i \sim N(x_i'\beta + z_i'\alpha, \tau^2)$, where $x_i$ and $z_i$ are known vectors, $\beta$ is an unknown vector of regression coefficients, and $\tau^2$ is an unknown variance. Furthermore, suppose that $\alpha \sim N(0, G)$, where $G$ depends on a vector $\theta$ of unknown variance components. Let $X$ and $Z$ be the matrices whose $i$th rows are $x_i'$ and $z_i'$, respectively. It is easy to see that this leads to the (Gaussian) linear mixed model (12.1) with $R = \tau^2 I$ (Exercise 12.4).

The two key elements in the above that define a Gaussian mixed model are (i) conditional independence (given the random effects) and a conditional distribution and (ii) the distribution of the random effects. We now use these basic elements to define a generalized linear mixed model, or GLMM. Suppose that, given a vector of random effects, $\alpha$, the responses $y_1, \ldots, y_n$ are (conditionally) independent such that the conditional distribution of $y_i$ given $\alpha$ is a member of the exponential family with pdf

$$f_i(y_i|\alpha) = \exp\left\{\frac{y_i\xi_i - b(\xi_i)}{a_i(\phi)} + c_i(y_i, \phi)\right\}, \tag{12.4}$$

where $b(\cdot)$, $a_i(\cdot)$, and $c_i(\cdot, \cdot)$ are known functions and $\phi$ is a dispersion parameter that may or may not be known. The quantity $\xi_i$ is associated with $\mu_i = E(y_i|\alpha)$, which, in turn, is associated with a linear predictor

$$\eta_i = x_i'\beta + z_i'\alpha, \tag{12.5}$$

where $x_i$ and $z_i$ are known vectors and $\beta$ is a vector of unknown parameters (the fixed effects), through a known link function $g(\cdot)$ such that

$$g(\mu_i) = \eta_i. \tag{12.6}$$

Furthermore, it is assumed that $\alpha \sim N(0, G)$, where the covariance matrix $G$ may depend on a vector $\theta$ of unknown variance components.

Note that according to the properties of the exponential family (see Appendix A.3), one has $b'(\xi_i) = \mu_i$. In particular, under the so-called *canonical link* function, one has

$$\xi_i = \eta_i;$$

that is, $g = h^{-1}$, where $h(\cdot) = b'(\cdot)$. Here, $h^{-1}$ represents the inverse function (not reciprocal) of $h$. A table of canonical links is given in McCullagh and Nelder (1989, p. 32). We consider some special cases.

*Example 12.3* (Mixed logistic model). Suppose that, given the random effects $\alpha$, binary responses $y_1, \ldots, y_n$ are conditionally independent *Bernoulli*. Furthermore, with $p_i = P(y_i = 1|\alpha)$, one has

$$\text{logit}(p_i) = x_i'\beta + z_i'\alpha,$$

where $\text{logit}(p) = \log\{p/(1-p)\}$ and $x_i$ and $z_i$ are as in the definition of GLMM. This is a special case of the GLMM, in which the (conditional) exponential family is Bernoulli and the link function is $g(\mu) = \text{logit}(\mu)$. Note that in this case the dispersion parameter $\phi = 1$.

*Example 12.4* (Poisson log-linear mixed model). The Poisson distribution is often used to model responses that are counts. Supposed that, given the random effects $\alpha$, the counts $y_1, \ldots, y_n$ are conditionally independent such that $y_i|\alpha \sim \text{Poisson}(\lambda_i)$, where

$$\log(\lambda_i) = x_i'\beta + z_i'\alpha$$

and $x_i$ and $z_i$ are as in the definition of GLMM. Again, this is a special case of GLMM, in which the (conditional) exponential family is Poisson and the link function is $g(\mu) = \log(\mu)$. The dispersion parameter $\phi$ in this case is again equal to 1.

The fact that the observations, or responses, are correlated makes it considerably more difficult to develop large-sample techniques for mixed model analysis. We first consider linear mixed models, for which the asymptotic theory is much more complete than for GLMMs. We focus on selected topics of interest. For a more complete coverage, see Jiang (2007).

## 12.2 REML: Restricted maximum likelihood

A main problem in mixed model analysis is estimation of the variance components. In many cases (e.g., quantitative genetics), the variance components are of main interest. In some other cases (e.g., longitudinal data analysis), the variance components themselves are not of main interest, but they need to be estimated in order to assess the variability of estimators of other quantities of interest, such as the fixed effects. Some of the earlier methods in mixed model analysis did not require the normality assumption. These include the analysis of variance (ANOVA) method, or Henderson's methods (Henderson 1953), and minimum norm quadratic unbiased estimation (MINQUE) method, proposed by C. R. Rao (e.g., Rao 1972). However, the ANOVA method is known to produce inefficient estimators of the variance components when the data are unbalanced. The MINQUE method, on the other hand, depends on some initial values of the variance components. Also, both ANOVA and MINQUE can result in estimators that fall outside the parameter space.

If normality is assumed, the efficient estimators of the variance components are the maximum likelihood estimators (MLEs). However, the latter had not been in serious use in linear mixed models, until Hartley and Rao (1967). The main reason was that, unlike the ANOVA estimator, the MLE under a linear mixed model was not easy to handle computationally in the earlier days. There was also an issue regarding the asymptotic behavior of the MLE, because, unlike the traditional i.i.d. case, the observations are correlated under a linear mixed model. Both issues were addressed by the Hartley–Rao paper. Asymptotic properties of the MLE were further studied by Miller (1977) for a wider class of models. On the other hand, the MLEs of the variance components are, in general, biased. Here, we are not talking about the finite-sample bias, which may vanish as the sample size increases. In fact, the following example due to Neyman and Scott (1948) shows that the bias can lead to inconsistent estimators of the variance components in a certain situation.

*Example 12.5* (The Neyman-Scott problem). Recall Example 3 in the Preface, where two measurements, $y_{ij}$, $j = 1, 2$, are taken from the $i$th patient. Write $y_{ij} = \mu_i + \epsilon_{ij}$, $i = 1, \ldots, m$, $j = 1, 2$, where $\mu_i$ is the unknown mean of the $i$th patient and $\epsilon_{ij}$ is the measurement error, whose variance is of main interest. Suppose that the $\epsilon_{ij}$'s are independent and distributed as $N(0, \sigma^2)$. It can be shown (Exercise 12.5) that the MLE of $\sigma^2$ is given by

$$\sigma_{\mathrm{ML}}^2 = \frac{1}{4m} \sum_{i=1}^{m} (y_{i1} - y_{i2})^2$$

$$= \frac{1}{4m} \sum_{i=1}^{m} (\epsilon_{i1} - \epsilon_{i2})^2. \tag{12.7}$$

Applying the law of large numbers to the right side of (12.7), we see that $\hat{\sigma}_{\mathrm{ML}}$ converges in probability to $\sigma^2/2$, not $\sigma^2$, as the number of patients, $m$, goes to infinity. Therefore, the MLE is inconsistent in this case.

The inconsistency of the MLE in Example 12.5 is due to the presence of many nuisance parameters—namely, the (unknown) means $\mu_i$, $1 \leq i \leq m$. Note that to do the maximum likelihood, one has to estimate all of the parameters, including the nuisance ones. There are a total of $m + 1$ unknown parameters (why?), whereas the total sample size is $2m$. Intuitively, this does not look like a very profitable enterprise. However, there is an easy way to get around, or get rid of, the nuisance parameters: by taking the differences

$z_i = y_{i1} - y_{i2}$, $\le i \le m$. Now, considering the $z_i$'s as the observations, there are $m$ observations and only one unknown parameter, $\sigma^2$ [note that the $z_i$'s are independent and distributed as $N(0, 2\sigma^2)$], so the situation has gotten much better. In fact, the MLE of $\sigma^2$ based on the $z_i$'s is given by

$$\hat{\sigma}^2_{\text{REML}} = \frac{1}{2m} \sum_{i=1}^{m} z_i^2$$

$$= \frac{1}{2m} \sum_{i=1}^{m} (\epsilon_{i1} - \epsilon_{i2})^2. \tag{12.8}$$

It follows, again, by the law of large numbers, that $\hat{\sigma}^2_{\text{REML}}$ converges in probability to $\sigma^2$ as $m \to \infty$, and therefore is consistent.

The "trick" used above is a special case of a method called restricted, or residual, maximum likelihood, or REML (and this is why the notation $\hat{\sigma}^2_{\text{REML}}$ is used). The method was proposed by Thompson (1962) and later put together on a broader basis by Patterson and Thompson (1971). It applies not only to the Neyman–Scott problem, where only the fixed effects are involved, but also to linear mixed models in general. Let the dimensions of $y$ and $\beta$ be $n$ and $p$, respectively. Without loss of generality, assume that $\text{rank}(X) = p$. Let $A$ be a $n \times (n - p)$ matrix of full rank such that $A'X = 0$. The REML estimators of the variance components are simply the MLEs based on $z = A'y$. It is seen from (12.1) that the distribution of $z$ does not depend on $\beta$. So by making the transformation $z = A'y$, the fixed effects have been removed. It can be shown that the REML estimators do not depend on the choice of $A$ (Exercise 12.6). Furthermore, several authors have argued that there is no loss of information in REML for estimating the variance components (e.g., Patterson and Thompson 1971; Harville 1977; Jiang 1996). For alternative derivations of REML, see Harville (1974), Barndorff-Nielsen (1983), Verbyla (1990), Heyde (1994), and Jiang (1996).

The REML estimators are typically derived under the normality assumption. However, the latter is likely to be violated in real-life problems. Due to such concerns, a quasilikelihood approach has been used in deriving the REML estimators without the normality assumption. The idea is to use the Gaussian REML estimators, even if the normality assumption does not hold (e.g., Richardson and Welsh 1994, Heyde 1994, 1997, Jiang 1996, 1997a). More specifically, the REML estimators are defined as the solution to the Gaussian REML equation. For example, for the mixed ANOVA model with the Hartley–Rao form of variance components, the REML equations are given by

$$y'Qy = \lambda(n - p),$$
$$y'QZ_iZ_i'Qy = \lambda\text{tr}(Z_i'QZ_i), \quad 1 \le i \le s, \tag{12.9}$$

where $Q = \Gamma^{-1} - \Gamma^{-1}X(X'\Gamma^{-1}X)^{-1}X'\Gamma^{-1}$ with $\Gamma = I_n + \sum_{i=1}^{s} \gamma_i Z_i Z_i'$ ($n$ is the dimension of $y$). See Jiang (2007, Section 1.4). In the sequel, such Gaussian REML estimators are simply called REML estimators, , even if normality is not assumed. An important question then is: How does the REML estimators behave (asymptotically) when normality does not hold? A related question is regarding the asymptotic superiority of REML over (straight) maximum likelihood (ML). It is seen in the Neyman–Scott problem (Example 12.5) that the REML estimator remains consistent as the number of nuisance parameters increases with the sample size, whereas the MLE fails to do so. Do we have such a difference in asymptotic behavior in general? Like REML, the MLEs are understood as the Gaussian MLEs when normality is not assumed.

To answer such questions, let us focus on the mixed ANOVA models defined by (12.1) and (12.2). Instead of normality, we assume that the components of $\alpha_i$ are i.i.d. with mean 0 and variance $\sigma_i^2$, $1 \leq i \leq s$; the components of $\epsilon$ are i.i.d. with mean 0 and variance $\tau^2$; and $\alpha_1, \ldots, \alpha_s, \epsilon$ are independent. We consider the Hartley–Rao form of variance components defined above Example 12.2. Based on these assumptions, Jiang (1996, 1997a) developed an asymptotic theory about REML estimation. Typically, a theorem requires some technical conditions. It is important that (i) the technical conditions make sense and (ii) ideally, only necessary assumptions are made. Regarding (i), Jiang (1996) set up the conditions so that they can be interpreted intuitively. For the most part, there are two conditions for the consistency of REML estimators. The first condition states that the variance components are asymptotically identifiable. To see what this means, let us forget about the asymptotic part, for now, and consider a simple example.

*Example 12.6.* Consider the following random effects model: $y_i = \mu + \alpha_i + \epsilon_i$, $i = 1, \ldots, m$, where $\mu$ is an unknown mean, the random effects $\alpha_1, \ldots, \alpha_n$ are independent and distributed as $N(0, \sigma^2)$, the errors $\epsilon_1, \ldots, \epsilon_n$ are independent and distributed as $N(0, \tau^2)$, and the random effects and errors are independent. It is clear that in this case, there is no way to "separate" the variance of the random effects from that of the errors. In other words, the $\alpha_i$'s and $\epsilon_i$'s could have the distributions $N(0, \sigma^2 + a)$ and $N(0, \tau^2 - a)$, respectively, for any $a$ such that $|a| \leq \sigma^2 \wedge \tau^2$, and the joint distribution of the $y_i$'s remains the same. Thus, in this case, the variance components are not (individually) identifiable (in fact, only $\sigma^2 + \tau^2$ is identifiable). If the variance components are not identifiable, they cannot be estimated consistently (why?). Note that the above model corresponds to the one-way random effects model of Example 12.2 with $n_i = 1, 1 \leq i \leq m$. On the other hand, if $n_i = 2, 1 \leq i \leq m$ (or, more generally, $n_i = k$ for some $k > 1$), then it is easy to see that the variance components $\sigma^2$ and $\tau^2$ are identifiable (intuitively, in this case, one can separate the two variances). Now, let us consider some cases in between. Suppose that all but a few $n_i$'s are equal to 1 and the rest of the $n_i$' are equal to 2; then, as one would expect, asymptotically, we will have an identifiability problem with the variance components (this is because, asymptotically, the roles of a few observations in model inference will be "washed out"; so the inference is essentially based on the observations corresponding to the unidentifiable variance components). On the other hand, if all but a few $n_i$'s are equal to

2 and the rest of the $n_i$' are equal to 1, then, asymptotically, we will be fine (for the same reason) in identifying of the variance components. This is what asymptotic identifiability means.

The second condition for the REML asymptotic theory states that the observations are infinitely informative. This is an extension of the simple concept that the sample size goes to infinity in the case of independent observations. However, there is a complication in extending this concept to linear mixed models, which we explain with an example.

*Example 12.7* (Two-way random effects model). Consider the random effects model $y_{ij} = \mu + u_i + v_j + e_{ij}$, $i = 1, \ldots, m_1$, $j = 1, \ldots, m_2$, where $u_i$'s and $v_j$'s are random effects and $e_{ij}$'s are errors, which are independent such that $u_i \sim N(0, \sigma_1^2)$, $v_j \sim N(0, \sigma_2^2)$, and $e_{ij} \sim N(0, \tau^2)$. The question is: What is the *effective* sample size? It turns out that the answer depends on which variance component one is interested in estimating. It can be shown that, asymptotically, the effective sample sizes for estimating $\sigma_1^2$, $\sigma_2^2$, and $\tau^2$ are $m_1$, $m_2$, and $m_1 m_2$, respectively. This can be seen from Theorem 12.1 below but, intuitively, it makes sense. For example, the random effect $u$ has $m_1$ (unobserved) realizations. Therefore, the effective sample size for estimating the variance of $u$ should be $m_1$ (not the total sample size $m_1 m_2$). In this special case, the infinitely informative assumption simply means that both $m_1$ and $m_2$ go to infinity, which is clearly necessary for consistently estimating all of the variance components.

In conclusion, under the assumptions that (a) the variance components are asymptotically identifiable and (b) the observations are infinitely informative, Jiang (1996) proved that the REML estimators are consistent, and this is true regardless of the normality assumption. Furthermore, the author established asymptotic normality of the REML estimators under the additional assumption that (c) the distributions of the random effects and errors are nondegenerate (i.e., they are not two-point distributions). Once again, normality is not needed for the asymptotic normality. To illustrate these results in further detail, we consider a special case of linear mixed models.

A linear mixed model is called a balanced mixed ANOVA model (or linear mixed model with balanced data; e.g., Searle et al. 1992, Section 4.6) if it can be expressed as (12.1) and (12.2), where

$$X = \otimes_{q=1}^{r+1} 1_{n_q}^{d_q}, \quad Z_i = \otimes_{q=1}^{r+1} 1_{n_q}^{i_q}, \quad i \in S,$$

with $d = (d_1, \ldots, d_{r+1}) \in S_{r+1} = \{0, 1\}^{r+1}$ [i.e., $d$ is a $(r+1)$-dimensionnal vector whose components are 0 or 1], $i = (i_1, \ldots, i_{r+1}) \in S \subset S_{r+1}$, $i \in S$, $1_k^1 = 1_k$ and $1_k^0 = I_k$ (recall $1_k$ and $I_k$ are the $k$-dimensional vector of 1's and identity matrix, respectively). Here, $r$ is the number of factors and $n_q$ is the number of levels for the $q$th factor, $1 \leq q \leq r + 1$, with $r + 1$ corresponding to number of replications for each cell (a cell is a combination of levels of different factors). Note that we now use the multiple index $i$ instead of the single index $i$. Similarly, the variance components are $\tau^2$ and $\sigma_i^2, i \in S$, or $\lambda$ and $\gamma_i, i \in S$ in the Hartley–Rao form. Example 12.2, with $n_i = k$, $1 \le i \le m$, and Example 12.7 are special cases of the balanced mixed ANOVA model. In fact, in the former case, (12.1) and (12.2) reduce to

$$y = 1_m \otimes 1_k \mu + I_m \otimes 1_k \alpha + \epsilon$$

with $y = (y_1', \ldots, y_m')'$, $y_i = (y_{ij})_{1 \le j \le k}$, $1 \le i \le m$, $\epsilon$ defined similarly, and $\alpha = (\alpha_i)_{1 \le i \le m}$. Similarly, in the latter case, (12.1) and (12.2) reduce to

$$y = 1_{m_1} \otimes 1_{m_2} + I_{m_1} \otimes 1_{m_2} u + 1_{m_1} \otimes I_{m_2} v + e$$

with $y = (y_1', \ldots, y_{m_1}')'$, $y_i = (y_{ij})_{1 \le j \le m_2}$, $1 \le i \le m_1$, $e$ defined similarly, $u = (u_i)_{1 \le i \le m_1}$, and $v = (v_j)_{1 \le j \le m_2}$. For another example, see Exercise 12.7. The balanced mixed ANOVA model is called *unconfounded* if (i) the fixed effects are not confounded with the random effects and errors [i.e., rank$(X, Z_i) > p$, $i \in S$ and $X \ne I_n$] and (ii) the random effects and errors are not confounded [i.e., the matrices $I_n, Z_i Z_i', i \in S$ are linearly independent]. Here, $n = \prod_{q=1}^{r+1} n_q$ is the total sample size. Also, the dimension of $\alpha_i$ is $m_i = \prod_{i_q=0} n_q$, $i \in S$.

For a general mixed ANOVA model (not necessarily balanced), the random effects and errors are said to be *nondegenerate* if the squares of them are not a.s. constants. We say the sequences of estimators $\hat\lambda, \hat\gamma_i, 1 < i \le s$, are asymptotically normal if there are sequences of positive numbers $p_i(n) \to \infty, 0 \le i \le s$, and a sequence of matrices $B_n$ satisfying

$$\limsup(\|B_n^{-1}\| \vee \|B_n\|) < \infty,$$

$$M_n \begin{bmatrix} p_0(n)(\hat\lambda - \lambda) \\ p_1(n)(\hat\gamma_1 - \gamma_1) \\ \vdots \\ p_s(n)(\hat\gamma_s - \gamma_s) \end{bmatrix} \xrightarrow{d} N(0, I_{s+1}),$$

where $\|B\| = \lambda_{\max}^{1/2}(B'B)$. Define the symmetric $(s+1) \times (s+1)$ matrix $\mathcal{I}_n$ whose $(i,j)$ element is $\operatorname{tr}(Z_i Z_i' Q Z_j Z_j' Q)/p_i(n)p_j(n)$, $1 \le i, j \le s$, the $(i,0)$ element is $\operatorname{tr}(Z_i Z_i' Q)/\lambda p_0(n)p_i(n)$, $1 \le i \le s$, and the $(0,0)$ element is $(n-p)/\lambda^2 p_0^2(n)$, where $Q$ is defined below (12.9). Furthermore, define $W = [I_n \ \sqrt{\gamma_1} Z_1 \ \cdots \ \sqrt{\gamma_s} Z_s]$ (block matrix), $Q_0 = W'QW$, and $Q_i = W'Q Z_i Z_i' QW$, $1 \le i \le s$. Let $\mathcal{K}_n$ be the $(s+1) \times (s+1)$ matrix whose $(i,j)$ element is

$$\frac{1}{p_i(n)p_j(n)} \sum_{l=1}^{n+m} \frac{\{E(\xi_l^4) - 3\}Q_{i,ll}Q_{j,ll}}{\lambda^{1_{(i=0)} + 1_{(j=0)}}},$$

$0 \le i, j \le s$, where $m = \sum_{i=1}^s m_i$, $Q_{i,kl}$ is the $(k,l)$ element of $Q_i$, $\xi_l = \epsilon_l/\sqrt{\lambda}$, $1 \le l \le n$, and $\xi_l = \alpha_{i,l-n-\sum_{k<i} m_k}/\sqrt{\lambda \gamma_i}$, $n + \sum_{k<i} m_k + 1 \le l \le n + \sum_{k \le i} m_k$, $1 \le i \le s$. Let $\mathcal{J}_n = 2\mathcal{I}_n + \mathcal{K}_n$. Jiang (1996) proved the following.

**Theorem 12.1.** Let the balanced mixed ANOVA model be unconfounded and the variance components $\tau^2$ and $\sigma_i^2, i \in S$, be positive. Then the following hold as $m_i \to \infty, i \in S$:

(i) There exist with probability tending to 1 REML estimators $\hat{\lambda}$ and $\hat{\gamma}_i, i \in S$, that are consistent, and the sequences $\sqrt{n-p}(\hat{\lambda} - \lambda), \sqrt{m_i}(\hat{\gamma}_i - \gamma_i), i \in S$, are bounded in probability.

(ii) If, in addition, the random effects and errors are nondegenerate, the REML estimators in (1) are asymptotically normal with $p_0(n) = \sqrt{n-p}$, $p_i(n) = \sqrt{m_i}, i \in S$, and $M_n = \mathcal{J}_n^{-1/2}\mathcal{I}_n$.

Note that $\mathcal{K}_n$ vanishes under the normality assumption, so that $\mathcal{J}_n = 2\mathcal{I}_n$ and $M_n = \mathcal{I}_n^{1/2}/\sqrt{2}$. The result of (i) is in the form of Cramér consistency (see Section 1.4). Later, Jiang (1997a) proved the Wald consistency of the REML estimators, in the sense that the maximizer of the restricted Gaussian likelihood (i.e., the likelihood of $z = A'y$ under normality) is consistent, under the same assumptions but without positiveness of the variance components. Note that if the variance components are indeed positive, then, asymptotically, the maximizer of the restricted Gaussian likelihood constitutes a root to the REML equations (why?), which, by definition, is the (vector of) REML estimators without the normality assumption.

The proof of (i) is based on the following basic argument. Suppose that $l_n(y, \theta)$ is a function that depends on both the observations $y$ and a $d$-dimensional vector $\theta$ of parameters. Suppose that there are sequences of positive numbers $p_i(n)$ and $q_i(n)$, $1 \leq i \leq d$, such that $p_i(n) \to \infty$, $p_i(n)q_i(n) \to \infty$, and the following hold:

$$\frac{1}{p_i(n)}\frac{\partial l_n}{\partial \theta_i} = O_P(1), \quad 1 \leq i \leq d, \tag{12.10}$$

$$\left[\frac{1}{p_i(n)p_j(n)}\frac{\partial^2 l_n}{\partial \theta_i \partial \theta_j}\right]_{1 \leq i,j \leq d} = G_n + o_P(1), \tag{12.11}$$

where $G_n$ is a sequence of matrices satisfying $\liminf \lambda_{\min}(G_n) > 0$, and

$$\frac{1}{p_i(n)p_j(n)p_k(n)}\sup_{\theta \in \Theta_n}\left|\frac{\partial^3 l_n(\tilde{\theta})}{\partial \theta_i \partial \theta_j \partial \theta_k}\right| = o_P(1), \quad 1 \leq i,j,k \leq d, \tag{12.12}$$

where $\Theta_n = \{\tilde{\theta} : |\tilde{\theta}_i - \theta_i| < q_i(n), 1 \leq i \leq d\}$. Let $a_n$ denote the vector whose $_i$th component is given by the left side of (12.10), and let $P_n = \text{diag}\{p_i(n), 1 \leq i \leq d\}$. The next thing we do is to choose a point $\theta_n$ that is "close enough" to $\theta$. This is defined by

$$P_n(\theta_n - \theta) = -\tilde{G}_n^{-1}a_n, \tag{12.13}$$

where $\tilde{G}_n$ is the left side of (12.11). It can be shown (Exercise 12.8) that

$$l_n(\tilde{\theta}) - l_n(\theta_n) = \frac{1}{2}\{P_n(\tilde{\theta} - \theta_n)\}'\tilde{G}_n\{P_n(\tilde{\theta} - \theta_n)\} + r_n, \qquad (12.14)$$

where $r_n$ is uniformly ignorable compared with the first term as $\tilde{\theta}$ varies on the boundary of the ellipsoid $E_n = \{\tilde{\theta} : |P_n(\tilde{\theta} - \theta)| \leq 1\}$; that is, $\bar{E}_n = \{\tilde{\theta} : |P_n(\tilde{\theta} - \theta)| = 1\}$. It follows that, asymptotically, we have $l(\tilde{\theta}) > l(\theta_n)$ for all $\tilde{\theta} \in \bar{E}_n$; hence, there is a solution to $\partial l_n / \partial \theta = 0$ in the interior of $E_n$ (why?). The above argument is due to Weiss (1971). The argument leads to the existence of the REML estimator by letting $l_n$ be the negative of the restricted Gaussian log-likelihood. The consistency and boundedness in probability of $P_n(\hat{\theta} - \theta)$, where $P_n = \text{daig}(\sqrt{n-p}, \sqrt{m_i}, i \in S)$ and $\theta = (\lambda, \gamma_i, i \in S)'$, follows from the closeness of $\theta_n$ to $\theta$ [i.e., (12.10) and (12.13)]. It turns out that the simple assumptions of Theorem 12.1 [above (i)] are all that is needed to carry out the above arguments, rigorously.

To prove (ii), note that, by (i) and the Taylor expansion, one can show

$$-a_n = G_n P_n(\hat{\theta} - \theta) + o_P(1), \qquad (12.15)$$

where $\hat{\theta}$ is the REML estimator that satisfies (i) (Exercise 12.8). The key step in proving the asymptotic normality of $\hat{\theta}$ is thus to argue that $a_n$ is asymptotically normal. With $l_n$ being the negative of the restricted Gaussian log-likelihood, the components of $a_n$ are quadratic forms of the random effects and errors (Exercise 12.9). Thus, the asymptotic normality follows from the CLT for quadratic forms that was established in Section 8.8 (as an application of the martingale central limit theorem). Note that the additional assumption that the random effects and errors are nondegenerate is necessary for the asymptotic normality of the REML estimators (Exercise 12.10).

The second part of the asymptotic theory on REML is regarding its comparison with ML. Again, we consider the special case of balanced mixed ANOVA models. For any $u, v \in S_{r+1} = \{0,1\}^{r+1}$, define $u \vee v = (u_1 \vee v_1, \ldots, u_{r+1} \vee v_{r+1})$ and $S_u = \{v \in S : v \leq u\}$, where $v \leq u$ if and only if $u_q \leq v_q, 1 \leq q \leq r+1$. Recall the expression $m_u = \prod_{u_q=0} n_q$. Furthermore, let $m_{u,S} = \min_{v \in S_u} m_v$ if $S_u \neq \emptyset$ and $m_{u,S} = 1$ otherwise. For two sequences of constants $b_n$ and $c_n$, $b_n \sim c_n$ means that $b_n/c_n = O(1)$ and $c_n/b_n = O(1)$. Jiang (1996) proved the following.

**Theorem 12.2.** Let the balanced mixed ANOVA model be unconfounded and the variance components $\tau^2$ and $\sigma_i^2, i \in S$ be positive. Then the following hold as $m_i \to \infty, i \in S$:

(i) There exist with probability tending to 1 the MLEs of $\lambda$ and $\gamma_i, i \in S$, that are consistent if and only if

$$\frac{p}{n} \to 0, \quad \frac{m_{i \vee d} m_{i \vee d, S}}{m_i^2} \to 0, \quad i \in S. \qquad (12.16)$$

(ii) If, in addition, the random effects and errors are nondegenerate, then there exist with probability tending to 1 the MLEs of $\lambda$ and $\gamma_i, i \in S$, that are asymptotically normal if and only if

$$p_0(n) \sim \sqrt{n-p}, \quad p_i(n) \sim \sqrt{m_i}, \quad i \in S \qquad (12.17)$$

and

$$\frac{p}{\sqrt{n}} \to 0, \quad \frac{m_{i\vee d}m_{i\vee d,S}}{m_i^{3/2}} \to 0, \quad i \in S. \qquad (12.18)$$

(iii) When (12.18) is satisfied, the MLEs are asymptotically normal with the same $p_i(n), i \in \{0\} \cup S$, and $M_n$ as for the REML estimators.

A comparison between Theorem 12.1 and Theorem 12.2 shows clearly the asymptotic superiority of REML over ML. Note that the overall assumptions of the two theorems are exactly the same, under which the REML estimators are consistent without any further assumption; whereas the MLE are consistent if and only if (12.16) holds. For the most part, this means that the rate at which the number of fixed effects increases must be slower than that at which the sample size increases. For example, in the Neyman–Scott problem (Example 12.5) we have $p/n = 1/2$; so (12.16) is violated. Similarly, under the same additional assumption that the random effects and errors are nondegenerate, the REML estimators are asymptotically normal without any further assumption; whereas the MLE are asymptotically normal if and only if (12.17) and (12.18) hold. Again, (12.18) fails, of course, in the Neyman–Scott problem.

Finally, when (12.18) holds, the REML estimators and MLEs are asymptotically equivalent, so neither has (asymptotic) superiority of over the other.

## 12.3 Linear mixed model diagnostics

Diagnostics or model checking has been a standard procedure for regression analysis (e.g., Sen and Srivastava 1990). There is a need for developing similar techniques for mixed models. For the most part, diagnostics include informal and formal model checking (McCullagh and Nelder 1989, p. 392). Informal model checking uses diagnostic plots for inspection of potential violations of model assumptions, whereas a standard technique for formal model checking is goodness-of-fit tests. To date, the diagnostic tools for linear mixed models are much more developed than for GLMMs. Therefore, we will only consider linear mixed model diagnostics.

A basic tool for regression diagnostics is the residual plots. Note that a linear regression model corresponds to (12.1) with $Z = 0$; so, in a way, the residuals may be viewed as the estimated errors (i.e., $\epsilon$). A Similar idea has been used for linear mixed model diagnostics, in which standard estimates of the random effects are the empirical best linear unbiased predictors (EBLUP). See, for example, Lange and Ryan (1989) and Calvin and Sedransk (1991). The BLUP for the random effects, $\alpha$, in (12.1) can be expressed as (e.g., Jiang 2007, Section 2.3)

$$\tilde{\alpha} = GZ'V^{-1}(y - X\tilde{\beta}), \tag{12.19}$$

where $V = \text{Var}(y) = ZGZ' + R$ and

$$\tilde{\beta} = (X'V^{-1}X)^{-1}X'V^{-1}y \tag{12.20}$$

is the best linear unbiased estimator (BLUE) of $\beta$. Here, it is assumed that $V$ is known; otherwise, the expressions are not computable. In practice, however, $V$ is unknown and typically depends on a vector, $\theta$, of variance components. If one replaces $\theta$ by $\hat{\theta}$, a consistent estimator, one obtains the EBLUP

$$\hat{\alpha} = GZ'\hat{V}^{-1}(y - X\hat{\beta}), \tag{12.21}$$

where $\hat{V}$ is $V$ with $\theta$ replaced by $\hat{\theta}$ and $\hat{\beta}$ is $\tilde{\beta}$ with $V$ replaced by $\hat{V}$. Thus, a natural idea is to use the plot of $\hat{\alpha}$ for checking for distributional assumptions regarding the random effects. Consider, for example, the one-way random effects model of Example 12.2. The EBLUPs for the random effects, $\alpha_i, 1 \leq i \leq m$, are given by

$$\hat{\alpha}_i = \frac{n_i\hat{\sigma}^2}{\hat{\tau}^2 + n_i\hat{\sigma}^2}(\bar{y}_{i\cdot} - \hat{\mu}), \quad i = 1, \ldots, m,$$

where $\hat{\sigma}^2$ and $\hat{\tau}^2$ are, say, the REML estimators of $\sigma^2$ and $\tau^2$, $\bar{y}_{i\cdot} = n_i^{-1}\sum_{j=1}^{n_i} y_{ij}$, and

$$\hat{\mu} = \left\{\sum_{i=1}^{m} \frac{n_i}{\hat{\tau}^2 + n_i\hat{\sigma}^2}\right\}^{-1} \sum_{i=1}^{m} \frac{n_i\bar{y}_{i\cdot}}{\hat{\tau}^2 + n_i\hat{\sigma}^2}.$$

One may use the EBLUPs for checking the normality of the random effects by making a Q-Q plot of the $\hat{\alpha}_i$'s. The Q-Q plot has the quantiles of the $\hat{\alpha}_i$'s plotted against those of the standard normal distribution. If the plot is close to a straight line, the normality assumption is reasonable.

However, empirical studies have suggested that the EBLUP is not accurate in checking the distributional assumptions about the random effects (e.g., Verbeke and Lesaffre 1996). Jiang (1998c) provided a theoretical explanation for the inaccuracy of EBLUP diagnostics. Consider, once again, the one-way random effects model of Example 12.2 and assume, for simplicity, that $n_i = k$, $1 \leq i \leq m$. Define the empirical distribution of the EBLUPs as

$$\hat{F}(x) = \frac{1}{m}\sum_{i=1}^{m} 1_{(\hat{\alpha}_i \leq x)}.$$

If the latter converges, in a certain sense, to the true underlying distribution of the random effects, say, $F(x)$, then the EBLUP is asymptotically accurate for the diagnostic checking. It can be shown that $\hat{F}(x) \xrightarrow{P} F(x)$ for every $x$ that is a continuity point of $F$ provided that $m \to \infty$ and $k \to \infty$. However, the latter assumption is impractical in most applications of linear mixed models. For example, in small area estimation (see the next chapter), $n_i$ represents the sample size for the $i$th small-area (e.g., a county), which is typically small. So it is not reasonable to assume that $n_i \to \infty$, or $k \to \infty$ (but it is reasonable to assume $m \to \infty$). In fact, one of the main motivations of introducing the random effects is that there is insufficient information for estimating the random effects individually (as for estimating a fixed parameter), but the information is sufficient for estimating the variance of the random effects (see the previous section). In other words, $m$ is large while the $n_i$'s are small.

For the goodness-of-fit tests, we first consider the mixed ANOVA model of (12.1) and (12.2). The hypothesis can be expressed as

$$H_0 : \quad F_i(\cdot|\sigma_i) = F_{0i}(\cdot|\sigma_i),\ 1 \leq i \leq s,\ G(\cdot|\tau) = G_0(\cdot|\tau), \qquad (12.22)$$

where $F_i(\cdot|\sigma_i)$ is the distribution of the components of $\alpha_i$, which depends on $\sigma_i$, $1 \leq i \leq s$, and $G(\cdot|\tau)$ is the distribution of the components of $\epsilon$, which depends on $\tau$. Here, $F_{0i}, 1 \leq i \leq s$, and $G_0$ are known distributions (such as the normal distribution with mean 0).

A special case of (12.22), in which $s = 2$, was considered in Section 2.6, where a $\chi^2$-test based on estimated cell frequencies was proposed (Jiang, Lahiri, and Wu 2001). The approach requires that the estimator of the model parameters, $\hat{\theta}$, be independent of the data used to compute the cell frequencies. Typically, such an independent estimator is obtained either from a different dataset or by splitting the data into two parts, with one part used in computing $\hat{\theta}$ and the other part used in computing the cell frequencies. The drawbacks of this approach are the following: (i) In practice there may not be another dataset available and (ii) splitting the data may result in loss of efficiency and therefore reduced power of the test. Jiang (2001) proposed a simplified $\chi^2$ goodness-of-fit test for the general hypothesis (12.22) that does not suffer from the above drawbacks. He noted that the denominator in Pearson's $\chi^2$-statistic [e.g., (2.23)] was chosen such that the limiting null distribution is $\chi^2$. However, except for a few special cases (such as binomial and Poisson distributions), the asymptotic null distribution of Pearson's $\chi^2$-test is not $\chi^2$ if the expected cell frequencies are estimated by the maximum likelihood method, no matter what denominators are used in place of the $\hat{E}_k$'s in (2.24) (see our earlier discussion in Section 2.6). Therefore, Jiang proposed to simply drop the $\hat{E}_k$'s in the denominator. This leads to the simplified test statistic

$$\hat{\chi}^2 = \frac{1}{a_n} \sum_{k=1}^{M} \left\{ N_k - E_{\hat{\theta}}(N_k) \right\}^2, \qquad (12.23)$$

where $a_n$ is a suitable normalizing constant, $N_k = \sum_{i=1}^{n} 1_{(y_i \in I_k)}$ (i.e., the observed frequency for the interval $I_k$), and $M$ is the number of intervals, or cells. Here, $E_{\theta}$ denotes the expectation given that $\theta$ is the parameter vector.

A key step in developing the latest goodness-of-fit test is to derive the asymptotic distribution of (12.23). This also involves the determination of

$a_n$, or the order of $a_n$ which is all we need. The main tool for deriving the asymptotic distribution is, again, the martingale central limit theorem. We illustrate the idea through an example.

*Example 12.8.* Consider the following extension of Example 12.7:

$$y_{ij} = x'_{ij}\beta + u_i + v_j + e_{ij},$$

where $x_{ij}$ is a $p$-dimensional vector of known covariates, $\beta$ is an unknown vector of fixed effects, and everything else is as in Example 12.7 except that the random effects and errors are not assumed normal. Instead, it is assumed that $u_i \sim F_1(\cdot|\sigma_1)$, $v_j \sim F_2(\cdot|\sigma_2)$, and $e_{ij} \sim G(\cdot|\tau)$. The null hypothesis is thus (12.22) with $s = 2$. Write $\xi_n = (\xi_{n,k})_{1\le k\le M}$, where $\xi_{n,k} = N_k - \mathrm{E}_{\hat\theta}(N_k)$. Then the test statistic (12.23) can be expressed as

$$\hat\chi^2 = a_n^{-1}|\xi_n|^2 = \left|a_n^{-1/2}T'_n\xi_n\right|^2$$

for any orthogonal matrix $T_n$. If we can find $T_n$ such that

$$a_n^{-1/2}T'_n\xi_n \xrightarrow{d} N(0, D), \tag{12.24}$$

where $D = \mathrm{diag}(\lambda_1, \ldots, \lambda_M)$, then by the continuous mapping theorem (Theorem 2.12), we have

$$\hat\chi^2 \xrightarrow{d} \sum_{k=1}^{M} \lambda_k Z_k^2, \tag{12.25}$$

where $Z_1, \ldots, Z_M$ are independent $N(0, 1)$ random variables. The distribution of the right side of (12.25) is known as a weighted $\chi^2$. To show (12.24), we need to show that for any $b \in R^M$, we have

$$b'a_n^{-1/2}T'_n\xi_n \xrightarrow{d} N(0, b'Db) \tag{12.26}$$

(Theorem 2.14). To show (2.26), we decompose the left side as

$$b'a_n^{-1/2}T'_n\xi_n = a_n^{-1/2}\sum_{k=1}^{M} b_{n,k}\{N_k - \mathrm{E}_\theta(N_k)\}$$

$$+ a_n^{-1/2}\sum_{k=1}^{M} b_{n,k}\{\mathrm{E}_{\hat\theta}(N_k) - \mathrm{E}_\theta(N_k)\}, \tag{12.27}$$

where $b_{n,k}$ is the $k$th component of $b_n = T_n b$ and $\theta$ denotes the true parameter vector. Suppress, for notation simplicity, the subscript $\theta$ in the first term of the right side of (12.27) (and also note that the expectation and probability are under the null hypothesis); we can write

$$N_k - \mathrm{E}(N_k) = \sum_{i=1}^{m_1} \sum_{j=1}^{m_2} \{1_{(y_{ij} \in I_k)} - \mathrm{P}(y_{ij} \in I_k)\}.$$

Note that the summand has mean 0, but we need more than this. Here, we use a technique called *projection*. Write (verify this)

$$
\begin{aligned}
&1_{(y_{ij} \in I_k)} - \mathrm{P}(y_{ij} \in I_k) \\
&= \mathrm{P}(y_{ij} \in I_k | u_i) - \mathrm{P}(y_{ij} \in I_k) \\
&\quad + \mathrm{P}(y_{ij} \in I_k | v) - \mathrm{P}(y_{ij} \in I_k) \\
&\quad + 1_{(y_{ij} \in I_k)} - \mathrm{P}(y_{ij} \in I_k | u, v) \\
&\quad + \mathrm{P}(y_{ij} \in I_k | u, v) - \mathrm{P}(y_{ij} \in I_k | u) - \mathrm{P}(y_{ij} \in I_k | v) + \mathrm{P}(y_{ij} \in I_k) \\
&= \zeta_{1,ijk} + \zeta_{2,ijk} + \delta_{1,ijk} + \delta_{2,ijk}.
\end{aligned}
\tag{12.28}
$$

In an exercise (Exercise 12.11), the reader is asked to show the following: (i) $\sum_{i=1}^{m_1} \sum_{j=1}^{m_2} \delta_{l,ijk} = O_{\mathrm{P}}(\sqrt{m_1 m_2})$, $l = 1, 2$; (ii)

$$\sum_{i=1}^{m_1} \sum_{j=1}^{m_2} \zeta_{1,ijk} = \sum_{i=1}^{m_1} \zeta_{1,ik}(u_i),$$

where $\zeta_{1,ik}(u_i) = \sum_{j=1}^{m_2} \zeta_{1,ijk}$ is a function of $u_i$; and, similarly, (iii)

$$\sum_{i=1}^{m_1} \sum_{j=1}^{m_2} \zeta_{2,ijk} = \sum_{j=1}^{m_2} \zeta_{2,jk}(v_j),$$

where $\zeta_{2,jk}(v_j) = \sum_{i=1}^{m_1} \zeta_{2,ijk}$ is a fundtion of $v_j$. It follows that

$$
\begin{aligned}
N_k - \mathrm{E}(N_k) &= \sum_{i=1}^{m_1} \zeta_{1,ik}(u_i) + \sum_{j=1}^{m_2} \zeta_{2,jk}(v_j) \\
&\quad + O_{\mathrm{P}}(\sqrt{m_1 m_2}).
\end{aligned}
\tag{12.29}
$$

Note that the first two terms on the right side of (12.29) are $O_{\mathrm{P}}(m_1^{1/2} m_2)$ and $O_{\mathrm{P}}(m_1 m_2^{1/2})$, respectively (Exercise 12.11).

Now, consider the difference $\mathrm{E}_{\hat{\theta}}(N_k) - \mathrm{E}_{\theta}(N_k)$. Consider $\mathrm{E}_{\theta}(N_k)$ as a function of $\theta$, say, $\psi_k(\theta)$. We have, by the Taylor expansion,

$$\psi_k(\hat{\theta}) - \psi_k(\theta) \approx \left( \frac{\partial \psi_k}{\partial \theta'} \right)(\hat{\theta} - \theta).$$

We now use an asymptotic expansion (see Jiang 1998c) that for any sequence of constant vectors $c_n$, we have

$$c_n'(\hat{\theta} - \theta) \approx \lambda_n' w + w' \Lambda_n w - \mathrm{E}(w' \Lambda_n w) + \text{a remaining term},$$

where $\lambda_n$ is a sequence of constant vectors, $\Lambda_n$ is a sequence of constant symmetric matrices, and $w = (e', u', v')'$. Here, $\approx$ means that the remaining term is of lower order. Note that both $\lambda'_n w$ and $w' \Lambda_n w - \mathrm{E}(w' \Lambda_n w)$ can be expressed as sums of martingale differences. Note that the components of $w$ may be denoted by $w_l, 1 \leq l \leq N = n + m_1 + m_2$, where $n = m_1 m_2$, $w_l, 1 \leq l \leq n$, correspond to the $e_{ij}$'s, $w_l, n + 1 \leq l \leq n + m_1$, to the $u_i$'s, and $w_l, n + m_1 + 1 \leq l \leq N$, to the $v_j$'s. For example, let $\lambda_{n,kl}$ be the $(k,l)$ element of $\Lambda_n$. Then we have (also see Example 8.3)

$$w' \Lambda_n w - \mathrm{E}(w' \Lambda_n w) = \sum_{k,l=1}^{N} \lambda_{n,kl} w_k w_l - \sum_{l=1}^{N} \lambda_{n,ll} \mathrm{E}(w_l^2)$$

$$= \sum_{l=1}^{N} \lambda_{n,ll} \{ w_l^2 - \mathrm{E}(w_l^2) \} + 2 \sum_{l=1}^{N} \sum_{k<l} \lambda_{n,kl} w_k w_l$$

$$= \sum_{l=1}^{N} \left[ \lambda_{n,ll} \{ w_l^2 - \mathrm{E}(w_l^2) \} + 2 \left( \sum_{k<l} \lambda_{n,kl} w_k \right) w_l \right].$$

The summands are a sequence of martingale differences with respect to the $\sigma$-fields $\mathcal{F}_l = \sigma(w_k, k \leq l), 1 \leq l \leq N$. Also, note that the sum of the first two terms on the right side of (12.29) can be expressed as $\sum_{l=n+1}^{N} \zeta_l(w_l)$ for some functions $\zeta_l(\cdot)$ (that depend on $k$) such that $\zeta_l(w_l), \mathcal{F}_l, n + 1 \leq l \leq N$, is a sequence of martingale differences. It follows that (12.27) can be expressed as a sum of martingale differences plus a term of lower order. The martingale central limit theorem (Theorem 8.7) is then applied.

Like Pearson's $\chi^2$-test, the above goodness-of-fit test depends on the number of intervals $M$ and how to choose the intervals $I_k, 1 \leq k \leq M$. It turns out that optimal choice of these intervals or cells is a difficult problem and there is no simple solution (e.g., Lehmann 1999, Section 5.7). Furthermore, the class of linear mixed models considered by Jiang (2001)—namely, the mixed ANOVA model—is a bit restrictive. In particular, the test does not apply to the problem of testing for multivariate normality of the random effects $\alpha_i$ in the longitudinal linear mixed model (12.3). For simplicity, suppose that the error $\epsilon_i$ is normal and we are interested in testing the hypothesis

$$\mathrm{H}_0 : \quad \alpha_i \sim N_d(\mu, \Sigma), \tag{12.30}$$

a $d$-dimensional multivariate normal distribution with (unknown) mean vector $\mu$ and covariance matrix $\Sigma$, where $d > 1$. Claeskens and Hart (2009) proposed an alternative approach to the resting of (12.30). For simplicity, consider the case $d = 2$. Write $\alpha_i = \mu + \Gamma u_i$, where $\Gamma \Gamma' = \Sigma$. Then the problem is equivalent to testing the hypothesis that $u_i \sim N_d(0, I)$, where $I$ is the $d$-dimensional identity matrix. Consider an Edgeworth expansion (see Section 4.3) of the density of $u_i$, $f$, around the standard bivariate normal density, $\phi$:

$$f(u) = \phi(u) \{ 1 + \kappa_3 H_3(u) + \kappa_4 H_4(u) + \cdots \},$$

where $\kappa_3, \kappa_4, \ldots$ are associated with the cumulants of $u_i$, and $H_3, H_4, \ldots$ are Hermite polynomials. The latter is defined by

$$H_j(u) = (-1)^j \phi^{-1}(u) \frac{d^j \phi(u)}{du^j}, \quad j = 1, 2, \ldots.$$

In particular, we have $H_3(u) = u^3 - 3u$ and $H_4(u) = u^4 - 6u^2 + 3$. By a reordering of the terms in the expansion, we may approximate the infinite series by a finite degree polynomial. This leads to the consideration of a density function of the form

$$f_M(u) = P_M^2(u)\phi(u), \tag{12.31}$$

where $P_M$ is a bivariate polynomial that can be expressed as

$$P_M(u) = \sum_{s+t \leq M} a_{st} u_1^s u_2^t \tag{12.32}$$

and the coefficients $a_{st}$ satisfy the constraint

$$\int f_M(u) \, du = 1. \tag{12.33}$$

For example, with $M = 2$, we have

$$P_2(u) = a_{00} + a_{10}u_1 + a_{01}u_2 + a_{20}u_1^2 + a_{11}u_1u_2 + a_{02}u_2^2.$$

The idea is that the distribution under the null hypothesis is a special case of (12.31) with $M = 0$. If there is evidence, provided by the data, suggesting that $M > 0$, then the null hypothesis should be rejected.

The question then is how to obtain the statistical evidence for $M > 0$. Claeskens and Hart proposed using Akaike's AIC (see Section 9.3). Let $\hat{l}_M$ denote the maximized log-likelihood function under $f_M$, and let $\hat{l}_0$ denote that under the null density. The AIC can be expressed as

$$\mathrm{AIC}(M) = -2\hat{l}_M + 2(N_M - 1), \tag{12.34}$$

where $N_M$ is the number of coefficients $a_{st}$ involved in $P_M$ and the subtraction of 1 is due to the integral constraint (12.33) (so $N_M - 1$ is the number of free coefficients). It can be shown that $N_M = 1 + M(M + 3)/2$ (Exercise 12.12). Thus, the null hypothesis is rejected if

$$\min_{1 \leq M \leq L} \{\mathrm{AIC}(M) - \mathrm{AIC}(0)\} < 0,$$

where $L$ is an upper bound for the $M$ under consideration, or, equivalently,

$$T_n = \max_{1 \leq M \leq L} \frac{2(\hat{l}_M - \hat{l}_0)}{M(M + 3)} > 1.$$

This is the test statistic proposed by Claeskens and Hart (2009). The authors stated that the asymptotic null distribution of $T_n$ is that of

$$\max_{r \geq 1} \frac{1}{r(r+3)} \sum_{j=1}^{r} \chi_{j+1}^2,$$

where $\chi_2^2, \chi_3^2, \ldots$ are independent random variables such that $\chi_j^2$ has a $\chi^2$-distribution with $j$ degrees of freedom ($j \geq 2$). In practice, a Monte Carlo method may be used to obtain the large-sample critical values for the test.

## 12.4 Inference about GLMM

Unlike linear mixed models, the likelihood function under a GLMM typically does not have an analytic expression. In fact, the likelihood function may involve high-dimensional intergrals that are difficult to evaluate even numerically. The following is an example.

*Example 12.9.* Suppose that, given the random effects $u_i, 1 \leq i \leq m_1$, and $v_j, 1 \leq j \leq m_2$, binary responses $y_{ij}$, $i = 1, \ldots, m_1$, $j = 1, \ldots, m_2$, are conditionally independent such that, with $p_{ij} = P(y_{ij} = 1 | u, v)$,

$$\text{logit}(p_{ij}) = \mu + u_i + v_j,$$

where $\mu$ is an unknown parameter, $u = (u_i)_{1 \leq i \leq m_1}$, and $v = (v_j)_{1 \leq j \leq m_2}$. Furthermore, the random effects $u_i$'s and $v_j$'s are independent such that $u_i \sim N(0, \sigma_1^2)$ and $v_j \sim N(0, \sigma_2^2)$, where the variances $\sigma_1^2$ and $\sigma_2^2$ are unknown. Thus, the unknown parameters involved in this model are $\psi = (\mu, \sigma_1^2, \sigma_2^2)'$. It can be shown (Exercise 12.13) that the likelihood function under this model for estimating $\psi$ can be expressed as

$$c - \frac{m_1}{2} \log(\sigma_1^2) - \frac{m_2}{2} \log(\sigma_2^2) + \mu y_{..}$$

$$+ \log \int \cdots \int \left[ \prod_{i=1}^{m_1} \prod_{j=1}^{m_2} \{1 + \exp(\mu + u_i + v_j)\}^{-1} \right]$$

$$\times \exp \left( \sum_{i=1}^{m_1} u_i y_{i.} + \sum_{j=1}^{m_2} v_j y_{.j} - \frac{1}{2\sigma_1^2} \sum_{i=1}^{m_1} u_i^2 - \frac{1}{2\sigma_2^2} \sum_{j=1}^{m_2} v_j^2 \right)$$

$$du_1 \cdots du_{m_1} dv_1 \cdots dv_{m_2}, \tag{12.35}$$

where $c$ is a constant, $y_{..} = \sum_{i=1}^{m_1} \sum_{j=1}^{m_2} y_{ij}$, $y_{i.} = \sum_{j=1}^{m_2} y_{ij}$, and $y_{.j} = \sum_{i=1}^{m_1} y_{ij}$. The multidimensional integral involved has dimension $m_1 + m_2$ (which increases with the sample size), and it cannot be further simplified.

Due to the numerical difficulties of computing the maximum likelihood estimators, some alternative methods of inference have been proposed. One approach is based on Laplace approximation to integrals (see Section 4.6). First, note that the likelihood function under the GLMM defined in Section 12.1 can be expressed as

$$L_q \propto |G|^{-1/2} \int \exp\left\{ -\frac{1}{2}\sum_{i=1}^n d_i - \frac{1}{2}\alpha'G^{-1}\alpha \right\} d\alpha,$$

where the subscript q indicates quasilikelihood and

$$d_i = -2 \int_{y_i}^{\mu_i} \frac{y_i - u}{a_i(\phi)v(u)} du,$$

known as the (quasi-) deviance. What it means is that the method to be developed does not require the full specification of the conditional distribution (12.4)—only the first two conditional moments are needed. Here, $v()$ corresponds to the variance function; that is,

$$\text{var}(y_i|\alpha) = a_i(\phi)v(\mu_i) \tag{12.36}$$

and $\mu_i = E(y_i|\alpha)$. In particular, if the underlying conditional distribution satisfies (12.4), then $L_q$ is the true likelihood. Using Laplace approximation (4.64), one obtains an approximation to the logarithm of $L_q$:

$$l_q \approx c - \frac{1}{2}\log|G| - \frac{1}{2}\log|q''(\tilde{\alpha})| - q(\tilde{\alpha}), \tag{12.37}$$

where $c$ does not depend on the parameters,

$$q(\alpha) = \frac{1}{2}\left(\sum_{i=1}^n d_i + \alpha'G^{-1}\alpha\right),$$

and $\tilde{\alpha}$ minimizes $q(\alpha)$. Typically, $\tilde{\alpha}$ is the solution to the equation

$$q'(\alpha) = G^{-1}\alpha - \sum_{i=1}^n \frac{y_i - \mu_i}{a_i(\phi)v(\mu_i)g'(\mu_i)} z_i = 0,$$

where $\mu_i = x_i'\beta + z_i'\alpha$. It can be shown that

$$q''(\alpha) = G^{-1} + \sum_{i=1}^n \frac{z_i z_i'}{a_i(\phi)v(\mu_i)\{g'(\mu_i)\}^2} + r, \tag{12.38}$$

where the remainder term $r$ has expectation 0 (Exercise 12.14). If we denote the term in the denominator of (12.38) by $w_i^{-1}$ and ignore the term $r$, assuming that it is in probability of lower order than the leading terms, then we have a further approximation

$$q''(\alpha) \approx Z'WZ + G^{-1},$$

where $Z$ is the matrix whose $i$th row is $z_i'$, and $W = \operatorname{diag}(w_1, \ldots, w_n)$. Note that the quantity $w_i$ is known as the GLM iterated weights (e.g., McCullagh and Nelder 1989, Section 2.5). By combining approximations (12.37) and (12.38), one obtains

$$l_q \approx c - \frac{1}{2}\left(\log|I + Z'WZG| + \sum_{i=1}^{n}\tilde{d}_i + \tilde{\alpha}'G^{-1}\tilde{\alpha}\right), \qquad (12.39)$$

where $\tilde{d}_i$ is $d_i$ with $\alpha$ replaced by $\tilde{\alpha}$. A further approximation may be obtained by assuming that the GLM iterated weights vary slowly as a function of the mean. Then because the first term inside the $(\cdots)$ in (12.39) depends on $\beta$ only through $W$, one may ignore this term and thus approximate $l_q$ by

$$l_{pq} \approx c - \frac{1}{2}\left(\sum_{i=1}^{n}\tilde{d}_i + \tilde{\alpha}'G^{-1}\tilde{\alpha}\right). \qquad (12.40)$$

Approximation (12.40) was first derived by Breslow and Clayton (1993), who called the procedure penalized quasilikelihood (PQL) by making a connection to the PQL of Green (1987).

It is clear that a number of approximations are involved in PQL. If the approximated log-likelihood, $l_{pq}$, is used in place of the true log-likelihood, we need to know how much the approximations affect inference about the GLMM, which include, in particular, estimation and testing problems. Let us first consider a testing problem. There is considerable interest, in practice, in testing for overdispersion, heteroscedasticity, and correlation among responses. In some cases, the problem is equivalent to testing for zero variance components. Lin (1997) considered a GLMM that has an ANOVA structure for the random effects so that $g(\mu) = [g(\mu_i)]_{1 \le i \le n}$ can be expressed as (12.2), where $\alpha_1, \ldots, \alpha_s$ are independent vectors of random effects such that the components of $\alpha_i$ are independent with distribution $F_i$ whose mean is 0 and variance is $\sigma_i^2$, $1 \le i \le s$. The null hypothesis is

$$H_0 : \sigma_1^2 = \cdots = \sigma_s^2 = 0. \qquad (12.41)$$

Note that under the null hypothesis, there are no random effects involved; so the GLMM become a GLM. In fact, let $\theta = (\sigma_1^2, \ldots, \sigma_s^2)'$ and $l(\beta, \theta)$ denote the second-order Laplace approximate quasi-log-likelihood. The latter is obtained in the similar way as PQL except using the second-order Taylor expansion in the Laplace approximation (see Section 4.6). A global score statistic for testing (12.41) is defined as

$$\chi_G^2 = U_\theta(\hat{\beta})'\tilde{I}(\hat{\beta})^{-1}U_\theta(\hat{\beta}),$$

where $\hat{\beta}$ is the MLE under the null hypothesis—that is, the MLE under the GLM, assuming independence of the responses—$U_\theta(\beta)$ is the gradient vector with respect to $\theta$ (i.e., $\partial l/\partial \theta$), and $\tilde{I}$ is the information matrix of $\theta$ evaluated under $H_0$, which takes the form

$$\tilde{I} = I_{\theta\theta} - I'_{\beta\theta}I_{\beta\beta}^{-1}I_{\beta\theta}$$

with $I_{\theta\theta} = \mathrm{E}\{(\partial l/\partial \theta)(\partial l/\partial \theta')\}$, $I_{\beta\theta} = \mathrm{E}\{(\partial l/\partial \beta)(\partial l/\partial \theta')\}$, and $I_{\beta\beta}$ is $I_{\theta\theta}$ with $\theta$ replaced by $\beta$. Note that given the estimator $\hat{\beta}$, under the null hypothesis, the information matrix can be estimated, using the properties of the exponential family (McCullagh and Nelder 1989, p. 350). In fact, Lin (1997) showed that the information matrix may be estimated when the exponential-family assumption is replaced by some weaker assumptions on the cumulants of the responses. Furthermore, the author showed that under some regularity conditions, the global score statistic $\chi_G^2$ follows a $\chi_s^2$-distribution asymptotically under (12.41). Some optimality of the test was also established. So, for the above testing problem, the PQL works fine. In fact, the second-order Laplace approximation is not essential for the asymptotic results to hold. What is essential is that the Laplace approximation (first or second order) becomes exactly accurate under the null hypothesis. Also, note that under the null hypothesis, the observations become independent. Therefore, the asymptotic distribution can be derived from the CLT for sums of independent random variables (see Section 6.4).

On the other hand, it is quite a different story for the estimation problems. First, let us complete the PQL for estimating the variance component parameters. Let $\theta$ denote the vector of variance components. So far in the derivation of PQL we have held $\theta$ fixed. Therefore, the maximizer of $l_{\mathrm{pq}}$ depends on $\theta$. Breslow and Clayton (1993) proposed substituting these "estimators" back to (12.39) and thus obtaining a profile quasi-log-likelihood function. Furthermore, the authors suggested further approximations that led to a similar form of REML in linear mixed models. See Breslow and Clayton (1993, pp. 11–12) for details. However, the procedure is known to lead to inconsistent estimators of the model parameters. Jiang (1999b) gave an example to demonstrate the inconsistency of PQL estimators, as follows.

*Example 12.10.* Consider a special case of the mixed logistic model of Example 12.3 that can be expressed as

$$\mathrm{logit}\{\mathrm{P}(y_{ij} = 1|\alpha)\} = x'_{ij}\beta + \alpha_i,$$

where $y_{ij}, 1 \leq i \leq m, 1 \leq j \leq n_i$, are binary responses that are conditionally independent given the random effects $\alpha = (\alpha_i)_{1 \leq i \leq m}$, $x_{ij} = (x_{ijk})_{1 \leq k \leq p}$ is a vector of covariates, and $\beta = (\beta_k)_{1 \leq k \leq p}$, a vector of unknown fixed effects. It is assumed that $\alpha_1, \ldots, \alpha_m$ are independent and distributed as $N(0, \sigma^2)$. For simplicity, let us assume that $\sigma^2$ is known and that $n_i, 1 \leq i \leq m$, are bounded. The $x_{ijk}$'s are assumed to be bounded as well. Let $\phi_i(t, \beta)$ denote the unique solution $u$ to the following equation:

$$\frac{u}{\sigma^2} + \sum_{j=1}^{n_i} h(x'_{ij}\beta + u) = t, \tag{12.42}$$

where $h(x) = e^x/(1 + e^x)$ (Exercise 12.15). Then the PQL estimator of $\beta$ is the solution to the following equation:

$$\sum_{i=1}^{m}\sum_{j=1}^{n_i}\{y_{ij} - h(x'_{ij}\beta + \tilde{\alpha}_i)\}x_{ijk} = 0, \quad 1 \le k \le p, \tag{12.43}$$

where $\tilde{\alpha}_i = \phi_i(y_{i\cdot}, \beta)$ with $y_{i\cdot} = \sum_{j=1}^{n_i} y_{ij}$. Denote this solution by $\hat{\beta}$.

Suppose that $\hat{\beta}$ is consistent; that is, $\hat{\beta} \xrightarrow{P} \beta$. Hereafter in this example $\beta$ denotes the true parameter vector. Let $\xi_{i,k}$ denote the inside summation in (12.43); that is, $\xi_{i,k} = \sum_{j=1}^{n_i}\{y_{ij} - h(x'_{ij}\beta + \tilde{\alpha}_i)\}x_{ijk}$. Then, by (12.42),

$$\frac{1}{m}\sum_{i=1}^{m}\xi_{i,k} = \frac{1}{m}\sum_{i=1}^{m}\sum_{j=1}^{n_i}\{\psi_{ij}(\hat{\beta}) - \psi_{ij}(\beta)\},$$

where $\psi_{ij}(\beta) = h(x'_{ij}\beta + \tilde{\alpha})$. Now, by the Taylor expansion, we have

$$\psi_{ij}(\hat{\beta}) - \psi_{ij}(\beta) = \sum_{k=1}^{p}\left.\frac{\psi_{ij}}{\partial \beta_k}\right|_{\tilde{\beta}}(\hat{\beta}_k - \beta_k),$$

where $\tilde{\beta}$ lies between $\beta$ and $\hat{\beta}$. It can be derived from (12.42) that

$$\frac{\partial \tilde{\alpha}_i}{\partial \beta_k} = -\frac{\sum_{j=1}^{n_i} h'(x'_{ij}\beta + \tilde{\alpha}_i)x_{ijk}}{\sigma^{-2} + \sum_{j=1}^{n_i} h'(x'_{ij}\beta + \tilde{\alpha}_i)} \tag{12.44}$$

(Exercise 12.15). Thus, it can be shown that (verify)

$$\left|\psi_{ij}(\hat{\beta}) - \psi_{ij}(\beta)\right| \le \frac{p}{4}\left(1 + \frac{\sigma^2}{4}n_i\right)\left(\max_{i,j,k}|x_{ijk}|\right)\max_{1 \le k \le p}|\hat{\beta}_k - \beta_k|. \tag{12.45}$$

It follows that $m^{-1}\sum_{i=1}^{m}\xi_{i,k} \xrightarrow{P} 0$, as $m \to \infty$.

On the other hand, $\xi_{1,k}, \dots, \xi_{m,k}$ are independent random variables; so it follows by the LLN that $m^{-1}\sum_{i=1}^{m}\{\xi_{i,k} - E(\xi_{i,k})\} \xrightarrow{P} 0$ as $m \to \infty$. If we combine the results, the conclusion is that

$$\frac{1}{m}\sum_{i=1}^{m} E(\xi_{i,k}) \longrightarrow 0, \quad 1 \le k \le p. \tag{12.46}$$

However, (12.46) is not true, in general (see Exercise 12.15). The contradiction shows that $\hat{\beta}$ cannot be consistent in general.

We have seen various applications of the Taylor expansion in statistical inference. In some cases, such as the delta method (Example 4.4), the expansion preserves good asymptotic properties of the estimators, such as consistency and asymptotic normality; in some other cases, such as PQL, the expansion leads to inconsistent estimators (Laplace expansion is derived from the Taylor expansion). The question is: Why is there such a difference? In the first case, the expansion is in the form $f(\hat{\theta}) \approx f(\theta) + f'(\theta)(\hat{\theta} - \theta) + \cdots$, where $\hat{\theta}$ is a consistent estimator of $\theta$. It follows that, as the sample size increases, the error of the expansion (truncated after a finite number of terms) vanishes. In the second case, the expansion is in the form $f(y) \approx f(x) + f'(x)(y - x) + \cdots$, where $y$ is the variable being integrated over. For any fixed $y \neq x$, the error of the expansion (again, truncated after a finite number of terms) will not vanish as the sample size increases, no matter how close $y$ is to $x$. In fact, as far as consistency is concerned, it does not matter whether one uses the first-order, second-order, or any fixed-order Laplace approximation in PQL, the resulting estimators would still be inconsistent.

Another alternative to maximum likelihood is based on the method of moments, one of the oldest methods of finding point estimators, dating back at least to Karl Pearson in the 1800s. It has the virtue of being conceptually simple to use. In the i.i.d. case, the method may be described as follows. Let $X_1, \ldots, X_n$ be i.i.d. observations whose distribution depends on a $p$-dimensional vector $\theta$ of parameters. To estimate $\theta$, we consider the first $p$ sample moments of the observations and let them equal their expectations (assumed exist), which are the moments of the distribution. This means that we solve the system of equations

$$\frac{1}{n} \sum_{i=1}^{n} X_i^k = \mu_k, \quad k = 1, \ldots, p, \tag{12.47}$$

where $\mu_k = \mathrm{E}(X_1^k)$. Note that the $\mu_k$'s are functions of $\theta$. The method can be extended in several ways. First, the observations do not have to be i.i.d. Second, the left side of (12.47) does not have to be the sample moments and may depend on the parameters as well. A nice property of the method of moments is that it almost always produces consistent estimators. To see this, write $\mu_k = \mu_k(\theta)$. Then the method of moments estimator of $\theta$, say $\hat{\theta}$, satisfies $\mu_k(\hat{\theta}) = \hat{\mu}_k, 1 \leq k \leq p$, where $\hat{\mu}_k$ denotes the $k$th sample moment. According to the law of large numbers, the left side of (12.47) converges, say a.s., to $\mathrm{E}(X_1^k) = \mu_k(\theta), 1 \leq k \leq p$, where $\theta$ is the true parameter vector. Thus, if the equations $\mu_k(\hat{\theta}) = \mu_k(\theta), 1 \leq k \leq p$, have a unique solution, one would have $\hat{\theta}$ equal to $\theta$ exactly. Now, because the SLLN does not have $\hat{\mu}_k = \mu_k$ exactly, $\hat{\theta}$ would not equal to $\theta$ exactly, but still be consistent.

Jiang (1998a) extended the method of moments to GLMM. Note that the observations are not i.i.d. under a GLMM; therefore, it may not make sense to use the sample moments. Instead, we consider *sufficient statistics* for the parameters of interest. Roughly speaking, these are statistics that

contain all of the information about the unknown parameters that we intend
to estimate (e.g., Lehmann 1983). Consider the GLMM with the ANOVA
structure [defined above (12.41)] and let $a_i(\phi) = \phi/w_i$, where $w_i$ is a (known)
weight. For example, $w_i = n_i$ for grouped data if the response is a group
average, where $n_i$ is the group size, and $w_i = 1/n_i$ if the response is a group
sum. Then a set of sufficient statistics for $\theta = (\beta', \sigma'_1, \ldots, \sigma'_s)'$ is

$$
\begin{aligned}
S_j &= \sum_{i=1}^{n} w_i x_{ij} y_i, & 1 \leq j \leq p, \\
S_{p+l} &= \sum_{i=1}^{n} w_i z_{i1l} y_i, & 1 \leq l \leq m_1, \\
&\;\;\vdots \\
S_{p+m_1+\cdots+m_{q-1}+l} &= \sum_{i=1}^{n} w_i z_{iql} y_i, & 1 \leq l \leq m_s,
\end{aligned}
\tag{12.48}
$$

where $Z_r = (z_{irl})_{1 \leq i \leq n, 1 \leq l \leq m_r}$, $1 \leq r \leq s$. Thus, a natural set of *estimating
equations* can be formulated as

$$
\sum_{i=1}^{n} w_i x_{ij} y_i = \sum_{i=1}^{n} w_i x_{ij} E_\theta(y_i), \; 1 \leq j \leq p, \tag{12.49}
$$

$$
\sum_{l=1}^{m_r} \left( \sum_{i=1}^{n} w_i z_{irl} y_i \right)^2 = \sum_{l=1}^{m_r} E_\theta \left( \sum_{i=1}^{n} w_i z_{irl} y_i \right)^2, \; 1 \leq r \leq s. \tag{12.50}
$$

Equations (12.50) are then modified to remove the squared terms. The reason
is that the expectation of the squared terms involve the additional dispersion
parameter $\phi$ (Exercise 12.16), which is not of main interest here. Suppose that
$Z_r, 1 \leq r \leq s$, are standard design matrices in the sense that each $Z_r$ consists
only of 0's and 1's, and there is exactly one 1 in each row and at least one 1
in each column. Then the modified equations are

$$
\sum_{(s,t) \in I_r} w_s w_t y_s y_t = \sum_{(s,t) \in I_r} w_s w_t E_\theta(y_s y_t), \qquad 1 \leq r \leq s, \tag{12.51}
$$

where $I_r = \{(s,t) : 1 \leq s \neq t \leq n, z'_{sr} z_{tr} = 1\} = \{(s,t) : 1 \leq s \neq t \leq n, z_{sr} = z_{tr}\}$ and $z'_{ir}$ is the $i$th row of $Z_r$. In other words, the combined estimating equations are (12.49) and (12.51). The expectations involved in these
equations are typically integrals of much lower dimension that those involved
in the likelihood function. Jiang (1998a) proposed to evaluate these expectations by a simple Monte Carlo method and therefore called the procedure the
method of simulated moments (MSM; e.g., McFadden 1989). Furthermore,
Jiang showed that under some regularity conditions, the solution to these
estimating equations is a consistent estimator of $\theta$, as expected.

A drawback of the method of moments is that the estimator may be inefficient. An estimator $\hat{\theta}$ is (asymptotically) efficient if its asymptotic variance,
or covariance matrix, is the smallest among a class of estimators. In the i.i.d.
case, this means that $\sqrt{n}(\hat{\theta} - \theta) \xrightarrow{d} N(0, \Sigma)$, where $\Sigma$ is a covariance matrix such that $\tilde{\Sigma} - \Sigma$ is nonnegative definite for any estimator $\tilde{\theta}$ satisfying

$\sqrt{n}(\tilde{\theta} - \theta) \xrightarrow{d} N(0, \tilde{\Sigma})$. For example, the MLEs are efficient under regularity conditions. On the other hand, the MSM estimators are inefficient, in general. The lack of efficiency is due to the fact that the estimating equations (12.49) and (12.51) are not optimal. To find the optimal estimating equation, let us consider a class of estimators of $\theta$ that are solutions to estimating equations of the following type:

$$B\{S - u(\theta)\} = 0, \qquad (12.52)$$

where $S = (S_j)_{1 \le j \le p+m}$, with $m = m_1 + \cdots + m_s$, is the vector of sufficient statistics given by (12.48), $B$ is a $(p + s) \times (p + m)$ matrix, and $u(\theta) = E_\theta(S)$. Here, $E_\theta$ denotes the expectation given that $\theta$ is the true parameter vector. It can be shown that, theoretically, the optimal $B$ is given by

$$B^* = U'V^{-1}, \qquad (12.53)$$

where $U = \partial u/\partial \theta'$ and $V = \text{Var}(S)$. To see this, let $Q(\theta)$ denote the left side of (12.52) and let $\tilde{\theta}$ be the solution to (12.52). By the Taylor expansion around the true $\theta$, we have $\tilde{\theta} - \theta \approx (BU)^{-1}Q(\theta)$. Thus, we have the approximation

$$\text{Var}(\tilde{\theta}) \approx \{(BU)^{-1}\}BVB'\{(BU)^{-1}\}', \qquad (12.54)$$

assuming that $BU$ is nonsingular. The right side of (12.54) is equal to $(U'VU)^{-1}$ when $B = B^*$. Now, it is an exercise of matrix algebra (Exercise 12.17) to show that the right side of (12.54) is greater than or equal to $(U'VU)^{-1}$, meaning that the difference is a nonnegative definite matrix.

Unfortunately, with the exception of some special cases, the optimal $B$ given by (12.53) is not computable because it involves the exact parameter vector $\theta$ that we intend to estimate. To solve this problem, Jiang and Zhang (2001) proposed the following two-step procedure. First, note that for any fixed $B$, the solution to (12.52) is consistent, even though it may be inefficient. The method of moments estimator of Jiang (1998a) is a special case corresponding to $B = \text{diag}(I_p, 1'_{m_1}, \ldots, 1'_{m_s})$. By using this particular $B$, we obtain a first-step estimator of $\theta$. We then plug in the first-step estimator into (12.53) to obtain the estimated $B^*$. The next thing we do is solve (12.52) with $B$ replaced by the estimated $B^*$. The result is what we call the second-step estimator. Jiang and Zhang showed that, subject to some regularity conditions, the second-step estimator not only is consistent but has the following *oracle* property: Its asymptotic covariance matrix is the same as that of the solution to (12.52) with the optimal $B$—that is, $B^*$ of (12.53) with the true parameter vector $\theta$. The following simulated example, taken from Jiang and Zhang (2001), illustrate the two-step procedure of estimation.

*Example 12.11.* Consider a special case of Example 12.10 with $x'_{ij}\beta = \mu$. Then the unknown parameters are $\mu$ and $\sigma$.

First, note that when $n_i = k$, $1 \le i \le m$, where $k \ge 2$ (i.e., when the data are balanced), the first-step estimators are the same as the second-step ones.

In fact, in this case, the estimating equations of Jiang (1998a), which is (12.52) with $B = \text{diag}(1, 1'_m)$, is equivalent to the optimal estimating equation—that is, (12.52) with $B = B^*$ (Exercise 12.18) given by (12.53). So, in the balanced case, the second-step estimators do not improve the first-step estimators.

Next, we consider the unbalanced case. A simulation study is carried out to compare the performance of the first- and second-step estimators. Here, we have $m = 100$, $n_i = 2$, $1 \leq i \leq 50$, and $n_i = 6$, $51 \leq i \leq 100$. The true parameters were chosen as $\mu = 0.2$ and $\sigma = 1.0$. The results based on 1000 simulations are summarized in Table 12.1, where SD represents the simulated standard deviation and the overall MSE is the MSE of the estimator of $\mu$ plus that of the estimator of $\sigma$. It is seen that the second-step estimators have about 43% reduction of the overall MSE over the first-step estimators.

**Table 12.1. Simulation results: mixed logistic model**

| Method of Estimation | Estimator of $\mu$ | | | Estimator of $\sigma$ | | | Overall MSE |
|---|---|---|---|---|---|---|---|
| | Mean | Bias | SD | Mean | Bias | SD | |
| 1st step | .21 | .01 | .16 | .98 | −.02 | .34 | .15 |
| 2nd step | .19 | −.01 | .16 | .98 | −.02 | .24 | .08 |

## 12.5 Mixed model selection

Model selection and model diagnostics, discussed previously in Section 12.3, are connected in that both are associated with the validity of the assumed model or models. However, unlike model diagnostics, in which a given model is checked for its appropriateness, model selection usually deals with a class of (more than one) potential, or candidate, models, in an effort to choose an "optimal" one among this class. For example, it is possible that a given model is found inappropriate by model diagnostics but is the optimal model by model selection among the class of candidate models. This simply means that no other candidate model is "more appropriate" than the current one, but it does not imply that the current one is "appropriate." Model selection is needed when a choice has to be made. On the other hand, there is much to say about how to make a good choice.

Earlier in Section 9.3, we introduced the information criteria for model selection in the case of a time series. As noted, the information criteria may be expressed in the general form of (9.25), where $\hat{D}_M$ is a measure of lack-of-fit by the candidate model $M$ with dimension $|M|$ and $\lambda_n$ is a penalty. Here, $n$ is supposed to be the "effective sample size." For example, in the case of i.i.d. observations, the effective sample size is equal to the (total) sample size. However, in the case of mixed effects models, it is less clear what the effective sample size is (although, in many cases, it is clear that the effective sample size is *not* equal to the sample size). Consider, for example, Example 12.8. As discussed in Example 12.7, the effective sample sizes for estimating $\sigma_1^2$, $\sigma_2^2$, and $\tau^2$ are $m_1$, $m_2$, and $m_1 m_2$, respectively, whereas the total sample size is $n = m_1 m_2$. Now, suppose that one wishes to select the fixed covariates, which are the components of $x_{ij}$, using the BIC. It is not clear what should be in place of $n$ in the $\lambda_n = \log(n)$ penalty. It does not seem to make sense to use the total sample size $n = m_1 m_2$.

Furthermore, in many cases, the (joint) distribution of the observations is not fully specified (up to some unknown parameters) under the assumed model. Thus, an information criteria that is dependent on the likelihood function may encounter difficulties. Once again, let us consider Example 12.8. Suppose that normality is not assumed. Instead, a non-Gaussian linear mixed model is considered (see Section 12.1). Now, suppose that one, again, wishes to select the fixed covariates using the AIC, BIC, or HQ (see Section 9.3). It is not clear how to do this because all three criteria require the likelihood function in order to evaluate $\hat{D}_M$.

Even in the i.i.d. case, there are still some practical difficulties in using the information criteria. For example, the BIC is known to have tendency of over-penalizing. In other words, the penalty $\lambda_n = \log(n)$ may be a little too much in some cases. On the other hand, the HQ criterion with $\lambda_n = c \log\{\log(n)\}$, where $c$ is a constant greater than 2, is supposed to have a lighter penalty than the BIC, but this is the case only if $n$ is sufficiently large. In a finite-sample situation, the story can be quite different. For example, for $n = 100$ we have $\log(n) = 4.6$ and $\log\{\log(n)\} = 1.5$; hence, if the constant $c$ in the HQ is chosen as 3, the BIC and HQ are almost the same. This raises another practical issue; that is, how to choose the constant $c$? In a large sample (i.e., when $n \to \infty$), the choice of $c$ does not make a difference in terms of consistency of model selection (see Section 9.3). However, in the case of a moderate sample size, the performance of the HQ may be sensitive to the choice of $c$.

Now, let us consider a different strategy for model selection. The basic idea of the information criteria may be viewed as trading off model fit with model complexity. Perhaps, we can do this in a different way. More specifically, the first term in (9.25), $\hat{D}_M$, is a measure of how good a model fits the data. If this is the only thing we have in mind, then "bigger" model always wins (why?). However, this is not going to happen so easily because the bigger model also gets penalized more due to the presence of the second term in (9.25), $\lambda_n |M|$. For simplicity, let us assume that there is a full model among the candidate models, say, $M_f$. Then this is the model that fits the best (why?). A model is called *optimal* if it is a true model with minimum dimension. For example, in the linear mixed model (12.1), a true model for $X\beta$ satisfies (12.1), but some of the components of $\beta$ may be zero. An optimal model for $X\beta$ is one that satisfies (12.1) with all the components of $\beta$ nonzero. Let $Q(M) = Q(M, y, \theta_M)$ denote a measure of lack-of-fit, where $y$ is the vector of observations, $\theta_M$ is the vector of parameters under $M$, and by measure of lack-of-fit it, means that $Q(M)$ satisfies the minimal requirement that $\mathrm{E}\{Q(M)\}$ is minimized when $M$

is a true model (but not necessarily optimal), and $\theta_M$ is the true parameter vector. We consider some examples.

*Example 12.12* (Negative log-likelihood). Suppose that the joint distribution of $y$ belongs to a family of parametric distributions $\{P_{M,\theta_M}, M \in \mathcal{M}, \theta_M \in \Theta_M\}$. Let $P_{M,\theta_M}$ have a (joint) pdf $f_M(\cdot|\theta_M)$ with respect to a $\sigma$-finite measure $\mu$. Consider

$$Q(M, y, \theta_M) = - \log\{f_M(y|\theta_M)\},$$

the negative log-likelihood. This is a measure of lack-of-fit (Exercise 12.19).

*Example 12.13* (Residual sum of squares). Consider the problem of selecting the covariates for a linear model so that $E(y) = X\beta$, where $X$ is a matrix of covariates whose columns are to be selected from a number of candidates $X_1, \ldots, X_K$ and $\beta$ is a vector of regression coefficients. A candidate model $M$ corresponds to $X_M \beta_M$, where the columns of $X_M$ are a subset of $X_1, \ldots, X_K$ and $\beta_M$ is a vector of regression coefficients of suitable dimension. Consider

$$Q(M, y, \beta_M) = |y - X_M \beta_M|^2,$$

which corresponds to the residual sum of squares (RSS). Here, again, we have a measure of lack-of-fit (Exercise 12.20).

Let $\hat{Q}(M) = \inf_{\theta_M \in \Theta_M} Q(M)$. Because $M_f$ is a full model, we must have $\hat{Q}(M_f) = \min_{M \in \mathcal{M}} \hat{Q}(M)$, where $\mathcal{M}$ denotes the set of candidate models. The value $\hat{Q}(M_f)$ is considered as a *baseline*. Intuitively, if $M$ is a true model, the difference $\hat{Q}(M) - \hat{Q}(M_f)$ should be (nonnegative but) sufficiently close to zero. This means that

$$\hat{Q}(M) - \hat{Q}(M_f) \leq c \tag{12.55}$$

for some threshold value $c$. The right side of (12.55) serves as a "fence" to confine the true models and exclude the incorrect ones. Once the fence is constructed, the optimal model is selected from those within the fence—that is, models that satisfy (12.55), according to the minimum dimension criterion. The procedure is therefore called *fence method* (Jiang et al. 2008). The minimum dimension criterion used in selecting models within the fence may be replaced by other criteria of optimality and thus gives flexibility to the fence to take scientific or economic considerations into account.

It is clear that the threshold value $c$ plays an important role here: It divides the true models and incorrect ones. Note that, in practice, we do not know which model is a true model and which one is incorrect (otherwise, model selection would not be necessary). Therefore, we need to know how much the difference on the left side of (12.55) is likely to be when $M$ is a true model and how much this is different when $M$ is incorrect. The hope is that the difference is of lower order for a true model than for an incorrect model; then, asymptotically, we would be able to separate these two groups. To be more specific, suppose that we wish to select the fixed covariates in the linear mixed model (12.1) using the RSS measure of Example 12.13. In this case, a model $M$ corresponds to a matrix $X$ of covariates. Therefore, we use $M$ and $X$ interchangeably. In particular, $M_f$ corresponds to $X_f$. Similarly, let $\beta$ and $\beta_f$ correspond to $X$ and $X_f$, respectively, for notation simplicity. Furthermore, we assume that there exists a true model among the candidate models. It follows that $M_f$ is a true model (why?). The minimizer of $Q(M)$ over all $\beta$ is the least squares estimator (see Section 6.7), $\hat{\beta} = (X'X)^{-1}X'y$. Here, for simplicity (and without loss of generality), we assume that all of the $X$'s are full rank. It follows that $\hat{Q}(M) = |y - X\hat{\beta}|^2 = y'P_{X\perp}y$, where $P_{X\perp} = I - P_X$ and $P_X = X(X'X)^{-1}X'$ is the projection matrix onto the linear space spanned by the columns of $X$ (and $P_{X\perp}$ the orthogonal projection). It follows that

$$\hat{Q}(M) - \hat{Q}(M_f) = y'(P_{X\perp} - P_{X_f^\perp})y$$
$$= y'(P_{X_f} - P_X)y. \tag{12.56}$$

First, assume that $M$ is a true model. This means that (12.1) holds; hence, $y - X\beta = Z\alpha + \epsilon \equiv \xi$, where $\beta$ is the true parameter vector corresponding to $X$. Thus, we have

$$(P_{X_f} - P_X)y = (P_{X_f} - P_X)(y - X\beta)$$
$$= (P_{X_f} - P_X)\xi \tag{12.57}$$

because $(P_{X_f} - P_X)X = 0$ (why?). Also, note that $P_{X_f} - P_X$ is idempotent— that is, $(P_{X_f} - P_X)^2 = P_{X_f} - P_X$ (Exercise 12.21). Therefore, by (12.56), we have (make sure that you follow every step of this derivation)

$$\begin{aligned} E\{\hat{Q}(M) - \hat{Q}(M_f)\} &= E\{y'(P_{X_f} - P_X)y\} \\ &= E\{y'(P_{X_f} - P_X)^2 y\} \\ &= E\{\xi'(P_{X_f} - P_X)^2 \xi\} \\ &= E\{\xi'(P_{X_f} - P_X)\xi\} \\ &= E[tr\{\xi'(P_{X_f} - P_X)\xi\}] \\ &= E[tr\{(P_{X_f} - P_X)\xi\xi'\}] \\ &= tr[E\{(P_{X_f} - P_X)\xi\xi'\}] \\ &= tr\{(P_{X_f} - P_X)E(\xi\xi')\} \\ &= tr\{(P_{X_f} - P_X)(ZGZ' + R)\}. \tag{12.58} \end{aligned}$$

Now, suppose that $M$ is incorrect. Then $y - X\beta = \xi$ no longer holds. However, because $M_f$ is a true model according to our assumption, we have $y - X_f\beta_f = \xi$, where $\beta_f$ is the true parameter vector corresponding to $X_f$. Thus, we can write $y = X_f\beta_f + y - X_f\beta_f = X_f\beta_f + \xi$, and, similar to (12.57), $(P_{X_f} - P_X)y = (P_{X_f} - P_X)\xi + (P_{X_f} - P_X)X_f\beta_f$,

$$y'(P_{X_f} - P_X)y = y'(P_{X_f} - P_X)^2 y$$
$$= \xi'(P_{X_f} - P_X)\xi$$
$$+ 2\beta_f' X_f'(P_{X_f} - P_X)\xi$$
$$+ |(X_f - P_X X_f)\beta_f|^2. \tag{12.59}$$

Note that $(P_{X_f} - P_X)X_f = X_f - P_X X_f$. The second term on the right side of (12.59) has mean 0, and the expectation of the first term is already computed as the right side of (12.58). It follows that

$$\mathrm{E}\{\hat{Q}(M) - \hat{Q}(M_f)\} = \mathrm{tr}\{(P_{X_f} - P_X)(ZGZ' + R)\}$$
$$+ |(X_f - P_X X_f)\beta_f|^2. \tag{12.60}$$

Comparing (12.58) and (12.60), we see that when $M$ is incorrect, the expectation of the left side of (12.55) (which is nonnegative) has an extra term compared to the expectation when $M$ is a true model. We now show, by an example, that this extra term makes a greater difference.

*Example 12.14.* Consider the following linear mixed model:

$$y_{ij} = x_{ij}'\beta + \alpha_i + \epsilon_{ij}, \quad i = 1,\ldots,m, j = 1,2,$$

where $\alpha_i$ and $\epsilon_{ij}$ are the same as in Example 12.2, $x_{ij}$ is a vector of known covariates whose components are to be selected, and $\beta$ is the corresponding vector of fixed effects. More specifically, there are three candidates for $x_{ij}$: 1, $1_{(j=2)}$, and $[1, 1_{(j=2)}]'$, which correspond to
(I) $y_{ij} = \beta_0 + \alpha_i + \epsilon_{ij}, i = 1,\ldots,m, j = 1,2$;
(II) $y_{i1} = \alpha_i + \epsilon_{i1}, y_{i2} = \beta_1 + \alpha_i + \epsilon_{i2}, i = 1,\ldots,m$;
(III) $y_{i1} = \beta_0 + \alpha_i + \epsilon_{i1}, y_{i2} = \beta_0 + \beta_1 + \alpha_i + \epsilon_{i2}, i = 1,\ldots,m$,
respectively (the values of the $\beta$ coefficients may be different even if the same notation is used). Suppose that model I is a true model; then it is the optimal model. Furthermore, model III is the full model, which is also a true model, but not optimal; and model II is an incorrect model, unless $\beta_0 = 0$ (why?). It is more convenient to use the notation and properties of the Kronecker products (see Appendix A.1). Recall that $1_2$, $I_2$, and $J_2$ represent the 2-dimensional vector of 1's, identity matrix and matrix of 1's, respectively. Also, let $\delta_2 = (0,1)'$. We refer some of the details to an exercise (Exercise 12.22). We have $P_{X_f} = m^{-1}J_m \otimes I_2$, $Z = I_m \otimes 1_2$, $G = \sigma^2 I_m$, and $R = \tau^2 I_m \otimes I_2$. Furthermore, if $X$ corresponds to model I, we have $P_X = (2m)^{-1}J_m \otimes J_2$. It follows that the right side of (12.58) is equal to $\tau^2$. On the other hand, if $X$ corresponds to model II, we have $P_X = m^{-1}J_m \otimes (\delta_2 \delta_2')$. Thus, the first term on the right side of (12.60) is $\sigma^2 + \tau^2$, and the second term is $\beta_0^2 m$. In conclusion, we have

$$\mathrm{E}\{\hat{Q}(M) - \hat{Q}(M_f)\} = \begin{cases} \tau^2 & \text{if } M = \text{model I} \\ \sigma^2 + \tau^2 + \beta_0^2 m & \text{if } M = \text{model II}, \end{cases}$$

where $\beta_0$ is the true fixed effect under model I and $\sigma^2$ and $\tau^2$ are the true variance components. Thus, if $m$ is large, there is a big difference in the expected value of the left side of (12.55) between a true model and an incorrect one, provided that $\beta_0 \neq 0$. On the other hand, if $\beta_0 = 0$, the difference between model I and model II is much less significant (intuitively, does it make sense?).

What Example 12.14 shows is something that holds quite generally. Let $d_M$ denote the left side of (12.55). We can expect that the order of $E(d_M)$ is lower when $M$ is a true model, and the order is (much) higher when $M$ is incorrect. For example, in Example 12.14, we have $E(d_M) = O(1)$ when $M$ is true and $E(d_M) = O(m)$ when $M$ is incorrect. This suggests that, perhaps, there is a cutoff in the middle. This is the main idea of the fence. Of course, there are some technical details and rigorous arguments, but let us first be clear about the main idea, as details can be filled in later. For example, part of the "details" is the following. We know that $E(d_M)$'s are of different orders, but what about $d_M$'s themselves? Note that we can write

$$d_M = E(d_M) + d_M - E(d_M)$$
$$= E(d_M) + \sqrt{\mathrm{var}(d_M)} \times \frac{d_M - E(d_M)}{\sqrt{\mathrm{var}(d_M)}},$$

and there is a reason to write it this way. Under regularity conditions, we have

$$\frac{d_M - E(d_M)}{\sqrt{\mathrm{var}(d_M)}} \xrightarrow{\mathrm{d}} N(0,1). \tag{12.61}$$

Again, details later, but let us say that (12.61) holds. The consequence is that

$$d_M = E(d_M) + O_P\{\mathrm{s.d.}(d_M)\}, \tag{12.62}$$

where $\mathrm{s.d.}(d_M) = \sqrt{\mathrm{var}(d_M)}$ (s.d. refers to standard deviation). We know that there is a difference in the order of the first term on the right side of (12.62). If the order of $\mathrm{s.d.}(d_M)$ "stays in the middle," then we would have a cutoff not only in terms of $E(d_M)$ but also in terms of $d_M$ between the true and incorrect models. To see that this is something we can expect, let us return to Example 12.14.

*Example 12.14 (continued).* We outline an argument that justifies the asymptotic normality (12.61). In the process, we also obtain the order of $\mathrm{s.d.}(d_M)$. Let $\mu$ denote the (true) mean vector of $y$. Then we can write $y = \mu + \eta$, where $\eta = Z\alpha + \epsilon = W\xi$ with $W = (Z\ I)$ and $\xi = (\alpha'\ \epsilon')'$. Note that $\xi$ is a $3m$-dimensional vector of independent random variables with mean 0. Thus, we can write, by (12.56),

$$d_M = \mu'(P_{X_f} - P_X)\mu + 2\mu'(P_{X_f} - P_X)\eta + \eta'(P_{X_f} - P_X)\eta$$
$$= \mu'(P_{X_f} - P_X)\mu + a'\xi + \xi'A\xi$$
$$= \mu'(P_{X_f} - P_X)\mu + E(\xi'A\xi)$$
$$\quad + a'\xi + \xi'A\xi - E(\xi'A\xi),$$

where $a = 2W'(P_{X_f} - P_X)\mu$ and $A = W'(P_{X_f} - P_X)W$. It follows that $E(d_M) = \mu'(P_{X_f} - P_X)\mu + E(\xi'A\xi)$. Therefore, with $\xi = (\xi_i)_{1 \le i \le 3m}$, $a = (a_i)_{1 \le i \le 3m}$, and $A = (a_{ij})_{1 \le i,j \le 3m}$, we have (see Example 8.3)

$$
\begin{aligned}
&d_M - E(d_M) \\
&= \sum_{i=1}^{3m} a_i \xi_i + \sum_{i=1}^{3m} a_{ii} \{\xi_i^2 - E(\xi_i^2)\} \\
&\quad + 2 \sum_{i=1}^{3m} \sum_{j<i} a_{ij} \xi_i \xi_j \\
&= \sum_{i=1}^{3m} \left[ a_i \xi_i + a_{ii} \{\xi_i^2 - E(\xi_i^2)\} + 2 \left( \sum_{j<i} a_{ij} \xi_j \right) \xi_i \right].
\end{aligned}
\tag{12.63}
$$

The summand in the last expression is a sequence of martingale differences with respect to the $\sigma$-fields $\mathcal{F}_i = \sigma(\xi_1, \ldots, \xi_i)$. Thus, the martingale central limit (Theorem 8.7) can be applied to establish (12.61). A suitable normalizing constant needs to be chosen in order to apply Theorem 8.7, which is asymptotically equivalent to, by (12.63),

$$
\text{var}(d_M) = \sum_{i=1}^{m} \text{var} \left[ a_i \xi_i + a_{ii} \{\xi_i^2 - E(\xi_i^2)\} + 2 \left( \sum_{j<i} a_{ij} \xi_j \right) \xi_i \right].
\tag{12.64}
$$

It can be shown that the right side of (12.64) is $O(m)$ (Exercise 12.23). It follows that s.d.$(d_M) = O(\sqrt{m})$. Therefore, by (12.62), $d_M = O_P(\sqrt{m})$ if $M$ is true and $d_M = O_P(m)$ is $M$ is incorrect. In fact, in this case, it can be shown that $d_M = O_P(1)$ if $M$ is true (Exercise 12.23).

Suppose that when $M$ is a true model, the second term on the right side of (12.62) dominates the first term. This means that the order of $E(d_M)$ is the same as or lower than that of s.d.$(d_M)$ if $M$ is true. Then the threshold value $c$ on the right side of (12.55) should be of the same order as s.d.$(d_M)$. For such a reason, Jiang et al. (2008) suggested that $c = c_1 \widehat{\text{s.d.}}(d_M)$, where $\widehat{\text{s.d.}}(d_M)$ is an estimator of s.d.$(d_M)$ and $c_1$ is a tuning constant. Note that now the threshold $c$ may depend on $M$, but $c_1$ is independent of the candidate models. It can be shown that if $c_1$ increases "slowly" as the overall sample size, $n$, increases, then the fence method is consistent in the sense that with probability tending to 1 the procedure will select the optimal model if it is among the candidates. The detailed condition on how slowly $c_1$ increases can be found in Jiang et al. (2008). Although in a large sample, the choice of $c_1$ does not make a difference in terms of the consistency, finite-sample performance of the fence may be influenced by the value of the tuning constant. To solve this problem, Jiang et al. (2008) developed the following strategy called the *adaptive fence*.

The idea is to let data determine the best tuning constant. First note that, ideally, one wishes to select $c_1$ that maximizes the probability of choosing the optimal model. Here, to be specific, the optimal model is understood as a true model that has the minimum dimension among all of the true models. This means that one wishes to choose $c_1$ that maximizes

$$P = \mathrm{P}(M_0 = M_{\mathrm{opt}}), \tag{12.65}$$

where $M_{\mathrm{opt}}$ represents the optimal model and $M_0 = M_0(c_1)$ is the model selected by the fence procedure with the given $c_1$. However, two things are unknown in (12.65): (i) Under what distribution should the probability P be computed? (ii) What is $M_{\mathrm{opt}}$? (If we knew the answer to the latter question, model selection would not be necessary.)

To solve problem (i), let us assume that there is a true model among the candidate models. It follows that the full model, $M_{\mathrm{f}}$, is a true model. Therefore, it is possible to bootstrap (see Chapter 14) under $M_{\mathrm{f}}$. For example, one may estimate the parameters under $M_{\mathrm{f}}$ and then use a model-based bootstrap to draw samples under $M_{\mathrm{f}}$. This allows us to approximate the probability distribution P on the right side of (12.65).

To solve problem (ii), we use the idea of maximum likelihood; namely, let $p^*(M) = \mathrm{P}^*(M_0 = M)$, where $\mathrm{P}^*$ denotes the empirical probability obtained by the bootstrapping. In other words, $p^*(M)$ is the sample proportion of times out of the total number of bootstrap samples that model $M$ is selected by the fence method with the given $c_1$. Let $p^* = \max_{M \in \mathcal{M}} p^*(M)$, where $\mathcal{M}$ denotes the set of candidate models. Note that $p^*$ depends on $c_1$. The idea is to choose $c_1$ that maximizes $p^*$. It should be kept in mind that the maximization is not without restriction. To see this, note that if $c_1 = 0$, then $p^* = 1$ (because when $c_1 = 0$, the procedure always chooses $M_{\mathrm{f}}$). Similarly, $p^* = 1$ for very large $c_1$ if there is a unique model, $M_*$, with the minimum dimension (because when $c_1$ is large enough, the procedure always chooses $M_*$). Therefore, what one looks for is "the peak in the middle" of the plot of $p^*$ against $c_1$.

Here is another look at the adaptive fence. Typically, the optimal model is the one from which the data are generated; then this model should be the most likely given the data. Thus, given $c_1$, one is looking for the model (using the fence procedure) that is most supported by the data or, in other words, one that has the highest (posterior) probability (see Chapter 15). The latter is estimated by bootstrapping. Note that although the bootstrap samples are generated under $M_{\mathrm{f}}$, they are almost the same as those generated under the optimal model. This is because, for example, the estimates corresponding to the zero parameters are expected to be close to zero, provided that the parameter estimation under $M_{\mathrm{f}}$ is consistent. One then pulls off the $c_1$ that maximizes the (posterior) probability and this is the optimal choice, say, $c_1^*$.

It can be shown that, like the fence method (with the nonadaptive $c_1$), the adaptive fence is consistent under suitable regularity conditions. This means that, with probability tending to 1, the fence procedure with $c = c_1^* \widehat{\mathrm{s.d.}}(d_M)$ in (12.55) will select $M_{\mathrm{opt}}$. One of the key conditions is that the bootstrap provides an aymptotically accurate approximation to the probability P in

(12.65). Some related topics will be discussed in Chapter 14. Furthermore, simulation results reported by Jiang et al. (2008) (also see Jiang, Nguyen, and Rao 2009) suggest that the adaptive fence significantly outperforms the information criteria as well as the nonadaptive fence in mixed model selection.

Another attractive feature of the fence method is its computational advantage, as compared to the information criteria; that is, one may not need to search for the entire model space (i.e., all of the candidate models) in order to find the optimal model. Once again, suppose that the minimum-dimension criterion is used in selecting models within the fence. Then one can always begin by checking the model that has the minimum dimension (for membership within the fence) and let the dimension gratually increase. Let $M_1$ be the first model that falls in the fence with $|M_1| = d_1$. At this point, one needs to check all of the models with dimension $d_1$ to see if there is any other model(s) that fall in the fence and then select the model with dimension $d_1$ that has the minimum $\hat{Q}(M)$, which is the optimal model (why?). A numerical procedure based on this idea, known as the *fence algorithm*, has been developed.

## 12.6 Exercises

*12.1.* Show that, in Example 12.1, the correlation between any two observations from the same individual is $\sigma^2/(\sigma^2 + \tau^2)$, whereas observations from different individuals are uncorrelated.

*12.2* (Two-way random effects model). For simplicity, let us consider the case of one observation per cell. In this case, the observations $y_{ij}$, $i = 1, \ldots, m$, $j = 1, \ldots, k$, satisfy $y_{ij} = \mu + u_i + v_j + e_{ij}$ for all $i$ and $j$, where $\mu$ is as in Example 1.1; $u_i$, $i = 1, \ldots, m$ and $v_j$, $j = 1, \ldots, k$, are independent random effects such that $u_i \sim N(0, \sigma_1^2)$, $v_j \sim N(0, \sigma_2^2)$; and $e_{ij}$'s are independent errors distributed as $N(0, \tau^2)$. Again, assume that the random effects and errors are independent. Show that this model is a special case of the mixed ANOVA model but not a special case of the longitudinal model.

*12.3.* Give an example of a special case of the longitudinal model that is not a special case of the mixed ANOVA model.

*12.4.* Suppose that, given a vector of random effects, $\alpha$, observations $y_1, \ldots, y_n$ are (conditionally) independent such that $y_i \sim N(x_i'\beta + z_i'\alpha, \tau^2)$, where $x_i$ and $z_i$ are known vectors, $\beta$ is an unknown vector of regression coefficients, and $\tau^2$ is an unknown variance. Furthermore, suppose that $\alpha$ is multivariate normal with mean 0 and covariance matrix $G$, which depends on a vector $\theta$ of unknown variance components. Let $X$ and $Z$ be the matrices whose $i$th rows are $x_i'$ and $z_i'$, respectively. Show that the vector of observations, $y = (y_1, \ldots, y_n)'$, has the same distribution as the Gaussian linear mixed model (1.1), where $\alpha \sim N(0, G)$, $\epsilon \sim N(0, \tau^2 I)$, and $\alpha$ and $\epsilon$ are independent.

*12.5.* Show that, in Example 12.5, the MLE of $\sigma^2$ is given by (12.7), and the REML estimator of $\sigma^2$ is given by (12.8).

*12.6.* Show that the REML estimators of the variance components (see Section 12.2) do not depend on the choice of $A$. This means that if $B$ is another $n \times (n - p)$ matrix of full rank such that $B'X = 0$, the MLEs of the variance components based on $A'y$ (assuming normality) are the same as those based on $B'y$.

*12.7.* Consider the following linear mixed models: $y_{ijk} = \beta_i + \alpha_j + \gamma_{ij} + \epsilon_{ijk}$, $i = 1, \ldots, a$, $j = 1, \ldots, b$, $k = 1, \ldots, c$, where $\beta_i$'s are fixed effects, the $\alpha_j$'s are random effects, the $\gamma_{ij}$'s are the interactions between the fixed and random effects, which are also considered as random effects, and the $\epsilon_{ijk}$'s are errors. Show that in this case, (12.1) and (12.2) can be expressed as

$$y = I_a \otimes 1_b \otimes 1_c \beta + 1_a \otimes I_b \otimes 1_c \alpha + I_a \otimes I_b \otimes 1_c \gamma + \epsilon.$$

*12.8.* This exercise is concerned with some of the arguments used in the proof of Theorem 12.1.

(i) Show that (12.14) holds under (12.10)–(12.12), where $\theta_n$ is defined by (12.13) and $r_n$ is uniformly ignorable compared with the first term as $\tilde{\theta}$ varies on $\bar{E}_n = \{\tilde{\theta} : |P_n(\tilde{\theta} - \theta)| = 1\}$.

(ii) Derive (12.15) using the Taylor expansion and the result of part (i) of the theorem.

*12.9.* Show that the components of $a_n$ in (12.15) [defined below (12.12)] are quadratic forms of the random effects and errors.

*12.10.* The additional assumption in part (ii) of Theorem 12.1 that the random effects and errors are nondegenerate is necessary for the asymptotic normality of the REML estimators. To see a simple example, suppose that $y_i = \mu + \epsilon_i$, $i = 1, \ldots, m$, where the $\epsilon_i$'s are independent such that $P(\epsilon_i = -1) = P(\epsilon_i = 1) = 1/2$. Note that in this case there are no random effects and the $\epsilon_i$'s are the errors.

(i) Show that the REML estimator of the variance of the errors, $\sigma^2 = 1$, is the sample variance, $\hat{\sigma}^2 = (m - 1)^{-1} \sum_{i=1}^m (y_i - \bar{y})^2$, where $\bar{y} = m^{-1} \sum_{i=1}^m y_i$.

(ii) Show that $\sqrt{m}(\hat{\sigma}^2 - 1)$ does not converge in distribution to a (nondegenerate) normal distribution.

*12.11.* This exercise is regarding the projection method that begins with the identity (12.28).

(i) Show that $E(\sum_{i=1}^{m_1} \sum_{j=1}^{m_2} \delta_{1,ijk})^2 = O(m_1 m_2)$. [Hint: Note that, given $u$ and $v$, the $\delta_{1,ijk}$'s are conditionally independent with $E(\delta_{1,ijk}|u, v) = 0$.]

(ii) Show that (a) if $i_1 \neq i_2, j_1 \neq j_2$, then $\delta_{2,i_1j_1k}$ and $\delta_{2,i_2j_2k}$ are independent; (b) if $j_1 \neq j_2$, then $E(\delta_{2,ij_1k}\delta_{2,ij_2k}|u) = 0$; and (c) if $i_1 \neq i_2$, then $E(\delta_{2,i_1jk}\delta_{2,i_2jk}|v) = 0$. It follows that $E(\sum_{i=1}^{m_1} \sum_{j=1}^{m_2} \delta_{2,ijk})^2 = O(m_1 m_2)$.

(iii) Show that $\sum_{i=1}^{m_1} \sum_{j=1}^{m_2} \zeta_{1,ijk} = \sum_{i=1}^{m_1} \zeta_{1,ik}(u_i)$, where $\zeta_{1,ik}(u_i) = \sum_{j=1}^{m_2} \zeta_{1,ijk}$ is a function of $u_i$. Therefore, we have $\sum_{i=1}^{m_1} \sum_{j=1}^{m_2} \zeta_{1,ijk} = O_P(m_1^{1/2} m_2)$.

(iv) Show that $\sum_{i=1}^{m_1} \sum_{j=1}^{m_2} \zeta_{2,ijk} = \sum_{j=1}^{m_2} \zeta_{2,jk}(v_j)$, where $\zeta_{2,jk}(v_j) = \sum_{i=1}^{m_1} \zeta_{2,ijk}$ is a function of $v_j$. Therefore, we have $\sum_{i=1}^{m_1} \sum_{j=1}^{m_2} \zeta_{2,ijk} = O_P(m_1 m_2^{1/2})$.

*12.12.* Show that the number of coefficents $a_{st}$ in the bivariate polynomial (12.32) is $N_M = 1 + M(M+3)/2$.

*12.13.* Show that the likelihood function in Example 12.9 for estimating $\psi$ can be expressed as (12.35).

*12.14.* Verify (12.38) and obtain an expression for $r$. Show that $r$ has expectation 0.

*12.15.* This exercise is related to Example 12.10.

(i) Show that (12.42) has a unique solution for $u$ when everything else is fixed.

(ii) Show that the PQL estimator of $\beta$ satisfies (12.43).

(iii) Derive (12.44).

(iv) Consider the following special case: $x'_{ij}\beta = \mu$, where $\mu$ is an unknown parameter, and $n_i = l, 1 \le i \le m$. Show that in this case, the left side of (12.46) is equal to $g(\mu) = \sigma^{-2}\sum_{k=0}^{l}\phi(k,\mu)P(y_1 = k)$, where $\phi(t,\mu)$ is the unique solution $u$ to the equation $\sigma^{-2}u + lh(\mu + u) = t$ and

$$P(y_1 = k) = \binom{l}{k}E\left[\frac{\{\exp(\mu + \eta)\}^k}{\{1 + \exp(\mu + \eta)\}^l}\right]$$

with the expectation taken with respect to $\eta \sim N(0, \sigma^2)$.

(v) Take $\sigma^2 = 1$. Make a plot of $g(\mu)$ against $\mu$ and show that $g(\mu)$ is not identical to zero (function). [Hint: You make use numerical integration or the Monte Carlo method to evaluate the expectations involved in (iv).]

*12.16.* Show that under a GLMM, $E(y_i^2)$ depends on $\phi$, the dispersion parameter in (12.4), but $E(y_i y_{i'})$ does not depend on $\phi$ if $i \ne i'$.

*12.17.* Show that for any matrices $B$, $U$, and $V$ such that $V > 0$ (positive definite), $U$ is full rank, and $BU$ is square and nonsingular, we have

$$\{(BU)^{-1}\}BVB'\{(BU)^{-1}\}' \ge (U'V^{-1}U)^{-1}$$

(i.e., the difference of the two sides is nonnegative definite). (Hint: The proof is very similar to that of Lemma 5.1.)

*12.18.* Consider a special case of Example 12.11 with $n_i = k$, $1 \le i \le m$, where $k \ge 2$. Show that in this case, the estimating equation of Jiang (1998a), which is (12.52) with $B = \text{diag}(1, 1'_m)$, is equivalent to the optimal estimating equation—that is, (12.52) with $B = B^*$ given by (12.53).

*12.19.* Show that the negative log-likelihood measure defined in Example 12.12 is a measure of lack-of-fit according to the definition above Example 12.12.

*12.20.* Show that the RSS measure defined in Example 12.13 is a measure of lack-of-fit according to the definition above Example 12.12.

*12.21.* This exercise involves some of the details in the derivations following (12.56).

(i) Show that $(P_{X_f} - P_X)X = 0$.

(ii) Show that $P_{X_f} - P_X$ is idempotent; that is, $(P_{X_f} - P_X)^2 = P_{X_f} - P_X$.

*12.22.* This exercise involves some details regarding Example 12.14.

(i) Show that $P_{X_f} = m^{-1} J_m \otimes I_2$.

(ii) Show that $P_X = (2m)^{-1} J_m \otimes J_2$ for $X$ corresponding to model I and $P_X = m^{-1} J_m \otimes (\delta_2 \delta_2')$ for $X$ corresponding to model II.

(iii) Show that for model I, the right side of (12.58) is equal to $\tau^2$, where $\tau^2$ is the true variance of the errors.

(iv) Show that for model II, the right side of (12.60) is equal to $\sigma^2 + \tau^2 + \beta_0^2 m$, where $\beta_0$ is the true fixed effect under model I and $\sigma^2$ and $\tau^2$ are the true variance components.

12.23. Show that the right side of (12.64) is $O(m)$. In fact, in this case, it can be shown that the order is $O(1)$ when $M$ is true [Hint: The most challenging part is to evaluate $\sum_{i,j} a_{ij}^2 = \text{tr}(A^2)$. Note that

$$\text{tr}(A^2) = \text{tr}\{(P_{X_f} - P_X)(ZZ' + I)(P_{X_f} - P_X)(ZZ' + I)\}.$$

Consider the two cases, model I and model II, separately, and use the results of the previous exercise.]

# 13

# Small-Area Estimation

It is seldom possible to have a large enough overall sample size to support reliable direct estimates for all the domains of interest. Therefore, it is often necessary to use indirect estimates that "borrow strength" by using values of the variables of interest from related areas, thus increasing the "effective" sample size.

**Rao (2003)**
*Small Area Estimation*

## 13.1 Introduction

The term "small area" typically refers to a population for which reliable statistics of interest cannot be produced due to certain limitations of the available data. Examples of domains include a geographical region (e.g., a state, county or municipality), a demographic group (e.g., a specific age × sex × race group), a demographic group within a geographic region, and so forth. Some of the groundwork in small-area estimation related to the population counts and disease mapping research has been done by the epidemiologists and the demographers. The history of small-area statistics goes back to 11th- century England and 17th- century Canada. See Brackstone (1987) and Marshall (1991).

There are various reasons for the scarcity of direct reliable data on the variables of interest for small areas. In a sample survey, a small-area estimation problem could arise simply due to the sampling design that aims to provide reliable data for larger areas and pays little or no attention to the small areas of interest. For example, in a statewide telephone survey of sample size 4300 in the state of Nebraska (USA), only 14 observations are available to estimate the prevalence of alcohol abuse in Boone county, a small county in Nebraska. The problem is even more severe for a direct survey estimation of the prevalence for white females in the age group 25–44 in this county since only one observation is available from the survey. See Meza et al. (2003) for details.

J. Jiang, *Large Sample Techniques for Statistics,*
DOI 10.1007/978-1-4419-6827-2_13, © Springer Science+Business Media, LLC 2010

Oversampling is often employed in surveys in order to increase sample sizes for some domains, but that leaves other domains with very few samples or even no sample since the total sample size is usually fixed due to the limitation of the survey budget. Various design strategies that incorporate factors influencing small-area data quality have been suggested in the literature (see Rao 2003, pp. 21–24 and the references therein). While these changes in a sampling design generally improve on the small-area estimation, the problem still remains due to the budget and other practical constraints. In some situations, administrative data can be used to produce small-area statistics. Such a statistic does not suffer from the sampling errors but is often subject to measurement errors, resulting in poor quality of small-area statistics. For example, law enforcement records are likely to underreport small-area crime statistics since many crimes are not reported to the law enforcement authorities. Even the count for certain subgroups (e.g., illegal immigrants and certain minority groups) of the population compiled from the census could be of poor quality because of nonresponse and issues related to hard-to-find populations. In a disease mapping problem, the small-area problem arises simply because of the small population size associated with the disease.

On the other hand, small-area statistics are needed in regional planning, apportioning congressional seats, and fund allocation in many government programs. Thus, the importance of producing reliable small-area statistics cannot be overemphasized. For example, in both developed and developing countries, governmental policies increasingly demand income and poverty estimates for small areas. In fact, in the United States more than $130 billion of federal funds are allocated annually based on these estimates. In addition, states utilize these small-area estimates to divide federal funds and their own funds to areas within the state. These funds cover a wide range of community necessities and services, such as education and public health. As another example, mapping a disease incidence over different small areas are useful in the allocation of government resources to various geographical areas and also to identify factors (such as existence of a nuclear reactor near a site) causing a disease. For such reasons, there have been a growing need to refine the manner in which these estimates are produced in order to provide an increased level of precision.

The problem of small-area estimation was previously visited in Section 4.8, where the idea of borrowing strength was introduced. For the most part, "strength" is borrowed by utilizing a statistical model. For example, consider the Fay–Herriot model of Example 4.18. It is assumed that (4.86) holds for all of the small areas $i = 1, \ldots, m$. Therefore, the unknown parameters $\beta$ and $A$ can be estimated using the data from all of the small areas. Also, under the normality model assumption, one is able to derive the best estimator (BP), or predictor, of the small-area means in terms of minimizing the mean squared prediction error (MSPE). The MSPE is an important measure of uncertainty in small-area estimation. Earlier in Section 4.8, a method of obtaining second-order unbiased estimator of the MSPE, known as the Prasad–Rao method, was discussed. The method is suitable, however, only to linear mixed models that are used for small-area estimation. On the other hand, survey data available for small-area estimation are often in the form of binary responses and counts, for which generalized linear mixed models (GLMM) are more appropriate. For example, Malec et al. (1997) discussed a small area estimation problem in the National Health Interview Survey (NHIS), in which the binary responses are indicators of whether the sampled individuals had made at least one visit to a physician within the past year. Therefore, as our first specific topic of the chapter, we discussed an extension of the Prasad–Rao method to small-area estimation with binary observations.

## 13.2 Empirical best prediction with binary data

As discussed in Section 4.8, the Prasad–Rao method is based on the Taylor series expansion. It is appealing to a practitioner for its simplicity of implementation. Following their approach, we consider the following mixed logistic model for binary data $y_{ij}$, the $j$th observation from the $i$th small area. Let $\alpha_i$ be a random effect associated with the $i$th small area, $1 \leq i \leq m$. It is assumed that given $\alpha = (\alpha_i)_{1 \leq i \leq m}$, $y_{ij}$, $i = 1, \ldots, m$, $j = 1, \ldots, n_i$, are conditionally independent binary such that

$$\text{logit}\{P(y_{ij} = 1|\alpha)\} = x'_{ij}\beta + \alpha_i, \tag{13.1}$$

where $x_{ij} = (x_{ijk})_{1 \leq k \leq p}$ is a vector of auxiliary variables that are available and $\beta = (\beta_k)_{1 \leq k \leq p}$ is a vector of unknown parameters. Furthermore, the random effects $\alpha_1, \ldots, \alpha_m$ are independent and distributed as $N(0, \sigma^2)$ with $\sigma^2$ being an unknown variance. It is clear that the mixed logistic model is a special case of GLMM discussed in the previous chapter (see Example 12.3).

Our main interest is to estimate the small-area means. Let $N_i$ be the population size for the $i$th small area. Then the small-area mean is equal to the proportion $p_i = N_i^{-1} \sum_{k=1}^{N_i} Y_{ik}$. The difference between $Y_{ik}$ and $y_{ij}$ is that $Y_{ik}$ is the $k$th value of interest in the population, whereas $y_{ij}$ is the $j$th sampled value (so $y_{ij} = Y_{ik}$ for some $k$). Now, consider (13.1) as a *superpopulation model*, with $j$ replaced by $k$ and $y$ by $Y$, in the sense that the finite population $Y_{ik}, k = 1, \ldots, N_i$, are realizations of random variables satisfying the model. Then we have

$$\frac{1}{N_i} \sum_{k=1}^{N_i} \{Y_{ik} - P(Y_{ik} = 1|\alpha_i)\} = O_P(N_i^{-1/2}) \tag{13.2}$$

(why?). By (13.1) (with $j$ replaced by $k$ and $y$ by $Y$), the left side of (13.2) is equal to $p_i - \zeta_i(\beta, \alpha_i)$, where $\zeta_i(\beta, \alpha_i) = N_i^{-1} \sum_{k=1}^{N_i} h(x'_{i,k}\beta + \alpha_i)$ with $h(x) = e^x/(1 + e^x)$. It follows that, to the order of $O_P(N_i^{-1/2})$, the small-area mean $p_i$ can be approximated by the mixed effect$\zeta_i(\beta, \alpha_i)$. Here, a mixed effect is a (possibly nonlinear) function of the fixed effects $\beta$ and random effect $\alpha_i$. The population size $N_i$ is usually quite large. Note that the small areas are not really "small" in terms of their population sizes. For example, a (U.S.) county is often considered a small area, even though there are tens of thousands of people living in the county; so $N_i$ is large. However, $N_i$ is much smaller compared to the population of the United States, and this is why $n_i$ is small (because the sample size is proportional to the population size). Therefore, the approximation error on the right side of (13.2) is often ignored, and we therefore consider $\xi_i(\beta, \alpha_i)$ as the small-area mean.

The estimation problem now becomes a prediction problem, and we can treat the problem more generally without restricting to a specific function $\zeta_i(\cdot, \cdot)$. It turns out that the functional form of $\zeta_i$ does not create much complication as long as it is smooth. Therefore, for the sake of simplicity, we focus on a special case—that is, the prediction of the area specific random effect

$$\zeta_i = \alpha_i. \tag{13.3}$$

If $\beta$ and $\sigma$ are known, the best predictor (BP) of $\alpha_i$, in the sense of minimum MSPE, is the conditional expectation $\tilde{\alpha}_i = \mathrm{E}(\alpha_i | y) = \mathrm{E}(\alpha_i | y_i)$, where $y_i = (y_{ij})_{1 \le j \le n_i}$. It can be shown that this can be expressed as

$$\begin{aligned}
\tilde{\alpha}_i &= \sigma \frac{\mathrm{E}[\xi \exp\{\phi_i(y_{i\cdot}, \sigma\xi, \beta)\}]}{\mathrm{E}[\exp\{\phi_i(y_{i\cdot}, \sigma\xi, \beta)\}]} \\
&= \psi_i(y_{i\cdot}, \theta), \tag{13.4}
\end{aligned}$$

where $y_{i\cdot} = \sum_{j=1}^{n_i} y_{ij}$, $\theta = (\beta', \sigma)'$, $\phi_i(k, u, v) = ku - \sum_{j=1}^{n_i} \log\{1 + \exp(x_{ij}'v + u)\}$, and the expectation is taken with respect to $\xi \sim N(0, 1)$ (Exercise 13.1). Unlike the linear mixed models (see Section 4.8), here the BP does not have a closed-form expression. Nevertheless, only one-dimensional integrals are involved in the expression, which can be evaluated numerically, either by numerical integration or by the Monte Carlo method. On the other hand, the lack of analytic expression does not keep us from studing the asymptotic behavior of the BP. For example, the following hold:

(i) $\tilde{\alpha}_i/\sigma \to \mathrm{E}(\xi) = 0$ as $\sigma \to 0$. In other words, $\tilde{\alpha}_i = o(\sigma)$ as $\sigma \to 0$.

(ii) For any $1 \le k \le n_i - 1$, as $\sigma \to \infty$,

$$\psi_i(k, \theta) \longrightarrow \frac{\int u \exp\{\phi_i(k, u, \beta)\}\, du}{\int \exp\{\phi_i(k, u, \beta)\}\, du}.$$

(iii) $\psi_i(0, \theta) \to -\infty$ and $\psi_i(n_i, \theta) \to \infty$ as $\sigma \to \infty$.

(iv) Suppose that $x_{ij} = x_i$ (i.e., the auxiliary variables are area level). Then, as $\sigma \to \infty$,

$$\psi_i(k, \theta) \longrightarrow \sum_{l=1}^{k-1} l^{-1} - \sum_{l=1}^{n_i-k-1} l^{-1} - x_i'\beta, \quad 1 \le k \le n_i - 1, n_i \ge 2,$$

where $\sum_1^0(\cdot)$ is understood as zero. Note that $\sum_{l=1}^{k-1} l^{-1} \sim \log(k) + C$, where $C$ is Euler's constant. Therefore, as $\sigma \to \infty$, we have

$$\tilde{\alpha}_i \approx \mathrm{logit}(\bar{y}_{i\cdot}) - x_i'\beta, \tag{13.5}$$

where $\bar{y}_{i\cdot} = n_i^{-1} y_{i\cdot}$, provided that $1 \leq y_{i\cdot} \leq n_i - 1$ and $n_i \geq 2$. Note that (13.5) holds even if $y_{i\cdot} = 0$ or $n_i$.

In practice, the parameters $\theta$ are usually unknown. It is then customary to replace $\theta$ by $\hat{\theta}$, a consistent estimator. The result is called empirical best predictor (EBP), given by

$$\hat{\alpha}_i = \psi_i(y_{i\cdot}, \hat{\theta}). \tag{13.6}$$

We first study the asymptotic behavior of the EBP when $m \to \infty$ and $n_i \to \infty$. The first limiting process is reasonable because the total number of small areas, $m$, is usually quite large. For example, in the NHIS problem discussed in the previous section, the number of small areas of interest is about 600. The second limiting process, however, is unreasonable because the sample size $n_i$ is typically small. On the other hand, it is important to know what would happen in the "ideal" situation, so that one can be sure that the approach is fundamentally sound, at least in large sample. Write

$$\hat{\alpha}_i - \alpha_i = |\psi_i(y_{i\cdot}, \hat{\theta}) - \psi_i(y_{i\cdot}, \theta)| + |\psi_i(y_{i\cdot}, \theta) - \alpha_i^*| + |\alpha_i^* - \alpha_i|, \tag{13.7}$$

where $\alpha_i^*$ is the maximizer of $\phi_i(y_{i\cdot}, u, \beta)$ over $u$. The idea is to show that the three terms on the right side of (13.7) have the orders $O_P(|\hat{\theta} - \theta|)$, $O_P(n_i^{-1})$, and $O_P(n_i^{-1/2})$, respectively. The order of the last term is derived using virtually the same technique as for the MLE (see Section 4.7; Exercise 13.4). The order of the second term is the result of Laplace approximation to integrals (see Section 4.6). More details about the approximation error in the Laplace approximation can be found in Chapter 4 of De Bruijn (1961). As for the first term, let us point out how the idea of the Laplace approximation is, again, useful here. By the Taylor expansion, it suffices to show that $\partial \psi_i / \partial \theta$ is bounded in probability. Write $g_i(v) = -n_i^{-1} \phi_i(y_{i\cdot}, \sigma v, \beta)$. For any function $f = f(y_{i\cdot}, v, \theta)$, define

$$T_i(f) = \frac{\int f \exp(-n_i g_i)\phi \, dv}{\int \exp(-n_i g_i)\phi \, dv},$$

where $\phi$ is the pdf of $N(0,1)$. Let $\varphi_i(y_{i\cdot}, \theta) = T_i(v)$, where $v$ represents the identity function; that is, $f(v) = v$. Then we have $\psi_i(y_{i\cdot}, \theta) = \sigma T_i(v)$. It is easy to show (Exercise 13.5) that

$$\frac{\partial \varphi_i}{\partial \theta_k} = n_i \left\{ T_i(v) T_i\left(\frac{\partial g_i}{\partial \theta_k}\right) - T_i\left(v \frac{\partial g_i}{\partial \theta_k}\right) \right\}, \tag{13.8}$$

$1 \leq k \leq p+1$, where $\theta_{p+1} = \sigma$. The expression on the right side might suggest that our goal is hopeless: As $n_i \to \infty$, how can the right side of (13.8) be bounded? However, the difference inside the curly brackets has a special form. By the Laplace approximation (e.g., De Bruijn 1961, §4), we have

$$T_i(f) = f(\tilde{v}_i) + O(n_i^{-1}),\tag{13.9}$$

where $\tilde{v}_i$ is the minimizer of $g_i$ with respect to $v$. Using (13.9), we have a cancellation of the leading term inside the curly brackets, or square brackets in the following expression, for the right side of (13.8) (verify this):

$$n_i^{-1}\left[\{\tilde{v}_i + O(n_i^{-1})\}\left\{\frac{\partial g_i}{\partial\theta_k}\Big|_{\tilde{v}_i} + O(n_i^{-1})\right\} - \left\{\tilde{v}_i\frac{\partial g_i}{\partial\theta_k}\Big|_{\tilde{v}_i} + O(n_i^{-1})\right\}\right]$$

$$= \frac{\partial g_i}{\partial\theta_k}\Big|_{\tilde{v}_i} O(1) + \tilde{v}_i O(1) - O(1) + O(n_i^{-1}).$$

The expression can now be expected to be bounded.

A consequence of (13.7) plus the orders of the terms is that the EBP is consistent in the sense that $\hat{\alpha}_i - \alpha_i \xrightarrow{P} 0$ as $n_i \to \infty$ (recall that $\hat{\theta}$ is a consistent estimator). However, as noted, the assumption $n_i \to \infty$ is impractical for small-area estimation. So from now on we will forget about the consistency and focus on the estimation of the MSPE of the EBP given that $n_i$ is bounded. It is easy to verify the following decomposition of the MSPE:

$$\mathrm{MSPE}(\hat{\alpha}_i) = \mathrm{E}(\hat{\alpha}_i - \alpha_i)^2$$
$$= \mathrm{E}(\hat{\alpha}_i - \tilde{\alpha}_i)^2 + \mathrm{E}(\tilde{\alpha}_i - \alpha_i)^2,\tag{13.10}$$

where $\tilde{\alpha}_i = \mathrm{E}(\alpha_i|y)$. Using expression (13.4) for $\tilde{\alpha}_i$, the second term has a closed-form expression—namely,

$$\mathrm{E}(\tilde{\alpha}_i - \alpha_i)^2 = \sigma^2 - b_i(\theta),\tag{13.11}$$

where $b_i(\theta) = \mathrm{E}\{\mathrm{E}(\alpha_i|y_i)\}^2 = \sum_{k=0}^{n_i}\psi_i^2(k,\theta)p_i(k,\theta)$ and

$$p_i(k,\theta) = \mathrm{P}(y_{i\cdot} = k)$$

$$= \sum_{z\in S(n_i,k)}\exp\left(\sum_{j=1}^{n_i}z_j x_{ij}'\beta\right)\mathrm{E}[\exp\{\phi_i(z_\cdot,\sigma\xi,\beta)\}]$$

with $S(n,k) = \{z = (z_1,\ldots,z_n) \in \{0,1\}^n : z_1 + \cdots + z_n = k\}$, $z_\cdot = \sum_{j=1}^{n_i}z_j$, and the expectation is taken with respect to $\xi \sim N(0,1)$ (Exercise 13.6). As for the first term, we use a technique introduced in Section 4.3, called the *formal derivation*. By the Taylor expansion, we have

$$\hat{\alpha}_i - \tilde{\alpha}_i = \psi_i(y_{i\cdot},\hat{\theta}) - \psi_i(y_{i\cdot},\theta)$$
$$= \frac{\partial\psi_i}{\partial\theta'}(\hat{\theta} - \theta) + \frac{1}{2}(\hat{\theta} - \theta)'\left(\frac{\partial^2\psi_i}{\partial\theta\partial\theta'}\right)(\hat{\theta} - \theta) + o_\mathrm{P}(|\hat{\theta} - \theta|^2).$$

Suppose that

$$\hat{\theta} - \theta = O_P(m^{-1/2}). \tag{13.12}$$

It follows that

$$E(\hat{\alpha}_i - \tilde{\alpha}_i)^2 = m^{-1} E\left\{\frac{\partial \psi_i}{\partial \theta'}\sqrt{m}(\hat{\theta} - \theta)\right\}^2 + o(m^{-1}). \tag{13.13}$$

To obtain a further approximation, let us first make it simpler by assuming that $\hat{\theta} = \hat{\theta}_{-i}$, an estimator that does not depend on $y_i$. We know that by making such an assumption, we may draw criticisms right the way: How practical is this assumption? Well, before we worry about it, let us first see how it works, if the assumption holds. The bottom line is that, as we will argue later, the special form $\hat{\theta} = \hat{\theta}_{-i}$ does not really matter. Note that if $\hat{\theta} = \hat{\theta}_{-i}$, then the estimator is independent of $y_i$. If we write $V_i(\theta) = mE(\hat{\theta}_{-i} - \theta)(\hat{\theta}_{-i} - \theta)'$, then we have (make sure that you follow the steps)

$$\begin{aligned}
& E\left\{\frac{\partial \psi_i}{\partial \theta'}\sqrt{m}(\hat{\theta}_{-i} - \theta)\right\}^2 \\
&= E[E\{(\cdots)^2|y_i. = k\}|_{k=y_i.}] \\
&= E\left[\left\{\frac{\partial}{\partial \theta'}\psi_i(k,\theta)\right\} V_i(\theta)\left\{\frac{\partial}{\partial \theta}\psi_i(k,\theta)\right\}\Big|_{k=y_i.}\right] \\
&= E\left\{\frac{\partial \psi_i}{\partial \theta'} V_i(\theta)\frac{\partial \psi_i}{\partial \theta}\right\} \\
&= \sum_{k=0}^{n_i}\left\{\frac{\partial}{\partial \theta'}\psi_i(k,\theta)\right\} V_i(\theta)\left\{\frac{\partial}{\partial \theta}\psi_i(k,\theta)\right\} p_i(k,\theta) \\
&\equiv a_i(\theta). \tag{13.14}
\end{aligned}$$

If we combined (13.10), (13.11), (13.13), and (13.14), we get

$$\text{MSPE}(\hat{\alpha}_{i,-i}) = \sigma^2 - b_i(\theta) + a_i(\theta)m^{-1} + o(m^{-1}). \tag{13.15}$$

Here, $\hat{\alpha}_{i,-i}$ indicates that $\theta$ is estimated by $\hat{\theta}_{-i}$ in the EBP.

To show that we can replace $\hat{\alpha}_{i,-i}$ by $\hat{\alpha}_i$, an EBP with $\theta$ estimated by $\hat{\theta}$, we need to show that the difference made by such a replacement is $o(m^{-1})$. Suppose that (13.12) holds and, in addition,

$$\hat{\theta}_{-i} - \hat{\theta} = o_P(m^{-1/2}). \tag{13.16}$$

The motivation for (13.16) is the following. Consider, for simplicity, the sample mean $\hat{\mu} = m^{-1}\sum_{j=1}^{m} X_j$ based on i.i.d. observations $X_1, \ldots, X_m$. Then $\hat{\mu}_{-i} = (m-1)^{-1}\sum_{j\neq i} X_j$; so we have

$$\hat{\mu}_{-i} - \hat{\mu} = \frac{1}{m(m-1)} \sum_{j \neq i} X_j - \frac{1}{m} X_i = O_P(m^{-1}),$$

provided that $E(|X_i|) < \infty$. Hence (13.16) holds. Later we consider a class of estimators of $\theta$ that satisfy (13.16). Suppose that (13.12) and (13.16) hold; then we have $\hat{\theta}_{-i} - \theta = \hat{\theta} - \theta + \hat{\theta}_{-i} - \hat{\theta} = O_P(m^{-1/2})$. Therefore, by the Taylor expansion, we have

$$\hat{\alpha}_{i,-i} - \tilde{\alpha}_i = \frac{\partial \psi_i}{\partial \theta'}(\hat{\theta}_{-i} - \theta) + o_P(|\hat{\theta}_{-i} - \theta|)$$

$$= O_P(m^{-1/2}),$$

$$\hat{\alpha}_i - \hat{\alpha}_{i,-i} = (\hat{\alpha}_i - \tilde{\alpha}_i) - (\hat{\alpha}_{i,-i} - \tilde{\alpha}_i)$$

$$= \left\{ \frac{\partial \psi_i}{\partial \theta'}(\hat{\theta} - \theta) + o_P(|\hat{\theta} - \theta|) \right\} - \left\{ \frac{\partial \psi_i}{\partial \theta'}(\hat{\theta}_{-i} - \theta) + o_P(|\hat{\theta}_{-i} - \theta|) \right\}$$

$$= \frac{\partial \psi_i}{\partial \theta'}(\hat{\theta} - \hat{\theta}_{-i}) + o_P(m^{-1/2})$$

$$= o_P(m^{-1/2}).$$

It follows that

$$\mathrm{MSPE}(\hat{\alpha}_i) = \mathrm{MSPE}(\hat{\alpha}_{i,-i}) + 2E(\hat{\alpha}_i - \hat{\alpha}_{i,-i})(\hat{\alpha}_{i,-i} - \alpha_i) + E(\hat{\alpha}_i - \hat{\alpha}_{i,-i})^2$$

$$= \mathrm{MSPE}(\hat{\alpha}_{i,-i}) + o(m^{-1}). \tag{13.17}$$

Note that $E(\hat{\alpha}_i - \hat{\alpha}_{i,-i})(\hat{\alpha}_{i,-i} - \alpha_i) = E(\hat{\alpha}_i - \hat{\alpha}_{i,-i})(\hat{\alpha}_{i,-i} - \tilde{\alpha}_i)$ (why?). To obtain the final approximation, note that, again by (13.12) and (13.16), the $V_i(\theta)$ involved in $a_i(\theta)$ [see (13.15)] can be approximated by

$$V_i(\theta) = m\{E(\hat{\theta} - \theta)(\hat{\theta} - \theta)' + E(\hat{\theta}_{-i} - \hat{\theta})(\hat{\theta} - \theta)'$$

$$+ E(\hat{\theta} - \theta)(\hat{\theta}_{-i} - \hat{\theta})' + E(\hat{\theta}_{-i} - \hat{\theta})(\hat{\theta}_{-i} - \hat{\theta})'\}$$

$$= m\{E(\hat{\theta} - \theta)(\hat{\theta} - \theta)' + o(m^{-1})\}$$

$$= V(\theta) + o(1),$$

where $V(\theta) = mE(\hat{\theta} - \theta)(\hat{\theta} - \theta)'$. It follows that

$$a_i(\theta) = \sum_{k=0}^{n_i} \left\{ \frac{\partial}{\partial \theta'} \psi_i(k, \theta) \right\} V(\theta) \left\{ \frac{\partial}{\partial \theta} \psi_i(k, \theta) \right\} p_i(k, \theta)$$

$$+ o(1). \tag{13.18}$$

Denoting the summation on the right side of (13.18) by $c_i(\theta)$, we have, by (13.17), (13.15), and (13.18),

$$\mathrm{MSPE}(\hat{\alpha}_i) = \sigma^2 - b_i(\theta) + c_i(\theta)m^{-1} + o(m^{-1}). \tag{13.19}$$

In other words, we have obtained a similar approximation as (13.15) without assuming that $\hat{\theta} = \hat{\theta}_{-i}$. Once again, the method of formal derivation (see

Section 4.3) is used here without justifying each step rigorously. For example, it is not necessarily true that $E\{o_P(m^{-1})\} = o(m^{-1})$. However, these steps can be justified, with rigor, under suitable regularity conditions (see Jiang and Lahiri 2001).

It remains to find estimators that satisfy (13.12) and (13.16). Recall the method of moments estimators for GLMM discussed in the previous section. It can be shown that, subject to some regularity conditions, these estimators satisfy (13.12) and (13.16). In the current case, the estimating equations (12.49) and (12.51) take the following form:

$$\sum_{i=1}^{m}\sum_{j=1}^{n_i} x_{ijk}y_{ij} = \sum_{i=1}^{m} x_{ijk}E_\theta(y_{ij}), \quad 1 \le k \le p, \tag{13.20}$$

$$\sum_{i=1}^{m}\sum_{j_1 \ne j_2} y_{ij_1}y_{ij_2} = \sum_{i=1}^{m}\sum_{j_1 \ne j_2} E_\theta(y_{ij_1}y_{ij_2}) \tag{13.21}$$

with $E_\theta(y_{ij}) = E\{h(x'_{ij}\beta + \sigma\xi)\}$ and $E_\theta(y_{ij_1}y_{ij_2}) = E\{h(x'_{ij_1}\beta + \sigma\xi)h(x'_{ij_2}\beta + \sigma\xi)\}$, $j_1 \ne j_2$ [recall $h(x) = e^x/(1+e^x)$], where the expectations are taken with respect to $\xi \sim N(0,1)$.

Based on the MSPE approximation (13.19), we can derive a second-order unbiased estimator of the MSPE. This means finding $\widehat{MSPE}(\hat\alpha_i)$ such that

$$E\{\widehat{MSPE}(\hat\alpha_i) - MSPE(\hat\alpha_i)\} = o(m^{-1}). \tag{13.22}$$

Write $d_i(\theta) = \sigma^2 - b_i(\theta)$. For the $c_i(\theta)$ in (13.19), we can simply replace $\theta$ by $\hat\theta$, a consistent estimator (e.g., the method of moment estimator). This is because the difference is $m^{-1}\{c_i(\hat\theta) - c_i(\theta)\} = o_P(m^{-1})$, provided that $c_i(\cdot)$ is continuous. However, we cannot simply replace $d_i(\theta)$ by $d_i(\hat\theta)$, because the difference is $d_i(\hat\theta) - d_i(\theta) = O_P(m^{-1/2})$, in the typical situations, provided that $\hat\theta$ satisfies (13.12). Therefore, we have to do a *bias correction* in order to reduce the bias to $o(m^{-1})$. Let $M(\theta)$ denote the vector of the difference between the two sides of (13.20) and (13.21) and let $\hat\theta$ be the solution to (13.20) and (13.21) [i.e., $M(\hat\theta) = 0$]. We have, by the Taylor expansion,

$$0 = M(\hat\theta)$$
$$= M(\theta) + \frac{\partial M}{\partial \theta'}(\hat\theta - \theta) + \frac{1}{2}\left[(\hat\theta - \theta)'\frac{\partial^2 M_k}{\partial\theta\partial\theta'}(\hat\theta - \theta)\right]_{1 \le k \le p+1}$$
$$+ \frac{1}{6}\left[\sum_{a,b,c}\frac{\partial^3 \tilde{M}_k}{\partial\theta_a\partial\theta_b\partial\theta_c}(\hat\theta_a - \theta_a)(\hat\theta_b - \theta_b)(\hat\theta_c - \theta_c)\right]_{1 \le k \le p+1},$$

where $M_k$ represents the $k$th component of $M$ and $\tilde{M}_k$ denotes $M_k$ evaluated at $\tilde\theta_k$, which lies between $\theta$ and $\hat\theta$ (but depends on $k$), $1 \le k \le p+1$. In typical situations, we have $(\partial M/\partial\theta')^{-1} = O_P(m^{-1})$, and the second and third derivatives of $M_k$ are $O_P(m)$. These, plus (13.12), imply

$$0 = M(\theta) + \frac{\partial M}{\partial \theta'}(\hat{\theta} - \theta) + \frac{1}{2}\left[(\hat{\theta} - \theta)'\frac{\partial^2 M_k}{\partial \theta \partial \theta'}(\hat{\theta} - \theta)\right]_{1 \le k \le p+1}$$

$$+ O_P(m^{-1/2}) \tag{13.23}$$

$$= M(\theta) + \frac{\partial M}{\partial \theta'}(\hat{\theta} - \theta) + O_P(1). \tag{13.24}$$

Equation (13.24) implies that

$$\hat{\theta} - \theta = -\left(\frac{\partial M}{\partial \theta'}\right)^{-1} M(\theta) + O_P(m^{-1}). \tag{13.25}$$

We now bring (13.25) back to (13.23) to replace $\hat{\theta} - \theta$ in the quadratic form. This leads to (verify this, especially the order of the remina term)

$$0 = M(\theta) + \frac{\partial M}{\partial \theta'}(\hat{\theta} - \theta) + \frac{1}{2}\hat{Q} + O_P(m^{-1/2}), \tag{13.26}$$

where $\hat{Q} = [M'(\theta)(\partial M'/\partial \theta)^{-1}(\partial^2 M_k/\partial \theta \partial \theta')(\partial M/\partial \theta')^{-1}M(\theta)]_{1 \le k \le p+1}$. This leads to the following expansion:

$$\hat{\theta} - \theta = -\left(\frac{\partial M}{\partial \theta'}\right)^{-1}\left\{M(\theta) + \frac{1}{2}\hat{Q}\right\} + O_P(m^{-3/2}) \tag{13.27}$$

$$= -\left(\frac{\partial M}{\partial \theta'}\right)^{-1} M(\theta) + O_P(m^{-1}). \tag{13.28}$$

The second equation is due to the fact that $\hat{Q} = O_P(1)$ (why?).

Now, use another Taylor expansion, this time for $d_i(\hat{\theta})$:

$$d_i(\hat{\theta}) = d_i(\theta) + \frac{\partial d_i}{\partial \theta'}(\hat{\theta} - \theta) + \frac{1}{2}(\hat{\theta} - \theta)'\frac{\partial^2 d_i}{\partial \theta \partial \theta'}(\hat{\theta} - \theta) + O_P(m^{-3/2}).$$

By plugging (13.27) into the first-order term and (13.28) into the second-order term (why using different expressions?), we get

$$d_i(\hat{\theta}) = d_i(\theta) - \frac{\partial d_i}{\partial \theta'}\left(\frac{\partial M}{\partial \theta'}\right)^{-1}\left\{M(\theta) + \frac{1}{2}\hat{Q}\right\} + \frac{1}{2}\hat{R} + O_P(m^{-3/2})$$

$$= d_i(\theta) + m^{-1}\hat{B}_i(\theta) + O_P(m^{-3/2}),$$

where $\hat{R}$ is the same as the components of $\hat{Q}$ except replacing $\partial^2 M_k/\partial \theta \partial \theta'$ by $\partial^2 d_i(\theta)/\partial \theta \partial \theta'$ and

$$\hat{B}_i(\theta) = -m\left[\frac{\partial d_i}{\partial \theta'}\left(\frac{\partial M}{\partial \theta'}\right)^{-1}\left\{M(\theta) + \frac{1}{2}\hat{Q}\right\} - \frac{1}{2}\hat{R}\right]$$

$$= -m\frac{\partial d_i}{\partial \theta'}\left(\frac{\partial M}{\partial \theta'}\right)^{-1} M(\theta) + O_P(1)$$

$$= -m\frac{\partial d_i}{\partial \theta'}\left\{\mathrm{E}\left(\frac{\partial M}{\partial \theta'}\right)\right\}^{-1} M(\theta) + O_P(1).$$

The second equation holds because $\hat{R} = O_P(m^{-1})$ (why?), $(\partial M/\partial\theta')^{-1} = O_P(m^{-1})$, and $\hat{Q} = O_P(1)$; the third equation is due to the facts that

$$\left(\frac{\partial M}{\partial\theta'}\right)^{-1} - \left\{\mathrm{E}\left(\frac{\partial M}{\partial\theta'}\right)\right\}^{-1}$$
$$= -\left\{\mathrm{E}\left(\frac{\partial M}{\partial\theta'}\right)\right\}^{-1}\left\{\frac{\partial M}{\partial\theta'} - \mathrm{E}\left(\frac{\partial M}{\partial\theta'}\right)\right\}\left(\frac{\partial M}{\partial\theta'}\right)^{-1}$$
$$= O(m^{-1})O_P(m^{1/2})O_P(m^{-1})$$
$$= O_P(m^{-3/2}),$$

and $M(\theta) = O_P(m^{1/2})$. Now, use the fact that $\mathrm{E}\{M(\theta)\} = 0$ to get

$$B_i(\theta) \equiv \mathrm{E}\{\hat{B}_i(\theta)\} = O(1) \tag{13.29}$$

[note that $M(\theta) = O_P(m^{1/2})$ is not enough to get (13.29)]. The left side of (13.29), multiplied by $m^{-1}$, is the leading term for the bias of $d_i(\hat{\theta})$. To bias-correct the latter, we subtract an estimator of the bias, $B_i(\hat{\theta})$ [note that $B_i(\hat{\theta})$ is different from $\hat{B}_i(\theta)$]. This time, the plugging in of $\hat{\theta}$ results in an overall difference of $o(m^{-1})$. The bias-corrected MSPE estimator is therefore

$$\widehat{\mathrm{MSPE}}(\hat{\alpha}_i) = d_i(\hat{\theta}) + m^{-1}\{c_i(\hat{\theta}) - B_i(\hat{\theta})\}. \tag{13.30}$$

In practice, $B_i(\hat{\theta})$ may be evaluated using a parametric bootstrap method; namely, the bootstrap samples are generated under the mixed logistic model with $\hat{\theta}$ treated as the true parameters. For each bootstrap sample, the expression $\hat{B}_i(\hat{\theta})$ is evaluated. The sample mean of these evaluations over the different bootstrap samples is then an approximation to $B_i(\hat{\theta})$. See the next chapter for more details.

It can be shown that the estimator given by (13.30) satisfies (13.22) (Exercise 13.7). Note that (13.30) has a form similar to the Prasad–Rao MSPE estimator (4.95). Therefore, the result obtained in this section may be viewed as an extension of the Prasad–Rao method to mixed logistic models for small-area estimation. Once again, the method of formal derivation (see Section 4.3) is used in deriving the second-order unbiased MSPE estimator. However, the end result [i.e., (13.22) holds for the MSPE estimator (13.30)], can be rigorously justified (see Jiang and Lahiri 2001).

## 13.3 The Fay–Herriot model

The Fay–Herriot model was first introduced in Example 4.18. It is one of the most popular models for small-area estimation that have been practically used. Yet, the model is surprisingly simple and therefore very useful in introducing the basic as well as advanced asymptotic techniques for small-area estimation. Recall that a Fay–Herriot model can be expressed as (4.86) with the assumptions following the expression. The small-area mean under the Fay–Herriot model can be expressed as $\theta_i = x_i'\beta + v_i$. Earlier, the EBLUP of $\theta_i$ was derived in Example 4.18 (continued). A slightly different expression is

$$\hat{\theta}_i = y_i - B_i(\hat{A})\{y_i - x_i'\tilde{\beta}(\hat{A})\}, \tag{13.31}$$

where $B_i(A) = D_i/(A + D_i)$ and

$$\tilde{\beta}(A) = \left(\sum_{i=1}^{m} \frac{x_i x_i'}{A + D_i}\right)^{-1} \sum_{i=1}^{m} \frac{x_i y_i}{A + D_i}$$

is the BLUE of $\beta$ given that $A$ is the true variance of the random effects $v_i$. Note that hereafter we make a switch of notation from $Y_i$ to $y_i$.

The $\hat{A}$ in (13.31) is supposed to be a consistent, or more precisely, $\sqrt{m}$-consistent estimator of $A$. Here, by $\sqrt{m}$-consistent it means $\hat{A} - A = O_P(m^{-1/2})$. There are several choices choices for $\hat{A}$. Prasad and Rao (1990) used the method of moments (P-R) estimator

$$\hat{A}_{\text{PR}} = \frac{y'P_{X^\perp}y - \text{tr}(P_{X^\perp}D)}{m - p}, \tag{13.32}$$

where $P_{X^\perp}$ is defined above (12.56) with $X = (x_i')_{1 \le i \le m}$. Without loss of generality, we assume that $X$ has full rank $p$. Datta and Lahiri (2000) considered ML and REML estimators of $A$, denoted by $\hat{A}_{\text{ML}}$ and $\hat{A}_{\text{RE}}$, respectively, which do not have closed-form expressions. Another estimator of $A$, proposed by Fay and Herriot (1979) in their original paper (in which they introduced the Fay–Herriot model), is given below. Let $Q = Q(A)$ be as defined below (12.9) with $\Gamma = \text{Var}(y) = \text{diag}(A + D_i, 1 \le i \le m)$. The Fay–Herriot (F-H) estimator of $A$, denoted by $\hat{A}_{\text{FH}}$, is obtained by solving iteratively the equation

$$\frac{y'Q(A)y}{m - p} = 1 \tag{13.33}$$

for $A$. It can be shown that (13.33) is unbiased in the sense that the expectation of the left side is equal to the right side if $A$ is the true variance (Exercise 13.8). Note that

$$y'Q(A)y = \sum_{i=1}^{m} \frac{\{y_i - x_i'\tilde{\beta}(A)\}^2}{A + D_i}. \tag{13.34}$$

All of the estimators, $\hat{A}_{\text{PR}}$, $\hat{A}_{\text{ML}}$, $\hat{A}_{\text{RE}}$, and $\hat{A}_{\text{FH}}$, are $\sqrt{m}$-consistent. The $\sqrt{m}$-consistency of $\hat{A}_{\text{ML}}$ and $\hat{A}_{\text{RE}}$ follows from the results of Jiang (1996), which do not require normality (see Section 12.2). The $\sqrt{m}$-consistency of $\hat{A}_{\text{PR}}$ is left as an exercise (Exercise 13.9). The corresponding property of $\hat{A}_{\text{FH}}$ may be implied by the general theory of estimating equations (e.g., Heyde 1997, §12) or shown directly (Exercise 13.10).

Furthermore, all of the estimators possess the following properties: (i) They are even functions of the data; that is, the estimators are unchanged when $y$ is replaced by $-y$; (ii) they are translation invariant; that is, the estimators are unchanged when $y$ is replaced by $y + Xb$ for any $b \in R^p$ (Exercise 13.11). Harville (1985) showed that for any estimator $\hat{A}$ that satisfies (i) and (ii), the MSPE of the corresponding EBLUP has the following decomposition:

$$\mathrm{MSPE}(\hat{\theta}_i) = \mathrm{E}(\tilde{\theta}_{\mathrm{B},i} - \theta_i)^2 + \mathrm{E}(\tilde{\theta}_i - \tilde{\theta}_{\mathrm{B},i})^2 + \mathrm{E}(\hat{\theta}_i - \tilde{\theta}_i)^2$$
$$= g_{1i}(A) + g_{2i}(A) + \mathrm{E}(\hat{\theta}_i - \tilde{\theta}_i)^2, \tag{13.35}$$

where $\tilde{\theta}_{\mathrm{B},i}$ is the BP of $\theta_i$ (in terms of minimizing the MSPE), given by (13.31) with $\hat{A}$ and $\tilde{\beta}(A)$ replaced by $A$ and $\beta$, which are the true parameters, and $\tilde{\theta}_i$ is the BLUP of $\theta_i$, given by (13.31) with $\hat{A}$ replaced by $A$, the true $A$. To compute the BP, both $A$ and $\beta$ have to be known; to compute the BLUP, $A$ has to be known. Both BP and BLUP are not available in typical situations; so the EBLUP is the only one among the three that is computable. However, decomposition (13.35) is very useful in deriving a second-order unbiased estimator of the MSPE of EBLUP. Further analytic expressions can be obtained for the first two terms—namely,

$$g_{1i}(A) = \frac{AD_i}{A + D_i}, \tag{13.36}$$

$$g_{2i}(A) = \left( \frac{D_i}{A + D_i} \right)^2 x_i' \left( \sum_{j=1}^m \frac{x_j x_j'}{A + D_j} \right)^{-1} x_i. \tag{13.37}$$

Since these terms have nothing to do with the estimator, they remain the same regardless of what estimator of $A$ is used as long as (i) and (ii) are satisfied. Also, note that $g_{1i}(A) = O(1)$ and $g_{2i}(A) = O(m^{-1})$, under regularity conditions.

On the other hand, the third term on the right side of (13.35) depends on the estimator of $A$. Under suitable regularity conditions, it can be shown (e.g., Datta and Lahiri 2000) that

$$\mathrm{E}(\hat{\theta}_i - \tilde{\theta}_i)^2 = \frac{D_i^2}{(A + D_i)^3} \mathrm{var}(\hat{A}) + o(m^{-1})$$
$$= g_{3i}(A) + o(m^{-1}), \tag{13.38}$$

where $\mathrm{var}(\hat{A})$ is the asymptotic variance of $\hat{A}$. It remains to evaluate $\mathrm{var}(\hat{A})$, and thus $g_{3i}(A)$, for the different estimators of $A$. Prasad and Rao (1990) showed that for $\hat{A}_{\mathrm{PR}}$,

$$g_{3i}(A) = \frac{2D_i^2}{(A + D_i)^3 m^2} \sum_{j=1}^m (A + D_j)^2. \tag{13.39}$$

Datta and Lahiri (2000) showed that for both $\hat{A}_{\mathrm{ML}}$ and $\hat{A}_{\mathrm{RE}}$,

$$g_{3i}(A) = \frac{2D_i^2}{(A+D_i)^3} \left\{ \sum_{j=1}^m (A+D_j)^{-2} \right\}^{-1} \qquad (13.40)$$

(here it is assumed that $p$, the rank of $X$, is bounded; otherwise, the asymptotic variance of $\hat{A}_{\mathrm{ML}}$ and $\hat{A}_{\mathrm{RE}}$ may be different; see Section 12.2). Finally, Datta et al. (2005) showed that for $\hat{A}_{\mathrm{FH}}$,

$$g_{3i}(A) = \frac{2D_i^2 m}{(A+D_i)^3} \left\{ \sum_{j=1}^m (A+D_j)^{-1} \right\}^{-2} . \qquad (13.41)$$

Despite the different expressions, it is seen from (13.39)–(13.41) that $g_{3i}(A) = O(m^{-1})$, which makes sense in view of (13.38). In conclusion, the leading term in the MSPE decomposition is $g_{1i}(A)$, which is $O(1)$, followed by two $O(m^{-1})$ terms, $g_{2i}(A)$ and $g_{3i}(A)$, so that

$$\mathrm{MSPE}(\hat{\theta}_i) = g_{1i}(A) + g_{2i}(A) + g_{3i}(A) + o(m^{-1}) \qquad (13.42)$$

with $g_{3i}(A)$ depending on the method of estimation of $A$.

It can be shown that (Exercise 13.12)

$$\text{right side of } (13.40) \leq \text{right side of } (13.41)$$
$$\leq \text{right side of } (13.39).$$

This means that in terms of the asymptotic (predictive) efficiency, ML and REML estimators are the best, followed by the F-H estimator, and then by the P-R estimator. These theoretical results are confirmed by the results of simulation studies carried out by Datta et al. (2005), who showed that the EBLUPs based on ML, REML, and F-H estimators of $A$ perform similarly in terms of the MSPE (ML and REML perform slightly better under normality), and all three perform significantly better than the EBLUP with the P-R estimator of $A$ when there is a large variability in the $D_i$'s among the small areas. Note that in the balanced case (i.e., $D_i = D, 1 \leq i \leq m$), the P-R, REML, and F-H estimators are identical (provided that the estimator is nonnegative), whereas the ML estimator is different, although the difference is expected to be small when $m$ is large (Exercise 13.13).

Based on the MSPE approximation (13.42) and using a similar bias-correction technique as in the previous section, second-order unbiased estimators of the MSPE can be obtained. Datta and Lahiri (2000) showed that if $b(A)$ is the asymptotic bias of $\hat{A}$ up to the order $o(m^{-1})$, then the estimator

$$\widehat{\mathrm{MSPE}}(\hat{\theta}_i) = g_{1i}(\hat{A}) + g_{2i}(\hat{A}) + 2g_{3i}(\hat{A})$$
$$- \left( \frac{D_i}{\hat{A}+D_i} \right)^2 b(\hat{A}) \qquad (13.43)$$

is second-order unbiased; that is,

$$E\{\widehat{MSPE}(\hat\theta_i) - MSPE(\hat\theta_i)\} = o(m^{-1}).  \tag{13.44}$$

The authors further showed that for $\hat A_{PR}$ and $\hat A_{RE}$, $b(A) = 0$, so that for EBLUP with P-R estimator of $A$, we have

$$\widehat{MSPE}(\hat\theta_i) = g_{1i}(\hat A_{PR}) + g_{2i}(\hat A_{PR}) + 2g_{3i,PR}(\hat A_{PR}),  \tag{13.45}$$

where $g_{3i,PR}(A)$ denotes the $g_{3i}(A)$ given by (13.39); similarly, for EBLUP with the REML estimator of $A$, $\widehat{MSPE}(\hat\theta_i)$ is given by (13.45) with PR replaced by RE, where $g_{3i,RE}(A)$ denotes the $g_{3i}(A)$ given by (13.40). As for EBLUP with ML estimator of $A$, Datta and Lahiri (2000) showed that

$$\widehat{MSPE}(\hat\theta_i)$$
$$= (13.45) \text{ with PR replaced by ML}$$
$$- \left(\frac{D_i}{\hat A_{ML} + D_i}\right)^2 \left\{\sum_{j=1}^{m}(\hat A_{ML} + D_j)^{-2}\right\}^{-1}$$
$$\times tr\left\{\left(\sum_{j=1}^{m}\frac{x_j x_j'}{\hat A_{ML} + D_j}\right)^{-1}\sum_{j=1}^{m}\frac{x_j x_j'}{(\hat A_{ML} + D_j)^2}\right\},  \tag{13.46}$$

where $g_{3i,ML}$ is the same as $g_{3i,RE}$. Finally, for EBLUP with the F-H estimator of $A$, Datta et al. (2005) showed that

$$\widehat{MSPE}(\hat\theta_i)$$
$$= (13.45) \text{ with PR replaced by FH}$$
$$-2\left(\frac{D_i}{\hat A_{FH} + D_i}\right)^2 \left\{\sum_{j=1}^{m}(\hat A_{FH} + D_j)^{-1}\right\}^{-3}$$
$$\times \left[m\sum_{j=1}^{m}(\hat A_{FH} + D_j)^{-2} - \left\{\sum_{j=1}^{m}(\hat A_{FH} + D_j)^{-1}\right\}^2\right],  \tag{13.47}$$

where $g_{3i,FH}(A)$ denotes $g_{3i}(A)$ given by (13.41).

The (unconditional) MSPE is one way to measure the uncertainty of the EBLUP. There are other measures of uncertainties associated with conditional expectations. First, note that one can write

$$MSPE(\hat\theta_i) = E[E\{(\hat\theta_i - \theta_i)^2|\theta\}],  \tag{13.48}$$

where the outside conditional expectation is with respect to the distribution of $\theta = (\theta_i)_{1\le i\le m}$ and the inside conditional expectation is with respect to the sampling distribution of $y$, known as the design-based (conditional) MSPE. If one considers the inside conditional expectation in (13.48), it leads to

$$\text{MSPE}_1(\hat{\theta}_i|\theta) = \text{E}\{(\hat{\theta}_i - \theta_i)^2|\theta\} \tag{13.49}$$

as a measure of uncertainty conditional on the small-area means. A good thing about $\text{MSPE}_1$ is that it has an exactly unbiased estimator. More generally, consider a predictor of the form $\hat{\theta}_i = y_i + h_i(y)$, where $h_i(\cdot)$ is a differentiable function. Rivest and Belmonte (2000) showed that

$$\text{MSPE}_1(\hat{\theta}_i|\theta) = D_i + 2\text{E}\{(y_i - \theta_i)h_i(y)|\theta\} + \text{E}\{h_i^2(y)|\theta\}. \tag{13.50}$$

Furthermore, applying the well-known Stein's identity (e.g., Casella and Berger 2002, p. 124), we have

$$\text{E}\{(y_i - \theta_i)h_i(y)|\theta\} = D_i\text{E}\left(\left.\frac{\partial h_i}{\partial y_i}\right|\theta\right). \tag{13.51}$$

Equations (13.50) and (13.51) lead to the expression

$$\text{MSPE}_1(\hat{\theta}_i|\theta) = D_i + 2D_i\text{E}\left(\left.\frac{\partial h_i}{\partial y_i}\right|\theta\right) + \text{E}\{h_i^2(y)|\theta\}$$

$$= \text{E}\left\{\left.D_i\left(1 + 2\frac{\partial h_i}{\partial y_i}\right) + h_i^2(y)\right|\theta\right\}. \tag{13.52}$$

Thus, an exactly (design-)unbiased estimator of $\text{MSPE}_1$ is

$$\widehat{\text{MSPE}}_1(\hat{\theta}_i) = D_i\left(1 + 2\frac{\partial h_i}{\partial y_i}\right) + h_i^2(y). \tag{13.53}$$

In particular, for the EBLUP, we have

$$h_i(y) = -\frac{D_i}{\hat{A} + D_i}\{y_i - x_i'\tilde{\beta}(\hat{A})\}, \tag{13.54}$$

where $\hat{A}$ is a specified estimator of $A$ and $\tilde{\beta}(A)$ is given below (13.31). Thus, for the EBLUP, we have (Exercise 13.14)

$$\frac{\partial h_i}{\partial y_i} = \frac{D_i}{(\hat{A} + D_i)^2}\{y_i - x_i'\tilde{\beta}(\hat{A})\}\frac{\partial \hat{A}}{\partial y_i}$$

$$- \frac{D_i}{\hat{A} + D_i}\left(1 - x_i'\frac{\partial\tilde{\beta}}{\partial y_i}\right), \tag{13.55}$$

$$\frac{\partial\tilde{\beta}}{\partial y_i} = \left(\sum_{j=1}^{m}\frac{x_j x_j'}{\hat{A} + D_j}\right)^{-1}$$

$$\times \left\{\frac{x_i}{\hat{A} + D_i} - \sum_{j=1}^{m}\frac{y_j - x_j'\tilde{\beta}(\hat{A})}{(\hat{A} + D_j)^2}x_j\frac{\partial\hat{A}}{\partial y_i}\right\} \tag{13.56}$$

(note that the latter expression is a vector). It remains to obtain expressions for $\partial \hat{A}/\partial y_i$ for the different estimators of $A$. For the P-R estimator, the expression is simple, because $\hat{A}_{PR}$ has the closed-form expression (13.32)—namely,

$$\frac{\partial \hat{A}_{PR}}{\partial y_i} = \frac{2}{m - p}\{y_i - x_i'(X'X)^{-1}X'y\} \tag{13.57}$$

(Exercise 13.11). For the other estimators of $A$, the expressions for $\partial \hat{A}/\partial y_i$ are more complicated and the well-known implicit function theorem in calculus needs to be used (Exercise 13.14; also see Datta et al. 2009). Despite the exact unbiasedness, the estimator (13.53) may be unstable. The reason is that, unlike $\widehat{\text{MSPE}}(\hat{\theta}_i)$, the estimator (13.53) involves the term $y_i$, the direct estimator from the $i$th small area. This term is subject to a large sampling variation compared to an estimator, such as $\hat{A}$, that is based on observations from all the small areas (why?).

Alternatively, we can express the MSPE as

$$\text{MSPE}(\hat{\theta}_i) = E[E\{(\hat{\theta}_i - \theta_i)^2|y_i\}]. \tag{13.58}$$

This leads to another measure of uncertainty:

$$\text{MSPE}_2(\hat{\theta}_i|y_i) = E\{(\hat{\theta}_i - \theta_i)^2|y_i\}. \tag{13.59}$$

Note that, unlike (13.49), here the conditioning is on $y_i$, the direct estimator from the $i$th small area (Fuller 1989). Also, note that if the direct estimator, $y_i$, is used instead of $\hat{\theta}_i$, then we have (Exercise 13.15)

$$\text{MSPE}_2(y_i|y_i) = \{y_i - E(\theta_i|y_i)\}^2 + \text{var}(\theta_i|y_i)$$
$$= g_{1i}(A) + \left(\frac{D_i}{A + D_i}\right)^2 (y_i - x_i'\beta)^2, \tag{13.60}$$

where $g_{1i}(A)$ is given by (13.36). The question then is whether one can do better than $y_i$ with the EBLUP $\hat{\theta}_i$. Datta et al. (2009) obtained the following approximation:

$$\text{MSPE}_2(\hat{\theta}_i|y_i) = g_{1i}(A) + g_{2i}(A) + \frac{(y_i - x_i'\beta)^2}{A + D_i}g_{3i}(A)$$
$$+ o_P(m^{-1}), \tag{13.61}$$

where $g_{ri}(A)$, $r = 1, 2, 3$, are the same as before. Comparing (13.60) and (13.61) and noting that both $g_{2i}(A)$ and $g_{3i}(A)$ are $O(m^{-1})$, we see that a term of order $O(1)$ is replaced by something of the order $O_P(m^{-1})$, when $y_i$ is replaced by $\hat{\theta}_i$. Thus, the EBLUP is doing better, as expected.

Using a similar bias-correction technique as in Section 13.2, Datta *et al.* (2009) obtained a second-order unbiased estimator of $\text{MSPE}_2$. The estimator has the general form

$$\widehat{\mathrm{MSPE}}_2(\hat{\theta}_i|y_i) = g_{1i}(\hat{A}) + g_{2i}(\hat{A}) + g_{3i}(\hat{A})\left[1 + \frac{\{y_i - x_i'\tilde{\beta}(\hat{A})\}^2}{\hat{A} + D_i}\right]$$

$$-\left(\frac{D_i}{\hat{A} + D_i}\right)^2 b_i(\hat{A}), \tag{13.62}$$

where, like $g_{3i}(A)$, $b_i(A)$ depends on which estimator of $A$ is used. For example,

$$b_i(A) = \left\{(A + D_i)\sum_{j=1}^{m}(A + D_i)^{-2}\right\}^{-1}$$

$$\times \left[\frac{\{y_i - x_i'\tilde{\beta}(A)\}^2}{A + D_i} - 1\right] \tag{13.63}$$

for the REML estimator of $A$, and

$$b_i(A) = 2\left\{\sum_{j=1}^{m}(A + D_j)^{-1}\right\}^{-3}$$

$$\times \left[m\sum_{j=1}^{m}(A + D_j)^{-2} - \left\{\sum_{j=1}^{m}(A + D_j)^{-1}\right\}^2\right]$$

$$+ \left\{\sum_{j=1}^{m}(A + D_j)^{-1}\right\}^{-1}\left[\frac{\{y_i - x_i'\tilde{\beta}(A)\}^2}{A + D_i} - 1\right] \tag{13.64}$$

for the F-H estimator of $A$.

Finally, we consider hierarchical Bayes estimation of the small-area means. Note that the Fay–Herriot model can be formulated as a hierarchical model, once a prior distribution, $p(\beta, A)$, is specified on $\beta$ and $A$. In this regard, we assume the following:

(i) Given the $\theta_i$'s, $y_i, i = 1, \ldots, m$, are independent such that $y_i \sim N(\theta_i, D_i)$.

(ii) Given $\beta$ and $A$, $\theta_i, i = 1, \ldots, m$, are independent such that $\theta_i \sim N(x_i'\beta, A)$.

(iii) The prior $p(\beta, A) \propto \pi(A)$, where $\pi(A)$ is a distribution for $A$ ($\propto$ means "proportional to").

Note that without (iii), this is the same as the Fay–Herriot model that we have been considering. The hierarchical Bayes estimator of $\theta_i$ is the posterior mean, $E(\theta_i|y)$; a measure of uncertainty for the Bayes estimator is the posterior variance, $\mathrm{var}(\theta_i|y)$. It can be shown (Exercise 13.16) that the conditional distribution of $\theta_i$ given $A$ and $y$ is normal with mean equal to the right side of (13.31) with $\hat{A}$ replaced by $A$ and variance equal to $g_{1i}(A) + g_{2i}(A)$. It follows that (see Appendix A.2)

$$E(\theta_i|y) = E\{E(\theta_i|A, y)|y\}$$
$$= y_i - E\left[\left(\frac{D_i}{A + D_i}\right)\{y_i - x_i'\tilde{\beta}(A)\}\bigg| y\right], \qquad (13.65)$$
$$\text{var}(\theta_i|y) = E\{\text{var}(\theta_i|A, y)|y\} + \text{var}\{E(\theta_i|A, y)\}$$
$$= E\{g_{1i}(A)|y\} + E\{g_{2i}(A)|y\}$$
$$+ \text{var}\left[\left(\frac{D_i}{A + D_i}\right)\{y_i - x_i'\tilde{\beta}(A)\}\bigg| y\right]. \qquad (13.66)$$

The conditional expectations involved in (13.65) and (13.66) do not have closed-form expressions. To obtain second-order approximations to (13.65) and (13.66), Datta et al. (2005) used the Laplace approximation (see Section 4.6). Let $c(A)$ denote a function for which $E\{c(A)|y\}$ exists. Define $h(A)$ by

$$\exp\{-mh(A)\} = |\Gamma(A)|^{-1/2}|X'\Gamma^{-1}(A)X|^{-1/2}\exp\left\{-\frac{1}{2}y'Q(A)y\right\},$$

where $\Gamma(A)$ and $Q(A)$ are the same as in (13.33) [where we used the notation $\Gamma$ instead of $\Gamma(A)$]. Note that $mh(A)$ is the same as the negative of the restrictive log-likelihood for estimating $A$ (see Section 12.2). The minimizer of $h(A)$ is therefore the REML estimator of $A$, $\hat{A}_{\text{RE}}$. To express the Laplace approximation to $E\{c(A)|y\}$ in a neat way, we need to introduce some notation. Write $\hat{c} = c(\hat{A}_{\text{RE}})$, $\hat{c}_r = \partial^r c(A)/\partial A^r|_{\hat{A}_{\text{RE}}}$, $r = 1, 2$, $\hat{h}_r = \partial^r h(A)/\partial A^r|_{\hat{A}_{\text{RE}}}$, $r = 2, 3$, and $\hat{\rho}_1 = \partial \log \pi(A)/\partial A|_{\hat{A}_{\text{RE}}}$. Then we have

$$E\{c(A)|y\} = \hat{c} + \frac{1}{2m\hat{h}_2}\left(\hat{c}_2 - \frac{\hat{h}_3}{\hat{h}_2}\hat{c}_1\right) + \frac{\hat{c}_1}{m\hat{h}_2}\hat{\rho}_1$$
$$+ O_P(m^{-2}). \qquad (13.67)$$

Similarly, provided that $\text{var}\{c(A)|y\}$ exists, we have

$$\text{var}\{c(A)|y\} = \frac{\hat{c}_1^2}{m\hat{h}_2} + O_P(m^{-2}). \qquad (13.68)$$

By applying (13.67) and (13.68) to (13.65) and (13.66), second-order approximations of $E(\theta_i|y)$ and $\text{var}(\theta_i|y)$ can be obtained. See Datta et al. (2005) for the results and D. D. Smith's Ph. D. dissertation (2001; University of Georgia) for the details of the proofs.

Note that, unlike the Laplace approximations we have previously seen [e.g., (12.37)], here we have the orders of the approximation errors. For example, the approximation error in (13.67) is $O_P(m^{-2})$ and $E(c(A)|y)$ is $O_P(1)$. We call such an approximation *asymptotically accurate*. In contrast, the Laplace approximation in (12.37) is not asymptotically accurate in that the approximation error does not go to zero (e.g., in probability) as the sample size increases. One may wonder why there is such a difference. For the most part, the Laplace approximation becomes accurate if, as the sample size increases, the underlying distribution becomes concentrated near the mode of the distribution. The posterior distribution of $A$ becomes concentrated near its mode as $m \to \infty$, as more information becomes available about $A$. On the other hand, in typical situations of GLMM discussed in Section 12.4, the distribution of the random effects does not become concentrated near its mode, even as the sample size increases, as the information about individual random effects remains limited. For example, in the mixed logistic model of Section 13.2, the $n_i$'s (which correspond to the information about the individual $\alpha_i$'s) remain bounded even as $m \to \infty$.

## 13.4 Nonparametric small-area estimation

Nonparametric models for small-area estimation have received much attention in recent literature. In particular, Opsomer et al. (2008) proposed a spline-based nonparametric model. For simplicity, let us first consider a nonparametric regression model, which can be expressed as

$$y_i = f(x_i) + \epsilon_i, \quad i = 1, \ldots, n, \tag{13.69}$$

where $f(\cdot)$ is an unknown function and the errors $\epsilon_i$ are independent with mean 0 and constant variance. The linear regression model (see Section 6.7) is a special case of (13.69) with $f(x) = x'\beta$, a vector of unknown regression coefficients. In the latter case, $f$ is unknown only up to a vector of parameters, $\beta$. In other words, the form of $f$ is known. It is clear that a nonparametric regression model offers more flexibility than the linear regression model in modeling the mean function of the observations. On the other hand, it is difficult to make an inference about the model if $f$ is completely unknown. A common strategy in practice is to approximate $f$ by a function that can be chosen from a rich family of parametric functions. One such family is called *P-splines*. For simplicity, consider the case of univariate $x$. Then a P-spline can be expressed as

$$\tilde{f}(x) = \beta_0 + \beta_1 x + \cdots + \beta_p x^p$$
$$+ \gamma_1 (x - \kappa_1)_+^p + \cdots + \gamma_q (x - \kappa_q)_+^p, \tag{13.70}$$

where $p$ is the degree of the spline, $q$ is the number of knots, $\kappa_j$, $1 \leq j \leq q$ are the knots, and $x_+ = x1_{(x>0)}$. Graphically, a P-spline is pieces of ($p$th degree) polynomials smoothly connected at the knots. The P-spline model, which is (13.69) with $f$ replaced by $\tilde{f}$, is fitted by *penalized least squares*—that is, by minimizing

$$|y - X\beta - Z\gamma|^2 + \lambda|\gamma|^2, \tag{13.71}$$

with respect to $\beta$ and $\gamma$, where $y = (y_i)_{1 \leq i \leq n}$, the $i$th row of $X$ is $(1, x_i, \ldots, x_i^p)$, the $i$th row of $Z$ is $[(x_i - \kappa_1)_+^p, \ldots, (x_i - \kappa_q)_+^p]$, $1 \leq i \leq n$, and $\lambda$ is a penalty, or smoothing, parameter. To determine $\lambda$, Wand (2003) used the following interesting connection to a linear mixed model (see the previous chapter). Suppose that the $\epsilon_i$'s are distributed as $N(0, \tau^2)$. Then if the $\gamma$'s are treated as independent random effects with the distribution $N(0, \sigma^2)$, the minimizer of (13.71) is the same as the best linear unbiased estimator (BLUE) for $\beta$ and the best linear unbiased predictor (BLUP) for $\gamma$, provided that $\lambda$ is identical to the ratio $\tau^2/\sigma^2$ (Exercise 13.17). Thus, the P-spline model is fitted the same way as the linear mixed model

$$y = X\beta + Z\gamma + \epsilon \qquad (13.72)$$

(e.g., Jiang 2007, §2.3.1). It should be pointed out that the connection is asymptotically valid only if the true underlying function $f$ is not a P-spline. To see this, consider the following example.

*Example 13.1.* Suppose that the true underlying function is a quadratic spline with two knots, given by

$$f(x) = 1 - x + x^2 - 2(x - 1)_+^2 + 2(x - 2)_+^2, \quad 0 \le x \le 3$$

(the shape is a half-circle between 0 and 1 facing up, a half-circle between 1 and 2 facing down, and a half-circle between 2 and 3 facing up). Note that this function is smooth in that it has a continuous derivative (Exercise 13.18). Obviously, the best approximating spline is $f$ itself, for which $q = 2$. However, if one uses the above linear mixed model connection, the ML (or REML) estimator of $\sigma^2$ is consistent only if $q \to \infty$ (i.e., the number of appearances of the spline random effects goes to infinity). This can be seen from the asymptotic theory of the previous chapter (see Section 12.2), but, intuitively, it make sense even without the theory (why?). The seeming inconsistency has two worrisome consequences: (i) The meaning of $\lambda$ may be conceptually difficult to interpret and (ii) the behavior of the estimator of $\lambda$ may be unpredictable.

Nevertheless, in most applications of P-splines, the unknown function $f$ is unlikely to be a P-spline, so, in a way, (13.70) is only used as an approximation, with the approximation error vanishing as $q \to \infty$. So, from now on this is the case on which we focus. Opsomer et al. (2008) incorporated the spline model with the small-area random effects. By making use of the spline–mixed-model connection, their approximating model simply has the term $W\alpha$ added to the right side of (13.72); that is,

$$y = X\beta + Z\gamma + W\alpha + \epsilon, \qquad (13.73)$$

where $\alpha$ is the vector of small-area random effects and $W$ is a known matrix. It is assumed that $\gamma$, $\alpha$, and $\epsilon$ are uncorrelated with means 0 and covariance matrices $\Sigma_\gamma$, $\Sigma_\alpha$, and $\Sigma_\epsilon$, respectively. The BLUE and BLUP are given by

$$\tilde{\beta} = (X'V^{-1}X)^{-1}X'V^{-1}y,$$
$$\tilde{\gamma} = \Sigma_\gamma Z'V^{-1}(y - X\tilde{\beta}),$$
$$\tilde{\alpha} = \Sigma_\alpha W'V^{-1}(y - X\tilde{\beta}),$$

where $V = \text{Var}(y) = Z\Sigma_\gamma Z' + W\Sigma_\alpha W' + \Sigma_\epsilon$ (e.g., Jiang 2007, §2.3.1). The EBLUE and EBLUP, denoted by $\hat{\beta}$, $\hat{\gamma}$, and $\hat{\alpha}$, are obtained by replacing $V$ by $\hat{V}$ in the corresponding expressions. Here, we assume that $V = V(\theta)$, where $\theta$ is a vector of variance components, so that $\hat{V} = V(\hat{\theta})$. The REML estimator is used for $\hat{\theta}$, as suggested by Opsomer et al. (2008).

Once again, the problem of main interest is the estimation of small-area means. As in Section 13.2, we may treat (13.37), approximately, as a super-population model, which can be expressed as

$$Y_{ij} = x'_{ij}\beta + z'_{ij}\gamma + w'_{ij}\alpha + \epsilon_{ij}, \tag{13.74}$$

$j = 1, \ldots, N_i$, where $N_i$ is the population size for the $i$th small area or sub-population. Then, as argued in Section 13.2, the population mean for the $i$th small area is approximately equal to the mixed effect

$$\theta_i = \bar{x}'_{i,\text{P}}\beta + \bar{z}'_{i,\text{P}}\gamma + \bar{w}'_{i,\text{P}}\alpha, \tag{13.75}$$

where $\bar{x}_{i,\text{P}}$, $\bar{z}_{i,\text{P}}$, and $\bar{w}_{i,\text{P}}$ are the population means of $x_{ij}$, $z_{ij}$, and $w_{ij}$, respectively, over $j = 1, \ldots, N_i$, and the approximation error is $O_\text{P}(N_i^{-1/2})$. Thus, by ignoring the approximation error, we may treat $\theta_i$ as the small-area mean. Also, note that, in most applications, we have $w'_{ij}\alpha = \alpha_i$, the random effect associated with the $i$th small area, so that $\bar{w}'_{i,\text{P}}\alpha = \alpha_i$. We will focus on this case in the sequel. The EBLUP of $\theta_i$ is then given by

$$\hat{\theta}_i = \bar{x}'_{i,\text{P}}\hat{\beta} + \bar{z}'_{i,\text{P}}\hat{\gamma} + \hat{\alpha}_i, \tag{13.76}$$

where $\hat{\alpha}_i$ is the $i$th component of $\hat{\alpha}$.

The MSPE of the EBLUP is, again, of interest. Here, however, we cannot apply the results of Prasad–Rao (see Section 4.8). The reason is that the latter results apply only to the case where the covariance matrix of $y$, $V$, is block-diagonal. It is easy to see that $V$ is not block-diagonal in this case (why?). Das et al. (2004) extended the Prasad–Rao method to general linear mixed models, which may not have a block-diagonal covariance structure. Their results are applicable to the current situation. First, note that the Kackar–Harville identity (4.90) holds for normal linear mixed models in general, with or without block-diagonal covariance structure, where $\eta$ is any mixed effects that can be expressed as $\eta = \phi'\beta + \psi'\nu$, where $\nu$ is the vector of all of the random effects involved. It is clear that (13.75) is a special case of $\eta$, with $\phi = \bar{x}_{i,\text{P}}$, $\psi = (\bar{z}'_{i,\text{P}}, \bar{w}'_{i,\text{P}})'$, and $\nu = (\gamma', \alpha')'$. Second, expression (4.92) also holds for normal linear mixed models in general; so this term has a closed-form expression. It remains to approximate the second term on the right side of (4.90) to the second order. Opsomer et al. (2008) used the following results due to Das et al. (2004) to obtain the second-order approximation to this term. Note that the results do not require normality (but the Kackar–Harville identity does). Let $l(\theta) = l(\theta, y)$ be a function, where $\theta = (\theta_i)_{1 \leq i \leq s}$ is a parameter with the parameter space $\Theta$ and $y$ is the vector of observations. For example, $l$ may be the restricted log-likelihood function that leads to the REML estimator of $\theta$, the vector of variance components (see Section 12.2).

**Theorem 13.1.** Suppose that the following hold:
(i) $l(\theta, y)$ is three-times continuously differentiable with respect to $\theta$;
(ii) the true $\theta \in \Theta^o$, the interior of $\Theta$;
(iii) $-\infty < \limsup_{n \to \infty} \lambda_{\max}(D^{-1}AD^{-1}) < 0$, where $A = \mathrm{E}\{\partial^2 l/\partial \theta^2\}$ with the second derivative evaluated at the true $\theta$ and $D = \mathrm{diag}(d_1, \ldots, d_s)$, with $d_i$'s being positive constants satisfying $d_* = \min_{1 \leq i \leq s} d_i \to \infty$, as $n \to \infty$;
(iv) the $g$th moments of the following are bounded for some $g > 0$:

$$\frac{1}{d_i}\left|\frac{\partial l}{\partial \theta_i}\right|, \quad \frac{1}{\sqrt{d_i d_j}}\left|\frac{\partial^2 l}{\partial \theta_i \partial \theta_j} - \mathrm{E}\left(\frac{\partial^2 l}{\partial \theta_i \partial \theta_j}\right)\right|, \quad \frac{d_*}{d_i d_j d_k}M_{ijk}, \quad 1 \leq i, j, k \leq s,$$

where the first and second derivatives are evaluated at the true $\theta$ and

$$M_{ijk} = \sup_{\tilde{\theta} \in S_\delta(\theta)} \left|\frac{\partial^3 l}{\partial \theta_i \partial \theta_j \partial \theta_k}\right|_{\theta = \tilde{\theta}}$$

with $S_\delta(\theta) = \{\tilde{\theta} : |\tilde{\theta}_i - \theta_i| \leq \delta d_*/d_i, 1 \leq i \leq s\}$ for some $\delta > 0$. Then the following results hold:
(I) There exists $\hat{\theta}$ such that for any $0 < \rho < 1$, there is an event set $B$ satisfying for large $n$ and on $B$, $\hat{\theta} \in \Theta$, $\partial l/\partial \theta|_{\theta = \hat{\theta}} = 0$, $|D(\hat{\theta} - \theta)| < d_*^{1-\rho}$, and

$$\hat{\theta} = \theta - A^{-1}a + r,$$

where $\theta$ is the true $\theta$, $a = \partial l/\partial \theta$, evaluated at the true $\theta$, and $|r| \leq d_*^{-2\rho}u$ with $\mathrm{E}(u^g) = O(1)$.
(II) $\mathrm{P}(B^c) \leq cd_*^{\tau g}$ with $\tau = (1/4) \wedge (1 - \rho)$ and $c$ being a constant.

Theorem 13.1 plays a key role in the proof of the following theorem. Write the BLUP of $\eta$ as $\tilde{\eta} = \tilde{\eta}(\theta, y) = \phi'\tilde{\beta} + \psi'\tilde{\nu}$, where $\tilde{\beta}$ is the BLUE of $\beta$ and $\tilde{\nu}$ the BLUP of $\nu$, both depending on the unknown true $\theta$. Define a truncated estimator $\hat{\theta}_t$ of $\theta$ as follows: $\hat{\theta}_t = \hat{\theta}$ if $|\hat{\theta}| \leq L_n$, and $\hat{\theta}_t = \theta^*$ otherwise, where $\hat{\theta}$ is the estimator in Theorem 13.1, $\theta^*$ is a known vector in $\Theta$, and $L_n$ is a sequence of positive numbers such that $L_n \to \infty$ as $n \to \infty$. Consider the EBLUP $\hat{\eta} = \tilde{\eta}(\hat{\theta}_t, y)$. Define $s_0 = \sup_{|\theta| \leq L_n} |\tilde{\eta}(\theta, y)|$ and $s_2 = \sup_{|\theta| \leq L_n} \|\partial^2 \tilde{\eta}/\partial \theta \partial \theta'\|$, where $\|M\| = \lambda_{\max}^{1/2}(M'M)$ is the spectral norm of matrix $M$ and $b = \partial \tilde{\eta}/\partial \theta$, evaluated at the true $\theta$.

**Theorem 13.2.** Suppose that the conditions of Theorem 13.1 are satisfied. Furthermore, supppose that there is $h > 0$, $g > 8$, and nonnegative constants $g_j$, $j = 0, 1, 2$, such that $\mathrm{E}(s_0^{2h}) = O(d_*^{g_0})$, $\mathrm{E}\{|b|^{2g/(g-2)}\} = O(d_*^{g_1})$, and $\mathrm{E}(s_2^2) = O(d_*^{g_2})$. If the inequalities

$$g_0 < \frac{g}{4}(h-1) - 2h, \ g_1 < \left(\frac{g}{g-2}\right)\left\{\frac{1}{2} \wedge \left(\frac{g}{4} - 2\right)\right\}, \ g_2 < \frac{1}{2}$$

hold, then we have

$$\mathrm{E}(\hat{\eta} - \tilde{\eta})^2 = \mathrm{E}(b'A^{-1}a)^2 + o(d_*^{-2}), \tag{13.77}$$

where $A$ and $a$ are the same as in Theorem 2.1.

For a mixed ANOVA satisfying (12.1) and (12.2), the first term on the right side of (13.77) can be specified up to the order $o(d_*^{-2})$. Note that in this case, $\theta = (\tau^2, \sigma_1^2, \ldots, \sigma_s^2)$. Define $S(\theta) = V^{-1}ZG\psi$, where $V = \mathrm{Var}(y) = \tau^2 I_n + \sum_{i=1}^s \sigma_i^2 Z_i Z_i'$, $Z = (Z_1 \ldots Z_s)$ and $G = \mathrm{Var}(\alpha) = \mathrm{diag}(\sigma_1^2 I_{m_1}, \ldots, \sigma_s^2 I_{m_s})$. Then [note that a minus sign is missing in front of the trace on the right side of (3.4) of Das et al. (2004)] we have

$$\mathrm{E}\{b'A^{-1}a\}^2 = -\mathrm{tr}\left(\frac{\partial S'}{\partial \theta}V\frac{\partial S}{\partial \theta}A^{-1}\right) + o(d_*^{-2}). \tag{13.78}$$

Combining (13.77) and (13.78) with the Kackar-Harville identity, we obtain a second-order approximation to the MSPE of EBLUP as

$$\mathrm{MSPE}(\hat{\eta}) = g_1(\theta) + g_2(\theta) + g_3(\theta) + o(d_*^{-2}), \tag{13.79}$$

where $g_1(\theta) = \psi'(G - GZ'V^{-1}ZG)\psi$, $g_2(\theta) = \{\phi - X'S(\theta)\}'(X'V^{-1}X)^{-1}\{\phi - X'S(\theta)\}$, and $g_3(\theta)$ is the first term on the right side of (13.78). By (13.79) and using a similar bias-correction technique as in Section 13.2, a second-order unbiased estimator of the MSPE can be obtained. See Das et al. (2004).

Unlike (13.42), the remaining term on the right side of (13.79) is expressed as $o(d_*^{-2})$. For mixed ANOVA models, Das et al. (2004) showed that $d_i^2$ may be chosen as $\mathrm{tr}\{(Z_i'QZ_i)^2\}$, $0 \le i \le s$, where $Q$ is defined below (12.9) with $\Gamma = V$. Thus, for (13.73), assuming $\Sigma_\gamma = \sigma_\gamma^2 I_q$, $\Sigma_\alpha = \sigma_\alpha^2 I_m$, and $\Sigma_\epsilon = \sigma_\epsilon^2 I_n$, where $\sigma_\gamma^2$ and $\sigma_\epsilon^2$ are positive, we have $d_0^2 = \mathrm{tr}(Q^2)$, $d_1^2 = \mathrm{tr}(QZZ'QZZ')$ and $d_2^2 = \mathrm{tr}(QWW'QWW')$. For (13.79) to be meaningful, we need to show that

$$d_*^2 = d_0^2 \wedge d_1^2 \wedge d_2^2 \to \infty \tag{13.80}$$

as $m \to \infty$, at the very least. We assume $W = \mathrm{diag}(1_{n_i}, 1 \le i \le m)$, as is often the case, where $n_i$ is the sample size for the $i$th small area, which are assumed to be bounded. Then we have $\lambda_{\max}(W\Sigma_\alpha W') = \sigma_\alpha^2 \lambda_{\max}(WW') = \sigma_\alpha^2 \lambda_{\max}(W'W) = \sigma_\alpha^2 \max_{1 \le i \le m} n_i$. It follows that

$$\begin{aligned} V &= \Sigma_\epsilon + Z\Sigma_\gamma Z' + W\Sigma_\alpha W' \\ &\le \sigma_\epsilon^2 I_n + \lambda_{\max}(W\Sigma_\alpha W')I_n + \sigma_\gamma ZZ' \\ &= bI_n + \sigma_\gamma^2 ZZ', \end{aligned} \tag{13.81}$$

where $b = \sigma_\epsilon^2 + \sigma_\alpha^2 \max_{1 \le i \le m} n_i$ is positive and bounded. Let $\lambda_1, \ldots, \lambda_q$ be the eigenvalues of $Z'Z$. Then the eigenvalues of $ZZ'$ are $\lambda_1, \ldots, \lambda_q, 0, \ldots, 0$ (there are $n - q$ zeros after $\lambda_q$). Therefore, the eigenvalues of $V^{-2}$ are $(b + \sigma_\gamma^2 \lambda_j)^{-2}$, $1 \le j \le q$, and $b^{-2}, \ldots, b^{-2}$ ($n - q$ $b^{-2}$'s). It follows that

$$\operatorname{tr}(V^{-2}) = \frac{n-q}{b^2} + \sum_{j=1}^{q} (b + \sigma_\gamma^2 \lambda_j)^{-2}. \tag{13.82}$$

Also, we have $V \ge \sigma_\epsilon^2 I_n$; hence, by (i) of §5.3.1, $V^{-1} \le \sigma_\epsilon^{-2} I_n$. It follows, by (iii) of Section 5.3.1, $\|V^{-1}\|^2 = \lambda_{\max}(V^{-2}) = \{\lambda_{\max}(V^{-1})\}^2 \le \{\lambda_{\max}(\sigma_\epsilon^{-2} I_n)\}^2 = \sigma_\epsilon^{-4}$. By (5.43), we have

$$\begin{aligned}
\|V^{-1}\|_2 &= \|Q + V^{-1}X(X'V^{-1}X)^{-1}X'V^{-1}\|_2 \\
&\le \|Q\|_2 + \|V^{-1}X(X'V^{-1}X)^{-1}X'V^{-1}\|_2 \\
&\le \|Q\|_2 + \|V^{-1}\|\sqrt{p+1} \\
&\le \|Q\|_2 + \sigma_\epsilon^{-2}\sqrt{p+1}
\end{aligned}$$

(Exercise 13.19). Therefore, by (13.82), we have $d_0 = \|Q\|_2 \ge \|V^{-1}\|_2 - \sigma_\epsilon^2\sqrt{p+1} \to \infty$, provided, for example, that $n - q \to \infty$.

Next, we consider $d_1$. Note that $d_1^2 = \|Z'QZ\|_2^2$, and $Z'QZ = Z'V^{-1}Z - Z'V^{-1}X(X'V^{-1}X)^{-1}X'V^{-1}Z$. Thus, by a similar argument, we have

$$\begin{aligned}
\|Z'V^{-1}Z\|_2 &= \|Z'QZ + Z'V^{-1}X(X'V^{-1}X)^{-1}X'V^{-1}Z\|_2 \\
&\le \|Z'QZ\|_2 + \|Z'V^{-1}X(X'V^{-1}X)^{-1}X'V^{-1}Z\|_2.
\end{aligned}$$

Again, by (i) and (ii) of Section 5.3.1 and that $V \ge \sigma_\epsilon^2 I_n + \sigma_\gamma^2 ZZ'$, we have

$$Z'V^{-1}Z \le Z'(\sigma_\epsilon^2 I_n + \sigma_\gamma^2 ZZ')^{-1}Z.$$

Note that the nonzero eigenvalues of $Z'(\sigma_\epsilon^2 I_n + \sigma_\gamma^2 ZZ')^{-1}Z$ are the same as the nonzero eigenvalues of $(\sigma_\epsilon^2 I_n + \sigma_\gamma^2 ZZ')^{-1/2} ZZ'(\sigma_\epsilon^2 I_n + \sigma_\gamma^2 ZZ')^{-1/2}$, which are $(\sigma_\epsilon^2 + \sigma_\gamma^2 \lambda_j)^{-1}\lambda_j$, $1 \le j \le q$, followed by $n - q$ zeros. It follows by (iii) of Section 5.3.1, that $\|Z'V^{-1}Z\| \le \sigma_\gamma^{-2}$. On the other hand, it can be shown (Exercise 13.19) that $\|Z'V^{-1}X(X'V^{-1}X)^{-1}X'V^{-1}Z\|_2 \le \|Z'V^{-1}Z\|\sqrt{p+1}$. Therefore, we have $\|Z'QZ\|_2 \ge \|Z'V^{-1}Z\|_2 - \sigma_\gamma^{-2}\sqrt{p+1}$. Now use, again, the inequality (13.81) and similar arguments as above to show that

$$\|Z'V^{-1}Z\|_2^2 \ge \sum_{j=1}^{q} \left( \frac{\lambda_j}{b + \sigma_\gamma^2 \lambda_j} \right)^2 \tag{13.83}$$

(Exercise 13.20). Thus, $d_1 \to \infty$, provided, for example, that the right side of (13.83) goes to infinity.

Finally, we consider $d_2$. By a similar argument, it can be shown that

$$\|W'QW\|_2 \ge \|W'V^{-1}W\|_2 - c\sqrt{p+1} \tag{13.84}$$

for some constant $c$ (Exercise 13.21). Now, again, using (13.81) and an inverse matrix identity (see Appendix A.1), we have

$$V^{-1} \geq (bI_n + \sigma_\gamma^2 ZZ')^{-1}$$
$$= b^{-1}\{I_n - \delta Z(I_q + \delta Z'Z)^{-1}Z'\},$$

where $\delta = \sigma_\gamma^2/b$. Thus, we have

$$W'V^{-1}W \geq b^{-1}\{W'W - \delta W'Z(I_q + \delta Z'Z)^{-1}Z'W\}.$$

Write $B = \delta Z(I_q + \delta Z'Z)^{-1}Z'$. Then we have $\lambda_{\max}(B) = \lambda_{\max}\{\delta(I_q + \delta Z'Z)^{-1/2}Z'Z(I_q + \delta Z'Z)^{-1/2}\} = \max_{1\leq j\leq q}\delta\lambda_j(1 + \delta\lambda_j)^{-1} \leq 1$. It follows that $\|W'BW\|_2 \leq (\max_{1\leq i\leq m} n_i)q$ (Exercise 13.21). Now, by (iii) of Section 5.3.1, we have $\|W'W\|_2 \leq \|bW'V^{-1}W + W'BW\|_2 \leq b\|W'V^{-1}W\|_2 + (\max_{1\leq i\leq m} n_i)q$. Thus, by (13.84) and the fact that $\|W'W\|_2 = (\sum_{i=1}^{m} n_i^2)^{1/2}$, we have

$$
\begin{aligned}
d_2 &= \|W'QW\|_2 \\
&\geq b^{-1}\left\{\left(\sum_{i=1}^{m} n_i^2\right)^{1/2} - \left(\max_{1\leq i\leq m} n_i\right)q\right\} - c\sqrt{p+1} \\
&\longrightarrow \infty,
\end{aligned}
$$

provided that, for example, $n_i \geq 1$, $1 \leq i \leq m$, and $q/\sqrt{m} \to 0$.

In the above arguments that lead to (13.80), there is a single assumption that is not very clear what it means. This is the assumption that the right side of (13.83) goes to $\infty$. Note that the assumption does not have to hold, even if $q \to \infty$. For the remaining part of this section, we consider a specific example and show that the assumption holds in this case, provided that $q \to \infty$ at a certain slower rate than $n$.

*Example 13.2.* Consider the following special case: $n_i = 1$, $1 \leq i \leq m$ (hence $n = m$), $x_i = i/n$, $1 \leq i \leq n = qr$, where $r$ is a positive integer, and $\kappa_u = (u-1)/q$, $1 \leq u \leq q$. We first show that for any fixed $q$, there is a positive integer $n(q)$ such that

$$\lambda_{\min}(Z'Z) \geq 1, \quad n \geq n(q). \tag{13.85}$$

Note that $Z'Z = [\sum_{i=1}^{n}(x_i - \kappa_u)_+^p(x_i - \kappa_v)_+^p]_{1\leq u,v\leq q}$. We have

$$\frac{1}{n}\sum_{i=1}^{n}(x_i - \kappa_u)_+^p(x_i - \kappa_v)_+^p \longrightarrow \int_0^1 (x - \kappa_u)_+^p(x - \kappa_v)_+^p \, dx$$
$$\equiv b_{uv}, \quad 1 \leq u, v \leq q,$$

as $n \to \infty$. The matrix $B = (b_{uv})_{1\leq u,v\leq q}$ is positive definite. To see this, let $\xi = (\xi_u)_{1\leq u\leq q} \in R^q$. Then

$$\xi' B \xi = \int_0^1 \left\{ \sum_{u=1}^q \xi_u (x - \kappa_u)_+^p \right\}^2 dx \geq 0$$

and the equality holds if and only if

$$\sum_{u=1}^q \xi_u (x - \kappa_u)_+^p = 0, \quad x \in [0,1]. \tag{13.86}$$

Let $x \in (0, \kappa_2]$; then by (13.86), we have $\xi_1 x^p = 0$ and hence $\xi_1 = 0$. Let $x \in (\kappa_2, \kappa_3]$; then by (13.86) and the fact that $\xi_1 = 0$, we have $\xi_2 (x - \kappa_2)^p = 0$, hence $\xi_2 = 0$; and so on. This implies $\xi_u = 0, 1 \leq u \leq q$. It follows, by Weyl's eigenvalue perturbation theorem [see (5.51)], that

$$\frac{1}{n} \lambda_{\min}(Z'Z) = \lambda_{\min}(B) > 0;$$

hence, there is $n(q) \geq 1$ such that (13.85) holds.

Without loss of generality, let $n(q)$, $q = 1, 2, \ldots$, be strictly increasing. Define the sequence $q(n)$ as $q(n) = 1, 1 \leq n < n(1)$, and $q(n) = j, n(j) \leq n < n(j+1), j \geq 1$. By the definition, it is seen that, with $q = q(n)$, (13.85) holds as long as $n \geq n(1)$. It follows that $\lambda_j \geq 1, 1 \leq j \leq q$; hence, the right side of (13.83) is bounded from below by $q/(b + \sigma_\gamma^2)^2$, for $q = q(n)$ with $n \geq n(1)$, which goes to infinity as $n \to \infty$.

## 13.5 Model selection for small-area estimation

As discussed in Section 13.1, the "strength" for the small-area estimation is borrowed by utilizing a statistical model. It is therefore not surprising that the choice of the model makes a difference. Although there is extensive literature on inference about small areas using statistical models, especially mixed effects models (see Rao 2003), model selection in small-area estimation has received much less attention. However, the importance of model selection in the small-area estimation has been noted by prominent researchers (e.g., Battese et al. 1988, Ghosh and Rao 1994). Datta and Lahiri (2001) discussed a model selection method based on computation of the frequentist's Bayes factor in choosing between a fixed effects model and a random effects model. They focused on a one-way balanced random effects model, which is Example 12.2 with $n_i = k$, $1 \leq i \leq m$, for the sake of simplicity and observed that the choice between a fixed effects model and a random effects one in this case is equivalent to testing the following one-sided hypothesis $H_0$: $\sigma^2 = 0$ vs. $H_1$: $\sigma^2 > 0$. Note that, however, not all model selection problems can be formulated as hypothesis testing problems. Meza and Lahiri (2005) demonstrated the limitations of Mallows' $C_p$ statistic in selecting the fixed covariates in a nested error regression model (see below). They showed by results of simulation studies that the $C_p$ method without modification does not work well in the current mixed model setting when the variance of the small-area random effects is large.

As noted in Section 12.5, the selection of a mixed effects model for the small-area estimation is one of the unconventional problems in that it is not easy to determine the effective sample size, which is used by traditional information criteria, such as the BIC and HQ, to calculate the penalty for model complexity. We use an example to illustrate.

*Example 13.3* (Nested-error regression). A well-known model for the small-area estimation was proposed by Battese et al. (1988), known as the nested error regression model. This may be regarded as an extension of the one-way random effects model of Example 12.2, expressed as

$$y_{ij} = x'_{ij}\beta + v_i + e_{ij}, \quad i = 1, \ldots, m, \ j = 1, \ldots, n_i,$$

where $x_{ij}$ is a vector of known auxiliary variables, $\beta$ is an unknown vector of regression coefficients, $v_i$ is a small-area specific random effect, $e_{ij}$ is a sampling error, and $n_i$ is the sample size for the $i$th small areas. It is assumed that the random effects and sampling errors are independent such that $v_i \sim N(0, \sigma_v^2)$ and $e_{ij} \sim N(0, \sigma_e^2)$, where $\sigma_v^2$ and $\sigma_e^2$ are unknown. Once again, the problem is what is the effective sample size $n$ that could be used to determine the penalty $\lambda_n$ in (9.25) (considered as a general criterion for model selection) for, say, the BIC. However, it is not clear at all what $n$ should be: The total sample size which is $\sum_{i=1}^{m} n_i$, the number of small areas, $m$, or something else? If of all the $n_i$ are equal, then it may be reasonable to assume that the effective sample size is proportional to $m$, but even this is very impractical. In typical situations of the small-area estimation, the $n_i$ are very different from small area to small area. For example, Jiang et al. (2007) gave a practical example, in which the $n_i$'s range from 4 to 2301. It is clearly difficult to determine the effective sample size in such a case.

There is a further issue that was not addressed in a general mixed model selection problem, discussed in Section 12.5. In the small-area estimation, the main interest is the estimation of small-area means, which may be formulated as a (mixed effects) prediction problem (see the previous sections). In other words, one needs to select the best model for predition, not for estimation. The information criteria, on the other hand, rely on the likelihood function that is for the estimation of parameters. To make our points, consider the Fay–Herriot model (Example 4.18; also see Section 13.3), expressed as

$$y_i = x'_i\beta + v_i + e_i, \quad i = 1, \ldots, m, \tag{13.87}$$

where $x_i$ is a known vector of known auxiliary variable, $\beta$ is a vector of unknown regression coefficients, $v_i$ is a small-area specific random effect, and $e_i$ is a sampling error. It is assumed that the $v_i$'s and $e_i$'s are independent such that $v_i \sim N(0, A)$ and $e_i \sim N(0, D_i)$. For simplicity, let $A > 0$ be known, for now. Let $\mathcal{M}$ denote the collection of candidate models. An important difference from Section 12.5 is that here we consider a situation where the true model is not a member of $\mathcal{M}$. Note that this is a scenario that is quite possible to occur in practice. The goal is to find an approximating model within $\mathcal{M}$ that best serves our main interest of prediction of mixed effects. The latter can be expressed as $\zeta = (\zeta_i)_{1 \leq i \leq m} = E(y|v) = \mu + v$, where $y = (y_i)_{1 \leq i \leq m}$, $\mu = (\mu_i)_{1 \leq i \leq m}$, and $v = (v_i)_{1 \leq i \leq m}$. Note the following useful expression:

$$\mu_i = E(y_i), \quad 1 \leq i \leq m. \tag{13.88}$$

Consider the problem of selecting the auxiliary variables. Then each $M \in \mathcal{M}$ corresponds to a matrix $X = (x_i')_{1 \leq i \leq m}$. We assume that $X$ is full rank. The information criteria are based on the log-likelihood function for estimating $\beta$, which, under $M$, can be expressed as (Exercise 13.22)

$$l(M) = -\frac{m}{2} \log(2\pi)$$
$$-\frac{1}{2} \sum_{i=1}^{m} \left\{ \log(A + D_i) + \frac{(y_i - x_i'\beta)^2}{A + D_i} \right\}. \tag{13.89}$$

The MLE for $\beta$ given by

$$\hat{\beta} = (X'V^{-1}X)^{-1}X'V^{-1}y$$
$$= \left( \sum_{i=1}^{m} \frac{x_i x_i'}{A + D_i} \right)^{-1} \sum_{i=1}^{m} \frac{x_i y_i}{A + D_i}, \tag{13.90}$$

where $V = \text{diag}(A + D_i, 1 \leq i \leq m)$. Note that the MLE is the same as the BLUE for $\beta$. Thus, the maximized log-likelihood is given by

$$\hat{l}(M) = c - \frac{1}{2}y'P(M)y, \tag{13.91}$$

where $c = -(1/2)\{m \log(2\pi) + \sum_{i=1}^{m} \log(A + D_i)\}$ and

$$P(M) = V^{-1} - V^{-1}X(X'V^{-1}X)^{-1}X'V^{-1}.$$

Now, consider a generalized information criterion (GIC) that has the form

$$\text{GIC}(M) = -2\hat{l}(M) + \lambda_m p, \tag{13.92}$$

where $p = \text{rank}(X)$ and $\lambda_m$ is a penalty. The AIC corresponds to (12.92) with $\lambda_m = 2$; whereas for the BIC, $\lambda_m = \log(m)$. We have (Exercise 13.22)

$$E\{\text{GIC}(M)\} = m - 2c + \mu'P(M)\mu + (\lambda_m - 1)p. \tag{13.93}$$

Since $m - 2c$ does not depend on $M$, the best model according to the GIC corresponds to the one that minimizes $C_2(M) = \mu'P(M)\mu + (\lambda_m - 1)p$.

On the other hand, the best predictor (BP) of $\zeta$ under $M$ in the sense of minimizing the MSPE is $\tilde{\zeta}_M = (\tilde{\zeta}_{M,i})_{1 \leq i \leq m}$, where

$$\tilde{\zeta}_{M,i} = \mathrm{E}_M(\zeta_i|y) = \frac{A}{A + D_i} y_i + \frac{D_i}{A + D_i} x_i'\beta \qquad (13.94)$$

(Exercise 13.22). Here, $\mathrm{E}_M$ represents expectation under $M$. By (13.88), the MSPE can be expressed as (verify)

$$
\mathrm{E}\left(\left|\tilde{\zeta}_M - \zeta\right|^2\right)
$$

$$
= \sum_{i=1}^m \mathrm{E}\left(\tilde{\zeta}_{M,i} - \zeta_i\right)^2
$$

$$
= \sum_{i=1}^m \mathrm{E}\left(\frac{A}{A + D_i} y_i - \zeta_i\right)^2 + 2\sum_{i=1}^m \frac{D_i}{A + D_i} x_i'\beta\mathrm{E}\left(\frac{A}{A + D_i} y_i - \zeta_i\right)
$$

$$
+ \sum_{i=1}^m \left(\frac{D_i}{A + D_i}\right)^2 (x_i'\beta)^2
$$

$$
= \sum_{i=1}^m \mathrm{E}\left(\frac{A}{A + D_i} y_i - \zeta_i\right)^2 - 2\sum_{i=1}^m \left(\frac{D_i}{A + D_i}\right)^2 x_i'\beta\mu_i
$$

$$
+ \sum_{i=1}^m \left(\frac{D_i}{A + D_i}\right)^2 (x_i'\beta)^2. \qquad (13.95)
$$

The first term on the right side of (13.95) is unknown but does not depend on $M$ or $\beta$. Let $S(M, \beta)$ denote the sum of the last two terms, or, in matrix expression, $S(M, \beta) = \beta'X'R^2X\beta - 2\mu'R^2X\beta$, where $R = \mathrm{diag}\{D_i/(A + D_i), 1 \leq i \leq m\}$. It is easy to see that the $\beta$ that minimizes $S(M, \beta)$ is $\beta^* = (X'R^2X)^{-1}X'R^2\mu$. It follows that $\inf_\beta S(M, \beta) = -\mu'R^2X(X'R^2X)^{-1}X'R^2\mu$. Thus, the best model in terms of minimizing the MSPE is the one that maximizes $C_1(M) = \mu'R^2X(X'R^2X)^{-1}X'R^2\mu$.

It is easy to see that if $M$ is a true model (i.e., if $\mu = X\beta$ for some $\beta$), then $\beta^* = \beta$. In fact, if there is a true model, say $M \in \mathcal{M}$, then a true model with minimal $p$ is the best model under both the GIC and BP (Exercise 13.23). However, here we are concerned with the situation where there is no true model among the candidate models. We now show, by a specific example, that in such a case these two criteria, the GIC and BP, can lead to completely different choices of optimal models.

*Example 13.4.* Let $A = 1$. Suppose that

$$
D_i = \begin{cases} 1, & 1 \leq i \leq m/4 \text{ or } m/2 + 1 \leq i \leq 3m/4 \\ 3, & m/4 + 1 \leq i \leq m/2 \text{ or } 3m/4 + 1 \leq i \leq m. \end{cases}
$$

Also, suppose that

$$\mu_i = \begin{cases} a, & 1 \le i \le m/4 \text{ or } m/2+1 \le i \le 3m/4 \\ -a, & m/4+1 \le i \le m/2 \text{ or } 3m/4+1 \le i \le m, \end{cases}$$

where $a$ is a positive constant. There are three candidate models under consideration. They are $M_1$: $x_i'\beta = \beta_1 x_{i,1}$, where

$$x_{i,1} = \begin{cases} 1, & 1 \le i \le m/4 \\ 2, & m/4+1 \le i \le m/2 \\ 0, & m/2+1 \le i \le m; \end{cases}$$

$M_2$: $x_i'\beta = \beta_2 x_{i,2}$, where

$$x_{i,2} = \begin{cases} 0, & 1 \le i \le m/2 \\ 9, & m/2+1 \le i \le 3m/4 \\ 4, & 3m/4+1 \le i \le m; \end{cases}$$

and $M_3$: $x_i'\beta = \beta_1 x_{i,1} + \beta_2 x_{i,2}$, where the $\beta$'s are unknown parameters whose values may be different under different models. Note that none of the candidates is a true model. It can be shown (Exercise 13.24) that the best model according to the GIC is $M_2$, as long as $\lambda_m > 1$. On the other hand, the best model in terms of the BP is $M_1$.

In Section 12.5 we introduced a new strategy, called the *fence*, for model selection. The idea is to build a statistical fence to isolate a subgroup of candidate models, known as correct models, from which an optimal model is chosen according to a criterion of optimality that can be flexible. Note that here we do not assume that there is a true model among the candidates; so the term "correct model" should be understood as a model that provides an approximation that is "good enough" (in fact, it should be always understood this way, as the "correct models" may not be the actual true models, even if the latter exist among the candidates). An apparent advantage of the fence is its flexibility in choosing a measure of lack-of-fit, $Q(M)$, used to build the statistical fence according to (12.55), and a criterion of optimality for selecting a model within the fence. Here, we consider model simplicity, in terms of minimal dimension of the parameter space, as the criterion of optimality, as usual. Furthermore, we explore the other flexibility of fence in choosing the measure of lack-of-fit by deriving a predictive measure of lack-of-fit. To do so, let us return to (13.95). Note that by (13.88), we can express the MSPE as

$$E\left(\left|\tilde{\zeta}_M - \zeta\right|^2\right) = E\left\{ I_1 + \sum_{i=1}^{m} \left(\frac{D_i}{A+D_i}\right)^2 (x_i'\beta)^2 \right.$$
$$\left. -2\sum_{i=1}^{m} \left(\frac{D_i}{A+D_i}\right)^2 x_i'\beta y_i \right\}, \tag{13.96}$$

where $I_1 = \sum_{i=1}^{m} E\{A(A+D_i)^{-1}y_i - \zeta_i\}^2$ does not depend on $M$. This naturally leads to the idea of minimizing the expression inside the expectation.

The rationale behind this idea is the same as indicated in our second Preface example (Example 2). Note that the expression inside the expectation can be expressed as a sum of independent random variables, say, $\sum_{i=1}^{m} \psi_i(\beta, y_i)$. Then, according to the CLT (Section 6.4), we have, under regularity conditions,

$$
\begin{aligned}
\sum_{i=1}^{m} \psi_i(\beta, y_i) &= \mathrm{E}\left\{ \sum_{i=1}^{m} \psi_i(\beta, y_i) \right\} + \sum_{i=1}^{m} [\psi_i(\beta, y_i) - \mathrm{E}\{\psi_i(\beta, y_i)\}] \\
&= \mathrm{E}\left\{ \sum_{i=1}^{m} \psi_i(\beta, y_i) \right\} + O_\mathrm{P}(m^{1/2}).
\end{aligned}
$$

Thus, provided $\sum_{i=1}^{m} \mathrm{E}\{\psi_i(\beta, y_i)\} = O(m)$, which can be reasonably assumed if $\mathrm{E}\{\psi_i(\beta, y_i)\} \neq 0$, $\mathrm{E}\{\sum_{i=1}^{m} \psi_i(\beta, y_i)\}$ is the leading term for $\sum_{i=1}^{m} \psi_i(\beta, y_i)$ and vice versa. Therefore, to the first-order approximation, we can simply drop the expectation sign if the expression inside the expectation is a sum of independent random variables. In fact, the idea can be generalized to some cases where inside the expectation is not necessarily a sum of independent random variables (but with caution; see Example 2, Example 2.1, and Section 3.1). By (13.96) and the fact that $I_1$ does not depend on $M$, we arrive at the following measure of lack-of-fit:

$$
\begin{aligned}
Q_1(M) &= \sum_{i=1}^{m} \left( \frac{D_i}{A + D_i} \right)^2 (x_i'\beta)^2 - 2 \sum_{i=1}^{m} \left( \frac{D_i}{A + D_i} \right)^2 x_i'\beta y_i \\
&= \beta' X' R^2 X \beta - 2y' R^2 X \beta.
\end{aligned}
\tag{13.97}
$$

The $\beta$ that minimizes (13.97) is given by

$$
\begin{aligned}
\tilde{\beta} &= (X' R^2 X)^{-1} X' R^2 y \\
&= \left\{ \sum_{i=1}^{m} \left( \frac{D_i}{A + D_i} \right)^2 x_i x_i' \right\}^{-1} \sum_{i=1}^{m} \left( \frac{D_i}{A + D_i} \right)^2 x_i y_i.
\end{aligned}
\tag{13.98}
$$

Note that (13.98) is different from the MLE (13.90). We call (13.98) the best predictive estimator, or BPE, in the sense that $\mathrm{E}(\tilde{\beta}) = \beta^*$, the (theoretically) best $\beta$ given below (13.95) that minimizes the MSPE. We call $Q_1$ a predictive measure of lack-of-fit. By plugging in the BPE, we obtain

$$
\begin{aligned}
\hat{Q}_1(M) &= \inf_\beta Q_1(M) \\
&= -y' R^2 X (X' R^2 X)^{-1} X' R^2 y,
\end{aligned}
\tag{13.99}
$$

which will be used in place of $\hat{Q}(M)$ in the fence inequality (12.55).

*Example 13.4 (continued).* A simulation study was carried out by Dr. Thuan Nguyen of Oregon Health and Science University (personal communication), in which the predictive fence method was compared with two of the information criteria, the AIC and BIC, as well as a nonpredictive (ML) fence. The latter is based on (12.55) with $\hat{Q}(M) = -\hat{l}(M)$, where $\hat{l}(M)$ is the maximized log-likelihood given by (13.91). Two different sample sizes are considered: $m = 50$ and $m = 100$. The value of $a$ (that is involved in the definition of $\mu_i$) is either 1 or 2. For both predictive and ML fence methods, the constant $c$ in (12.55) is chosen adaptively (see Section 12.5), with the bootstrap sample size $B = 100$. A total of $N = 100$ simulations were run. Table 13.1 shows the empirical (or simulated) MSPE, obtained by averaging the difference $|\hat{\zeta} - \zeta|^2$ over all $N$ simulations. Here, $\hat{\zeta}$ is the empirical best predictor (EBP). For the three likelihood-based methods, the AIC, BIC, and ML fence, the EBP is obtained by replacing $\beta$ by $\hat{\beta}$, the MLE, in the expression of the BP (13.94); for the predictive fence, the EBP is obtained by replacing $\beta$ by $\tilde{\beta}$, the BPE, in (13.94). The numbers in Table 13.1 show two apparent

### Table 13.1. Simulated MSPE

| $a$ $m$ | AIC | BIC | ML Fence | Predictive Fence |
|---------|-------|-------|----------|------------------|
| 1  50   | 53.5  | 53.6  | 53.6     | 49.3             |
| 2  50   | 117.7 | 117.2 | 117.0    | 95.6             |
| 1 100   | 110.0 | 109.5 | 109.3    | 97.9             |
| 2 100   | 246.5 | 246.3 | 245.0    | 197.2            |

"clusters," with the AIC, BIC, and ML fence in one cluster and the predictive fence in the other. One may wonder why there is such a difference. Table 13.2 shows another set of summaries. Here are reported the empirical (or simulated) probabilities, in terms of percentages, that each model is selected. It is seen that the three likelihood-based methods have the highest probabilities of selecting $M_2$, whereas the predictive fence has the highest probabilities of selecting $M_1$. Recall that, theoretically, $M_2$ is the model most favored by the GIC (of which the AIC and BIC are special cases), whereas $M_1$ is the best model in terms of the BP. Given the way that the predictive fence is developed (and also the quite different focus of the ML method), one would not be surprised to see such a difference.

The method of deriving a predictive measure of lack-of-fit can be extended to more general situations. In the case of the Fay–Herriot model with unknown $A$, the variance of $v_i$, the MSPE can be expressed as (Exercise 13.25)

$$E(|\tilde{\zeta}_M - \zeta|^2)$$
$$= E\left\{ (y - X\beta)'R^2(y - X\beta) + 2A\mathrm{tr}(R) - \mathrm{tr}(D) \right\}, \tag{13.100}$$

where $D = \mathrm{diag}(D_i, 1 \le i \le m)$. Equation (13.100) suggests the measure

$$Q_1(M) = (y - X\beta)'\Gamma^2(y - X\beta) + 2A\mathrm{tr}(\Gamma). \tag{13.101}$$

Given $A$, (13.101) is minimized by $\tilde{\beta}$ given by (13.98). Thus, we have (verify) $\hat{Q}(M) = \inf_{A \ge 0} \tilde{Q}_1(M)$, where $\tilde{Q}_1(M) = y'RP_{(RX)\perp}Ry + 2A\,\mathrm{tr}(R)$ with

**Table 13.2. Percentage of selected model**

| $a$ $m$ | Model | AIC | BIC | Fence ML | Predictive Fence |
|---|---|---|---|---|---|
| 1 50 | 1 | 18 | 21 | 16 | 63 |
| | 2 | 74 | 78 | 72 | 17 |
| | 3 | 8 | 1 | 12 | 20 |
| 2 50 | 1 | 0 | 0 | 2 | 93 |
| | 2 | 85 | 95 | 97 | 0 |
| | 3 | 15 | 5 | 1 | 7 |
| 1 100 | 1 | 6 | 8 | 6 | 82 |
| | 2 | 84 | 90 | 85 | 7 |
| | 3 | 10 | 2 | 9 | 11 |
| 2 100 | 1 | 0 | 0 | 0 | 100 |
| | 2 | 85 | 94 | 100 | 0 |
| | 3 | 15 | 6 | 0 | 0 |

$P_{(RX)^\perp} = I_m - P_{RX}$ and $P_{RX} = RX(X'R^2X)^{-1}X'R$. Another simulation study shows similar performance of the predictive fence compared to the other three methods (details omitted).

Finally, we derive a predictive measure under the nested error regression model (see Example 13.3). More specifically, we assume that a nested error regression model holds for a superpopulation of finite populations in a way similar to Section 13.2. Let $Y_k, k = 1, \ldots, N_i$, $i = 1, \ldots, m$, represent the finite populations (small areas). We assume that auxiliary data $X_{ilk}, k = 1, \ldots, N_i, l = 1, \ldots, p$, are available for the finite populations and so are the population sizes $N_i$'s. The superpopulation model can be expressed as

$$Y_{ik} = X'_{ik}\beta + v_i + e_{ik}, \quad i = 1, \ldots, m, k = 1, \ldots, N_i, \qquad (13.102)$$

where $X_{ik} = (X_{ilk})_{1 \le l \le p}$ and other assumptions are the same as in Example 13.3. We are interested in the small-area means $\mu_i = N_i^{-1} \sum_{k=1}^{N_i} Y_{i,k}$, $i = 1, \ldots, m$. We consider a *model-assisted* method using the BP method. On the other hand, the performance of the model-assisted method will be evaluated using a design-based MSPE—that is, MSPE with respect to the sampling distribution within the finite populations. This is practical because, in surveys, one always samples from a finite population (although the model-based MSPE is often used as an approximation in the sense described in Section 13.2).

As in Example 13.2, let $y_{ij}$, $i = 1, \ldots, m, j = 1, \ldots, n_i$, represent the sampled $Y$'s. We use the notation $y_i = (y_{ij})_{1 \le j \le n_i}$, $y = (y_i)_{1 \le i \le m}$, $\bar{y}_{i\cdot} = n_i^{-1} \sum_{j=1}^{n_i} y_{ij}$, and $\bar{X}_{i,P} = N_i^{-1} \sum_{k=1}^{N_i} X_{i,k}$. Also, let $I_i$ denote the set of sampled indexes, so that $Y_{ik}$ is sampled if and only if $k \in I_i, i = 1, \ldots, m$. Under the nested error regression model (13.102), the BP for $\mu_i$ is

$$\tilde{\mu}_{M,i} = E_M(\mu_i|y)$$

$$= \frac{1}{N_i} \sum_{k=1}^{N_i} \mathrm{E}_M(Y_{i,k}|y_i)$$

$$= \frac{1}{N_i} \left\{ \sum_{j=1}^{n_i} y_{ij} + \sum_{k \notin I_i} \mathrm{E}_M(Y_{i,k}|y_i) \right\}$$

$$= \bar{X}'_{i,\mathrm{P}}\beta + \left\{ \frac{n_i}{N_i} + \left(1 - \frac{n_i}{N_i}\right) \frac{n_i \sigma_v^2}{n_0 + n_i \sigma_i^2} \right\} (\bar{y}_{i\cdot} - \bar{x}'_{i\cdot}\beta),$$

where $\mathrm{E}_M$ denotes the model-based (conditional) expectation. The design-based MSPE has the following expression:

$$\mathrm{MSPE} = \mathrm{E}_{\mathrm{d}}(|\tilde{\mu}_M - \mu|^2) = \sum_{i=1}^{m} \mathrm{E}_{\mathrm{d}}(\tilde{\mu}_{M,i} - \mu_i)^2,$$

where $\mathrm{E}_{\mathrm{d}}$ represents the design-based expectation. Furthermore, we have

$$\mathrm{E}_{\mathrm{d}}(\tilde{\mu}_{M,i} - \mu_i)^2 = \mathrm{E}_{\mathrm{d}}(\tilde{\mu}_{M,i}^2) - 2\mu_i \mathrm{E}_{\mathrm{d}}(\tilde{\mu}_{M,i}) + \mu_i^2. \qquad (13.103)$$

We now use $I_i$ to write $\bar{y}_{i\cdot} = n_i^{-1} \sum_{k=1}^{N_i} Y_{i,k} 1_{(k \in I_i)}$. Thus, we have

$$\mathrm{E}_{\mathrm{d}}(\bar{y}_{i\cdot}) = \frac{1}{n_i} \sum_{k=1}^{N_i} Y_{i,k} \mathrm{P}_{\mathrm{d}}(k \in I_i)$$

$$= \frac{1}{N_i} \sum_{k=1}^{N_i} Y_{i,k} = \mu_i.$$

Note that the $Y_{i,k}$'s are nonrandom under the sampling distribution and that the design-based probability $\mathrm{P}_{\mathrm{d}}(k \in I_i) = n_i/N_i$, assuming equal probability sampling. It follows that

$$\mathrm{E}_{\mathrm{d}}(\tilde{\mu}_{M,i})$$
$$= \bar{X}'_{i,\mathrm{P}}\beta + \left\{ \frac{n_i}{N_i} + \left(1 - \frac{n_i}{N_i}\right) \frac{n_i \sigma_v^2}{\sigma_e^2 + n_i \sigma_v^2} \right\} (\mu_i - \bar{X}'_{i,\mathrm{P}}\beta)$$
$$= \left(1 - \frac{n_i}{N_i}\right) \frac{\sigma_e^2 \bar{X}'_{i,\mathrm{P}}\beta}{\sigma_e^2 + n_i \sigma_v^2} + \left\{ \frac{n_i}{N_i} + \left(1 - \frac{n_i}{N_i}\right) \frac{n_i \sigma_v^2}{\sigma_e^2 + n_i \sigma_v^2} \right\} \mu_i.$$

If we bring the latest expression of $\mathrm{E}_{\mathrm{d}}(\tilde{\mu}_{M,i})$ to (13.103), a term $\mu_i^2$ will show up, which is unknown. The idea is to find a design-unbiased estimator of $\mu_i^2$, because then we can write (13.103), and therefore the MSPE, as the expectation of something, which is the "trick" we are using here. It can be shown (Exercise 13.26) that

$$\hat{\mu}_i^2 = \frac{1}{n_i} \sum_{j=1}^{n_i} y_{ij}^2 - \frac{N_i - 1}{N_i(n_i - 1)} \sum_{j=1}^{n_i} (y_{ij} - \bar{y}_{i\cdot})^2 \qquad (13.104)$$

is a design-unbiased estimator for $\mu_i^2$; that is, $E_d(\hat{\mu}_i^2) = \mu_i^2$. Write

$$a_i(\sigma_v^2, \sigma_e^2) = \left(1 - \frac{n_i}{N_i}\right) \frac{\sigma_e^2}{\sigma_e^2 + n_i \sigma_v^2},$$

$$b_i(\sigma_v^2, \sigma_e^2) = 1 - 2\left\{\frac{n_i}{N_i} + \left(1 - \frac{n_i}{N_i}\right)\frac{n_i \sigma_v^2}{\sigma_e^2 + n_i \sigma_v^2}\right\}.$$

Then we can express the MSPE as

$$\text{MSPE} = E_d\left[\sum_{i=1}^{m} \{\tilde{\mu}_{M,i}^2 - 2a_i(\sigma_v^2, \sigma_e^2)\bar{X}_{i,P}'\beta\bar{y}_{i\cdot} + b_i(\sigma_v^2, \sigma_e^2)\hat{\mu}_i^2\}\right].$$

Thus, a predictive measure of lack-of-fit is obtained by removing the expectation sign. This leads to

$$Q(M) = \sum_{i=1}^{m} \{\tilde{\mu}_{M,i}^2 - 2a_i(\sigma_v^2, \sigma_e^2)\bar{X}_{i,P}'\beta\bar{y}_{i\cdot} + b_i(\sigma_v^2, \sigma_e^2)\hat{\mu}_i^2\}.$$

## 13.6 Exercises

*13.1.* This exercise is associated with the mixed logistic model of (13.1).
(i) Show that $E(\alpha_i|y) = E(\alpha_i|y_i)$, where $y_i = (y_{ij})_{1 \leq j \leq n_i}$.
(ii) Verify (13.3) for $\tilde{\alpha}_i = E(\alpha_i|y_i)$.
*13.2.* Verify the limiting behaviors (i)–(iii) below (13.4).
*13.3.* Verify the limiting behavior (iv) below (13.4) and also (13.5). [Hint: The following formulas might be useful. For $1 \leq k \leq n - 1$, we have

$$\int_0^\infty \frac{x^{k-1}}{(1+x)^n} dx = \frac{(k-1)!(n-k-1)!}{(n-1)!},$$

$$\int_0^\infty \log(x) \frac{x^{k-1}}{(1+x)^n} dx = \frac{(k-1)!(n-k-1)!}{(n-1)!} \left(\sum_{l=1}^{k-1} l^{-1} - \sum_{l=1}^{n-k-1} l^{-1}\right).]$$

*13.4.* Show that the last term on the right side of (13.7) has the order $O_P(n_i^{-1/2})$. You may recall the argument of showing a similar property of the MLE (see Section 4.7).
*13.5.* Verify (13.8).
*13.6.* Verify expression (13.11), where $\tilde{\alpha}_i = E(\alpha_i|y)$ has expression (13.4).
*13.7.* Show, by formal derivation, that the estimator (13.30) satisfies (13.22). [Hint: You may use the fact that $E\{c_i(\hat{\theta}) - c_i(\theta)\} = o(1)$ and $E\{B_i(\hat{\theta}) - B_i(\theta)\} = o(1)$.]
*13.8.* Show that (13.33) is unbiased in the sense that the expectation of the left side is equal to the right side if $A$ is the true variance of the random effects. Also verify (13.34).

*13.9.* Show that the estimator $\hat{A}_{\mathrm{PR}}$ defined by (13.32) is $\sqrt{m}$-consistent.

*13.10.* Here is a more challenging exercise that the previous one: Show that the Fay–Herriot estimator $\hat{A}_{\mathrm{FH}}$, defined as the solution to (13.33), is $\sqrt{m}$-consistent.

*13.11.* Show that the estimators $\hat{A}_{\mathrm{PR}}$, $\hat{A}_{\mathrm{ML}}$, $\hat{A}_{\mathrm{RE}}$, and $\hat{A}_{\mathrm{FH}}$ in Section 13.3 possess the following properties: (i) They are even functions of the data— that is, the estimators are unchanged when $y$ is replaced by $-y$ and (ii) they are translation invariant—that is, the estimators are unchanged when $y$ is replaced by $y + Xb$ for any $b \in R^p$.

*13.12.* Show that the right side of (13.40) $\leq$ the right side of (13.41) $\leq$ the right side (13.39) for any $A \geq 0$.

*13.13.* Show that, in the balanced case (i.e., $D_i = D, 1 \leq i \leq m$), the P-R, REML, and F-H estimators of $A$ in the Fay–Herriot model are identical (provided that the estimator is nonnegative), whereas the ML estimator is different, although the difference is expected to be small when $m$ is large.

*13.14.* This exercise involves some calculus derivations.

(i) Verify expressions (13.55) and (13.56).

(ii) Verify expression (13.57).

(iii) Obtain an expression for $\partial \hat{A}_{\mathrm{FH}}/\partial y_i$. You man use the well-known result in calculus on differentiation of implicit functions.

*13.15.* Verify (13.60); that is, the expression of (13.59) when $\hat{\theta}_i$ is replaced by $y_i$.

*13.16.* Show that, under the hierarchical Bayes model near the end of Section 13.3, the conditional distribution of $\theta_i$ given $A$ and $y$ is normal with mean equal to the right side of (13.31) with $\hat{A}$ replaced by $A$ and variance equal to $g_{1i}(A) + g_{2i}(A)$, where $g_{1i}(A)$ and $g_{2i}(A)$ are given by (13.36) and (13.37), respectively.

*13.17.* Show that the minimizer of (13.71) is the same as the best linear unbiased estimator (BLUE) for $\beta$ and the best linear unbiased predictor (BLUP) for $\gamma$ in the linear mixed model $y = X\beta + Z\gamma + \epsilon$, where $\gamma \sim N(0, \sigma^2 I_q)$, $\epsilon \sim N(0, \tau^2 I_n)$, and $\gamma$ and $\epsilon$ are indepedent, provided that $\lambda$ is identical to the ratio $\tau^2/\sigma^2$. For the definition of BLUE and BLUP, see Section 5.6.

*13.18.* This exercise is related to the quadratic spline with two knots in Example 13.1.

(i) Plot the quadratic spline.

(ii) Show that the function is smooth in that it has a continuous derivative on $[0, 3]$, including at the knots.

*13.19.* Establish the following inequalities.

(i) $\|V^{-1}X(X'V^{-1}X)^{-1}X'V^{-1}\|_2^2 \leq \|V^{-1}\|^2(p+1)$.

(ii) $\|Z'V^{-1}X(X'V^{-1}X)^{-1}X'V^{-1}Z\|_2^2 \leq \|Z'V^{-1}Z\|^2(p+1)$.

*13.20.* Establish ineqaulity (13.83).

*13.21.* This exercise is related to the arguments in Section 13.4 that show $d_2 \to \infty$.

(i) Show that (13.84) holds for some constant $c > 0$ and determine this constant.

(ii) Show that $\|W'BW\|_2 \leq (\max_{1 \leq i \leq m} n_i)q$.

13.22. This exercise involves some details in Section 13.5.

(i) Verify that the likelihood function under $M$ is given by (13.89).

(ii) Verify (13.93).

(iii) Show that the BP of $\zeta$ under $M$, in the sense of minimizing the MSPE, is $\tilde{\zeta}_M = (\tilde{\zeta}_{M,i})_{1 \leq i \leq m}$, where $\tilde{\zeta}_{M,i}$ is given by (13.94).

13.23. Continue with the previous exercise. Show that if there is a true model $M \in \mathcal{M}$, then a true model with minimal $p$ is the optimal model under both the GIC and BP.

13.24. Consider Example 13.4.

(i) Recall that the best model in terms of BP is the one that maximizes $C_1(M)$, defined below (13.95). Show that

$$
C_1(M) = \begin{cases} (49/640)a^2m, & M = M_1 \\ 0, & M = M_2 \\ (49/640)a^2m, & M = M_3. \end{cases}
$$

Thus, the best model under BP is $M_1$, because it has the same (maximum) $C_1(M)$ as $M_3$ but is simpler.

(ii) Recall the best model under GIC is the one that minimizes $C_2(M)$, defined below (13.93). Show that

$$
C_2(M) = \begin{cases} s + \lambda_m - 1, & M = M_1, \\ s + \lambda_m - 1 - (49/712)a^2m, & M = M_2, \\ s + 2(\lambda_m - 1)1 - (49/712)a^2m, & M = M_3, \end{cases}
$$

where $s = \sum_{i=1}^m (A + D_i)^{-1}\mu_i^2$. Thus, $M_2$ is the best model under the GIC.

13.25. Derive the expression of MSPE (13.100) in the case that $A$ is unknown. [Hint: First, note that $\tilde{\zeta}_M = y - R(y - X\beta)$. Also note that $E(e'Ry) = E\{e'R(\mu + v + e)\} = E(e'Re) = \text{tr}(RD)$.]

13.26. Show that the estimator given by (13.104) is design-unbiased for $\mu_i^2$; that is, $E_d(\hat{\mu}_i^2) = \mu_i^2$. [Hint: Use the index set $I_i$; see a derivation below (13.103).]

# 14

# Jackknife and Bootstrap

... the bootstrap, and the jackknife, provide approximate *frequency* statements, not approximate *likelihood* statements. Fundamental inference problems remain, no matter how well the bootstrap works.

**Efron (1979)**
*Bootstrap methods: Another look at the jackknife*

## 14.1 Introduction

This chapter deals with statistical methods that, in some way, avoid mathematical difficulties that one would be facing using traditional approaches. The traditional approach of mathematical statistics is based on analytic expressions, or formulas, so avoiding these might seem itself a formidable task, especially in view of the chapters that so far have been covered. It should be pointed out that we have no objection of using mathematical formulas—in fact, some of these are pleasant to use. However, in many cases, such formulas are simply not available or too complicated to use. In a landmark paper, Efron (1979) showed why the bootstrap, to which the jackknife may be thought of as a linear approximation, is useful in solving a variety of inference problems that are otherwise intractable analytically. For simplicity, let us first consider the case where $X_1, \ldots, X_n$ are i.i.d. observations from an unknown distribution $F$. Let $X = (X_1, \ldots, X_n)$ and let $R = R(X, F)$ be a random variable, possibly depending on both $X$ and $F$. The goal is to estimate the distribution of $R$, called the *sampling distribution*. In many cases, one is interested in quantities or characteristics associated with the sampling distribution, such as the mean, variance, and percentiles, rather than the sampling distribution itself. For example, suppose that $\hat{\theta} = \hat{\theta}(X)$ is an estimator of a population parameter $\theta$ and that $R = \hat{\theta} - \theta$. Then the mean of $R$ is what we call the bias of $\hat{\theta}$, and the variance of $R$ is the squared standard deviation of $\hat{\theta}$ (an estimate of the standard deviation is what we call the standard error). Furthermore,

J. Jiang, *Large Sample Techniques for Statistics*,
DOI 10.1007/978-1-4419-6827-2_14, © Springer Science+Business Media, LLC 2010

the percentiles of $R$ are often used to construct confidence intervals for $\theta$. In some cases, these quantities, even the sampling distribution, have simple analytic expressions. Below is a classic example.

*Example 14.1* (The sample mean). Suppose that $\theta = \int x \, dF(x)$, the mean of $F$, and $\hat{\theta} = \bar{X}$, the sample mean. Let $R = \hat{\theta} - \theta$. Then we have $\mathrm{E}(R) = 0$; in other words, $\bar{X}$ is an unbiased estimator of $\theta$, which is a well-known fact. Furthermore, we have $\mathrm{var}(R) = \mathrm{var}(\bar{X}) = \sigma^2/n$, where $\sigma^2 = \int (x-\theta)^2 \, dF(x)$ is the variance of $F$. Finally, if $X_i$ is normal, then $R \sim N(0, \sigma^2/n)$. In fact, even if the $X_i$'s are not normal, according to the CLT, we have $\sqrt{n}R \xrightarrow{\mathrm{d}} N(0, \sigma^2)$; therefore the distribution of $R$ can be approximated by $N(0, \sigma^2/n)$.

If a simple analytic expression is available, it is often straightforward to make statistical inference. For example, in Example 14.1, the standard error of $\hat{\theta}$ is $\hat{\sigma}/\sqrt{n}$, where $\hat{\sigma}$ is an estimate of the standard deviation of $F$. Typically, the latter is estimated by the sqaure root of $s^2 = (n-1)^{-1} \sum_{i=1}^n (X_i - \bar{X})^2$, known as the sample variance, or $\hat{\sigma}^2 = n^{-1} \sum_{i=1}^n (X_i - \bar{X})^2$, which is the MLE of $\sigma^2$ under normality. Note that the only difference between $s^2$ and $\hat{\sigma}^2$ is the divisor—$n-1$ for $s^2$ and $n$ for $\hat{\sigma}^2$. Furthermore, suppose that the $X_i$'s are normal. Then the $100(1-\alpha)$th percentile of $R$ is $z_\alpha \sigma/\sqrt{n}$, where $0 < \alpha < 1$ and $z_\alpha$ is the $100(1-\alpha)$th percentile of $N(0,1)$; that is, $\Phi(z_\alpha) = 1 - \alpha$, where $\Phi(\cdot)$ is the cdf of $N(0,1)$. Thus, a $100(1-\alpha)\%$ confidence interval for $\theta$ is $[\bar{X} - z_{\alpha/2}\sigma/\sqrt{n}, \bar{X} + z_{\alpha/2}\sigma/\sqrt{n}]$ if $\sigma$ is known. A "little" complication occurs when $\sigma$ is unknown (which is practically the case), but the problem is solved due to a celebrated result of William Sealy Gosset, who showed in 1908 that the distribution of $R/(s/\sqrt{n})$ is $t_{n-1}$, the Student's $t$-distribution with $n-1$ degrees of freedom. Thus, in this case, a $100(1-\alpha)\%$ confidence interval for $\theta$ is $[\bar{X} - t_{n-1,\alpha/2}s/\sqrt{n}, \bar{X} + t_{n-1,\alpha/2}s/\sqrt{n}]$, where $t_{n-1,\alpha}$ is the $100(1-\alpha)$th percentile of $t_{n-1}$. A further complication is encountered when the $X_i$'s are not normal, because the result of $t$-distribution only applies to the normal case. Fortunately, there is the CLT that comes to our aid when the sample size is large. The CLT states that regardless of the distribution of the $X_i$'s, as $n$ goes to infinity, the distribution of $\sqrt{n}R/\sigma$ converges to $N(0,1)$. Then with a simple argument using Slutsky's theorem [see Theorem 2.13(ii)], we have $\sqrt{n}R/s \xrightarrow{\mathrm{d}} N(0,1)$ as $n \to \infty$ (note that $s$ is a consistent estimator of $\sigma$). It follows that in a large sample, an approximate $100(1-\alpha)\%$ confidence interval for $\theta$ is $[\bar{X} - z_{\alpha/2}s/\sqrt{n}, \bar{X} + z_{\alpha/2}s/\sqrt{n}]$, and the same is true with $s$ replaced by $\hat{\sigma}$.

The situation in the above example is the simplest that one can possibly imagine. Still, when there is a "minor" departure from the ideal we have to look for help, either from some clever mathematical derivation, or from the large-sample theory, and we are lucky to have such help around. However, in many cases, we are not so lucky in getting the help, as in the following examples.

*Example 14.2* (The sample median). The sample median is often used as a robust alternative to the sample mean. If $n$ is an odd number, say, $n = 2m-1$, then the sample median is $X_{(m)}$, where $X_{(1)} < \cdots < X_{(n)}$ are the order statistics. If $n$ is an even number, say $n = 2m$, then the sample median is defined as $[X_{(m)} + X_{(m+1)}]/2$. See Example 1.5. Unlike the sample mean, the sample median is not very sensitive to a few outliers. For example, suppose that a random sample of annual incomes from a company's employees are, in U.S. dollars, $26,000, 31,000, 42,000, 28,000$, and 2 million (the last one happens to be that of a high-ranking manager). If we use the sample mean as an estimate of the mean annual income of the company, the number is $425,400, which suggests that people are making big money in this company. However, the sample median is $31,000, which seems a more reasonable estimate of the mean annual income. The sample median is also more tolerable to missing values than the sample mean. For example, the following are life spans of five insects (in days), born at the same time: 17.4, 24.1, 13.9, *, 20.2 (the * corresponds to an insect that is still alive by the end of the experiment). Obviously, there is no way to compute the sample mean, but the sample median is 20.2, because the * is going to be the largest number anyway.

Suppose, for simplicity, that $n = 2m - 1$. Consider $R = X_{(m)} - \theta$, where $\theta$ is the population median, defined as $F^{-1}(0.5)$, where $F^{-1}$ is defined by (7.4). Being a special case of the order statistics, the distribution of $R$ can be derived (Exercise 14.1). However, with the exception of some special cases (see Exercise 14.2), there is no analytic expressions for the mean and variance of $R$. Bickel (1967) showed that both $E(R)$ and $\text{var}(R)$ are $O(n^{-1})$. Furthermore, if the underlying distribution, $F$, has a pdf $f$ that is continuous and positive at $\theta$, then, similar to Exercise 6.16, it can be shown that $\sqrt{n}R \xrightarrow{d} N(0, \sigma^2)$ with $\sigma^2 = \{2f(\theta)\}^{-2}$. In fact, Bickel (1967) showed that the limit of $n \, \text{var}(R)$ actually agrees with the asymptotic variance $\sigma^2$ (recall that convergence in distribution does not necessarily imply convergence of the variance; see Example 2.1). However, the asymptotic distribution offers little help for evaluating $E(R)$ (why?). Also, the asymptotic normality requires that $F$ has a continuous density that is positive at $\theta$. In many cases, one is dealing with an underlying distribution that does not have a density, such as a discrete distribution. In such a case, even the result of asymptotic variance may not apply.

*Example 14.3.* Recall that, in Section 13.2, we derived a second-order unbiased MSPE estimator for the EBP of a random effect associated with the small area—namely, (13.30). Although the expression might appear simple, it actually involves some tedious mathematical derivations if one intends to "spell it out." In fact, Jiang et al. (2002a, p. 1808) showed that the detailed expression is not simple even for the simplest case with $x'_{ij}\beta = \mu$. The problem with complicated analytic expressions is twofold. First, the derivation of such an expression requires mathematical skills, patience (to carry on the derivation rather than being intimidated), and carefulness (not to make mistakes!). Second, the programming of such a sophisticated formula into computer codes, as part of the modern-time scientific computing, is not trivial, and mistakes are often made in the process. These together may prove to be a daunting task for someone hoping for a routine operation of applied statistics.

In summary, some of the major difficulties of the traditional approaches are the following: (i) They rely on analytical expressions that may be difficult to derive. The derivation of such formulas or expressions is often tedious, requiring advanced mathematical skills, patience, and carefulness, the combination of which may be beyond the capability of an ordinary practitioner. (ii) The programming of a sophisticated analytical expression, even if it is available, is not trivial, and errors often occur at this stage. It may take a significant amount of time to find the errors, only if one is lucky enough—sometimes, the errors are found many years after the software package is commercialized. (iii) The analytic formulas are often case-by-case. A theoretician may help to derive a formula for one case, as William Gosset did with the $t$-distribution, but he/she cannot be there all the time whenever there is a new problem. (iv) The theoretical formulas derived under certain distributional assumptions may not be robust against violation of these assumptions. (v) Sometimes it requires a very large sample in order for the asymptotic results (e.g., the asymptotic variance) to be accurate as an approximation (e.g., to the true normalized variance). As will be seen, the methods discussed in the present chapter do not suffer, or at least suffer much less, from these difficulties.

We begin with an introduction to the jackknife, followed by an extension of the method to a nonconventional situation. We then discuss the classical bootstrap method and its extensions to two major areas: time series analysis and mixed models.

## 14.2 The jackknife

The *jackknife*, also known as the Quenouille–Tukey jackknife, was proposed by Quenouille (1949) as a way of estimating the bias of an estimator. Later, Tukey (1958) discovered that the jackknife can also be used to estimate the variance of an estimator. He coined the name "jackknife" for the method to imply that the method is an all-purpose tool for statistical analysis.

Consider, once again, the case of i.i.d. observations, say, $X_1, \ldots, X_n$. Let $\hat{\theta} = \hat{\theta}(X_1, \ldots, X_n)$ be an estimator of a parameter $\theta$. The interest is to estimate the bias of $\hat{\theta}$; that is,

$$\text{bias}(\hat{\theta}) = \text{E}(\hat{\theta}) - \theta. \tag{14.1}$$

Quenouille's proposal was the following. Define the $i$th *jackknife replication* of $\hat{\theta}$, $\hat{\theta}_{-i}$, as the estimator of $\theta$ that is computed the same way as $\hat{\theta}$ except using the data $X_1, \ldots, X_{i-1}, X_{i+1}, \ldots, X_n$ (i.e., the original data with the $i$th observation removed). For example, suppose that $\hat{\theta} = \bar{X} = n^{-1} \sum_{j=1}^{n} X_j$, the sample mean, and $\theta$ is the population mean. Then $\hat{\theta}_{-i}$ is the sample mean based on the data without the $i$th observation; that is, $\hat{\theta}_{-i} = (n - 1)^{-1} \sum_{j \neq i} X_j$. We then take the average of the jackknife replications, $\bar{\theta}_J = n^{-1} \sum_{i=1}^{n} \hat{\theta}_{-i}$. The jackknife estimator of the bias (14.1) is defined as

$$\widehat{\text{bias}}_J(\hat{\theta}) = (n - 1)(\bar{\theta}_J - \hat{\theta}). \tag{14.2}$$

The bias estimator leads to a bias-corrected estimator of $\theta$,

$$\hat{\theta}_J = \hat{\theta} - \widehat{\text{bias}}_J(\hat{\theta}) = n\hat{\theta} - (n - 1)\bar{\theta}_J. \tag{14.3}$$

From the definition, it is not easy to see the motivation of (14.2) and (14.3). How did Quenouille come up with the idea? Although this might never be known, it is a very useful strategy, in general, to start with a simple case. Experience tells us that the bias of a consistent estimator is often in the order of $O(n^{-1})$ if it is not unbiased. For example, suppose that $X_i \sim N(\mu, \sigma^2)$. The MLE of $\sigma^2$ is $\hat{\sigma}^2 = n^{-1} \sum_{i=1}^{n} (X_i - \bar{X})^2$, which is not unbiased. However, the bias of $\hat{\sigma}^2$ is $-\sigma^2/n$ (why?). Let us assume that the bias (14.1) can be expressed as

$$\text{bias}(\hat{\theta}) = \frac{a_1}{n} + \frac{a_2}{n^2} + O(n^{-3}), \tag{14.4}$$

where $a_1$ and $a_2$ are some constants. Then the bias of the $i$th jackknife replication can be expressed as

$$\text{bias}(\hat{\theta}_{-i}) = \frac{a_1}{n - 1} + \frac{a_2}{(n - 1)^2} + O(n^{-3}), \tag{14.5}$$

$1 \leq i \leq n$. It follows that $\text{bias}(\bar{\theta}_J)$ has the same expression. Thus, we have

$$\begin{aligned} \text{E}(\bar{\theta}_J - \hat{\theta}) &= \text{bias}(\bar{\theta}_J) - \text{bias}(\hat{\theta}) \\ &= \frac{a_1}{n(n - 1)} + O(n^{-3}). \end{aligned}$$

This leads to the expression

$$\text{E}\{(n - 1)(\bar{\theta}_J - \hat{\theta})\} = \frac{a_1}{n} + O(n^{-2}). \tag{14.6}$$

This suggests that the estimator (14.2) has the expectation whose leading term is the same as that of $\text{bias}(\hat{\theta})$. Thus, we can use (14.2) to "correct" the bias of $\hat{\theta}$ in that

$$\begin{aligned} \text{E}\{\hat{\theta} - (n - 1)(\bar{\theta}_J - \hat{\theta})\} &= \text{E}(\hat{\theta}) - \text{E}\{(n - 1)(\bar{\theta}_J - \hat{\theta})\} \\ &= \theta + \text{bias}(\hat{\theta}) - \text{E}\{(n - 1)(\bar{\theta}_J - \hat{\theta})\} \\ &= \theta + O(n^{-2}). \end{aligned}$$

Thus, if we define (14.3) as the bias-corrected estimator, we have

$$\text{bias}(\hat{\theta}_\text{J}) = O(n^{-2}). \tag{14.7}$$

Comparing (14.4) with (14.7), we see that the jackknife does the job of reducing the order of bias, from $O(n^{-1})$ to $O(n^{-2})$.

In general, one may not have a simple expression like (14.4), but the bias-reduction property of jackknife can often be (rigorously) justified in a way that is motivated by the simple case (see below).

The jackknife can also be used to estimate the variance of $\hat{\theta}$. This was first noted by Tukey (1958), whose variance estimator has the expression

$$\widehat{\text{var}}_\text{J}(\hat{\theta}) = \frac{n-1}{n} \sum_{i=1}^{n} (\hat{\theta}_{-i} - \bar{\theta}_\text{J})^2. \tag{14.8}$$

Expression (14.8) looks a lot like the sample variance $s^2$ (see Example 14.1) except that the factor in front is $(n-1)/n$ instead of $(n-1)^{-1}$. The reason for this is twofold. On the one hand, the estimator $\hat{\theta}_{-i}$, which is based on $n-1$ observations, tends to be less variable than a single observation $X_i$ (why?). Thus, the sum $\sum_i (\hat{\theta}_{-i} - \bar{\theta}_\text{J})^2$ is expected to be (much) smaller than $\sum_i (X_i - \bar{X})^2$. On the other hand, the target of our estimation is also smaller—the variance of an estimator tends to be smaller than that of a single observation. Therefore, the factor $(n-1)^{-1}$ for the sample variance is adjusted (in this case, "amplified") to make (14.8) a suitable estimator for the target. For example, in the case of the sample mean, the right side of (14.8) is equal to $s^2/n$, which is an unbiased estimator of $\text{var}(\hat{\theta}) = \sigma^2/n$ (Exercise 14.3). Thus, in this case, the adjusted factor, $(n-1)/n$ is "just right." This simple explanation actually motivated at least some of the rigorous justifications. See below.

So far, we have restricted ourself to the i.i.d. situation. There have been extensive studies on extending the jackknife to non-i.i.d. cases. In particular, in a series papers, Wu (1986) and Shao and Wu (1987), among others, proposed the delete-$d$ and weighted jackknife for regression analysis with heteroscedastic errors. In a classical linear regression model (see Section 6.7), the errors are assumed to have the same variance, known as homoscedastic errors. However, such an assumption may not hold in many situations. A heteroscedastic linear regression model is the same as (6.79) except that the variance of the error may depend on $i$. To be consistent with Wu's notation, we write this as

$$y_i = x_i'\beta + e_i, \tag{14.9}$$

where $y_i$, $x_i$, and $\beta$ are the same as $Y_i$, $x_i$, and $\beta$ in (6.79), but the $e_i$'s are assume to be independent with mean 0 and $\text{var}(e_i) = \sigma_i^2$, $1 \le i \le n$. A problem of interest is to estimate the covariance matrix of the ordinary least squares (OLS) stimator, denoted by $\hat{\beta}$. Although it might seem more reasonable to use a weighted least squares (WLS) estimator in case of heteroscedastic errors, the optimal weights are known to depend on the $\sigma_i^2$'s, which are unknown (see Lemma 5.1). Thus, the OLS is often more convenient to use, especially when the degree of heteroscedasticity is unknown. When $\sigma_i^2 = \sigma^2$, $1 \leq i \leq n$, a straightforward extension of the jackknife variance estimator for $\mathrm{Var}(\hat{\beta})$ is

$$\widehat{\mathrm{Var}}(\hat{\beta})_{\mathrm{J}} = \frac{n-1}{n} \sum_{i=1}^{n} (\hat{\beta}_{-i} - \bar{\beta}_{\mathrm{J}})(\hat{\beta}_{-i} - \bar{\beta}_{\mathrm{J}})', \qquad (14.10)$$

where $\hat{\beta}_{-i}$ is the OLS estimator without using $y_i$ and $x_i$, $1 \leq i \leq n$, and $\bar{\beta}_{\mathrm{J}} = n^{-1} \sum_{i=1}^{n} \hat{\beta}_{-i}$ (Miller 1974). Hinkley (1977) pointed out a number of shortcomings of (14.10). He suggested a weighted jackknife estimator,

$$\widehat{\mathrm{Var}}(\hat{\beta})_{\mathrm{H}} = \frac{n}{n-p} \sum_{i=1}^{n} (1 - h_i)^2 (\hat{\beta}_{-i} - \hat{\beta})(\hat{\beta}_{-i} - \hat{\beta})', \qquad (14.11)$$

where $h_i = x_i'(X'X)^{-1}x_i$, with $X = (x_i')_{1 \leq i \leq n}$, and $p$ is the rank of $X$. Here, we assume, for simplicity, that $X'X$ is nonsingular. Wu (1986) argued that Hinkley's estimator may be improved by using a different weighting scheme and/or allowing more than one observations to be removed in each jackknife replication. More precisely, Wu's proposal is the following. Let $d$ be an integer such that $1 \leq d \leq n-1$ and $r = n - d$. Let $S_r$ denote the collection of all subsets of $\{1, \ldots, n\}$ with size $r$. For $s = \{i_1, \ldots, i_r\} \in S_r$, let $X_s$ be the submatrix consisting of the $i_1$th, ..., $i_r$th rows of $X$ and let $y_s = (y_{i_1}, \ldots, y_{i_r})'$. Let $\hat{\beta}_s = (X_s'X_s)^{-1}X_s'y_s$ be the OLS estimator of $\beta$ based on $y_i, x_i, i \in s$. Again, for simplicity, we assume that $X_s'X_s$ is nonsingular for all $s \in S_r$. The weighted delete-$d$ jackknife estimator of $\mathrm{Var}(\hat{\beta})$ is defined as

$$\widehat{\mathrm{Var}}(\hat{\beta})_{\mathrm{J},d} = \binom{n-p}{d-1}^{-1} \sum_{s \in S_r} w_s (\hat{\beta}_s - \hat{\beta})(\hat{\beta}_s - \hat{\beta})', \qquad (14.12)$$

where $w_s = |X_s'X_s|/|X'X|$. One may wonder why the weights $w_s$ are chosen this way. Wu interpreted this by noting the following representation of the OLS estimator. First, consider a simple case, $y_i = \alpha + \beta x_i + e_i, i = 1, \ldots, n$, where $x_i$ is a scalar. In this case, it can be shown (Exercise 13.4) that

$$\hat{\beta} = \sum_{i<j} w_{ij} \hat{\beta}_{ij}, \qquad (14.13)$$

where $\hat{\beta}_{ij} = (y_i - y_j)/(x_i - x_j)$ is the OLS estimator of $\beta$ based on the pair of observations $(y_i, x_i)$ and $(y_j, x_j)$ and $w_{ij} = (x_i - x_j)^2 / \sum_{i'<j'} (x_{i'} - x_{j'})^2$. Note that here the weight $w_{ij}$ is proportional to

$$(x_i - x_j)^2 = \begin{vmatrix} 1 & x_i \\ 1 & x_j \end{vmatrix}.$$

In general, Wu (1986, Theorem 1) showed that the OLS estimator of $\beta$ has the following representation: For any $r \geq p$,

$$\hat{\beta} = \frac{\sum_{s \in S_r} |X_s' X_s| \hat{\beta}_s}{\sum_{s \in S_r} |X_s' X_s|}. \qquad (14.14)$$

We may interpret (14.14) as that the OLS estimator based on the full data is a weighted average of the OLS estimators based on all subsets of the data with size $r$, where the weights are proportional to $|X_s' X_s|$. Alternatively, (14.14) can be written as

$$\sum_{s \in S_r} |X_s' X_s| (\hat{\beta}_s - \hat{\beta}) = 0. \qquad (14.15)$$

The implication is that the weighted mean of $\hat{\beta}_s - \hat{\beta}$ is equal to zero, where the weights are proportional to $|X_s' X_s|$. Thus, as a second-order analogue, Wu proposed the weighted delete-$d$ jackknife estimator (14.12), where the weights $w_s$ are proportional to $|X_s' X_s|$. As in the i.i.d. case [see the discussion following (14.8)], a constant factor is adjusted to make it "just right." One consideration is to make sure that the estimator is (exactly) unbiased under homoscedastic errors (Wu 1986, Theorem 3). See below for a further consideration in terms of asymptotic unbiasedness. In particular, note that the sum of the weights in (14.12) [i.e., $\binom{n-p}{d-1}^{-1} \sum_{s \in S_r} w_s$] is equal to $(n - p - d + 1)/d$, not 1 [see Wu (1986), Lemma 1(ii)]. An important special case is the weighted delete-1 jackknife estimator of $\text{Var}(\hat{\beta})$. This can be expressed as (Exercise 14.5)

$$\widehat{\text{Var}}(\hat{\beta})_{\text{J},1} = \sum_{i=1}^{n} (1 - h_i)(\hat{\beta}_{-i} - \hat{\beta})(\hat{\beta}_{-i} - \hat{\beta})'. \qquad (14.16)$$

As noted, the weighted delete-$d$ jackknife estimator of $\text{Var}(\hat{\beta})$ is exactly unbiased under homoscedastic errors. What happens under heteroscedastic errors? To answer this question, let us define $h = \max_{1 \leq i \leq n} h_i$, where $h_i$ is defined below (14.11). The value of $h$ is used as an imbalanced measure (e.g., Shao and Wu 1987). An estimator of $\text{Var}(\hat{\beta})$, say $\hat{v}$, is called asymptotically unbiased (AU) if $n\{\text{E}(\hat{v}) - \text{Var}(\hat{\beta})\} \to 0$ as $n \to \infty$. The reason for multiplying the difference by $n$ is that, typically, $\text{Var}(\hat{\beta})$ is $O(n^{-1})$. So, the concept of AU only makes sense if the bias $\text{E}(\hat{v}) - \text{Var}(\hat{\beta})$ is $o(n^{-1})$; that is, $n\{\text{E}(\hat{v}) - \text{Var}(\hat{\beta})\} \to 0$. Shao and Wu (1987) proved the following theorem. Note that here both $d$ and $h$ are considered dependent on $n$.

**Theorem 14.1.** Suppose that the $\sigma_i^2$'s are bounded and $(X'X)^{-1} = O(n^{-1})$. If $\sup_n (dh) < 1$, then $\text{E}\{\widehat{\text{Var}}(\hat{\beta})_{\text{J},d}\} = \text{Var}(\hat{\beta}) + O(dh/n)$. Therefore, $\widehat{\text{Var}}(\hat{\beta})_{\text{J},d}$ is AU provided that $dh \to 0$.

The proof of Theorem 14.1, which we outline below, has pieces of the major techniques that have been used in developing jackknife estimators and justifying its properties. It is clear that, somehow, we have to evaluate $\text{E}\{(\hat{\beta}_s - \hat{\beta})(\hat{\beta}_s - \hat{\beta})'\} = \text{Var}(\hat{\beta}_s - \hat{\beta})$ (why?). We can write $\hat{\beta}_s - \hat{\beta}$ as

$$(X_s'X_s)^{-1}X_s'y_s - (X'X)^{-1}X'y$$
$$= \{(X_s'X_s)^{-1} - (X'X)^{-1}\}X_s'y_s - (X'X)^{-1}(X'y - X_s'y_s). \tag{14.17}$$

Why do this? Well, what happens is that the two terms on the right side of (14.17) are uncorrelated. Note that $X'y - X_s'y_s = X_{\bar{s}}'y_{\bar{s}}$, where $\bar{s}$ is the complement of $s$ with respect to $\{1, \ldots, n\}$. From this observation, we get

$$\mathrm{Var}(\hat{\beta}_s - \hat{\beta}) = \{(X_s'X_s)^{-1} \quad (X'X)^{-1}\}X_s'V_sX_s\{(X_s'X_s)^{-1} - (X'X)^{-1}\}$$
$$+(X'X)^{-1}X_{\bar{s}}'V_{\bar{s}}X_{\bar{s}}(X'X)^{-1}, \tag{14.18}$$

where $V_s = \mathrm{Var}(e_s)$ and $V_{\bar{s}}$ is defined similarly. It follows that $\mathrm{E}\{\widehat{\mathrm{Var}}(\hat{\beta}_{J,d})\}$ can be expressed as $S_1 + S_2$, with $S_1$ and $S_2$ corresponding to the two terms on the right side of (14.18). The arguments to follow show that $S_1$ is a term of $O(dh/n)$. Therefore, the main player is $S_2$.

Next, we look at $S_2$ more carefully and separate a main player within $S_2$. The determination of this main player also tells us what appropriate constant factor (see above) should be used. Note that

$$S_2 = \binom{n-p}{d-1}^{-1} \sum_{s \in S_r} w_s (X'X)^{-1} X_{\bar{s}}' V_{\bar{s}} X_{\bar{s}} (X'X)^{-1}.$$

To see what would be a main player in $S_2$, let us first consider a special case. Suppose that $w_s = 1$; that is, all of the weights are equal to 1. Then the summation is equal to $(X'X)^{-1}(\sum_{s \in S_r} X_{\bar{s}}'V_{\bar{s}}X_{\bar{s}})(X'X)^{-1}$. Note that

$$\sum_{s \in S_r} X_{\bar{s}}'V_{\bar{s}}X_{\bar{s}} = \sum_{s \in S_r} \sum_{i \in s} \sigma_i^2 x_i x_i'$$
$$= \sum_{i=1}^{n} \sum_{s \ni i} \sigma_i^2 x_i x_i'$$
$$= \sum_{i=1}^{n} \sigma_i^2 x_i x_i' |\{s \in S_r, s \ni i\}|$$
$$= \binom{n-1}{d-1} \sum_{i=1}^{n} \sigma_i^2 x_i x_i'$$

(why?). Thus, in this special case, we have

$$S_2 = \binom{n-p}{d-1}^{-1} \binom{n-1}{d-1} (X'X)^{-1} \left(\sum_{i=1}^{n} \sigma_i^2 x_i x_i'\right) (X'X)^{-1}$$
$$= \binom{n-p}{d-1}^{-1} \binom{n-1}{d-1} \mathrm{Var}(\hat{\beta}). \tag{14.19}$$

It can be shown (Exercise 14.6) that the factor in the front is $1 + O(d/n)$. Thus, in this special case, $S_2 = \{1 + O(d/n)\}\mathrm{Var}(\hat{\beta}) = \mathrm{Var}(\hat{\beta}) + O(d/n^2)$, by the assumptions of the theorem. Of course, the weights are not necessarily equal to 1, but we can write $S_2$ as $S_{21} - S_{22}$, where $S_{21}$ is $S_2$ when the weights are equal [i.e., the right side of (14.19)] and $S_{22}$ is the difference between the special and general cases; that is,

$$S_{22} = \binom{n-p}{d-1}^{-1} \sum_{s \in S_r} (1 - w_s)(X'X)^{-1}X_{\bar{s}}'V_{\bar{s}}X_{\bar{s}}(X'X)^{-1}. \quad (14.20)$$

It can be shown (see below) that $S_{22} = O(dh/n)$. Therefore, in view of the argument for the special case of $w_s = 1$, it is clear that $S_{21}$ is the main player, and $S_{21} = \text{Var}(\hat{\beta}) + O(d/n^2)$. It follows that $\text{E}\{\widehat{\text{Var}}(\hat{\beta})_{\text{J},d}\} = S_1 + S_2 = S_1 + S_{21} - S_{22} = \text{Var}(\hat{\beta}) + O(dh/n)$.

It remains to show that both $S_1$ and $S_{22}$ are $O(dh/n)$. We show the latter as an example and leave the former as an exercise (Exercise 14.7). We use a technique introduced in Section 3.5, known as the unspecified $c$; namely, let $c$ denote a positive constant that may have different values in different places (e.g., Shao and Wu 1987, p. 1566). Also, note that $w_s \leq 1$ for all $s$ (Exercise 14.7). Therefore, we have, by (14.20) and the assumptions of the theorem,

$$\text{tr}(S_{22}) = \binom{n-p}{d-1}^{-1} \sum_{s \in S_r} (1 - w_s)\text{tr}\{(X'X)^{-1}X_{\bar{s}}'V_{\bar{s}}X_{\bar{s}}(X'X)^{-1}\}$$

$$\leq \frac{c}{n}\binom{n-p}{d-1}^{-1} \sum_{s \in S_r} (1 - w_s)\text{tr}\{(X'X)^{-1/2}X_{\bar{s}}'X_{\bar{s}}(X'X)^{-1/2}\}$$

$$= \frac{c}{n}\binom{n-p}{d-1}^{-1} \sum_{s \in S_r} (1 - w_s)\sum_{i \in \bar{s}} \text{tr}\{(X'X)^{-1/2}x_i x_i'(X'X)^{-1/2}\}$$

$$= \frac{c}{n}\binom{n-p}{d-1}^{-1} \sum_{s \in S_r} (1 - w_s)\sum_{i \in \bar{s}} h_i \quad [h_i \text{ is defined below (14.11)}]$$

$$\leq c\frac{dh}{n}\binom{n-p}{d-1}^{-1} \sum_{s \in S_r} (1 - w_s).$$

Now, use an identity [Wu 1986, Lemma 1(ii)] $\sum_{s \in S_r} w_s = \binom{n-p}{d}$ to get

$$\text{tr}(S_{22}) \leq c\frac{dh}{n}\binom{n-p}{d-1}^{-1} \left\{\binom{n}{d} - \binom{n-p}{d}\right\} = O(dh/n)$$

(Exercise 14.6), which implies $\|S_{22}\| = \lambda_{\max}(S_{22}) = O(dh/n)$.

Another extension of the jackknife is considered in the next section.

## 14.3 Jackknifing the MSPE of EBP

The jackknife was proposed in the context of estimating the bias and variance of an estimator. So far, the development and extension have been around the

*Example 14.5* (James–Stein estimator and naive jackknife). Let the observations $y_1, \ldots, y_m$ be independent such that $y_i \sim N(\theta_i, 1)$, $1 \leq i \leq m$. In the context ofsimultaneous estimation of $\theta = (\theta_1, \ldots, \theta_m)'$, it is well known that, for $m \geq 3$, the James–Stein estimator dominates the maximum likelihood estimator, which is simply $y = (y_1, \ldots, y_m)'$, in terms of the frequentist risk under a sum of squares error loss function (e.g., Lehmann 1983, p. 302). Efron and Morris (1973) provided an empirical Bayes justification of the James–Stein estimator. Their Bayesian model may be equivalently written as a simple random effects model, $y_i = v_i + e_i, i = 1, \ldots, m$, where the random effects $v_i$ and sampling errors $e_i$ are independent with $v_i \sim N(0, A)$ and $e_i \sim N(0, 1)$, where $A$ is unknown. Write $B = (1 + A)^{-1}$. Then the James–Stein estimator can be interpreted as an EBP under the random effects model. For example, the BP of $\zeta = v_1$ is $\tilde{\zeta} = (1 - B)y_1$. The estimator of $B$ proposed by Efron and Morris (1973) is $\hat{B} = (m - 2)/\sum_{i=1}^{m} y_i^2$. Alternatively, the MLE of $B$ is $\hat{B} = m/\sum_{i=1}^{m} y_i^2$. By plugging in $\hat{B}$, we get the EBP, $\hat{\zeta} = (1 - \hat{B})y_1$.

A straightforward extension of the jackknife for estimating the MSPE of $\hat{\zeta}$ would define the $i$th jackknife replication of $\hat{\zeta}$ as one derived the same way except without the observation $y_i$. Denote this naive jackknife replication by $\hat{\zeta}^*_{-i}$. The question is: What is $\hat{\zeta}^*_{-1}$? To be more specific, suppose that the MLE of $B$ is used. If we follow the derivation of of the EBP, then we must have $\tilde{\zeta}_{-1} = 0$ (why?); hence, $\hat{\zeta}^*_{-1} = 0$. On the other hand, for $i \geq 2$, we have $\tilde{\zeta}_{-i} = (1 - B)y_1$ (same as $\tilde{\zeta}$) and $\hat{B}_{-i} = (m-1)/\sum_{j \neq i} y_j^2$; hence, $\hat{\zeta}^*_{-i} = (1 - \hat{B}_{-i})y_1$. It can be shown (Exercise 14.9) that $\mathrm{MSPE}(\hat{\zeta}) = 1 - B + 2B/m + o(m^{-1})$. On the other hand, we have $\mathrm{E}(\hat{\zeta}^*_{-1} - \hat{\zeta})^2 = \mathrm{E}(\hat{\zeta}^2) = A(1 - B) + o(1)$ and $\mathrm{E}(\hat{\zeta}^*_{-i} - \hat{\zeta})^2 = 2B/m^2 + o(m^{-2})$, $i \geq 2$. With these, it can be shown that the expectation of the right side of (14.27), with $\hat{\zeta}_{-i}$ replaced by $\zeta^*_{-i}, 1 \leq i \leq m$, is equal to $\mathrm{MSPE}(\hat{\zeta}) + A(1 - B) + o(1)$ (Exercise 14.9). In other words, the bias of the naive jackknife MSPE estimator does not even go to zero as $m \to \infty$.

The problem with the naive jackknife can be seen clearly from the above example. The observation $y_1$ plays a critical role in the prediction of $\zeta = v_1$. This observation cannot be removed no matter what. Any jackknife replications should only be with respect to $\hat{\psi}$, not $y_S$ [see (14.22)]. Therefore, we define the $i$th jackknife replication of $\hat{\zeta}$ as

$$\hat{\zeta}_{-i} = \pi(y_S, \hat{\psi}_{-i}), \tag{14.28}$$

where $y_S$ is the same as in (14.22) and $\hat{\psi}_{-i}$ is the $i$th jackknife replication of $\hat{\psi}$, described below.

So far, the presence of $\hat{\psi}$ is, more or less, just a notation—we have not given the specific form of $\hat{\psi}$. In many applications, the estimator $\hat{\psi}$ belongs to a class of M-estimators. Here, an M-estimator is associated with a solution, $\hat{\psi}$, to the following equation:

$$F(\psi) = \sum_{j=1}^{m} f_j(\psi, y_j) + a(\psi) = 0, \tag{14.29}$$

where $f_j(\cdot, \cdot)$ are vector-valued functions satisfying $\mathrm{E}\{f_j(\psi, y_j)\} = 0, 1 \leq j \leq m$, if $\psi$ is the true parameter vector, and $a(\cdot)$ is a vector-valued function which may depend on the joint distribution of $y_1, \ldots, y_m$. When $a(\psi) \neq 0$, it plays the role of a modifier, or penalizer. We consider some examples.

*Example 14.6* (ML estimation). Consider the longitudinal linear mixed model (12.3). Let $\psi = (\beta', \theta')'$. Under regularity conditions, the MLE of $\psi$ satisfies (14.29) with $a(\psi) = 0$, $f_j(\psi, y_j) = [f_{j,k}(\psi, y_j)]_{1 \leq k \leq p+q}$, where $p$ is the dimension of $\beta$ and $q$ the dimension of $\theta$;

$$[f_{j,k}(\psi, y_j)]_{1 \leq k \leq p} = X_j' V_j^{-1}(\theta)(y_j - X_j\beta);$$

$$f_{j,p+l}(\psi, y_j) = (y_j - X_j\beta)' V_j^{-1}(\theta) \frac{\partial V_j}{\partial \theta_l} V_j^{-1}(\theta)(y_j - X_j\beta)$$

$$-\mathrm{tr}\left\{ V_j^{-1}(\theta) \frac{\partial V_j}{\partial \theta_l} \right\}, \quad 1 \leq l \leq q,$$

where $V_j(\theta) = R_j + Z_j G_j Z_j'$ (Exercise 14.10).

*Example 14.7* (REML estimation). Continue with the previous example. Similarly, the REML estimator of $\theta$ is defined as a solution to the REML equation (see Section 12.2). The REML estimator of $\beta$ is defined as $\hat{\beta} = \{\sum_{j=1}^{m} X_j' V_j^{-1}(\hat{\theta}) X_j\}^{-1} \sum_{j=1}^{m} X_j' V_j^{-1}(\hat{\theta}) y_j$, where $\hat{\theta}$ is the REML estimator of $\theta$. It can be shown (Exercise 14.11) that the REML estimator of $\psi$ satisfies (14.29), where the $f_j$'s are the same as in Example 14.6; $a(\psi) = [a_k(\psi)]_{1 \leq k \leq p+q}$ with $a_k(\psi) = 0, 1 \leq k \leq p$, and

$$a_{p+l}(\psi) = \sum_{j=1}^{m} \mathrm{tr}\left[ V_j^{-1}(\theta) X_j \{X' V^{-1}(\theta) X\}^{-1} X_j' V_j^{-1}(\theta) \frac{\partial V_j}{\partial \theta_l} \right],$$

$1 \leq l \leq q$. Here, $X = (X_j)_{1 \leq j \leq m}$ and $V(\theta) = \mathrm{diag}\{V_j(\theta), 1 \leq j \leq m\}$.

Consider a jackknife replication of (14.29); that is,

$$F_{-i}(\psi) = \sum_{j \neq i} f_j(\psi, y_j) + a_{-i}(\psi) = 0, \tag{14.30}$$

$1 \leq i \leq m$. The $i$th jackknife replication of $\hat{\psi}$, $\hat{\psi}_{-i}$, is defined as a solution to (14.30). Sometimes, a solution to (14.29) may not exist, or exist but not within the parameter space (e.g., negative values for variances). Therefore, we define an M-estimator of $\psi$ as $\hat{\psi} = \dot{\psi}$ if the solution to (14.29) exists within the parameter space, and $\hat{\psi} = \psi^*$ otherwise, where $\psi^*$ is a known vector within the parameter space. Similarly, we define $\hat{\psi}_{-i} = \check{\psi}_{-i}$ if the solution to (14.30) exists within the parameter space, and $\hat{\psi}_{-i} = \psi^*$ otherwise.

Jiang et al. (2002a,b) showed that the jackknife MSPE estimator, defined by (14.27) with $\hat{\psi}, \hat{\psi}_{-i}, 1 \leq i \leq m$, being the M-estimator and its jackknife replications, has a similar bias reduction property, known as asymptotic unbiasedness, as the jackknife variance estimator considered in Section 14.2. For simplicity, below we consider a very special case. Let $\psi$ be a scalar parameter and $S - \{1\}$ in (14.22). Also, assume that $f_i = f$ (i.e., not dependent on $j$) and $a(\psi) = a_{-i}(\psi) = 0$ in (14.29) and (14.30). Let $y_1, \ldots, y_m$ be independent with the same distribution as $Y$. Before we study the asymptotic unbiasedness of the jackknife MSPE estimator, let us first consider an important property of the M-estimator. For notation convenience, write $\hat{\psi}_{-0} = \hat{\psi}$ and $F_{-0} = F$. The M-estimators $\hat{\psi}_{-i}, 0 \leq i \leq m$, are said to be consistent uniformly (c.u.) at rate $m^{-d}$ if for any $b > 0$, there is a constant $B$ (possibly depend on $b$) such that $\mathrm{P}(A_{i,b}^c) \leq Bm^{-d}, 0 \leq i \leq m$, where $A_{i,b} = \{F_{-i}(\hat{\psi}_{-i}) = 0, |\hat{\psi}_{-i} - \psi| \leq b\}$ and $\psi$ is the true parameter. The M-estimating equations are said to be *standard* if $f(\psi, Y) = (\partial/\partial\psi)l(\psi, Y)$ for some function $l(\psi, u)$ that is three times continuously differentiable with respect to $\psi$ and satisfies

$$\mathrm{E}\left\{\frac{\partial^2}{\partial\psi^2}l(\psi, Y)\right\} > 0.$$

The ML and REML equations (see Examples 14.6 and 14.7) are multivariate extensions of the standard M-estimating equations. The following theorem is given in Jiang et al. (2002a) as a proposition.

**Theorem 14.2.** Suppose that the M-estimating equations are standard and the $2d$th moments of

$$\left|\frac{\partial^r}{\partial\psi^r}l(\psi, Y)\right|, \quad r = 1, 2, \quad \sup_{|\psi' - \psi| \leq b_0}\left|\frac{\partial^3}{\partial\psi^3}l(\psi', Y)\right|$$

are finite for some $d \geq 1$ and $b_0 > 0$, where $\psi$ is the true parameter. Then there exist M-estimators $\hat{\psi}_{-i}, 0 \leq i \leq m$, that are c.u. at rate $m^{-d}$.

The proof is left as an exercise (Exercise 14.12). The next thing we do is to establish the asymptotic unbiasedness properties for the jackknife estimators of $\mathrm{MSAE}(\hat{\zeta})$ and $\mathrm{MSPE}(\tilde{\theta})$ separately and then combine them [see (14.24) and (14.27)]. To do so we also need some regularity conditions on the EBP (14.22) (note that now $S = \{1\}$). We assume that

$$|\pi(Y_1, \psi)| \leq \omega(Y_1)(1 \vee |\psi|^\lambda) \tag{14.31}$$

for some constant $\lambda > 0$ and measurable function $\omega(\cdot)$ such that $\omega(\cdot) \geq 1$ [$a \vee b = \max(a, b)$]. The c.u. property can now be generalized. Let $\mathcal{A} = \sigma(y_1, \ldots, y_m)$, the $\sigma$-field generated by the $y_i$'s. Define a measure $\mu_\omega$ on $\mathcal{A}$ as

$$\mu_\omega(A) = E\{\omega^2(Y)1_A\}, \quad A \in \mathcal{A}. \tag{14.32}$$

The M-estimators $\hat{\psi}_{-i}, 0 \le i \le m$, are said to be c.u. with respect to $\mu_\omega$ (c.u. $\mu_\omega$) at rate $m^{-d}$ if for any $b > 0$, there is a constant $B$ (possibly dependent on $b$) such that $\mu_\omega(A_{i,b}^c) \le Bm^{-d}, 0 \le i \le m$, where $A_{i,b}$ is the same as above. Below are some remarks regarding c.u. $\mu_\omega$ and its connection to c.u.

*Remark 1.* Because $\omega(\cdot) \ge 1$, that the M-estimators are c.u. $\mu_\omega$ at rate $m^{-d}$ implies that they are c.u. at rate $m^{-d}$. Conversely, if there is $\tau > 2$ such that $E(|\omega(Y)|^\tau) < \infty$, then that the M-estimators are c.u. at rate $m^{-d}$ implies that they are c.u. $\mu_\omega$ at rate $m^{-d(1-2/\tau)}$ (Exercise 14.13). In particular, if $E\{\omega^4(Y)\} < \infty$, then if the M-estimators are c.u. at rate $m^{-2d}$, they are c.u. $\mu_\omega$ at rate $m^{-d}$. This is useful in checking the c.u. $\mu_\omega$ property because, under a suitable moment condition, it reduces to checking the c.u. property.

*Remark 2.* In practice, the function $\omega$ may be chosen in the following way: Find a positive number $\lambda$ such that $\omega(y) = \sup_\psi\{|\pi(y,\psi)|/(1 \vee |\psi|^\lambda)\} < \infty$ for every $y$ and use this $\omega$.

The following theorem states the asymptotic unbiasedness of the jackknife MSAE estimator.

**Theorem 14.3.** Suppose that (i) $E\{(\partial/\partial\psi)f(\psi,Y)\} \ne 0$; (ii) for some constants $d > 2$ and $b_0 > 0$, the expectations of the following are finite, where $\psi$ is the true parameter:

$$|f(\psi,Y)|^{2d}, \quad \left|\frac{\partial}{\partial\psi}f(\psi,Y)\right|^{2d}, \quad \sup_{|\psi'-\psi|\le b_0}\left|\frac{\partial^2}{\partial\psi^2}f(\psi',Y)\right|^{2d},$$

$$\sup_{|\psi'-\psi|\le b_0}\left\{\frac{\partial^3}{\partial\psi^3}f(\psi',Y)\right\}^4;$$

$$\omega^4(Y), \quad \left\{\frac{\partial^2}{\partial\psi^2}\pi(Y,\psi)\right\}^4, \quad \sup_{|\psi'-\psi|\le b_0}\left\{\frac{\partial^3}{\partial\psi^3}\pi(Y,\psi')\right\}^2,$$

$$\sup_{|\psi'-\psi|\le b_0}\left|\frac{\partial}{\partial\psi}\pi(Y,\psi')\right|^{2d};$$

and (iii) $\hat{\psi}_{-i}, 0 \le i \le m$ are c.u. $\mu_\omega$ at rate $m^{-d}$. Then we have

$$E\{\widehat{\mathrm{MSAE}}(\hat{\zeta})_J\} - \mathrm{MSAE}(\hat{\zeta}) = o(m^{-1-\epsilon})$$

for any $0 < \epsilon < (d-2)/(2d-1)$.

The next theorem states the asymptotic unbiasedness of the jackknife bias-corrected estimator for the MSPE of $\tilde{\zeta}$.

**Theorem 14.4.** Suppose that (i) $E\{(\partial/\partial\psi)f(\psi,Y)\} \ne 0$; (ii) for some $d > 2$ and $b_0 > 0$, the $2d$th moments of the following are finite:

$$|f(\psi, Y)|, \quad \left|\frac{\partial^r}{\partial \psi^r} f(\psi, Y)\right|, \quad r = 1, 2, \quad \sup_{|\psi' - \psi| \le b_0} \left|\frac{\partial^3}{\partial \psi^3} f(\psi, Y)\right|,$$

and $\sup_{|\psi' - \psi| \le b_0} |b^{(4)}(\psi')|$ is bounded, where $\psi$ is the true parameter and $b^{(4)}$ represents the fourth derivative; and (iii) $\hat{\psi}_{-i}, 0 \le i \le m$ are c.u. at rate $m^{-d}$. Then we have

$$E\{\widehat{\mathrm{MSPE}}(\tilde{\zeta})_J\} \quad \mathrm{MSPE}(\tilde{\zeta}) - o(m^{-1-\epsilon})$$

for any $0 < \epsilon < (d - 2)/(2d + 1)$.

By combining Theorems 14.3 and 14.4 and in view of (14.24) and (14.27), we obtain the asymptotic unbiasedness of the jackknife MSPE estimator.

**Theorem 14.5.** Suppose that (i)–(iii) of Theorem 14.3 and (ii) of Theorem 14.4 hold; then we have

$$E\{\widehat{\mathrm{MSPE}}(\hat{\zeta})_J\} - \mathrm{MSPE}(\hat{\zeta}) = o(m^{-1-\epsilon})$$

for any $0 < \epsilon < (d - 2)/(2d + 1)$.

At this point, we would like to discuss some of the ideas used in the proofs. A basic technique is Taylor series expansions (see below). This allows us to approximate the EBP by something simpler. Typically, the approximation error in the Taylor expansion is expressed in terms of $O_P$ or $o_P$. We need to "convert" such a result to convergence in expectation. In this regard, an earlier result, Lemma 3.17, is found very useful. We now give an outline of the Taylor expansions. The details can be found in Jiang et al. (2002b). First, consider expansions of the M-estimators. It is fairly straightforward to do it for $\hat{\psi} - \psi$. On the other hand, it is more challenging to obtain an expansion for $\hat{\psi}_{-i} - \hat{\psi}$, the main reason being that we need to carry out the expansion for $\hat{\psi}_{-i} - \psi$ to a higher order (than for $\hat{\psi} - \psi$) [because we need to consider the sum of $(\hat{\psi}_{-i} - \hat{\psi})^2$ over $i$]. The idea is to do the Taylor expansion twice, or do it in two steps, first obtaining a rough approximation and then using it for a more accurate result. Write $f_j = f_j(\psi, y_j)$, $g_j = (\partial/\partial \psi) f_j(\psi, y_j)$, $f. = \sum_j f_j$, $f._{-i} = \sum_{j \neq i} f_j$, $\bar{f} == m^{-1} f.$, and so forth. Define $D_i = \{\hat{\psi}_{-i}$ satisfies $F_{-i} = 0$ and $|\hat{\psi}_{-i} - \psi| \le m^{\delta - 1/2}\}, 0 \le i \le m$, and $G = \{|\bar{g} - E(\bar{g})| \le m^{-\Delta}|E(\bar{g})|\}$ for some $\delta, \Delta > 0$. Let $C_i = D_0 \cap D_i, i \ge 1$. For expanding $\hat{\psi} - \psi$, we have

$$f_j(\hat{\psi}, y_j) = f_j + g_j(\hat{\psi} - \psi) + \frac{1}{2} h_j(\hat{\psi} - \psi)^2 + r_j,$$

where $h_j = (\partial^2/\partial \psi^2) f_j(\psi_j^*, y_j)$ and $\psi_j^*$ lies between $\psi$ and $\hat{\psi}$ [note that $\psi_j^*$ depends on $j$ (why?)], and $|r_j| \le m^{3\delta - 3/2} u$ and $E(u^d)$ is bounded. Hereafter, you do not need to verify the bounds for the $r$'s, as the goal here is to illustrate the main idea, but you may want to think about why, as always. Also, the $r$'s are not necessarily the same, even if the same notation is used (this is similar to the unspecified $c$; see, for example, the end of Section 14.2). If we sum over $j$, then we have, on $C_0$ (why does the first equation hold?),

$$0 = \sum_j f_j(\hat{\psi}, y_j)$$

$$= f_\cdot + g_\cdot(\hat{\psi} - \psi) + \frac{1}{2}h_\cdot(\hat{\psi} - \psi)^2 + r. \tag{14.33}$$

$$= f_\cdot + E(g_\cdot)(\hat{\psi} - \psi) + r_1 + r, \tag{14.34}$$

where $r_1 = \{g_\cdot - E(g_\cdot)\}(\hat{\psi} - \psi)$, $|r| \leq m^{2\delta}u$, and $E(u^d)$ is bounded. This implies that

$$\hat{\psi} - \psi = -\{E(g_\cdot)\}^{-1}f_\cdot + r \tag{14.35}$$

with $|r| \leq m^{2\delta-1}u$ and $E(u^d)$ bounded.

The two equation numbers in (14.33) and (14.34) are not left by mistake (normally one would need just one equation number for such a series of equations), as it will soon become clear. For expanding $\hat{\psi}_{-i} - \hat{\psi}$, we have, on $C_i$,

$$f_j(\hat{\psi}_{-i}, y_j) = f_j + g_j(\hat{\psi}_{-i} - \psi) + \frac{1}{2}h_{-i,j}(\hat{\psi}_{-i} - \psi)^2 + r_{-i,j},$$

where $h_{-i,j} = (\partial^2/\partial\psi^2)f_j(\psi^*_{-i,j}, y_j)$, $\psi^*_{-i,j}$ is between $\psi$ and $\hat{\psi}_{-i}$, and $|r_{-i,j}| \leq m^{3\delta-3/2}u_j$ with $E(u_j^d)$ bounded. We then sum over $i$ to have, on $C_i$,

$$0 = \sum_{j \neq i} f_j(\hat{\psi}_{-i}, y_j)$$

$$= f_{\cdot-i} + g_{\cdot-i}(\hat{\psi}_{-i} - \psi) + \frac{1}{2}h_{-i,\cdot-i}(\hat{\psi}_{-i} - \psi)^2 + r_{-i,\cdot-i} \tag{14.36}$$

$$= f_\cdot + E(g_\cdot)(\hat{\psi}_{-i} - \psi) + r_i \tag{14.37}$$

with $|r_i| \leq m^{2\delta}u_i$ and $E(u_i^d)$ bounded. Again, the two equations numbers, (14.36) and (14.37) are not left without a purpose. We now subtract (14.37) from (14.34) to get $0 = E(g_\cdot)(\hat{\psi} - \hat{\psi}_{-i}) + r_i$, where $|r_i| \leq m^{2\delta}u_i$ and $E(u_i^d)$ is bounded. This impies a (rough) bound

$$|\hat{\psi}_{-i} - \hat{\psi}| \leq m^{2\delta-1}u_i \tag{14.38}$$

with $E(u_i^d)$ bounded. Using (14.38), it can be shown that

$$h_{-i,\cdot-i}(\hat{\psi}_{-i} - \psi)^2 = E(h_{-i,\cdot-i})(\hat{\psi} - \psi)^2 + r_i \tag{14.39}$$

with $|r_i| \leq m^{3\delta-1/2}u_i$ and $E(u_i^d$ bounded. We now subtract (14.36) from (14.33) and observe that $h_\cdot(\hat{\psi} - \psi)^2 = E(h_\cdot)(\hat{\psi} - \psi)^2 + r$ with $|r| \leq m^{2\delta-1/2}u$

and $E(u^d)$ bounded, to get $0 = f_i + g_.(\hat{\psi} - \hat{\psi}_{-i}) + r_i$ with $|r_i| \leq m^{3\delta-1/2}u_i$ and $E(u_i^d)$ bounded. This gives us a more accurate expansion:

$$
\begin{aligned}
\hat{\psi}_{-i} - \hat{\psi} &= g_.^{-1}(f_i + r_i) \\
&= \{E(g_.)\}^{-1}f_i + r_i
\end{aligned}
\tag{14.40}
$$

with $|r_i| \leq m^{-1-\epsilon}u_i$ for $\epsilon = (1/2 - 3\delta) \wedge \Delta$ and $E(u_i^d)$ bounded.

The expansions for $\hat{\psi}_{-i} - \hat{\psi}$ lead to those for $h(\hat{\psi}_{-i}) - h(\hat{\psi}), 1 \leq i \leq m$ [see (14.26)], and similar techniques are used in obtaining expansions for the EBPs, $\hat{\zeta}_{-i} - \hat{\zeta}, 1 \leq i \leq m$, which are the keys to the proofs.

To conclude this section, we would like to make a note on the extension of Theorem 14.5 beyond the simple case that we have considered. This has much to do with the role of the $a$'s in (14.29) and (14.30). In the simple case we assumed that the $a$'s are zero (functions), so there is no such a problem. In general, Jiang et al. (2002a) had the following restriction on the $a$'s in order to obtain the asymptotic unbiasedness of the jackknife MSPE estimator,

$$
\sum_{i=1}^{m} \{a(\psi) - a_{-i}(\psi)\} = O(m^{-\nu})
\tag{14.41}
$$

for some constant $\nu > 0$, where $\psi$ is the true parameter vector. To see that (14.41) is not a serious restriction, we illustrate it with an example.

*Example 14.7 (continued).* Write $\Delta_i = a(\psi) - a_{-i}(\psi)$, $V_j = V_j(\theta)$, and $V = V(\theta)$. From the definition of $a$ we have $a_{-i,k}(\psi) = 0, 1 \leq k \leq p$, and $a_{-i,p+l}(\psi)$ is defined the same way as $a_{p+l}(\psi)$ except with $\sum_{j=1}^{m}$ replaced by $\sum_{j\neq i}, 1 \leq l \leq q$. Note that $X'V^{-1}X = \sum_{j=1}^{m} A_j$, with $A_j = X_j'V_j^{-1}X_j$, also needs to be adjusted. Thus, $\Delta_{i,k} = 0, 1 \leq k \leq p$, and, for $1 \leq l \leq q$,

$$
\Delta_{i,p+l} = \text{tr}(A_.^{-1}B_{i,l}) + \sum_{j\neq i}\text{tr}\{(A_.^{-1} - A_{.-i}^{-1})B_{j,l}\},
$$

where $B_{j,l} = X_jV_j^{-1}(\partial V_j/\partial\theta_l)V_j^{-1}X_j$, $A_. = A_{.-0} = \sum_{j=1}^{m} A_j$, and $A_{.-i} = \sum_{j\neq i} A_j$ (verify the expression for $\Delta_{i,p+l}$). Under regularity conditions, we have $A_.^{-1} - A_{.-i}^{-1} = -A_.^{-1}A_iA_.^{-1} + O(m^{-3})$. Thus, we have

$$
\begin{aligned}
\sum_{i=1}^{m}\Delta_{i,p+l} &= \sum_{i=1}^{m}\text{tr}(A_.^{-1}B_{i,l}) - \sum_{i=1}^{m}\sum_{j\neq i}\{\text{tr}(A_.^{-1}A_iA_.^{-1}B_{j,l}) + O(m^{-3})\} \\
&= \sum_{i=1}^{m}\text{tr}(A_.^{-1}B_{i,l}) - \sum_{i=1}^{m}\sum_{j=1}^{m}\text{tr}(A_.^{-1}A_iA_.^{-1}B_{j,l}) \\
&\quad + \sum_{i=1}^{m}\text{tr}(A_.^{-1}A_iA_.^{-1}B_{i,l}) + O(m^{-1})
\end{aligned}
$$

$$= \sum_{i=1}^{m} \text{tr}(A_.^{-1}B_{i,l}) - \sum_{j=1}^{m} \text{tr}\left\{ A_.^{-1}\left(\sum_{i=1}^{m} A_i\right) A_.^{-1}B_{j,l} \right\} + O(m^{-1})$$
$$= O(m^{-1}).$$

Thus, (14.41) is satisfied with $\nu = 1$.

## 14.4 The bootstrap

When Efron (1979) introduced the bootstrap, he did not spend much time developing theory, as he wrote that he would proceed (the bootstrap) "by a series of examples, with little offered in the way of general theory." Not by coincidence, the book, *An Introduction to the Bootstrap*, that he coauthored with Tibshirani begins with a quote from *Faust* (von Goethe 1808):

> Dear friend, theory is all gray,
> and the golden tree of life is green.

Nevertheless, these by no mean intend to undermine the importance of theory. The bootstrap method has a sound theoretical basis, and this was certainly in the mind of Efron when he proposed the method. As indicated in Section 14.1, the problem of interest is the sampling distribution of $R = R(X_1, \ldots, X_n, F)$, where $X_1, \ldots, X_n$ are independent observations with the common distribution $F$. Efron proposed to approximate the distribution of $R$ by that of $R^* = R(X_1^*, \ldots, X_n^*, F_n)$, where $F_n$ is the empirical distribution of $X_1, \ldots, X_n$ defined by (7.1) and $X_1^*, \ldots, X_n^*$ are i.i.d. samples from $F_n$. Practically—and this is perhaps the best known features of the bootstrap—the distribution of $R^*$ can be approximated using the Monte Carlo method, by drawing $X_1^*, \ldots, X_n^*$ with replacement from the box that consists of $X_1, \ldots, X_n$. One apparent advantage of the bootstrap is that it is conceptually simple and intuitive, especially if one has a good sense of the concepts of estimation. For example, if one expects that $F_n$ is a good estimator of $F$ (which it is; see Chapter 7), then it makes sense to approximate the distribution of $R$ by $R^*$. Of course, this is not (yet) a rigorous justification. It took some years for the theory of bootstrap to develop, as in many other cases of methodology developments. One of the early theoretical studies of the bootstrap, in terms of its asymptotic properties, was carried out by Bickel and Freedman (1981). As indicated by these authors, the bootstrap "would probably be used in practice only when the distributions could not be estimated analytically. However, it is of some interest to check that the bootstrap approximation is valid in situations which are simple enough to handle analytically." Some of the situations are considered below, where the results are mainly based on Bickel and Freedman's studies.

The first simple case to look at is bootstrapping the mean. Suppose that $F$ has finite mean $\mu$ and variance $\sigma^2$. A standard *pivotal quantity* is the t-statistic, defined by $t_n = \sqrt{n}(\mu_n - \mu)/\sigma_n$, where $\mu_n = \bar{X} = n^{-1}\sum_{i=1}^{n} X_i$ and

$\sigma_n^2 = \hat{\sigma}^2 = n^{-1} \sum_{i=1}^n (X_i - \mu_n)^2$. Here, a pivotal quantity is a random variable whose distribution is free of parameters. The change of notation (from $\bar{X}$ to $\mu_n$, etc.) is for technical convenience, because we are going to allow the bootstrap sample size, $m$, to be different from $n$. Let $X_1^*, \ldots, X_m^*$ be i.i.d. samples from $F_n$; define $\mu_m^*$ and $\sigma_m^*$ similarly, and let $t_{m,n}^* = \sqrt{m}(\mu_m^* - \mu_n)/\sigma_m^*$ and $t_n^* = t_{n,n}^*$. To make the case even simpler, let us first consider the $t$'s without the denominators. In the case $m = n$, the bootstrap distribution of $\sqrt{n}(\mu_n^* - \mu_n)$ is used to approximate the sampling distribution of $\sqrt{n}(\mu_n - \mu)$. We may think of this approximation as making two changes at the same time: (i) $\mu_n$ to $\mu_n^*$ and (ii) $\mu$ to $\mu_n$. If one only looks at change (ii), one would expect an error in the order of $O(n^{-1/2})$ (why?), which likely may affect the asymptotic distribution, right? [Recall $\sqrt{n}(\mu_n - \mu) \xrightarrow{\mathrm{d}} N(0, \sigma^2)$; so any change of $O(n^{-1/2})$ may result in a shift of the mean of the asymptotic distribution.] However, there is also an error due to change (i). These two errors somehow cancel each other, to a large extent. In fact, the same thing happens even if $m$ and $n$ go to $\infty$ independently, as the following theorem states.

**Theorem 14.6.** Let $X_1, X_2, \ldots$ be i.i.d. with positive variance $\sigma^2$. For almost all sample sequences $X_1, X_2, \ldots$ we have, as $m, n \to \infty$, the following:
(i) The conditional distribution of $\sqrt{m}(\mu_m^* - \mu_n)$ given $X_1, \ldots, X_n$ converges weakly to $N(0, \sigma^2)$.
(ii) $s_m^* \to \sigma$ in the conditional probability; that is, for every $\epsilon > 0$,

$$P(|s_m^* - \sigma| > \epsilon | X_1, \ldots, X_n) \longrightarrow 0.$$

Recall that, by the CLT, the distribution of $\sqrt{n}(\mu_n - \mu)$ converges weakly to $N(0, \sigma^2)$ as $n \to \infty$. Thus, (i) of Theorem 14.6 states that the asymptotic distribution is unchanged when we consider the bootstrap version of $\sqrt{n}(\mu_n - \mu)$, even if the bootstrap sample size may be different from the sample size. This, combined with (ii) of Theorem 14.6, implies that the asymptotic distribution of the bootstrap pivot $t_{m,n}^*$, and hence $t_n^*$, coincides with the classical one, which is $N(0, 1)$ (why?). The proof of Theorem 14.6 is based on the following argument that can be easily extended to more general situations (therefore, the result of Theorem 14.6 is certainly extendable). Let $\Gamma_2$ be the set of all distributions $G$ satisfying $\int x^2 \, dG(x) < \infty$. Define convergence of a sequence, $G_\alpha \in \Gamma_2$, where $\alpha$ represents a certain index, as (see Section 2.4)

$$G_\alpha \Rightarrow G \text{ iff } G_\alpha \xrightarrow{\mathrm{w}} G \text{ and } \int x^2 \, dG_\alpha(x) \longrightarrow \int x^2 \, dG(x). \quad (14.42)$$

It can be shown that convergence in the sense of (14.42) is equivalent to convergence in a metric, $d_2$, on $\Gamma_2$ defined as $d_2(G, H) = \sqrt{E\{(X - Y)^2\}}$, $G, H \in \Gamma_2$, where the infimum is over all joint distribution of $(X, Y)$ such that $X$ and $Y$ have marginal distributions $G$ and $H$, respectively. (One may compare this result with Theorem 2.17 or §2.7.9.) The metric $d_2$ satisfies the following nice inequality, due to Mallows (1972). Let $Z_1(G), \ldots, Z_m(G)$ denote a sequence of independent random variables with common distribution $G$, and let $G^{(m)}$ be the distribution of $m^{-1/2} \sum_{j=1}^{m} [Z_j(G) - E\{Z_j(G)\}]$. Then we have

$$d_2[G^{(m)}, H^{(m)}] \leq d_2(G, H), \quad G, H \in \Gamma_2. \tag{14.43}$$

With this notation, the distribution of $\sqrt{m}(\mu_m^* - \mu_n)$ is simply $F_n^{(m)}$, and the distribution of $\sqrt{m}(\mu_m - \mu)$ is $F^{(m)}$. By (14.43), $F_n^{(m)}$ is at least as close to $F^{(m)}$ as $F_n$ is to $F$, and $d_2(F_n, F)$ goes to zero by the CLT.

As noted, Theorem 14.6 implies that the asymptotic distribution of $t_n^*$ is the same as that of $t_n$. As far as this result is concerned, the bootstrap approximation is correct but does not seem to have any advantage over the classical approximation. On the other hand, intuitively, one would expect the bootstrap distribution, which is based on the empirical distribution of the data, to better approximate the true distribution of the pivot than the (classical) asymptotic distribution, because the latter requires that $n$ goes to infinity. This intuition is correct, at least for some simple cases. For example, consider the case that $\sigma$ is known. In such a case, it is customary to consider $z_n = \sqrt{n}(\mu_n - \mu)/\sigma$ instead of $t_n$. Singh (1981) showed that, in this case, the bootstrap distribution of $\mu_n^*$ better approximates the distribution of $z_n$ than the classical asymptotic distribution, $N(0, 1)$. More specifically, we have

$$\sqrt{n} \sup_x |P(z_n \leq x) - P(t_n^* \leq x | X_1, \ldots, X_n)| \xrightarrow{\text{a.s.}} 0. \tag{14.44}$$

Compare (14.44) with the Edgeworth expansion (4.27), which implies that

$$\sqrt{n}|P(z_n \leq x) - \Phi(x)| = \frac{\kappa_3}{6}(1 - x^2)\phi(x) + O(n^{-1/2}),$$

where $\kappa_3 = E(X_1 - \mu)^3/\sigma^3$ and $\Phi(\cdot), \phi(\cdot)$ are the cdf and pdf of $N(0, 1)$, respectively. It is clear that as long as $\kappa_3$ is not equal to zero, the bootstrap distribution is asymptotically closer to $P(z_n \leq x)$ than $\Phi(x)$, which is the classical approximation.

Next, we consider the problem of bootstrapping von Mises functionals. Let $X_1, \ldots, X_n$ be i.i.d. $p$-dimensional observations. Many pivots of interest can be expressed in the form

$$\frac{\sqrt{n}\{h(S_n/n) - h(\mu)\}}{v(U_n/n)} \tag{14.45}$$

and have asymptotic normal distributions, where $S_n = \sum_{i=1}^{n} s(X_i)$, $U_n = \sum_{i=1}^{n} u(X_i)$, $h$, $s$, and $u$ are vector-valued functions, and $v$ is a real-valued function. The traditional method of deriving the asymptotic distribution is the delta method (see Example 4.4); namely, write

$$\sqrt{n}\left\{h\left(\frac{S_n}{n}\right) - h(\mu)\right\} = \frac{\partial h}{\partial \mu'}\sqrt{n}\left(\frac{S_n}{n} - \mu\right) + o_P(1)$$

and go from there. Here, we assume that all of the functions involved are differentiable. In the bootstrap, we replace $S_n$, $U_n$, and $\mu$ in (14.45) by $S_n^*$, $U_n^*$, and $S_n/n$, respectively, where $S_n^*$ and $U_n^*$ are $S_n$ and $U_n$, respectively, with the $X_i$'s replaced by $X_i^*$'s, and $X_1^*, \ldots, X_n^*$ are i.i.d. samples from $F_n$, the empirical distribution of $X_1, \ldots, X_n$ that puts mass $1/n$ on each $X_i$. To study the asymptotic behavior of the current bootstrap procedure, we use the idea of the delta method, also known as *linearization*, in a functional space. Recall that the delta method is based on the Taylor expansion. Let $h : \mathcal{F} \to R$ be a functional, where $\mathcal{F}$ is a convex set of probability measures on $R^p$ that includes all point masses and $F$, the true distribution of the $X_i$'s. Suppose that there is an expansion of $h$ at $F$ such that for $G$ close to $F$, we have

$$h(G) - h(F) = \int \psi(x, F) \, d(G - F)(x) + R,$$

where $R \to 0$ at a faster rate than the distance between $G$ and $F$, as the latter goes to zero. The function $\psi$ can be treated as a derivative. If we impose the regularity condition $\int \psi(x, F) \, dF(x) = 0$, then the expansion can be simply expressed as

$$h(G) - h(F) = \int \psi(x, F) \, dG(x) + R. \tag{14.46}$$

A functional that satisfies (14.46) is often called a von Mises functional. A more rigorous treatment of this subject, known as Gâteaux derivative, can be given (e.g., Serfling 1980). However, it is often more convenient to use (14.46) in another way, without having to go through the rigorous justification. First, we use (14.46) to motivate an asymptotic expansion and, in particular, obtain the leading term of the expansion; then we use the expansion to obtain the asymptotic distribution and directly justify our arguments. An intuitive way of obtaining the first term on the right side of (14.46) is the following:

$$\int \psi(x, F) \, dG(x) = \frac{\partial}{\partial \epsilon} h[F + \epsilon(G - F)] \Big|_{\epsilon=0}. \tag{14.47}$$

To illustrate the method, we consider the special case

$$h(G) = \int \int \omega(x, y) \, dG(x) \, dG(y), \tag{14.48}$$

where $\omega(x, y) = \omega(y, x)$, assuming that the functional is well defined. It can be shown (Exercise 14.14) that

$$\psi(x, F) = 2 \left\{ \int \omega(x, y) \, dF(y) - h(F) \right\}. \tag{14.49}$$

Now, let $G = F_n^*$, the empirical distribution of $X_1^*, \ldots, X_n^*$, and $F = F_n$. Then, we can (always) write, in view of (14.46),

$$h(F_n^*) - h(F_n) = \int \psi(x, F_n)\, dF_n^*(x) + R_n$$

with $\int \psi(x, F_n)\, dF_n^*(x) = n^{-1} \sum_{i=1}^{n} \psi(X_i^*, F_n)$. It can be shown (rigorously) that for almost all $X_1, X_2, \ldots$, we have $\sqrt{n} R_n \to 0$ in conditional probability, and the conditional distribution of $n^{-1/2} \sum_{i=1}^{n} \psi(X_i^*, F_n)$ converges weakly to the normal distribution with mean 0 and variance

$$\tau^2 = \int \psi^2(x, F)\, dF(x)$$

$$= 4 \left[ \int \left\{ \int \omega(x, y)\, dF(x) \right\}^2 dF(x) - h^2(F) \right]. \qquad (14.50)$$

This leads to the following theorem.

**Theorem 14.7.** Let $h$ be given by (14.48). If $\int \omega^2(x, y)\, dF(x)\, dF(y) < \infty$ and $\int \omega^2(x, x)\, dF(x) < \infty$, then, for almost all $X_1, X_2, \ldots$, as $n \to \infty$, the conditional distribution of $\sqrt{n}\{h(F_n^*) - h(F_n)\}$ converges weakly to $N(0, \tau^2)$ with $\tau^2$ given by (14.50).

The idea of the proof is pretty much explained in the discussion leading to the theorem. On the other hand, according to a classical result (von Mises 1947), under the same conditions, the distribution of $\sqrt{n}\{h(F_n) - h(F)\}$ converges weakly to $N(0, \tau^2)$ as $n \to \infty$. Thus, again, the asymptotic distribution of the bootstrap matches the classical one. We use an example to illustrate an application.

*Example 14.8.* Recall the one-sample Wilcoxon statistic (see Example 11.8), $U_n = \binom{n}{2}^{-1} \sum_{1 \leq i < j \leq n} 1_{(X_i + X_j > 0)}$. Consider the functional $\theta = \theta(F) = P(X_1 + X_2 > 0)$. Note that this is a special case of (14.48) with $\omega(x, y) = 1_{(x+y>0)}$. Then we have $\theta(F_n) = n^{-2} \sum_{i=1}^{n} \sum_{j=1}^{n} 1_{(X_i + X_j > 0)}$, which is a special case of the $V$-statistics (see Exercise 14.15, which also involves some of the results below). It can be shown that

$$|\sqrt{n}(U_n - \theta) - \sqrt{n}\{\theta(F_n) - \theta(F)\}| \leq \frac{2}{\sqrt{n}} \qquad (14.51)$$

and, similarly,

$$|\sqrt{n}(U_n^* - U_n) - \sqrt{n}\{\theta(F_n^*) - \theta(F_n)\}| \leq \frac{2}{\sqrt{n}}, \qquad (14.52)$$

where $U_n^*$ is $U_n$ with the $X_i$'s replaced by $X_i^*$'s, the latter being i.i.d. samples from $F_n$, and $F_n^*$ is the empirical d.f. of $X_1^*, \ldots, X_n^*$. According to Theorem 14.7, $\sqrt{n}\{\theta(F_n^*) - \theta(F_n)\}$ has the same asymptotic (conditional) distribution as $\sqrt{n}\{\theta(F_n) - \theta(F)\}$. Thus, by (14.51) and (14.52), $\sqrt{n}(U_n^* - U_n)$ has the same asymptotic (conditional) distribution as $\sqrt{n}(U_n - \theta)$. Therefore, it is valid to bootstrap $\sqrt{n}(U_n^* - U_n)$ in order, for instance, to obtain the power of the one-sample Wilcoxon test, also known as Wilcoxon signed-rank test, that is related to the distribution of $\sqrt{n}(U_n - \theta)$ (see Sections 11.2 and 11.3).

One of the examples that Efron (1979) used in his seminal paper to demonstrate the bootstrap method is estimating the median. In fact, this is one of the many statistics that are associated with the quantile process, defined as

$$Q_n(t) = \sqrt{n}\{F_n^{-1}(t) - F^{-1}(t)\}, \quad 0 < t < 1, \tag{14.53}$$

where $F^{-1}$ is defined by (7.4). The quantile process is a stochastic process. It is known (e.g., Bickel 1966) that if $F$ has continuous and positive density $f$, then $Q_n$ converges weakly to $B/(f \circ F^{-1})$ in the space of probability measures on $D[a, b]$, where $B$ is the Brownian bridge on $[0, 1]$ (see Section 10.5), and $f \circ F^{-1}(t) = f\{F^{-1}(t)\}$, $t \in (0, 1)$. Here, $0 < a \leq b < 1$, $D[a, b]$ denotes the space of functions on $[a, b]$ that are right-continuous and possess left-limit at each point, and the weak convergence is defined the same way as in Section 7.3. To bootstrap the quantile process, we consider

$$Q_n^*(t) = \sqrt{n}\{(F_n^*)^{-1}(t) - F_n^{-1}(t)\}, \quad 0 < t < 1.$$

Given this result, the validity of bootstrapping the quantile process is justified by the following theorem.

**Theorem 14.8.** If $F$ has continuous and positive density $f$, then for almost all sample sequences $X_1, X_2, \ldots$, $Q_n^*$ converges weakly, given $X_1, \ldots, X_n$, to $B/(f \circ F^{-1})$ in the space of probability measures on $D[a, b]$.

The proof amounts to writing $Q_n^* = \sqrt{n}\{F \circ (F_n^*)^{-1} - F \circ F_n^{-1}\}/R_n$, where

$$R_n = \frac{\{F \circ (F_n^*)^{-1} - F \circ F_n^{-1}\}}{(F_n^*)^{-1} - F_n^{-1}},$$

and then arguing that $\sqrt{n}\{F \circ (F_n^*)^{-1} - F \circ F_n^{-1}\}$ converges weakly to $B$ and $\|R_n - f \circ F^{-1}\| \to 0$ (make sense?) in probability, where $\|\cdot\|$ denotes the supremum norm in $D[a, b]$ [similar to (7.7)]. See Bickel and Freedman (1981, pp. 1206–1207) for details. A consequence of Theorem 14.8 is the following result on bootstrapping the median. Note that the median is $\nu = F^{-1}(1/2)$. Let $\nu_n = F_n^{-1}(1/2)$, the sample median (the definition is slightly different from that, e.g., in Example 14.2, when $n$ is even, but is asymptotically equivalent). It is well known that $\sqrt{n}(\nu_n - \nu)$ converges weakly to $N(0, \sigma^2)$, where $\sigma^2 = \{4f^2(\nu)\}^{-1}$, provided that $f(\nu) > 0$ (see Exercise 6.16 for a special case, but it can be proved more generally; see Exercise 14.16). Let $\nu_n^* = (F_n^*)^{-1}(1/2)$ denote the median of $F_n^*$.

**Corollary 14.1.** Suppose that $F$ has a unique median. Then, under the condition of Theorem 14.8, we have for almost all sample sequences $X_1, X_2, \ldots$, that $\sqrt{n}(\nu_n^* - \nu_n)$ converges weakly to $N(0, \sigma^2)$ for the same $\sigma^2$.

In fact, for the conclusion of Corollary 14.1 to hold, all one needs is that $\nu$ is unique and $F$ has a positive derivative at $\nu$.

So far, some special cases have been looked at and the bootstrap method is valid in those cases. In general, one needs to know when the bootstrap works and also if there are situations where it does not work. Regarding the first question, Bickel and Freedman (1981) provided the following general guidelines. Roughly speaking, the bootstrap works if the following hold:

(i) For all $G$ in a "neighborhood" of $F$, into which $F_n$ falls eventually with probability 1, the distribution of $R(X_1, \ldots, X_n, G)$ converges weakly to a distribution $\mathcal{L}_G$.

(ii) The map $G \mapsto \mathcal{L}_G$ is continuous.

Regarding the second question, it is known that there are situations where the bootstrap fails. A well-known counterexample, also given by Bickel and Freedman (1981), is the following.

*Example 14.9.* Let $F$ be the Uniform$(0, \theta]$ distribution, where $\theta > 0$ is unknown. The usual pivot for $\theta$ is $\xi_n = n\{\theta - X_{(n)}\}/\theta$, which has a limiting Exponential(1) distribution (Exercise 14.17). It is natural to bootstrap $\xi_n$ by $\xi_n^* = n\{X_{(n)} - X_{(n)}^*\}/X_{(n)}$, where $X_{(1)}^* \leq \cdots \leq X_{(n)}^*$ are the order statistics of $X_1^*, \ldots, X_n^*$. Note that $\theta$ is the upper end of the support of $F$ [i.e., $F(\theta) = 1$ and $F(x) < 1, \forall x < \theta$] and $X_{(n)}$ is the upper end of the support of $F_n$ $[F_n\{X_{(n)}\} = 1$ and $F_n(x) < 1, \forall x < X_{(n)}]$. Therefore, the distribution of $\xi_n^*$ is the bootstrap distribution of $\xi_n$ according to the *principle of bootstrap* described at the beginning of this section. However, the bootstrap distribution does not converge weakly. In fact, it can be shown (see below) that

$$\limsup n\{X_{(n)} - X_{(n-1)}\} = \infty \quad \text{a.s.} \tag{14.54}$$

[note that this implies $\limsup n\{X_{(n)} - X_{(n-k+1)}\} = \infty$ for every $k \geq 2$; compare (14.55) below] and, for any fixed $k \geq 1$,

$$\liminf n\{X_{(n)} - X_{(n-k+1)}\} = 0 \quad \text{a.s.} \tag{14.55}$$

Now, suppose that, with probability 1, the conditional distribution of $\xi_n^*$ given $X_1, \ldots, X_n$ converges weakly to a distribution, say $G$. Then for any continuity point $x > 0$ of $G$ (see Section 2.4), we have

$$\lim_{n \to \infty} \mathrm{P}(\xi_n^* > x | X_1, \ldots, X_n) = 1 - G(x).$$

Now, see what happens. On the one hand, by (14.54), we have with probability 1 that there is a subsequence of $n$ so that $\Delta_n = n\{X_{(n)} - X_{(n-1)}\}/X_{(n)} > x$ holds for the subsequence (Exercise 14.17). It follows that, with probability 1 along the subsequence,

$$1 - G(x) = \lim \mathrm{P}(\xi_n^* > x | X_1, \ldots, X_n)$$
$$\geq \lim \mathrm{P}(\xi_n^* > \Delta_n | X_1, \ldots, X_n)$$

$$= \lim P\{X^*_{(n)} < X_{(n-1)}|X_1,\ldots,X_n\}$$

$$= \lim \left(1 - \frac{2}{n}\right)^n$$

$$= e^{-2} > 0 \tag{14.56}$$

(Exercise 14.17). On the other hand, by (14.55), for any fixed $k \geq 1$ we have, with probability 1, that there is a subsequence of $n$ so that $\delta_n = n\{X_{(n)} - X_{(n-k+1)}\}/X_{(n)} < x$ holds for the subsequence (Exercise 14.17). It follows that, with probability 1 along the subsequence,

$$\begin{aligned}
1 - G(x) &= \lim P(\xi^*_n > x|X_1,\ldots,X_n) \\
&\leq \lim P(\xi^*_n > \delta_n|X_1,\ldots,X_n) \\
&= \lim P\{X^*_{(n)} < X_{(n-k+1)}|X_1,\ldots,X_n\} \\
&= \lim \left(1 - \frac{k}{n}\right)^n \\
&= e^{-k} \tag{14.57}
\end{aligned}$$

(Exercise 14.17). Because (14.57) holds for every $k \geq 1$, it contradicts (14.56) (when $k > 2$). Thus, the conditional distribution of $\xi^*_n$ must not converge.

It remains to show (14.54) and (14.55). We give an outline of the proof below and leave the details to another exercise (Exercise 14.18). The idea is to use the Borel–Cantelli lemma—namely, part (ii) of Lemma 2.5. However, this requires pairwise independent of the events, whereas the sequence $\eta_n = n\{X_{(n)} - X_{(n-k+1)}\}, n \geq 1$, are not pairwise independent. Therefore, the first thing we do is to construct a sequence of independent random variables that is asymptotically identical to a subsequence of $\eta_n$ (this kind of technique is sometimes called *coupling*). Let $a_n, n \geq 0$, be a sequence of positive integers that is strictly increasing and satisfies

$$\sum_{n=n_k}^{\infty} \frac{a_{n-1}}{a_n - k} < \infty \tag{14.58}$$

for every $k \geq 0$, where $n_k$ is the first integer $n$ such that $a_n > k$. Why chose $a_n$ this way? Well, see below. Let $\zeta_{n,k}$ be the $k$th largest value of $X_i, a_{n-1} < i \leq a_n$ (so $\zeta_{n,1} = \max_{a_{n-1} < i \leq a_n} X_i$). Let $A_{n,k}$ be the event that $X_{(a_n-k+1)} = X_i$ for some $1 \leq i \leq a_{n-1}$. Then we have for any $k \geq 1$,

$$\begin{aligned}
P(A_{n,k}) &\leq \sum_{i=1}^{a_{n-1}} P\{X_{(a_n-k+1)} = X_i\} \\
&= \frac{a_{n-1}}{a_n - k + 1}.
\end{aligned}$$

For fixed $k \geq 2$, write $A_n = A_{n,1} \cup A_{n,k}, n \geq n_k$. Then we have, by (14.58),

$$\sum_{n=n_k}^{\infty} P(A_n) \leq \sum_{n=n_k}^{\infty} \left( \frac{a_{n-1}}{a_n} + \frac{a_{n-1}}{a_n - k + 1} \right) < \infty.$$

Therefore, by the Borel–Cantelli lemma [part (i) of Lemma 2.5; note that this part does not require pairwise independence], we have $P(A_n \text{ i.o.}) = 0$. It follows that, with probability 1, we have $A_n^c$ for large $n$, which implies $\eta_{a_n} = a_n \{ \zeta_{n,1} - \zeta_{n,k} \} \equiv \zeta_n$, for large $n$.

Note that $\zeta_n, n \geq 1$, are independent. Thus, to use part (ii) of the Borel–Cantelli lemma, we need to show that for any $b > 0$,

$$\sum_{n=1}^{\infty} P(\zeta_n > b) = \infty \text{ and } \sum_{n=1}^{\infty} P(\zeta_n < b) = \infty, \tag{14.59}$$

and this should complete the proof. The divergences of the two series in (14.59) follow from an evaluation of the summands; namely, the distribution of $\zeta_n$ is the same as that of $a_n(X_{n,a_n-a_{n-1}} - X_{n-k+1,a_n-a_{n-1}})$, where $X_{r,n}$ is the $r$th-order statistic of $X_1, \ldots, X_n$. It can be shown that, for any $r < s$, the distribution of $n(X_{s,n} - X_{r,n})$ weakly converges to the Gamma$(s-r, 1)$ distribution as $n \to \infty$ and (14.58) implies $a_n - a_{n-1} \sim a_n$ [i.e., $(a_n - a_{n-1})/a_n \to 1$]. Therefore, the summand probabilites in (14.59) are bounded away from zero for large $n$; hence, the corresponding series must diverge.

The failure of the bootstrap in this particular example is due to the lack of uniformity in the convergence of $F_n$ to $F$. In other words, the conditions (i) and (ii) below Corollary 14.1 do not hold in this case. Efron and Tibshirani (1993, Section 7.4) discussed the same example, where the authors used a simulation to illustrate what happens. In the simulation, the authors generated 50 Uniform$[0, 1]$ random variables, $X_1, \ldots, X_{50}$. Then 2000 bootstrap samples, each of size 50 and drawn with replacement from $X_1, \ldots, X_{50}$, were generated. For each bootstrap sample, the estimator $\hat{\theta}^* = X_{(n)}^*$ was computed, and the histogram based on the 2000 $\hat{\theta}^*$'s was made. It was evident that the histogram was a poor approximation to the true distribution of $\hat{\theta}$ (Exercise 14.19). For example, the (conditional) probability that $\hat{\theta}^* = \hat{\theta}$ is $1 - (1 - 1/n)^n \to 1 - e^{-1} \approx 0.632$ (why?). However, we know for sure that the probability that $\hat{\theta}$ equals any given value is zero.

So far, the discussions on the bootstrap have been limited to the classical situation (i.e., the case of i.i.d. observations). Although this is the case in which the bootstrap was originally proposed, extensive studies have been carried out in efforts to extend the method beyond the i.i.d. case. In the next two sections we consider some of these extensions.

## 14.5 Bootstrapping time series

As we have seen, the i.i.d. assumption can be relaxed in many ways. A case that is slightly different from i.i.d. is independent but not identically distributed observations. This includes the important special case of regression (see Section 6.7). Efron (1979) proposed to bootstrap the residuals under a regression model. Note that although the observations $Y_i$ in (6.79) are not i.i.d., the errors $\epsilon_i$ are. On the other hand, the residuals may be viewed as estimates of the errors. Thus, by analogy with the i.i.d. case, it seems rational to bootstrap the residuals. In other words, one approximates the distribution of the errors by the empirical distribution of the residuals, and this was Efron's proposal. His idea was extended by Freedman (1984) to stationary linear models. Consider a dynamic model that can be expressed as

$$Y_t = Y_t A + Y_{t-1} B + X_t C + \epsilon_t, \tag{14.60}$$

$t = 0, \pm 1, \pm 2 \ldots$, where $X_t$ and $Y_t$ are $1 \times a$ and $1 \times b$ vectors of observations, respectively, $A$, $B$, and $C$ are coefficient matrices, and $\epsilon_t$ is a $1 \times b$ vector of errors satisfying $E(\epsilon_t) = 0$. It is assumed that $(X_t, \epsilon_t)$ are i.i.d. for $t = 0, \pm 1, \pm 2, \ldots$ with finite fourth moments. Note that we can write, from (14.60), $Y_t(I - A) = Y_{t-1}B + X_tC + \epsilon_t$. Thus, if $I - A$ is invertible, we have

$$\begin{aligned}
Y_t &= Y_{t-1}B(I - A)^{-1} + (X_tC + \epsilon_t)(I - A)^{-1} \\
&= Y_{t-2}\{B(I - A)^{-1}\}^2 + (X_{t-1}C + \epsilon_{t-1})(I - A)^{-1}\{B(I - A)^{-1}\} \\
&\quad + (X_tC + \epsilon_t)(I - A)^{-1} \\
&\cdots \\
&= Y_{t-k}\{B(I - A)^{-1}\}^k + \sum_{s=0}^{k-1} \xi_{t,s},
\end{aligned} \tag{14.61}$$

where $\xi_{t,s} = (X_{t-s}C + \epsilon_{t-s})(I - A)^{-1}\{B(I-A)^{-1}\}^s$. If we let $k$ go to infinity on the right side of (14.61), assuming that $\|B(I-A)^{-1}\| < 1$, where $\|\cdot\|$ represents the spectral norm defined above (5.11), we have the series expansion

$$Y_t = \sum_{s=0}^{\infty} \xi_{t,s}, \tag{14.62}$$

which holds, say, almost surely. This implies that $Y_t, t = 0, \pm 1, \pm 2, \ldots$, is strictly stationary (see Section 9.1). The matrices $A$, $B$, and $C$ can be estimated by the LS method (see Section 6.7). For example, let $y_t$ be the first component of $Y_t$ and $e_t$ the first component of $\epsilon_t$. Then the first-component equation of (14.60) can be written as $y_t = U_t\beta + e_t$, where $U_t$ is a vector that involves $Y_t, Y_{t-1}$, and $X_t$ and $\beta$ is the vector that combines the first columns of $A, B$, and $C$. Let $\hat{\beta}$ be the LS estimator of $\beta$. Then the residuals for fitting the first-component equation are $\hat{e}_t = y_t - U_t\hat{\beta}$. We then combine the residuals for fitting the equations of different components of (14.60) and denote the combined residual vector by $\hat{\epsilon}_t$. Now, suppose that $(Y_t, X_t)$ is observed for $t = 0, \ldots, n$. Let $\hat{F}$ be the empirical distribution of $(X_t, \hat{\epsilon}_t), t = 1, \ldots, n$—that is, a multivariate probability distribution that puts mass $1/n$ on each point $(X_t, \hat{\epsilon}_t)$. Let $(X_s^*, \epsilon_s^*)$ be independent with the common distribution $\hat{F}$

for $s = 0, \pm 1, \pm 2, \ldots$ The bootstrap samples $Y_t^*$ are then generated in the same way as (14.62); that is,

$$Y_t^* = \sum_{s=0}^{\infty} \xi_{t,s}^*, \qquad (14.63)$$

where $\xi_{t,s}^*$ is $\xi_{t,s}$ with $X$ and $\epsilon$ replaced by $X^*$ and $\epsilon^*$, respectively (and everything else is unchanged).

A nice feature of the bootstrap is that the extension of the method to various non-i.i.d. situations is often intuitively simple (although this is not always the case) as long as one sticks with a few basic principles. The most important of those is the *plug-in principle* that was introduced in the previous section, although not given the name; namely, the distribution of $R(X, F)$ is approximated by that of $R(X^*, \hat{F})$, where $X$ is a sample from $F$, $\hat{F}$ is an estimator of $F$, and $X^*$ is a sample from $\hat{F}$. Here, $X$ is a vector of observations that do not have to be i.i.d. and $F$ is the joint distribution (Exercise 14.20). As we will see, one can propose and develop variations of the plug-in principle according to the problem of interest.

After all, the extensions do need to be justified. Freedman (1984) listed two kinds evidence that are needed to show that the (extension of) bootstrap "works": (i) A showing that the bootstrap gives the right answers with large samples, so it is at least as sound as the conventional asymptotics; and (ii) a showing that in finite samples, the bootstrap actually outperforms the conventional asymptotics. For example, Freedman (1984) provided evidence for (i). Also, see the previous section, where the bootstrap was justified asymptotically in several cases. On the other hand, in many cases, a bootstrap method is used when there is no alternative method that could be used, not even by the asymptotics. As a result, one does not know the answer provided by the asymptotic theory, conventional or otherwise. However, a combination of (i) and (ii) may allow us to study large-sample behavior of the bootstrap through finite samples. We add this later approach as (iii) A showing of improved finite sample performance of the bootstrap as the sample size increases via simulation studies. This latest evidence that bootstrap works is relatively easy to get for practitioners and researchers, especially those who are not sophisticatedly equipped with the large-sample theory. We demonstrate this in various occasions in the sequel.

In a way, Freedman's method may be viewed as model-based or parametric bootstrap. The idea is to first estimate the unknown parameters under the assumed model, which leads to an estimate of the underlying distribution, from which the bootstrap samples will be drawn. Extension of this idea to other parametric time series models, such as ARMA models (see Section 9.4), is fairly straightforward. However, the scope of such parametric models is rather limited—a time series in real life may not satisfy, for example, an ARMA model. In such a case, a nonparametric bootstrap method—that is, one that does not rely on a specific parametric model assumption—is needed.

There have been two major approaches to bootstrapping time series non-parametrically. The first is based on an idea of approximating a (nonparametric) time series by a sequence of parametric time series. This kind of methods have been proposed in the statistical literature. In particular, Grenander (1981) called it the *method of sieves*. Bühlmann (1997) used the latter method in bootstrapping a time series. It is known that many stationary time series can be represented, or approximated, by the so-called $AR(\infty)$ process (e.g., Bickel and Bühlmann 1997). Here, an $AR(\infty)$ process is defined as a stationary process $X_t, t \in Z = \{0, \pm 1, \pm 2, \ldots\}$, satisfying

$$\sum_{j=0}^{\infty} \phi_j (X_{t-j} - \mu_X) = \epsilon_t, \quad t \in Z, \tag{14.64}$$

where $\mu_X = E(X_t)$, the $\epsilon_t$'s are uncorrelated with mean 0, and the $\phi_j$'s satisfy $\phi_0 = 1$ and $\sum_{j=0}^{\infty} \phi_j^2 < \infty$. If we write $\tilde{X}_t = X_t - \mu_X$, then the left side of (14.64) may be viewed as the left side of (9.1), with $X_t$ replaced by $\tilde{X}_t$ and $b_j = -\phi_j, 1 \leq j \leq p$, as $p \to \infty$. Thus, for example, even if the time series is not ARMA (or we do not know if it is ARMA), we may still approximate it with an $AR(p)$ with perhaps a large $p$. This naturally leads to the following strategy, called *sieve bootstrap*. Given a sample $X_1, \ldots, X_n$ of the time series, first fit an $AR(p)$ model, where $p$ is supposed to increase with $n$ (at a suitable rate; see below). For example, the AR model may be fitted by solving the Yule–Walker equation (9.33). This leads to estimates of the AR coefficients, say, $\hat{\phi}_1, \ldots, \hat{\phi}_p$. We then compute the residuals of the fit,

$$\hat{\epsilon}_t = \sum_{j=0}^{p} \hat{\phi}_j (X_{t-j} - \bar{X}), \quad t = p+1, \ldots, n,$$

where $\hat{\phi}_0 = 1$ and $\bar{X}$ is the sample mean. We then centralize the residuals by

$$\tilde{\epsilon}_t = \hat{\epsilon}_t - \frac{1}{n-p} \sum_{t=p+1}^{n} \hat{\epsilon}_t.$$

Denote the empirical cdf of the centralized residuals by

$$\hat{F}(x) = \frac{1}{n-p} \sum_{t=p+1}^{n} 1_{(\tilde{\epsilon}_t \leq x)}.$$

Then we resample $\epsilon_t^*, t \in Z$, independently from $\hat{F}$ and define the bootstrapped time series, $X_t^*, t \in Z$, by the AR model

$$\sum_{j=0}^{p} \hat{\phi}_j (X_{t-j}^* - \bar{X}) = \epsilon_t^*, \quad t \in Z.$$

Here, again, the plug-in principle is in play, although it might be (much) easier to view it as a variation of Efron's plug-in principle (Exercise 14.22). According to (14.64), the process $X_t, t \in Z$ is a function of $\epsilon = (\epsilon_t)_{t \in Z}$, and the parameters $\mu_X$ and $\phi_j, j = 0, 1, \ldots$. For example, if the function $\Phi(x) = \sum_{j=0}^{\infty} \phi_j z^j$ of a complex variable $z$ is nonzero for $|z| \leq 1$, then $X_t$ can be expressed as $X_t = \mu_X + \sum_{j=0}^{\infty} \psi_j \epsilon_{t-j}$, where the $\psi_j$'s depend on the $\phi_j$'s with $\psi_0 = 1$ (e.g., Anderson 1971; also see Section 4.4.3). Let $\theta$ denote the vector of all of the parameters and let $G$ be the (joint) distribution of $\epsilon$. Then we have $(X_t, t \in Z) = R(\epsilon, \theta)$, where $\epsilon \sim G$ and $R$ is the functional corresponding to (14.64). The plug-in principle here is to approximate the distribution of $R(\epsilon, \theta)$ by that of $R(\epsilon^*, \hat{\theta})$, where $\epsilon^* \sim \hat{G}$, an estimator of $G$, and $\hat{\theta}$ is an estimator of $\theta$ ($R$ is the same). Here, $\hat{G}$ and $\hat{\theta}$ are chosen, respectively, as the empirical distribution of the residuals and the vector of $\hat{\mu}_X = \bar{X}$ and, say, the Yule–Walker estimators of the $\phi$'s. Another difference between the sieve bootstrap and Efron's bootstrap is that the sieve bootstrap samples $X_t^*, t \in Z$, are not a subset of the original data $X_1, \ldots, X_n$ (this is also true for Freedman's model-based bootstrap).

Now, the sieve bootstrap has been introduced, so far as a proposal. It has the lead-off for being intuitive—and this is important. Although there are seemingly exceptions, good statistical methods are almost always intuitive. The next step is to justify the method rigorously. This brings in the second stage of the development. The justification was given by Bühlmann (1997) from two aspects that provided the evidences of (i) and (iii) following Freedman's remark (see above). For evidence of (i), we consider the simple case of bootstrapping the mean. First, note that in order to approximate the time series by an AR($p$) process, the order $p$ needs to increase with $n$. This makes sense if the time series is an AR($\infty$) process or can be approximated by an AR($\infty$) process. The question is how fast should $p$ increase with $n$? Bühlmann (1997) required that

$$p = o\left(\left\{\frac{n}{\log n}\right\}^{1/4}\right). \tag{14.65}$$

Under (14.65) and some regularity conditions, he showed that the sieve bootstrap has the following asymptotic properties. Let P$^*$ denote the conditional probability given $X_1, \ldots, X_n$, and var$^*$ denote the variance under P$^*$. Then, as $n \to \infty$, we have

$$\text{var}^*\left(\frac{1}{\sqrt{n}}\sum_{t=1}^{n} X_t^*\right) - \text{var}\left(\frac{1}{\sqrt{n}}\sum_{t=1}^{n} X_t\right) = o_P(1). \tag{14.66}$$

Note that (14.66) can be written as $n \, \text{var}^*(\bar{X}^*) - n \, \text{var}(\bar{X}) = o_P(1)$. Recall that $n \, \text{var}(\bar{X})$ corresponds to the asymptotic variance of $\bar{X}$. Thus, (14.66) may be interpreted as that the asymptotic bootstrap variance is the same as the asymptotic variance. Furthermore, suppose that

$$\frac{1}{\sqrt{n}} \sum_{t=1}^{n} (X_t - \mu_X) \xrightarrow{\text{d}} N(0, \tau^2), \tag{14.67}$$

where $\tau^2 = \sum_{k=-\infty}^{\infty} r(k)$ with $r(k) = \text{cov}(X_0, X_k)$; then we have

$$\sup_{x \in R} \left| \text{P}^* \left\{ \frac{1}{\sqrt{n}} \sum_{t=1}^{n} (X_t^* - \bar{X}) \leq x \right\} - \text{P} \left\{ \frac{1}{\sqrt{n}} \sum_{t=1}^{n} (X_t - \mu_X) \leq x \right\} \right|$$
$$= o_P(1) \tag{14.68}$$

as $n \to \infty$ (Exercise 14.23).

For evidence of (iii), Bühlmann (1997) carried out a simulation study. The goal is to study finite sample performance of the bootstrap. Note that although, in theory, (14.65) is what we need, there are still (infinitely) many choices of $p$ that satisfy this assumption. In order to choose a $p$ for practical use, Bühlmann proposed to use the AIC (see Section 9.3). An interesting question is: Does the $p$ chosen by the AIC satisfy (14.65) in some sense? Suppose that the coefficients $\phi_j$ in (14.64) satisfy $\phi_j \sim c_1 j^{-a}$ as $j \to \infty$, where $c_1$ and $a$ are constants with $a > 1$. Shibata (1980) showed that the $p$ chosen by the AIC satisfies $p \sim c_2 n^{1/2a}$ with probability tending to 1 (i.e., $p/c_2 n^{1/2a} \xrightarrow{\text{P}}$ 1 as $n \to \infty$) for some constant $c_2$. Comparing this result with (14.65), we see that the latter holds as long as $a > 2$. However, it should be noted that the $p$ selected by the AIC is a random variable depending on the observed data, whereas in (14.65), $p$ is supposed to be nonrandom. In Section 9.3 we discussed the consistency of the AIC, BIC, and other information criteria for selecting the order $p$ for AR. The story was that the BIC is consistent in the sense that the probability of selecting the true order of the AR process goes to 1 as $n \to \infty$, whereas the AIC is inconsistent in this regard. However, this result does not apply to the current case. The reason is that the underlying process here is not a finite-order AR (or at least we do not know if it is). In other words, the true AR order $p$ is $\infty$. In this case, it may be argued that the AIC is consistent and the BIC is not. This is why the AIC is prefered here. So, given the data, we first use the AIC to select an appropriate order $p$ and then use the sieve bootstrap with the selected $p$ as the order of the AR to generate bootstrap samples.

The following models were considered in the simulation study. The first is an AR(48) process, $X_t = \sum_{j=1}^{48} \phi_j X_{t-j} + \epsilon_t$, where $\phi_j = (-1)^{j+1} 7.5/(j+1)^3$ and the $\epsilon_t$'s are i.i.d. $\sim N(0,1)$. The second is an ARMA(1,1) process, $X_t = 0.8 X_{t-1} + \epsilon_t - 0.5 \epsilon_{t-1}$, where the $\epsilon_t$'s are i.i.d. $\sim 0.95 N(0,1) + 0.05 N(0,100)$. The third is the same as the second except that the coefficient 0.8 is replaced by $-0.8$. The last one is a self-exciting threshold autoregressive (SETAR; Moeanaddin and Tong 1990), SETAR(2; 1, 1), with $X_t = 1.5 - 0.9 X_{t-1} + \epsilon_t$ if $X_{t-1} \leq 0$, and $X_t = -0.4 - 0.6 X_{t-1} + \epsilon_t$ if $X_{t-1} > 0$, where the $\epsilon_t$'s are i.i.d. $\sim N(0.4)$. These four models are denoted by M1, M2, M3, and M4, respectively. The statistic $T_n =$ sample median of $X_1, \ldots, X_n$ is considered, and the normalized variance of $T_n$, $\sigma_n^2 = n \operatorname{var}(T_n)$ is of interest. The performance of the bootstrap variance was studied under two different sample sizes, $n = 64$ and $n = 512$. The results are presented in Table 14.1. Reported are the true values of $\sigma_n^2$, computed based on 1000 simulations, the mean (E), and standard derivation (sd) of $(\sigma_n^2)^* = n \operatorname{var}^*(T_n^*)$ over 100 simulation runs, where $T_n^*$ is the sample median of $X_1^*, \ldots, X_n^*$, the number of bootstrap replicates (that were used to compute var*) is 300, and the relative mean squared error (RMSE), defined as the MSE of $(\sigma_n^2)^*$ over the simulations [i.e., the average of $\{(\sigma_n^2)^* - \sigma_n^2\}^2$ over the 100 simulation runs] divided by $\sigma_n^4$. An estimated standard error of the RMSE is given in parentheses. It can be seen that the performance of $(\sigma_n^2)^*$, the bootstrap estimator of $\sigma_n^2$, improves in virtually all cases and aspects as $n$ increases. The only exception might be $E\{(\sigma_n^2)^*\}$ under M4, which does not seem to get closer to $\sigma_n^2$ as $n$ increases. However, since the true value of $\sigma_n^2$ also changes with $n$, a more reliable measure of performance is the RESE (and the corresponding standard error), which improves in all of the cases as the sample size increases.

**Table 14.1. Sieve bootstrap variance estimation**

|  | $\sigma_n^2$ | $E\{(\sigma_n^2)^*\}$ | $sd\{(\sigma_n^2)^*\}$ | RMSE |
|---|---|---|---|---|
| $n = 64$ | | | | |
| M1 | 16.4 | 13.1 | 8.6 | 0.31 (0.061) |
| M2 | 14.1 | 8.1 | 8.2 | 0.52 (0.063) |
| M3 | 3.1 | 5.0 | 5.2 | 3.09 (0.891) |
| M4 | 8.9 | 7.8 | 2.0 | 0.07 (0.008) |
| $n = 512$ | | | | |
| M1 | 16.7 | 16.1 | 4.3 | 0.07 (0.009) |
| M2 | 14.2 | 12.5 | 6.5 | 0.22 (0.046) |
| M3 | 2.6 | 2.9 | 0.7 | 0.08 (0.020) |
| M4 | 9.8 | 8.0 | 1.1 | 0.05 (0.004) |

Prior to Bühlmann's proposal of sieve bootstrap, Künsch (1989) introduced another strategy of bootstraping a stationary time series nonparametrically, called the *block bootstrap*. The idea is to resample "blocks" of successive observations rather than individual observations. Why blocks? First, the intention was to mimic Efron's bootstrap, in which the resampled observations are a subset of the original ones. The question is how to do this. Note that the successive observations in a time series contain important information about the dynamics of the series. Such information would be lost if one resamples the individual observations independently. Thus, the idea of block-sampling arises naturally. On the other hand, if the blocks are resampled independently, which is the case here (see below), it raises a concern on whether the blocks in the real-life series are actually independent, because, otherwise, the resampled blocks cannot mimic the real-life blocks. However, as we have seen (throughout this book), the independence assumption may not be as critical as one might have thought, depending what one wishes to do. For example, the ergodic theorem (see Section 7.6) assumes no independence, but the result is very similar to the classical SLLN, which assumes i.i.d. After all, the target of Künsch's proposal is stationary processes with short-range dependence. Roughly speaking, if the process has short-range dependence, then the observations are approximately independent if they are far apart. Künsch showed that the idea of blocks works for such processes, at least asymptotically, provided that the block size increases with the sample size.

Suppose that we observe $X_1, \ldots, X_N$ from a stationary process. For a given positive integer $m$, define $Y_t = (X_t, \ldots, X_{t-m+1})$. The empirical $m$-dimensional marginal distribution is defined as $p_N = n^{-1} \sum_{t=1}^{n} \delta(Y_t)$, where $n = N - m + 1$, and $\delta(y)$ denotes the point mass at $y \in R^m$. Note that $\rho_N$ depends on $m$, although the latter is suppressed for notation simplicity. Our interest is the distribution of a statistic that can be expressed as $T_N = T(p_N)$, where $T$ is a real-valued functional on the set of all probability measures on $R^m$. In order to bootstrap $T_N$, we select blocks of length $l$ at random. For simplicity, let $n = kl$, where $k$ is a positive integer. The bootstrap $m$-dimensional marginal distribution is

$$p_N^* = \frac{1}{n} \sum_{j=1}^{k} \sum_{t=I_j+1}^{I_j+l} \delta(Y_t), \qquad (14.69)$$

where $I_1, \ldots, I_k$ are i.i.d. and uniformly distributed over $0, 1, \ldots, n - l$. Note that (14.69) is a random (conditional) probability measure on $R^m$, and it has the following equivalent expressions (Exercise 14.24):

$$p_N^* = \frac{1}{n} \sum_{t=1}^{n} f_t \delta(Y_t), \qquad (14.70)$$

where $f_t$ is the number of $j$'s such that $t - l \leq I_j \leq t - 1$, and

$$p_N^* = \frac{1}{n} \sum_{t=1}^{n} \delta(Y_t^*), \qquad (14.71)$$

where the $k$ blocks $(Y_1^*, \ldots, Y_l^*), (Y_{l+1}^*, \ldots, Y_{2l}^*), \ldots, (Y_{n-l+1}^*, \ldots, Y_n^*)$ are i.i.d. with the distribution

$$p_{Y,n}^l = \frac{1}{n - l + 1} \sum_{t=0}^{n-l} \delta\{(Y_{t+1}, \ldots, Y_{t+l})\}.$$

We then form a bootstrap statistic $T_N^* = T(p_N^*)$. Some regularity conditions are imposed, as follows.

(A1) $T_N \xrightarrow{a.s.} T(F^m)$ as $N \to \infty$, where $F^m$ is the distribution of $(X_m, \ldots, X_1)$.

(A2) The influence function, defined as

$$\text{IF}(y, F^m) = \lim_{\epsilon \downarrow 0} \epsilon^{-1}[T\{(1-\epsilon)F^m + \epsilon\delta(y)\} - T(F^m)]$$

($\epsilon \downarrow 0$ means that $\epsilon$ approaches zero from the positive side), exists for all $y \in R^m$, so that $n^{-1/2}\sum_{t=1}^{n}\text{IF}(Y_t, F^m) \overset{d}{\longrightarrow} N(0, \sigma^2)$ as $N \to \infty$, where

$$\sigma^2 = \sum_{k=-\infty}^{\infty} \text{E}\{\text{IF}(Y_0, F^m)\text{IF}(Y_k, F^m)\} \qquad (14.72)$$

(Exercise 14.25).

(A3) The remaining term $R_N$ in the linearization

$$T(p_N) = T(F^m) + \frac{1}{n}\sum_{t=1}^{n}\text{IF}(Y_t, F^m) + R_N,$$

is of the order $o_P(n^{-1/2})$.

As indicated by (A3), the influence function allows one to linearize the statistical functional $T(p_N)$. The technique is useful, for example, in deriving the asymptotic distribution of the statistical functional.

To further illustrate the method, we focus below on the special case of bootstrapping—again, the sample mean. In this case, we have $m = 1$, $T(F) = \int xF(dx)$ (with $F = F^1$) (Exercise 14.26) and $\text{IF}(x, F) = x - \mu$, where $\mu = \text{E}(X_t)$. The statistic $T_N = N^{-1}\sum_{t=1}^{N} X_t$ and the bootstrap statistic can be expressed as $T_N^* = k^{-1}\sum_{j=1}^{k} U_{n,j}$, where the $U_{n,j}$'s are i.i.d. uniformly distributed among the points $u_i = l^{-1}(X_{i+1} + \cdots + X_{i+l})$, $i = 0, \ldots, n-l$. Thus, in particular, we have

$$\text{E}(T_N^*|X_1, \ldots, X_N) = \text{E}(U_{n,1})$$

$$= \frac{1}{l(n-l+1)}\sum_{i=0}^{n-l}\sum_{t=1}^{l} X_{i+t}. \qquad (14.73)$$

Furthermore, the bootstrap variance estimator, $\hat{\sigma}_{\text{B}}^2$, is defined as the conditional variance of $T_N^*$,

$$\text{var}(T_N^*|X_1, \ldots, X_N) = \frac{\text{var}(U_{n,1})}{k}$$

$$= \frac{1}{kl^2(n-l+1)}\sum_{i=0}^{n-l}\left[\sum_{t=1}^{l}\{X_{i+t} - \text{E}(U_{n,1})\}\right]^2 \qquad (14.74)$$

(Exercise 14.26). Künsch (1989) gave a theoretical justification of the block bootstrap in this case, under some additional conditions. The first two conditions may be intuitively interpreted as that the conditional first and second moments of $\sqrt{n}(T_N^* - T_N)$ agree asymptotically with those (unconditional moments) of $\sqrt{n}(T_N - \mu)$:

(B1) $\sqrt{n}\{E(T_N^*|X_1,\ldots,X_N) - T_N\} \xrightarrow{\text{a.s.}} 0$;

(B2) $n\hat{\sigma}_B^2 \xrightarrow{\text{a.s.}} \sigma^2$, as $N \to \infty$, where $\sigma^2$ is given in (A2) above.

The third condition is similar to the Lindeberg condition for the CLT for triangular arrays of independent random variables [see (6.36)]:

(B3) With probability 1, $\max_{0 \le i \le n-l} |\sum_{t=1}^{l}(X_{i+t} - \mu)| = o(\sqrt{n})$.

Suppose that (A1), (A2), and (B1)–(B3) hold, then we have

$$\sup_x |P(T_N^* - T_N \le x|X_1,\ldots,X_N) - P(T_N - \mu \le x)| \xrightarrow{\text{a.s.}} 0$$

as $N \to \infty$. Thus, at least in this case, the validity of the block bootstrap is justified in terms of Freedman's evidence (i) (see above). A remaining question is how to verify condition (B3). Künsch (1989) suggests an approach via strong approximation—namely, by showing that there is a Brownian motion (see Section 10.5) $B(t)$ such that, with probability 1,

$$\sum_{s \le t}(X_s - \mu) - B(t) = o\left(\sqrt{t}\right).$$

With such an approximation, (B3) can be established by using the properties of Brownian motion, provided that $l = O(n^\alpha)$ for some $\alpha < 1$. More specifically, Künsch showed that the following conditions are sufficient for (B3): $E(|X_t|^p) < \infty$ and $l = o(n^{1/2-1/p})$ some $p > 2$.

In addition, Künsch (1989) provided simulation results as Freedman's type (ii) evidence that supports the block bootstrap.

Bühlmann (2002) discussed on advantages and disadvantages of the block bootstrap and sieve bootstrap as they compare to each other. The advantages of the block bootstrap include that it is the most general method, so far, in bootstrapping a time series; its implementation of resampling is no more difficult than Efron's i.i.d. bootstrap. Note that, unlike the sieve bootstrap, the resampled data in the block bootstrap is a subset of the original data. In this regard, the procedure is more similar to Efron's than the sieve bootstrap. The disadvantages of the block bootstrap include that the resampled data are not viewed as a reasonable sample mimicking the data-generating process; it is not stationary and it exhibits artifacts where resampled blocks are linked together. The latter implies that the plug-in principle for bootstrapping an estimator is not appropriate. The advantages of the sieve bootstrap include that it resamples from a reasonable time series model; the plug-in rule is employed for defining and computing the bootstrapped estimator; and the method is easy to implement due to the simplicity of fitting an AR model. In fact, the sieve bootstrap discussed above is considered the best if the data-generating process is a linear time series representable as an AR($\infty$).

## 14.6 Bootstrapping mixed models

Mixed effects models (see Chapter 12) is another case of correlated data. Unlike the time series, it is usually inappropriate to make the assumption of stationarity for mixed models. Therefore, some of the approaches of the previous section no longer work. Nevertheless, the model-based or parametric bootstrap is still a useful strategy, in many cases effectively, for mixed models.

We begin with the problem of the linear mixed model prediction, discussed in most parts of Chapter 13. More specifically, we are interested in constructing highly accurate prediction intervals for mixed effects of interest. Consider the linear mixed model

$$y = X\beta + Zu + e, \qquad (14.75)$$

where $y$ is an $n \times 1$ vector of observations, $\beta$ is a $p \times 1$ vector of unknown parameters (the fixed effects), $u$ is an $m \times 1$ vector of random effects, and $e$ is an $n \times 1$ vector of errors. The matrices $X$ and $Z$ are observed or known; the random effects $u$ and errors $e$ are assumed to have means 0 and covariance matrices $G$ and $R$, respectively; and $u$ and $e$ are uncorrelated. Our problem of interest is the prediction of a linear mixed effect, expressed as $\zeta = c'(X\beta + Zu)$ for some known constant vector $c$. Such mixed effects include the small-area means, say, under a Fay–Herriot model (see Section 13.3) and genetic merits of breeding animals (e.g., Jiang 2007, Section 2.6). Under the normality assumption—that is, $u \sim N(0, G)$, $e \sim N(0, R)$, and $u$ and $e$ are uncorrelated—the conditional distribution of $\zeta$ given the data $y$ is $N(\tilde{\mu}, \tilde{\sigma}^2)$, where

$$\tilde{\mu} = c'\{X\beta + ZGZ'V^{-1}(y - X\beta)\}, \qquad (14.76)$$

$$\tilde{\sigma}^2 = c'Z(G - GZ'V^{-1}ZG)Z'c, \qquad (14.77)$$

and $V = \mathrm{Var}(y) = R + ZGZ'$. Therefore, for any $\alpha \in (0, 1)$, we have

$$P\left(\left|\frac{\zeta - \tilde{\mu}}{\tilde{\sigma}}\right| \le z_{\alpha/2} \,\middle|\, y\right) = 1 - \alpha,$$

where $z_\alpha$ is the $\alpha$-critical value of $N(0, 1)$ [i.e., $\Phi(\alpha) = 1 - \alpha$, where $\Phi$ is the cdf of $N(0, 1)$]. It follows that the probability is $1 - \alpha$ that the interval $I = [\tilde{\mu} - z_{\alpha/2}\tilde{\sigma}, \tilde{\mu} + z_{\alpha/2}\tilde{\sigma}]$ covers $\zeta$ (why?). Of course, the latter is not a practical solution for the prediction interval, because there are unknown parameters involved. In addition to $\beta$, the covariance matrices $G$ and $R$ typically also depend on some unknown dispersion parameters, or variance components. Suppose that $G = G(\psi)$ and $R = R(\psi)$, where $\psi$ is a $q$-dimensional vector of variance components, and $\theta = (\beta', \psi')'$. It is customary to replace $\theta$ by an estimator, $\hat{\theta} = (\hat{\beta}', \hat{\psi}')'$. However, with the replacement of $\theta$ by $\hat{\theta}$, the resulting (prediction) interval no longer has the coverage probability $1 - \alpha$.

To illustrate the problem more explicitly, let us consider prediction under the Fay–Herriot model (see Section 13.3). The model may be thought of equivalently as having two levels. In Level 1, we have, conditional on $\zeta_1, \ldots, \zeta_n$, that

$y_1, \ldots, y_n$ are independent such that $y_i \sim N(\zeta_i, D_i)$, where $D_i > 0$ are known. In Level 2, we have $\zeta_i = x_i'\beta + u_i$, $1 \leq i \leq n$, where the $x_i$'s are observed vectors of covariates, $\beta$ is a $p \times 1$ vector of unknown regression coefficients, the $u_i$'s are small-area-specific random effects that are independent and distributed as $N(0, A)$, and $A > 0$ is an unknown variance. A prediction interval for $\zeta_i$ based on the Level 1 model only is given by $I_{1,i} = [y_i - z_{\alpha/2}\sqrt{D_i}, y_i + z_{\alpha/2}\sqrt{D_i}]$. This interval has the right coverage probability $1 - \alpha$ but is hardly useful, because its length is too large due to the high variability of the predictor based on a single observation $y_i$. Alternatively, one may construct a prediction interval based on the Level 2 model only. We consider a special case.

*Example 14.10.* Consider the special case of the Fay–Herriot model with $D_i = 1$ and $x_i'\beta = \beta$. At Level 2, the $\zeta_i$'s are i.i.d. with the distribution $N(\beta, A)$. If $\beta$ and $A$ were known, a Level 2 prediction interval that has the coverage probability $1 - \alpha$ would be $I_2 = [\beta - z_{\alpha/2}\sqrt{A}, \beta + z_{\alpha/2}\sqrt{A}]$. Note that the interval is not area-specific (i.e., not dependent on $i$). Bacause $\beta$ and $A$ are unknown, we replace them by estimators. The standard estimators are $\hat{\beta} = \bar{y} = n^{-1}\sum_{i=1}^{n} y_i$ and $\hat{A} = \max(0, s^2 - 1)$, where $s^2 = (n-1)^{-1}\sum_{i=1}^{n}(y_i - \bar{y})^2$. This leads to the interval $\tilde{I}_2 = [\hat{\beta} - z_{\alpha/2}\sqrt{\hat{A}}, \hat{\beta} + z_{\alpha/2}\sqrt{\hat{A}}]$, but, again, the coverage probability is no longer $1 - \alpha$. However, because $\hat{\beta}$ and $\hat{A}$ are consistent estimators, the coverage probability is $1 - \alpha + o(1)$ (Exercise 14.27). We can do better than this with a simple Level 2 bootstrap: First, draw independent samples $\zeta_1^*, \ldots, \zeta_n^*$ from $N(\hat{\beta}, \hat{A})$; then draw $y_i^*, i = 1, \ldots, n$, independently such that $y_i^* \sim N(\zeta_i^*, 1)$. Compute $\hat{\beta}^*$ and $\hat{A}^*$ the same way as $\hat{\beta}$ and $\hat{A}$ except using the $y_i^*$'s; then determine the cutoff points $t_1$ and $t_2$ such that

$$P^* \left( \hat{\beta}^* - t_1\sqrt{\hat{A}^*} \leq \zeta_i^* \leq \hat{\beta}^* + t_2\sqrt{\hat{A}^*} \right) = 1 - \alpha,$$

where $P^*$ denotes the bootstrap probability, evaluated using a large number of bootstrap replications. Note that, at least approximately, $t_1$ and $t_2$ do not depend on $i$ (why?). We then define a (Level 2) bootstrap prediction interval as $\hat{I}_2 = [\hat{\beta} - t_1\sqrt{\hat{A}}, \hat{\beta} + t_2\sqrt{\hat{A}}]$. It can be shown that $\hat{I}_2$ has a coverage probability of $1 - \alpha + o(n^{-1/2})$. It is an improvement but still not accurate enough.

Intuitively, one should do better by combining both levels of the model. In fact, Cox (1975) proposed the following empirical Bayes interval for $\zeta_i$ using information of both levels:

$$I_i^C : \quad (1 - \hat{B}_i)y_i + \hat{B}_i x_i'\hat{\beta} \pm z_{\alpha/2}\sqrt{D_i(1 - \hat{B}_i)}, \quad (14.78)$$

where $\hat{B}_i$ and $\hat{\beta}$ are (consistent) estimators of $B_i = D_i/(A + D_i)$ and $\beta$, respectively. It can be seen that the center of the interval (14.78) is a weighted average of the centers of the Level 1 and Level 2 intervals. If $D_i$ is much larger than $A$, then $1 - \hat{B}_i$ is expected to be close to zero. In this case, the center of $I_i^C$ is close to the Level 2 center, but the length will be close to zero and, in particular, much smaller than that of the Level 1 interval (which is $\sqrt{D_i}$). On the other hand, if $A$ is much larger than $D_i$, then the center of $I_i^C$ is expected to be close to the Level 1 center, and the length is close to $\sqrt{D_i}$, much smaller than that of the Level 2 interval (which would be $\sqrt{A}$ if $A$ were known). Thus, the empirical Bayes interval does seem to have the combined strength of both levels. It can be shown that, under regularity conditions, the coverage probability of $I_i^C$ is $1 - \alpha + O(n^{-1})$. Can one do better than $I_i^C$? In the following, we consider a parametric bootstrap procedure, proposed and studied by Chatterjee et al. (2008), that leads to a prediction interval with coverage probability $1-\alpha+O(n^{-3/2})$, provided that the number of parameters under model (14.75) is bounded.

Let $d = p+q$, the total number of fixed parameters under model (14.75). It is not assumed that $d$ is fixed or bounded; so it may increase with the sample size $n$. This is practical in many applications, where the number of fixed parameters may be comparable to the sample size. See, for example, Section 12.2. According to the discussion below (14.77), the key is to approximate the distribution of $T = (\zeta - \hat{\mu})/\hat{\sigma}$, where $\hat{\mu}$ and $\hat{\sigma}$ are estimators of $\tilde{\mu}$ and $\tilde{\sigma}$, respectively, defined by (14.76) and (14.77). For simplicity, assume that $X$ is of full rank and the estimator $\hat{\beta} = (X'X)^{-1}X'y$, which is the ordinary least squares estimator of $\beta$. Consider a bootstrap version of (14.75),

$$y^* = X\hat{\beta} + Zu^* + e^*, \qquad (14.79)$$

where $u^* \sim N(0, \hat{G})$, $e^* \sim N(0, \hat{R})$ with $\hat{G} = G(\hat{\psi})$ and $\hat{R} = R(\hat{\psi})$, and $u^*$ and $e^*$ are independent. Here, $\hat{\psi}$ may be chosen as the REML estimator (see Section 12.2). From (14.79) we generate $y^*$ and then obtain $\hat{\beta}^*$ and $\hat{\psi}^*$ the same way as $\hat{\beta}$ and $\hat{\psi}$, except using $y^*$. Next, obtain $\hat{\mu}^*$ and $\hat{\sigma}^*$ using $\hat{\beta}^*$ and $\hat{\psi}^*$, (14.76) and (14.77). Define $\zeta^* = c'(X\hat{\beta} + Zu^*)$. The distribution of $T^* = (\zeta^* - \hat{\mu}^*)/\hat{\sigma}^*$ conditional on $y$ is the parametric bootstrap approximation to the distribution of $T$. Under some regularity conditions, Chatterjee et al. (2008) showed that

$$\sup_x |\mathrm{P}^*(T^* \leq x) - \mathrm{P}(T \leq x)| = O_\mathrm{P}(d^3 n^{-3/2}), \qquad (14.80)$$

where $\mathrm{P}^*(T^* \leq x) = \mathrm{P}(T^* \leq x|y) \equiv F^*(x)$. As a consequence, if $d^2/n \to 0$ and for any $\alpha \in (0,1)$, let $q_1$ and $q_2$ be the real numbers such that

$$F^*(q_2) - F^*(q_1) = 1 - \alpha,$$

then we have (Exercise 14.28)

$$\mathrm{P}(\hat{\mu} + q_1\hat{\sigma} \leq \zeta \leq \hat{\mu} + q_2\hat{\sigma}) = 1 - \alpha + O(d^3 n^{-3/2}). \qquad (14.81)$$

Going back to the special case of the Fay–Herriot model, consider the problem of constructing prediction intervals for the small-area means $\zeta_i$. Suppose that the maximum of the diagonal elements of $P_X = X(X'X)^{-1}X'$ is

$O(p/n)$ (make sense?) and the $D_i$'s are bounded from above as well as away from zero. Then, under some regularity conditions on $\hat{A}$, we have

$$P\left\{\zeta_i \in \left[\hat{\zeta}_i + q_{i1}\sqrt{D_i(1-\hat{B}_i)},\ \hat{\zeta}_i + q_{i2}\sqrt{D_i(1-\hat{B}_i)}\right]\right\}$$
$$= 1 - \alpha + O(p^3 n^{-3/2}), \tag{14.82}$$

where $\hat{\zeta}_i = (1-\hat{B}_i)y_i + \hat{B}_i x_i'\hat{\beta}$, $\hat{B}_i = D_i/(\hat{A}+D_i)$, and $q_{i1}$ and $q_{i2}$ satisfy

$$P^*\left\{\zeta_i^* \in \left[\hat{\zeta}_i^* + q_{i1}\sqrt{D_i(1-\hat{B}_i^*)},\ \hat{\zeta}_i^* + q_{i2}\sqrt{D_i(1-\hat{B}_i^*)}\right]\right\}$$
$$= 1 - \alpha + O_P(p^3 n^{-3/2}) \tag{14.83}$$

with $\hat{\zeta}_i^* = (1-\hat{B}_i^*)y_i^* + \hat{B}_i^* x_i'\hat{\beta}^*$ and $\hat{B}_i^* = D_i/(\hat{A}^*+D_i)$. For example, the term $O_P(p^3 n^{-3/2})$ in (14.83) may be taken as 0 in order to determine $q_{i1}$ and $q_{i2}$. Thus, in particular, if $p^3/\sqrt{n} \to 0$, the prediction interval in (14.82) is asymptotically more accurate than $I_i^C$ of (14.78). Regarding finite-sample performance, Chatterjee et al. (2008) carried out a simulation study under the special case with $m = 15$ and $x_i'\beta = 0$. The 15 small areas are divided into 5 groups with 3 areas in each group. The $D_i$'s are the same within each group, but vary between groups according to the one of the following patterns: (a) 0.2, 0.4, 0.5, 0.6, 4.0 or (b) 0.4, 0.8, 1.0, 1.2, 8.0. Pattern (a) was considered by Datta et al. (2005; see Section 13.3). Pattern (b) is simply pattern (a) multiplied by 2. The true value of $A$ is 1 for pattern (a), and $A = 2$ for pattern (b). Note that the choice of $q_{i1}$ and $q_{i2}$ that satisfy (14.83) is not unique. Two most commonly used choices are (1) equal-tail, in which $q_{i1}$ and $q_{i2}$ are chosen such that the tail probabilities on both sides are $\alpha/2$ and (2) shortest-length, in which $q_{i1}$ and $q_{i2}$ are chosen to minimize the length of the prediction interval (while maintaining the coverage probability $1 - \alpha$). The parametric bootstrap (PB) prediction intervals corresponding to (1) and (2) are denoted by PB-1 and PB-2, respectively. The number of bootstrap replications used to evaluate (14.83) is 1000. In the PB procedure, the unknown variance $A$ is estimated by the Fay–Herriot method [F-H; see Section 13.3, in particular, (13.33)], and $\beta$ is estimated by the EBLUE, given below (13.31) with $A$ replaced by its F-H estimator. The PB intervals are compared with three other competitors. The first is the Cox empirical Bayes interval (Cox) (14.78), with the Prasad–Rao (P-R) estimator of $A$ (13.32). The second (PR) and third (FH) are both in the form of EBLUP $\pm$ 1.96$\sqrt{\widehat{\text{MSPE}}}$. In the second case, $\widehat{\text{MSPE}}$ is the Prasad–Rao MSPE estimator (13.45) with the P-R estimator of $A$; in the third case, $\widehat{\text{MSPE}}$ is the MSPE estimator proposed by Datta et al. (2005) [i.e., (13.47)] with the F-H estimator of $A$. The results, based on 10,000 simulation runs and reported by Chatterjee et al. (2008) in their Table 1 and Table 2, are summarized in Table 14.2, where A1–A5 correspond to the five cases of pattern (a) and B1–B5 correspond to those of pattern (b). The numbers in the main body of the table are empirical coverage probabilities, in terms of percentages, and lengths of the prediction intervals (in parentheses). The nominal coverage probability is 0.95. It is seen that, with the exception of one case, the PB methods outperform their competitors (the way to look at the table is to first consider coverage probabilities; if the latter are similar, then compare the corresponding lengths). The differences are more significant for pattern (b), where all of the competing methods have relatively poor coverage probabilities, whereas the PB intervals appear to be very accurate. These results are consistent with the theoretical findings discussed above.

**Table 14.2. Coverage probabilities and lengths of prediction intervals**

| Pattern | Cox | PR | FH | PB-1 | PB-2 |
|---------|-----|-----|-----|------|------|
| A1 | 83.1 (3.12) | 92.4 (3.82) | 90.4 (3.57) | 96.1 (4.50) | 95.7 (4.42) |
| A2 | 85.4 (2.14) | 98.0 (3.19) | 93.7 (2.50) | 96.2 (2.83) | 95.9 (2.79) |
| A3 | 85.8 (2.02) | 98.0 (3.08) | 93.9 (2.36) | 96.0 (2.65) | 95.6 (2.61) |
| A4 | 86.1 (1.89) | 98.2 (2.93) | 94.3 (2.19) | 96.1 (2.43) | 95.7 (2.39) |
| A5 | 89.7 (1.12) | 97.3 (1.87) | 95.2 (1.23) | 95.7 (1.28) | 95.3 (1.26) |
| B1 | 85.5 (4.87) | 89.3 (5.35) | 89.5 (5.18) | 95.7 (6.55) | 95.4 (6.47) |
| B2 | 83.6 (2.68) | 87.3 (2.93) | 86.0 (2.82) | 95.2 (3.90) | 94.9 (3.75) |
| B3 | 83.4 (2.49) | 86.8 (2.71) | 85.7 (2.60) | 95.2 (3.53) | 94.9 (3.49) |
| B4 | 82.9 (2.27) | 86.2 (2.46) | 85.0 (2.36) | 95.0 (3.22) | 94.5 (3.18) |
| B5 | 83.0 (1.21) | 84.8 (1.29) | 84.0 (1.23) | 94.9 (1.72) | 94.6 (1.70) |

Our next example of bootstrapping mixed models involves the fence method for mixed model selection, discussed earlier in Section 12.5. Recall that the fence is built via the inequality

$$d_M \leq c_1 \widehat{\text{s.d.}}(d_M), \tag{14.84}$$

where $d_M = \hat{Q}(M) - \hat{Q}(M_f)$ and $M_f$ denotes the full model. An adaptive fence procedure is described near the end of Section 12.5, which involves bootstrap (in order to evaluate $p^*$). The bootstrap is typically done parametrically in a similar way as Chatterjee et al. (2008). Jiang et al. (2009) proposed a simplified version of the adaptive fence. The motivation is that, in most cases, the calculation of $\hat{Q}(M)$ is fairly straightforward, but the evaluation of $\widehat{\text{s.d.}}(d_M)$ can be quite challenging. Sometimes, even if an expression can be obtained for $\widehat{\text{s.d.}}(d_M)$, its accuracy as an estimate of the standard deviation cannot be guaranteed in a finite-sample situation. In the simplified version, the estimator $\widehat{\text{s.d.}}(d_M)$ is absorbed into $c_1$, which is then chosen adaptively. In other words, one considers the fence inequality (12.55), where $c$ is chosen adaptively in the same way as described near the end of Section 12.5. Under suitable regularity conditions, the authors showed consistency of the simplified adaptive fence, as in Jiang et al. (2008) for the adaptive fence. For the most part, the result states the following: (i) There is a $c^*$ that is at least a local maximum of $p^*$ and an approximate global maximum in the sense that the $p^*$ at $c^*$ goes to

1 in probability and (ii) the probability that $M_0^* = M_{\mathrm{opt}}$, where $M_0^*$ is the model selected by the fence (12.55) with $c = c^*$ and $M_{\mathrm{opt}}$ is the optimal model [see (12.65)], goes to 1 as both the sample size and the number of bootstrap replications increase.

We demonstrate the consistency empirically through a simulation study. Jiang et al. (2009) used a simulation design that mimic the well-known Iowa crops data (Battese et al. 1988). The original data were obtained from 12 Iowa, USA, counties in the 1978 June Enumerative Survey of the U.S. Department of Agriculture as well as from land observatory satellites on crop areas involving corn and soybeans. The objective was to predict mean hectares of corn and soybeans per segment for the 12 counties using the satellite information. Battese et al. (1988) proposed the following nested error regression model:

$$y_{ij} = x'_{ij}\beta + v_i + e_{ij}, \quad i = 1, \ldots, m, j = 1, \ldots, n_i, \tag{14.85}$$

where $i$ represents county and $j$ represents the segment within the county; $y_{ij}$ is the number of hectares of corn (or soybeans), $v_i$ is a small-area-specific random effect, and $e_{ij}$ is the sampling error. It is assumed that the random effects are independent and distributed as $N(0, \sigma_v^2)$, the sampling errors are independent and distributed as $N(0, \sigma_e^2)$, and the random effects and sampling errors are uncorrelated. For the Iowa crops data, $m = 12$ and the $n_i$'s ranged from 1 to 6. Battese et al. (1988) used $x'_{ij}\beta = \beta_0 + \beta_1 x_{ij1} + \beta_2 x_{ij2}$, where where $x_{ij1}$ and $x_{ij2}$ are the number of pixels classified as corn and soybeans, respectively, according to the satellite data. The authors did discuss, however, various model selection problems associated with the nested error regression, such as whether or not to include quadratic terms in the model. The latter had motivated a model selection problem described below.

In the simulation study, the number of clusters, $m$, is either 10 or 15, setting up a situation of increasing sample size. Note that the $n_i$'s are typically small and therefore not expected to increase. Here, the $n_i$'s are generated from a Poisson(3) distribution and fixed throughout the simulations. The random effects, $v_i$, and errors $e_{ij}$, are both generated independently from the $N(0,1)$ distribution. The components of the covariates, $x_{ij}$, are to be selected from $x_{ijk}$, $k = 0, 1, \ldots, 5$, where $x_{ij0} = 1$; $x_{ij1}$ and $x_{ij2}$ are generated independently from $N(0,1)$ and then fixed throughout; $x_{ij3} = x_{ij1}^2$, $x_{ij4} = x_{ij2}^2$, and $x_{ij5} = x_{ij1}x_{ij2}$. The simulated data are generated under two models:

Model I. The model that involves the linear terms only; that is, $x_{kij}$, $k = 0, 1, 2$, with all of the regression coefficients equal to 1;

Model II. The model that involves both the linear and the quadratic terms; that is, $x_{kij}$, $k = 0, \ldots, 5$, with all of the regression coefficients equal to 1.

The measure $Q(M)$ is chosen as the negative log-likelihood function. The number of bootstrap replications is 100. Results based on 100 simulation runs are reported in Table 14.3, which shows empirical probabilities (%) of selection of the optimal model (i.e., the model from which the data is generated) for the simplified adaptive fence. The results show that even in these cases of fairly small sample size, the performance of the simplified adaptive fence is quite satisfactory. Note the improvement of the results when $m$ increases, which is in line with the consistency result.

**Table 14.3. Simplified adaptive fence: simulation results**

| Optimal Model | # of Clusters, $m$ | Empirical Probability (in %) |
|:---:|:---:|:---:|
| Model I | 10 | 82 |
| Model I | 15 | 99 |
| Model II | 10 | 98 |
| Model II | 15 | 100 |

So far, we have been talking about the parametric bootstrap. A question of interest is how to bootstrap nonparametrically in mixed model situations. It seems obvious that Efron's i.i.d. bootstrap cannot be applied directly to the data, neither do the strategies such as sieve or block bootstraps, discussed in the previous section, which require stationarity. In fact, under a mixed model, linear or generalized linear, the vector $y$ of all of the observations is viewed as a single $n$-dimensional observation and there is no other replication or stationary copy of this observation (in other words, one has an i.i.d. sample of size 1). On the other hand, there are some i.i.d. random variables "inside" the model, at least for most mixed models that are practically used. For example, in most cases, the random effects are assumed to be i.i.d. However, one cannot bootstrap the random effects directly, because they are not observed—and this is a major difference from the i.i.d. situation. Hall and Maiti (2006) had a clever idea on how to bootstrap the random effects nonparametrically, at least in some cases.

Consider, once again, the nested error regression model (14.85). Suppose that the random effects are i.i.d. but not normal, or at least we do not know if they are normal. Then, what can we do in order to generate the random effects? The parametric bootstrap discussed above usually requires normality, or at least a parametric distribution of the random effects, so this strategy encounters a problem. On the other hand, one cannot use Efron's (nonparametric) bootstrap because the random effects are not observed, as mentioned. The answer by Hall and Maiti: Depending on what you want. In many cases, the quantity of interest does not involve every piece of information about the distribution of the random effects. For example, Hall and Maiti observed that the MSPE of EBLUP (e.g., Section 4.8; Chapter 13) involves only the second and fourth moments of the random effects and errors, up to the order of $o(m^{-1})$. This means that for random effects and errors from any distributions with the same second and fourth moments, the MSPE of EBLUP, with some suitable estimators of the variance components, are different only by a term of $o(m^{-1})$ (Exercise 14.29). This observation leads to a seemingly simple strategy: First, estimate the second and fourth moments of the random effects and errors; then draw bootstrap samples of the random effects and errors from distributions that match the first (which is 0) and estimated second and fourth moments; given the bootstrap random effects and errors, use (14.85) to generate bootstrap data, and so on. The hope is that this leads to an MSPE estimator whose bias is $o(m^{-1})$ (i.e., second-order unbiased).

To be more precise, for any $\mu_2, \mu_4 \geq 0$ such that $\mu_2^2 \leq \mu_4$, let $D(\mu_2, \mu_4)$ denote the distribution of a random variable $\xi$ such that $E(\xi) = 0$ and $E(\xi^j) = \mu_j, j = 2, 4$. Hall and Maiti (2006) suggested using moment estimators, $\hat{\sigma}_v^2$, $\hat{\sigma}_e^2$, $\hat{\gamma}_v$, and $\hat{\gamma}_e$, of $\sigma_v^2$, $\sigma_e^2$, $\gamma_v$, and $\gamma_e$, where $\gamma_v, \gamma_e$ are the fourth moments of $v_i$ and $e_{ij}$, respectively, so that they satisfy the constraints $\hat{\sigma}_v^4 \leq \hat{\gamma}_v, \hat{\sigma}_e^4 \leq \hat{\gamma}_e$ (why do we need these constraints?). Suppose that we are interested in the prediction of the mixed effects

$$\zeta_i = X_i'\beta + u_i, \tag{14.86}$$

where $X_i$ is a known vector of covariates, such as the population mean of the $x_{ij}$'s. Denote the EBLUP of $\zeta_i$ by $\hat{\zeta}_i$. To obtain a bootstrap estimator of the MSPE of $\hat{\zeta}_i$, $MSPE_i = E(\hat{\zeta}_i - \zeta_i)^2$, we draw samples $v_i^*, i = 1, \ldots, m$, independently from $D(\hat{\sigma}_v^2, \hat{\gamma}_v)$, and $e_{ij}^*, i = 1, \ldots, m, j = 1, \ldots, n_i$, independently, from $D(\hat{\sigma}_e^2, \hat{\gamma}_e)$. Then, mimicking (14.85), define

$$y_{ij}^* = x_{ij}'\hat{\beta} + v_i^* + e_{ij}^*, \quad i = 1, \ldots, m, j = 1, \ldots, n_i,$$

where $\hat{\beta}$ is the EBLUE of $\beta$ [i.e., (4.88) with $\sigma_v^2$ and $\sigma_e^2$ (in $V$) replaced by their estimators]. Let $\hat{\zeta}_i^*$ be the bootstrap version of $\hat{\zeta}_i$, obtained the same way except using $y^*$ instead of $y$. We then define a bootstrap MSPE estimator

$$\widehat{MSPE}_i = E\{(\hat{\zeta}_i^* - \zeta_i^*)^2 | y\}, \tag{14.87}$$

where $\zeta_i^* = X_i'\hat{\beta} + v_i^*$ [see (14.86)], and the conditional expectation is evaluated, as usual, by replications of $y^*$. Euation (14.87) produces an MSPE estimator whose bias is $O(m^{-1})$, not $o(m^{-1})$ (see Hall and Maiti 2006, Section 4).

To obtain an MSPE estimator whose bias is $o(m^{-1})$, we bias-correct $\widehat{MSPE}_i$ using a *double bootstrap* as follows. The first bootstrap is done above. Conditional on $v^*$ and $e^*$, draw samples $v_i^{**}, i = 1, \ldots, m$ and $e_{ij}^{**}, i = 1, \ldots, m, j = 1, \ldots, n_i$ independently from $D(\hat{\sigma}_v^{*2}, \hat{\gamma}_v^*)$ and $D(\hat{\sigma}_e^{*2}, \hat{\gamma}_e^*)$, respectively, where $\hat{\sigma}_v^{*2}$, and so on are computed the same way as $\hat{\sigma}_v^2$, and so on. except using $y^*$ instead of $y$. Then, similarly, define

$$y_{ij}^{**} = x_{ij}'\hat{\beta}^* + v_i^{**} + e_{ij}^{**}, \quad i = 1, \ldots, m, j = 1, \ldots, n_i,$$

and compute the double-bootstrap version of $\hat{\zeta}_i$ and $\hat{\zeta}_i^{**}$. After that, compute

$$\widehat{MSPE}_i^* = E\{(\hat{\zeta}_i^{**} - \zeta_i^{**})^2 | y^*\}, \tag{14.88}$$

where $\zeta_i^{**} = X_i'\hat{\beta}^* + v_i^{**}$ and the conditional expectation is evaluated by replications of $y^{**}$. Equation (14.88) is the bootstrap analogue of (14.87) [the way to understand this is to forget that (14.87) is computed via bootstrap; then (14.88) is just a bootstrap analogue of (14.87)]. According to one of the classical usages of the bootstrap, we estimate the bias of $\widehat{\mathrm{MSPE}}_i$ by

$$\widehat{\mathrm{bias}}_i = \mathrm{E}(\widehat{\mathrm{MSPE}}_i^* | y) - \widehat{\mathrm{MSPE}}_i,$$

where the conditional expectation is evaluated by replications of $y^*$ [note that, after taking the conditional expectation in (14.88), $\widehat{\mathrm{MSPE}}_i^*$ is a function of $y^*$]. This leads to a bias-corrected MSPE estimator

$$\widehat{\mathrm{MSPE}}_i^{\mathrm{bc}} = \widehat{\mathrm{MSPE}}_i - \widehat{\mathrm{bias}}_i = 2\widehat{\mathrm{MSPE}}_i - \mathrm{E}(\widehat{\mathrm{MSPE}}_i^* | y). \qquad (14.89)$$

Hall and Miati (2006) showed that, under regularity conditions, we have

$$\mathrm{E}(\widehat{\mathrm{MSPE}}_i^{\mathrm{bc}}) = \mathrm{MSPE}_i + o(m^{-1}), \qquad (14.90)$$

so the bootstrap bias-corrected MSPE estimator is second-order unbiased.

Regarding the distribution $D(\mu_2, \mu_4)$, clearly, the choice is not unique. It remains a question of what is the optimal choice of such a distribution. The simplest example is, perhaps, constructed from a three-point distribution depending on a single parameter $p \in (0, 1)$, defined as $\mathrm{P}(\xi = 0) = 1 - p$ and

$$\mathrm{P}\left(\xi = \frac{1}{\sqrt{p}}\right) = \mathrm{P}\left(\xi = -\frac{1}{\sqrt{p}}\right) = \frac{p}{2}.$$

It is easy to verify that $\mathrm{E}(\xi) = 0, \mathrm{E}(\xi^2) = 1$, and $\mathrm{E}(\xi^4) = 1/p$. Thus, if we let $p = \mu_2^2/\mu_4$, the distribution of $\sqrt{\mu_2}\xi$ is $D(\mu_2, \mu_4)$. Note that the inequality $\mu_2^2 \leq \mu_4$ always hold with the equality if and only if $\xi^2$ is degenerate (i.e., a.s. a constant). Another possibility is the rescaled Student $t$-distribution whose degrees of freedom $\nu > 4$ and is not necessarily an integer. The distribution has first and third moments 0, $\mu_2 = 1$, and $\mu_4 = 3(\nu - 2)/(\nu - 4)$, which is always greater than 3, meaning that the tails are heavier than those of the normal distribution.

The moment-matching, double-bootstrap procedure may be computationally intensive, and so far there have not been published comparisons of the method with other exsiting methods that also produce second-order unbiased MSPE estimators, such as the Prasad–Rao (see Section 4.8) and jackknife (see Section 14.3). The latter is also considered a resampling method. On the other hand, the idea of Hall and Maiti is potentially applicable to mixed models with more complicated covariance structure, such as mixed models with crossed random effects. We conclude this section with an illustrative example.

*Example 14.11.* Consider a two-way random effects model

$$y_{ij} = \mu + u_i + v_j + e_{ij},$$

$i = 1, \ldots, m_1$, $j = 1, \ldots, m_2$, where $\mu$ is an unknown mean; the $u_i$'s are i.i.d. with mean 0, variance $\sigma_1^2$, and an unknown distribution $F_1$; the $v_j$'s are i.i.d. with mean 0, variance $\sigma_2^2$, and an unknown distribution $F_2$; the $e_{ij}$'s are i.i.d. with mean 0, variance $\sigma_0^2$, and an unknown distribution $F_0$; and $u$, $v$, and $e$ are independent. Note that the observations are *not clustered* under this model. As a result, the jackknife method of Jiang et al. (2002; see Section 14.3) may not apply. Consider a mixed effect that can be expressed as $\zeta - a_0\mu + a_1'u + a_2'v$. The BLUP of $\zeta$ can be expressed as $\tilde{\zeta} = a_0 y_{..} + a_1'\tilde{u} + a_2'\tilde{v}$, where $y_{..} - (m_1 m_2)^{-1} \sum_{i=1}^{m_1} \sum_{j=1}^{m_2} y_{ij}$ and $\tilde{u}$ and $\tilde{v}$ are the BLUPs of $u$ and $v$, respectively (e.g., Section 4.8). According to the general result of Das et al. (2004; see Theorem 3.1 therein), the MSPE of the EBLUP of $\zeta$ can be expressed as

$$\text{MSPE} = g_1(\theta) + g_2(\theta) + g_3 + o(m_*^{-1}),$$

where $\theta = (\sigma_0^2, \sigma_1^2, \sigma_2^2)'$; $g_1(\theta) = a'(G - GZ'V^{-1}ZG)a$ with $a = (a_1', a_2')'$, $G = \text{diag}(\sigma_1^2 I_{m_1}, \sigma_2^2 I_{m_2})$, $Z = (I_{m_1} \otimes 1_{m_2}\ 1_{m_1} \otimes I_{m_2})$, and $V = \sigma_0^2 I_{m_1 m_2} + ZGZ'$; $g_2(\theta) = \{a_0 - 1_{m_1 m_2}'V^{-1}ZGa\}^2/(1_{m_1 m_2}'V^{-1}1_{m_1 m_2})$ (you may simplify these expressions considerably in this simple case; see Exercise 14.30); $g_3 = \text{E}(h'A^{-1}b)^2$ with $h = \partial\tilde{\zeta}/\partial\theta$, $A = \text{E}(\partial^2 l/\partial\theta\partial\theta')$, $l$ being the Gaussian restricted log-likelihood function, $b = \partial l/\partial\theta$; and $m_* = m_1 \wedge m_2$, provided that the Gaussian REML estimator of $\theta$ is used for the EBLUP (see Section 12.2). Note that $g_3$ is not necessarily a function of $\theta$ unless the data are normal [and this is why the notation $g_3$, not $g_3(\theta)$, is used]. However, it is anticipated that $g_3$ can be expressed as $s + o(m_*^{-1})$, where $s$ is a term that depends only on the fourth moments of $u$, $v$, and $e$, in addition to $\theta$. If this conjecture turns out to be true (Exercise 14.30), then it is understandable that a similar idea of the moment-matching bootstrap should apply to this case as well.

## 14.7 Exercises

*14.1.* Let $X_1, \ldots, X_n$ be an i.i.d. sample from a distribution with cdf $F$ and pdf $f$, and $X_{(1)} < \cdots < X_{(n)}$ be the order statistics. Then the cdf of $X_{(i)}$ is given by

$$G_{(i)}(x) = \sum_{k=i}^{n} \binom{n}{k}\{F(x)\}^k\{1 - F(x)\}^{n-k},$$

and the pdf of $X_i$ is given by

$$g_{(i)}(x) = \frac{n!}{(i-1)!(n-i)!}\{F(x)\}^{i-1}\{1 - F(x)\}^{n-i}f(x).$$

Using these results, derive the cdf and pdf of $R$ in Example 14.2, assuming $n = 2m - 1$ for simplicity.

*14.2.* Continue with the previous exercise. Suppose that $X_i$ has the Uniform$[0,1]$ distribution. Show that $X_{(i)} \sim \text{Beta}(i, n-i+1)$, $1 \leq i \leq n$, and therefore obtain the mean and variance of $R$ for $n = 2m - 1$.

*14.3.* Consider the example of sample mean discussed below (14.1). Showed that, in this case, the right side of (14.8) is equal to $s^2/n$, where $s^2 = (n-1)^{-1} \sum_{i=1}^{n} (X_i - \bar{X})^2$ is the sample variance. Therefore, we have $E\{\widehat{\text{var}}_J(\hat{\theta})\} = \text{var}(\hat{\theta})$; in other words, the jackknife variance estimator is unbiased.

*14.4.* Verify the representation (14.13) and show that $\hat{\beta}_{ij}$ is the OLS estimator of $\beta$ based on the following regression model:

$$y_i = \alpha + \beta x_i + e_i,$$
$$y_j = \alpha + \beta x_j + e_j.$$

*14.5.* Show that when $d = 1$, (14.12) reduces to (14.16), which is the weighted delete-1 jackknife estimator of $\text{Var}(\hat{\beta})$.

*14.6.* Regarding the outline of the proof of Theorem 14.1 (near the end of Section 14.2), show the following:

$$\binom{n-p}{d-1}^{-1} \binom{n-1}{d-1} = O\left(\frac{d}{n}\right);$$
$$\binom{n-p}{d-1}^{-1} \left\{ \binom{n}{d} - \binom{n-p}{d} \right\} = O(1).$$

*14.7.* This exercise involves some further details regarding the outline of the proof of Theorem 14.1.

(i) Show that $w_s \leq 1$, $s \in S_r$.

(ii) Use the technique of unspecified $c$ (see Section 3.5) to show that $\text{tr}(S_1) = O(dh/n)$, hence $\|S_1\| = O(dh/n)$.

*14.8.* Show that, in Example 14.4 [continued following (14.22)], the MSPE of $\tilde{\zeta}$ is equal to $\sigma_v^2 \sigma_e^2 / (\sigma_e^2 + n_i \sigma_v^2)$, which is the same as $\text{var}(\zeta|y_i)$.

*14.9.* This exercise involves some details in Example 14.5.

(i) Show that $\text{MSPE}(\tilde{\zeta}) = 1 - B$.

(ii) Show that $\text{MSPE}(\hat{\zeta}) = 1 - B + 2B/m + o(m^{-1})$.

(iii) Show that $E(\hat{\zeta}_{-1}^* - \hat{\zeta})^2 = A(1 - B) + o(1)$.

(iv) Show that $E(\hat{\zeta}_{-i}^* - \hat{\zeta})^2 = 2B/m^2 + o(m^{-2})$, $i \geq 2$.

(v) Show that, with $\hat{\zeta}_{-i}$ replaced by $\hat{\zeta}_{-i}^*$, $1 \leq i \leq m$, the expectation of the right side of (14.27) is equal to $\text{MSPE}(\hat{\zeta}) + A(1 - B) + o(1)$.

*14.10.* Regarding Example 14.6, show that the MLE of $\psi$ satisfies (14.29) (under regularity conditions that you may need to specify) with $a(\psi) = 0$ and $f_j(\psi, y_j)$, $1 \leq j \leq p + q$, specified in the example.

*14.11.* Regarding Example 14.7, show that the REML estimator of $\psi$ satisfies (14.29) (according to the definition under non-Gaussian linear mixed models; see Section 12.2), where the $f_j$'s are the same as in Example 14.6 and $a(\psi)$ is given in Example 14.7. [Hint: You may use the identity

$$V^{-1} = V^{-1}X(X'V^{-1}X)^{-1}X'V^{-1} + A(A'VA)^{-1}A',$$

where $V = V(\theta)$ and $A$ is any $n \times (n-p)$ matrix of full rank ($n$ is the dimention of $y = (y_j)_{1 \le j \le m}$) such that $A'X = 0$ (e.g., Searle et al. 1992, p. 451).]

14.12. Prove Theorem 14.2. [Hint: Consider a neighborhood of $\psi$ and show that the values of the function $l$ at the boundary of the neighborhood are greater than that at the center with high probability.]

14.13. Regarding Remark 1 below (11.32), show the following.

(i) That the M-estimators $\psi_{-i}, 0 \le i \le m$, are c.u. $\mu_\omega$ at rate $m^{-d}$ implies that they are c.u. at rate $m^{-d}$.

(ii) Conversely, if there is $\tau > 2$ such that $E(|\omega(Y)|^\tau) < \infty$, then the M-estimators $\hat\psi_{-i}, 0 \le i \le m$, are c.u. at rate $m^{-d}$ implies that they are c.u. $\mu_\omega$ at rate $m^{-d(1-2/\tau)}$.

14.14. Show that, with the functional $h$ defined by (14.48), the derivative $\psi$ is given by (14.49). [Hint: Use (14.47). You do not need to justify it rigorously.]

14.15. This exercise involves some of the details in Example 14.8.

(i) Show that the functional $\theta$ is a special case of (14.48).

(ii) Verify that $\theta(F_n) = n^{-2} \sum_{i=1}^n \sum_{j=1}^n 1_{(X_i+X_j>0)}$, which is a $V$-statistic.

(iii) Verify (14.51) and (14.52) and thus conclude that the asymptotic conditional distribution of $\sqrt{n}(U_n^* - U_n)$ given $X_1, \ldots, X_n$ is the same as the asymptotic distribution of $\sqrt{n}(U_n - \theta)$.

(iv) What is the asymptotic distribution of $\sqrt{n}(U_n - \theta)$?

14.16. Let $X_1, \ldots, X_n$ be i.i.d. with cdf $F$ and pdf $f$. For any $0 < p < 1$, let $\nu_p$ be such that $F(\nu_p) = p$. Suppose that $f(\nu_p) > 0$. Show that for any sequence $m = m_n$ such that $m/n = p + o(n^{-1/2})$, we have $\sqrt{n}\{X_{(m)} - \nu_p\} \xrightarrow{d} N(0, \sigma_p^2)$, where $X_{(i)}$ is the $i$th order statistic, and $\sigma_p^2 = p(1-p)/f^2(\nu_p)$.

14.17. This exercise is related to some of the details in Example 14.9.

(i) Show that $\xi_n = n\{\theta - X_{(n)}\}/\theta$ converges weakly to Exponential(1).

(ii) Show that for any $k \ge 1$,

$$P\{X_{(n)}^* < X_{(n-k+1)}|X_1, \ldots, X_n\} = \left(1 - \frac{k}{n}\right)^n.$$

(iii) Show that (14.45) and (14.55) imply $\limsup \Delta_n = \infty$ a.s. and $\liminf \delta_n = 0$ a.s.

14.18. Continue with the previous exercise.

(i) Give an example of a sequence $a_n, n \ge 0$, of positive integers that is strictly increasing and satisfies (14.58) for every $k \ge 0$.

(ii) Show that for every $k \ge 1$ and $1 \le i \le n$,

$$P\{X_{(a_n-k+1)} = X_i\} = \frac{1}{a_n - k + 1}.$$

(iii) Argue that $P(A_n \text{ i. o.}) = 0$ implies that, with probability 1, $\eta_{a_n} = \zeta_n$ for large $n$.

(iv) Using the above result, argue that (14.54) and (14.55) follow if (14.59) holds for any $b > 0$.

(v) Show that the distribution of $\zeta_n$ is the same as that of $a_n(X_{n,a_n-a_{n-1}} - X_{n-k+1,a_n-a_{n-1}})$, where $X_{r,n}$ is $r$th order statistic of $X_1, \ldots, X_n$. Furthermore, show that for any $r < s$, the distribution of $n(X_{s,n} - X_{r,n})$ weakly converges to the Gamma$(s - r, 1)$ distribution as $n \to \infty$.

*14.19.* In this exercise you are encouraged to study the large-sample behavior of the bootstrap through some simulation studies. Two cases will be considered, as follows. In each case, consider $n = 50, 100$, and 200. In both cases, a sample $X_1, \ldots, X_n$ are drawn independently from a distribution, $F$. Then 2000 bootstrap samples are drawn, given the values of $X_1, \ldots, X_n$. A parameter, $\theta$, is of interest. Let $\hat{\theta}$ be the estimator of $\theta$ based on $X_1, \ldots, X_n$ and $\hat{\theta}^*$ be the bootstrap version of $\hat{\theta}$.

(i) $F = $ Uniform$[0, 1]$, $\theta = $ the median of $F$ (which is $1/2$), $\hat{\theta} = $ the sample median of $X_1, \ldots, X_n$, and $\hat{\theta}^* = $ the sample median of $X_1^*, \ldots, X_n^*$, the bootstrap sample. Make a histogram based on the 2000 $\hat{\theta}^*$'s.

(ii) $F = $ Uniform$[0, \theta]$, $\hat{\theta} = X_{(n)} = \max_{1 \le i \le n} X_i$, and $\hat{\theta}^* = X_{(n)}^* = \max_{1 \le i \le n} X_i^*$. Make a histogram based on the 2000 $\hat{\theta}^*$'s.

(iii) Make a histogram of the true distribution of $\hat{\theta}$ for case (i). This can be done by drawing 2000 sample of $X_1, \ldots, X_n$ and computing $\hat{\theta}$ for each sample. Compare this histogram with that of (i). What do you conclude?

(iv) Make a histogram of the true distribution of $\hat{\theta}$ for case (ii). This can be done by drawing 2000 sample of $X_1, \ldots, X_n$, and compute $\hat{\theta}$ for each sample. Compare this histogram with that of (ii). What do you conclude?

*14.20.* Regarding the plug-in principle of bootstrap summarized below (14.63), what are $X, F, R(X, F)$ for bootstrapping under the dynamic model (14.60)? What are $X^*, \hat{F}$, and $R(X^*, \hat{F})$ in this case?

*14.21.* Show that the coefficients $\phi_j, j = 0, 1, \ldots$, in (14.64) are functions of $F$, the joint distribution of the $X$'s and $\epsilon$'s. (You may impose some regularity conditions, if necessary.)

*14.22.* Is the plug-in principle used in the sieve bootstrap [see Section 14.5, below (14.64)] the same as Efron's plug-in principle [see Section 14.5, below (14.63)]? Why?

*14.23.* This exercise has three parts.

(i) Interpret the expression of the asymptotic variance $\tau^2$ in (14.66) given below the equation.

(ii) Show that (14.67) is equivalent to

$$P^* \left\{ \frac{1}{\sqrt{n}} \sum_{t=1}^{n} (X_t^* - \bar{X}) \le x \right\} - P \left\{ \frac{1}{\sqrt{n}} \sum_{t=1}^{n} (X_t - \mu_X) \le x \right\} = o_P(1)$$

for every $x$.

(iii) Interpret (14.67) by the plug-in principle.

*14.24.* Verify the two equivalent expressions of (14.69)—that is, (14.70) and (14.71).

*14.25.* Interpret expression (14.72) of the asymptotic variance $\sigma^2$.

*14.26.* This exercise is regarding the special case of block-bootstrapping the sample mean, discussed in Section 14.5.

(i) Show that the influence function IF defined in (A2) [above (14.72)] is given by $\mathrm{IF}(x, F) = x - \mu$, where $\mu = \mathrm{E}(X_t)$.

(ii) Show that the bootstrap statistic can be expressed as

$$T_N^* = \frac{1}{k} \sum_{j=1}^{k} U_{n,j},$$

where the $U_{n,j}$'s are i.i.d. uniformly distributed among the points $u_i = l^{-1}(X_{i+1} + \cdots + X_{i+l})$, $i = 0, \ldots, n - l$. Thus, in particular, the conditional mean and variance of $T_N^*$ given $X_1, \ldots, X_N$ are given by (14.73) and (14.74), respectively.

*14.27.* Show that, in Example 14.10, the coverage probability of $\tilde{I}_{2,i}$ is $1 - \alpha + o(1)$; that is, as $n \to \infty$,

$$\mathrm{P}\left(\hat{\mu} - z_{\alpha/2}\sqrt{\hat{A}} \leq \zeta_i \leq \hat{\mu} + z_{\alpha/2}\sqrt{\hat{A}}\right) = 1 - \alpha + o(1).$$

*14.28.* Show that (14.80) and $d^2/n \to 0$ imply (14.81).

*14.29.* Consider a special case of the nested error regression model (14.85) with $x_{ij}'\beta = \mu$ and $n_i = k$, $1 \leq i \leq m$, where $\mu$ is an unknown mean and $k \geq 2$. Suppose that the random effects $v_i$ are i.i.d. with an unknown distribution $F$ that has mean 0 and finite moment of any order; the errors $e_{ij}$ are also i.i.d. with mean 0 and finite moment of any order. The REML estimators of the variances $\sigma_v^2, \sigma_e^2$, which do not require normality (see Section 12.2), are given by $\hat{\sigma}_v^2 = (\mathrm{MSA} - \mathrm{MSE})/k$ and $\hat{\sigma}_e^2 = \mathrm{MSE}$, where $\mathrm{MSA} = \mathrm{SSA}/(m - 1)$ with $\mathrm{SSA} = k \sum_{i=1}^{m}(\bar{y}_{i\cdot} - \bar{y}_{\cdot\cdot})^2$, $\bar{y}_{i\cdot} = k^{-1}\sum_{j=1}^{k} y_{ij}$, $\bar{y}_{\cdot\cdot} = (mk)^{-1}\sum_{i=1}^{m}\sum_{j=1}^{k} y_{ij}$, and $\mathrm{MSE} = \mathrm{SSE}/m(k - 1)$ with $\mathrm{SSE} = \sum_{i=1}^{m}\sum_{j=1}^{k}(y_{ij} - \bar{y}_{i\cdot})^2$, if one ignores the nonnegativity constraint on $\hat{\sigma}_v^2$, which you may throughout this exercise for simplicity. The EBLUP for the random effect $v_i$ is given by

$$\hat{v}_i = k\hat{\sigma}_v^2(\hat{\sigma}_e^2 + k\hat{\sigma}_v^2)^{-1}(\bar{y}_{i\cdot} - \bar{y}_{\cdot\cdot}).$$

Suppose that $m \to \infty$ while $k$ is fixed. Show that $\mathrm{MSPE}(\hat{v}_i) = \mathrm{E}(\hat{v}_i - v_i)^2$ can be expressed as $a + o(m^{-1})$, where $a$ depends only on the second and fourth moments of $F$ and $G$.

*14.30.* This exercise is related to Example 14.11.

(i) Simplify the expressions for $g_1(\theta)$ and $g_2(\theta)$.

(ii) It is conjectured that $g_3$ can be expressed as $s + o(m_*^{-1})$, where $s$ depends only on the second and fourth moments of $u$, $v$, and $e$. Is it true? (This may involve some tedious derivations but is doable).

# 15

# Markov-Chain Monte Carlo

## 15.1 Introduction

There are various things named after Monte Carlo, almost all of which originated from the Monte Carlo Casino in Monaco. In the mid-1940s, mathematicians John von Neumann and Stanislaw Ulam were working on a secret (nuclear) project at the Los Alamos National Laboratory in New Mexico. The project involved such calculations as the amount of energy that a neutron is likely to give off following a collision with an atomic nucleus. It turned out that the calculations could not be carried out analytically, so the two scientists suggested to solve the problem by using a random number-generating computer. Due to the secrecy of their project, they code-named their method *Monte Carlo*, referring to the Monaco casino, where Ulam's uncle would borrow money to gamble (Ulam was born in Europe). The one thing that Monte Carlo, as a method of scienctific computating, and casino gambling have in common is that both play with chances, either large (in terms of convergence probability) or small (in terms of winning a big prize at the Casino) chances.

The idea of the simplest Monte Carlo method came from the law of large numbers. Suppose that one wishes to evaluate the integral $\int f(x)\,dx$. Suppose that the integrand $f$ can be expressed as $f(x) = g(x)p(x)$, where $f$ is a pdf. Then we have

$$\int f(x)\,dx = \int g(x)p(x)\,dx = \mathrm{E}\{g(X)\}, \tag{15.1}$$

where $X$ is a random variable whose pdf is $p$. Now, suppose that $p$ is known, so that we can draw i.i.d. samples $X_1, \ldots, X_n$ from $p$. Then by the SLLN and (15.1), we have

$$\frac{1}{n}\sum_{i=1}^{n} g(X_i) \longrightarrow \int f(x)\,dx \tag{15.2}$$

almost surely, as $n \to \infty$. Here, we assume, of course, that the integral or expectation in (15.1) is finite, so that the SLLN applies (see Section 6.3). This

J. Jiang, *Large Sample Techniques for Statistics,*
DOI 10.1007/978-1-4419-6827-2_15, © Springer Science+Business Media, LLC 2010

means that we can approximate the integral (15.1) by the left side of (15.2) with a large $n$. The procedure practically requires drawing a large number of random samples using a computer. One small problem, at this point, with the SLLN is that the convergence in (15.2) can be very slow. For example, in Table 11.2 we presented some numerical values for $\rho_w, \rho_s$, and the asymptotic significance levels. Try to evaluate some of them using the SLLN (Exercise 15.1). A bigger problem occurs when the dimension of the integral is high. Here, the integral in (15.1) is understood as one-dimensional. What if it is two-dimensional? Well, the SLLN still applies—all we have to do is to draw $X_1, \ldots, X_n$ independently from the bivariate pdf, $p$, and then take the average. Of course, this should still work out. What if the integral is 80-dimensional? Theoretically, the SLLN still works the same way, but, practically, there may be a problem. Here is an example.

*Example 15.1.* An example of a high-dimensional integral is given in Example 12.9. Note that, subject to the multiplicative factor

$$c = (2\pi\sigma_1^2)^{m_1/2}(2\pi\sigma_2^2)^{m_2/2}\exp(-\mu y_{..}),$$

the integral in (12.35) can be written as (15.1) with $x = (u', v')'$, $\int = \int \cdots \int$, whose dimension is $m_1 + m_2$, $dx = \prod_{i=1}^{m_1} du_i \prod_{j=1}^{m_2} dv_j$, and

$$g(x) = \prod_{i=1}^{m_1}\prod_{j=1}^{m_2}\{h(\mu + u_i + v_j)\}^{y_{ij}}\{1 - h(\mu + u_i + v_j)\}^{1-y_{ij}},$$

$$p(x) = \frac{1}{(2\pi\sigma_1^2)^{m_1/2}(2\pi\sigma_2^2)^{m_2/2}}\exp\left(-\frac{1}{2\sigma_1^2}\sum_{i=1}^{m_1}u_i^2 - \frac{1}{2\sigma_2^2}\sum_{j=1}^{m_2}v_j^2\right),$$

where $h(x) = e^x/(1 + e^x)$ (verify). So, if $m_1 = m_2 = 40$, we have an 80-dimensional integral, as mentioned above. By the way, this example is related to the infamous salamander mating data, first reported in McCullagh and Nelder (1989, Section 14.5), where the $u_i$ and $v_j$ represent random effects corresponding to female and male animals (i.e., salamanders), respectively, and $y_{ij}$ is the indicator of a successful mating (1 = Yes, 0 = No). As noted in Example 12.9, there is no analytic expression for the integral involved here. However, noticing that $p(x)$ is the multivariate pdf of $X = (u_1, \ldots, u_{m_1}, v_1, \ldots, v_{m_2})'$, it looks like one could use the same strategy [i.e., (15.2)] to approximate the integral, right? Not quite. To see this, let us assume that $m_1 = m_2 = 40$. The problem is that $g(x)$ is a product of 1600 terms with each term less than 1. It is very possible that such a term is numerically zero. For example, suppose that each term in the product is 0.5; then $g(x) = 0.5^{1600}$, which is mathematically positive, of course. However, when the value is entered in a computer using the statistical programming language **R**, the returned value is 0. If one takes the summation, or average, of such terms on the left side of (15.2), one is not going to get anything but 0 without a huge $n$!

There is another difficulty that often occurs with the Monte Carlo method: Sometimes, the pdf $p$ is not completely known—so how can we sample from it? This happens, for example, with Bayesian analysis. The idea of Bayesian inference may be interpreted as updating ones knowledge about the unknowns using the data information. The knowledge is expressed in terms of the distribution of the unknown (multidimensional) parameter, say, $\theta$. The knowledge prior to seeing the data is the *prior* density, $\pi_0(\theta)$. The conditional pdf of the data $x$ given $\theta$ is $f(x|\theta)$. The knowledge is updated by computing the posterior—that is, the conditional density of $\theta$ given the observed data $x$—

$$p(\theta|x) = \frac{f(x|\theta)\pi_0(\theta)}{\int f(x|\theta)\pi_0(\theta) \, d\theta}. \tag{15.3}$$

Although (15.3) is expressed in terms of integration with respect to Lebesgue measure, the posterior is easily extended to discrete distributions, with the integration replaced by summation (or integration with respect to the counting measure). In many cases, the integral in the denominator of (15.3) may be unknown. In fact, because (15.3) is a ratio, one only needs to know the product $f(x|\theta)\pi_0(\theta)$ up to a constant, known or otherwise. In particular, the prior $\pi_0$ only needs to be specified proportionally (see Exercise 15.3). However, this may not help much with regard to the integral in the denominator. So, in this case, the posterior is known only up to a multiplicative constant. Note that, given the data $x$, the denominator of (15.3) is a constant. However, unless this constant is known, one cannot sample directly from the posterior, as is often desired in Bayesian analysis. Sometimes, a strategy called *rejection sampling* may help to solve the problem. Suppose that one wishes to sample from a pdf $p(\theta)$ but cannot do so directly. Instead, there is a function $q(\theta)$ in hand with the following properties:

(i) $q(\theta) \geq 0$ and $q(\theta) > 0$ for all $\theta$ such that $p(\theta) > 0$;

(ii) $\int q(\theta) \, d\theta < \infty$, and one is able to draw random samples from a pdf proportional to $q$;

(iii) the *importance ratio* $p(\theta)/q(\theta) \leq b$ for all $\theta$ and some constant $b > 0$, which is known.

Given a $q$ with properties (i)–(iii), we proceed as follows:

1. sample $\theta$ at random from the pdf in (ii) that is proportional to $q$;

2. sample $u$ from Uniform$[0,1]$ independent of $\theta$;

3. accept $\theta$ as a draw from $p(\cdot)$ if $u \leq p(\theta)/bq(\theta)$; otherwise, reject the drawn and return to step 1.

It can be shown (Exercise 15.4) that the accepted $\theta$ has the pdf $p(\theta)$; that is, the pdf of the drawn $\theta$, conditional on $u \leq p(\theta)/bq(\theta)$, is $p(\theta)$. The upper bound $b$ for the importance ratio plays an important role, practically, in rejection sampling. It is preferable that the pdf in (ii) is a close approximation to $p(\theta)$; so that one can choose $b$ as small as possible. If this is not the case that one has to choose a very large $b$ in order to bound the importance ratio, the sample from step 1 will almost always be rejected. As a result, the sampling will proceed very slowly, if at all.

Geman and Geman (1984) discussed an alternative approach, called *Gibbs sampler*, for sampling iteratively from approximate distributions. The samples drawn is a Markov chain (see Section 10.2) whose unique stationary distribution is identical to the posterior $p(\theta|x)$, which is our *target* distribution (i.e., the distribution from which we wish to sample). Thus, according to the Markov-chain convergence theorem (Theorem 10.2), the distribution of the drawn converges to its limiting distribution which is the target distribution as the number of iterations increases. The term "Gibbs" originated from image processing, in which context the posterior of interest is a Gibbs distribution, expressed as

$$p(\theta|x) \propto \exp\left\{-\frac{E(\theta_1,\ldots,\theta_d)}{\lambda T}\right\},$$

where $\lambda$ is a positive constant, $T$ is the temperature of the system, $E$ is the energy (function) of the system, and $\theta_j$ is the characteristic of interest for the $j$th component of the system. The Gibbs sampler belongs to a class of iterative sampling strategies called Markov-chain Monte Carlo, or MCMC. The class shares the same property that the samples iteratively drawn from a Markov chain, which explains the name, and they are useful when it is not possible, or computationally inefficient, to sample directly from $p(\theta|x)$. The convergence property of Markov chains plays an important role and, in fact, is "at the core of the motivation" for various MCMC algorithms (Robert and Casella 2004, p. 231). Thus, the focus of the current chapter is MCMC. We begin naturally with the Gibbs sampler.

## 15.2 The Gibbs sampler

For some reason, the Gibbs sampler was largely unknown to the statistical community, after its proposal, until Gelfand and Smith (1990) pointed out that Geman and Geman's sampling scheme could in fact be useful for other posterior sampling problems. In fact, the method is not restricted to posterior sampling and Bayesian inference. Let $f(x)$ denote a target density function with respect to a $\sigma$-finite measure $\nu$, where $x = (x_i)_{1 \le i \le k} \in R^s$ with $x_i$ being a $s_i$-dimensional subvector, $1 \le i \le k$, such that $s_1 + \cdots + s_k = s$. Let $x_{-i}$ denote the vector $(x_1',\ldots,x_{i-1}',x_{i+1}',\ldots,x_k')'$ and $f(x_i|x_{-i})$ denote the conditional density induced by $f$. The problem of interest is to draw samples from $f(x)$, but the situation is such that this cannot be done directly due to some computational difficulties. For example, the dimension $s$ may be too high or $f(x)$ is intractable. However, it is fairly easy to sample from the conditional distributions $f(x_i|x_{-i}), 1 \le i \le k$. Thus, we proceed with successive drawings from the conditionals, as follows. Starting with $x^0 = (x_i^0)_{1 \le i \le k}$, draw

$$
\begin{aligned}
x_1^1 \quad &\text{from} \quad f(x_1|x_2^0,\ldots,x_k^0), \\
\text{then} \quad x_2^1 \quad &\text{from} \quad f(x_2|x_1^1,x_3^0,\ldots,x_k^0),
\end{aligned}
$$

then $x_3^1$  from  $f(x_3|x_1^1, x_2^1, x_4^0, \ldots, x_k^0)$,

$$\vdots$$

then $x_k^1$  from  $f(x_k|x_1^1, \ldots, x_{k-1}^1)$.

This completes the first cycle of the drawings. We then update the initial value $x^0$ by $x^1$, and repeat the process. This procedure is called the Gibbs sampler.

A question that one first wants answered is why the Gibbs sampler works. Before going into the mathematical details, we would like to make a couple of notes. The first is that it is completely possible to recover the joint distribution, $f$, from the conditional distributions, which is what the Gibbs sampler is trying to do. To see this, consider the special case of $k = 2$ and $s_1 = s_2 = 1$. Denote the joint, marginal, and conditional pdf's of $X$ and $Y$ by $f(x, y)$, $f_X(x)$, $f_Y(y)$, $f_{X|Y}(x|y)$, and $f_{Y|X}(y|x)$, respectively. Then since $f(x, y) = f_{Y|X}(y|x)f_X(x) = f_{X|Y}(x|y)f_Y(y)$, we have

$$
\begin{aligned}
f(x, y) &= f_{Y|X}(y|x)f_X(x) \\
&= \frac{f_{Y|X}(y|x)f_X(x)}{\int f_Y(y)\, dy} \\
&= f_{Y|X}(y|x)\left\{\int \frac{f_Y(y)}{f_X(x)}\, dy\right\}^{-1} \\
&= f_{Y|X}(y|x)\left\{\int \frac{f_{Y|X}(y|x)}{f_{X|Y}(x|y)}\, dy\right\}^{-1}.
\end{aligned}
$$

So, at least from a theoretical point of view, the joint is fully recoverable from the conditionals. Note that with the exception of special cases such as independence, it is not possible to recover the joint from the marginals.

The second note is that the idea of having something done componentwisely and iteratively is not new. For example, the Gauss–Seidel algorithm was well known to numerical analysis more than 100 years before Gibbs sampler. See, for example, Young (1971). Suppose that one wishes to solve a large system of equations expressed as

$$L(x_1, \ldots, x_m) = 0, \tag{15.4}$$

where $L$ is a $R^m$-valued function. Such a problem can be numerically challenging even if $L$ is a linear function if the dimension $m$ is high (see Jiang 2000b for a nonlinear version). The Gauss–Seidel method solves a system of univariate equations iteratively. Note that it is often trivial to solve a univariate equation, either analytically or numerically. Let $L_i$ be the $i$th component of $L$, $1 \le i \le m$. Given the current value $x^0 = (x_1^0, \ldots, x_m^0)$, solve

$$x_1^1 \quad \text{from} \quad L_1(x_1, x_2^0, \ldots, x_m^0) = 0,$$
$$\text{then} \quad x_2^1 \quad \text{from} \quad L_2(x_1^1, x_2, x_3^0, \ldots, x_m^0) = 0,$$

$$\text{then } x_3^1 \quad \text{from} \quad L_3(x_1^1, x_2^1, x_3, x_4^0, \ldots, x_m^0) = 0,$$

$$\vdots$$

$$\text{then } x_m^1 \quad \text{from} \quad L_m(x_1^1, \ldots, x_{m-1}^1, x_m) = 0.$$

This completes one cycle of the iteration. We then update the current value $x^0$ by $x^1$, and repeat the process. A nice feature of the algorithm is that it is totally intuitive, and there is much more. Under mild conditions, the algorithm has the property of *global convergence*, which means that starting with any initial value $x^0$, the sequence $x^l, l = 0, 1, 2, \ldots$, generated by the iterative procedure converges to the (unique) solution to (15.4).

The Gibbs sampler shares a similar intuitiveness to the Gauss–Seidel algorithm; so the next thing we want to make sure is that it converges. Because our goal is to draw samples from the target distribution, it is desirable that the distribution of the drawn converge to the target distribution. As noted, the latter is indeed true as a result of the Markov-chain convergence theorem. We show this, again, for the special case $k = 2$ and $s_1 = s_2 = 1$, with a note given at the end on obvious extensions of the results. More general treatments of the topic can be found in Tierney (1994) and Robert and Casella (2004, Chapter 10). To be more specific, we also assume that $\nu$ is the Lebesgue measure on $R$. Let $x$ and $y$ denote the two components (instead of $x_1$ and $x_2$). Let the initial values be $X_0 = x_0$ and $Y_0 = y_0$. Then, at iteration $t$, the Gibbs sampler draws $X_t$ from $f_{X|Y}(x|Y_{t-1})$ and then $Y_t$ from $f_{Y|X}(y|X_t)$, $t = 1, 2, \ldots$. First, we show that $X_t$ is a Markov chain. Note that here we are dealing with a chain whose state-space may be an arbitrary subset of an Euclidean space. Therefore, we need to extend the definition in Section 10.2 a little bit, but very much along the same line. A function $K(u, v)$, where $u, v \in R^a$ for some positive integer $a$ is called a *transition kernel* if (i) for any $u \in R^a$, $K(u, \cdot)$ is an $a$-variate pdf and (ii) for any $B \in \mathcal{B}$, $K(\cdot, B)$ is measurable, where $\mathcal{B}$ denotes the Borel $\sigma$-field on $R^a$. The process $X_t, t = 0, 1, 2, \ldots$, is a Markov chain with transition kernel $K$ if

$$P(X_t \in B|X_0, \ldots, X_{t-1}) = \int_B K(X_{t-1}, v) \, dv$$
$$= P(X_t \in B|X_{t-1}). \tag{15.5}$$

Note that the second equality in (15.5) is implied by the first one. For the $X_t$ induced by the Gibbs sampler, we have

$$P(X_t \in B|X_0, \ldots, X_{t-1}) = \int_B f_{X_t|X_0, \ldots, X_{t-1}}(x|X_0, \ldots, X_{t-1}) \, dx$$
$$= \int_B \int f_{X_t|X_0, \ldots, X_{t-1}, Y_{t-1}}(x|X_0, \ldots, X_{t-1}, y)$$
$$\times f_{Y_{t-1}|X_0, \ldots, X_{t-1}}(y|X_0, \ldots, X_{t-1}) \, dy \, dx.$$

According to the definition of the Gibbs sampler, we have

$$f_{X_t|X_0,\ldots,X_{t-1},Y_{t-1}}(x|X_0,\ldots,X_{t-1},y) = f_{X|Y}(x|y),$$
$$f_{Y_{t-1}|X_0,\ldots,X_{t-1}}(y|X_0,\ldots,X_{t-1}) = f_{Y|X}(y|X_{t-1}).$$

Thus, if we let

$$K(u,v) = \int f_{Y|X}(y|u)f_{X|Y}(v|y)\,dy, \tag{15.6}$$

we have, continuing with the derivation, that

$$P(X_t \in B|X_0,\ldots,X_{t-1}) = \int_B \int f_{X|Y}(x|y)f_{Y|X}(y|X_{t-1})\,dy\,dx$$
$$= \int_B K(X_{t-1},v)\,dv.$$

The argument shows that $X_t$ is a Markov chain with transition kernel given by (15.6). Similarly, it can be shown that $Y_t$ is a Markov chain (Exercise 15.5). The chain that we are most interested in is $Z_t = (X_t, Y_t)'$. This, too, can be shown to be a Markov chain with transition kernel

$$K(w,z) = f_{X|Y}(x|v)f_{Y|X}(y|x) \tag{15.7}$$

for $w = (u,v)'$ and $z = (x,y)'$ (Exercise 15.5).

As seen in Section 10.2, there are four regularity conditions associated with the convergence properties of Markov chains. These are irreducibility, aperiodicity, recurrence, and the existence of a stationary distribution. Once again, we need to extend the definitions in Section 10.2 to general state-space Markov chains. A stationary distribution for the Markov chain is a probability distribution $\pi$ that satisfies

$$\pi(B) = \int P(X_1 \in B|X_0 = x)\pi(x)\,dx$$
$$= \int K(x,B)\pi(x)\,dx \tag{15.8}$$

for every $B \in \mathcal{B}$, where $K(x,B) = \int_B K(x,v)\,dv$ and $K$ is the transition kernel of the Markov chain. The Markov chain, or $K$, is $\pi$-*irreducible* if for all $x \in D = \{x : \pi(x) > 0\}$, $\pi(B) > 0$ implies that $P(X_t \in B|X_0 = x) > 0$ for some $t \geq 1$. A $\pi$-irreducible $K$ is periodic if there is an integer $d \geq 2$ and a collection of nonempty disjoint sets $E_0,\ldots,E_{d-1}$ such that for $i = 0,\ldots,d-1$ and all $x \in E_i$, $K(x,E_t) = 1$, $t = i+1$ mod $d$; otherwise, $K$ is aperiodic. Here, $a = b$ mod $d$ if $a - b$ can be divided by $d$. The definition of recurrence can also be extended, but for the reason below we do not need it at this point. The reason is that these conditions are connected so that there is no need to assume all of them. See Theorem 10.2, for example, in the discrete state-space case. For the Markov chains we are dealing with in this section, it is fairly straightforward to find the stationary distributions. In fact, this is exactly how these chains are constructed so that they have the desired stationary distribtions as potential limiting distributions. We consider some examples.

*Example 15.2.* For the Gibbs Markov chain $X_t$ above, it is easy to verify that $f_X(x)$, or $\pi_X(B) = \int_B f_X(x)\, dx$, is a stationary distribution. Similarly, $f_Y(y)$, or $\pi_Y(B) = \int_B f_Y(y)\, dy$, is a stationary distribution for the Gibbs Markov chain $Y_t$ (Exercise 15.6).

*Example 15.3.* For the Gibbs Markov chain $Z_t = (X_t, Y_t)'$, it is easy to verify that $f(x, y)$, or $\pi(B) = \int_B f(x, y)\, dx\, dy$, is a stationary distribution (Exercise 15.6), which is the target distribution from which we wish to sample.

Given the existence of the stationary distribution, convergence of the Markov chain is implied by irreducibility and aperiodicity, as the following theorem states (see Roberts and Smith 1994). Let $K^t(x, \cdot)$ denote the kernel for $X_t$ given $X_0 = x$; that is,

$$P(X_t \in B | X_0 = x) = \int_B K^t(x, v)\, dv \tag{15.9}$$

for any $B \in \mathcal{B}$. Note that when $t = 1$, (15.9) reduces to (15.5).

**Theorem 15.1.** Suppose that the Markov chain has transition kernel $K$ and stationary distribution $\pi$ so that $K$ is $\pi$-irreducible and aperiodic. Then, for all $x \in D = \{x : \pi(x) > 0\}$, the following hold:
(i) $\int |K^t(x, v) - \pi(v)|\, dv \to 0$ as $t \to \infty$;
(ii) for any real-valued, $\pi$-integrable function $g$,

$$\frac{1}{n} \sum_{t=1}^{n} g(X_t) \xrightarrow{\text{a.s.}} \int g(x)\pi(x)\, dx \quad \text{as } n \to \infty.$$

The result of (i) implies that $P(X_t \in B | X_0 = x) \to \pi(B)$ as $t \to \infty$, for every $B \in \mathcal{B}$ regardless of the starting point $x$ [as long as $\pi(x) > 0$]. So, in a way, this is similar to the global convergence of the Gauss–Seidel algorithm (see our earlier discussion). To see this, note that, by (15.9), we have

$$|P(X_t \in B | X_0 = x) - \pi(B)| = \left| \int_B K^t(x, v)\, dv - \int_B \pi(v)\, dv \right|$$

$$\leq \int_B |K^t(x, v) - \pi(v)|\, dv$$

$$\leq \int |K^t(x, v) - \pi(v)|\, dv.$$

The conclusion of (i) also implies that the stationary distribution is unique up to a set of Lebesgue measure 0 (why?). The result of (ii) may be regarded as a SLLN for the Markov chain.

Theorem 15.1 provides us with a tool to find sufficient conditions for the convergence of Gibbs sampler. We already know the existence of the stationary distribution (see Example 15.3); so all we have to do is to verify that the kernel (15.7) is $F$-irreducible and aperiodic, where $F$ is the cdf of the target distribution, $f$. Roberts and Smith (1994) found that the following conditions are sufficient for the Gibbs sampler to have these two properties. A function $h: R^2 \to [0, \infty)$ is said to be lower semicontinuous at zero if for any $z = (x, y)'$ with $h(z) > 0$, there exists an open neighborhood $S(z)$ of $z$ and $\epsilon > 0$ such that $h(w) \geq \epsilon$ for all $w \in S(z)$. A function $h: R \to [0, \infty)$ is said to be locally bounded if for any $x$, there are constants $\delta, c > 0$ (which may depend on $x$) such that $h(u) \leq c$ for all $u$ such that $|u - x| < \delta$. The subset $D$ is said to be connected if for any $w, z \in D$, there is a continuous map $h: [0, 1] \to D$ such that $h(0) = w$ and $h(1) = z$.

**Theorem 15.2.** The conditions of Theorem 15.1 hold for the Gibbs sampler provided that $F$ is lower semicontinuous at zero, $D$ is connnected, and both $f_X(\cdot)$ and $f_Y(\cdot)$ are locally bounded.

Note that although Theorems 15.1 and 15.2 are stated for the case $s = 2$, or two-stage Gibbs sampler, they actually apply to general $s \geq 2$ or the multistage Gibbs sampler as well, with obvious modifications. We consider some applications of Theorem 15.2.

*Example 15.4* (Bivariate-normal Gibbs). Consider sampling from the bivariate normal distribution with means 0, variances 1 and correlation coefficient $\rho$. Although it is straightforward to sample directly in this case (so that the Gibbs sampler is not needed), it makes a trivial special case to illustrate the conditions of Theorem 15.2. The Gibbs sampler draws $X_{t+1}$ from $N(\rho y_t, 1 - \rho^2)$, where $y_t$ is the realized value of $Y_t$, and then $Y_{t+1}$ from $N(\rho x_{t+1}, 1 - \rho^2)$, where $x_{t+1}$ is the realized value of $X_{t+1}$. The cdf $F$ is

$$F(x, y) = \int_{-\infty}^{x} \int_{-\infty}^{y} \frac{1}{2\pi\sqrt{1 - \rho^2}} \exp\left\{ -\frac{u^2 - 2\rho u v + v^2}{2(1 - \rho^2)} \right\} du\, dv.$$

The function is continuous on $R^2$, which implies that it is lower semicontinuous at zero. Furthermore, $D = R^2$, which is obviously connected. Finally, $f_X$ and $f_Y$ are both the pdf of $N(0, 1)$, which is (locally) bounded by $1/\sqrt{2\pi}$.

*Example 15.5* (Autoexponential Gibbs). A 3-dimensional auto-exponential model (Besag 1974) is defined as the distribution of $X = (X_1, X_2, X_3)'$ such that $X_i > 0, i = 1, 2, 3$, and the pdf of $X$ is

$$f(x_1, x_2, x_3) \propto \exp\{-(x_1 + x_2 + x_3 + ax_1x_2 + bx_2x_3 + cx_3x_1)\},$$

where $a$, $b$, and $c$ are known positive constants. It is easy to show that

$$X_1|x_2, x_3 \sim \text{Exponential}(1 + ax_2 + cx_3),$$
$$X_2|x_3, x_1 \sim \text{Exponential}(1 + bx_3 + ax_1),$$
$$X_3|x_1, x_2 \sim \text{Exponential}(1 + cx_1 + bx_2)$$

(Exercise 15.8). The cdf of $f$, $F(x_1, x_2, x_3)$, is the integral of $f(y_1, y_2, y_3)$ over $(0, x_1] \times (0, x_2] \times (0, x_3]$, which is continuous on $D = (0, \infty)^3$. It follows that $F$ is lower semicontinuous at zero. The set $D$ is obviously connnected. Finally, it can be shown (Exercise 15.8) that

$$f_{X_1, X_2}(x_1, x_2) \propto \frac{\exp\{-(2x_1 + x_2 + ax_1x_2)\}}{1 + bx_2 + cx_1}$$
$$\times \int_0^\infty \frac{\exp\{-(x_2 + ax_1x_2)\}}{1 + bx_2 + cx_1} \, dx_2$$
$$\leq c \int_0^\infty e^{-x_2} \, dx_2 = c,$$

where $c$ is a normalizing constant. Thus, $f_{X_1, X_2}$ is (locally) bounded. Similarly, it can be shown that $f_{X_2, X_3}$ and $f_{X_3, X_1}$ are locally bounded.

It should be pointed out that for Markov chains with a discrete state-space, the result of Theorem 10.2 applies, whose key conditions can be verified according to the definitions and results of Section 10.2 (see Exercise 15.9).

## 15.3 The Metropolis–Hastings algorithm

Prior to Gibbs sampler, a MCMC algorithm was given by Metropolis et al. (1953) in a statistical physics context. The algorithm was later extended by Hastings (1970) to what is now called the Metropolis–Hastings algorithm. In fact, the Gibbs sampler may be viewed as a special case of the Metropolis–Hastings algorithm (e.g., Robert and Casella 2004, Section 10.2.2). However, the former is simpler and more intuitive for a beginner and is therefore seen as a better introductary subject. This is why we choose to present the Gibbs sampler first, and not as an implication of the Metropolis–Hastings algorithm. The original algorithm proposed by Metropolis et al. (1953), known as the Metropolis algorithm, involves a *jumping distribution* $q(x, y)$ such that $q(x, \cdot)$ is a pdf, with respect to $\nu$, for every $x$ and $q(x, y) = q(y, x)$. Given the current value $X_t = x$, a candidate value for $X_{t+1}$, say $y$, is sampled from $q(x, \cdot)$. A decision is then made on whether or not to "jump" (to the candidate), based on a chance process. To do so, first compute the ratio $r = f(y)/f(x)$, where, as in Section 15.2, $f$ is the target distribution and, for now, we assume that $f(x) > 0$. Then we set $X_{t+1} = y$ (jumping to the candidate value) with probability $\alpha = r \wedge 1$ and $X_{t+1} = X_t$ (staying at the current place) with probability $1 - \alpha$, where $a \wedge b = \min(a, b)$.

Although the Metropolis algorithm may not seem as intuitive as the Gibbs sampler, it, indeed, generates a Markov chain. To see this, let $Y_t$ denote the value sampled from the jumping distribution given $X_t$. For simplicity, let the target density $f$ be with respect to the Lebesgue measure and assume that it is positive everywhere. Write, as in the previous section [following (15.5)],

$$P(X_{t+1} \in B | X_0, \ldots, X_t) = \int_B f_{X_{t+1}|X_0,\ldots,X_t}(x|X_0,\ldots,X_t)\, dx$$

$$= \int_B \int f_{X_{t+1}|X_0,\ldots,X_t,Y_t}(x|X_0,\ldots,X_t,y) f_{Y_t|X_0,\ldots,X_t}(y|X_0,\ldots,X_t)\, dy\, dx$$

and note that $f_{Y_t|X_0,\ldots,X_t}(y|X_0,\ldots,X_t) = q(X_t, y)$. Thus, we have

$$P(X_{t+1} \in B | X_0, \ldots, X_t)$$

$$= \int_B \int f_{X_{t+1}|X_0,\ldots,X_t,Y_t}(x|X_0,\ldots,X_t,y) q(X_t,y)\, dy\, dx$$

$$= \int \int_B f_{X_{t+1}|X_0,\ldots,X_t,Y_t}(x|X_0,\ldots,X_t,y)\, dx\, q(X_t,y)\, dy$$

$$= \int P(X_{t+1} \in B | X_0, \ldots, X_t, Y_t = y) q(X_t, y)\, dy.$$

It is easy to verify that

$$P(X_{t+1} \in B | X_0, \ldots, X_t, Y_t = y) = \alpha 1_{(y \in B)} + (1 - \alpha) 1_{(X_t \in B)},$$

where $r = f(y)/f(X_t)$. Thus, continuing with the derivation, we have

$$P(X_{t+1} \in B | X_0, \ldots, X_t)$$

$$= \int \{\alpha 1_{(y \in B)} + (1 - \alpha) 1_{(X_t \in B)}\} q(X_t, y)\, dy$$

$$= \int_B \alpha q(X_t, y)\, dy + 1_{(X_t \in B)} \left\{ 1 - \int \alpha q(X_t, y)\, dy \right\}$$

$$= \int_B K(X_t, y)\, dy,$$

where, with $r = f(y)/f(x)$, $p(x) = \int \alpha q(x, y)\, dy$ and $\delta_x$ being the Dirac (or point) mass at $x$, the transition kernel can be expressed as

$$K(x, y) = \alpha q(x, y) + \{1 - p(x)\} \delta_x(y). \tag{15.10}$$

The main motivation is that not only does the Metropolis algorithm generate a Markov chain, but also that the chain has $f$ as its stationary distribution. To see this, note that it is straightforward to verify that

$$\left\{ \frac{f(y)}{f(x)} \wedge 1 \right\} q(x, y) f(x) = \left\{ \frac{f(x)}{f(y)} \wedge 1 \right\} q(y, x) f(y).$$

Here, we have used the fact that $q(x, y)$ is symmetric in $x$ and $y$. On the other hand, it is obvious that

$$\{1 - p(x)\}\delta_x(y)f(x) = \{1 - p(y)\}\delta_y(x)f(y)$$

[note that $\delta_x(y) = 0$ and $\delta_y(x) = 0$ unless $x = y$]. With these, it is easy to see that the transition kernel $K$ is *reversible* with respect to $f$ in the sense that

$$K(x, y)f(x) = K(y, x)f(y), \qquad (15.11)$$

which implies that $f$ is the stationary distribution (Exercise 15.10).

Equation (15.11) has, in fact, motivated a series of MCMC algorithms: Construct a Markov chain so that its transition kernel is reversible with respect to the target distribution! Hastings (1970) proposed his construction of such Markov chains in a way that extends the Metropolis algorithm. A key restriction that Hastings was able to remove is that $q(x, y)$ be symmetric. Now, $q(x, y) = q(y, x)$ is no longer needed. To ensure that (15.11) still holds, we modify the acceptance probability $\alpha$ in the Metropolis algorithm by

$$\alpha = \left\{ \frac{f(y)q(y, x)}{f(x)q(x, y)} \right\} \wedge 1. \qquad (15.12)$$

Here, we assume that the denominator $f(x)q(x, y)$ is positive; otherwise, $\alpha$ is defined to be 1. Note that, as before, $\alpha$ is a function of $x$ and $y$; therefore, the complete notation is $\alpha = \alpha(x, y)$. Also note that when $q(x, y)$ is symmetric, (15.12) reduces to the $\alpha$ in the Metropolis algorithm. With the new definition of $\alpha$ and everything else the same as in the Metropolis algorithm, the procedure is called the Metropolis–Hastings algorithm, or M-H algorithm. It can be shown that the sequence $X_t$ generated by the M-H algorithm remains a Markov chain with the transition kernel $K$ still given by (15.10) (with the new definition of $\alpha$, of course), and (15.11) continues to hold (Exercise 15.11). It follows, by the same argument, that $f$ is the stationary distribution of the Markov chain generated by the M-H algorithm, known as the M-H chain. By choosing different jumping distributions $q$, one obtains different MCMC algorithms as special cases of the M-H algorithm. Below are some examples.

*Example 15.6* (Random walk chain). Suppose that given $X_t = x$, $X_{t+1} = x + \xi$, where $\xi$ is sampled independently from a density $g$, so that $q(x, y) = g(y - x)$. This is known as the random walk chain (see Example 10.2 for a case of a discrete state-space Markov chain). Natural choices for $g$ include the uniform and normal distributions.

*Example 15.7* (Autoregressive chain). An extension of the random walk chain is the autoregressive chain, or AR chain. This can be expressed as that, given $X_t = x$, $X_{t+1} = a + b(x - a) + \xi$, where $\xi$ is the same as in Example 15.6. Thus, in this case, $q(x, y) = g\{y - a - b(x - a)\}$.

*Example 15.8* (Rejection sampling chain). The sampling of $X_{t+1}$ given $X_t = x$ can be done via rejection sampling (see Section 15.1). This strategy may be useful if direct sampling from a certain distribution is not possible or inconvenient. Recall that earlier we had assumed the existence of a (known) constant $b$ such that $f(x) \leq bg(x)$, where $g(x)$ is the density from which we actually sample. Sometimes, this restriction leads to choosing an excessively large $b$, and hence an inefficient algorithm with many rejections, not to mention that, in some cases, one simply does not know what $b$ is. The M-H algorithm provides a simple remedy to this difficulty. Let $b$ be a given positive constant. Given $X_t = x$, the rejection sampling of $X_{t+1}$ is done equivalently as follows. Pairs $(Y, U)$ are generated such that $Y \sim g$ and $U|Y \sim \text{Uniform}[0, bg(Y)]$ until a pair satisfying $U < f(Y)$ is obtained, whose value of $Y$, say, $y$, is then the candidate value. It can be shown that the jumping distribution corresponding to this sampling scheme satisfies

$$q(x, y) \propto f(y) \wedge \{bg(y)\} \qquad (15.13)$$

(Exercise 15.12). Furthermore, let $B = \{x : f(x) \leq bg(x)\}$. Then, the acceptance probability (15.12) of the M-H algorithm reduces to

$$\alpha = \begin{cases} 1, & x \in B \\ bg(x)/f(x), & x \notin B, y \in B \\ \{f(y)g(x)/f(x)g(y)\} \wedge 1, & x \notin B, y \notin B. \end{cases} \qquad (15.14)$$

Note that the jumping distribution $q$, by definition, is a (Markov chain) transition kernel, which we call the jumping kernel. On the other hand, it is easy to see that $\delta_x(y)$ in (15.10) is also a transition kernel, which we call the Dirac kernel. Thus, (15.10) suggests that the transition kernel of the M-H chain may be viewed as a weighted average of the jumping kernel and Dirac kernel [note that $1 - p(x) = \int (1 - \alpha)\mu_x(dy)$, where $\mu_x(dy) = q(x, y)\,dy$]. Therefore, with regard to the regularity conditions of Theorem 15.1, it is not surprising that irreducibility (aperiodicity) of the jumping kernel has an impact on that of the M-H kernel. On the other hand, such properties of the jumping kernel alone cannot guarantee the corresponding properties of the M-H kernel. This is because the rejection of the candidate (in the M-H algorithm) also plays a role and therefore affects the behavior of the M-H chain. More precisely, the following theorem, given by Roberts and Smith (1994), states the connections.

**Theorem 15.3.** (i) If the jumping kernel is aperiodic, or $\text{P}(X_t = X_{t-1}) > 0$ for some $t \geq 1$, then the M-H kernel is aperiodic.

(ii) If the jumping kernel is $f$-irreducible, and $q(x, y) = 0$ if and only if $q(y, x) = 0$, then the M-H kernel is $f$-irreducible.

Note that the condition $\text{P}(X_t = X_{t-1}) > 0$ for some $t$ simply means that there is a positive probability that the M-H chain does not move, either due to the rejection of the candidate or that the candidate coincides with the current state. Because we already know that the stationary distribution of the

M-H chain is $f$, by Theorem 15.1, the conditions of (i) and (ii) of Theorem 15.3 are sufficient for the conclusions (i) and (ii) of Theorem 15.1 for the M-H chain. More explicit conditions are given, for example, by Roberts and Tweedie (1996), as follows.

**Theorem 15.4.** Let $f$ be bounded and positive on every compact set of its support $\mathcal{E}$ and there be constants $\epsilon, \delta > 0$ such that $x, y \in \mathcal{E}, |x - y| \leq \delta$ implies $q(x, y) \geq \epsilon$. Then the M-H chain is $f$-irreducible and aperiodic.

Here, the support $\mathcal{E}$ means that $f(x) > 0$ if and only if $x \in \mathcal{E}$. In fact, the conditions imply that the M-H chain is $\nu$-irreducible, where $\nu$ is the Lebesgue measure, which implies $f$-irreducibility (why?). We consider some examples.

*Example 15.7 (continued).* Suppose that $f$ is continuous and has support $(a - 1, a + 1)$ for some $a \in R$. Let $g(x) = 1_{(-1<x<1)}/2$ [i.e., the pdf of Uniform$(-1, 1)$]. Also, let $0 < b < 1$. Clearly, $f$ is bounded and positive on every compact subset of $(a - 1, a + 1)$. Furthermore, let $\delta = b/2$. For any $x, y \in (a - 1, a + 1)$ such that $|x - y| \leq \delta$, we have

$$|y - a - b(x - a)| = |y - x + (1 - b)(x - a)|$$
$$\leq |y - x| + (1 - b)|x - a|$$
$$< \delta + 1 - b = 1 \quad b/2 < 1;$$

hence, $q(x, y) = g\{y - a - b(x - a)\} = 1/2$. Therefore, the conditions of Theorem 15.4 are satisfied. It follows that the conclusions of Theorem 15.1 hold for the AR chain and, hence, the random walk chain (Example 15.6) as a special case.

*Example 15.8 (continued).* Suppose that $f$ is the pdf of a *truncated distribution* whose pdf is continuous and positive everywhere. In other words, $f(x) = h(x), x \in [A, B]$, and $f(x) = 0$ elsewhere for some $A < B$, where $h(x)$ is continuous pdf such that $h(x) > 0$ for all $x$. Let $g$ be a pdf that is also continuous and positive everywhere. Clearly, $f$ is bounded and positive on every compact subset of its support $\mathcal{E} = [A, B]$. Furthermore, let $a_1 = \inf_{x \in [A, B]} f(x)$ and $a_2 = \inf_{x \in [A, B]} g(x)$, which are both positive. By (15.13), there is a constant $c > 0$ such that $q(x, y) = c[f(y) \wedge \{bg(y)\}]$. Thus, for any $x, y \in [A, B]$, we have $q(x, y) \geq c\{a_1 \wedge (ba_2)\} = \epsilon > 0$; so the conditions of Theorem 15.4 are satisfied for the $\epsilon$ and any $\delta > 0$. It follows that the convergence results of Theorem 15.1 hold for the rejection sampling chain in this special case.

## 15.4 Monte Carlo EM algorithm

The EM (Expectation–Maximization) algorithm was originally proposed by Dempster et al. (1977) to overcome computational difficulties associated with the maximum likelihood method. A key element of the EM algorithm is the so-called incomplete data. Usually, these consist of the observed data, $y$, and some unobserved random variables, $v$. For example, $v$ may be a vector of missing observations, or random effects in a linear mixed model, or a GLMM. The idea is to choose $v$ appropriately so that maximum likelihood becomes easy for the complete data. Let $w = (y', v')'$ denote the complete data. Suppose that $w$ has a pdf $f(w|\theta)$ that depends on a vector $\theta$ of unknown parameters. In the E-step of the algorithm, one computes the conditional expectation

$$Q\{\theta|\theta^{(k)}\} = \mathrm{E}\{\log f(w|\theta)|y, \theta^{(k)}\},$$

where $\theta^{(k)}$ is the estimated $\theta$ at the current iteration $k$. Then, in the M-step, one maximizes $Q\{\theta|\theta^{(k)}\}$ with respect to $\theta$ to obtain an updated estimator $\theta^{(k+1)}$. The procedure is iterated until convergence (e.g., Wu 1983). Here, we are particularly interest in using the EM algorithm for mixed model analysis. Laird and Ware (1982) applied the EM algorithm to the estimation of the variance components in a Gaussian mixed model (see Section 12.2). They focused on a special class of Gaussian mixed models for longitudinal data analysis (e.g., Diggle et al. 1996), but the same idea applies to Gaussian mixed models with multiple random effect factors that can be expressed as

$$y = X\beta + Z_1 v_1 + \cdots + Z_s v_s + e, \tag{15.15}$$

where $\beta$ is the vector of fixed effects; each $v_r$ is a vector of random effects that are independent and distributed as $N(0, \sigma_r^2)$, $1 \leq r \leq s$; $e$ is a vector of errors that are independent and distributed as $N(0, \tau^2)$; $X, Z_1, \ldots, Z_s$ are known matrices; and $v_1, \ldots, v_s$, and $e$ are independent. For simplicity, let $X$ be of full rank. If we treat the random effects as missing data, we have $v = (v_1', \ldots, v_s')'$. The unknown parameters are $\theta = (\beta', \tau^2, \sigma_1^2, \ldots, \sigma_s^2)'$. The log-likelihood based on the complete data has the expression

$$l = c - \frac{1}{2}\left(n\log\tau^2 + \sum_{r=1}^{s} m_r \log\sigma_r^2 + \sum_{r=1}^{s} \frac{|v_r|^2}{\sigma_r^2}\right.$$
$$\left. + \frac{1}{\tau^2}\left|y - X\beta - \sum_{r=1}^{s} Z_r v_r\right|^2\right), \tag{15.16}$$

where $c$ does not depend on the data or parameters, $n$ is the dimension of $y$, and $m_r$ is the dimension of $v_r$ (Exercise 15.13). To complete the E-step, we need expressions for $\mathrm{E}(v_r|y)$ and $\mathrm{E}(|v_r|^2|y)$, $1 \leq r \leq s$. It can be shown that

$$\mathrm{E}(v_r|y) = \sigma_r^2 Z_r' V^{-1}(y - X\beta),$$
$$\mathrm{E}(|v_r|^2|y) = \sigma_r^4 (y - X\beta)' V^{-1} Z_r Z_r' V^{-1}(y - X\beta)$$
$$+ \sigma_r^2 m_r - \sigma_r^4 \mathrm{tr}(Z_r' V^{-1} Z_r), \quad 1 \leq r \leq s, \tag{15.17}$$

where $V = \tau^2 I_n + \sum_{r=1}^{s} \sigma_r^2 Z_r Z_r'$ (Exercise 15.13). Once the E-step is done, the M-step is easier, because the maximizer of $\mathrm{E}\{l|y, \theta^{(k)}\}$ is given by $\theta^{(k+1)} = [\{\beta^{(k+1)}\}', \{\tau^2\}^{(k+1)}, \{\sigma_1^2\}^{(k+1)}, \ldots, \{\sigma_s^2\}^{(k+1)}]'$, where, in particular,

$$\beta^{(k+1)} = (X'X)^{-1}X'\left\{y - \sum_{q=1}^{s} Z_q \mathrm{E}(v_q|y)|_{\theta=\theta^{(k)}}\right\},$$

$$\{\sigma_r^2\}^{(k+1)} = \frac{1}{m_r}\mathrm{E}(|v_r|^2|y)|_{\theta=\theta^{(k)}}, \quad 1 \leq r \leq s, \qquad (15.18)$$

and there is also a closed-form expression for $\{\tau^2\}^{(k+1)}$ in terms of $y$ and $\theta^{(k)}$.

However, the strategy encounters some difficulties when dealing with GLMMs. The problem is the E-step. There are no closed-form expressions for the conditional expectations such as (15.17). In some relatively simple cases, the conditional expectations may be evaluated using numerical integration, but for more complicated cases, such as GLMMs with crossed random effects (see Example 15.1), numerical integration may be intractable. McCulloch (1994) proposed using the Gibbs sampler to approximate the conditional expectations in the E-step. He considered a special case of the GLMM that can be expressed as a threshold model. Let $u_i, i = 1, \ldots, n$, be some continuous random variables that are unobservable. Instead, one observes $y_i = 1_{(u_i>0)}$; that is, whether or not $u_i$ exceeds a threshold which, without loss of generality, is set to zero. Such models have been used in economics to describe, for example, the mechanisms that drive particular choices of consumers (e.g., Manski and McFadden 1981). To incorporate with the covariate information and random effects, a linear mixed model is assumed for $u = (u_i)_{1\leq i\leq n}$ so that it satisfies (15.15) with $y$ replaced by $u$, and everything else being the same. McCulloch showed that, in this case, the Gibbs sampler for the E-step is equivalent to the following:

1. Compute $\mu_i = \mathrm{E}(u_i|u_j, j \neq i) = x_i'\beta + c_i'(u_{-i} - X_{-i}\beta)$, where $x_i'$ is the $i$th row of $X$, $c_i = \mathrm{Cov}(u_i, u_{-i})$, $u_{-i} = (u_j)_{j\neq i}$, and $X_{-i}$ is $X$ without its $i$th row; also compute $\sigma_i^2 = \mathrm{var}(u_i|u_j, j \neq i)$.

2. Simulate $u_i$ from a truncated normal distribution with mean $\mu_i$ and standard deviation $\sigma_i$ so that $u_i$ is simulated from the distribution truncated above zero if $y_i = 1$ and from the distribution truncated below zero if $y_i = 0$.

Using the Gibbs sampler, McCulloch (1994) analyzed the salamander mating data (see Example 15.1) via the EM algorithm and obtained the MLEs of the parameters under the GLMM with crossed random effects.

In general, an EM algorithm that involves Monte Carlo approximations in the E-step is called a Monte Carlo EM (MCEM) algorithm. The MCEM by the Gibbs sampler described above has certain limitations. For example, it only applies to the *probit* link (e.g., McCullagh and Nelder 1989) with normally distributed random effects; otherwise, the sampling from the conditional distribution of $u_i$ given $u_{-i}$ may encounter a problem. To extend the MCEM to more general GLMMs, McCulloch (1997) used the M-H algorithm instead of the Gibbs sampler for the E-step. Consider a GLMM with a *canonical link* (McCullagh and Nelder 1989) such that the conditional density of $y_i$ given the random effects $v$ can be expressed as

$$f(y_i|v, \beta, \phi) = \exp\left\{\frac{y_i\eta_i - b(\eta_i)}{a(\phi)} + c(y_i, \phi)\right\}, \tag{15.19}$$

where $\eta_i = x_i'\beta + z_iv$, $x_i$ and $z_i$ are known vectors, $a(\cdot)$, $b(\cdot)$, and $c(\cdot, \cdot)$ are known functions, $\beta$ is a vector of fixed effects, and $\phi$ is an additional dispersion parameter. Furthermore, it is assumed that $v \sim g(\cdot|\theta)$, where $\theta$ is a vector of variance components. Such GLMMs were considered in Section 12.4, where it was indicated that maximum likelihood estimation of the model parameters may be computationally challenging. McCullogh (1997) suggested evaluating the MLEs using the MCEM. Note that the log-likelihood function based on the complete data can be expressed as

$$l = \log f(y|v, \beta, \phi) + \log g(v|\theta).$$

Thus, the E-step involves evaluations of the conditional expectations

$$\text{E}\{\log f(y|v, \beta, \phi)|y\} \quad \text{and} \quad \text{E}\{\log g(v|\theta)|y\}. \tag{15.20}$$

To obtain Monte Carlo approximations to (15.20), we need to draw samples from the conditional distribution $f_{v|y}$, which is the target distribution $f$ by the notation of Section 15.3. Let $v = (v_i)_{1\leq i\leq m}$. McCullogh (1997) suggested using the M-H algorithm in an $m$-step cycle at each iteration, with one component being sampled in each step. A variation is the following, which is simpler for illustration purpose. Let $v$ be be previous draw. We need a jumping distribution to sample a candidate from for the current draw. A natural choice is the (marginal) distribution of $v$—that is, $q = g(\cdot|\theta)$ (note that, in the E-step, all of the parameters, $\beta$, $\phi$, and $\theta$, are evaluated at the current value). There is another advantage for this choice of $q$: The acceptance probability for the M-H algorithm has the neat form

$$\alpha = \left\{\frac{\prod_{i=1}^n f(y_i|v^*, \beta, \phi)}{\prod_{i=1}^n f(y_i|v, \beta, \phi)}\right\} \wedge 1, \tag{15.21}$$

where $v^*$ is the candidate sampled from $g(\cdot|\theta)$ (Exercise 15.14). Note that the expression only involves the conditional density of $y$ given $v$, which is given by (15.19) in closed form. We consider an example.

*Example 15.9.* Consider the mixed logistic model (13.1). Denote the random effects by $v$ instead. Then we have

$$f(y_{ij}|v, \beta, \sigma) = \frac{\exp\{y_{ij}(x_{ij}'\beta + v_i)\}}{1 + \exp(x_{ij}'\beta + v_i)}.$$

Furthermore, we have $g(v|\sigma) = (2\pi\sigma)^{-m/2}\exp(-|v|^2/2\sigma^2)$, where $|v|^2 = \sum_{i=1}^m v_i^2$. Thus, the sampling from $q$ is done by drawing $v_1^*, \ldots, v_m^*$ independently from $N(0, \sigma^2)$. Furthermore, we have (verify)

$$\alpha = \left\{ \prod_{i=1}^{m} \prod_{j=1}^{n_i} \frac{1 + \exp(x'_{ij}\beta + v_i)}{1 + \exp(x'_{ij}\beta + v_i^*)} \right\} \exp \left\{ \sum_{i=1}^{m} y_{i\cdot}(v_i^* - v_i) \right\}.$$

It should be pointed out that MCEM does not have to involve MCMC, as above. Other Monte Carlo methods, such as i.i.d. sampling, may also be used, depending on the situation. For example, Booth and Hobert (1999) used a strategy called *importance sampling* for the E-step, which allowed them to draw i.i.d. samples for the Monte Carlo approximation. They noted that the E-step is all about computing $Q\{\psi|\psi^{(k)}\} = \mathrm{E}\{\log f(y, v|\psi)|y, \psi^{(k)}\}$, where $\psi = (\beta', \phi, \theta')'$ and $k$ represents the current step. The expected value is computed under the conditional distribution of $v$ given $y$, which has density

$$f\{v|y, \psi^{(k)}\} \propto f\{y|v, \beta^{(k)}, \phi^{(k)}\} f\{v|\theta^{(k)}\}, \tag{15.22}$$

where $f(v|\theta)$ represents the (marginal) density of $v$ when $\theta$ is the parameter vector, and so on. There is a normalizing constant involved in (15.22), which is the (marginal) density $f\{y|\psi^{(k)}\}$. This constant is difficult to evaluate; however, as noted by Booth and Hobert, the constant does not play a role in the next M-step, because it only depends on $\psi^{(k)}$, which is held fixed during the maximization. The idea of importance sampling is the following. Suppose that one wishes to evaluate an integral of the form $I(f) = \int f(x) \, dx$ for some function $f \geq 0$. We can write the integral in an alternative expression:

$$I(f) = \int \frac{f(x)}{h(x)} h(x) \, dx = \mathrm{E} \left\{ \frac{f(\xi)}{h(\xi)} \right\},$$

where $\xi$ has pdf $h$. Here, $h$ is chosen such that $f(x) > 0$ implies $h(x) > 0$; in other words, the support of $f$ is a subset of the support of $h$. Let $\xi_1, \ldots, \xi_L$ be i.i.d. samples drawn from $h$; then the integral $I(f)$ can be approximated by $L^{-1} \sum_{l=1}^{L} f(\xi_l)/h(\xi_l)$ according to the SLLN, if $\mathrm{E}\{f(\xi)/h(\xi)\} < \infty$. For the evaluation of the conditional expectation in the E-step, Booth and Hobert suggested to use a multivariate $t$-distribution as $h$. The latter depends on the dimension $d$, mean vector $\mu$, covariance matrix $\Sigma$, and degrees of freedom $\nu$, so that it has the joint pdf

$$h(x) = \frac{\Gamma\{(d+\nu)/2\}}{(\pi\nu)^{d/2}\Gamma(\nu/2)} |\Sigma|^{-1/2} \left\{ 1 + \frac{1}{\nu}(x - \mu)'\Sigma^{-1}(x - \mu) \right\}^{-(d+\nu)/2},$$

$x \in R^d$, where $\Gamma$ is the gamma function. Let $v_1^*, \ldots, v_L^*$ be i.i.d. samples generated from $h$; then we have the approximation

$$Q\{\psi|\psi^{(k)}\} \approx \frac{1}{L} \sum_{l=1}^{L} w_{kl} \log f(y, v_l^*|\psi), \tag{15.23}$$

where $w_{kl} = f\{v_l^*|y, \psi^{(k)}\}/h(v_l^*)$, known as the *importance weights*. The right side of (15.23) is then maximized with respect to $\psi$ to get $\psi^{(k+1)}$. As noted, the right side of (15.23) is not a completely known function of $\psi$ due to the unknown constant $f\{y|\psi^{(k)}\}$, but the latter makes no difference in the M-step and therefore can simply be ignored (i.e., replaced by 1). Note that the importance weights do not depend on $\psi$ and therefore are treated as constants during the M-step (see Exercise 15.15).

There have been some good advances in computing the MLEs in the GLMM using MCEM algorithms. Although the procedures are still computationally intensive, with the fast developments of computer hardware and technology, it is very probable that computation of the exact MLEs in the GLMM will eventually become a routine operation. On the other hand, one important theoretical problem regarding the MLEs in the GLMM remains unsolved. To make a long story short, recall the salamander mating data that have had a significant impact to the development of GLMMs as well as other fields even since its original publication by McCullagh and Nelder (1989, Section 14.5). See, for example, Karim and Zeger (1992), Lin and Breslow (1996), Jiang (1998a), McCulloch (1994), and Booth and Hobert (1999). In particular, the authors of the last two papers were able to compute the MLEs of a GLMM that fits the data. However, one fundamental question has yet to be answered: Are the MLEs consistent as the number of female and male salamanders involved in the experiments grow? To make this an even simpler question, consider Example 15.1 (or earlier Example 12.9). Suppose that $\sigma_1^2 = \sigma_2^2 = 1$; so $\mu$ is the only unknown parameter. Suppose that $m_1, m_2 \to \infty$. Is the MLE of $\mu$ consistent? Even for this seemingly trivial case, the answer is not known but is expected to be anything but trivial (see Exercise 15.16)!

## 15.5 Convergence rates of Gibbs samplers

The key to the success of a MCMC strategy is the convergence of the Markov chain to its stationary, or *equilibrium*, distribution. A important practical issue, however, is not just whether or not the chain converges, but how fast it converges, for it certainly would not help if the Markov chain takes years to converge. In fact, the convergence rate has been a main factor in designing more efficient MCMC algorithms, although it is not the only factor.

Consider, for example, a random variable $X = (X_1, X_2, X_3)$ with three components. Suppose that it is desired to draw samples from the distribution of $X$, $f(\cdot)$. For notation simplicity, let $f(\xi|\eta)$ denote a conditional density, with respect to a certain measure $\nu$, of $\xi$ given $\eta$, and so on. Several Gibbs sampler strategies are in consideration. The *direct* Gibbs sampler draws samples from $f(x_i|x_{-i})$, $i = 1, 2, 3$, in turn, where $x_{-i}$ is the vector $(x_j)_{j\neq i}$. Now, suppose that we are able to draw $x_2$ and $x_3$ together from $f(x_2, x_3|x_1)$. This will be the case if we are able to draw $x_2$ from $f(x_2|x_1)$ and then $x_3$ from $f(x_3|x_1, x_2)$. This leads to an alternative strategy, called *grouping*, by drawing from $f(x_1|x_2, x_3)$ and $f(x_2, x_3|x_1)$ in turn. In other words, we are running a two-stage Gibbs sampler that involves $x_1$ and $(x_2, x_3)$. Furthermore, suppose that we can draw $x_1$ from $f(x_1|x_2)$ and $x_2$ from $f(x_2|x_1)$ (i.e., with $x_3$ integrated out). Then the Gibbs sampler can be restricted to $x_1$ and $x_2$. This strategy is called *collapsing*. It makes sense, in particular, if the problem of interest only involves the joint distribution of $X_1$ and $X_2$. In conclusion, we have, at least, three different strategies: direct, grouping, and collapsing. Which one should we use?

As noted, one way to investigate the problem is to study the convergence rates of the corresponding (Gibbs) Markov chains. Traditionally, such studies were done by making use of the (advanced) theory of Markov chains (e.g., Geman and Geman 1984; Nummelin 1984; Tierney 1991). In a series of papers, Liu (1994) and Liu et al. (1994, 1995) provided an "elementary" approach to the problem using simple functional analysis and inequalities. Let $X_0, X_1, \ldots$ be consecutive samples of a stationary Markov chain with stationary distribution $f$ and transition kernel $K$. For the rest of the section, E( ) and var( ) denote the expectation and variance under $f$, whereas $E_p(\ )$ is the expectation under a probability measure $p$, if it is different from $f$. The Hilbert space of all mean 0, square-integrable, complex-valued functions of $X$ is denoted by

$$L_0^2(f) = \{g(X) : E\{g(X)\} = 0 \text{ and } E\{|g(X)|^2\} < \infty\}.$$

The inner product in the Hilbert space is defined by

$$\langle g(X), h(X) \rangle = E\{g(X)\overline{h(X)}\}, \quad g(X), h(X) \in L_0^2(f),$$

where $\bar{c}$ denotes the complex conjugate of $c$ (and also $|c|$ denotes the modulus of the complex number $c$). The variance of an element $g(X)$ in $L_0^2(f)$ is then defined as $\|g(X)\|^2 = \langle g(X), g(X) \rangle = E\{|g(X)|^2\}$. Define a operator $\mathbf{F}$ on $L_0^2(f)$, called the *forward operator*, by

$$\mathbf{F}g(X) = E\{g(X_1)|X_0 = X\} = \int g(y)K(X, y)\, dy. \qquad (15.24)$$

It is clear that $\mathbf{F}$ is an operator from $L_0^2(f)$ to itself. Define the norm of $\mathbf{F}$ as

$$\|\mathbf{F}\| = \sup_{g \in L_0^2(f), \|g\|=1} \|\mathbf{F}g\|,$$

where $g$ is a short form of $g(X)$. Let $p$ be a probability measure. The Pearson $\chi^2$-discrepancy between $p$ and $f$, denoted by $d_f(p, f)$, is defined as

$$d_f^2(p, f) = \int \left\{ \frac{p^2(x)}{f(x)} \right\} dx - 1. \qquad (15.25)$$

It is easy to see that $d_f^2(p, f) = \mathrm{var}\{p(X)/f(X)\}$ and, therefore, nonnegative (which explains the use of square). However, $d_f$ is not a distance (Exercise 15.17). Nevertheless, it can be shown that $d_f$ is a stronger measure of discrepancy than the $L^1$-distance, and $d_f^2$ is a stronger measure than a certain kind of Kullback–Leibler information distance (Exercise 15.17).

Let $f_t$ denote the pdf of $X_t$. Note that, by the Markovian property, we have the following expressions (Exercise 15.18):

$$\mathrm{E}_{f_t}\{g(X)\} = \int\int g(y)K(x,y)f_{t-1}(x)\,dx\,dy, \qquad (15.26)$$

$$\mathrm{E}\{g(X)\} = \int\int g(y)K(x,y)f(x)\,dx\,dy \qquad (15.27)$$

for every $g \in L_0^2(f)$. It follows that

$$\mathrm{E}_{f_t}\{g(X)\} - \mathrm{E}\{g(X)\} = \int\int g(y)K(x,y)\left\{\frac{f_{t-1}(x)}{f(x)} - 1\right\}f(x)\,dx\,dy$$

$$= \int \mathbf{F}g(x)\left\{\frac{f_{t-1}(x)}{f(x)} - 1\right\}f(x)\,dx$$

according to the definition of $\mathbf{F}$, (5.24) (verify the last expression). It follows, by the Cauchy–Schwarz inequality [see (5.60)], that

$$|\mathrm{E}_{f_t}\{g(X)\} - \mathrm{E}\{g(X)\}| \leq \|\mathbf{F}g(X)\|d_f(f_{t-1},f)$$
$$\leq \|\mathbf{F}\| \cdot \|g(X)\|d_f(f_{t-1},f). \qquad (15.28)$$

Note that (Exercise 15.18)

$$\int\left\{\frac{f_{t-1}(x)}{f(x)} - 1\right\}^2 f(x)\,dx = d_f^2(f_{t-1},f). \qquad (15.29)$$

Note that (15.28) holds for every $g(X) \in L_0^2(f)$. Now, consider $g(X) = f_t(X)/f(X) - 1$. We have $\mathrm{E}\{g(X)\} = 0$ and

$$\mathrm{E}\{g(X)^2\} = \int\left\{\frac{f_t(x)}{f(x)} - 1\right\}^2 f(x)\,dx = d_f^2(f_t,f) \qquad (15.30)$$

[see (15.29)], which we assume to be finite. Thus, $g(X) \in L_0^2(f)$ and therefore satisfies (15.28). However, for this particular $g(X)$, the left side of (15.28) is equal to $d_f^2(f_t,f)$, whereas the right side is equal to $\|\mathbf{F}\|d_f(f_t,f)d_f(f_{t-1},f)$. Thus, again, assuming that (15.30) is finite, we obtain the inequality

$$d_f(f_t,f) \leq \|\mathbf{F}\|d_f(f_{t-1},f). \qquad (15.31)$$

The inequality applies to any $t$ so long as (15.30) is finite. So, if we assume that the $\chi^2$-discrepancy is finite for some $t = t_0$, then we have, for any $t \geq t_0$,

$$d_f(f_t,f) \leq \|\mathbf{F}\|d_f(f_{t-1},f)$$
$$\leq \|\mathbf{F}\|^2 d_f(f_{t-2},f)$$
$$\cdots$$
$$\leq \|\mathbf{F}\|^{t-t_0}d_f(f_{t_0},f)$$
$$= c_0\|\mathbf{F}\|^t, \qquad (15.32)$$

where $c_0 = d_f(f_{t_0}, f)/\|\mathbf{F}\|^{t_0} < \infty$.

Inequalities (15.32) suggest that the convergence rate of the Markov chain is closely related to $\|\mathbf{F}\|$, the norm of the forward operator. Furthermore, Liu et al. (1995) showed that, under mild conditions, we have $\|\mathbf{F}\| < 1$ (it is easy to show $\|\mathbf{F}\| \leq 1$; see Exercise 15.19). Thus, the Pearson $\chi^2$-discrepancy between $f_t$ and $f$ decreases at a geometric rate, and the rate is determined by $\|\mathbf{F}\|$. Let $X \sim f$; then for any $B \in \mathcal{B}$, the Borel $\sigma$-field, we have

$$|\mathrm{P}(X_t \in B) - \mathrm{P}(X \in B)| = \left| \int_B f_t(x) \, dx - \int_B f(x) \, dx \right|$$
$$\leq \int |f_t(x) - f(x)| \, dx$$
$$= \|f_t - f\|_{L^1}$$
$$\leq d_f(f_t, f)$$

[see Exercise 15.17, part (ii)]. Thus, $\sup_{B \in \mathcal{B}} |\mathrm{P}(X_t \in B) - \mathrm{P}(X \in B)|$ also converges to zero at the above geometric rate.

Returning to the three Gibbs sampler schemes discussed earlier. Let $\mathbf{F}_d$, $\mathbf{F}_g$, and $\mathbf{F}_c$ denote the forward operators of the direct, grouped, and collapsed Gibbs samplers. Liu et al. (1994) proved the following elegant inequalities

$$\|\mathbf{F}_c\| \leq \|\mathbf{F}_g\| \leq \|\mathbf{F}_d\|. \tag{15.33}$$

The results are intuitive, in view of the above result on the convergence rate. For example, with grouping, one is running a two-stage cycle rather than three-stage one; so would expect a faster convergence rate; that is, $\|\mathbf{F}_g\| \leq \|\mathbf{F}_d\|$. Similarly, since one additional variable, $x_3$, has to be sampled and updated jointly with $x_2$, one would expect the collapsed chain to converge faster than the grouped chain; that is, $\|\mathbf{F}_c\| \leq \|\mathbf{F}_g\|$. To establish these inequalities, we introduce another concept, called *maximum correlation*, and relate it to the norm of the forward operator. Given two vector-valued random variables $\xi$ and $\eta$, their maximum correlation is defined as

$$\rho(\xi, \eta) = \sup_{g,h:\mathrm{var}\{g(\xi)\}<\infty, \mathrm{var}\{h(\eta)\}<\infty} \mathrm{cor}\{g(\xi), h(\eta)\}, \tag{15.34}$$

where $\mathrm{cor}(U, V) = \mathrm{cov}(U, V)/\sqrt{\mathrm{var}(U)\mathrm{var}(V)}$. A useful alternative expression is given by (Exercise 15.20)

$$\{\rho(\xi, \eta)\}^2 = \sup_{g:\mathrm{var}\{g(\xi)\}=1} \mathrm{var}[\mathrm{E}\{g(\xi)|\eta\}]. \tag{15.35}$$

A connection between the forward operator and the maximum correlation is the following, which can be derived directly from the definitions of $\mathbf{F}$ [see (15.24)] as well as its norm, and (15.35) (Exercise 15.20):

$$\|\mathbf{F}\| = \rho(X_0, X_1). \tag{15.36}$$

Consider a stationary Markov chain $M_t = (X_t, U_t), t = 0, 1, \ldots$, generated by a two-stage Gibbs sampler that cycles between $X$ and $U$, where $U_t$ may be multidimensional. Liu et al. (1994, Theorem 3.2) showed the following connection regarding the maximum correlation between consecutive samples of the chain and that between $X$ and $U$ within the same cycle,

$$\rho(M_{t-1}, M_t) = \rho(X_t, U_t), \quad t = 1, 2, \ldots. \tag{15.37}$$

Now, use the notation $X$, $Y$, and $Z$ for $X_1$, $X_2$, and $X_3$, respectively. To establish the left-side inequality in (15.33), first consider $U_t = Y_t$. By (15.36) and (15.37), we have $\|\mathbf{F}_c\| = \rho(M_{t-1}, M_t) = \rho(X_t, Y_t)$, which is

$$\sup_{g:\mathrm{var}\{g(X_t)\}=1} \mathrm{var}[\mathrm{E}\{g(X_t)|Y_t\}]. \tag{15.38}$$

Now, consider $U_t = (Y_t, Z_t)$. Again, by (15.36) and (15.37), we have $\|\mathbf{F}_g\| = \rho(M_{t-1}, M_t) = \rho\{X_t, (Y_t, Z_t)\}$, which is

$$\sup_{g:\mathrm{var}\{g(X_t)\}=1} \mathrm{var}[\mathrm{E}\{g(X_t)|Y_t, Z_t)\}]. \tag{15.39}$$

That (15.38) is bounded by (15.39) follows from the facts that, with $\xi = \mathrm{E}\{g(X_t)|Y_t, Z_t\}$, $\eta = Y_t$, and $\zeta = \mathrm{E}\{g(X_t)|Y_t\}$, we have $\mathrm{var}(\xi) = \mathrm{var}\{\mathrm{E}(\xi|\eta)\} + \mathrm{E}\{\mathrm{var}(\xi|\eta)\} \geq \mathrm{var}\{\mathrm{E}(\xi|\eta)\} = \mathrm{var}(\zeta)$.

The proof of the right-side inequality is more technical and involves the properties of the *backward operator*, defined as

$$\mathbf{B}g(X) = \mathrm{E}\{g(X_0)|X_1 = X\} = \int g(x) \frac{f(x)K(x, X)}{f(X)} dx. \tag{15.40}$$

The backward operator also satisfies (15.36) (with $\mathbf{F}$ replaced by $\mathbf{B}$). See Liu et al. (1994) for more details.

Inequalities (15.32) and (15.33) suggest that, in terms of convergence rate, collapsing is better than grouping, which, in turn, is better than the direct Gibbs sampler. However, one cannot simply conclude that collapsing is always the best strategy, followed by grouping, and so on, based on these facts alone. The reason is that, for example, sometimes it is much easier to sample $X_1$ given $X_2$ and $X_3$ than to sample $X_1$ given $X_2$, and this would give an advantage to grouping in terms of sampling. As noted by Liu (1994), a good Gibbs sampler must meet two conflicting criteria: (i) drawing one component conditioned on the others must be computationally simple and (ii) the Markov chain induced by the Gibbs sampler must converge reasonably fast to its stationary distribution. Given the variables $X_1, \ldots, X_s$, there are many ways to group them. How to do so is a decision that the users have to make, "providing an opportunity to their ingenuity" (Liu 1994, p. 963). The results about the convergence rates should be helpful to the users in making a compromise decision. For example, it seems to be a reasonable strategy to "group" or "collapse" whenever it is computationally feasible. We consider an example.

*Example 15.10.* Liu (1994) considered the following Bayesian missing value problem. Recall that the posterior distribution is calculated by (15.3), where $x$ represents the vector of observations. In many applications, however, the data contain missing values. Let $x_i = (y_i, z_i), i = 1, \ldots, n$ denote the data, where $y_i$ is the observed part and $z_i$ is the missing part, for the $i$th subject. Write $X = (Y, Z)$, where $Y = (y_1, \ldots, y_n)$ and $Z = (z_1, \ldots, z_n)$. The posterior based on the observed data can be expressed as

$$p(\theta|Y) = \int p(\theta|Y, Z) f(Z|Y) \, dZ, \tag{15.41}$$

where $p(\theta|Y, Z)$ is the posterior based on the complete data and $f(Z|Y)$ is the conditional density of $Z$ given $Y$, known as the predictive distribution. If we are able to draw samples $Z^{(1)}, \ldots, Z^{(K)}$ from the latter, then, by the ergodic theorem (see Section 7.6) and (15.41), we have

$$p(\theta|Y) \approx \frac{1}{K} \sum_{k=1}^{K} p\{\theta|Y, Z^{(k)}\},$$

if $K$ is large. The problem is that it is often impossible to draw $Z$ directly from the predictive distribution. Tanner and Wong (1987) proposed a strategy, called data augmentation (DA), to handle the problem. The DA may be regarded as a two-stage Gibbs sampler, in which one draws $\theta$ from $p(\theta|Y, Z)$ and then $Z$ from $f(Z|\theta, Y)$, and iterates. As we have seen, this generates a Markov chain whose stationary distribution is $f(\theta, Z|Y)$, the joint conditional distribution of $\theta$ and $Z$ given $Y$, which is, of course, more than we are asking for because $f(Z|Y)$ can be evaluated from $f(\theta, Z|Y)$.

If our main interest is the predictive distribution, this can be done by collapsing down $\theta$—namely, from a Gibbs sampler with $n + 1$ components, $\theta, z_1, \ldots, z_n$ (which is what DA does), to that with $n$ components, $z_1, \ldots, z_n$ (note that everything is conditional on $Y$, of course). To see that this is something often convenient to do, consider a special case. Suppose that $X_1$, $X_2$, and $X_3$ are random variables such that $X_1$ and $X_3$ are independent given $X_2$,

$$P(X_2 = 1|X_1 = 0) = P(X_2 = 0|X_1 = 1) = \alpha,$$
$$P(X_3 = 1|X_2 = 0) = P(X_3 = 0|X_2 = 1) = \beta,$$

where $\alpha, \beta \in (0, 1)$ are parameters. This is a simple case of the so-called *graphical model*. Suppose that the incomplete data for $(X_1, X_2, X_3)$ are $(1, 1, 0)$, $(1, z_2, 1)$, and $(1, z_3, 1)$, where $z_2$ and $z_3$ are missing values (note that there is no $z_1$). Here, $Y = (x_{11}, x_{12}, x_{13}, x_{21}, x_{23}, x_{31}, x_{33}) = (1, 1, 0, 1, 1, 1, 1)$. Write $u = (x_{11}, x_{12}, x_{13})$ and $v = (x_{21}, x_{23}, x_{31}, x_{33})$. Below, let $f(\xi|\eta)$ denote the conditional pmf of $\xi$ given $\beta$ and refer the derivations to an exercise (Exercise 15.21). It is can be shown that

$$f(v, z_2, z_3|u, \alpha, \beta) = \alpha^{2-z_2-z_3}(1-\alpha)^{1+z_2+z_3}\beta^{3-z_2-z_3}(1-\beta)^{z_2+z_3}. \tag{15.42}$$

By integrating out $\alpha$ and $\beta$ over $[0,1] \times [0,1]$ from (15.42), an expression for $f(v, z_2, z_3|u)$ is obtained. Then since

$$f(z_3|z_2, Y) = \frac{f(z_2, z_3|Y)}{f(z_2|Y)} = \frac{f(v, z_2, z_3|u)}{f(z_2|Y)f(v|u)},$$

it can be shown that

$$\frac{P(z_3 = 0|z_2, Y)}{P(z_3 = 1|z_2, Y)} = \frac{(2 - z_2)(3 - z_2)}{(1 + z_2)(2 + z_2)}. \tag{15.43}$$

From (15.43), the conditional distribution of $z_3$ given $z_2$ and $Y$ can be easily derived, and similarly for the conditional distribution of $z_2$ given $z_3$ and $Y$.

Another implication of (15.32) and (15.36) is that convergence rate of the Gibbs sampler is closely related to the autocorrelation, or autocovariance, of the chain generated by the Gibbs sampler—the closer the autocovariance to zero, the faster the chain is converging. This important observation suggests a way of monitoring the convergence by plotting the autocovariance (or autocorrelation) function. For example, Figure 15.1, which is kindly provided by Professor J. S. Liu, shows an autocovariance plot comparing the convergence rates of the direct and collapsed Gibbs samplers for analyzing Murray's data (Murry 1977). The data consist of 12 bivariate observations with some of the components missing. The observations are assumed drawn from a bivariate normal distribution with means zero and unknown covariance matrix. Liu (1994) compared the direct and collapsed Gibbs sampling schemes in DA (see Example 15.10). In this case, the collapsed Gibbs sampler involves drawing from the noncentral $t$-distribution, which is easy to implement. The plot shows two groups of autocovariance curves, each with eight curves for eight missing values. The curves are estimated from simulations of 100 independent chains with 100 iterations for each chain. The estimated autocovariance for the direct scheme is about 2.5 times larger than that for the collapsed scheme, indicating a (much) faster convergence rate for the collapsed Gibbs sampler.

## 15.6 Exercises

*15.1.* Using the simple Monte Carlo method based on the SLLN [i.e., (15.2)] numerically evaluate $\rho_w, \rho_s$, and the asymptotic significance level in Table 11.2 for $\alpha = 0.05$ for the case $F = t_3$. Try $n = 1000$ and $n = 10,000$. Do the numerical values seem to stabilize (you can find out by repeating the calculation with the same $n$ and comparing the results to see how much they vary)? What about $n = 100,000$.

*15.2.* This exercise is in every book, chapter, or section about Bayesian inference. Suppose that the distribution of $y$ depends on a single parameter, $\theta$. The conditional distribution of the observation $y$ given $\theta$ is $N(\theta, \sigma^2)$, where

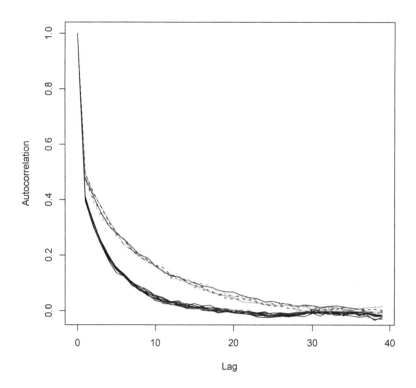

**Fig. 15.1.** *A plot from Liu (1994): The upper group of curves correspond to the direct, while the lower group to the collapsed, Gibbs samplers.*

$\sigma^2$ is assumed known. Let the prior for $\theta$ also be normal, say, $\theta \sim N(\mu, \tau^2)$. Derive the posterior of $\theta$.

*15.3.* Now, let us consider a discrete situation. Suppose that the observation $y$ follows a Poisson distribution with mean $\theta$. Furthermore, the prior for $\theta$ is proportional to $\exp(\nu \log \theta - \eta \theta)$; that is,

$$\pi(\theta) \propto \exp\{\nu \log(\theta) - \eta \theta\},$$

where $\eta, \nu$ are some hyperparameters. What is the posterior of $\theta$?

*15.4.* Regarding *rejection sampling*, introduced in Section 15.1, show that the pdf of the drawn $\theta$ conditional on $u \le p(\theta)/bq(\theta)$ is $p(\theta)$. (Hint: Use Bayes' theorem; see Appendix A.3).

*15.5.* This exercise is regarding the Markovian properties of the Gibbs sampler for the special case described above (15.5).

(i) Show that $Y_t$ is a Markov chain and derive its transition kernel.

(ii) Show that $Z_t = (X_t, Y_t)'$ is a Markov chain with the transition kernel (15.7).

15.6. Regarding Examples 15.2 and 15.3, show the following:

(i) $f_X(x)$, or $\pi_X(B) = \int_B f_X(x)\, dx$, is a stationary distribution for the Markov chain $X_t$.

(ii) $f_Y(y)$, or $\pi_Y(B) = \int_B f_Y(y)\, dy$, is a stationary distribution for the Markov chain $Y_t$.

(iii) $f(x, y)$, or $\pi(B) = \int_B f(x, y)\, dx\, dy$, is a stationary distribution for the Markov chain $Z_t$.

15.7. Consider the bivariate-normal Gibbs sampler of Example 15.4.

(i) Show that the marginal chain $X_t$ has transition kernel

$$K(u, v) = \frac{1}{2\pi(1 - \rho^2)} \int \exp\left\{ -\frac{(v - \rho y)^2}{2(1 - \rho^2)} - \frac{(y - \rho u)^2}{2(1 - \rho^2)} \right\} dy.$$

(ii) Show that the stationary distribution of the chain $X_t$ is $N(0, 1)$.

(iii) Show that $X_{t+1} = \rho^2 X_t + \epsilon_t, t = 1, 2, \ldots$, where $\epsilon_t$ are i.i.d. with distribution $N(0, 1 - \rho^4)$. Using this relation, show that $X_t$ is a (scalar) Markov chain with stationary distribution $N(0, 1)$ (here you are not supposed to use the results of the previous parts, but rather to derive it directly).

15.8. Verify the expressions for the conditional density of $X_1$ given $X_2 = x_2$ and $X_3 = x_3$, as well as the joint density of $X_1, X_2$ in Example 15.5.

15.9. This exercise provides a very simple special case of the Gibbs sampler that was considered by Casella and George (1992). Let the state-space be $S = \{0, 1\}^2$. A probability distribution $\pi$ defined over $S$ has $\pi(\{(i, j)\}) = \pi_{ij}$, $i, j = 0, 1$. The Gibbs sampler draws $\theta = (\theta_1, \theta_2)$ according to the following:

1. Given $\theta_2 = j$, draw $\theta_1$ such that

$$\pi_1(0|j) = P(\theta_1 = 0|\theta_2 = j) = \frac{\pi_{0j}}{\pi_{\cdot j}}, \quad \pi_1(1|j) = P(\theta_1 = 1|\theta_2 = j) = \frac{\pi_{1j}}{\pi_{\cdot j}},$$

where $\pi_{\cdot j} = \pi_{1j} + \pi_{2j}, j = 0, 1$.

2. Given $\theta_1 = i$, draw $\theta_2$ such that

$$\pi_2(0|i)) = P(\theta_2 = 0|\theta_1 = i) = \frac{\pi_{i0}}{\pi_{i\cdot}}, \quad \pi_2(1|i) = P(\theta_2 = 1|\theta_1 = i) = \frac{\pi_{i1}}{\pi_{i\cdot}},$$

where $\pi_{i\cdot} = \pi_{i0} + \pi_{i1}, i = 0, 1$.

Given the initial value $\theta_0 = (\theta_{01}, \theta_{02})$, let $\theta_t = (\theta_{t1}, \theta_{t2})'$ be the drawn of iteration $t, t = 1, 2, \ldots$.

(i) Show that $\theta_t$ is a (discrete state-space) Markov chain with transition probability

$$p\{(i, j), (k, l)\} = \frac{\pi_{kj} \pi_{kl}}{\pi_{\cdot j} \pi_{k\cdot}}.$$

(ii) Show that $\pi$ is the stationary distribution of the chain.

(iii) Using the result of Section 10.2 (in particular, Theorem 10.2), find simple sufficient conditions (or condition) for the convergence of the Markov chain. Are the sufficient conditions (or condition) that you find also necessary, and why? [Hint: The results of Section 10.2 apply to vector-valued discrete space Markov chains in obvious ways. Or, if you wish, you may consider a scalar chain that assumes values in correspondence with $\theta_t$ (e.g., $\psi_t = 10\theta_{t1} + \theta_{t2}$). This is always possible for a discrete state-space chain.]

15.10. Show that (15.11) implies that $f$ is the stationary distribution with respect to $K$.

15.11. Show that the M-H algorithm, described below (15.12), generates a Markov chain $X_t$ whose transition kernel is given by (15.10) with $\alpha$ defined by (15.12) if $f(x)q(x, y) > 0$, and $\alpha = 1$ otherwise. Furthermore, the transition kernel is reversible with respect to $f$, the target density, in the sense of (15.11). [For simplicity, you may assume that $f(x)q(x, y) > 0$ for all $x, y$.]

15.12. This exercise is related to some of the details of the rejection sampling chain of Example 15.8 as a special case of the M-H algorithm.

(i) Show that the rejection sampling scheme described above (15.13) is equivalent to that described in Section 15.1 that involves sampling of a $u$ from Uniform$[0, 1]$.

(ii) Show that the jumping distribution $q(x, y)$ satisfies (15.13).

(ii) Show that the acceptance probability of (15.12) reduces to (15.14).

15.13. Verify that under the Gaussian mixed model (15.15), the log-likelihood function can be expressed as (15.16), where $c$ does not depend on the data or parameters. Also verify the expressions of the conditional expectations (15.17).

15.14. Regarding the M-H algorithm described below (15.20), show that the acceptance probability of the M-H algorithm [i.e., (15.12)] can be expressed as (15.21).

15.15. Specify the right side of (15.23) for the mixed logistic model of Example 15.9. Also specify the importance weights $w_{kl}$ by replacing $f\{y|\psi^{(k)}\}$ by 1, which does not affect the result of the M-step.

15.16. Recall the open question given at the end of Section 15.4 regarding the consistency of MLE in a very simple case of the GLMM. The conjectured answer to the question is yes. Here is evidence (and you can surely find more). Let $m = m_1 \wedge m_2$. Consider a subset of the data $X_k = y_{kk}, k = 1, \ldots, m$. Note that $X_1, \ldots, X_m$ are i.i.d. Let $\tilde{\mu}$ denote the MLE of $\mu$ based on $X_1, \ldots, X_m$. Show that $\tilde{\mu}$ is a consistent estimator of $\mu$; that is, $\tilde{\mu} \xrightarrow{P} \mu$ as $m \to \infty$. The point is that even the MLE based on a subset of the data is consistent. If one uses the MLE based on the full data (and therefore more information), one should do better, right? The challenge (not necessarily for this exercise) is to provide a rigorous justification.

15.17. This exercise is regarding the measure $d_f$ defined by the square root of (15.25).

(i) Show that $d_f$ is not a distance [the definition of a distance is given by the requirements 1–4 below (6.63)].

(ii) Show that $d_f$ is a stronger measure of descrepancy than the $L^1$-distance in the sense that $\|p - f\|_{L^1} \leq d_f(p, f)$, where

$$\|p - f\|_{L^1} = \int |p(x) - f(x)|\, dx$$

is the $L^1$-distance between $p$ and $f$.

(iii) Show that $d_f^2$ is a stronger measure of descrepancy than the Kullback Leibler information distance defined as

$$d_{\mathrm{KL}}(p, f) = \mathrm{E}_p\left[\log\left\{\frac{p(X)}{f(X)}\right\}\right];$$

that is, we have $d_{\mathrm{KL}}(p, f) \leq d_f^2(p, f)$.

15.18. Continue with the previous exercise.

(i) Verify expressions (15.26) and (15.27).

(ii) Verify the identity (15.29).

15.19. Recall the definition of $\|\mathbf{F}\|$ below (15.24). Show that $\|\mathbf{F}\| \leq 1$.

15.20. This exercise is related to the maximum correlation introduced in Section 15.5 [see (15.34)].

(i) Verify the alternative expression (15.35).

(ii) Derive the identity (15.36).

15.21. This exercise is related to Example 15.10.

(i) Verify expression (15.42).

(ii) Derive an expression for $f(v, z_2, z_3|u)$ by integrating out $\alpha$ and $\beta$ over $[0, 1] \times [0, 1]$ from (15.42).

(iii) Verify (15.43); then derive $f(z_3|z_2, Y)$.

(iv) Similarly, derive $f(z_2|z_3, Y)$.

# A

## Appendix

## A.1 Matrix algebra

### A.1.1 Numbers associated with a matrix

Let $A = (a_{ij})_{1\leq i\leq m, 1\leq j\leq n}$ be an $m \times n$ matrix of real numbers. The spectral norm of $A$ is defined as $\|A\| = \{\lambda_{\max}(A'A)\}^{1/2}$, and the 2-norm of $A$ is defined as $\|A\|_2 = \{\mathrm{tr}(A'A)\}^{1/2}$, where $\lambda_{\max}$ denotes the largest eigenvalue (see below). The following inequalities hold (see Lemma 3.7):

$$\|A\| \leq \|A\|_2 \leq \sqrt{m \wedge n}\|A\|.$$

If $m = n$, $A$ is called a square matrix. The trace of a square matrix $A$ is defined as the sum of the diagonal elements of $A$; that is, $\mathrm{tr}(A) = \sum_{i=1}^{n} a_{ii}$. The trace has the following properties:

(i) $\mathrm{tr}(A) = \mathrm{tr}(A')$, where $A'$ denotes the transpose of $A$.

(ii) $\mathrm{tr}(A + B) = \mathrm{tr}(A) + \mathrm{tr}(B)$ for any square matrices $A$ and $B$ of the same dimension.

(iii) $\mathrm{tr}(cA) = c\,\mathrm{tr}(A)$ for any square matrix $A$ and real number $c$.

(iv) $\mathrm{tr}(AB) = \mathrm{tr}(BA)$ for any matrices $A$ and $B$, provided that $AB$ and $BA$ are well-defined square matrices.

Let $\pi = (\pi_1, \ldots, \pi_n)$ denote an arbitrary permutation of $(1, \ldots, n)$. The number $\#(\pi)$ of inversions of $\pi$ is the number of exchanges of pairs of integers $\pi$ to bring them to the natural order $1, \ldots, n$. Let $A = (a_{ij})_{1\leq i,j\leq n}$ be a square matrix. The determinant of $A$ is defined as

$$|A| = \sum_{\text{all } \pi} (-1)^{\#(\pi)} \prod_{i=1}^{n} a_{\pi_i, i}.$$

The determinant has the following properties:

(i) $|A| = |A'|$.

(ii) If the row (or column) of $A$ is multiplied by a number $c$, $|A|$ is multiplies by $c$. It follows that $|cA| = c^n|A|$ for $n \times n$ matrix $A$.

J. Jiang, *Large Sample Techniques for Statistics*,
DOI 10.1007/978-1-4419-6827-2, © Springer Science+Business Media, LLC 2010

(iii) If two rows (or columns) of $A$ are interchanged, the sign of $|A|$ changes. It follows that if two rows (or columns) of $A$ are identical, then $|A| = 0$.

(iv) The value of $|A|$ is unchanged if to its $i$th row (column) is added $c$ times the $j$th row (column), where $c$ is a real number. Thus, in particular, if the rows (or columns) of $A$ are not linearly independent, then $|A| = 0$.

(v) If $A = \mathrm{diag}(a_1, \ldots, a_n)$, then $|A| = a_1 \cdots a_n$.

(vi) $|AB| = |A| \cdot |B|$ if the determinants are well defined.

(vii) $|AA'| \geq 0$ and $|A'A| \geq 0$ for any matrix $m \times n$ matrix $A$.

(viii) $\left| \begin{pmatrix} A & C \\ 0 & B \end{pmatrix} \right| = |A| \cdot |B|$.

(ix) $|I_m + AB| = |I_n + BA|$ for any $m \times n$ matrix $A$ and $n \times m$ matrix $B$.

### A.1.2 Inverse of a matrix

The inverse of a matrix $A$ is defined as a matrix $B$ such that $AB = BA = I$, the identity matrix. The inverse of $A$, if it exists, is unique and denoted by $A^{-1}$. The following are some basic properties of the inverse:

(i) $A^{-1}$ exists if and only if $|A| \neq 0$.

(ii) $(A')^{-1} = (A^{-1})'$.

(iii) $(cA)^{-1} = c^{-1}A^{-1}$, where $c$ is a nonzero real number.

(iv) $(AB)^{-1} = B^{-1}A^{-1}$, if $A^{-1}$ and $B^{-1}$ both exist.

(v) $\mathrm{diag}(a_1, \ldots, a_n)^{-1} = \mathrm{diag}(a_1^{-1}, \ldots, a_n^{-1})$. More generally, if $A$ is a block-diagonal matrix, $A = \mathrm{diag}(A_1, \ldots, A_k)$, where the diagonal blocks are nonsingular, which is equivalent to $|A_j| \neq 0, 1 \leq j \leq k$, then $A^{-1} = \mathrm{diag}(A_1^{-1}, \ldots, A_k^{-1})$.

An inverse-matrix identity that is very useful is the following. For any $n \times n$ nonsingular matrix $A$, $n \times q$ matrix $U$, and $q \times n$ matrix $V$, we have

$$(P + UV)^{-1} = P^{-1} - P^{-1}U(I_q + VP^{-1}U)^{-1}VP^{-1}. \tag{A.1}$$

One of the applications of identity (A.1) is the following. Denote the $n \times 1$ vector of 1's by $1_n$ (and recall that $I_n$ is the $n \times n$ identity matrix). Let $J_n = 1_n 1_n'$. Then, by (A.1) it is easy to show that for any real numbers $a$ and $b$ such that $a \neq 0$ and $a + nb \neq 0$, we have

$$(aI_n + bJ_n)^{-1} = \frac{1}{a}\left(I_n - \frac{b}{a+nb}J_n\right);$$

we also have $|aI_n + bJ_n| = a^{n-1}(a + nb)$.

For any matrix $A$, whether it is nonsingular or not, there always exists a matrix $A^-$ satisfying $AA^-A = A$. Such an $A^-$ is called a generalized inverse of $A$. Note that here we use the term "a generalized inverse" instead of "the generalized inverse" because such an $A^-$ may not be unique. Two special kinds of generalized inverse are often of interest.

Any matrix $A^-$ satisfying

$$AA^- A = A \quad \text{and} \quad A^- AA^- = A^-$$

is called a reflexible generalized inverse of $A$. Given a generalized inverse $A^-$ of $A$, one can produce a generalized inverse that is reflexible by $A_r^- = A^- AA^-$.

If the generalized inverse is required to satisfy the conditions, known as the Penrose conditions, (i) $AA^- A = A$, (ii) $A^- AA^- = A^-$, (iii) $AA^-$ is symmetric, and (iv) $A^- A$ is symmetric, it is called the Moore–Penrose inverse. In other words, a reflexible generalized inverse that satisfies the symmetry conditions (iii) and (iv) is the Moore–Penrose inverse. It can be shown that for any matrix $A$, its Moore–Penrose inverse exists and is unique.

### A.1.3 Kronecker products

Let $A = (a_{ij})_{1 \leq i \leq m, 1 \leq j \leq n}$ be a matrix. Then for any matrix $B$, the Kronecker product $A \otimes B$ is defined as the partitioned matrix $(a_{ij}B)_{1 \leq i \leq m, 1 \leq j \leq n}$. For example, if $A = I_m$ and $B = 1_n$, then $A \otimes B = \text{diag}(1_n, \ldots, 1_n)$. Below are some well-known and useful properties of the Kronecker products:

(i) $(A_1 + A_2) \otimes B = A_1 \otimes B + A_2 \otimes B$.

(ii) $A \otimes (B_1 + B_2) = A \otimes B_1 + A \otimes B_2$.

(iii) $c \otimes A = A \otimes c = cA$, where $c$ is a real number.

(iv) $A \otimes (B \otimes C) = (A \otimes B) \otimes C$.

(v) $(A \otimes B)' = A' \otimes B'$.

(vi) If $A$ is partitioned as $A = [A_1 \ A_2]$, then $[A_1 \ A_2] \otimes B = [A_1 \otimes B \ A_2 \otimes B]$. However, if $B$ is partitioned as $[B_1 \ B_2]$, then $A \otimes [B_1 \ B_2] \neq [A \otimes B_1 \ A \otimes B_2]$.

(vii) $(A_1 \otimes B_1)(A_2 \otimes B_2) = (A_1 A_2) \otimes (B_2 B_2)$.

(viii) If $A$ and $B$ are nonsingular, so is $A \otimes B$, and $(A \otimes B)^{-1} = A^{-1} \otimes B^{-1}$.

(ix) $\text{rank}(A \otimes B) = \text{rank}(A)\text{rank}(B)$.

(x) $\text{tr}(A \otimes B) = \text{tr}(A)\text{tr}(B)$.

(xi) If $A$ is $m \times m$ and $B$ is $k \times k$, then $|A \otimes B| = |A|^m |B|^k$.

(xii) The eigenvalues of $A \otimes B$ are all possible products of an eigenvalue of $A$ and an eigenvalue of $B$.

### A.1.4 Matrix differentiation

If $A$ is a matrix whose elements are functions of $\theta$, a real-valued variable, then $\partial A / \partial \theta$ represents the matrix whose elements are the derivatives of the corresponding elements of $A$ with respect to $\theta$. For example, if

$$A = \begin{pmatrix} a_{11} & a_{12} \\ a_{21} & a_{22} \end{pmatrix}, \quad \text{then} \quad \frac{\partial A}{\partial \theta} = \begin{pmatrix} \partial a_{11}/\partial \theta & \partial a_{12}/\partial \theta \\ \partial a_{21}/\partial \theta & \partial a_{22}/\partial \theta \end{pmatrix}.$$

If $a = (a_i)_{1 \leq i \leq k}$ is a vector whose components are functions of $\theta = (\theta_j)_{1 \leq j \leq l}$, a vector-valued variable, then $\partial a / \partial \theta'$ is defined as the matrix $(\partial a_i / \partial \theta_j)_{1 \leq i \leq k, 1 \leq j \leq l}$. Similarly, $\partial a' / \partial \theta$ is defined as the matrix $(\partial a / \partial \theta')'$.

The following are some useful results.

(i) (Innerproduct) If $a$, $b$, and $\theta$ are vectors, then

$$\frac{\partial(a'b)}{\partial\theta} = \left(\frac{\partial a'}{\partial\theta}\right) b + \left(\frac{\partial b'}{\partial\theta}\right) a.$$

(ii) (Quadratic form) If $x$ is a vector and $A$ is a symmetric matrix, then

$$\frac{\partial}{\partial x} x' A x = 2Ax.$$

(iii) (Inverse) If the matrix $A$ depends on a vector $\theta$ and is nonsingular, then, for any component $\theta_i$ of $\theta$,

$$\frac{\partial A^{-1}}{\partial\theta_i} = -A^{-1}\left(\frac{\partial A}{\partial\theta_i}\right) A^{-1}.$$

(iv) (Log-determinant) If the matrix $A$ above is also positive definite, then, for any component $\theta_i$ of $\theta$,

$$\frac{\partial}{\partial\theta_i} \log(|A|) = \mathrm{tr}\left(A^{-1}\frac{\partial A}{\partial\theta_i}\right).$$

## A.1.5 Projection

For any matrix $X$, the matrix $P_X = X(X'X)^{-1}X'$ is called the projection matrix to $\mathcal{L}(X)$, the linear space spanned by the columns of $X$. Here, it is assumed that $X'X$ is nonsingular; otherwise, $(X'X)^{-1}$ should be replaced by $(X'X)^-$, the generalized inverse (see Section A.1.2).

To see why $P_X$ is given such a name, note that any vector in $\mathcal{L}(X)$ can be expressed as $v = Xb$, where $b$ is a vector of the same dimension as the number of columns of $X$. Then we have $P_X v = X(X'X)^{-1}X'Xb = Xb = v$; that is, $P_X$ keeps $v$ unchanged.

The orthogonal projection to $\mathcal{L}(X)$ is defined as $P_{X^\perp} = I - P_X$, where $I$ is the identity matrix. Then, for any $v \in \mathcal{L}(X)$, we have $P_{X^\perp}v = v - P_X v = v - v = 0$. In fact, $P_{X^\perp}$ is the projection matrix to the orthogonal space of $X$, denoted by $\mathcal{L}(X)^\perp$.

If we define the projection of any vector $v$ to $\mathcal{L}(X)$ as $P_X v$, then if $v \in \mathcal{L}$, the projection of $v$ is itself; if $v \in \mathcal{L}(X)^\perp$, the projection of $v$ is zero (vector). In general, we have the orthogonal decomposition $v = v_1 + v_2$, where $v_1 = P_X v \in \mathcal{L}(X)$ and $v_2 = P_{X^\perp}v \in \mathcal{L}(X)^\perp$ such that $v_1'v_2 = v'P_X P_{X^\perp}v = 0$, because $P_X P_{X^\perp} = P_X(1 - P_X) = P_X - P_X^2 = 0$.

The last equation recalls an important property of a projection matrix; that is, any projection matrix is idempotent (i.e., $P_X^2 = P_X$). For example, if $X = 1_n$ (see Section A.1.2), then $P_X = 1_n(1_n'1_n)^{-1}1_n' = n^{-1}J_n = \bar{J}_n$. The orthogonal projection is thus $I_n - \bar{J}_n$. It is easy to verify that $\bar{J}_n^2 = \bar{J}_n$ and $(I_n - \bar{J}_n)^2 = I_n - \bar{J}_n$.

Another useful result involving projections is the following. Suppose that $X$ is $n \times p$ such that $\text{rank}(X) = p$ and that $V$ is $n \times n$ and positive definite. For any $n \times (n - p)$ matrix $A$ such that $\text{rank}(A) = n - p$ and $A'X = 0$, we have

$$A(A'VA)^{-1}A' = V^{-1} - V^{-1}X(X'V^{-1}X)^{-1}X'V^{-1}. \tag{A.2}$$

Equation (A.2) may be expressed in a different way:

$$P_{V^{1/2}A} = I - P_{V^{-1/2}X},$$

where $V^{1/2}$ and $V^{-1/2}$ are the square root matrix of $V$ and $V^{-1}$, respectively (see the next section). In particular, if $V = I$, we have $P_A = I - P_X = P_{X^\perp}$. If $X$ is not of full rank, (A.2) still holds with $(X'V^{-1}X)^{-1}$ replaced by $(X'V^{-1}X)^-$ (see Section A.1.2).

## A.1.6 Decompositions of matrices and eigenvalues

There are various decompositions of a matrix satisfying certain conditions. Two of them are most relevant to this book.

The first is Choleski's decomposition. Let $A$ be a nonnegative definite matrix. Then there exists an upper-triangular matrix $U$ such that $A = U'U$. An application of the Choleski decomposition is the following. For any $k \times 1$ vector $\mu$ and $k \times k$ covariance matrix $V$, one can generate a $k$-variate normal random vector with mean $\mu$ and covariance matrix $V$. Simply let $\xi = \mu + U'\eta$, where $\eta$ is a $k \times 1$ vector whose components are independent $N(0, 1)$ random variables and $U$ is the upper-triangular matrix in the Choleski decomposition of $V$.

Another decomposition is the eigenvalue decomposition. For any $n \times n$ symmetric matrix $A$, there exists an orthogonal matrix $T$ such that $A = TDT'$, where $D = \text{diag}(\lambda_1, \ldots, \lambda_n)$, and $\lambda_1, \ldots, \lambda_n$ are the eigenvalues of $A$. In particular, if $A \geq 0$ (i.e., nonnegative definite, in which case the eigenvalues are nonnegative), we define $D^{1/2} = \text{diag}(\sqrt{\lambda_1}, \ldots, \sqrt{\lambda_n})$ and $A^{1/2} = TD^{1/2}T'$, called the square root matrix of $A$. It follows that $(A^{1/2})^2 = A$. If $A$ is positive definite, then we write $A^{-1/2} = (A^{1/2})^{-1}$, which is identical to $(A^{-1})^{1/2}$. Thus, for example, an alternative way of generating the $k$-variate normal random vector (see above) is to let $\xi = \mu + V^{1/2}\eta$. The definition of $A^{1/2}$ can be extended $A^r$ for any $A \geq 0$ and $r \in [0, 1]$; that is, $A^r = T\text{diag}(\lambda_1^r, \ldots, \lambda_n^r)$. For example, the Löwner–Heinz inequality states that for any matrices $A$ and $B$ such that $A \geq B \geq 0$ and $r \in [0, 1]$, we have $A^r \geq B^r$.

The eigenvalue decomposition is one way of diagonalizing a matrix $A$ such that $T'AT = D$, where $T$ is orthogonal and $D$ is diagonal. It follows that any symmetric matrix $A$ can be diagonalized by an orthogonal matrix such that the diagonal elements are the eigenvalues of $A$. Furthermore, if symmetric matrices $A_1, \ldots, A_k$ are commuting—that is, if $A_j A_j = A_j A_i, 1 \leq i \neq j \leq k$—then they are simultaneously diagonalizable in the sense that there is an orthogonal matrix $T$ that $T'A_jT = D_j, 1 \leq j \leq k$, where $D_j$ is a diagonal matrix whose diagonal elements are the eigenvalues of $A_j$, $1 \leq j \leq k$.

The largest and smallest eigenvalues of a symmetric matrix, $A$, are denoted by $\lambda_{\max}(A)$ and $\lambda_{\min}(A)$, respectively. The following properties hold. Let $\lambda_1, \ldots, \lambda_n$ be the eigenvalues of $A$.

(i) For any positive integer $p$, the eigenvalues of $A^p$ are $\lambda_1^p, \ldots, \lambda_n^p$. Thus, if $A \geq 0$, then $\lambda_{\max}(A^p) = \{\lambda_{\max}(A)\}^p$, $\lambda_{\min}(A^p) = \{\lambda_{\min}(A)\}^p$.

(ii) $\operatorname{tr}(A) = \lambda_1 + \cdots + \lambda_n$.

(iii) $|A| = \lambda_1 \cdots \lambda_n$.

(iv) For any matrices $A$ and $B$ (not necessarily symmetric), the nonzero eigenvalues of $AB$ are the same as the nonzero eigenvalues of $BA$. This implies, in particular, that $\lambda_{\max}(AB) = \lambda_{\max}(BA)$ and $\lambda_{\min}(AB) = \lambda_{\min}(BA)$, if the eigenvalues of $AB, BA$ are all positive. Another consequence is that $\lambda_{\max}(A'A) = \lambda_{\max}(AA')$ for any matrix $A$.

Finally, if $A$ and $B$ are symmetric matrices, whose eigenvalues, arranged in decreasing orders, are $\lambda_1 \geq \cdots \geq \lambda_k$ and $\mu_1 \geq \cdots \geq \mu_k$, respectively, then Weyl's perturbation theorem states that

$$\max_{1 \leq i \leq k} |\lambda_i - \mu_i| \leq \|A - B\|.$$

An application of Weyl's theorem is the following. If $A_n$ is a sequence of symmetric matrices such that $\|A_n - A\| \to 0$ as $n \to \infty$, where $A$ is a symmetric matrix, then the eigenvalues of $A_n$ converge to those of $A$ as $n \to \infty$.

## A.2 Measure and probability

Let $\Omega$ denote the space of all elements of interest. In our case, $\Omega$ is typically the space of all possible outcomes so that the probability of $\Omega$ is equal to one. This said, the mathematical definition of probability has yet to be given, even although the concept might seem straightforward as a common sense.

### A.2.1 Measures

First, we need to define what is a collection of "reasonable outcomes". Let $\mathcal{F}$ be a collection of subsets of $\Omega$ satisfying the following three properties:

(i) The empty set $\emptyset \in \mathcal{F}$.

(ii) $A \in \mathcal{F}$ implies the complement $A^c \in \mathcal{F}$.

(iii) $A_i \in \mathcal{F}, i \in I$, where $I$ is a discrete set of indexes, implies $\cup_{i \in I} A_i \in \mathcal{F}$.

Then $\mathcal{F}$ is called a $\sigma$-field. The pair $(\Omega, \mathcal{F})$ is then called a measurable space. The elements of $\mathcal{F}$ are called measurable sets with respect to $\mathcal{F}$, or simply measurable sets, when the context is clear. Let $\mathcal{S}$ be a collection of subsets of $\Omega$. The smallest $\sigma$-field that contains $\mathcal{S}$, denoted by $\sigma(\mathcal{S})$, is called the $\sigma$-field generated by $\mathcal{S}$. The smallest $\sigma$-field does exist. To see this, let $\mathcal{F}$ be the collection of all sets obtained by taking complements, union, or intersection of elements of $\mathcal{S}$ in any order and possibly multiple times, plus the empty set (which may be understood as taking the union of no elements). It is easy to see that $\mathcal{F}$ is a $\sigma$-field and, in fact, $\mathcal{F} = \sigma(\mathcal{S})$. We consider some examples.

*Example A.1.* Let $A$ be a nonempty proper subset of $\Omega$ (i.e., $A \neq \Omega$). Then $\sigma(\{A\}) = \{\emptyset, A, A^c, \Omega\}$.

*Example A.2* (Borel $\sigma$-field). Let $\mathcal{O}$ be the collection of all open sets on $R$, the real line. The $\sigma$-field $\mathcal{B} = \sigma(\mathcal{O})$ is called the Borel $\sigma$-field on $R$. It can be shown that $\mathcal{B} = \sigma(\mathcal{I})$, where $\mathcal{I}$ is the collection of all finite open intervals.

Let $(\Omega, \mathcal{F})$ be a measurable space. A function $\nu\colon \mathcal{F} \mapsto [0, \infty]$ is called a *measure* if the following conditions are satisfied:
(i) $\nu(\emptyset) = 0$;
(ii) if $A_i \in \mathcal{F}, i = 1, 2, \ldots$ are disjoint (i.e., $A_i \cap A_j = \emptyset, i \neq j$), then $\nu\left(\bigcup_{i=1}^{\infty} A_i\right) = \sum_{i=1}^{\infty} \nu(A_i)$.
In the special case, in which $\nu(\Omega) = 1$, $\nu$ is called a *probability measure*. The triple $(\Omega, \mathcal{F}, \nu)$ is then called a measure space; when $\nu(\Omega) = 1$, this is called a *probability space*. Although probability measures are what we are dealing with most of the time, the following nonprobability measures are often used in order to define a probability mass function, or a probability density fuction (see the next subsection).

*Example A.3* (Counting measure). Let $\mathcal{F}$ be the collection of all subsets of $\Omega$. It is easy to verify that $\mathcal{F}$ is a $\sigma$-field. Now, define, for any $A \in \mathcal{F}$, $\nu(A) = $ the number of elements in $A$ [so $\nu(A) = \infty$ if $A$ contains infinitely many elements]. It is easy to verify that $\nu$ is a measure on $(\Omega, \mathcal{F})$, known as the *counting measure*. In particular, if $\Omega$ is *countable* in the sense that there is a one-to-one correspondence between $\Omega$ and the set positive integers, we may let $\mathcal{F}$ be the collection of all subsets of $\Omega$. This is a $\sigma$-field, known as the *trivial $\sigma$-field*. A counting measure is then defined on $(\Omega, \mathcal{F})$.

*Example A.4* (Lebesgue measure). There is a unique measure $\nu$ on $(R, \mathcal{B})$ that satisfies $\nu([a, b]) = b - a$ for any finite interval $[a, b]$. This is called the *Lebesgue measure*. In particular, if $\Omega = [0, 1]$, the Lebesgue measure is a probability measure.

Some basic properties of a measure are the following. Let $(\Omega, \mathcal{F}, \nu)$ be a measure space and assume that all the sets considered are in $\mathcal{F}$.
(i) (Monotonicity) $A \subset B$ implies $\nu(A) \leq \nu(B)$.
(ii) (Subadditivity) For any collection of sets $A_i, i \in I$, in $\mathcal{F}$, we have $\nu\left(\bigcup_{i \in I} A_i\right) \leq \sum_{i \in I} \nu(A_i)$.
(iii) (Continuity) If $A_1 \subset A_2 \subset \cdots$, then

$$\nu(\lim_{n \to \infty} A_n) = \lim_{n \to \infty} \nu(A_n), \tag{A.3}$$

where $\lim_{n\to\infty} A_n = \cup_{i=1}^{\infty} A_i$. Similarly, if $A_1 \supset A_c \supset \cdots$ and $\nu(A_1) < \infty$, then (A.3) holds with $\lim_{n\to\infty} A_n = \cap_{i=1}^{\infty} A_i$.

A measure $\nu$ on $(\Omega, \mathcal{F})$ is called $\sigma$-finite if there is a countable collection of measurable sets $A_i, i \in I$, such that $\nu(A_i) < \infty, i \in I$, and $\Omega = \cup_{i\in I} A_i$.

Let $P$ denote a probability measure on $(R, \mathcal{B})$. The *cumulative distribution function* (cdf) of $P$ is defined as

$$F(x) = P((-\infty, x]), \quad x \in R. \tag{A.4}$$

The cdf has the following properties:

(i) $F(-\infty) \equiv \lim_{x\to-\infty} F(x) = 0$, $F(\infty) \equiv \lim_{x\to\infty} F(x) = 1$;
(ii) $F$ is nondecreasing; that is, $F(x) \leq F(y)$ if $x < y$;
(iii) $F$ is right-continuous; that is, $\lim_{y\to x, y>x} F(y) = F(x)$.

It can be shown that for any real-valued function $F$ that satisfies the above properties (i)–(iii), there is a unique probability measure $P$ such that $F$ can be expressed as (A.4).

The concept of cdf can be extended to multivariate case. Let $(\Omega_i, \mathcal{F}_i), i = 1, \ldots, k$, be measurable spaces. The product $\sigma$-field on the product space $\prod_{i=1}^{k} \Omega_i$ is defined as $\sigma(\prod_{i=1}^{k} \mathcal{F}_i)$, where $\prod_{i=1}^{k} \Omega_i = \{(x_1, \ldots, x_k) : x_i \in \Omega_i, 1 \leq i \leq k\}$ and $\prod_{i=1}^{k} \mathcal{F}_i = \{A_1 \times \cdots \times A_k : A_i \in \mathcal{F}_i, 1 \leq i \leq k\}$ with $A_1 \times \cdots \times A_k = \{(a_1, \ldots, a_k) : a_i \in A_i, 1 \leq i \leq k\}$. Note that $\prod_{i=1}^{k} \mathcal{F}_i$ is not necessarily a $\sigma$-field. If $(\Omega_i, \mathcal{F}_i, \nu_i), 1 \leq i \leq k$, are measure spaces, where $\nu_i, 1 \leq i \leq k$, are $\sigma$-finite, there is a unique $\sigma$-finite measure $\nu$ on $\{\prod_{i=1}^{k} \Omega_i, \sigma(\prod_{i=1}^{k} \mathcal{F}_i)\}$ such that for all $A_i \in \mathcal{F}_i, 1 \leq i \leq k$,

$$\nu(A_1 \times \cdots \times A_k) = \nu_1(A_1) \cdots \nu_k(A_k).$$

This is called the product measure, denoted by $\nu = \nu_1 \times \cdots \times \nu_k$. In particular, if $\Omega_i = R, \mathcal{F}_i = \mathcal{B}, 1 \leq i \leq k$, the corresponding product space and $\sigma$-field are denoted by $R^k$ and $\mathcal{B}^k$, respectively. Let $P$ be a probability measure on $(R^k, \mathcal{B}^k)$. The joint cdf of $P$ is defined as

$$F(x_1, \ldots, x_k) = P\{(-\infty, x_1] \times \cdots \times (-\infty, x_k]\}, \tag{A.5}$$

$x_1, \ldots, x_k \in R$. In the special case where $P = P_1 \times P_k$, where $P_i$ is a probability measure on $(\Omega_i, \mathcal{F}_i), 1 \leq i \leq k$, (A.5) becomes

$$F(x_1, \ldots, x_k) = F_1(x_1) \cdots F_k(x_k), \tag{A.6}$$

$x_1, \ldots, x_k \in R$, where $F_i$ is the cdf of $P_i, 1 \leq i \leq k$.

### A.2.2 Measurable functions

Let $f$ be a map from $\Omega$ to $\Lambda$, an image space. Suppose that there is a $\sigma$-field $\mathcal{G}$ on $\Lambda$ such that

$$f^{-1}(B) \equiv \{\omega \in \Omega : f(\omega) \in B\} \in \mathcal{F}, \quad \forall B \in \mathcal{G}; \tag{A.7}$$

then $f$ is said to be a measurable map from $(\Omega, \mathcal{F})$ to $(\Lambda, \mathcal{G})$. In particular, if $\Lambda = R^k$ for some $k$, $f$ is called a measurable function. If, in addition, $\mathcal{G} = \mathcal{B}^k$, $f$ is called *Borel measurable*.

Now, suppose that $(\Omega, \mathcal{F}, P)$ is a probability space. Any measurable function from $(\Omega, \mathcal{F})$ to $(R, \mathcal{B})$ is called a *random variable*. Similarly, any measurable function from $(\Omega, \mathcal{F})$ to $(R^k, \mathcal{B}^k)$ $(k > 1)$ is called a random vector, or vector-valued random variable. Let $X$ be a random variable. Define a probability measure on $(R, \mathcal{B})$ by

$$PX^{-1}(B) = P\{X^{-1}(B)\}, \quad B \in \mathcal{B}, \tag{A.8}$$

where $X^{-1}(B)$ is defined by (A.7) with $f$ replaced by $X$. $PX^{-1}$ is called the distribution of $X$. The cdf of $X$ is defined as

$$F(x) = PX^{-1}\{(-\infty, x]\} = P(X \le x), \quad x \in R. \tag{A.9}$$

Note that (A.9) is the same as (A.4) with $P$ replaced by $PX^{-1}$; in other words, the cdf of $X$ is the same as the cdf of $PX^{-1}$. Note that a random variable, by definition, must be finite, whereas a measurable function may take infinite values, depending on the definition of $\Lambda$. We consider an example.

*Example A.5.* If $X$ is a random variable, then for any $\epsilon > 0$, there is $b > 0$ such that $P(|X| \le b) > 1 - \epsilon$. This is because, by continuity [see (A.3)],

$$
\begin{aligned}
1 &= PX^{-1}\{(-\infty, \infty)\} \\
&= PX^{-1}\left(\lim_{n \to \infty} [-n, n]\right) \\
&= \lim_{n \to \infty} PX^{-1}([-n, n]) \\
&= \lim_{n \to \infty} P(|X| \le n).
\end{aligned}
$$

Therefore, there must be some $b = n$ such that $P(|X| \le b) > 1 - \epsilon$.

The following are some basic facts related to Borel-measurable functions.

(i) $f$ is Borel measurable if and only if $f^{-1}\{(-\infty, b)\} \in \mathcal{F}$ for any $b \in R$.

(ii) If $f$ and $g$ are Borel measurable, so are $fg$ and $af + bg$ for any real numbers $a$ and $b$; and $f/g$ is Botel measurable if $g(\omega) \ne 0$ for any $\omega \in \Omega$.

(iii) If $f_1, f_2, \ldots$ are Borel measurable, so are $\sup_n f_n, \inf_n f_n, \limsup_n f_n$, and $\liminf_n f_n$.

(iv) If $f$ is measurable from $(\Omega, \mathcal{F})$ to $(\Lambda, \mathcal{G})$ and $g$ is measurable from $(\Lambda, \mathcal{G})$ to $(\Gamma, \mathcal{H})$, then the composite function, $g \circ f(\omega) = g\{f(\omega)\}, \omega \in \Omega$, is measurable from $(\Omega, \mathcal{F})$ to $(\Gamma, \mathcal{H})$. In particular, if $\Gamma = R$ and $\mathcal{H} = \mathcal{B}$, then $g \circ f$ is Borel measurable.

(iv) If $\Omega$ is a Borel subset of $R^k$, where $k \ge 1$, then any continuous function from $\Omega$ to $R$ is Borel measurable.

A class of noncontinuous measurable functions plays an important role in the definition of integrals (see below). These are called *simple measurable functions*, or simply *simple functions*, defined as

$$f(\omega) = \sum_{i=1}^{s} a_i I_{A_i}(\omega), \tag{A.10}$$

where $a_1, \ldots, a_s$ are real numbers, $A_1, \ldots, A_s$ are measurable sets, and $I_A$ is the indicator function defined as

$$I_A(\omega) = \begin{cases} 1, & \omega \in A \\ 0, \text{ otherwise} \end{cases}$$

### A.2.3 Integration

Let $(\Omega, \mathcal{F}, \nu)$ be a measure space. If $f$ is a simple function defined by (A.10), its integral with respect to $\nu$ is defined as

$$\int f \, d\nu = \sum_{i=1}^{s} a_i \nu(A_i).$$

From the definition, it follows that $\int I_A \, d\nu = \nu(A)$. Thus, the integral may be regarded as an extension of the measure.

Next, if $f$ is a nonnegative Borel-measurable function, let $S_f$ denote the collection of all nonnegative simple functions $g$ such that $g(\omega) \le f(\omega)$ for all $\omega \in \Omega$. The integral of $f$ with respect to $\nu$ is defined as

$$\int f \, d\nu = \sup_{g \in S_f} \int g \, d\nu.$$

Finally, any Borel-measurable function $f$ can be expressed as $f = f^+ - f^-$, where $f^+(\omega) = f(\omega) \vee 0$ and $f^-(\omega) = -f(\omega) \wedge 0$. Because both $f^+$ and $f^-$ are nonnegative Borel measurable (why?), the above definition of integral applies to both $f^+$ and $f^-$. If at least one of the integrals $\int f^+ \, d\nu$ and $\int f^- \, d\nu$ is finite, the integral of $f$ with respect to $\nu$ is defined as

$$\int f \, d\nu = \int f^+ \, d\nu - \int f^- \, d\nu.$$

*Example A.3 (continued).* Let $\Omega$ be a countable set and $\nu$ be the counting measure. Then, for any Borel-measurable function $f$, we have

$$\int f \, d\nu = \sum_{\omega \in \Omega} f(\omega).$$

So, in this case, the integral is the summation.

*Example A.4 (continued).* If $\Omega = R$ and $\nu$ is the Lebesgue measure, the integral is called the *Lebesgue integral*, which is usually denoted by

$$\int f \, d\nu = \int_{-\infty}^{\infty} f(x) \, dx.$$

The definition extends to the multidimensional case in an obvious way. The connection between the Lebesgue integral and the Riemann integral (see §1.5.4.32) is that the two are equal when the latter is well defined. Below are some basic properties of the integrals and well-known results. Although some were already given in Section 1.5, they are listed again for completeness.

(i) For any Borel-measurable functions $f$ and $g$, as long as the integrals involved exist, we have

$$\int (af + bg) \, d\nu = a \int f \, d\nu + b \int g \, d\nu$$

for any real numbers $a, b$.

(ii) For any Borel-measurable functions $f$ and $g$ such that $f \leq g$ a.e. $\nu$ [which means that $\nu(\{\omega : f(\omega) > g(\omega)\}) = 0$], we have $\int f \, d\nu \leq \int g \, d\nu$, provided that the integrals exist.

(iii) If $f$ is Borel measurable and $f \geq 0$ a.e. $\nu$, then $\int f d\nu = 0$ implies $f = 0$ a.e. $\nu$.

(iv) (Fatou's lemma) Let $f_1, f_2, \ldots$ be a sequence of Borel-measurable functions such that $f_n \geq 0$ a.e. $\nu$, $n \geq 1$; then

$$\int \left( \liminf_{n \to \infty} f_n \right) d\nu \leq \liminf_{n \to \infty} \left( \int f_n \, d\nu \right).$$

(v) (Monotone convergence theorem) If $f_1, f_2, \ldots$ are Borel measurable such that $0 \leq f_1 \leq f_2 \leq \cdots$ and $\lim_{n \to \infty} f_n = f$ a.e. $\nu$, then

$$\int \left( \lim_{n \to \infty} f_n \right) d\nu = \lim_{n \to \infty} \left( \int f_n \, d\nu \right). \tag{A.11}$$

(vi) (Dominated convergence theorem) If $f_1, f_2, \ldots$ are Borel measurable such that $\lim_{n \to \infty} f_n = f$ a.e. $\nu$ and there is an integrable function $g$ (i.e., $\int |g| \, d\nu < \infty$) such that $f_n \leq g$ a.e. $\nu$, $n \geq 1$, then (A.11) holds.

(vii) (Differentiation under the integral sign) Suppose that for each $\theta \in (a, b) \subset R$, $f(\cdot, \theta)$ is Borel measurable, $\partial f/\partial \theta$ exists, and $\sup_{\theta \in (a,b)} |\partial f/\partial \theta|$ is integrable. Then, for each $\theta \in (a, b)$, $\partial f/\partial \theta$ is integrable and

$$\frac{\partial}{\partial \theta} \int f(\omega, \theta) \, d\nu = \int \frac{\partial}{\partial \theta} f(\omega, \theta) \, d\nu.$$

(viii) (Change of variable) Let $f$ be measurable from $(\Omega, \mathcal{F}, \nu)$ to $(\Lambda, \mathcal{G})$ and $g$ be Borel measurable on $(\Lambda, \mathcal{G})$. Then we have

$$\int_{\Omega} (g \circ f) \, d\nu = \int_{\Lambda} g \, d(\nu \circ f^{-1}), \tag{A.12}$$

provided that either integral exists. Here, the measure $\nu \circ f^{-1}$ is defined similarly as (A.8); that is,

$$\nu \circ f^{-1}(B) = \nu\{f^{-1}(B)\}, \quad B \in \mathcal{G},$$

where $f^{-1}(B)$ is defined by (A.7).

(ix) (Fubini's theorem) Let $\nu_i$ be a $\sigma$-finite measure on $(\Omega_i, \mathcal{F}_i), i = 1, 2$, and $f$ be Borel measurable on $\{\Omega_1 \times \Omega_2, \sigma(\mathcal{F}_1 \times \mathcal{F}_2)\}$ whose integral with respect to $\nu_1 \times \nu_2$ exists. Then $\int_{\Omega_1} f(\omega_1, \omega_2) \, d\nu_1$ is Borel measurable on $(\Omega_2, \mathcal{F}_2)$ whose integral with respect to $\nu_2$ exists, and

$$\int_{\Omega_1 \times \Omega_2} f(\omega_1, \omega_2) \, d\nu_1 \times \nu_2 = \int_{\Omega_2} \left\{ \int_{\Omega_1} f(\omega_1, \omega_2) \, d\nu_1 \right\} d\nu_2.$$

(x) (Radon-Nikodym derivative) Let $\mu$ and $\nu$ be two measures on $(\Omega, \mathcal{F})$ and $\mu$ be $\sigma$-finite. $\nu$ is said to be *absolutely continuous* with respect to $\mu$, denoted by $\nu \ll \mu$, if for any $A \in \mathcal{F}$, $\mu(A) = 0$ implies $\nu(A) = 0$. The Radon–Nikodym theorem states that if $\nu \ll \mu$, there exists a nonnegative Borel-measurable function $f$ such that

$$\nu(A) = \int_A f \, d\mu, \quad \forall A \in \mathcal{F}. \tag{A.13}$$

The $f$ is unique a.e. $\mu$ in the sense that if $g$ is another Borel-measurable function that satisfies (A.13), then $f = g$ a.e. $\mu$. The function $f$ in (A.13) is called the Radon–Nikodym derivative, or density, of $\nu$ with respect to $\mu$, denoted by $f = d\nu/d\mu$. The following result holds, which is, again, similar to the change-of-variables rule in calculus: If $\nu \ll \mu$, then for any nonnegative Borel-measurable function $f$, we have

$$\int f \, d\nu = \int f \left( \frac{d\nu}{d\mu} \right) d\mu. \tag{A.14}$$

### A.2.4 Distributions and random variables

The pdf of a random variable $X$ on $(\Omega, \mathcal{F}, P)$ is defined as $F(x) = P(X \leq x), x \in R$. Similarly, the joint pdf of random variables $X_1, \ldots, X_k$ on $(\Omega, \mathcal{F}, P)$ is defined as $F(x_1, \ldots, x_k) = P(X_1 \leq x_1, \ldots, X_k \leq x_k), x_1, \ldots, x_k \in R$. This definition is consistent with (A.9), whose multivariate version is

$$F(x_1, \ldots, x_k) = PX^{-1}\{(-\infty, x]\}, \quad x = (x_1, \ldots, x_k) \in R^k,$$

where $(-\infty, x] = (-\infty, x_1] \times \cdots \times (-\infty, x_k]$.

The random variable $X$ is said to be *discrete* if its possible values are a finite or countable subset of $R$. Let $\mu$ be the counting measure (see Example A.3) and $\nu = PX^{-1}$. It is clear that $\nu \ll \mu$ (why?). The Radon–Nikodym

derivative $f = d\nu/d\mu$ is called the *probability mass function*, or pmf, of $X$.
The definition of pmf can be easily extended to the multivariate case.

A random variable $X$ on on $(\Omega, \mathcal{F}, P)$ is said to be *continuous* if $\nu = PX^{-1} \ll \mu$, the Lebesgue measure. In such a case, the Radon–Nikodym derivative $f = d\nu/d\mu$ is called the *probability density function*, or pdf, of $X$. Again, the definition can be easily extended to the multivariate case.

A list of commonly used random variables and their pmf's or pdf's can be found, for example, in Casella and Berger (2002, pp. 621–626).

The mean, or expected value, of a random variable $X$ on $(\Omega, \mathcal{F}, P)$ is defined as $E(X) = \int_\Omega X \, dP$. Usually, the expected value is calculated via a pmf or pdf. Let $\mu$ be a $\sigma$-finite measure on $(R, \mathcal{B})$ and assume that $PX^{-1} \ll \mu$. Let $f = dPX^{-1}/d\mu$. Then, by (A.12) and (A.14), we have

$$E(X) = \int_\Omega x \circ X \, dP$$

$$= \int_R x \, dPX^{-1}$$

$$= \int_R x \left( \frac{dPX^{-1}}{d\mu} \right) d\mu$$

$$= \int xf \, d\mu.$$

In particular, if $\mu$ is the counting measure, we have

$$E(X) = \sum_{x \in S} xf(x) = \sum_{x \in S} xP(X = x),$$

where $S$ is the set of possible values for $X$ (note that in this case, $X$ must be discrete—why?); if $\mu$ is the Lebesgue measure on $(a, b)$, where $a$ and $b$ can be finite or infinite, we have

$$E(X) = \int_a^b xf(x) \, dx.$$

The variance of $X$ is defined as $\mathrm{var}(X) = E(X - \mu_X)^2$, where $\mu_X = E(X)$. Another expression of the variance is $\mathrm{var}(X) = E(X^2) - \{E(X)\}^2$. The covariance and correlation between two random variables $X$ and $Y$ on the same probability space are defined as $\mathrm{cov}(X, Y) = E(X - \mu_X)(Y - \mu_Y)$ and $\mathrm{cor}(X, Y) = \mathrm{cov}(X, Y)/\sqrt{\mathrm{var}(X)\mathrm{var}(Y)}$, respectively. Similar to the variance, an alternative expression for the covariance is $\mathrm{cov}(X, Y) = E(XY) - E(X)E(Y)$. If $X = (X_1, \ldots, X_k)'$ is a random vector, its expected value is defined as $E(X) = [E(X_1), \ldots, E(X_k)]'$ and its covariance matrix is defined as $\mathrm{Var}(X) = [\mathrm{cov}(X_i, X_j)]_{1 \leq i, j \leq k}$. If $Y = (Y_1, \ldots, Y_l)'$ is another random vector, where $l$ need not be the same as $k$, the covariance matrix between $X$ and $Y$ is defined as $\mathrm{Cov}(X, Y) = [\mathrm{cov}(X_i, Y_j)]_{1 \leq i \leq k, 1 \leq j \leq l}$. Some basic properties of the expected values and variances are the following, assuming their existence.

(i) $E(aX + bY) = aE(X) + bE(Y)$ for any constants $a$ and $b$.

(ii) $\text{var}(X + Y) = \text{var}(X) + 2\,\text{cov}(X, Y) + \text{var}(Y)$. In particular, if $X$ and $Y$ are *uncorrelated* in that $\text{cov}(X, Y) = 0$, then $\text{var}(X, Y) = \text{var}(X) + \text{var}(Y)$.

(iii) $\text{Cov}(X, Y) = E\{X - E(X)\}\{Y - E(Y)\}'$, where the expected value of a random matrix, $\xi = (\xi_{ij})_{1 \le i \le k, 1 \le j \le l}$, is defined as $E(\xi) = [E(\xi_{ij})]_{1 \le i \le k, 1 \le j \le l}$. This gives an equivalent definition of the covariance matrix between $X$ and $Y$, which is often more convenient in derivations. In particular, $\text{Cov}(X, X) = \text{Var}(X) = E\{X - E(X)\}\{X - E(X)\}'$. Thus, we have $\text{Cov}(Y, X) = \text{Cov}(X, Y)'$, and, similar to (ii), $\text{Var}(X + Y) = \text{Var}(X) + \text{Cov}(X, Y) + \text{Cov}(X, Y)' + \text{Var}(Y)$.

(iv) $\text{var}(aX) = a^2 \text{var}(X)$ for any constant $a$; $\text{Var}(AX) = A\,\text{Var}(X)A'$ for any constant matrix $A$; and $\text{Cov}(AX, BY) = A\,\text{Cov}(X, Y)B'$ for any constant matrices $A$ and $B$.

(v) $\text{var}(X) \ge 0$ and $\text{var}(X) = 0$ if and only if $X$ is a.s. a constant; that is, $P(X = c) = 1$ for some constant $c$. Similarly, $\text{Var}(X) \ge 0$ (i.e., positive definite); and $|\text{Var}(X)| = 0$ if and only if there is a constant vector $a$ and a constant $c$ such that $a'X = c$ a.s.

(vi) (The covariance inequality) $|\text{cov}(X, Y)| \le \sqrt{\text{var}(X)\text{var}(Y)}$. This implies, in particular, that the correlation between $X$ and $Y$ is always between $-1$ and $1$; when $\text{cor}(X, Y) = -1$ or $1$, there are constants $a$, $b$, and $c$ such that $aX + bY = c$ a.s.; in other words, one is (a.s.) a linear function of the other.

Other quantities associated with the expected values include the moments and central moments. The $p$th *moment* of a random variable $X$ is defined as $E(X^p)$; the $p$th *absolute moment* is $E(|X|^p)$; and the $p$th *central moment* is $\gamma_p = E\{(X - \mu_X)^p\}$, assuming existence, of course, in each case. The skewness and kurtosis of $X$ are defined as $\kappa_3 = \gamma_3/\sigma^3$ and $\kappa_4 = (\gamma_4/\sigma^4) - 3$, respectively, where $\sigma = \sqrt{\text{var}(X)}$, known as the *standard deviation* of $X$.

The random variables $X_1, \ldots, X_k$ are said to be independent if

$$P(X_1 \in B_1, \ldots, X_k \in B_k) = P(X_1 \in B_1) \cdots P(X_k \in B_k)$$

for any $B_1, \ldots, B_k \in \mathcal{B}$. Equivalently, $X_1, \ldots, X_k$ are independent if their joint pdf is the product of their individual pdf's; that is,

$$F(x_1, \ldots, x_k) = F_1(x_1) \cdots F_k(x_k)$$

for all $x_1, \ldots, x_k \in R$, where $F(x_1, \ldots, x_k) = P(X_1 \le x_1, \ldots, X_k \le x_k)$ and $F_j(x_j) = P(X_j \le x_j), 1 \le j \le k$. If the joint pdf $f$ of $X_1, \ldots, X_k$ with respect to a product measure, $\mu = \mu_1 \times \cdots \times \mu_k$, exists, where the $\mu_j$'s are $\sigma$-finite, independence of $X_1, \ldots, X_k$ is also equivalent to

$$f(x_1, \ldots, x_k) = f_1(x_1) \cdots f_k(x_k)$$

for all $x_1, \ldots, x_k$, where $f_j$ is the pdf of $X_j$ with respect to $\mu_j$, which can be derived by integrating the joint pdf; that is,

$$f_j(x_j) = \int f(x_1, \ldots, x_k) d\mu_1 \cdots d\mu_{j-1} d\mu_{j+1} \cdots d\mu_k.$$

The random variables $X_1, \ldots, X_k$ are said to be independent and identically distributed, or i.i.d., if they are independent and have the same distribution, that is, $F_1 = \cdots = F_k$.

### A.2.5 Conditional expectations

An extension of the expected values is conditional expectations. Let $X$ be an integrable random variable on $(\Omega, \mathcal{F}, P)$. Let $\mathcal{A}$ be a sub-$\sigma$-field of $\mathcal{F}$. The *conditional expectation* of $X$ given $\mathcal{A}$, denoted by $E(X|\mathcal{A})$, is the a.s. unique random variable that satisfies the following conditions:

(i) $E(X|\mathcal{A})$ is measurable from $(\Omega, \mathcal{A})$ to $(R, \mathcal{B})$.

(ii) $\int_A E(X|\mathcal{A})\, dP = \int_A X\, dP$ for every $A \in \mathcal{A}$.

The *conditional probability* is a special case of the conditional expectation, with $X = 1_B$, the indicator function of $B \in \mathcal{F}$, denoted by $P(B|\mathcal{A}) = E(1_B|\mathcal{A})$.

The definition leads to the following result, which is useful in checking if a random variable is the conditional expectation of another random variable: Suppose that $\xi$ is integrable and $\eta$ is $\mathcal{A}$-measurable. Then $\eta = E(\xi|\mathcal{A})$ if and only if $E(\xi\zeta) = E(\eta\zeta)$ for every $\zeta$ that is bounded and $\mathcal{A}$-measurable.

The conditional expectation has the same properties as the expected value—all one has to do is to keep the notation $|\mathcal{A}$ during the operations; however, there are also some important differences. A main difference is that, unlike the expected value which is a constant, the conditional expectation is a random variable, unless in some special cases, such as when $\mathcal{A} = \{\emptyset, \Omega\}$.

If $Y$ is another random variable on $(\Omega, \mathcal{F}, P)$, the conditional expectation of $X$ given $Y$ is defined as $E(X|Y) = E\{X|\sigma(Y)\}$, where $\sigma(Y)$ is the $\sigma$-field generated by $Y$, defined as $\sigma(Y) = Y^{-1}(\mathcal{B}) = \{Y^{-1}(B) : B \in \mathcal{B}\}$, where $Y^{-1}(B)$ is defined by (A.7) with $f = Y$. Note that $E(X|Y)$ is a function of $Y$ so let $E(X|Y) = h(Y)$, where $h$ is a Borel-measurable function. Then the conditional expectation of $X$ given $Y = y$ is defined as

$$E(X|Y = y) = h(y).$$

Given the definitions of the conditional expectations, the conditional variances of $X$ given $\mathcal{A}$, or $X$ given $Y$, are defined as

$$\mathrm{var}(X|\mathcal{A}) = E[\{X - E(X|\mathcal{A})\}^2|\mathcal{A}],$$
$$\mathrm{var}(X|Y) = E[\{X - E(X|Y)\}^2|Y],$$

respectively. Similar to the variance, the identities

$$\mathrm{var}(X|\mathcal{A}) = E(X^2|\mathcal{A}) - \{E(X|\mathcal{A})\}^2,$$
$$\mathrm{var}(X|Y) = E(X^2|Y) - \{E(X|Y)\}^2$$

hold. The latter gives an easier way to define $\mathrm{var}(X|Y = y)$ as

$$E(X^2|Y = y) - \{E(X|Y = y)\}^2.$$

Two of the most useful properties of the conditional expectations (variances) are the following, assuming the existence of those throughout.

(i) If $\mathcal{A}_1 \subset \mathcal{A}_2$, then

$$E(X|\mathcal{A}_1) = E\{E(X|\mathcal{A}_2)|\mathcal{A}_1\} \quad \text{a.s.} \tag{A.15}$$

In particular, because the trivial $\sigma$-field, $\{\emptyset, \Omega\}$, is a sub-$\sigma$-field of any $\sigma$-field, we have, by letting $\mathcal{A}_1 = \{\emptyset, \Omega\}$ and $\mathcal{A}_2 = \mathcal{A}$ in (A.15),

$$E(X) = E\{E(X|\mathcal{A})\}.$$

Similarly, if $X$, $Y$, and $Z$ are random variables, we have

$$E(X|Y) = E\{E(X|Y, Z)|Y\} \quad \text{a.s.,}$$
$$E(X) = E\{E(X|Y)\}.$$

Here, $E(X|Y, Z) = E\{X|\sigma(Y, Z)\}$ and $\sigma(Y, Z)$ is defined the same way as $\sigma(Y)$, treating $(Y, Z)$ as a vector-valued random variable.

(ii) The following identity holds for the conditional variance (see above):

$$\text{var}(X) = E\{\text{var}(X|Y)\} + \text{var}\{E(X|Y)\}.$$

In addition, the conditional expectation $E(X|Y)$ is the minimizer of

$$E\{X - g(Y)\}^2$$

over all Borel-measurable functions $g$ such that $E\{g^2(Y)\} < \infty$. Here is another useful result: Let $X$ be a random variable such that $E(|X|) < \infty$, and let $Y$ and $Z$ be random vectors. If $(X, Y)$ and $Z$ are independent, then

$$E(X|Y, Z) = E(X|Y) \quad \text{a.s.} \tag{A.16}$$

We say that, given $Y$, $X$ and $Z$ are *conditionally independent* if

$$P(A|Y, Z) = P(A|Y) \quad \text{a.s.} \quad \forall A \in \sigma(X), \tag{A.17}$$

where the conditional probability $P(A|\xi)$ is defined as $E\{1_A|\sigma(\xi)\}$ for any random vector $\xi$. Given the definition, a similar result to (A.16) is the following. If $(X, Y)$ and $Z$ are independent, then given $Y$, $X$ and $Z$ are conditionally independent. The conclusion may not be true if $Z$ is independent of $X$ but not of $(X, Y)$.

## A.2.6 Conditional distributions

Let $X$ be a random vector on a probability space $(\Omega, \mathcal{F}, P)$ and let $\mathcal{A}$ be a sub-$\sigma$-field of $\mathcal{F}$. There exists a function $P(\cdot, \cdot)$ defined on $\mathcal{B}^k \times \Omega$, where $k$ is the dimension of $X$, such that the following hold:

(i) $P(B, \omega) = P\{X^{-1}(B)|\mathcal{A}\}$ a.s. for any fixed $B \in \mathcal{B}^k$ [see (A.7) for the definition of $X^{-1}(B)$].

(ii) $P(\cdot, \omega)$ is a probability measure on $(R^k, \mathcal{B}^k)$ for any fixed $\omega \in \Omega$.

If $Y$ is measurable from $(\Omega, \mathcal{F})$ to $(\Lambda, \mathcal{G})$. Then there exists a function defined on $\mathcal{B}^k \times \Lambda$, denoted by $P_{X|Y}(\cdot|\cdot)$, such that the following hold:

(i) $P_{X|Y}(B|y) = P\{X^{-1}(B)|Y = y\} \equiv E\{1_{X^{-1}(B)}|Y = y\}$ (see above) a.s. $PY^{-1}$ [see (A.8)] for any fixed $B \in \mathcal{B}^k$.

(ii) $P_{X|Y}(\cdot|y)$ is a probability measure on $(R^k, \mathcal{B}^k)$ for any fixed $y \in \Lambda$. The following holds. If $g(\cdot, \cdot)$ is a Borel function such that $E|g(X, Y)| < \infty$, then

$$E\{g(X, Y)|Y = y\} = E\{g(X, y)|Y = y\}$$
$$= \int_{R^k} g(x, y) \, dP_{X|Y}(x|y) \text{ a.s. } PY^{-1}.$$

Let $(\Lambda, \mathcal{G}, P_1)$ be a probability space. Suppose that $P_2$ is a function from $\mathcal{B}^k \times \Lambda$ to $R$ such that

(i) $P_2(B, \cdot)$ is Borel measurable for any $B \in \mathcal{B}^k$; and

(ii) $P_2(\cdot|y)$ is a probability measure on $(R^k, \mathcal{B}^k)$ for any $y \in \Lambda$.

Then there is a unique probability measure on $\{R^k \times \Lambda.$ $\sigma(\mathcal{B}^k \times \mathcal{G})\}$ such that

$$P(B \times C) = \int_C P_2(B, y) \, dP_1(y) \tag{A.18}$$

for any $B \in \mathcal{B}^k$ and $C \in \mathcal{G}$. In particular, if $(\Lambda, \mathcal{G}) = (R^l, \mathcal{B}^l)$ and define $X(x, y) = x$ and $Y(x, y) = y$ for $(x, y) \in R^k \times R^l$, then $P_1 = PY^{-1}$ and $P_2(\cdot, y) = P_{X|Y}(\cdot|y)$ (see above), and the probability measure (A.18) is the joint distribution of $(X, Y)$ that has the joint cdf

$$F(x, y) = \int_{(-\infty, y]} P_{X|Y}\{(-\infty, x]|v\} \, dPY^{-1}(v), \quad x \in R^k, y \in R^l,$$

where $(-\infty, a]$ is defined above (see the beginning of Section A.2.4). $P_{X|Y}(\cdot|y)$, denoted by $P_{X|Y=y}$, is called the conditional distribution of $X$ given $Y = y$. If $P_{X|Y=y}$ has a pdf with respect to $\nu$, a $\sigma$-finite measure on $(R^k, \mathcal{B}^k)$, the pdf is denoted by $f_{X|Y}(\cdot|y)$, known as the conditional pdf of $X$ given $Y = y$.

## A.3 Some results in statistics

### A.3.1 The multivariate normal distribution

A random vector $\xi$ is said to have a $k$-dimensional multivariate normal distribution with mean vector $\mu$ and covariance matrix $\Sigma$, or $\xi \sim N(\mu, \Sigma)$, if the joint pdf of $\xi$ is given by

$$f(x) = \frac{1}{(2\pi)^{k/2}|\Sigma|^{1/2}} \exp\left\{-\frac{1}{2}(x-\mu)'\Sigma^{-1}(x-\mu)\right\}, \qquad x \in R^k.$$

When $k = 1$, the pdf reduces to that of $N(\mu, \sigma^2)$ with $\sigma^2 = \Sigma$. Below are some useful results associated with the multivariate normal distribution. Here, we assume that all of the matrix products are well defined.

1. If $\xi \sim N(\mu, \Sigma)$, then for any constant matrix $A$, $A\xi \sim N(A\mu, A\Sigma A')$.

2. If $\xi \sim N(\mu, \Sigma)$, then for any constant matrices $A$ and $B$, $A\xi$ and $B\xi$ are independent if and only if $A\Sigma B' = 0$. If $\xi$ is multivariate normal, the components of $\xi$ are independent if and only if they are uncorrelated; that is, $\text{cov}(\xi_i, \xi_j) = 0$, $i \neq j$, where $\xi_i$ is the $i$th component of $\xi$.

3. If $\xi \sim N(\mu, \Sigma)$ and $\xi, \mu$ and $\Sigma$ are partitioned accordingly as

$$\xi = \begin{pmatrix} \xi_1 \\ \xi_2 \end{pmatrix}, \qquad \mu = \begin{pmatrix} \mu_1 \\ \mu_2 \end{pmatrix}, \qquad \Sigma = \begin{pmatrix} \Sigma_{11} & \Sigma_{12} \\ \Sigma_{21} & \Sigma_{22} \end{pmatrix},$$

then the conditional distribution of $\xi_1$ given $\xi_2$ is multivariate normal with mean vector $\mu_1 + \Sigma_{12}\Sigma_{22}^{-1}(\xi_2 - \mu_2)$ and covariance matrix $\Sigma_{11} - \Sigma_{12}\Sigma_{22}^{-1}\Sigma_{21}$. Note that $\Sigma_{21} = \Sigma_{12}'$.

4. Let $\xi$ be a random vector such that $E(\xi) = \mu$ and $\text{Var}(\xi) = \Sigma$. Then, for any constant symmetric matrix $A$, we have

$$E(\xi'A\xi) = \mu'A\mu + \text{tr}(A\Sigma).$$

In particular, if $\xi \sim N(0, \Sigma)$, then $\xi'A\xi$ is distributed as $\chi_r^2$ if and only if $A\Sigma$ is idempotent (see Section A.1.5) and $r = \text{rank}(A)$. In particular, if $\xi = (\xi_i)_{1 \leq i \leq k} \sim N(0, I_k)$, then $|\xi|^2 = \xi_1^2 + \cdots + \xi_k^2 \sim \chi_k^2$.

5. If $\xi \sim N(\mu, \Sigma)$, $a$ is a constant vector, and $A$ and $B$ are constant symmetric matrices, then $a'\xi$ and $\xi'A\xi$ are independent if and only if $b'\Sigma A = 0$; $\xi'A\xi$ and $\xi'B\xi$ are independent if and only if $A\Sigma B = 0$. Also, we have

$$\text{cov}(\xi'A\xi, b'\xi) = 2b'\Sigma A\mu,$$
$$\text{cov}(\xi'A\xi, \xi'B\xi) = 4\mu'A\Sigma B\mu + 2\,\text{tr}(A\Sigma B\Sigma).$$

6. If $\xi \sim N(0, 1)$, $\eta \sim \chi_d^2$, and $\xi$ and $\eta$ are independent, then

$$t = \frac{\xi}{\sqrt{\eta/d}} \sim t_d,$$

the $t$-distribution with $d$ degrees of freedom, which has the pdf

$$f(x) = \frac{\Gamma\{(d+1)/2\}}{\sqrt{d\pi}\,\Gamma(d/2)}\left(1 + \frac{x^2}{d}\right)^{-(d+1)/2}, \qquad -\infty < x < \infty.$$

An extension of the $t$-distribution is the multivariate $t$-distribution. A $k$-dimensional multivariate $t$-distribution with mean vector $\mu$, covariance matrix $\Sigma$, and degrees of freedom $d$ has the joint pdf

$$\frac{\Gamma\{(d+k)/2\}}{(d\pi)^{k/2}\Gamma(d/2)}|\Sigma|^{-1/2}\left\{1+\frac{1}{d}(x-\mu)'\Sigma^{-1}(x-\mu)\right\}^{-(d+k)/2}, \quad x \in R^d.$$

7. If $\xi_j \sim \chi^2_{d_j}$, $j = 1, 2$, and $\xi_1$ and $\xi_2$ are independent, then

$$F = \frac{\xi_1/d_1}{\xi_2/d_2} \sim F_{d_1,d_2},$$

the $F$-distribution with $d_1$ and $d_2$ degrees of freedom, which has the pdf

$$f(x) = \frac{\Gamma\{(d_1+d_2)/2\}}{\Gamma(d_1/2)\Gamma(d_2/2)}\left(\frac{d_1}{d_2}\right)^{d_1/2} x^{d_1/2-1}\left\{1+\left(\frac{d_1}{d_2}\right)x\right\}^{-(d_1+d_2)/2},$$

$-\infty < x < \infty$.

## A.3.2 Maximum likelihood

Let $X$ be a vector of observations and let $f(\cdot|\theta)$ the pdf of $X$, with respect to a $\sigma$-finite measure $\mu$—that is dependent on a vector of parameters, $\theta$. Let $x$ denote the observed value of $X$. The notation $f(x|\theta)$ can be viewed in two ways. For a fixed $\theta$, it is the pdf of $X$ when considered as a function of $x$; for the fixed $x$ (as the observed $X$), it is viewed as a function of $\theta$, known as the *likelihood function*. In the latter case, a different notation is often used, $L(\theta|x) = f(x|\theta)$. The *log-likelihood* is the logarithm of the likelihood function, denoted by $l(\theta|x) = \log\{L(\theta|x)\}$.

A widely used method of estimation is the *maximum likelihood*; namely, the parameter vector $\theta$ is estimated by the maximizer of the likelihood function. More precisely, let $\Theta$ be the *parameter space*—that is, the space of possible values of $\theta$. Suppose that there is $\hat\theta \in \Theta$ such that

$$L(\hat\theta|x) = \sup_{\theta\in\Theta} L(\theta|x);$$

then $\hat\theta$ is called the maximum likelihood estimator, or MLE, of $\theta$.

Under widely existing regularity conditions, the maximum likelihood is carried out by differentiating the log-likelihood with respect to $\theta$ and solving the equations that equate the derivatives to zero; that is,

$$\frac{\partial}{\partial\theta}l(\theta|x) = 0. \tag{A.19}$$

Equation (A.19) is known as the ML equation. It should be noted that a solution to the ML equation is not necessarily the MLE. However, under some more restrictive conditions, the solution indeed coincides with the MLE. For example, if (A.19) has a unique solution and it can be made sure that the maximum of $L(\theta|x)$ does not occur on the boundary of $\Theta$, then the MLE is identical to the solution of the ML equation.

Associated with the log-likelihood function is the *information matrix*, or Fisher information matrix, defined as

$$I(\theta) = E_\theta \left( \frac{\partial l}{\partial \theta} \frac{\partial l}{\partial \theta'} \right), \tag{A.20}$$

where $\partial l/\partial \theta = (\partial/\partial\theta)l(\theta|X)$ and $E_\theta$ stands for expectation with $\theta$ being the true parameter vector. It should be noted—and this is important—that the $\theta$ in $E_\theta$ must be the same as the $\theta$ in $\partial l/\partial \theta$ on the right side of (A.20).

Under some regularity conditions, the following nice properties hold for the log-likelihood function.

(i) [An integrated version of (A.19)]

$$E_\theta \left\{ \frac{\partial}{\partial \theta} l(\theta|X) \right\} = 0.$$

(ii) [Another expression of (A.20)]

$$I(\theta) = -E_\theta \left\{ \frac{\partial^2 l}{\partial\theta\partial\theta'} \right\},$$

where $\partial^2 l/\partial\theta\partial\theta' = (\partial^2/\partial\theta\partial\theta')l(\theta|X)$. Properties (i) and (ii) lead to a third expression for $I(\theta)$:

$$I(\theta) = \text{Var}_\theta \left( \frac{\partial l}{\partial \theta} \right),$$

where $\partial l/\partial \theta = (\partial/\partial\theta)l(\theta|X)$ and $\text{Var}_\theta$ stands for covariance matrix with $\theta$ being the true parameter vector.

A well-known result involving the Fisher information matrix is the Cramér–Rao lower bound. For simplicity, let $\theta$ be a scalar parameter. Let $\hat{\delta}$ be an unbiased estimator of $\delta = g(\theta)$; that is, $E_\theta(\hat{\delta}) = g(\theta)$ for all $\theta$, where $g$ is a differentiable function. Under regularity conditions, we have

$$\text{var}_\theta(\hat{\delta}) \geq \frac{\{g'(\theta)\}^2}{I(\theta)}.$$

For a multivariate version, see, for example, Shao (2003, p. 169).

An important and well-known property of the MLE is its asymptotic efficiency in the sense that, under regularity conditions, the asymptotic covariance matrix of the MLE is equal to the Cramér–Rao lower bound. For example, in the i.i.d. case, we have, as $n \to \infty$,

$$\sqrt{n}(\hat{\theta} - \theta) \xrightarrow{d} N\{0, I^{-1}(\theta)\},$$

where the right side is the multivariate normal distribution with mean vector 0 and covariance matrix $I^{-1}(\theta) = I(\theta)^{-1}$ (see Section A.3.1).

Many testing problems involve the *likelihood ratio*. This is defined as

$$\frac{\sup_{\theta \in \Theta_0} L(\theta|x)}{\sup_{\theta \in \Theta} L(\theta|x)}, \tag{A.21}$$

where $\Theta_0$ is the parameter space under the null hypothesis, $H_0$, and $\Theta$ is the parameter space without assuming $H_0$.

### A.3.3 Exponential family and generalized linear models

The concept of generalized linear models, or GLMs, is closely related to that of the exponential family. The distribution of a random variable $Y$ is a member of the exponential family if its pdf or pmf can be expressed as

$$f(y; \theta) = \exp\left\{ \frac{y\theta - b(\theta)}{a(\phi)} + c(y, \phi) \right\}, \tag{A.22}$$

where $a(\cdot)$, $b(\cdot)$, and $c(\cdot, \cdot)$ are known functions, $\theta$ is an unknown parameter, and $\phi$ is an additional dispersion parameter, which may or may not be known. Many of the well-known distributions are members of the exponential family. These include normal, Gamma, binomial, and Poisson distributions.

An important fact regarding the exponential family is the following relationship between the mean of $Y$ and $\theta$:

$$\mu = \mathrm{E}(Y) = b'(\theta).$$

In many cases, this establishes an 1–1 correspondence between $\mu$ and $\theta$. Another relationship among $\theta$, $\phi$, and the variance of $Y$ is

$$\mathrm{var}(Y) = b''(\theta)a(\phi).$$

The following is an example.

*Example A.6.* Suppose that $Y \sim \text{Binomial}(n, p)$. Then the pmf of $Y$ can be expressed as (A.22) with

$$\theta = \log\left(\frac{p}{1-p}\right), \quad b(\theta) = n\log(1 + e^\theta), \quad \text{and} \quad a(\phi) = \log\binom{n}{y}.$$

Note that in this case, $\phi = 1$. It follows that $b'(\theta) = ne^\theta/(1+e^\theta) = np = \mathrm{E}(Y)$ and $b''(\theta) = ne^\theta/(1 + e^\theta)^2 = np(1-p) = \mathrm{var}(Y)$.

McCullagh and Nelder (1989) introduced the GLM as an extension of the classical linear models. Suppose the following:

(i) The observations $y_1, \ldots, y_n$ are independent.

(ii) The distribution of $y_i$ is a member of the exponential family, which can be expressed as

$$f_i(y) = \exp\left\{ \frac{y\theta_i - b(\theta_i)}{a_i(\phi)} + c_i(y, \phi) \right\}.$$

(iii) The mean of $y_i$, $\mu_i$, is associated with a linear predictor $\eta_i = x_i'\beta$ through a link function; that is,

$$\eta_i = g(\mu_i),$$

where $x_i$ is a vector of known covariates, $\beta$ is a vector of unknown regression coefficients, and $g(\cdot)$ is a link function.

Assumptions (i)–(iii) define a GLM. By the properties of the exponential family mentioned above, $\theta_i$ is associated with $\eta_i$. In particular, if

$$\theta_i = \eta_i,$$

the link function $g(\cdot)$ is called *canonical*.

The function $a_i(\phi)$ typically takes the form $a_i(\phi) = \phi/w_i$, where $w_i$ is a weight. For example, if the observation $y_i$ is the average of $k_i$ observations (e.g., a binomial proportion, where $k_i$ is the number of Bernoulli trials), then $w_i = k_i$; if the observation is the sum of $k_i$ observations (e.g., a binomial or sum of Bernoulli observations), then $w_i = 1/k_i$.

### A.3.4 Bayesian inference

Suppose that $Y$ is a vector of observations and $\theta$ is a vector of parameters that are not observable. Let $f(y|\theta)$ represent the probability density function (pdf) of $Y$ given $\theta$ and let $\pi(\theta)$ represent a *prior* pdf for $\theta$. Then the *posterior* pdf of $\theta$ is given by

$$p(\theta|y) = \frac{f(y|\theta)\pi(\theta)}{\int f(y|\theta)\pi(\theta)\,d\theta}. \tag{A.23}$$

Obtaining the posterior is often the goal of Bayesian inference. In particular, some numerical summaries may be obtained from the posterior. For example, a Bayesian point estimator of $\theta$ is often obtained as the *posterior mean*:

$$E(\theta|y) = \int \theta p(\theta|y)\,d\theta$$
$$= \frac{\int \theta f(y|\theta)\pi(\theta)\,d\theta}{\int f(y|\theta)\pi(\theta)\,d\theta};$$

the *posterior variance*, $\text{var}(\theta|y)$, on the other hand, is often used as a Bayesian measure of uncertainty. The notation $d\theta$ in the above, which corresponds to the Lebesgue measure, can be replaced by $d\mu$, where $\mu$ is a $\sigma$-finite measure with respect to which $\pi(\cdot)$ is defined.

A discrete probabilistic version of (A.23) is called the *Bayes rule*, which is often useful in computing the conditional. Suppose that there are a number of events, $A_1, \ldots, A_k$, such that $A_i \cap A_j = \emptyset, i \neq j$, and $A_1 \cup \cdots \cup A_k = \Omega$, the sample space. Then, for any event $B$, we have

$$P(A_i|B) = \frac{P(B|A_i)P(A_i)}{\sum_{j=1}^{k} P(B|A_j)P(A_j)}, \tag{A.24}$$

$1 \le i \le k$. To see the connection between (A.23) and (A.24), suppose that $\pi$ is a discrete distribution over $\theta_1, \ldots, \theta_k$, and the distribution of $Y$ is also discrete. Then, for any possible value $y$ of $Y$, we have, by (A.24) with $A_i = \{\theta = \theta_i\}, 1 \le i \le k$, and $B = \{Y = y\}$,

$$P(\theta = \theta_i|Y = y) = \frac{\Gamma(Y - y|\theta - \theta_i)P(\theta = \theta_i)}{\sum_{j=1}^{k} P(Y = y|\theta = \theta_j)P(\theta = \theta_j)},$$

which is the discrete version of (A.23).

The posterior can be used to obtain a *posterior predictive distribution* of a future observation, $\tilde{Y}$. Suppose that $Y$ and $\tilde{Y}$ are conditionally independent given $\theta$ (see Section A.2.5); then, by (A.17), we have $f(\tilde{y}|\theta, y) = f(\tilde{y}|\theta)$. Therefore, the posterior predictive pdf is given by

$$p(\tilde{y}|y) = \int p(\tilde{y}, \theta|y) \, d\theta$$

$$= \int f(\tilde{y}|\theta, y)p(\theta|y) \, d\theta$$

$$= \int f(\tilde{y}|\theta)p(\theta|y) \, d\theta$$

(here, as usual, $f$ and $p$ denote the pdf's, and the rule of notation is that $p$ is used whenever the conditioning involves $y$ only; otherwise, $f$ is used).

Similar to Section A.3.2, the pdf $f(y|\theta)$, considered as a function of $\theta$, is called the *marginal likelihood*, or simply likelihood. The ratio of the posterior $p(\theta|y)$ evaluated at the points $\theta_1$ and $\theta_2$ under a given model is called the *posterior odds* for $\theta_1$ compared to $\theta_2$—namely,

$$\frac{p(\theta_1|y)}{p(\theta_2|y)} = \frac{\pi(\theta_1)f(y|\theta_1)/f(y)}{\pi(\theta_2)f(y|\theta_2)/f(y)}$$

$$= \frac{\pi(\theta_1)}{\pi(\theta_2)} \cdot \frac{f(y|\theta_1)}{f(y|\theta_2)},$$

according to the Bayes rule (A.23). In other words, the posterior odds is simply the prior odds multiplied by the *likelihood ratio*, $f(y|\theta_1)/f(y|\theta_2)$ [see (A.21)]. The concept of (posterior) odds is most familiar when $\theta$ takes two possible values, with $\theta_2$ being the complement of $\theta_1$.

A similar concept is the *Bayesian factor*. This is used, for example, when a discrete set of competing models is proposed for model selection. The Bayesian factor is the ratio of the marginal likelihood under one model to that under another model. If we label the two competing models by $M_1$ and $M_2$, respectively, then the ratio of their posterior probabilities is

$$\frac{p(M_1|y)}{p(M_2|y)} = \frac{\pi(M_1)}{\pi(M_2)} \times \text{Bayesian factor}(M_1, M_2),$$

which defines the Bayesian factor—namely,

$$\text{Bayesian factor}(M_1, M_2) = \frac{f(y|M_1)}{f(y|M_2)}$$

$$= \frac{\int \pi(\theta_1|M_1)f(y|\theta_1, M_1)\, d\theta_1}{\int \pi(\theta_2|M_2)f(y|\theta_2, M_2)\, d\theta_2}.$$

### A.3.5 Stationary processes

Many important processes have the *stationarity* properties, in one way or the other. For simplicity, consider a process $X(t), t \geq 0$, taking values in $R$.

The process is said to be *strongly stationary* if $[X(t_1), \ldots, X(t_n)]$ and $[X(t_1 + h), \ldots, X(t_n + h)]$ have the same joint distribution for all $t_1, \ldots, t_n$ and $h > 0$. Note that if $X(t), t \geq 0$, is strongly stationary, then, in particular, the $X(t)$'s are identically distributed. However, strong stationarity is a much stronger property than identical distribution. The process is said to be *weakly* (or *second-order*) stationary if $\mathrm{E}\{X(t_1)\} = \mathrm{E}\{X(t_2)\}$ and $\mathrm{cov}\{X(t_1), X(t_2)\} = \mathrm{cov}\{X(t_1 + h), X(t_2 + h)\}$ for all $t_1, t_2$ and $h > 0$.

The terms used here might suggest that a strongly stationary process must be weakly stationary. However, this is not implied by the definition, unless the second moment of $X(t)$ is finite for all $t$ (in which case, the claim is true). On the other hand, a weakly stationary process may not be strongly stationary, of course, unless the process is Gaussian, as in the first example below.

*Example A.7* (Gaussian process). A real-valued process $X(t), t \geq 0$, is said to be *Gaussian* if each finite-dimensional vector $[X(t_1), \ldots, X(t_n)]'$ has a multivariate normal distribution. Now, suppose that the Gaussian process is weakly stationary. Then the vectors $U = [X(t_1), \ldots, X(t_n)]'$ and $V = [X(t_1 + h), \ldots, X(t_n + h)]'$ have the same mean vector [which is $(\mu, \ldots, \mu)$, where $\mu = \mathrm{E}\{X(0)\}$]. Furthermore, the weak stationarity property implies that $\mathrm{Var}(U) = \mathrm{Var}(V)$ (see Section A.2.4). Thus, by the properties of multivariate normal distribution (see Section A.3.1), $U$ and $V$ have the same joint distribution. In other words, the Gaussian process is strongly stationary.

*Example A.8* (Markov chains). Let $X(t), t \geq 0$, be an irreducible Markov chain taking values in a countable subset $S$ of $R$ and with a unique stationary distribution $\pi$ (see Section 10.2). The finite-dimensional distributions of the process depend on the initial distribution $p_0$ of $X(0)$, and it is not generally true that $X(t), t \geq 0$, is stationary in either sense. However, if $p_0 = \pi$—that is, the initial distribution is the same as the stationary distribution—then the distribution $p_t$ of $X(t)$ satisfies

$$p_t(j) = P\{X(t) = j\}$$
$$= \sum_{i \in S} P\{X(0) = i\}P\{X(t) = j | X(0) = i\}$$
$$= \sum_{i \in S} p_0(i)p^{(t)}(i,j)$$
$$= \sum_{i \in S} \pi(i)p^{(t)}(i,j) \;\; = \;\; \pi(j), \quad j \in S,$$

the last identity implied by (10.7) and (10.17). In other words, the distribution of $X(t)$ does not depend on $t$. By a similar argument, it can be shown that the joint distribution of $X(t_1 + h), \ldots, X(t_n + h)$ does not depend on $h$ for every $n > 1$. Thus, the process $X(t), t \geq 0$, is strongly stationary.

A fundamental theory associated with weak stationary processes is called the spectral theorem, or spectral representation. Define the *autocovariance function* of a weakly stationary process $X(t), -\infty < t < \infty$, by

$$c(t) = \text{cov}\{X(s), X(s+t)\}$$

for any $s, t \in R$. Thus, in particular, $c(0) = \text{var}\{X(t)\}$. The *autocorrelation function* is defined as $\rho(t) = c(t)/c(0), t \in R$. The spectral theorem for autocorrelation functions states that if $c(0) > 0$ and $\rho(t)$ is continuous at $t = 0$, then $\rho(t)$ is the cf (see Section 2.4) of some distribution $F$; that is,

$$\rho(t) = \int_{-\infty}^{\infty} e^{it\lambda} \, dF(\lambda), \quad t \in R.$$

The distribution $F$ is called the *spectral distribution function* of the process. If $X(n), n = 0, \pm 1, \ldots$, is a discrete-time process such that $\sum_{n=-\infty}^{\infty} |\rho(n)| < \infty$, then $F$ has a density $f$, called the *spectral density function*, given by

$$f(\lambda) = \frac{1}{2\pi} \sum_{n=-\infty}^{\infty} e^{-in\lambda} \rho(n), \quad \lambda \in [-\pi, \pi].$$

Not only does the autocorrelation function of a weakly stationary process have the spectral representation, but the process itself also enjoys a nice spectral representation. Suppose that $X(t), -\infty < t < \infty$, is weakly stationary with $E\{X(t)\} = 0$ and that $\rho(t)$ is continuous. Then there exists a complex-valued process $S(\lambda), -\infty < \lambda < \infty$, such that

$$X(t) = \int_{-\infty}^{\infty} e^{it\lambda} \, dS(\lambda), \quad -\infty < t < \infty$$

(see Section 10.6 for the definition of a stochastic integral). The process $S$ is called the *spectral process* of $X$.

As for strongly stationary processes, a well-known result is the ergodic theorem. This may be regarded as an extension of the SLLN. The theorem is usually stated for a discrete-time process, $X_n, n = 1, 2, \ldots$. If the latter is strongly stationary such that $E(|X_1|) < \infty$, then there is a random variable, $Y$, with the same mean as the $X$'s such that

$$\frac{1}{n} \sum_{i=1}^{n} X_i \xrightarrow{\text{a.s.}} Y$$

and $E(X_n) \to E(Y)$ as $n \to \infty$. The random variable $Y$ can be expressed as a conditional expectation. Let $(\Omega, \mathcal{F}, P)$ be a probability space. A measurable map $T\colon \Omega \to \Omega$ is said to be measure-preserving if $P(T^{-1}A) = P(A)$ for $A \in \mathcal{F}$, where $T^{-1}A = \{\omega \in \Omega, T(\omega) \in A\}$. Any stationary process $X_n, n = 0, 1, \ldots$, can be thought of as being generated by a measure-preserving transformation $T$ in the sense that there exists a random variable $X$ defined on a probability space $(\Omega, \mathcal{F}, P)$ and a map $T\colon \Omega \to \Omega$ such that the process $XT^n, n \geq 0$, has the same joint distribution as $X_n, n \geq 0$, where $XT^n(\omega) = X\{T^n(\omega)\}$, $\omega \in \Omega$, and $XT^0 = X$. The process $X_n, n \geq 0$, is said to be *ergodic* if the transformation $T$ satisfies the following: For any $A \in \mathcal{F}$, $T^{-1}(A) = A$ implies $P(A) = 0$ or $1$. The ergodic theorem can now be restated as that if $T$ is measure-preserving and $E(|X|) < \infty$, then

$$\frac{1}{n} \sum_{i=0}^{n-1} XT^i \xrightarrow{\text{a.s.}} E(X|\mathcal{I}),$$

where $\mathcal{I}$ is the invariant $\sigma$-field defined as $\mathcal{I} = \{A \in \mathcal{F} : T^{-1}A = A\}$. In particular, if the process $X_n, n \geq 0$, is ergodic, then $Y = E(X|\mathcal{I}) = E(X)$.

## A.4 List of notation and abbreviations

The list is in alphabetical order, although the actual letters that appear in different places in the text may be different:

$a \wedge b$: $= \min(a, b)$.

$a \vee b$: $= \max(a, b)$.

a.s.: almost surely.

$a'$: the transpose of vector $a$.

$\dim(a)$: the dimension of vector $a$.

$A \leq B$, where $A$ and $B$ are symmetric matrices: This means $B - A$ is nonnegative definite.

$A < B$, where $A$ and $B$ are symmetric matrices: This means $B - A$ is positive definite.

$A^c$: the complement of set $A$.

$|A|$: the determinant of matrix $A$.

$A'$: the transpose of matrix $A$.

$\lambda_{\min}(A)$: the smallest eigenvalue of matrix $A$.

$\lambda_{\max}(A)$: the largest eigenvalue of matrix $A$.

$\mathrm{tr}(A)$: the trace of matrix $A$.

$\|A\|$: the spectral norm of matrix $A$ defined as $\|A\| = \{\lambda_{\max}(A'A)\}^{1/2}$.

$\|A\|_2$: the 2-norm of matrix $A$ defined as $\|A\|_2 = \{\mathrm{tr}(A'A)\}^{1/2}$.

$\mathrm{rank}(A)$: the (column) rank of matrix $A$.

$A^{1/2}$: the square root of a nonnegative definite matrix $A$ (see Section A.1.6).

If $A$ is a set, $|A|$ represents the cardinality of $A$.

ACR: autocorrelation.

ACV: autocovariance.

AIC: Akaike's information criterion.

ANOVA: analysis of variance.

AR: autoregressive process.

ARMA: autoregressive moving average process.

ARE: asymptotic relative efficiency.

$a_n = O(b_n)$: This means that the sequence $a_n/b_n, n = 1, 2, \ldots$, is bounded.

$a_n = o(b_n)$: This means that the sequence $a_n/b_n \to 0$ as $n \to \infty$.

$a_n \sim b_n$, where both $a_n$ and $b_n$ are sequences of real numbers: This means $a_n/b_n \to 1$ as $n \to \infty$.

AU: asymptotically unbiased.

$\mathcal{B}$: the Borel $\sigma$-field.

BIC: Bayesian information criterion.

BLUE: best linear unbiased estimator.

BLUP: best linear unbiased predictor.

BP: best predictor.

$\mathcal{C}$: the space of continuous functions with the uniform metric.

cdf: cumulative distribution function.

cf: characteristic function.

CLT: central limit theorem.

$C_k^m$: the binomial coefficient equal to the number of ways of choosing $k$ items from $m$ items without considering the order; this is also denoted by $\binom{m}{k}$.

$\mathrm{Cov}(\xi, \eta)$: the covariance matrix between random vectors $\xi$ and $\eta$ (see Section A.2.4).

$\xrightarrow{\text{a.s.}}$: almost sure convergence.

$\xrightarrow{\text{d}}$: convergence in distribution.

$\xrightarrow{\text{P}}$: convergence in probability.

$\xrightarrow{L^p}$: convergence in $L^p$.

c.u.: continuous uniformly.

$\mathcal{D}$: the space of functions that are right continuous and possess left-limit at each point, with the uniform metric.

$\mathrm{diag}(A)$: for $A$ being a square matrix, this is the vector of diagonal elements of $A$.

diag$(A_1, \ldots, A_k)$: the block-diagonal matrix with $A_1, \ldots, A_k$ on its diagonal; the definition also includes the diagonal matrix, when $A_1, \ldots, A_k$ are numbers.

Distributions: Binomial$(n, p)$ — binomial distribution with $n$ independent trials and probability $p$ of success for each trial; Cauchy$(\mu, \sigma)$ — Cauchy distribution with pdf $f(x|\mu, \sigma) = (\pi\sigma[1 + \{(x - \mu)/\sigma\}^2])^{-1}$, $-\infty < x < \infty$; DE$(\mu, \sigma)$ — double exponential distribution with pdf $f(x|\mu, \sigma) = (2\sigma)^{-1} \exp(-|x - \mu|/\sigma)$, $-\infty < x < \infty$; $\chi_\nu^2$ — $\chi^2$-distribution with $\nu$ degrees of freedom; Exponential$(\lambda)$ — exponential distribution with mean $\lambda$; $N(\mu, \sigma^2)$ — normal distribution with mean $\mu$ and variance $\sigma^2$, or $N(\mu, \Sigma)$ — multivariate normal distribution with mean vector $\mu$ and covariance matrix $\Sigma$; NM$(p, \tau)$ — normal mixture distribution with cdf $(1 - p)\phi(x) + p\Phi(x/\tau)$, where $\Phi$ is the cdf of $N(0, 1)$; Poisson$(\lambda)$ — Poisson distribution with mean $\lambda$; $t_\nu$ — $t$-distribution with $\nu$ degrees of freedom; Uniform$[a, b]$ — uniform distribution over $[a, b]$.

$\nabla$: the gradient operator.

$\delta_x(y)$: the Dirac (or point) mass at $x$, which $= 1$ if $y = x$ and $0$ otherwise.

$E_\theta$: This notation is often used for expectation under the distribution with $\theta$ being the true parameter (vector).

$E_M$: This notation is sometimes used for model-based expectation; or expectation under model $M$.

$E_d$: This notation is sometimes used for design-based expectation.

$E(\xi|\eta)$: conditional expectation of $\xi$ given $\eta$.

EBLUE: empirical best linear unbiased estimator.

EBLUP: empirical best linear unbiased predictor.

EBP: empirical best predictor.

EM: Expectation–Maximization (algorithm).

$\emptyset$: empty set.

$E_\theta$: expectation when $\theta$ is the true parameter (vector).

$f \circ g$: $f \circ g(x) = f(g(x))$ for functions $f, g$.

$F^{-1}(t)$: If $F$ is a cdf, this is defined as $\inf\{x : F(x) \geq t\}$.

$f(x) = O\{g(x)\}$: This means $f(x)/g(x)$ is bounded for all $x$.

$f(x) = o\{g(x)\}$: This means $f(x)/g(x) \to 0$ as $x \to \infty$ (or $x \to 0$).

$f(x) \sim g(x)$: This means $f(x)/g(x) \to 1$ as $x \to \infty$ (or $x \to 0$).

$f(x|y)$: the conditional density function.

$\mathcal{F}$: This notation is usually used for a $\sigma$-field.

$\mathcal{F}_n$: a sequence of $\sigma$-fields such that $\mathcal{F}_n \subset \mathcal{F}_{n+1}, n = 1, 2, \ldots$.

$F_n$: This notation is often (but not always) used for the empirical distribution of observations $X_1, \ldots, X_n$.

$F'_-(x)$ $(F'_+(x))$: the left (right) derivative of $F$ at $x$.

$\Gamma(\cdot)$: the gamma function.

GLM: generalized linear model.

GLMM: generalized linear mixed model.

HQ: Hannan–Quinn criterion.

$i$: in the definition of cf (see above), for example, this represents $\sqrt{-1}$.

iff: if and only if.

i.i.d.: independent and identically distributed.

inf: infimum.

$I_n$: the $n$-dimensional identity matrix.

$I(\theta)$ (or $\mathcal{I}(\theta)$): the Fisher information (matrix).

$J_n$: the $n \times n$ matrix of 1's, or $J_n = 1_n 1_n'$ (see below).

$\bar{J}_n := n^{-1} J_n$.

$\kappa_3$: the skewness (parameter).

$\kappa_4$: the kurtosis (parameter).

$\mathcal{L}(A)$: the linear space spanned by the columns of matrix $A$.

LIL: law of the iterated logarithm.

log: logarithm of base $e$, or natural logarithm.

logit: the logit function defined as $\mathrm{logit}(p) = \log\{p/(1-p)\}$, $0 < p < 1$.

$L^p$: the $L^p$ space of functions or random variables.

$\liminf x_n$: the smallest limit point of the sequence $x_n$.

$\limsup x_n$: the largest limit point of the sequence $x_n$.

$\limsup A_n$, where $A_1, A_2, \ldots$ is a sequence of events: This is defined as $\cap_{N=1}^{\infty} \cup_{n=N}^{\infty} A_n$.

LLN: law of large numbers.

LSE: least squares estimator.

MA: moving average process.

MC: Markov chain.

MCEM: Monte Carlo EM (algorithm).

MCMC: Markov-chain Monte Carlo.

$M_f$: This notation is often used to denote a full model.

mgf: moment generating function.

MINQUE: minimum norm quadratic unbiased estimation.

ML: maximum likelihood.

MLE: maximum likelihood estimator.

MM: method of moments.

$M_{\mathrm{opt}}$: This notation is often used to denote an optimal model.

MSA: for balanced data $y_{ij}$, $1 \le i \le m$, $1 \le j \le k$, MSA $=$ SSA$/(k-1)$.

MSE: mean squared error (or, see below).

MSE: for balanced data $y_{ij}$, $1 \le i \le m$, $1 \le j \le k$, MSE $=$ SSE$/m(k-1)$.

MSM: method of simulated moments.

MSPE: mean squared prediction error.

$\#A$, where $A$ is a set: This represents the cardinality of set $A$.

$N(\mu, \Sigma)$: The multivariate normal distribution with mean vector $\mu$ and covariance matrix $\Sigma$.

OLS: ordinary least squares.

$1_A$, where $A$ is an event: This represents the indicator of event $A$.

$1_n$, where $n$ is a positive integer: the $n$-dimensional vector of 1s.

$1_n^0 := I_n$.

$1_n^1 := 1_n$.

$\Omega$: This usually represents a probability space.

$O_P, o_P$: big O and small o in probability (see Section 3.4).

$\otimes$: Kronecker product.

$P(A|B)$: conditional probability of $A$ given $B$.

$P_A$: the projection matrix to $\mathcal{L}(A)$ defined as $P_A = A(A'A)^- A'$, where $A^-$ is the generalized inverse of $A$ (see §A.1.2).

$P_{A\perp}$: the projection matrix with respect to the linear space orthogonal to $\mathcal{L}(A)$, defined as $P_A = I - P_A$, where $I$ is the identity matrix.

$\partial A$, the boundary of set $A$.

$\partial \xi/\partial \eta'$: When $\xi = (\xi_i)_{1\leq i\leq a}$, $\eta = (\eta_j)_{1\leq j\leq b}$, this notation means the matrix $(\partial \xi_i/\partial \eta_j)_{1\leq i\leq a,1\leq j\leq b}$.

$\partial^2 \xi/\partial \eta \partial \eta'$: When $\xi$ is a scalar, $\eta = (\eta_j)_{1\leq j\leq b}$, this notation means the matrix $(\partial^2 \xi/\partial \eta_j \partial \eta_k)_{1\leq j,k\leq b}$.

pdf: probability density function.

pmf: probability mass function.

PQL: penalized quasi-likelihood.

$\propto$: proportional to.

r.c.: relatively compact.

$R^d$: the $d$-dimensional Euclidean space; in particular, $R^1 = R$ represents the real line.

REML: restricted maximum likelihood.

RSS: residual sum of squares.

s.d.: standard deviation.

SDE: stochastic differential equation.

$S_\delta(a)$: the $\delta$-neighborhood of $a$; that is, $\{x : |x - a| < \delta\}$.

SLLN: strong law of large numbers.

SSA: for balanced data $y_{ij}$, $1 \leq i \leq m, 1 \leq j \leq k$, SSA $= k \sum_{i=1}^m (\bar{y}_{i\cdot} - \bar{y}_{\cdot\cdot})^2$.

SSE: for balanced data $y_{ij}$, $1 \leq i \leq m, 1 \leq j \leq k$, SSE $= \sum_{i=1}^m \sum_{j=1}^k (y_{ij} - y_{i\cdot})^2$.

sup: supremum.

TMD: two-parameter martingale differences.

Var($\xi$): covariance matrix of the random vector $\xi$.

$\text{var}_\theta$: variance when $\theta$ is the true parameter (vector).

WLLN: weak law of large numbers.

WLS: weighted least squares.

WN: white noise process.

w.r.t.: with respect to.

$[x]$ for real number $x$: This is the largest integer less than or equal to $x$.

$|x|$ for $x \in R^d$: This is defined as $(\sum_{i=1}^d x_i^2)^{1/2}$, where $x_i$ is the $i$th component of $x$, $1 \leq i \leq d$.

$x^+$: defined as $x$ is $x > 0$ and 0 otherwise.

$x^-$: defined as $-x$ if $x < 0$ and 0 otherwise.

$\bar{X}$: the sample mean of $X_1, \ldots, X_n$.

$X_{(i)}$: the $i$th order statistic of $X_1, \ldots, X_n$ such that $X_{(1)} \leq \cdots \leq X_{(n)}$.

Var($\xi$): the covariance matrix of random vector $\xi$ (see §A.2.4).

$(X_i)_{1 \leq i \leq m}$: When $X_1, \ldots, X_m$ are matrices with the same number of columns, this is the matrix that combines the rows of $X_1, \ldots, X_m$, one after the other.

$(y_i)_{1 \leq i \leq m}$: When $y_1, \ldots, y_m$ are column vectors, this notation means the column vector $(y_1', \ldots, y_m')'$.

$(y_{ij})_{1 \leq i \leq m, 1 \leq j \leq n_i}$: In the case of clustered data, where $y_{ij}$, $j = 1, \ldots, n_i$, denote the observations from the $i$th cluster, this notation represents the vector $(y_{11}, \ldots, y_{1n_1}, y_{21}, \ldots, y_{2n_2}, \ldots, y_{m1}, \ldots, y_{mn_m})'$.

$y_i.$, $\bar{y}_i.$, $y._j$, $\bar{y}._j$, $y_{..}$ and $\bar{y}_{..}$: In the case of clustered data $y_{ij}$, $i = 1, \ldots, m$, $j = 1, \ldots, n_i$, $y_i. = \sum_{j=1}^{n_i} y_{ij}$, $\bar{y}_i. = n_i^{-1} y_i.$, $y_{..} = \sum_{i=1}^{m} \sum_{j=1}^{n_i} y_{ij}$, $\bar{y}_{..} = (\sum_{i=1}^{m} n_i)^{-1} y_{..}$; in the case of balanced data $y_{ij}$, $1 \leq i \leq a$, $j = 1, \ldots, b$, $y_i. = \sum_{j=1}^{b} y_{ij}$, $\bar{y}_i. = b^{-1} y_i.$, $y._j = \sum_{i=1}^{a} y_{ij}$, $\bar{y}._j = a^{-1} y._j$, $y_{..} = \sum_{i=1}^{a} \sum_{j=1}^{b} y_{ij}$, $\bar{y}_{..} = (ab)^{-1} y_{..}$.

$y | \eta \sim$: the distribution of $y$ given $\eta$ is ...; note that here $\eta$ may represent a vector of parameters or random variables, or a combination of both.

Y-W: Yule–Walker.

# References

Akaike, H. (1973), Information theory as an extension of the maximum likelihood principle, in *Second International Symposium on Information Theory* (B. N. Petrov and F. Csaki eds.), Akademiai Kiado, Budapest, 267–281.

Akaike, H. (1974), A new look at the statistical model identification, *IEEE Trans. Automatic Control* 19, 716–723.

An Hong-Zhi, Chen Zhao-Guo, and Hannan, E. J. (1982), Autocorrelation, autoregression and autoregressive approximation, *Ann. Statist.* 10, 926–936.

Anderson, T. W. (1971), *The Statistical Analysis of Time Series*, Wiley, New York.

Anderson, T. W., and Darling, D. A. (1954), A test of goodness of fit, *J. Amer. Statist. Assoc.* 49, 765–769.

Arnold, L. (1974), *Stochastic Differential Equations: Theory and Applications*, Wiley, New York.

Atwood, L. D., Wilson, A. F., Bailey-Wilson, J. E., Carruth, J. N., and Elston, R. C. (1996), On the distribution of the likelihood ratio test statistic for a mixture of two normal distributions, *Commun. Statist. - Simulation* 25, 733–740.

Bachman, G., Narici, L., and Beckenstein, E. (2000), *Fourier and Wavelet Analysis*, Springer, New York.

Bamber, D. (1975), The area above the ordinal dominance graph and the area below the receiver operating characteristic graph, *J. Math. Psych.* 12, 387–415.

Barndorff-Nielsen, O. (1983), On a formula for the distribution of the maximum likelihood estimator, *Biometrika* 70, 343–365.

Barndorff-Nielsen, O. E., and Cox, D. R. (1989), *Asymptotic Techniques for Use in Statistics*, Chapman & Hall, London.

Battese, G. E., Harter, R. M., and Fuller, W. A. (1988), An error-components model for prediction of county crop areas using survey and satellite data, *J. Amer. Statist. Assoc.* 80, 28–36.

Beran, R. (1984), Bootstrap methods in statistics, *Jber. Dt. Math. Verein.* 86, 24–30.

Bernstein, S. N. (1937), On several modifications of Chebyshev's inequality, *Doklady Akad. Nauk SSSR* 17, 275–277.

Berry, A. C. (1941), The accuracy of the Gaussian approximation to the sum of independent variates, *Trans. Amer. Math. Soc.* 49, 122–136.

Besag, J. (1974), Spatial interaction and the statistical analysis of lattice systems (with discussion), *J. Roy. Statist. Soc. B* 36, 192–236.

Bickel, P. J. (1966), Some contributions to the theory of order statistics, *Proc. 5th Berkeley Symp. Math. Statist. Probab.* 1, 575–592.

Bickel, P. J. (1967), Some contributions to the theory of order statistics, *Proc. 5th Berkeley Symp. Math. Statist. Probab.* 1, 575–591.

Bickel, P. J., and Bühlmann, P. (1997), Closure of linear processes, *J. Theoret. Probab.* 10, 445–479.

Bickel, P. J., and Freedman, D. A. (1981), Some asymptotic theory for the bootstrap, *Ann. Statist.* 9, 1196–1217.

Bickel, P. J., Klaassen, C. A. J., Ritov, Y., and Wellner, J. A. (1993), *Efficient and Adaptive Estimation for Semiparametric Models*, Springer, New York.

Billingsley, P. (1968), *Convergence of Probability Measure*, Wiley, New York.

Billingsley, P. (1995), *Probability and Measure*, 3nd ed., Wiley, New York.

Birnbaum, Z. W. and McCarty, R. C. (1958), A distribution-free upper confidence bound for $P(Y < X)$ based on independent samples of $X$ and $Y$, *Ann. Math. Statist.* 29, 558–562.

Black, F., and Sholes, M. (1973), The pricing of options and corporate liabilities, *J. Politi. Econ.* 81, 637–659.

Bollerslev, T. (1986), Generalized autoregressive conditional heteroskedasticity, *J. Econometrics* 31, 307–327.

Booth, J. G., and Hobert, J. P. (1999), Maximum generalized linear mixed model likelihood with an automated Monte Carlo EM algorithm, *J. Roy. Statist. Soc. B* 61, 265–285.

Bozdogan, H. (1994), Editor's general preface, in *Engineering and Scientific Applications*, Vol. 3 (H. Bozdogan ed.), *Proceedings of the First US/Japan Conference on the Frontiers of Statistical Modeling: An Informational Approach*, ix–xii, Kluwer Academic Publishers, Dordrecht.

Brackstone, G. J. (1987), Small area data: policy issues and technical challenges, in *Small Area Statistics* (R. Platek, J. N. K. Rao, C. E. Sarndal, and M. P. Singh eds.) 3–20, Wiley, New York.

Breslow, N. E., and Clayton, D. G. (1993), Approximate inference in generalized linear mixed models, *J. Amer. Statist. Assoc.* 88, 9–25.

Brockwell, P. J., and Davis, R. A. (1991), *Time Series: Theory and Methods*, 2nd ed., Springer, New York.

Brown, L. D., Wang, Y., and Zhao, L. H. (2003), Statistical equivalence at suitable frequencies of GARCH and stochastic volatility models with the corresponding diffusion model, *Statist. Sinica* 13, 993–1013.

Bühlmann, P. (1997), Sieve bootstrap for time series, *Bernoulli* 3, 123–148.

Bühlmann, P. (2002), Bootstraps for time series, *Statist. Sci.* 17, 52–72.

Burkholder, D. L. (1966), Martingale transforms, *Ann. Math. Statist.* 37, 1494–1504.

Calvin, J. A., and Sedransk, J. (1991), Bayesian and frequentist predictive inference for the patterns of care studies, *J. Amer. Statist. Assoc.* 86, 36–48.

Casella, G. and Berger, R. L. (2002), *Statistical Inference*, 2nd ed., Duxbury, Thomson Learning, Pacific Grove, CA.

Casella, G., and George, E. I. (1992), Explaining the Gibbs sampler, *Amer. Statist.* 46, 167–174.

Chan, N. N., and Kwong, M. K. (1985), Hermitian matrix inequalities and a conjecture, *Amer. Math. Monthly* 92, 533–541.

Chatterjee, S., Lahiri, P., and Li, H. (2008), Parametric bootstrap approximation to the distribution of EBLUP, and related prediction intervals in linear mixed models, *Ann. Statist.* 36, 1221–1245.

Chernoff, H. and Lehmann, E. L. (1954), The use of maximum-likelihood estimates in $\chi^2$ tests for goodness of fit, *Ann. Math. Statist.* 25, 579–586.

Chiang, T.-P. (1987), On Markov models of random fields, *Acta Math. Appl. Sinica* (English) 3, 328–341.

Chiang, T.-P. (1991), Stationary random field: Prediction theory, Markov models, limit theorems, *Contemp. Math.* 118, 79–101.

Chow, Y. S. (1960), A martingale inequality and the law of large numbers, *Proc. Amer. Math. Soc.* 11, 107–111.

Chow, Y. S. (1965), Local convergence of martingales and the law of large numbers, *Ann. Math. Statist.* 36, 552–558.

Chow, Y. S., and Teicher, H. (1988), *Probability Theory*, Springer, New York.

Chung, K.-L. (1948), On the maximum partial sums of sequence of independent random variables, *Trans. Amer. Math. Soc.* 64, 205–233.

Chung, K.-L. (1949), An estimate concerning the Kolmogoroff limit distribution, *Trans. Amer. Math. Soc.* 67, 36–50.

Claeskens, G., and Hart, J. D. (2009), Goodness-of-fit tests in mixed models (with discussion), *TEST* 18, 213–239.

Cox, D. R. (1975), Prediction intervals and empirical Bayes confidence intervals, in *Perspectives in Probability and Statistics, Papers in Honor of M. S. Bartlett*

(J. Gani, ed.), Applied Probability Trust, University of Sheffield, Sheffield, 47–55.

Cramér, H. (1936), Über eine Eigenschaft der normalen Verteilungsfunction, *Math. Zeitschr.* 41, 405–414.

Cramér, H. (1946), *Mathematical Mathods of Statistics*, Princeton University Press, Princeton, N.J.

Csörgő, M., and Révész, R. (1975), A new method to prove Strassen-type laws of invariance principle, I and II, *Z. Wahrsch. verw. Geb.* 31, 255–269.

Das, K., Jiang, J., and Rao, J. N. K. (2004), Mean squared error of empirical predictor, *Ann. Statist.* 32, 818–840.

DasGupta, A. (2008), *Asymptotic Theory of Statistics and Probability*, Springer, New York.

Datta, G. S., and Lahiri, P. (2000), A unified measure of uncertainty of estimated best linear unbiased predictors in small area estimation problems, *Statist. Sinica* 10, 613–627.

Datta, G. S., and Lahiri, P. (2001), Discussions on a paper by Efron and Gous, *Model Selection*, IMS Lecture Notes/Monograph 38, P. Lahiri ed., Institute of Mathematical Statistics, Beachwood, OH.

Datta, G. S., Kubokawa, T., Rao, J. N. K., and Molina, I. (2009), Estimation of mean squared error of model-based small area estimators, *TEST*, to appear.

Datta, G. S., Rao, J. N. K., and Smith, D. D. (2005), On measuring the variability of small area estimators under a basic area level model, *Biometrika* 92, 183–196.

David, H. A. and Nagaraja, H. N. (2003), *Order Statistics*, 3rd ed., Wiley, New York.

De Bruijn, N. G. (1961), *Asymptotic Methods in Analysis*, North-Holland, Amsterdam.

Dehling, H., Mikosch, T., and Sørensen, M. (2002), *Empirical Process Techniques for Dependent Data*, Birkhäuser, Boston.

Dehling, H., and Philipp, W. (2002), Empirical process techniques for dependent data, in *Empirical Process Techniques for Dependent Data* (H. Dehling, T. Mikosch and M. Sørensen eds.), Birkhäuser, Boston.

de Leeuw, J. (1992), Introduction to Akaike (1973) information theory and an extension of the maximum likelihood principle, in *Breakthroughs in Statistics* (S. Kotz and N. L. Johnson eds.), Springer, London, Vol. 1, 599–609.

Dempster, A., Laird, N., and Rubin, D. (1977), Maximum likelihood from incomplete data via the EM algorithm (with discussion), *J. Roy. Statist. Soc. B* 39, 1–38.

Diggle, P. J., Liang, K. Y., and Zeger, S. L. (1996). *Analysis of Longitudinal Data*. Oxford University Press, Oxford.

Donsker, M. (1951), An invariance principle for certain probability limit theorems, *Mem. Amer. Math. Soc.* 6, .

Donsker, M. (1952), Justification and extension of Doob's heuristic approach to the Kolmogorov-Smirnov theorems, *Ann. Math. Statist.* 23, 277-i-281.

Doob, J. L. (1949), Heuristic approach to the Kolmogorov-Smirnov theorems, *Ann. Math. Statist.* 20, 393–403.

Doob, J. L. (1953), *Stochastic Processes*, Wiley, New York.

Doob, J. L. (1960), Notes on martingale theory, *Proc. Fourth Berkeley Symp. Math. Statist. Prob.* 2, 95–102.

Durbin, J. (1960), The fitting of time series models, *Int. Statist. Rev.* 28, 233–244.

Durrett, R. (1991), *Probability: Theory and Examples*, Wadsworth, Pacific Grove, CA.

Durrett, R. (1996), *Stochastic Calculus: A Practical Introduction*, CRC Press, Boca Raton, FL.

Dvoretzky, A., Erdös, P., and Kakutani, S. (1961), Nonincrease everywhere of the Brownian motion process, *Proc. Fourth Berkeley Symposium* II, 103–116.

Dvoretzky, A., Keifer, J., and Wolfowitz, J. (1956), Asymptotic minimax character of the sample distribution functions and of the classical multinomial estimator, *Ann. Math. Statist.* 27, 642–669.

Dynkin, E. B. (1957), Inhomogeneous strong Markov processes, *Dokl. Akad. Nauk SSSR* 113, 261–263.

Engle, R. F. (1982), Autoregressive conditional heteroscedasticity with estimates of the variance of United Kingdom inflation, *Econometrica* 50, 987–1007.

Efron, B. (1979), Bootstrap method: Another look at the jackknife, *Ann. Statist.* 7, 1–26.

Efron, B., and Morris, C. (1973), Stein's estimation rule and its competitors: An empirical Bayes approach, *J. Amer. Statist. Assoc.* 68, 117–130.

Efron, B., and Tibshirani, R. J. (1993), *An Introduction to the Bootstrap*, Chapman & Hall/CRC, New York.

Esseen, C. G. (1942), On the Liapounoff limit of error in the theory of probability, *Ark. Math. Astr. och Fysik* 28A, 1–19.

Faber, G. (1910), Über stetige Funktionen II, *Math. Ann.* 69, 372–443.

Fan, J., and Yao, Q. (2003), *Nonlinear Time Series*, Springer, New York.

Fay, R. E., and Herriot, R. A. (1979), Estimates of income for small places: an application of James-Stein procedures to census data, *J. Amer. Statist. Assoc.* 74, 269–277.

Federá Serio, G., Manara, A., and Sicoli, P. (2002), Giuseppe Piazzi and the Discovery of Ceres, in W. F. Bottke Jr., A. Cellino, P. Paolicchi, and R. P.

Binzel, *Asteroids III*, Tucson, Arizona: Univ. of Arizona Press, pp. 17–24, Retrieved 2009-06-25.

Feller, W. (1968), *An introduction to Probability Theory and Its Applications*, Wiley, New York, Vol. I.

Feller, W. (1971), *An introduction to Probability Theory and Its Applications*, Wiley, New York, Vol. II.

Ferguson, T. S. (1996), *A Course in Large Sample Theory*, Chapman & Hall, London.

Finkelstein, H. (1971), The law of iterated logarithm for empirical distributions, *Ann. Math. Statist.* 42, 607–615.

Fisher, R. A. (1922a), On the interpretation of chi-square from contingency tables, and the calculation of P, *J. Roy. Statist. Soc.* 85, 87–94.

Fisher, R. A. (1922b), On the mathematical foundations of the theoretical statistics, *Phil. Trans. R. Soc.* 222, 309–368.

Foderà Serio, G., Manara, A., and Sicoli, P. (2002), Giuseppe Piazzi and the discovery of Ceres, in *Asteroids* III (W. F. Bottke Jr., A. Cellino, P. Paolicchi and R. P. Binzel, eds.), University of Arizona Press, Tucson, 17–24.

Forrester, J., and Ury, H. K. (1969), The signed-rank (Wilcoxon) test in the rapid analysis of biological data, *Lancet* 1, 239–241.

Fox, R., and Taqqu, M. S. (1985), Noncentral limit theorems for quadractic forms in random variables having long-range dependence, *Ann. Probab.* 13, 428–446.

Freedman, D. (1984), On bootstrapping two-stage least squares estimates in stationary linear models, *Ann. Statist.* 12, 827–842.

Fuller, W. A. (1989), Prediction of true values for the measurement error model, in *Conference on Statistical Analysis of Measurement Error Models and Applications*, Humbolt State University, Arcata, CA.

Gelfand, A. E., and Smith, A. F. M. (1990), Sampling-based approaches to calculating marginal densities, *J. Amer. Statist. Assoc.* 85, 398–409.

Gelman, A., Carlin, J. B., Stern, H. S., and Rubin, D. B. (1995), *Bayesian Data Analysis*, Chapman & Hall London.

Geman, S., and Geman, D. (1984), Stochastic relaxation, Gibbs distributions and the Bayesian restoration of images, *IEEE Trans. Patterns. Anal. Mach. Intell.* 6, 721–741.

Ghosh, M., and Rao, J. N. K. (1994), Small area estimation: An appraisal (with discussion), *Statist. Sci.* 9, 55–93.

Goethe, J. W. von (1808), *Faust*, Part One.

Grenander, U. (1981), *Abstract Inference*, Wiley, New York.

Green, P. J. (1987), Penalized likelihood for general semi-parametric regression models, *Int. Statist. Rew.* 55, 245–259.

Guttorp, P., and Lockhart, R. A. (1988), On the asymptotic distribution of quadratic forms in uniform order statistics, *Ann. Statist.* 16, 433–449.

Hall, P. (1992), *The Bootstrap and Edgeworth Expansion*, Springer, New York.

Hall, P., and Heyde, C. C. (1980), *Martingale Limit Theory and Its Application*, Academic Press, New York.

Hall, P., and Malti, T. (2006), Nonparametric estimation of mean-squared prodiction error in nested-error regression models, *Ann. Statist.* 34, 1733–1750.

Hannan, E. J. (1970), *Multiple Time Series*, Wiley, New York.

Hannan, E. J. (1980), The estimation of the order of an ARMA process, *Ann. Statist.* 8, 1071–1081.

Hannan, E. J., and Heyde, C. C. (1972), On limit theorems for quadratic functions of discrete time series, *Ann. Math. Statist.* 43, 2058–2066.

Hannan, E. J., and Quinn, B. G. (1979), The determination of the order of an autoregression, *J. Roy. Statist. Soc. B* 41, 190–195.

Hanna, E. J., and Rissanen, J. (1982), Recursive estimation of mixed autoregressive-moving average order, *Biometrika* 69, 81–94.

Hardy, G., Littlewood, J. E. and Pólya, G. (1934), *Inequalities*, Cambridge Univ. Press.

Hartigan, J. A. (1985), A failure of likelihood asymptotics for normal mixtures, *Proceedings of the Berkeley Conference in Honor of Jerzy Neyman and Jack Kiefer* (L. M. Le Cam and R. A. Olshen eds.), Vol. II, 807–810, Wadsworth, Belmont, CA.

Hartley, H. O., and Rao, J. N. K. (1967), Maximum likelihood estimation for the mixed analysis of variance model, *Biometrika* 54, 93–108.

Hartman, P., and Wintner, A. (1941), On the law of the iterated logarithm, *Amer. J. Math.* 63, 169–176.

Harville, D. A. (1974), Bayesian inference for variance components using only error contrasts, *Biometrika* 61, 383–385.

Harville, D. A. (1977), maximum likelihood approaches to variance components estimation and related problems, *J. Amer. Statist. Assoc.* 72, 320–340.

Harville, D. A. (1985), Decomposition of prediction error, *J. Amer. Statist. Assoc.* 80, 132–138.

Hastings, W. K. (1970), Monte Carlo sampling methods using Markov chains and their applications, *Biometrika* 57, 97–109.

Henderson, C. R. (1953), Estimation of variance and covariance components, *Biometrics* 9, 226–252.

Heyde, C. C. (1994), A quasi-likelihood approach to the REML estimating equations, *Statist. Probab. Lett.* 21, 381–384.

Heyde, C. C. (1997), *Quasi-Likelihood and Its Application*, Springer, New York.

Heyde, C. C., and Scott, D. J. (1973), Invariance principle for the law of the iterated logarithm for martingales and processes with stationary increments, *Ann. Probab.* 1, 428–436.

Hinkley, D. V. (1977), Jackknifing in unbalanced situations, *Technometrics* 19, 285–292.

Hjort, N. L., and Jones, M. C. (1996), Locally parametric nonparametric density estimation, *Ann. Statist.* 24, 1619–1647.

Hoeffding, W. (1956), On the distribution of the number of successes in independent trials, *Ann. Math. Statist.* 27, 713–721.

Hoeffding, W. (1961), The strong law of large numbers for U-statistics, *Inst. Mimeo Ser.* 302, 1–10.

Hsieh, F., and Turnbull, B. W. (1996), Non-parametric and semi-parametric estimation of the receiver operating characteristic curve, *Ann. Statist.* 24, 25–40.

Hu, I. (1985), A uniform bound for the tail probability of Kolmogorov-Smirnov statistics, *Ann. Statist.* 13, 821–826.

Huang, D. (1988a), Convergence rate of sample autocorrelations and autocovariances for stationary time series, *Sci. Sinica* XXXI, 406–424.

Huang, D. (1988b), Recursive method for ARMA model estimation (I), *Acta Math. Appl. Sinica* 4, 169–192.

Huang, D. (1989), Recursive method for ARMA model estimation (II), *Acta Math. Appl. Sinica* 5, 332–354.

Huang, D. (1992), Central limit theorem for two-parameter martingale differences with application to stationary random fields, *Sci. Sinica* XXXV, 413–425.

Hunt, G. (1956), Some theorems concerning Brownian motion, *Trans. Amer. Math. Soc.* 81, 294–319.

Hurvich, C. M., and Tsai, C. L. (1989), Regression and time series model selection in small samples, *Biometrika* 76, 297–307.

Ibragimov, I. A., and Linnik, Y. V. (1971), *Independent and Stationary Sequence of Random Variables*, Wolters-Noordhoff, Groningen.

James, F. (2006), *Statistical Methods in Experimental Physics*, 2nd ed., World Scientific Singapore.

Jiang, J. (1989), Uniform convergence rate of sample autocovariances and autocorrelations for linear spatial series, *Adv. Math. (Chinese)* 18, 497–499.

Jiang, J. (1991a), Uniform convergence rate of sample ACV and ACR for linear spatial series under more general martingale condition, *Adv. Math. (Chinese)* 20, 39–50.

Jiang, J. (1991b), Parameter estimation of spatial AR model, *Chinese Ann.*

*Math.* 12B, 432–444.

Jiang, J. (1993), Estimation of spatial AR models, *Acta Math. Appl. Sinica* 9, 174–187.

Jiang, J. (1996), REML estimation: Asymptotic behavior and related topics, *Ann. Statist.* 24, 255–286.

Jiang, J. (1997a), Wald consistency and the method of sieves in REML estimation, *Ann. Statist.* 25, 1781–1803.

Jiang, J. (1997b), Sharp upper and lower bounds for asymptotic levels of some statistical tests, *Statist. Probab. Lett.* 35, 395–400.

Jiang, J. (1998a), Consistent estimators in generalized linear mixed models, *J. Amer. Statist. Assoc.* 93, 720–729.

Jiang, J. (1998b), On unbiasedness of the empirical BLUE and BLUP, *Statist. Probab. Lett.* 41, 19–24.

Jiang, J. (1998c), Asymptotic properties of the empirical BLUP and BLUE in mixed linear models, *Statist. Sinica* 8, 861–885.

Jiang, J. (1999a), Some laws of the iterated logarithm for two parameter martingales, *J. Theoret. Probab.* 12, 49–74.

Jiang, J. (1999b), On maximum hierarchical likelihood estimators, *Commun. Statist.: Theory Methods* 28, 1769–1776.

Jiang, J. (2000a), A matrix inequality and its statistical application, *Linear Algebra Applic.* 307, 131–144.

Jiang, J. (2000b), A nonlinear Gauss-Seidel algorithm for inference about GLMM, *Comput. Statist.* 15, 229–241.

Jiang, J. (2001), On actual significance levels of combined tests, *Nonparametric Statist.* 13, 763–774.

Jiang, J. (2007), *Linear and Generalized Linear Mixed Models and Their Applications*, Springer, New York.

Jiang, J., Jia, H., and Chen, H. (2001), Maximum posterior estimation of random effects in generalized linear mixed models, *Statist. Sinica* 11, 97–120.

Jiang, J., and Lahiri, P. (2001), Empirical best prediction for small area inference with binary data, *Ann. Inst. Statist. Math.* 53, 217–243.

Jiang, J., and Lahiri, P. (2006), Mixed model prediction and small area estimation (with discussion), *TEST* 15, 1–96.

Jiang, J., Lahiri, P., and Wan, S. (2002a), A unified jackknife theory for empirical best prediction with M-estimation, *Ann. Statist.* 30, 1782–1810.

Jiang, J., Lahiri, P., and Wan, S. (2002b), Jackknifing the mean squared error of empirical best predictor: A theoretical synthesis, Tech. Report, Dept. Statistics, University of California, Davis, CA.

Jiang, J., Lahiri, P., and Wu, C.-H. (2001), A generalization of the Pearson's

$\chi^2$ goodness-of-fit test with estimated cell frequencies, *Sankhyā, Ser. A* 63, 260–276.

Jiang, J., Rao, J. S., Gu, Z., and Nguyen, T. (2008), Fence methods for mixed model selection, *Ann. Statist.* 36, 1669–1692.

Jiang, J., Nguyen, T., and Rao, J. S. (2009), A simplified adaptive fence procedure, *Statist. Probab. Lett.* 79, 625–629.

Jiang, J., and Zhang, W. (2001), Robust estimation in generalized linear mixed models, *Biometrika* 88, 753–765.

Jones, M. C., Marron, J. S., and Sheather, S. J. (1996), A brief survey of bandwidth selection for density estimation, *J. Amer. Statist. Assoc.* 91, 401–407.

Kackar, R. N., and Harville, D. A. (1981), Unbiasedness of two-stage estimation and prediction procedures for mixed linear models, *Commun. Statist.: Theory Methods* 10, 1249–1261.

Kackar, R. N., and Harville, D. A. (1984), Approximations for standard errors of estimators of fixed and random effects in mixed linear models, *J. Amer. Statist. Assoc.* 79, 853–862.

Karim, M. R., and Zeger, S. L. (1992), Generalized linear models with random effects: Salamander mating revisited, *Biometrics* 48, 631–644.

Katz, M. (1968), A note on the weak law of large numbers, *Ann. Math. Statist.* 39, 1348–1349.

Khan, L. A., and Thaheem, A. B. (2000), On the equivalence of the Heine-Borel and the Bolzano-Weierstrass theorems, *Int. J. Math. Edu. Sci. Technol.* 31, 620–622.

Khintchine, A. (1924), Über einen Satz der Wahrscheinlichkeitsrechnung, *Fund. Math.* 6, 9–20.

Kolmogorov, A. (1929), Über das Gesetz des iterierten Logarithmus, *Math. Ann.* 101, 126–135.

Koroljuk, V. S., and Borovskich, Yu. V. (1994), *Theory of U-Statistics*, Kluwer, Dordrecht, The Netherlands.

Kosorok, M. P. (2008), *Introduction to Empirical Processes and Semiparametric Inference*, Springer, New York.

Künsch, H. R. (1989), The jackknife and the bootstrap for general stationary observations, *Ann. Statist.* 17, 1217–1241.

Kutoyants, Y. A. (2004), *Statistical Inference for Ergodic Diffusion Processes*, Springer, London.

Łagodowski, Z. A., and Rychlik, Z. (1986), Rate of convergence in the strong law of large numbers for martingales, *Probab. Theory Rel. Fields* 71, 467–476.

Lai, T. L., Robbins, H., and Wei, C. Z. (1979), Strong consistency of least squares estimates in multiple regression II, *J. Multivariate Anal.* 9, 343–361.

Lai, T. L., and Wei, C. Z. (1982), A law of the iterated logarithm for double arrays of independent random variables with applications to regression and time series models, *Ann. Probab.* 10, 320–335.

Lai, T. L., and Wei, C. Z. (1984), Moment inequalities with applications to regression and time series models, *Inequalities in Statistics and Probability*, IMS Lecture Notes - Monograph Ser. 5, Institute of Mathematical Statistics, Hayward, CA, 165–172.

Laird, N. M., and Ware, J. M. (1982), Random effects models for longitudinal data, *Biometrics* 38, 963–974.

Lange, N., and Ryan, L. (1989), Assessing normality in random effects models, *Ann. Statist.* 17, 624–642.

Langyintuo, A., and Mekuria, M. (2008), Assessing the influence of neighborhood effects on the adoption of improved agricultural technologies in developing agriculture, *African J. Agric. Resource Econ.* 2, 151–169.

Le Cam, L. (1986), *Asymptotic Methods in Statistical Decision Theory*, Springer, Berlin.

Le Cam, L., and Yang, G. (1990), *Asymptotics in Statistics: Some Basic Concepts*, Springer, New York.

Lee, A. J. (1990), *U-Statistics*, Marcel Dekker, New York.

Lehmann, E. L. (1975), *Nonparametrics*, Holden-Day, San Francisco.

Lehmann, E. L. (1983), *Theory of Point Estimation*, Wiley, New York.

Lehmann, E. L. (1986), *Testing Statistical Hypotheses* 2nd. ed., Chapman & Hall, London.

Lehmann, E. L. (1999), *Elements of Large-Sample Theory*, Springer, New York.

Lichstein, J. W., Simons, T. R., Shriner, S. A., and Franzreb, K. E. (2002), Spatial autocorrelation and autoregressive models in ecology, *Ecol. Monogr.* 72, 445–463.

Lin, X. (1997), Variance components testing in generalized linear models with random effects, *Biometrika* 84, 309–326.

Lin, X., and Breslow, N. E. (1996), Bias correction in generalized linear mixed models with multiple components of dispersion, *J. Amer. Statist. Assoc.* 91, 1007–1016.

Liu, J. S. (1994), The collapsed Gibbs sampler in Bayesian computations with applications to a gene regulation problem, *J. Amer. Statist. Assoc.* 89, 958–966.

Liu, J. S., Wong, W. H., and Kong, A. (1994), Covariance structure of the Gibbs sampler with applications to the comparisons of estimators and augmentation schemes, *Biometrika* 81, 27–40.

Liu, J. S., Wong, W. H., and Kong, A. (1995), Covariance structure and convergence rate of the Gibbs sampler with various scans, *J. R. Statist. Soc. B* 57, 157–169.

Malec, D., Sedransk, J., Moriarity, C. L., and LeClere, F. B. (1997), Small area inference for binary variables in the National Health Interview Survey, *J. Amer. Statist. Assoc.* 92, 815–826.

Mallows, C. L. (1972), A note on asymptotic joint normality, *Ann. Math. Statist.* 43, 508–515.

Manski, C. F., and McFadden, D. (1981), *Structural Analysis of Discrete Data with Econometric Applications*, MIT Press, Cambridge, MA.

Marcinkiewicz, J., and Zygmund, A. (1937a), Sur les fonctions indepdendantes, *Fundam. Math.* 29, 60–90.

Marcinkiewicz, J., and Zygmund, A. (1937b), Remarque sur la loi du logarithme itéré, *Fundam. Math.* 29, 215–222.

Marron, J. S., and Nolan, D. (1988), Canonical kernels for density estimation, *Statist. Probab. Lett.* 7, 195–199.

Marshall, R. J. (1991), A review of methods for the statistical analysis of spatial patterns of disease, *J. Roy. Statist. Soc. A* 154, 421–441.

Mason, D. M., Shorack, G. R., and Wellner, J. A. (1983), Strong limit theorems for oscillation moduli of the uniform empirical process, *Z. Wahrsch. verw. Geb.* 65, 83–97.

Massart, P. (1990), The tight constant in the Dvoretzky-Kiefer-Wolfowitz inequality, *Ann. Probab.* 18, 1269–1283.

McCullagh, P., and Nelder, J. A. (1989). *Generalized Linear Models*, 2nd ed., Chapman & Hall, London.

McCulloch, C. E. (1994), Maximum likelihood variance components estimation for binary data, *J. Amer. Statist. Assoc.* 89, 330–335.

McCulloch, C. E. (1997), Maximum likelihood algorithms for generalized linear mixed models, *J. Amer. Statist. Assoc.* 92, 162–170.

McFadden, D. (1989), A method of simulated moments for estimation of discrete response models without numerical integration, *Econometrika* 57, 995–1026.

Metropolis, N., Rosenbluth, A., Rosenbluth, M., Teller, A., and Teller, E. (1953), Equations of state calculations by fast computing machines, *J. Chem. Phys.* 21, 1087–1092.

Meza, J., and Lahiri, P. (2005), A note on the Cp statistic under the nested error regression model, *Survey Methodology* 31, 105–109.

Meza, J., Chen, S., and Lahiri, P. (2003), Estimation of lifetime alcohol abuse for Nebraska counties, unpublished manuscript.

Miller, J. J. (1977), Asymptotic properties of maximum likelihood estimates in the mixed model of analysis of variance, *Ann. Statist.* 5, 746–762.

Miller, R. G. (1974), An unbalanced jackknife, *Ann. Statist.* 2, 880–891.

Mises, R. von (1947), On the asymptotic distribution of differentiable statistical functions, *Ann. Math. Statist.* 18, 309–348.

Moeanaddin, R., and Tong, H. (1990), Numerical evaluation of distributions in nonlinear autoregression, *J. Time Series Anal.* 11, 33–48.

Moore, D. S. (1978), Chi-square tests, in *Studies in Statistics* (R. V. Hogg, ed.), Mathematical Society of America, Providence, RI.

Móricz, F. (1976), Moment inequalities and the strong laws of large numbers, *Z. Wahrsch. verw. Geb.* 35, 299–314.

Morris, C. N. (1983), Parametric empirical Bayes inference: theory and applications, *J. Amer. Statist. Assoc.* 78, 47–59.

Murray, G. D. (1977), Comment on "Maximum likelihood from incomplete data via the EM algorithm" by A. P. Dempster, N. Laird, and D. B. Rubin, *J. Roy. Statist. Soc. B* 39, 27–28.

Nelson, D. (1990), ARCH models as diffusion approximations, *J. Econometrics* 45, 7–38.

Neyman, J., and Scott, E. (1948), Consistent estimates based on partially consistent observations, *Econometrika* 16, 1–32.

Nishii, R. (1984), Asymptotic properties of criteria for selection of variables in multiple regression, *Ann. Statist.* 12, 758–765.

Nummelin, E. (1984), *General Irreducible Markov Chains and Non-negatice Operators*, Cambridge University Press, Cambridge.

Opsomer, J. D., Claeskens, G., Ranalli, M. G., Kauermann, G., and Breidt, F. J. (2008), Non-parametric small area estimation using penalized spline regression, *J. R. Statist. Soc. B* 70, 265–286.

Owen, D. B. (1962), *Handbook of Statistical Tables*, Addison-Wesley, Reading, MA.

Paley, R. E. A. C., Wiener, N., and Zygmund, A. (1933), Notes on random functions, *Math. Z.* 37, 647–668.

Patterson, H. D., and Thompson, R. (1971), Recovery of interblock information when block sizes are unequal, *Biometrika* 58, 545–554.

Pearson, K. (1900), On a criterion that a given system of deviations from the probable in the case of a corrected system of variables is such that it can be reasonably supposed to have arisen from random sampling, *Philos. Mag. 5th Series* 50, 157–175.

Peng, L., and Zhou, X.-H. (2004), Local linear smoothing of receiver operating characteristic (ROC) curves, *J. Statist. Planning Inference* 118, 129–143.

Petrov, V. V. (1975), *Sums of Independent Random Variables*, Springer, Berlin.

Prasad, N. G. N., and Rao, J. N. K. (1990), The estimation of mean squared errors of small area estimators, *J. Amer. Statist. Assoc.* 85, 163–171.

Quenouille, M. (1949), Approximation tests of correlation in time series, *J. Roy.*

*Statist. Soc., Ser. B* 11, 18–84.

Rao, C. R. (1972), Estimation of variance and covariance components in linear models, *J. Amer. Statist. Assoc.* 67, 112–115.

Rao, C. R., and Kleffe, J. (1988), *Estimation of Variance Components and Applications*, North-Holland, Amsterdam.

Rao, J. N. K. (2003), *Small Area Estimation*, Wiley, Hoboken, NJ.

Rice, J. A. (1995), *Mathematical Statistics and Data Analysis*, 2nd ed., Duxbury Press, Belmont, CA.

Richardson, A. M., and Welsh, A. H. (1994), Asymptotic properties of restricted maximum likelihood (REML) estimates for hierarchical mixed linear models, *Austral. J. Statist.* 36, 31–43.

Rivest, L.-P., and Belmonte, E. (2000), A conditional mean squared error of small area estimators, *Survey Methodology* 26, 67–78.

Robert, C. P., and Casella, G. (2004), *Monte Carlo Statistical Methods*, 2nd ed., Springer, New York.

Roberts, G. O., and Smith, A. F. M. (1994), Simple conditions for the convergence of the Gibbs sampler and Metropolis-Hastings algorithms, *Stoch. Proc. Appl.* 49, 207–216.

Roberts, G. O., and Tweedie, R. L. (1996), Geometric convergence and central limit theorem for multidimensional Hastings and Metropolic algorithms, *Biometrika* 83, 95–110.

Robinson, G. K. (1991), That BLUP is a good thing: The estimation of random effects (with discussion), *Statist. Sci.* 6, 15–51.

Rosenblatt, M. (1952), Limit theorems associated with variants of the von Mises statistic, *Ann. Math. Statist.* 23, 617–623.

Rosenthal, H. P. (1970), On the subspaces $L^p$ $(p > 2)$ spanned by sequences of independent random variables, *Israel J. Math.* 8, 273–303.

Ross, S. M. (1983), *Stochastic Processes*, Wiley, New York.

Samuels, M. L., and Witmer, J. A. (2003), *Statistics for the Life Sciences*, 3rd ed., Pearson Education, Upper Saddle River, NJ.

Schmidt, W. H., and Thrum, R. (1981), Contributions to asymptotic theory in regression models with linear covariance structure, *Math. Operationsforsch. Statist. Ser. Statist.* 12, 243–269.

Schwarz, G. (1978), Estimating the dimension of a model, *Ann. Statist.* 6, 461–464.

Scott, D. W. (1992), *Multivariate Density Estimation*, Wiley, New York.

Searle, S. R., Casella, G., and McCulloch, C. E. (1992), *Variance Components*, Wiley, New York.

Sen, A., and Srivastava, M. (1990), *Regression Analysis*, Springer, New York.

Serfling, R. J. (1980), *Approximation Theorems of Statistics*, Wiley, New York.

Shao, J. (2003), *Mathematical Statistics*, Springer, New York.

Shao, J., and Wu, C. F. J. (1987), Heteroscedasticity-robustness of jackknife variance estimators in linear models, *Ann. Statist.* 15, 1563–1579.

Shibata, R. (1980), Asymptotically efficient selection of the order of the model for estimating parameters of a linear process, *Ann. Statist.* 8, 147–164.

Shibata, R. (1984), Approximate efficiency of a selection procedure for the number of regression variables, *Biometrika* 71, 43–49.

Shorack, G. R., and Wellner, J. A. (1986), *Empirical Processes with Applications to Statistics*, Wiley, New York.

Singh, K. (1981), On the asymptotic accuracy of Efron's bootstrap, *Ann. Statist.* 9, 1187–1195.

Slepian, D. (1962), The one-sided barrier problem for Gaussian noise, *Bell System Tech. J.* 41, 463–501.

Smirnov, N. V. (1936), Sur la distribution de $\omega^2$ (Critérium de M. R. v. Mises), *C. R. Acad. Sci. Paris* 202, 449–452.

Smirnov, N. V. (1944), Approximating the distribution of random variables by empirical data (in Russian), *Usp. Mat. Nauk* 10, 179–206.

Smythe, R. T. (1973), Strong laws of large numbers for $r$-dimensional arrays of random variables, *Ann. Probab.* 1, 164–170.

Stout, W. F. (1974), *Almost Sure Convergence*, Academic Press, New York.

Strassen, V. (1964), An invariance principle for the law of the iterated logarithm, *Z. Warsch. verw. Geb.* 3, 211–226.

Strassen, V. (1966), A converse to the law of the iterated logarithm, *Z. Warsch. verw. Geb.* 4, 265–268.

Stroock, D. W., and Varadhan, S. R. S. (1979), *Multidimensional Diffusion Processes*, Springer, Berlin.

Swets, J. A., and Pickett, R. M. (1982), *Evaluation of Diagnostic Systems: Methods from Signal Detection Theory*, Academic Press, New York.

Tanner, M. A., and Wong, W. H. (1987), The calculation of posterior distribution by data augmentation (with discussion), *J. Amer. Statist. Assoc.* 82, 528–550.

Thompson, W. A., Jr. (1962), The problem of negative estimates of variance components, *Ann. Math. Statist.* 33, 273–289.

Tierney, L. (1991), Exploring posterior distributions using Markov chains, in *Computer Science and Statistics: Proc. 23rd Symp. Interface* (E. M. Keramidas ed.), Interface Foundation, Fairfax Station, VA, 563–570.

Tierney, L. (1994), Markov chains for exploring posterior distributions (with discussion), *Ann. Statist.* 22, 1701–1786.

Tjøstheim, D. (1978), Statistical spatial series modelling, *Adv. Appl. Probab.* 10, 130–154.

Tjøstheim, D. (1983), Statistical spatial series modelling II: Some further results on unilateral lattice processes, *Adv. Appl. Probab.* 15, 562–584.

Tong, Y. L. (1980), *Probability Inequalities in Multivariate Distributions*, Academic Press, New York.

Tukey, J. (1958), Bias and confidence in not quite large samples, *Ann. Math. Statist.* 29, 614.

van der Vaart, A. W. (1998), *Asymptotic Statistics*, Cambridge University Press, Cambridge.

van der Vaart, A. W., and Wellner, J. A. (1996), *Weak Convergence and Empirical Processes with Applications to Statistics*, Springer, New York.

Varadhan, S. R. S. (1966), Asymptotic probabilities and differential equations, *Commun. Pure Appl. Math.* 19, 261–286.

Verbeke, G., and Lesaffre, E. (1996), A linear mixed-effects model with heterogeneity in the random-effects population, *J. Amer. Statist. Assoc.* 91, 217–221.

Verbyla, A. P. (1990), A conditional derivation of residual maximum likelihood, *Austral. J. Statist.* 32, 227–230.

Ville, J. (1939), *Etude Critique de la Notion de Collectif*, Gauthier-Villars, Paris.

Wald, A. (1949), Note on the consistency of the maximum likelihood estimate, *Ann. Math. Statist.* 20, 595–601.

Wand, M. (2003), Smoothing and mixed models, *Comput. Statist.* 18, 223–249.

Wang, Y. (2002), Asymptotic nonequivalence of GARCH models and diffusions, *Ann. Statist.* 30, 754–783.

Weiss, L. (1955), The stochastic coverage of a function of sample successive differences, *Ann. Math. Statist.* 26, 532–535.

Weiss, L. (1971), Asymptotic properties of maximum likelihood estimators in some nonstandard cases, *J. Amer. Statist. Assoc.* 66, 345–350.

Wichura, M. J. (1973), Some Strassen-type laws of the iterated logarithm for multiparameter stochastic processes with independent increments, *Ann. Probab.* 1, 272–296.

Wieand, S., Gail, M. H., James, B. R., and James, K. L. (1989), A family of nonparametric statistics for comparing diagnostic markers with paired and unpaired data, *Biometrika* 76, 585–592.

Wiener, N. (1923), Differential spaces, *J. Math. Phys.* 2, 131–174.

Wittmann, R. (1985), A general law of iterated logarithm, *Z. Wahrsch. verw. Geb.* 68, 521–543.

Wold, H. (1938), *A Study in the Analysis of Stationary Time Series*, Almqvist and Wiksell, Uppsala.

Wolfe, J. H. (1971), A Monte Carlo study of the sampling distribution of the likelihood ratio for mixture of multinomial distributions, *Tech. Bull. STB* 72-2, Naval Research and Training Laboratory, San Diego, CA.

Wolfowitz, J. (1949), On Wald's proof of the consistency of the maximum likelihood estimate, *Ann. Math. Statist.* 20, 601–602.

Wood, C. L., and Altavela, M. M. (1978), Large-sample results for Kolmogorov-Smirnov statistics for discrete distributions, *Biometrika* 65, 239–240.

Wu, C. F. J. (1083), On the convergence properties of the EM algorithm, *Ann. Statist.* 11, 95–103.

Wu, C. F. J. (1986), Jackknife, bootstrap and other resampling methods in regression analysis (with discussion), *Ann. Statist.* 14, 1261–1350.

Wu, W. B. (2007), Strong invariance principles for dependent random variables, *Ann. Probab.* 35, 2294–2320.

Yaglom, A. M. (1957), Some classes of random fields in n-dimensional space, related to stationary processes, *Theory Probab. Appl.* 2, 273–320.

Young, D. (1971), *Iterative Solutions of Large Linear Systems*, Academic Press, New York.

Zhan, X. (2002), *Matrix inequalities*, Lecture Notes in Mathematics No. 1790, Springer, New York.

Zygmund, A. (1951), An individual ergodic theorem for non-commutative transformations, *Acta Math. Szeged* 14, 103–110.

# Index